EMBRYOLOGY

EMBRYOLOGY

CONSTRUCTING THE ORGANISM

Edited by Scott F. Gilbert and Anne M. Raunio

With illustrations by Nancy J. Haver

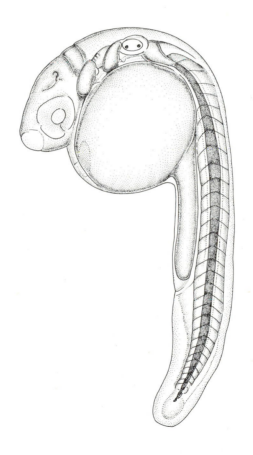

SINAUER ASSOCIATES, *Inc. Publishers*
Sunderland, MA 01375 U.S.A.

The Cover
Front Cover: Bipinnaria larva of a starfish.
Back cover: Larva of sipunculid.
Photographs by Mary E. Rice, Smithsonian Marine Station, Fort Pierce, Florida.

Embryology: Constructing the Organism

Copyright © 1997 by Sinauer Associates, Inc. All rights reserved.

This book may not be reproduced in whole or in part without permission.

Address inquiries and orders to:

Sinauer Associates, Inc., P.O. Box 407,

23 Plumtree Road, Sunderland, MA 01375 U.S.A.

FAX: 413-549-1118

email: publish@sinauer.com

Library of Congress Cataloging-in-Publication Data

Embryology: constructing the organism / edited by Scott F. Gilbert and Anne M. Raunio.

     p.   cm.

Includes bibliographical references and index.

ISBN 0-87893-237-2 (cloth : alk. paper)

1. Embryology.   I. Gilbert, Scott F.   II. Raunio, Anne M.

QL955.E43   1997

571.8'61—dc21   97-13045

                                                                    CIP

Printed in Canada

5   4   3   2   1

Dedicated to N. J. Berrill (1903–1996)
and Michael E. Somers,
powerful inducers who never lost focus on the whole embryo.

Contents

Preface

WHY, IN THE LAST YEARS OF THE TWENTIETH CENTURY, would one want a new embryology book? Didn't developmental biology replace embryology? Wasn't the phenotypic science of embryonic anatomy put on a cellular and molecular basis during the last quarter of this century?

There are several ways to answer the question as to why a new embryology book is needed. The first answer is that the replacement of embryology by modern developmental biology has been very selective. Only a few organisms—the fly *Drosophila melanogaster*, the nematode *Caenorhabditis elegans*, the frog *Xenopus laevis*, the mouse *Mus musculus*, and, increasingly, the fish *Danio rerio*, the chick *Gallus domesticus*, and the primate *Homo sapiens*—have been privileged to be studied extensively on the molecular level. In other words, of the millions of insect species, only one is known on a significant developmental level; of the thousands of amphibians, only one species represents that whole remarkably diverse class.

The second answer, contingent on the first, is that without knowledge of many other species, one cannot do significant comparative studies. There is much new material in this volume. Indeed, we are beginning to see a renaissance of embryological studies. Where new techniques are being utilized, major changes are being wrought in our knowledge of embryos.

The third answer is that the study of evolution now demands knowledge of the development of many species. Changes in developmental processes are increasingly being seen as causes of macroevolution. This renewed interest in the causal mechanisms of macroevolution is fueling a new generation of studies in evolutionary embryology.

The fourth answer is that textbooks of developmental biology are now being used for upper-level courses in which students already have some knowledge of cell biology and genetics. This has created a vacuum at the sophomore level. Students come to our developmental biology classes knowing about transcription and translation but knowing nothing about teloblasts or neural crest cells. Students have to know the "what" before they can ask the "how." Döllinger, the mentor of von Baer, noted that students should first be shown the embryo. Then they will ask the appropriate questions as to how changes in the developing embryo can possibly occur. We hope that this book will inspire students to learn the "what" and ask those "how" questions that are addressed in the developmental biology texts.

The fifth answer is that textbooks in developmental biology report the success stories. They discuss the embryos about which we know the most. This embryology book discusses the things about which we do not know so much. While developmental biology books highlight the few organisms that have been extensively studied, this book reports about those organisms whose development is not as well known. The organisms of most animal classes have not had their fate maps drawn nor their patterns of gene expression probed. The student of developmental biology should discover whole new worlds to explore.

The sixth answer is aesthetic. Many of today's developmental biologists entered the field because they are fascinated by the beauty, order, and interactions of real developing organisms. Whole embryos and larvae are wonder-full forms of life and should be brought back into the curriculum. Embryology is a science of becoming. It is a science of emerging complexity; hence it may be the science of the twenty-first century. And when one realizes that the adult *Cecropia* moth flies only once—to find its mate—and then dies, one realizes that the larval stages are not a small portion of the animal's life.

This book sees a continuity between the changing anatomy emphasized in embryology courses and changing patterns of cells and gene expression emphasized in developmental biology courses. One approach cannot be complete without the other. Alfred North Whitehead (a philosopher of process and change much admired by embryologists of the 1930s and 1940s) wrote that there is no true distinction between the reality of an object and the processes that cause its being. This is consistent with our studies in development.

This is a "timeless" book. No contributor had the time to write it. But they did. It is to them that we owe a big thank you. Each chapter is an act of love from its author(s) to the reader. Each author wanted to write his or her chapter to introduce students to organisms that are so worthwhile to know. Scott's job was to round up these people and convince them that if they didn't write these chapters, students wouldn't be introduced to the incredible organisms they study. Anne's job was to read the chapters and to make certain that someone with a basic biological knowledge (but not a specialist's insight) could understand and enjoy them.

After the first two introductory chapters, each chapter can stand alone. There is no requirement to read this book in any particular order, and students may be more comfortable starting with the more familiar embryological organisms—echinoderms, amphibians, and birds—before starting in on bryozoans, myriapods, and dicyemid mesozoans. And since in recent years the development of plants and animals has been linked through the studies of gene expression and signal transduction, we are particularly pleased to at least represent this kingdom with a concluding chapter that integrates the physiology and anatomy of angiosperm development with the new genetic analyses that are revolutionizing this field.

The authors were asked to write for college sophomores, not for their colleagues, and certainly not for posterity. It is our hope that this book will stimulate the formation of new embryology courses focusing on embryonic and larval structures and functions. There is no substitute for seeing the real creature.

This volume owes much to its predecessors. F. M. Balfour's *A Treatise on Comparative Embryology* (1881) was probably the last of the English-language comparative embryology texts comprehensive enough to include both mesozoans and vertebrates. It was written explicitly for the integration of embryology and evolutionary biology, and we hope to accomplish with more than twenty authors what Balfour did alone. In the early years of this century, the *Textbook of Embryology* attempted to provide a comprehensive synthesis of embryology and was actually two separate volumes: E. W. MacBride's *Invertebrata* (1914) and J. G. Kerr's *Vertebrates with the Exception of Mammalia* (1919). Since then most comparative embryology textbooks have remained on one side or the other of the vertebral divide; it has only been since the discovery of the genetic similarities uniting insect and mammalian development that the gap between the invertebrates and vertebrates has seemed artificial. Comparative invertebrate embryology especially has had several excellent textbooks, including G. Reverberi's *Experimental Embryology of Marine and Fresh-water Invertebrates* (1971), G. Brusca's *General Patterns of Invertebrate Development* (1975), and the classic reference textbook, *Invertebrate Embryology*, edited by M. Kumé and K. Dan and translated into English in 1968. The beautifully illustrated Russian textbook, *The Comparative Embryology of Invertebrate Animals* by O. M. Ivanova-Kazas (1975), remains untranslated into English but has been an excellent source of references and figures. We hope this volume lives up to the standards set by the above-mentioned books.

Acknowledgments

The haunting, sometimes eerie beauty of larvae and embryos have been captured by many fine scientists and microscopists. The photographs in this volume are gratefully reproduced with thanks both to the chapter contributors for finding and collecting the photographic material and to the many scientists who supplied prints and permission so we could show their work.

This project could not have been started or completed without the commitment from Andy Sinauer that it was time for a new embryology book to see print. It could not have achieved the visual standards necessary for the appropriate presentation of its subject without the fine hand of biological artist Nancy Haver, whose intricate and lovely stipple art adorns the chapters. The editors and contributors alike express thanks and admiration for her work.

The hard work of Jefferson Johnson of the Sinauer Associates production department resulted in an elegant design and the clear presentation of a lot of complicated material. He and production manager Christopher Small put together pieces from many different sources to produce a uniform and flowing text. And project editor Carol Wigg once again oversaw with patience and fortitude the mechanics of turning 23 manuscripts into a book.

Contributors

Gary J. Brusca, Department of Biology, Humboldt State University, Arcata, CA 95521 U.S.A.

Richard C. Brusca, Grice Marine Biological Laboratory, University of Charleston, Charleston, SC 85721 U.S.A.

J. R. Collier, Biology Department, Brooklyn College of The City University of New York, Brooklyn, NY 11210 U.S.A.

Yolanda P. Cruz, Department of Biology, Oberlin College, Oberlin, OH 44074 U.S.A.

Richard P. Elinson, Department of Zoology, University of Toronto, Toronto, Ontario M5S 1A1 CANADA

Charles H. Ellis, Jr., Department of Biology, Northeastern University, Boston, MA 02115 U.S.A.

Anne Fausto-Sterling, Division of Biology and Medicine, Brown University, Providence, RI 02912 U.S.A.

Paul E. Fell, Department of Zoology, Connecticut College, New London, CT 06320 U.S.A.

Scott F. Gilbert, Department of Biology, Swarthmore College, Swarthmore, PA 19081 U.S.A.

Jonathan Henry, Department of Cell and Structural Biology, University of Illinois, Urbana, IL 61801 U.S.A.

Piroschka Horvath, Department of Biology, Texas A&M University, College Station, TX 77843 U.S.A.

William R. Jeffery, Department of Biology, Pennsylvania State University, University Park, PA 16801 U.S.A.

Charles B. Kimmel, Department of Biology, University of Oregon, Eugene, OR 97403 U.S.A.

James A. Langeland, Biology Department, Kalamazoo College, Kalamazoo, MI 49006 U.S.A.

Vicki J. Martin, Department of Biological Sciences, University of Notre Dame, Notre Dame, IN 46556 U.S.A.

Mark Q. Martindale, Department of Organismal Biology and Anatomy, University of Chicago, Chicago, IL 60637 U.S.A.

John F. Pilger, Department of Biology, Agnes Scott College, Decatur, GA 30030 U.S.A.

Robert M. Savage, Department of Biology, Williams College, Williamstown, MA 01267 U.S.A.

Einhard Schierenberg, Zoological Institute, University of Cologne, D-50923 Köln GERMANY

Gary C. Schoenwolf, Department of Neurobiology and Anatomy, University of Utah School of Medicine, Salt Lake City, UT 84132 U.S.A.

Fritz E. Schwalm, Biology Department, Texas Woman's University, Denton, TX 76204 U.S.A.

Marty Shankland, Department of Zoology, University of Texas, Austin, TX 78712 U.S.A.

Susan R. Singer, Department of Biology, Carleton College, Northfield, MN 55057 U.S.A.

Billie J. Swalla, Department of Biology, Pennsylvania State University, University Park, PA 16802 U.S.A.

J. R. Whittaker, Department of Biology, University of New Brunswick, Fredericton, NB E3B 6E1 CANADA

Gregory A. Wray, Department of Ecology and Evolution, The State University of New York, Stony Brook, NY 11794 U.S.A.

Russel L. Zimmer, Department of Biological Sciences, University of Southern California, Los Angeles, CA 90089 U.S.A.

SECTION I

Introduction

CHAPTER 1

Characteristics of Metazoan Development

Gary J. Brusca, Richard C. Brusca, and Scott F. Gilbert

S OPPOSED TO THE PROTOZOA, WHICH ARE USUALLY viewed as being unicellular, the metazoa are multicellular. This distinction is sometimes blurred, however, because there are a number of protozoa that form rather complex colonies with some division of labor among different cell types. The metazoa possess certain qualities that must be considered in concert with the basic idea of multicellularity. The cells of metazoa are organized into functional units, generally as tissues and organs, with specific roles that support the life of the whole animal. These cell types are interdependent and their activities are coordinated into predictable patterns and relationships. Structurally, the cells of metazoan animals are organized as layers, which develop through a series of events early in an organism's embryogeny. These embryonic tissues, or germ layers, are the framework upon which the metazoan body is constructed. *Thus, the cells of metazoa are specialized, interdependent, coordinated in function, and develop through layering during embryogeny.* This combination of features is absent from protozoa.

Eggs and Embryos

The attributes that distinguish the metazoa are the result of metazoan embryonic development. To put it another way, adult phenotypes result from specific sequences of developmental stages—evolutionary patterns reveal themselves, in part, through species' ontogenesis. Indeed, embryogenesis is the key feature distinguishing metazoa from protozoa. Bearing this in mind, it is not surprising to discover that both animal unity and diversity are as evident in patterns of development as they are in the architecture of adults. The patterns of development discussed in this chapter reflect this unity and diversity, and serve as a basis for understanding the embryogeny of specific phyla in later chapters.

Eggs

Biological processes are generally cyclical. The production of one generation after another through reproduction exemplifies this generality, as the term *life cycle* implies. At what point one begins describing such cycles is a matter of convenience. For our purposes in this chapter we choose to begin with the **egg**, or **ovum**—a remarkable cell capable of developing into a new individual. Once the egg is fertilized, all the different cell types of an adult metazoan are derived during embryogenesis from this single totipotent cell. A fertilized egg contains not only the information necessary to direct development, but also some quantity of nutrient material, called **yolk**, that sustains the early stages of life.

Animals that simply deposit their fertilized eggs, either freely or in capsules, are said to be **oviparous**. A great number of invertebrates as well as some vertebrates (amphibians, fishes, reptiles, and birds) are oviparous. Animals that brood their embryos internally and nourish them directly, such as placental mammals, are described as **viviparous**. **Ovoviviparous** animals brood their embryos internally but rely on the yolk within the eggs to nourish their developing young. Most internally-brooding invertebrates are ovoviviparous.

Eggs tend to be polarized along what is called their **animal-vegetal axis**. This polarity may be apparent in the egg itself, or it may be recognizable only as development proceeds. The vegetal pole is commonly associated with the formation of nutritive organs (e.g., the digestive system), whereas the animal pole tends to produce other regions of the embryo. These and many other manifestations of the egg's polarity will be explored more completely throughout this chapter.

Metazoan ova are categorized primarily by the amount and location of yolk within the cell (Figure 1.1), two factors that greatly influence certain aspects of development.

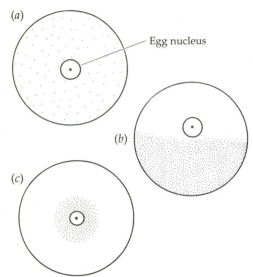

(a)

Egg nucleus

(b)

(c)

Figure 1.1   Types of ova. The stippling denotes the distribution and relative concentration of yolk within the cytoplasm. *(a)* An isolecithal ovum has a small amount of yolk distributed evenly. *(b)* The yolk in a telolecithal ovum is concentrated toward the vegetal pole. The amount of yolk in such eggs varies greatly. *(c)* A centrolecithal ovum has yolk concentrated at the center of the cell. (From Brusca and Brusca 1990.)

Isolecithal eggs contain a relatively small amount of yolk that is more-or-less evenly distributed throughout the cell. Ova in which the yolk is concentrated at one end (the vegetal pole) are termed **telolecithal eggs**; those in which the yolk is concentrated in the center are called **centrolecithal eggs**. The actual amount of yolk in telolecithal and centrolecithal eggs is highly variable. Yolk production (**vitellogenesis**) is typically the longest phase of egg production, and its duration varies by an order of magnitude among species. Rates of yolk production depend on the specific vitellogenic mechanism used. In general, so-called *r*-selected (opportunistic) species have evolved vitellogenic pathways for the rapid conversion of food into egg production, whereas so-called *K*-selected (specialized) species utilize slower pathways.

Cleavage

The stimulus that initiates development in an egg is usually provided at **fertilization** by the entry into the egg of a sperm cell and the subsequent fusion of the male and female nuclei to produce a fertilized egg, or **zygote**. The initial process of cell division of a zygote is called **cleavage**, and the resulting cells are called **blastomeres**. Certain aspects of the patterns of early cleavage are determined by the amount and placement of yolk, whereas other features are apparently inherent in the genetic programming of the particular organism. Isolecithal and weakly to moderately telolecithal ova generally undergo **holoblastic cleavage**. That is, the cleavage planes pass completely through the cell, producing blastomeres that are separated from one another by thin cell membranes (Figure 1.2*a*). Whenever very large amounts of yolk are present (e.g., as in strongly telolecithal eggs), the cleavage planes do not pass readily through the dense yolk, so the blastomeres are not fully separated by cell membranes. This pattern of early cell division is called **meroblastic cleavage** (Figure 1.2*b*). The pattern of cleavage in centrolecithal eggs is dependent on the amount of yolk and varies from holoblastic to various modifications of meroblastic (for examples, see the descriptions of arthropod development in Chapter 13).

ORIENTATION OF CLEAVAGE PLANES.   A number of terms are used to describe the relationship of the planes of cleavage to the animal-vegetal axis of the egg and the relationships of the resulting blastomeres to each other (Figure 1.3). Cell divisions during cleavage are often referred to as either **equal** or **unequal**, the terms indicating the comparative sizes of groups of blastomeres. The term **subequal** is used when blastomeres are only slightly different in size. When cleavage is distinctly unequal, the larger cells are called **macromeres** and usu-

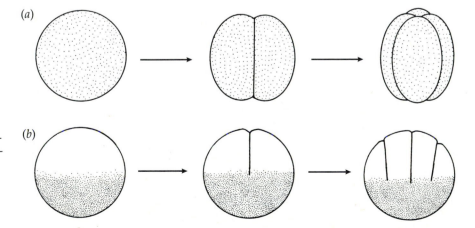

(a)

(b)

Figure 1.2   Types of early cleavage in developing zygotes. *(a)* Holoblastic cleavage. The cleavage planes pass completely through the cytoplasm. *(b)* Meroblastic cleavage. The cleavage planes do not pass completely through the yolky cytoplasm. (From Brusca and Brusca 1990.)

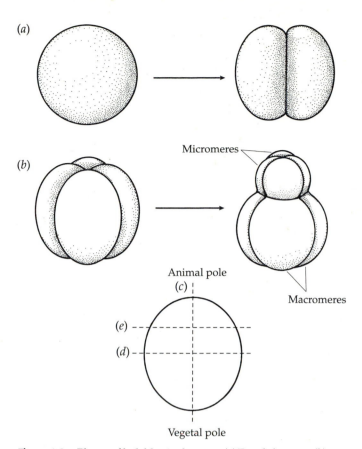

RADIAL AND SPIRAL CLEAVAGE.   Most animals display one of two basic cleavage patterns defined on the basis of the orientation of the blastomeres about the animal-vegetal axis. These patterns are called **radial cleavage** and **spiral cleavage** and are illustrated in Figure 1.4. Radial cleavage involves strictly meridional and transverse divisions. Thus, the blastomeres are arranged in rows either parallel or perpendicular to the animal-vegetal axis. The placement of the blastomeres shows a radially symmetrical pattern in polar view.

Spiral cleavage is quite another matter. Although not inherently complex, it can be difficult to describe. The first

Figure 1.3   Planes of holoblastic cleavage. (*a*) Equal cleavage. (*b*) Unequal cleavage produces micromeres and macromeres. (*c–e*) Planes of cleavage relative to the animal-vegetal axis of the egg or zygote. (*c*) Longitudinal (= meridional) cleavage parallel to the animal-vegetal axis. (*d*) Equatorial cleavage perpendicular to the animal-vegetal axis and bisecting the zygote into equal animal and vegetal halves. (*e*) Latitudinal cleavage perpendicular to the animal-vegetal axis but not passing along the equatorial plane. (From Brusca and Brusca 1990.)

ally lie at the vegetal pole. The smaller cells are called **micromeres** and are usually located at the animal pole.

Cleavage planes that pass parallel to the animal-vegetal axis produce **longitudinal** (= **meridional**) divisions; those that pass at right angles to the axis produce **transverse** divisions. Transverse divisions may be either **equatorial**, when the embryo is separated equally into animal and vegetal halves, or simply **latitudinal** when the division plane does not pass through the "equator" of the embryo.

Figure 1.4   Comparison of radial versus spiral cleavage through the 8-cell stage. During radial cleavage, the cleavage planes all pass either perpendicular or parallel to the animal-vegetal axis of the embryo. Spiral cleavage involves a tilting of the mitotic spindles, commencing with the division from four to eight cells. The resulting cleavage planes are neither perpendicular nor parallel to the axis. The polar views of the resulting 8-cell stages illustrate the differences in blastomere orientation. (From Brusca and Brusca 1990.)

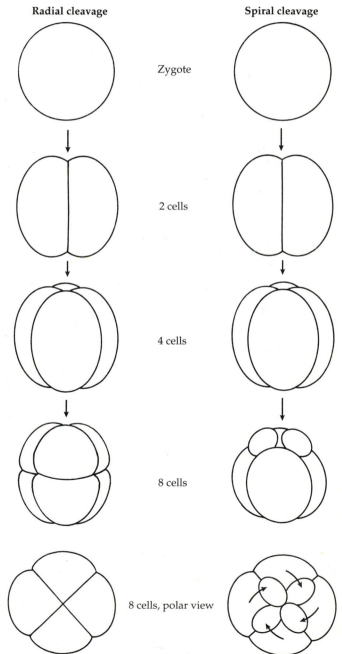

two divisions are meridional and usually equal to subequal. Subsequent divisions, however, result in the displacement of blastomeres in such a way that they lie in the furrows between one another. This condition is a result of the formation of the mitotic spindles at acute angles rather than parallel to the axis of the embryo; hence the cleavage planes are neither perfectly meridional nor perfectly transverse. The division from 4 to 8 cells involves a displacement of the cells near the animal pole in a clockwise (**dextrotropic**) direction (viewed from the animal pole). The next division, from 8 to 16 cells, occurs with a displacement in a counterclockwise (**levotropic**) direction; the next is clockwise, and so on, "twisting" back and forth until approximately the 64-cell stage. (In reality, divisions are frequently nonsynchronous; not all of the cells divide at the same rate. Thus, a particular embryo may not proceed from 4 cells to 8, to 16, and so on as neatly as in our generalized example.)

An elaborate coding system for spiral cleavage was developed by E. B. Wilson (1892) during his extensive studies on the polychaete worm *Neanthes succinea* conducted at Woods Hole Biological Station. Wilson's system is usually applied to spiral cleavage in order to trace cell fates and compare development among various taxa. The following account of spiral cleavage is a general one, but it provides a point of reference for considering the patterns seen in different animal groups. Although this coding system may seem a bit confusing at first, it will quickly become evident that it is a rather simple and elegant means by which one may follow the developmental lineage of each and every cell in a developing embryo.

At the 4-cell stage, following the initial meridional divisions, the cells are given the codes A, B, C, and D and are labeled clockwise in that order when viewed from the animal pole (Figure 1.5*a*). These four cells are referred to as a quartet of macromeres, and may be collectively coded as simply Q. The D cell commonly is slightly larger than the others, providing a starting point for the coding process. The next division is more-or-less unequal, with the four cells nearest the animal pole being displaced in a dextrotropic fashion, as explained above. These four smaller cells are called the first quartet of micromeres (collectively the 1q cells) and are given the individual codes of 1a, 1b, 1c, and 1d. The numeral "1" indicates a member of the first micromere quartet to be produced; the letters correspond to their respective macromere origins. The capital letters designating the macromeres are now preceded by the numeral "1" to indicate that they have divided once and produced a first micromere set (Figure 1.5*b*). We may view this 8-cell embryo as four pairs of daughter cells that have been produced by the divisions of the four original macromeres, as follows:

$$A \nearrow^{1a}_{\searrow 1A} \qquad B \nearrow^{1b}_{\searrow 1B} \qquad C \nearrow^{1c}_{\searrow 1C} \qquad D \nearrow^{1d}_{\searrow 1D}$$

One "rule" that will aid you in tracing the cells and their codes through spiral cleavage is that the *only* code numbers that are changed through subsequent divisions are the prefix numbers of the macromeres. These are changed to indicate the number of times these individual macromeres have divided, and to correspond to the number of micromere quartets thus produced. So, at the 8-cell stage, we can designate the existing blastomeres as the 1Q (= 1A, 1B, 1C, 1D) and the 1q (= 1a, 1b, 1c, 1d).

It should be mentioned that although the macromeres and micromeres are sometimes similar in size, these terms are nonetheless always used in describing spiral cleavage. Much of the size discrepancy depends upon the amount of yolk present at the vegetal pole in the original egg; this yolk tends to be retained primarily in the larger macromeres.

The division from 8 to 16 cells occurs levotropically and involves cleavage of each macromere and micromere. The macromeres (1Q) divide to produce a second quartet of micromeres (2q = 2a, 2b, 2c, 2d), and 4 daughter macromeres, whose prefix numeral is changed to "2." The first micromere quartet also divides and now comprises eight cells, each of which is identifiable not only by the letter corresponding to its parent macromere but now

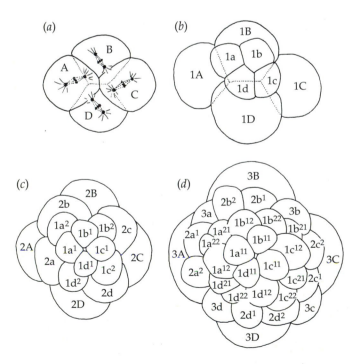

Figure 1.5 Spiral cleavage from 4 to 32 cells (assumed synchronous) labeled with E. B. Wilson's coding system. All diagrams are viewed from the animal pole surface. (From Brusca and Brusca 1990.)

by the addition of superscript numerals. For example, the 1a micromere (of the 8-cell embryo) divides to produce two daughter cells coded the $1a^1$ and the $1a^2$ cells. The cell that is physically nearer the animal pole of the embryo receives the superscript "1," the other cell the superscript "2." Thus, the 16-cell stage (Figure 1.5c) includes the following cells:

$$\text{Derivatives of the 1q} \begin{cases} 1a^1 \quad 1b^1 \quad 1c^1 \quad 1d^1 \\ 1a^2 \quad 1b^2 \quad 1c^2 \quad 1d^2 \end{cases}$$

$$\text{Derivatives of the 1Q} \begin{cases} 2q = 2a \quad 2b \quad 2c \quad 2d \\ 2Q = 2A \quad 2B \quad 2C \quad 2D \end{cases}$$

The next division (from 16 to 32 cells) involves dextrotropic displacement. The third micromere quartet (3q) is formed, the daughter macromeres are now given the prefix "3" (3Q), and all of the 12 existing micromeres divide. Superscripts are added to the derivatives of the first and second micromere quartets according to the rule of position as stated above. Thus, the $1b^1$ cell divides to yield the $1b^{11}$ and $1b^{12}$ cells; the $1a^2$ cell yields the $1a^{21}$ and $1a^{22}$ cells; the 2c yields the $2c^1$ and $2c^2$, and so on. Do not think of these superscripts as double-digit numbers (e.g., "twenty-one" and "twenty-two") but rather as two-digit sequences reflecting the precise lineage of each cell ("two-one" and "two-two").

The elegance of Wilson's system is that each code tells the history as well as the position of the cell in the embryo. For instance, the code $1b^{11}$ indicates that the cell is a member (derivative) of the first quartet of micromeres; that its parent macromere is the B cell; that the original 1b micromere has divided twice since its formation; and that this particular cell rests uppermost in the embryo relative to its sister cells. The 32-cell stage (Figure 1.5d) is composed of the following:

$$\text{Derivatives of the 1q} \begin{cases} 1a^{11} \quad 1b^{11} \quad 1c^{11} \quad 1d^{11} \\ 1a^{12} \quad 1b^{12} \quad 1c^{12} \quad 1d^{12} \\ 1a^{21} \quad 1b^{21} \quad 1c^{21} \quad 1d^{21} \\ 1a^{22} \quad 1b^{22} \quad 1c^{22} \quad 1d^{22} \end{cases}$$

$$\text{Derivatives of the 2q} \begin{cases} 2a^1 \quad 2b^1 \quad 2c^1 \quad 2d^1 \\ 2a^2 \quad 2b^2 \quad 2c^2 \quad 2d^2 \end{cases}$$

$$\text{Derivatives of the 2Q} \begin{cases} 3q = 3a \quad 3b \quad 3c \quad 3d \\ 3Q = 3A \quad 3B \quad 3C \quad 3D \end{cases}$$

The division to 64 cells follows the same pattern, with appropriate coding changes and additions of superscripts. The displacement is levotropic and results in the following cells:

$$\text{Derivatives of the 1q} \begin{cases} 1a^{111} \quad 1b^{111} \quad 1c^{111} \quad 1d^{111} \\ 1a^{112} \quad 1b^{112} \quad 1c^{112} \quad 1d^{112} \\ 1a^{121} \quad 1b^{121} \quad 1c^{121} \quad 1d^{121} \\ 1a^{122} \quad 1b^{122} \quad 1c^{122} \quad 1d^{122} \\ 1a^{211} \quad 1b^{211} \quad 1c^{211} \quad 1d^{211} \\ 1a^{212} \quad 1b^{212} \quad 1c^{212} \quad 1d^{212} \\ 1a^{221} \quad 1b^{221} \quad 1c^{221} \quad 1d^{221} \\ 1a^{222} \quad 1b^{222} \quad 1c^{222} \quad 1d^{222} \end{cases}$$

$$\text{Derivatives of the 2q} \begin{cases} 2a^{11} \quad 2b^{11} \quad 2c^{11} \quad 2d^{11} \\ 2a^{12} \quad 2b^{12} \quad 2c^{12} \quad 2d^{12} \\ 2a^{21} \quad 2b^{21} \quad 2c^{21} \quad 2d^{21} \\ 2a^{22} \quad 2b^{22} \quad 2c^{22} \quad 2d^{22} \end{cases}$$

$$\text{Derivatives of the 3q} \begin{cases} 3a^1 \quad 3b^1 \quad 3c^1 \quad 3d^1 \\ 3a^2 \quad 3b^2 \quad 3c^2 \quad 3d^2 \end{cases}$$

$$\text{Derivatives of the 3Q} \begin{cases} 4q = 4a \quad 4b \quad 4c \quad 4d \\ 4Q = 4A \quad 4B \quad 4C \quad 4D \end{cases}$$

Notice that no two cells share precisely the same code, so exact identification of individual blastomeres and their lineages is always possible.

The Problem of Cell Fates

Tracing the fates of cells through development has been a popular and productive endeavor of embryologists for over a century. Such studies have played a major role in enabling researchers not only to describe development but also to establish homologies among attributes in different animals. Although the cells of embryos eventually become established as functional parts of tissues or organs, there is a great deal of variation in the timing of the establishment of cell fates and in how firmly fixed the fates eventually become. Even in the adult stages of some animals (e.g., sponges) the cells retain the ability to change their structure and function, although under normal conditions they are relatively specialized. Furthermore, many groups of animals have remarkable powers to regenerate lost parts, wherein cells may dedifferentiate and then generate new tissues and organs. But in other cases, cell fates are quite firmly fixed and cells are able only to produce more of their own kind.

By carefully watching the development of any animal, one can predict that certain cells are going to form certain structures. In some cases, cell fates are determined very early during cleavage—as early as the 2- or 4-cell stage. If one experimentally removes a blastomere from the early embryo of such an animal, that embryo will fail to develop normally; the fates of the cells have already become fixed, and the missing cell cannot be replaced. Animals whose cell fates are established very early are said to have **determinate cleavage**. On the other hand, the blastomeres of some animals can be separated at the 2-cell, 4-cell, or even later stages, and each separate cell will develop normally; in such cases the fates of the cells are not fixed until relatively late in development. Such animals are said to have **indeterminate cleavage**. Eggs that undergo determinate cleavage are often called **mosaic ova**, because the fates of regions of undivided cells can be mapped. Eggs that undergo indeterminate cleavage are often called **regulative ova**, in that they can "regulate" to accommodate lost blastomeres.

In any case, formation of an animal's basic body plan is in most cases complete by the time the embryo comprises about 10^4 cells (usually after 1 or 2 days). By this time, all available embryonic material has been apportioned into specific cell groups, or "founder regions." These regions are few in number and are large, each becoming a territory within which still-more intricate developmental patterns unfold. As these zones of undifferentiated tissue are established, the unfolding genetic code drives them to develop into their "preassigned" body tissues, organs, or other structures. Graphic representations of these regions are called **fate maps**.

In the past it has been a general practice to equate mosaic eggs and determinate cleavage with spirally cleaving embryos, and to equate regulative ova and indeterminate cleavage with radially cleaving embryos. However, surprisingly few actual tests for determinacy have been performed, and what evidence is available suggests that there are many exceptions to this generalization. That is, some embryos with spiral cleavage appear indeterminate, and some with radial cleavage appear determinate. Much more work remains to be done on these matters, and for the present the relationships among these features of early development are questionable.

In spite of the variations and exceptions, there is a remarkable underlying consistency in the fates of blastomeres among embryos that develop by typical spiral cleavage. Many examples of these similarities are discussed in later chapters, but we illustrate the point by noting that the germ layers of spirally cleaving embryos tend to arise from the same groups of cells. The first three quartets of micromeres and their derivatives give rise to ectoderm (the outer germ layer), the 4a, 4b, 4c, and 4Q cells to endoderm (the inner germ layer), and the 4d cell to mesoderm (the middle germ layer). Many students of embryology view this uniformity of cell fates as strong evidence that taxa sharing this pattern are related to one another in some fundamental way and that they share a common evolutionary heritage.

Blastula Types

The product of early cleavage is called the **blastula**, which may be defined developmentally as the embryonic form preceding the formation of embryonic germ layers. Several types of blastulae are recognized among invertebrates. A **coeloblastula** frequently results from radial cleavage of ova with relatively small amounts of yolk (Figure 1.6a). This blastula is a hollow ball of cells, the wall of which is usually one cell-layer thick. The space within the sphere of cells is the **blastocoel**, or **primary body cavity**. Spiral cleavage often results in a solid ball of cells called a **stereoblastula** (Figure 1.6b); obviously there is no blastocoel at this stage. Meroblastic cleavage typically results in a cap or disc of cells (the **blastodisc**) at the animal pole above an uncleaved mass of yolk. This arrangement is appropriately termed a **discoblastula** (Figure 1.6c). Some centrolecithal ova undergo odd cleavage patterns to form a **periblastula**, similar in some respects to a coeloblastula, that is centrally filled with noncellular yolk (Figure 1.6d).

Gastrulation and Germ Layer Formation

Through one or more of several methods, the blastula develops toward a multilayered form, a process called **gastrulation** (Figure 1.7). The structure of the blastula dictates to some degree the nature of the process and the form of the resulting embryo, the **gastrula**. Gastrulation is the formation of the embryonic germ layers, the tissues on which all subsequent development eventually depends. In fact, we may view gastrulation as the embryonic analogue of the transition from protozoan to metazoan grades of complexity. It achieves separation of those cells that must interact directly with the environment (i.e., locomotory, sensory, and protective functions) from those that process materials ingested from the environment (i.e., nutritive functions).

The initial inner and outer sheets of cells are the **endoderm** and **ectoderm**, respectively; a third germ layer, the **mesoderm**, is produced in most animals between the ecto-

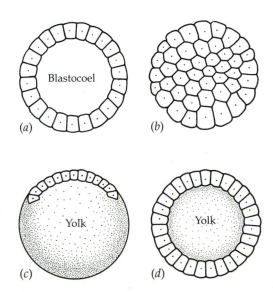

Figure 1.6 Types of blastulae. These diagrams represent sections along the animal–vegetal axis. (a) Coeloblastula. The blastomeres form a hollow sphere with a wall one cell layer thick. (b) Stereoblastula. Cleavage results in a solid ball of blastomeres. (c) Discoblastula. Cleavage has produced a cap of blastomeres that lies at the animal pole, above a solid mass of yolk. (d) Periblastula. Blastomeres form a single cell layer enclosing an inner yolky mass. (From Brusca and Brusca 1990.)

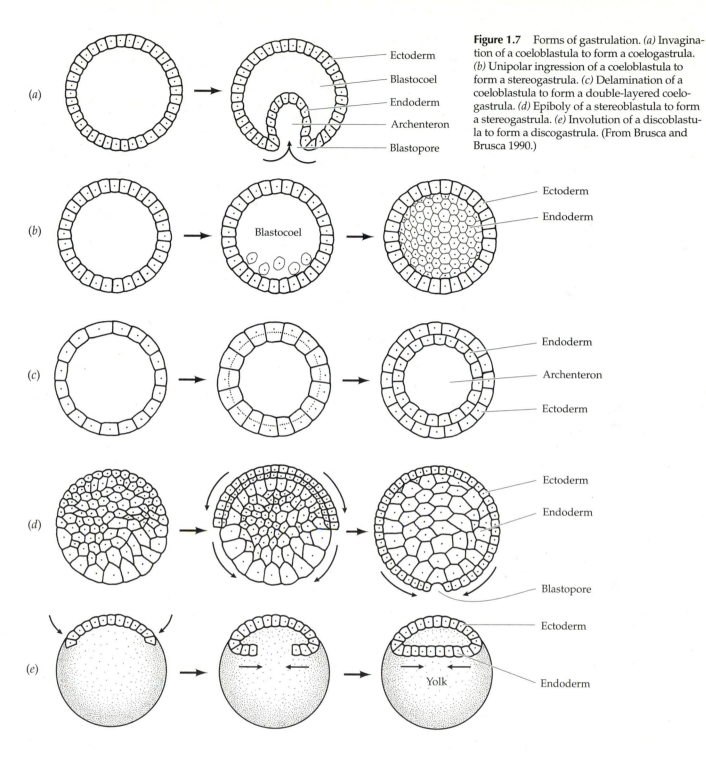

Figure 1.7 Forms of gastrulation. *(a)* Invagination of a coeloblastula to form a coelogastrula. *(b)* Unipolar ingression of a coeloblastula to form a stereogastrula. *(c)* Delamination of a coeloblastula to form a double-layered coelogastrula. *(d)* Epiboly of a stereoblastula to form a stereogastrula. *(e)* Involution of a discoblastula to form a discogastrula. (From Brusca and Brusca 1990.)

derm and the endoderm. One of the principle examples of the unity among the metazoa is the consistency of the fates of these germ layers. For example, ectoderm always forms the nervous system and the outer skin and its derivatives; endoderm the main portion of the gut and associated structures; and mesoderm the coelomic lining, the circulatory system, most of the internal support structures, and the musculature. The process of gastrulation, then, is a critical one in establishing the basic materials and their locations for the building of the whole organism.

Coeloblastulae often gastrulate by **invagination**, a process commonly used to illustrate gastrulation in general zoology classes. The cells in one area of the surface of the blastula (frequently at or near the vegetal pole) grow inward as a sac within the blastocoel (Figure 1.7a). These invaginated cells are now called the endoderm, the sac thus formed is the embryonic gut, or **archenteron**, and the opening to the outside is the **blastopore**. The outer cells are now called ectoderm, and a double-layered, hollow **coelogastrula** has been formed.

The coeloblastulae of many cnidarians undergo gastrulation processes that result in a solid **stereogastrula**. Usually the cells of the blastula divide such that the cleavage planes are perpendicular to the surface of the embryo. Some of the cells detach from the wall and migrate into the blastocoel, eventually filling it with a solid mass of endoderm. This process is called **ingression** (Figure 1.7b) and may occur only at the vegetal pole (unipolar ingression) or over virtually the whole blastula (multipolar ingression). In a few instances (e.g., certain hydroids) the cells of the blastula divide with cleavage planes that are parallel to the surface, a process called **delamination** (Figure 1.7c). Delamination also produces a layer or mass of endoderm surrounded by a layer of ectoderm.

Stereoblastulae that result from holoblastic cleavage generally undergo gastrulation by **epiboly**. Because stereoblastulae have no blastocoel into which the presumptive endoderm can migrate by any of the above methods, gastrulation involves a rapid growth of presumptive ectoderm around the presumptive endoderm (Figure 1.7d). Cells of the animal pole proliferate rapidly, growing down and over the vegetal cells to enclose them as endoderm. The archenteron typically forms secondarily as a space within the endoderm.

Figure 1.7e illustrates gastrulation by **involution**, a process that usually follows the formation of a discoblastula. The cells around the edge of the disc divide rapidly and grow beneath the disc, thus forming a double-layered gastrula with ectoderm on the surface and endoderm below. There are several other types of gastrulation, mostly variations or combinations of the above processes.

Mesoderm and Body Cavities

Some time following gastrulation, a middle layer forms between the ectoderm and the endoderm. This middle layer may be derived from ectoderm, as it is in members of the diploblastic phyla Cnidaria and Ctenophora, or from endoderm, as it is in all of the triploblastic phyla. In the first case the middle layer is said to be **ectomesoderm**, and in the latter case **endomesoderm**, or "true" mesoderm. Thus, the triploblastic condition, by definition, includes entomesoderm. In most texts, the term *mesoderm* in a general sense refers to endomesoderm rather than ectomesoderm.

In diploblastic and certain triploblastic phyla (the acoelomates), the middle layer does not form thin sheets of cells; rather, it produces a fairly solid but loosely organized **mesenchyme** consisting of a gel matrix (the **mesoglea**) that contains various cellular and fibrous inclusions. In a few cases (e.g., the hydrozoans) a virtually noncellular mesoglea lies between the ectoderm and endoderm.

One of the major trends in the evolution of the triploblastic metazoa has been the development of a fluid-filled cavity between the outer body wall and the digestive tube—that is, between the derivatives of the the ectoderm and the endoderm. The evolution of this structural device created a radically new architecture, a tube-within-a-tube design in which the inner tube (the gut and its associated organs) was freed from the constraint of being attached to the outer tube (the body wall) except at the ends. The fluid-filled cavity not only served as a mechanical buffer between these two largely independent tubes, but it also allowed for the development and expansion of new structures within the body, served as storage chambers for various body products (such as gametes), provided a medium for circulation, and was in itself an incipient hydroskeleton.

Three major grades of construction are recognized among the triploblastic metazoa: **acoelomate**, **pseudocoelomate**, and **eucoelomate**. The acoelomate grade (Greek *a*, "without," and *coel*, "cavity") occurs in two triploblastic phyla, Platyhelminthes and Nematoda. In these animals, the mesoderm forms a more or less solid mass of tissue between the gut and the body wall (Figure 1.8a). In most other triploblastic phyla, an actual space develops as a fluid-filled cavity between the body wall and the gut. In many phyla

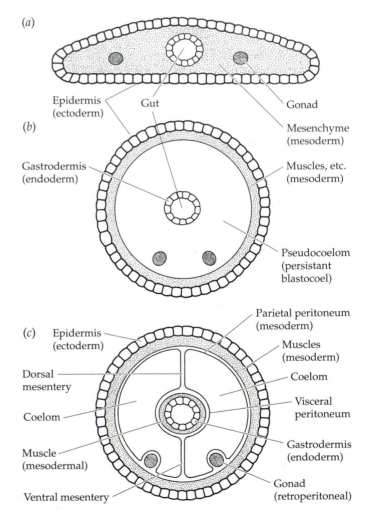

Figure 1.8 Principal body plans of triploblastic metazoa (diagrammatic cross sections). (*a*) Acoelomate. (*b*) Pseudocoelomate. (*c*) Eucoelomate. (From Brusca and Brusca 1990.)

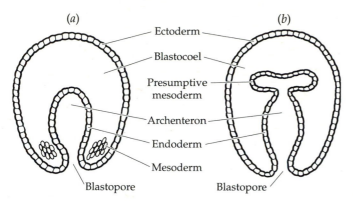

Figure 1.9 Methods of mesoderm formation in late gastrulae (frontal sections). *(a)* Mesoderm formed from derivatives of a mesentoblast. *(b)* Mesoderm formed by archenteric pouching. (From Brusca and Brusca 1990.)

(e.g., annelids, molluscs, and echinoderms), this cavity arises within the mesoderm itself and is completely enclosed within a thin cellular lining called the **peritoneum**, which is derived from the mesoderm. Such a cavity is called a true **coelom**, or eucoelom. Notice that the organs of the body are not actually free within the coelomic space, but are separated from it by the peritoneum (Figure 1.8*c*).

Several phyla of triploblastic metazoa (nematodes, rotifers, and others) possess a body cavity that is neither formed from the mesoderm nor fully lined by peritoneum or any other form of mesodermally derived tissue. Such a cavity is called a **pseudocoelom** (Greek *pseudo*, "false") (Figure 1.8*b*). Unlike the enclosed organs of true coelomates, the organs of pseudocoelomate animals lie free within the body cavity and are bathed directly in its fluid. The embryonic origin of the body spaces in pseudocoelomate animals is unclear. In some cases it appears to be a persistent blastocoel; in other cases its origin is quite unknown.

In eucoelomate animals, mesoderm generally originates in one of two basic ways. In most phyla that undergo spiral cleavage (e.g., annelids, molluscs), a single micromere—the 4d cell, known as the **mesentoblast**—proliferates as mesoderm between the walls of the developing archenteron (endoderm) and the body wall (ectoderm) (Figure 1.9*a*). The other cells of the 4q (the 4a, 4b, and 4c cells) contribute to endoderm. In some other taxa (e.g., echinoderms and chordates), the mesoderm arises from the wall of the archenteron itself (that is, from preformed endoderm), either as a solid sheet or as pouches (Figure 1.9*b*).

The formation and subsequent development of mesoderm is intimately associated with the formation of the body cavity in coelomate metazoa. In those instances where mesoderm has been produced as solid masses derived from a mesentoblast, the body cavity arises through a process called **schizocoely**. Normally in such cases, paired pockets of mesoderm gradually enlarge and become hollow, eventually becoming thin-walled coelomic spaces (Figure 1.10*a,b*). The number of such paired coeloms varies among animal phyla and is frequently associated with segmentation, as it is in annelid worms (Figure 1.10*c*).

The other general method of coelom formation is called **enterocoely**; it accompanies the process of mesoderm formation from the archenteron. In the most direct sort of enterocoely, mesoderm production and coelom formation are one and the same process. Figure 1.11*a* illustrates this process, called **archenteric pouching**. A pouch or pouches form in the gut wall. Each pouch eventually pinches off from the gut as a complete coelomic compartment. The walls of these pouches are defined as mesoderm. In some cases the mesoderm arises from the walls of the archen-

Figure 1.10 Coelom formation by schizocoely (frontal sections). *(a)* Precoelomic conditions with paired packets of mesoderm. *(b)* Hollowing of the mesodermal packets to produce a pair of coelomic spaces. *(c)* Progressive proliferation of serially arranged pairs of coelomic spaces. The process as depicted is characteristic of metameric annelids. (From Brusca and Brusca 1990.)

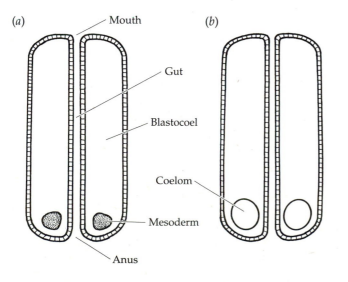

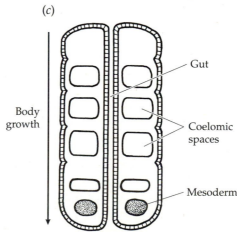

(a)

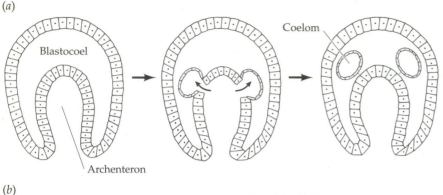

(b)

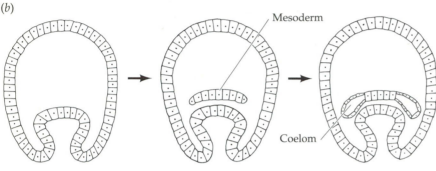

(c)

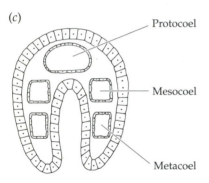

Figure 1.11   Coelom formation by enterocoely (frontal sections). *(a)* Archenteric pouching. *(b)* Proliferation and subsequent hollowing of a plate of mesoderm from the archenteron. *(c)* The typical tripartite arrangement of coeloms in a deuterostome embryo. (From Brusca and Brusca 1990.)

teron as a solid sheet or plate that later becomes bilayered and hollow (Figure 1.11*b*). Some authors consider this process to be a form of schizocoely (because of the "splitting" of the mesodermal plate), but it is in fact a modified form of enterocoely. Enterocoely frequently results in a tripartite arrangement of the body cavities, which are designated **protocoel**, **mesocoel**, and **metacoel** (Figure 1.11*c*).

Body Symmetry

A fundamental aspect of an animal's body plan is its overall shape or geometry. So in order to discuss anatomical architecture and function, we must acquaint ourselves with a basic aspect of body form: symmetry. **Symmetry** refers to the arrangement of body structures relative to some axis of the body. Animals that can be bisected or split along at least one plane, so that the resulting halves are similar to one another, are said to be **symmetrical**. For example, a shrimp can be bisected vertically through its midline, head to tail, to produce right and left halves that

are mirror images of one another. Animals that have no **plane of symmetry** are said to be **asymmetrical**. Many sponges, for example, have an irregular growth form and so lack any clear plane of symmetry.

Most animals possess some kind of symmetry, but within this context a great variety of body design has evolved. The simplest form of symmetry is **spherical symmetry**; it is seen in an animal that assumes the form of a sphere, with its parts arranged concentrically around, or radiating from, a central point (Figure 1.12). A sphere has an infinite number of planes of symmetry that can pass through its center to divide the organism into like halves. Spherical symmetry is rare in nature and, in the strictest sense, is found only in certain protozoa. Creatures with spherical symmetry share an important functional attribute with asymmetrical organisms: both groups lack **polarity**. That is, no clear differentiation along an axis exists, other than from the center of the body toward its surface. In all other forms of symmetry, some level of polarity has been achieved, and with polarity comes specialization of body regions and structures.

A body displaying **radial symmetry** has the general form of a cylinder, with one main axis around which the various body parts are arranged (Figure 1.13). In a body displaying perfect radial symmetry, the body parts are arranged equally around the axis, and any plane of sectioning that passes along that axis results in similar halves (rather like a cake being divided and subdivided into equal halves and quarters). Nearly perfect radial symmetry occurs in the simplest sponges, in many cnidarian polyps, and in oocytes (Figure 1.13). Perfect radial sym-

Figure 1.12 Spherical symmetry in animals. *(a)* In spherical symmetry, any plane passing through the center divides the organism into like halves. *(b)* A radiolarian protozoan. (From Brusca and Brusca 1990.)

(a)

(b)

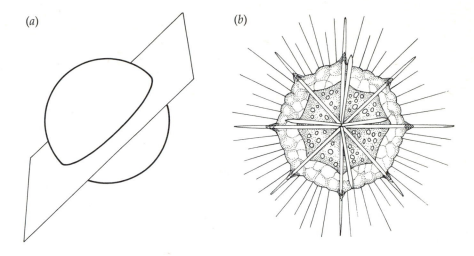

metry is relatively rare, however, and most radially symmetrical animals have evolved modifications on this theme. **Biradial symmetry**, for example, occurs where portions of the body are specialized and only two planes of sectioning can divide the animal into similar halves. Common examples of biradial organisms are sea anemones and ctenophores.

One adaptively significant feature of radial symmetry is that such animals can confront their environment in numerous directions. A radially symmetrical animal has no front or back end; it is organized about an axis that passes through the center of its body, like an axle through a wheel. When a gut is present, this axis passes through the mouth-bearing (**oral**) surface to the opposite (**aboral**) surface. Radial symmetry is most common in sessile species (e.g., sponges and sea anemones) and drifting pelagic species (e.g., jellyfishes and ctenophores). Given these lifestyles, it is clearly advantageous to be able to confront

the environment equally well from a variety of directions. In such creatures one generally finds feeding structures and sensory receptors placed in such a way that they contact the environment more-or-less equally in all directions from the body axis. Furthermore, many fundamentally bilaterally symmetrical animals have become "functionally" radial in certain ways associated with sessile lifestyles. For example, their feeding structures may be in the form of a whorl of radially arranged tentacles, an arrangement allowing more efficient contact with their surroundings.

The body parts of **bilaterally symmetrical** animals are oriented about an axis that passes from the front (**anterior**) to the back (**posterior**) end. There is a single plane of symmetry, passing along the axis of the body to separate right and left sides, the **midsagittal plane** (or median sagittal plane). A longitudinal plane passing perpendicular to the sagittal plane and separating the backside (**dorsal**) from the underside (**ventral**) is called a **frontal plane**. Any plane that cuts across the body, from side to side, is called a **transverse plane** (or, more simply, a cross section) (Figure 1.14). In bilaterally symmetrical animals, the term **lateral** refers to the sides of the body, or to structures away from and to the right and left of the midsagittal plane. The term **medial** refers to the midline of the body, or to structures on or near the midsagittal plane.

Whereas spherical and radial symmetry are typically associated with sessile or drifting animals, bilaterality is generally found in animals with controlled mobility. In these animals, the anterior end of the body confronts the environment first. With the evolution of bilateral symmetry and unidirectional movement, one finds an associated concentration of feeding and sensory structures at the anterior end. The formation of a head end is called **cephal-**

(a)

(b)

Figure 1.13 In radial symmetry, the body parts are arranged radially around a central oral-aboral axis. *(a)* A representation of perfect radial symmetry. *(b)* Sea urchin oocyte. *(a* from Brusca and Brusca 1990.)

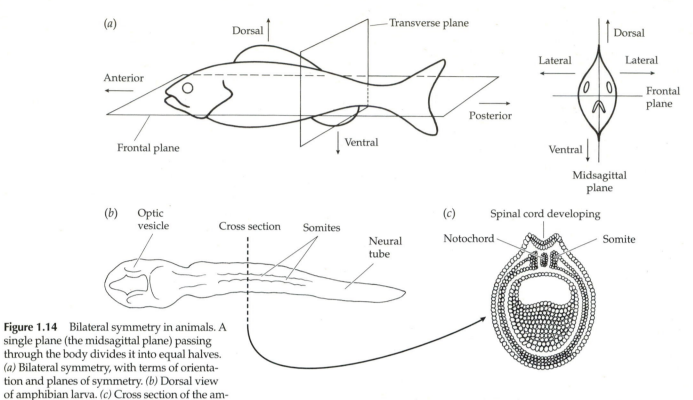

Figure 1.14 Bilateral symmetry in animals. A single plane (the midsagittal plane) passing through the body divides it into equal halves. (a) Bilateral symmetry, with terms of orientation and planes of symmetry. (b) Dorsal view of amphibian larva. (c) Cross section of the amphibian larva at the transverse plane indicated.

ization. Furthermore, with the differentiation of dorsal and ventral surfaces, the ventrum often becomes locomotory and the dorsum often is specialized for protection. A variety of secondary asymmetrical modifications of bilateral (and radial) symmetry have occurred, for example, the spiral coiling of gastropods and hermit crabs. Most asymmetrical modifications result from the displacement of certain body parts because of the habit of being fastened on one side rather than on the ventral surface, or from unequal development of certain paired structures.

Development and Evolution

Homology

Evolution consists of hereditable changes that alter development. When we say that the one-toed horse is derived from a five-toed ancestor, we are saying that changes have occurred in the development of the limb cartilage cells. Some genes involved in chondrocyte growth, placement, or differentiation have been altered. The origin of any new structure or structural modification does not take place in an adult, but by modifications of embryonic or larval development.

Some of the best evidence for evolution is provided by featues that are homologous between embryos of different organisms. **Homologies** are seen when unity of form persists in a diversity of structures that share a common origin—structures inherited from a common ancestor. Be-

cause evolution involves *descent with modification*, we should not necessarily expect structures in related organisms to be identical, only similar. The bones of a human hand, a bat wing, a bird wing, and a seal flipper are all composed of the same fundamental elements, but the elements are arranged differently. The concept of homology is one of the most crucial ideas in biology today. As David Wake (1995) has written:

> Homology is the central concept for *all* biology. Whenever we say that a mammalian hormone is the "same" as a fish hormone, that a human sequence is the "same" as a sequence in a chimp or a mouse, that a HOX gene is the "same" in a mouse, a fruit fly, a frog, and a human—even when we argue that discoveries about a roundworm, a fruit fly, a frog, a mouse, or a chimp have relevance to the human condition—we have made a bold and direct statement about homology.

Homology was originally defined as "the same organ in different animals under every variety of form and function" (Owen, 1843). In recent years, this definition has been defined more carefully. First, modern definitions of homology have extended the levels at which homology may be ascertained. In addition to the organs and skeletal parts recognized by Owen, one can now seek homologies between different organisms by looking at their genes and proteins. Second, modern definitions stress that homologous features can only exist between organisms derived from a common ancestor. Thus, one such definition (Futuyma, 1986) proposes that homology is the possession in two or more species of a trait derived, with or without modification, from the same trait in their common ancestor.

Similar structures need not be homologues. Characteristics are homologous only if they have "continuity of information" from their common origin. Such continuity means that there is a genealogical relationship. Thus, characteristics of two organisms can be homologous only if the same character (or its precursor) was present in their common ancestor. To give a cultural example, one could discuss whether the pyramids of Egypt and Mexico are homologous (van Valen 1982). If they share a common ancestor, they would be considered homologous structures (each diverging from the same ancestral pyramid design). If they do not share a common ancestor, they would be seen an example of **convergence**. Convergent structures have similar characteristics but are not derived from a common ancestor. The bivalved shells of clams and the bivalved shells of brachiopods (lamp shells) evolved independently and are therefore not homologous. Such products of convergence are often referred to as **analogues**. The wing of a mosquito and the wing of a bird are another pair of analogous structures. They perform the same functions and look similar, but their parts are not linked by continuity with a common ancestor.

Distinguishing convergent structures (which evolve independently in different groups to perform similar functions) from homologous structures (which are derived from a common ancestor) is a major task in determining which organisms are closely related. Identifying homologies provides a powerful means of unraveling evolutionary history and classifying organisms.

Synapomorphy and Symplesiomorphy

Synapomorphies are shared homologues inherited from an immediate common ancestor. These "shared derived traits" separate one group from all others. The mammary glands of different species of mammals may differ from one another, but the presence of mammary glands themselves is a synapomorphy that characterizes the mammals as a group distinguishable from all other vertebrates. Mammary glands constitute a shared derived trait. This particular trait did not arise *de novo*. Rather, the mammary gland evolved from other types of glands found in reptiles. Thus, at some deeper level mammals and reptiles also share a homology (the glandular precursor of a mammary gland). Another example is the presence of hair, also a defining trait (synapomorphy) of mammals. Hair is derived from evolutionary modifications of reptilian scales. All mammals have hair, and *no other group* of animals has it.

A related term, **symplesiomorphy**, is used for homologous features inherited from ancestors more remote than the immediate common ancestor. Thus, hair and mammary glands are synapomorphic characters that define the emergence of mammals among vertebrates, but they are symplesiomorphic traits shared by species groups *within* the class Mammalia. Because symplesiomorphies are shared ancestral traits, they tell us nothing about relationships among the groups that share them. Thus, among the vertebrate classes, the mammals are distinguished by the synapomorphies of hair and mammary glands. Similarly, the jawed vertebrates are shown to be more closely related to each other than they are to the agnathans (jawless fishes) because they share the derived trait (synapomorphy) of articulating jaws. However, *within* the class Mammalia, the presence of hair and mammary glands are shared ancestral traits—symplesiomorphies—and are of no use in evaluating relationships among the orders of mammals.

Synapomorphies in embryonic and larval stages provide some of the best evidence for evolutionary descent from a common ancestor. Although the adult barnacle looks nothing like a crab or lobster, the presence of a nauplius larva in barnacle development indicates that barnacles should be grouped with the crustaceans, because the nauplius larva is a unique synapomorphy defining the taxon Crustacea. Similarly, although an adult tunicate doesn't look like any vertebrate (or like anything else on this planet), the presence of a notochord—a synapomorphy that defines the phylum Chordata—in the larvae of both groups shows the affinity of tunicates and vertebrates.

Homology (synapomorphies and symplesiomorphies) can also be seen at the molecular level. Such relationships can be observed in the homologous genes whose protein products control cell division. Even between very divergent species (such as humans and yeasts), these genes share very similar nucleotide sequences. Some of these genes are so similar that the gene for the human protein can replace the homologous gene in yeast and allow the yeast cells to divide. The presence in numerous phyla of a large number of very similar genes involved in cell division strongly suggests that all of these genes were derived from a common ancestor and are thus homologous.

Although these homologous genes are usually common to many phyla, there are often small differences in their nucleotide sequences. These differences may have little or no effect on the mechanics of cell function, but they can be used as synapomorphies to separate the living kingdoms into smaller groups. For instance, if there is a particular base pair change in the gene encoding the small subunit (18S) ribosomal RNA, and that base pair substitution is found only in a certain group of animals, then that group of animals can be hypothesized to be united by that substitution. By cataloging many of these structural and molecular synapomorphies, one can attempt to reconstruct the relationships among the various taxa. One such reconstruction is shown in Figure 1.15.

Researchers can also compare the rates of these nucleotide sequence changes with the fossil record in an attempt to calculate "clocks" of gene evolution. On the basis of protein and gene homologies, one extensive study recently concluded that prokaryotes and eukaryotes diverged about 2 billion years ago, and that animals, plants, and fungi did not diverge from each other until nearly a billion years later (Doolittle et al. 1996). The plants may have diverged first, leaving a common ancestor for animals and fungi

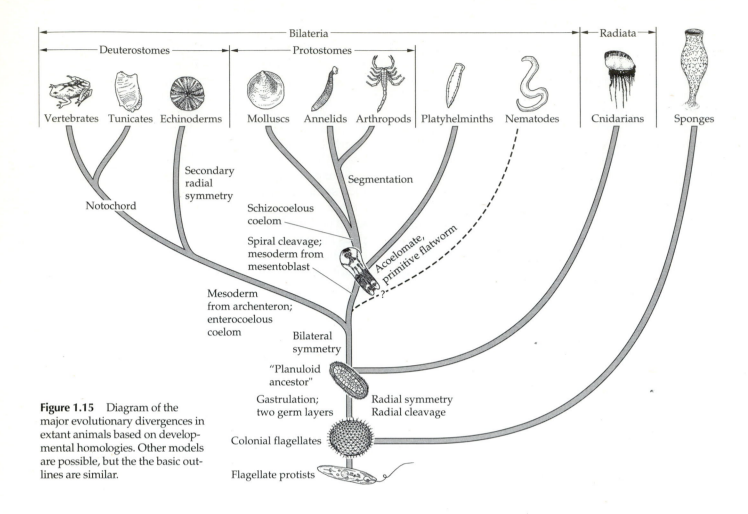

Figure 1.15 Diagram of the major evolutionary divergences in extant animals based on developmental homologies. Other models are possible, but the the basic outlines are similar.

(Figure 1.16a). Most studies have implicated the choanoflagellates as the protists most likely to have shared a common ancestor with the animals. One recently published hypothesis of evolutionary divergence within animals, based on paleontological data (Valentine et al. 1996), is shown in Figure 1.16b. These studies are not without their critics. Assumptions have been made in reading the fossil record, in choosing which changes are important and which are not, in deciding which changes are the result of evolutionary convergence and which show common ancestry, and in choosing the mathematical algorithms that integrate the different sets of data. Still, one can get a sense of the outline of how the major living groups evolved. For more information on how these relationships are detected and depicted, see Brusca and Brusca (1990) and Raff (1996).

Levels of Homology

Any assignment of homology must specify the *level* at which this assignment is being made (see Bolker and Raff 1996). For example, bat wings and bird wings are homologous if they are viewed as tetrapod forelimbs. That is to say, the bony structures of these forelimbs are similar because they are derived from the common ancestor of all tetrapods. However, bird and bat wings are not homolo-

gous when viewed as functional wings. Wings emerged in the birds, and bat wings emerged *independently* in one group of mammals. On the other hand, the gliding wing of the albatross and the paddle-like wing of the penguin are homologous as wings, since both are derived from an ancestral avian winged structure.

The level of homology is of critical concern when assessing the relationship between molecular and structural homologies. For example, the Pax6 protein may be present in every photoreceptor in the animal kingdom. Thus, it has been claimed that, on the molecular level, all photoreceptors share a homology, derived from a primordial photoreceptor containing the Pax6 protein (Quiring et al. 1994). However, on the structural level, the compound eye of the fly, the "inverted" cephalopod eye, and the mammalian eye are definitely analogous structures, each formed in a very different manner and having very different anatomies. Similarly, the fly leg and the chick leg are obviously analogous. The arthropod appendage develops from the ectodermal cells of an imaginal disc that everts during metamorphosis and secretes a cuticular exoskeleton. The chick leg forms from interactions between a particular region of trunk ectoderm with a particular region of mesoderm, which form a cartilagenous endoskeleton

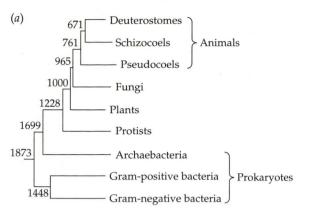

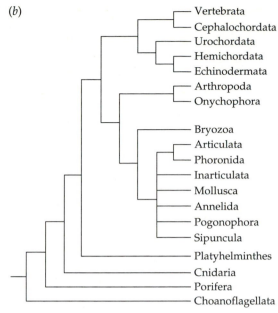

Figure 1.16 Ancestral relationships within the animal kingdom proposed on the basis of molecular homologies. (*a*) Relationship of all life calculated on the basis of 57 different enzyme amino acid sequences shared among organisms. The numbers indicate the proposed time (in millions of years) when these groups separated from one another. (*b*) Patterns of phylogenetic branching inferred by the analysis of 18S ribosomal RNA sequences. (The branches here are not scaled to represent any time intervals.) (*a* from Doolittle et al. 1996; *b* from Valentine et al. 1996.)

that gradually is replaced by bone. The anatomical development and adult anatomy of fly and chick legs have hardly anything in common. However, the chick and the fly apparently use the same sets of genes to specify both the placement of the limbs along the body axis and the polarity of the limb axes. Some key regulatory genes are homologous, whereas the structures are analogous.

Developmental Homologies

Homologies can relate embryology and evolution, and biologists have used embryological homologies in several ways. One way embryological homology can be used is to show the relationships among taxa. As mentioned earlier, synapomorphies (derived homologues) can sometimes be seen in embryonic or larval states more readily than in adults. The larvae of tunicates demonstrate their affinity to vertebrates, and the spirally cleaving embryos of annelids show their affinity to molluscs. Also, adult structures that do not appear to be derived from a common ancestor may be found to have the same embryological origins. The most celebrated of these homologies is that between the middle ear bones of mammals (the stapes, malleus, and incus bones) and the articulation region of the reptilian jaw. Although the adult structures differ enormously, the embryonic development of the mammalian ear bones parallels the fossil record of cartilage changes in the reptilian jaw apparatus.

Homologies also link development and evolution by showing how developmental changes in identical structures can produce evolutionary novelties. One mechanism that produces variety among homologous structures is **allometric growth**. Allometric growth is seen when different

growth rates develop in different parts of the organism. In the male fiddler crab, *Uca pugnax*, the mass of the waving claw increases six times faster than the mass of the rest of its body. Thus, unlike other crabs (and unlike the females of its own species), this animal's claw becomes enormous, although the structure is homologous to the normal claw. Allometric growth is important in distinguishing the different morphologies of ants, giving the workers ("soldiers") larger jaws.

Another way to produce an evolutionary change is by **heterochrony**, the phenomenon wherein animals change the rate of their development or the relative time of appearance of features inherited from their ancestors. For instance, some salamanders have accelerated the production of gonads relative to the rest of the embryo, causing the animal to become a sexually mature adult at a much earlier stage. In *Bolitoglossa occidentalis*, the adult stage is reached while this amphibian still has webbed feet and is very small. Its minute body size and feet that can be used to produce suction enable *B. occidentalis* to climb trees, a niche rarely occupied by salamanders. This species is thought to have evolved from a related species that passes through a juvenile stage similar to the adult stage of *B. occidentalis*.

Another way embryonic homologies have been used is to look at the expression patterns of homologous genes during development. As mentioned earlier, the *Pax6* gene appears to be expressed in the developing eye throughout the animal kingdom. Instead of having developed independently in several groups of organisms, it is possible that all the different types of animal eyes each evolved from a common ancestor that expressed the *Pax6* gene (Quiring et al. 1994). Similarly, a gene called *tinman* (because the mutant lacks a heart) appears to be expressed in heart primordia in both flies and vertebrates (Scott 1994). Not only are these genes homologous, but the patterns of their expression appear to be similar. This level of homology has important implications for the mechanisms of evolution, and it implies powerful conservatism within the animal kingdom.

A variation of this way of using homologies in embryonic gene expression to study evolution is to consider certain pathways as homologous. Here the "characteristic" being compared is not a structure or a molecule, but a biochemical pathway (De Robertis and Sasai 1996; Gilbert et al. 1996). For instance, the neural tube in vertebrates (deuterostomes) is specified by certain proteins blocking the action of a protein called bone morphogenesis protein 4 (BMP4). Interestingly, the same proteins appear to specify the neural tube in flies (protostomes)—also by blocking BMP4. The pathways by which the neural tube is constructed in the ventral portion of the fly embryo and the dorsal portion of the vertebrate embryo appear to be remarkably constant despite the 500 million years of divergence between these two groups. This similarity suggests that there may have been only one original way of making a neural tube, and that it was not invented separately in protostomes and deuterostomes.

Another way homologous genes have been used to examine evolution is in studies that indicate how, in related organisms, homologous structures may take different forms because the expression of homologous genes has been altered during development. For instance, there have been enormous debates about the origin of the amphibian leg from the fish fin. As Richard Owen pointed out, there is considerable homology between the bones of the fish fin and the tetrapod limb, the pectoral and pelvic fins of the fish being homologous to the fore and hindlimbs, respectively. While specific homologies were able to be made between the proximal elements (zeugopod; tibia and fibula) of the fin and limb, homologies between the autopod of the limb (the hand or foot at the distal end) and the rays of the fins "did not hold water." While there seems to be homology for the proximal elements of the limb, the autopod seems to be something new.

Recent studies have strongly suggested that the expression of a particular cluster of genes called the *Hox* genes may be crucial in the change from fin to limb. In the early limb bud of chicks and mice, certain *Hox* genes are expressed only in the posterior end of the limb bud. This is similar to the situation in the zebrafish fin bud (Figure 1.17; Sordino et al. 1995). However, in the development of the foot, there is a second phase, where the expression of these genes changes. Instead of being restricted to the posterior of the limb bud, the *Hox* genes are expressed across the distal mesoderm. This band of expression is coincident with the "digital arch" from which the digits form. These studies show that, whereas the *Hox* gene expression pattern is "homologous" between fish fin and chick leg in the proximal regions, the expression in the late bud distal mesenchyme is new. It also confirms the paleontological-developmental studies of Shubin and Alberch (1986), who proposed that the path of digit formation was not (as previously believed) through the fourth digit (making the fin rays homologous the other digits), but through an arch of distal wrist condensations (metapterygia) that begins posteriorly and turns anteriorly across the distal mesenchyme. Thus, the border of *Hox* gene expression follows the metapterygial axis that Shu-

Figure 1.17 Differences in the expression of the *Hox* gene *Hoxd-11* in fish fin and tetrapod leg. (*a*) Regions of *Hoxd-11* expression in the mouse hindlimb from an early bud stage to a later one. During the later stages, the *Hoxd-11* expression pattern extends across the anterior-posterior border of the progress zone. (*b*) In the zebrafish pectoral fin, *Hoxd-11* expression continues posteriorly, but does not extend anteriorly. (*c, d*) Origin of digits as an evolutionary novelty. (*c*) Diagrammatic representation of a primitive fish fin showing a central axis (black) with rays radiating anteriorly (light gray) and posteriorly (dark gray). (*d*) Current view of autopod formation. The axis originally extends posteriorly, but then curves anteriorly across the metapterygial cartilage. The tibia is considered to branch anteriorly, but the digits are not homologous to any rays. (*a, b* after Sordino et al. 1995; *c, d* after Nelson and Tabin 1995.)

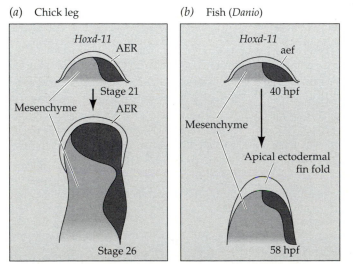
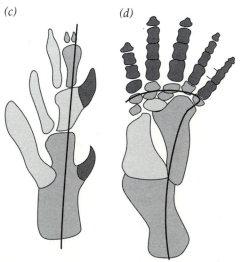

(*a*) Chick leg   (*b*) Fish (*Danio*)   (*c*)   (*d*)

bin and Alberch hypothesized as being the origin of digits. The reoriented, distal, phase of this *Hox* gene expression pattern represents a new and "derived" condition (i.e., a synapomorphy). This, in turn, might have evolved due to changes in the regulation of these *Hox* genes. We are finally coming to a point in biology where changes in gene expression on the molecular level can be linked to evolutionary change at the phylogenetic level.

A deeper understanding of evolutionary processes will emerge as we come to understand how changes in developmental processes create new structures and new combinations of old structures. We will not have this knowledge until we know the developmental processes occurring in the various animal phyla. In 1894, Wilhelm Roux, one of the founders of experimental embryology, predicted that once the physiology of development was known, comparisons between the phyla could delineate the paths by which life evolved on earth. We are finally at the point of being able to fulfill his prediction. But before we can do so, we must learn more about the various ways that embryos have evolved. Fortunately, the study of embryology is among the world's more fascinating subjects. There are few real "crimes against nature," but making embryology boring must count as one of them.

Literature Cited

Bolker, J. A. and R. A. Raff. 1996. Developmental genetics and traditional homology. *BioEssays* 18: 489–494.

Brusca, R. C. and G. J. Brusca. 1990. *Invertebrates*. Sinauer Associates, Sunderland, MA.

De Robertis, E. M. and Y. Sasai. 1996. A common plan for dorsoventral patterning in Bilateria. *Nature* 380: 37–40.

Doolittle, R. F., D.-F. Feng, S. Tsang, G. Cho and E. Little. 1996. Determining divergence times of the major kingdoms of living organisms with a protein clock. *Science* 271: 470–477.

Futuyma, D. J. 1986. *Evolutionary Biology*, 2nd Ed. Sinauer Associates, Sunderland, MA.

Gilbert, S. F. 1997. *Developmental Biology*, 5th Ed. Sinauer Associates, Sunderland, MA.

Gilbert, S. F., J. M. Opitz and R. A. Raff. 1996. Resynthesizing evolutionary and developmental biology. *Dev. Biol.* 173: 357–372.

Nelson, C. E. and C. Tabin. 1995. Footnote on limb evolution. *Nature* 375: 630–631.

Owen, R. 1843. *Lectures on Comparative Anatomy and Physiology of the Invertebrate Animals, delivered at the Royal College of Surgeons in 1843.* Longman, Brown, Green and Longman, London.

Quiring, R., U. Waldorf, U. Kotler and W. J. Gehring. 1994. Homology of the *eyeless* gene of *Drosophila* to the *Small eye* gene of mice and aniridia in humans. *Science* 265: 785–789.

Raff, R. A. 1996. *The Shape of Life: Genes, Development, and the Evolution of Animal Form.* University of Chicago Press, Chicago.

Roth, V. L. 1988. The biological basis of homology. In *Ontogeny and Systematics.* C. J. Humphries (ed.), Columbia University Press, New York, pp. 1–26.

Roux, W. 1894. The problems, method, and scope of developmental mechanics. Trans. W. H. Wheeler. In *Biological Lectures of the Marine Biological Laboratory, Woods Hole,* Ginn, Boston, pp. 149–190.

Scott, M. P. 1994. Intimations of a creature. *Cell* 79: 1121–1124.

Shubin, N. H. and P. Alberch. 1986. A morphogenetic approach to the origin and basic organization of the tetrapod limb. *Evol. Biol.* 20: 319–387

Sordino, P., F. van der Hoeven and D. Duboule. 1995. *Hox* gene expression in teleost fins and the origin of the vertebrate digits. *Nature* 375: 678–681.

Valentine, J. W., D. H. Erwin and D. Jablonski. 1996. Developmental evolution of metazoan body plans: The fossil evidence. *Dev. Biol.* 173: 373–381.

van Valen, L. 1982. Homology and causes. *J. Morphol.* 173: 305–312.

Wilson, E. B. 1892. The cell lineage of *Nereis*. *J. Morphol.* 6: 361–480.

CHAPTER 2

The Concept of Larvae

Paul E. Fell

D EVELOPMENT CONSISTS OF A SERIES OF COORDINATED stages from egg through adult (Figure 2.1). The fertilized egg, or **zygote**, undergoes a series of mitotic divisions called **cleavage**, during which it is converted into a population of many cells. Then, in the process of **gastrulation**, the cells of the embryo become rearranged, giving rise to the basic organization of the body. After organs develop in the embryonic stage, the organism is often not a mature form capable of reproduction. Rather, it is an immature stage of that organism. The immature stage of some animals generally resembles the adult organism, in which case it is called a **juvenile**. The juvenile progressively develops, or matures, into the adult. On the other hand, many animals possess a free-living immature form unlike the adult. Such an immature stage is a **larva**. In some cases there is a series of larval stages in the life history of an animal. The larva undergoes a more or less dramatic transformation into the adult, which frequently involves new organogenesis and sometimes reorganization of the basic body plan. The development of a larva into an adult is called **metamorphosis**.

Since larval forms are unlike adults, they frequently carry out special functions in the life cycles of organisms. Included among these functions are feeding, dispersal, and habitat selection. Some larvae, especially among the insects, become dormant and may be the only form in which a population survives adverse environmental conditions.

Feeding

Some larvae, including sponge *parenchymellas*, some coelenterate *planulae*, and tunicate *tadpoles*, do not feed. They spend only a short time in the plankton before settling and metamorphosing into adults. The larvae of certain other organisms are relatively long-lived and are the only feeding stage in the life history; the adults reproduce and die within a short time following metamorphosis. For example, this situation is found in a number of insects and the brook lamprey. The *caterpillar* of the cecropia silkworm moth eats leaves, increasing in mass 5000-fold during its six-week existence. The *ammocete* larva of the brook lamprey burrows into the sediments of streams and feeds on diatoms. The larval period may last for as long as 6.5 years. In both cases, the larva gives rise to a complex adult with all the nutritive reserves needed for producing gametes and completing reproduction before it dies.

In most cases both the larvae and adults feed, but the diets of these different stages may be very different. For example, the *pluteus* larvae of sea urchins feed on phytoplankton, whereas the adults eat macroalgae (seaweeds) and other larger foods. Although the larva of the sea lamprey, like that of the brook lamprey, feeds on diatoms, the adult is an ectoparasite/predator on other fish. During the course of metamorphosis, the *tadpole* of the leopard frog is transformed from an aquatic herbivore into a semiterrestrial carnivore that preys on a variety of animals, including insects. Such situations eliminate competition between larvae and adults for food and may allow each stage to exploit appropriate seasonally abundant food sources.

Many animals, especially marine invertebrates, produce large numbers of small eggs containing relatively little yolk. An oyster may produce 60 million eggs during a reproductive season, and some sea urchins produce up to 400 million eggs per female each year. The nutrient reserves of the eggs are adequate for the development of very simple feeding larvae, but not for the development of complex larvae or adults. These simple larvae must begin to feed immediately since they rapidly deplete their yolk stores. As they feed, they grow and increase in complexity (Figure 2.2). The more complex late-stage larvae then meta-

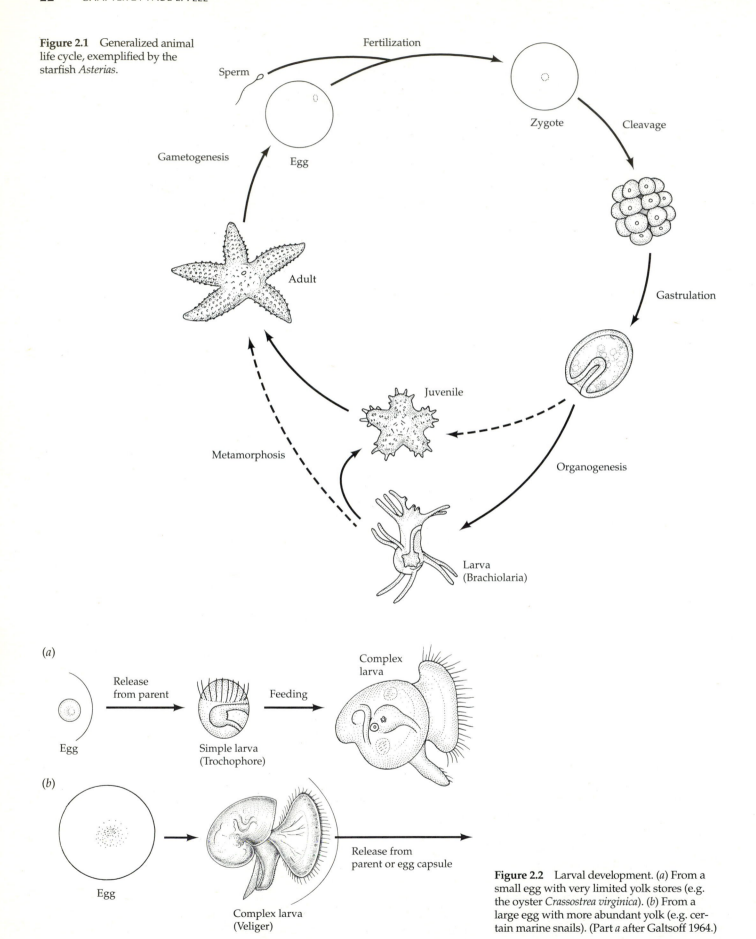

Figure 2.1 Generalized animal life cycle, exemplified by the starfish *Asterias*.

Fertilization

Sperm

Egg

Gametogenesis

Zygote

Cleavage

Gastrulation

Adult

Juvenile

Organogenesis

Metamorphosis

Larva (Brachiolaria)

(a)

Egg

Release from parent

Simple larva (Trochophore)

Feeding

Complex larva

(b)

Egg

Complex larva (Veliger)

Release from parent or egg capsule

Figure 2.2 Larval development. (*a*) From a small egg with very limited yolk stores (e.g. the oyster *Crassostrea virginica*). (*b*) From a large egg with more abundant yolk (e.g. certain marine snails). (Part *a* after Galtsoff 1964.)

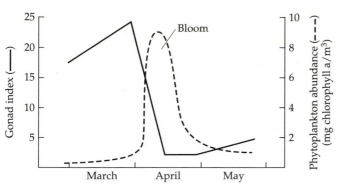

Figure 2.3 Spawning by the green sea urchin *Strongylocentrotus droebachiensis* in relation to the spring phytoplankton bloom. As spawning occurs, the gonad index (wet weight of the gonads expressed as a percentage of the total wet weight of the animal) rapidly declines. (After Himmelman 1975.)

morphose into small juveniles or adults. The eggs and larvae of such animals typically experience a high rate of mortality, often becoming the food of other organisms.

Other animals produce fewer larger eggs that may develop directly into complex larvae which then begin to feed. These larvae may be able to survive for at least short periods when food is scarce or absent. Development up to the feeding stage sometimes takes place within special protective capsules or parental **brood pouches**, and the embryos generally exhibit a lower rate of mortality compared to those developing from very small eggs.

For animals that produce feeding larvae, it is important that the larval period coincides with an abundance of larval food. In aquatic environments, the larval food is often phytoplankton. The spawning of gametes by some animals with plankton-feeding (**planktotrophic**) larvae and the release of such larvae by other animals that brood embryos have been shown to be highly correlated with phytoplankton blooms (Figure 2.3) and to be stimulated by phytoplankton. For example, when sea urchins and mussels, collected prior to their normal spawning period, are maintained in the laboratory without phytoplankton, they either do not spawn or exhibit at most a low incidence of spawning. However, the addition of phytoplankton stimulates spawning by many individuals within a few days. The spawning response depends upon the concentration of phytoplankton. Thus in these instances, phytoplankton directly signals the abundance of food. This situation is advantageous because the occurrence of phytoplankton blooms depends upon a number of interacting factors and cannot be reliably predicted on the basis of any one. The actual spawning inducer apparently is a phenolic compound released by phytoplankton (Starr et al. 1990, 1992).

In barnacles, which brood embryos, first-stage *nauplius* larvae (Figure 2.4) hatch within the mantle cavity of the parent and escape into the surrounding water in response to phytoplankton. However, phytoplankton does not act directly on the larvae. In the spring when the adults re-

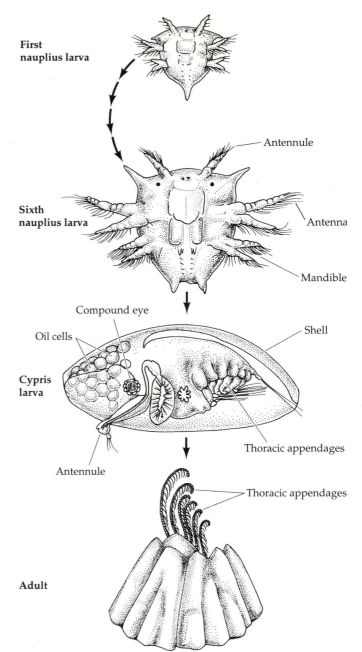

Figure 2.4 Larval stages and metamorphosis in the common rock barnacle *Semibalanus balanoides*. (After Costlow and Bookhout 1957 and Walley 1969.)

sume feeding on plankton, they secrete a fatty acid compound into the mantle cavity. This substance stimulates vigorous muscular activity in the larvae which results in their breaking out of the egg capsules (Barnes 1957; Clare and Walker 1986; Song et al. 1990).

Dispersal and Settlement

Although the adults of some organisms travel over much greater distances than their larvae, the larvae of many ani-

Figure 2.5 Factors influencing patterns of larval dispersal and settlement.

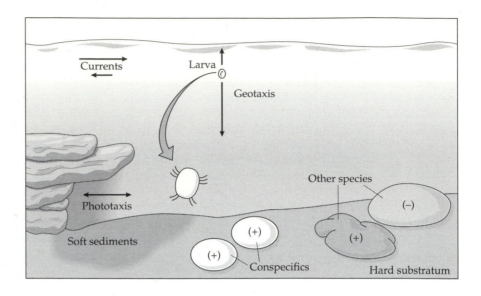

mals play an important role in dispersal. Most aquatic larvae are small and their rates of movement are frequently very slow compared to those of water currents. Therefore, to a large extent, they are passively distributed within the environment. Although this is true, larval behavior may be a significant factor in determining the final distribution pattern. Many larvae exhibit a **phototaxis**, moving toward or away from light. They may also respond positively or negatively to the force of gravity (**geotaxis**) (Figure 2.5). The larvae of many organisms possess eyes, statocysts, and other sense organs. Various stimuli may interact to influence larval behavior, and responses to individual stimuli may change during the larval period (see Crisp 1974; Chia and Rice 1978). For example, some larvae move toward the bottom when they are exposed to light but move upward when they are carried under an object that shields them from the light. Larvae may swim near the surface of the water during early larval life and move along the bottom when they are older.

Behavior, which regulates the position of larvae in the water column, may determine which currents will be used for dispersal. At different depths, the currents may flow with different velocities and/or in different directions. By riding different currents that move in opposite directions, at various times during the larval period, aquatic larvae may be retained within a limited area rather than being dispersed into potentially unfavorable habitats. For example, the larvae of estuarine animals tend to be flushed from an estuary with the outward flowing river water. Certain crab larvae reduce this tendency by rising in the water column during the flooding tide that carries them up river and then descending during the ebbing tide to a position where the seaward currents are relatively slow (Cronin and Forward 1979).

In some other aquatic animals, many aspects of the larval period have been highly modified to minimize the risk of dispersal into unfavorable situations. Freshwater lampsilid mussels provide a striking example. These mussels produce a *glochidium* larva (Figure 2.6a) that develops from the fertilized egg within a brood pouch formed by portions of its mother's gills. Following expulsion from the brood pouch, the glochidium clamps onto the gills of a fish with its two shell valves. The tissues of the fish grow progressively over the glochidium, forming a cyst in which the larva completes development (Figure 2.6b). At the end of the parasitic phase, during which the larva ap-

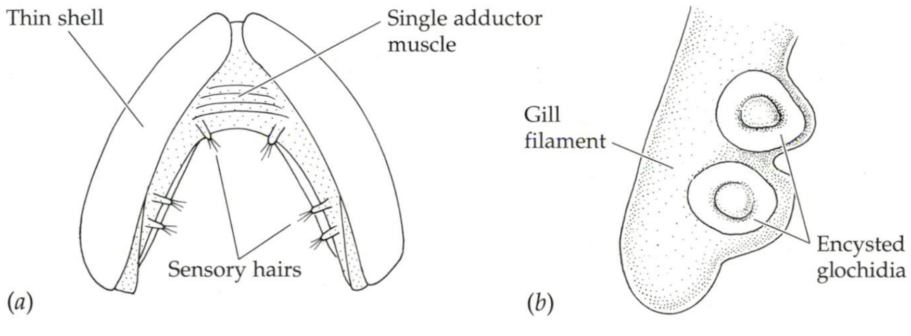

Figure 2.6 Glochidium larvae of a freshwater mussel. (*a*) Free-living larva with gaping shell valves. (*b*) Glochidia encysted in the gills of a fish. (After Coker et al. 1919.)

parently absorbs nutrients from its host, a juvenile mussel breaks out of the cyst and falls to the bottom. If attachment to a fish does not occur, the glochidium dies within a short period of time. Contact between glochidia and fish is promoted by structural and behavioral adaptations of the brooding female mussel. Adult females of some species possess fishlike lures that develop from the posterior mantle folds. These lures exhibit eyelike spots at one end and a rhythmically waving tail at the other. When a predatory fish is attracted to the lure, the female mussel discharges masses of glochidia that are eaten by the fish. Many of the larvae evidently are moved into the gut and digested, but some become attached to the gills of the fish. Such transfer to a fish host prevents the larval mussels from being carried downstream by currents (Welsh 1969).

Other freshwater mussels release glochidia that fall to the bottom of the river or stream. These larvae attach to the fins or other body regions of fish that come into chance contact with them. The glochidia of many such species possess a prominent hook at the free margin of each shell valve, and these facilitate attachment to the fish host (Welsh 1969).

A phenomenon related to larval dispersal is **site selection**. This is especially important for nonmotile and sedentary organisms. After a period of dispersal and development, larvae of numerous species acquire the ability to settle and initiate metamorphosis. Such larvae are said to be **competent**. Competent larvae, which have reached a stage preceding metamorphosis, explore surfaces with which they come in contact. When a larva encounters a favorable substratum, it may settle almost immediately. On the other hand, if a larva fails to locate a suitable site, settling may be delayed or prevented. Such postponement of settling increases the likelihood that a favorable site will be found. For example, the larvae of the red abalone (*Haliotis rufescens*), which have a very specific substratum requirement for settling, become competent for settling at 7 days following fertilization; then, at any time for up to a month, they rapidly settle out of the water column upon making contact with an appropriate substratum. (Also see the discussion later in this section on *Haliotis rufescens* larval settlement.) However, beyond this period, they exhaust their energy stores and are no longer capable of settling (Morse 1991). In selecting a place for settling, larvae may respond to a number of factors including texture, chemical composition, light, and currents.

Barnacles, such as *Semibalanus*, develop a series of six *nauplius* larvae that are characterized by the possession of three pairs of head appendages: antennules, antennae, and mandibles (Figure 2.4). These larvae feed on phytoplankton. The last nauplius molts and gives rise to a non-feeding *cypris* larva that functions in site selection and settlement. The cypris possesses a bivalve shell, compound eyes, antennules, cement glands, and six pairs of thoracic appendages that are used for swimming. When the larva has found a favorable place for settling, it adheres to the substratum by means of its antennules and releases attachment cement. The shell and compound eyes are shed, and the larva metamorphoses into a small adult-like juvenile (Walley 1969).

Barnacle larvae settle gregariously. That is, they settle in clusters. The settlement of a few larvae on a surface greatly increases the chances of other larvae settling there. The protein, arthropodin, present in the cuticle of settled barnacles promotes larval settlement; and this effect is species-specific (Figure 2.7). Slate panels soaked in an aqueous extract of barnacles, containing arthropodin, strongly stimulate cypris larvae to settle in contrast to untreated panels (Crisp and Meadows 1963). Gregarious larval settling is important for these nonmotile organisms, which are cross-fertilizing **contemporaneous hermaphrodites** (adults simultaneously produce both spermatozoa and eggs), because fertilization is internal. An acting male deposits spermatozoa in the mantle cavity of a receptive individual by means of a highly extensible penis. Obviously, if two individuals are not within a certain limited distance of each other, breeding can not take place. Even if the problem of fertilization did not exist, the presence of barnacles on a surface would serve as a cue to the larvae that the site is a favorable one for settling, but such a cue is not infallible.

Larvae of the honeycomb worm *Phragmatopoma* also settle gregariously. These worms construct tubes of sand grains and debris held together by a proteinaceous cement, and large reefs are formed as new recruits settle on the tubes of established adults. The larvae are induced to settle by the cementing protein (Morse 1991).

The larvae of certain species of the annelid *Spirorbis*, which possesses a calcareous tube cemented to a surface, exhibit a high degree of substratum selectivity during settlement. Closely related species of *Spirorbis* occupy differ-

Figure 2.7 Experiment demonstrating gregarious larval settlement in barnacles. (After Knight-Jones 1954.)

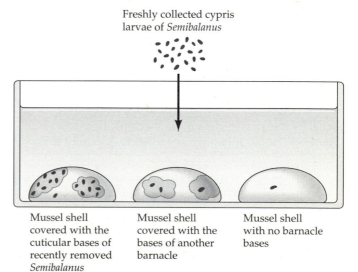

Freshly collected cypris larvae of *Semibalanus*

Mussel shell covered with the cuticular bases of recently removed *Semibalanus*

Mussel shell covered with the bases of another barnacle

Mussel shell with no barnacle bases

ent substrata even when they occur together in the same general area. *Spirorbis borealis* is found on the brown alga *Fucus serratus,* whereas *Spirorbis corallinae* occurs on the red coralline alga, *Corallina officinalis.* A third species, *Spirorbis tridentatus,* is attached to rocks. When the larvae of *S. borealis* are given a choice of settling on *Fucus* and/or *Corallina* in the laboratory, they settle predominantly on *Fucus.* In the same situation, the larvae of *S. corallinae* settle almost exclusively on *Corallina.* The larvae of *S. tridentatus* do not settle on macroalgae but readily colonize stones (de Silva 1962). Exposure of panels to an aqueous extract of *Fucus* greatly increases the number of *S. borealis* larvae settling on them compared to untreated control panels (Williams 1964).

The larvae of a number of animals, besides *Spirorbis corallinae,* are induced to settle by contact with coralline red algae. One of these is the red abalone *Haliotis rufescens.* The larvae of this mollusc respond to a small algal peptide that has properties similar to those of gamma-aminobutyric acid (GABA), a neurotransmitter; and GABA is effective in inducing larval settlement. The newly metamorphosed juveniles eat coralline red algae, but older abalones have a diverse diet (Morse 1991).

In these cases, a substance(s) produced by another organism makes a substratum specifically attractive for larval settlement. In other cases, it appears that a potential settling site may be made specifically unattractive by a similar mechanism. The bryozoan *Bugula* is often overgrown and smothered by the colonial tunicate, *Diplosoma;* and it would be beneficial if the larvae of *Bugula* could avoid settling close to this competitor. Interestingly, *Bugula* larvae can be reversibly prevented from settling by placing them into sea water in which *Diplosoma* has been previously kept (Young and Chia 1981).

Reproduction

Although larvae cannot reproduce sexually (see the discussion of neoteny below), **asexual reproduction** may occur at any stage in the life cycle of an organism, even in the larva. For example, the oceanic larvae of certain starfish produce embryo-like to larva-like **embryoids** that develop from various regions of the larval body (Figure 2.8). These "embryos" detach from the primary larva and take up an independent planktonic existence. Such asexual propagation by larvae may promote long-distance dispersal by currents and increase the probability that some larvae will find a suitable place to settle (Jaeckle 1994).

Larval Structure and Function

Larvae, like all stages in the life history of an organism, are subject to evolutionary change. Over time larvae may become more efficient in their feeding, acquire increased capacity for locomotion, become better adapted to stressful environments, and/or change in a variety of other ways (see Emlet and Ruppert 1994). The larvae of some animals undergo dramatic transformations during metamorphosis; and within certain limits, changes in the larvae may have little direct influence on the adults. For example, in certain aquatic larvae that have a long planktonic life, structures related to swimming, suspension feeding and/or the reduction of sinking velocity may be highly developed and then they are absorbed or discarded at the time of metamorphosis. The *veliger* larva of the mud snail *Ilyanassa obsoleta,* possesses a large bilobed structure associated with its head known as the **velum** (Figure 2.9a). This structure, which is ciliated, functions in feeding and locomotion. During metamorphosis, the velum becomes severed from the rest of the body and the young snail creeps away, leaving this larval organ behind (Scheltema 1962).

In other cases, changes in larval structure/function are carried over to the adult stage. Such changes are selected for or against depending upon how they affect

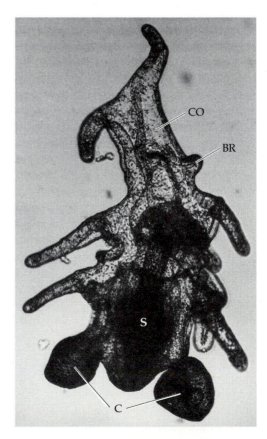

Figure 2.8 Asexual budding by a sea star brachiolaria larva. Note the secondary larvae (C), and the coelom (CO), stomach (S), and brachiolar arm (BR) of the primary larva. (From Jaeckle 1994.)

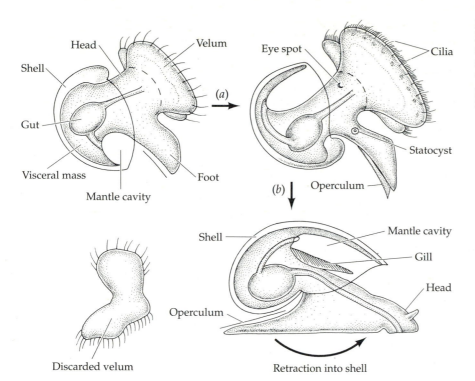

Figure 2.9 Development of a gastropod mollusc. (*a*) Torsion in the veliger larva, shifting the mantle cavity to a dorsal position behind the head. (*b*) Metamorphosis of the advanced veliger larva into a juvenile snail. During this process, the larval velum is shed. (Modified from various sources.)

both the larva and the adult. For example, the caterpillar of the tiger moth *Utethesia ornatrix* feeds on plants containing toxic alkaloids. These alkaloids accumulate within its body and are transferred during metamorphosis to the adult stage, protecting both the larva and the adult from predators. Furthermore, the newly laid eggs of this moth contain alkaloids derived not only from the female adult, but also from male semen. Thus chemicals ingested by the larva protect all other stages of development as well (Dussourd et al. 1988). This defense mechanism of the moth obviously depends upon its resistance to the toxic effects of the alkaloids from the egg through the adult. Another example is torsion in gastropod molluscs (Garstang 1928). In the early veliger larva, the mantle cavity is in a ventral position behind the foot. However, as development continues, the visceral mass is rotated 180° in relation to the head and foot, bringing the mantle cavity into a dorsal position behind the head (Figure 2.9*b*). This rotation is brought about by contraction of an asymmetrically positioned retractor muscle and/or differential growth. At the end of the torsion, the head and velum can be drawn into the shell, followed by the foot, which often bears a horny operculum on its dorsal surface. It has been suggested that torsion is an adaptation for protecting the head of the larva and the adult. Another possible advantage of torsion for the adult is that the osphradium, a chemoreceptor situated in the mantle cavity, is exposed to a current of water coming from in front of the animal.

Neoteny

Neoteny is the phenomenon brought about by the maturation of the reproductive system in a larval form that fails to undergo metamorphosis. As a result, the former adult stage is lost. For example, a number of salamander species are neotenic. These include the mudpuppy *Necturus maculosus*, the Mexican axolotl *Ambystoma mexicanum*, and the Texas salamander *Eurycea neotenes*. Since metamorphosis is normally induced by the thyroid hormones, triiodothyronine and thyroxine, there are several different levels at which mutations could result in neoteny: the hypothalamus, the adenohypophysis, the thyroid gland, and the target tissues of the larva. Depending upon where the block to metamorphosis occurs, some neotenic amphibians can be experimentally stimulated to undergo metamorphosis.

In *Ambystoma mexicanum*, the adenohypophysis apparently does not release thyroid stimulating hormone that promotes the synthesis of thyroid hormones by the thyroid gland. This salamander can be induced to undergo metamorphosis by transplanting the adenohypophysis of another species into it (Blount 1950) or by treatment with thyroid hormones (Huxley 1920; Prahlad and De Lanney 1965). On the other hand, the tissues of *Necturus* are unresponsive to thyroid hormones (Frieden 1981).

The northwestern salamander *Ambystoma gracile* inhabits moist woodlands along the Pacific coast at altitudes ranging from sea level to approximately 3500 m. At sea level most of the larvae undergo metamorphosis at one year of age and become sexually mature one year later. Nearly all of the remaining larvae metamorphose during their second year, but a few become neotenic. However, in montane populations, metamorphosis is rare and most individuals become neotenic (Snyder 1956). Neoteny appears to be advantageous for populations where the aquatic environment is normally more hospitable than the terrestrial one (Wilbur and Collins 1973). Predation, competition and/or climatic conditions may be important factors.

Summary

The necessity of larval forms for feeding, dispersal, and/or site selection has produced a wealth of intricate and beautiful transitory phases in animal life cycles. These stages link the embryo with the adult and have their own anatomical and physiological properties that make them fascinating to study. The larvae of some animals are specialized for feeding and growth; those of others function primarily in dispersal and site selection. However, many larval forms perform all of these functions. In some cases larvae occupy the same general habitat as the adults, whereas in other cases they live in a very different environment. The larvae of some animals may be the only stage in the life cycle to survive seasonally occurring adverse conditions.

Literature Cited

Barnes, H. 1957. Processes of restoration and synchronization in marine ecology. The spring diatom increase and the spawning of the common barnacle, *Balanus balanoides* (L.). *Ann. Biol.* 33: 67–85.

Blount, B. F. 1950. The effects of heteroplastic hypophyseal grafts upon the axolotl, *Ambystoma mexicannum. J. Exp. Zool.* 113: 717–739.

Chia, F.-S. and M. E. Rice (eds.). 1978. *Settlement and Metamorphosis of Marine Invertebrate Larvae.* Elsevier/North-Holland Biomedical Press, New York.

Clare, A. S. and G. Walker. 1986. Further studies on the control of the hatching process in *Balanus balanoides* (L.). *J. Exp. Mar. Biol. Ecol.* 97: 295–304.

Coker, R. E., A. R. Shira, H. W. Clark and A. D. Howard. 1919. Natural history and propagation of fresh-water mussels. *Bull. Bureau Fish.* 37: 75–181.

Costlow, J. D., Jr. and C. G. Bookhout. 1957. Larval development of *Balanus eburneus* in the laboratory. *Biol. Bull. Woods Hole* 112: 313–324.

Crisp, D.J. 1974. Factors influencing the settlement of marine invertebrate larvae. In *Chemoreception in Marine Organisms* P.T. Grant and A.M.Mackie (eds.). pp. 177–265. Academic Press, New York.

Crisp, D. J. and P. S. Meadows. 1963. Absorbed layers; The stimulus to settlement in barnacles. *Proc. R. Soc. London* B 158: 364–387.

Cronin, T. W. and R. B. Forward, Jr. 1979. Tidal vertical migration: An endogenous rhythm in estuarine crab larvae. *Science* 205: 1020–1022.

de Silva, P. H. D. H. 1962. Experiments on choice of substrate by *Spirorbis* larvae. *J. Exp. Biol.* 39: 483–490.

Dussourd, D. E., K. Ubik, C. Harvis, J. Resch, J. Meinwald and T. Eisner. 1988. Biparental defensive endowment of eggs with acquired plant alkaloids in the moth *Utethesia ornatrix. Proc. Natl. Acad. Sci. USA* 85: 5992–5996.

Emlet, R. B. and E. E. Ruppert (eds.). 1994. Symposium: Evolutionary morphology of marine invertebrate larvae and juveniles. *Amer. Zool.* 34: 479–585.

Frieden, E. 1981. The dual role of thyroid hormones in vertebrate development and calorigenesis. In *Metamorphosis: A Problem in Developmental Biology*, 2nd Ed. L.I. Gilbert and E. Frieden (eds.), pp. 545–563. Plenum Press, New York.

Galtsoff, P. S. 1964. The American oyster, *C. virginica* (Gmelin). U.S. Fish and Wildlife Service, Fish Bull. 64: 480 pp.

Garstang, W. 1928. Presidential address to the British Association for the Advancement of Science, Section D. Reprinted in *Larval Forms and Other Zoological Verses*, 1985, pp. 77–98. University of Chicago Press, Chicago.

Himmelman, J. H. 1975. Phytoplankton as a stimulus for spawning in three marine invertebrates. *J. Exp. Mar. Biol. Ecol.* 20: 199–214.

Huxley, J. 1920. Metamorphosis of axolotl caused by thyroid feeding. *Nature* 104: 435.

Jaeckle, W. B. 1994. Multiple modes of asexual reproduction by tropical and subtropical sea star larvae: An unusual adaptation for genet dispersal and survival. *Biol. Bull. Woods Hole* 186: 62–71.

Knight-Jones, E. W. 1954. Laboratory experiments on gregariousness during settling in *Balanus balanoides* and other barnacles. *J. Exp. Biol.* 30: 584–598.

Morse, A. N. L. 1991. How do planktonic larvae know where to settle? *Amer. Sci.* 79: 154–167.

Prahlad, K. V. and L. E. De Lanney. 1965. A study of induced metamorphosis in the axolotl. *J. Exp. Zool.* 160: 137–146.

Scheltema, R. A. 1962. Pelagic larvae of New England intertidal gastropods. I. *Nassarius obsoletus* (Say) and *Nassarius vibex* (Say). *Trans. Amer. Microscop. Soc.* 81: 1–11.

Snyder, R. C. 1956. Comparative features of the life histories of *Ambystoma gracile* (Baird) from populations at low and high altitudes. *Copeia* 41–50.

Song, W.-C., D. L. Holland and E. M. Hill. 1990. The production of eicosanoids with egg-hatching activity in barnacles. *Proc. R. Soc. London* B 241: 9–12.

Starr, M., J. H. Himmelman and J. C. Therriault. 1990. Direct coupling of marine invertebrate spawning with phytoplankton blooms. *Science* 247: 1071–1074.

Starr, M., J. H. Himmelman and J. C. Therriault. 1992. Isolation and properties of a substance from the diatom *Phaeodactylum tricornutum* which induces spawning in the sea urchin *Strongylocentrotus droebachiensis. Mar. Ecol. Prog. Ser.* 79: 275–287.

Walley, L. J. 1969. Studies on the larval structure and metamorphosis of *Balanus balanoides* (L.). *Phil. Trans. R. Soc. London* B 256: 237–280.

Welsh, J. H. 1969. Mussels on the move. *Natl. Hist.* 78: 56–59.

Wilbur, H. M. and J. P. Collins. 1973. Ecological aspects of amphibian metamorphosis. *Science* 182: 1305–1314.

Williams, G. B. 1964. The effect of extracts of *Fucus serratus* in promoting the settlement of larvae of *Spirorbis borealis* [Polychaeta]. *J. Mar. Biol. Assoc. UK* 44: 397–414.

Young, C. M. and F.-S. Chia. 1981. Laboratory evidence for delay of larval settlement in response to a dominant competitor. *Int. J. Inv. Reprod. Dev.* 3: 221–226.

SECTION II

Primitive Multicellular Organisms

CHAPTER 3

Dicyemid Mesozoans

Piroschka Horvath

SPECIES IN THE PHYLUM DICYEMIDA HAVE BODY PLANS THAT are among the simplest of any multicellular organism. This phylum was originally termed Mesozoa because these organisms are intermediate in complexity to the metazoans and the protozoans: they are multicellular, have few cells, no body cavities, and no differentiated organs (van Benéden 1876). Many unrelated groups were added to the phylum Mesozoa but were later removed. More recently, the phylum Mesozoa encompassed only two unrelated groups, class Dicyemida and class Orthonectida. Today, most authorities acknowledge that these two groups are actually distinct and separate phyla, phylum Dicyemida and phylum Orthonectida (Noble et al. 1989).

The life cycle of the Diycemida possesses some features in common with other phyla as well as some unique ones. Features unique to the dicyemid life cycle include: (1) the body plans of the life cycle stages contain fewer cells than any multicellular animal; and (2) there exists a nested cell-within-a-cell arrangement wherein embryos and certain cells develop inside of other cells. Features shared with other groups include: (1) both asexual and sexual reproduction occur during the dicyemid life cycle; and (2) sexual reproduction occurs via self-fertilization.

The Dicyemid Life Cycle: Three Stages

There are three morphologically distinct stages in the dicyemid life cycle: the vermiform stage (Figure 3.1), the infusorigen stage (Figure 3.2), and the infusoriform stage (Figure 3.3). Most of the dicyemid life cycle takes place on the surface of the kidneys of a benthic cephalopod host such as an octopus. Cephalopod kidneys are special structures called renal appendages; they are enclosed inside renal sacs filled with urine. During the cephalopod's early development, the walls of the renal appendages are de-

rived from both the epithelial wall of the vena cava and the epithelium of the renal sac (Schipp and von Boletzky 1975). At the microscopic level, the surfaces of the renal appendages are filled with crypts.

Vermiform Stage

Vermiform dicyemids are especially adapted for surviving on the cephalopod renal appendages (Hochberg 1982). A mature vermiform body consists of one huge, long, and slender axial cell that is totally surrounded by a single external layer of 20–40 ciliated cells; the number of these ciliated cells is constant within a species but varies in different species (Hochberg 1983). Between one and over one hundred cells called axoblasts (Figure 3.4), are engulfed inside of the axial cell (Lapan and Morowitz 1975). Some of the external ciliated cells at the anterior end of the vermiform form a swollen structure called the **calotte**. The heavily ciliated calotte is inserted into a crevice on the surface of the host's kidney and acts as an anchor, while most of the vermiform's long thin body hangs free in the host's urine (Figure 3.5). An external ciliated cell of the vermiform has many convolutions, called **ruffles**, in its cell membrane (Ridley 1968). These ruffles increase the outer surface area of the vermiform, adapting it to absorb nutrients from the urine more efficiently (Bresciani and Fenchel 1965).

The first sign that a cephalopod's kidney is being colonized by a dicyemid species is the presence of the vermiform **stem nematogen** (Lameere 1916; Nouvel 1947; McConnaughey 1951). This vermiform subtype is rare because, although it is the first type of vermiform generated, it is quickly outnumbered by the vermiform embryos it produces. The two types of vermiform dicyemid can be distinguished because the stem nematogen (Figure 3.1c) has two or three axial cells, while the vermiform embryos contain only one axial cell.

An axoblast cell of the vermiform stem nematogen un-

Figure 3.1 The vermiform. (*a*) Differential interference contrast light micrograph of a live unstained vermiform of dicyemid species *Dicyemennea californica* from cephalopod host *Octopus bimaculoides*. Ciliated cells at the swollen anterior end of the vermiform form the calotte. Engulfed within the single giant axial cell of the vermiform are two vermiform embryos (arrowheads) and numerous round single-celled axoblasts. The posterior third of the internal axial cell extends all the way to the end but is not visible because it is not in the plane of focus of this micrograph. Scale bar = 80 μm. (*b*) Scanning electron micrograph of the anterior end of a very young vermiform of a dicyemid species from cephalopod host *Octopus bimaculoides*. The external surface of the vermiform is covered with cilia. The cilia of the swollen calotte are more densely packed than the cilia of the trunk. Scale bar = 8 μm. (*c–f*). Schematic diagrams of the two vermiform subtypes. External ciliated cells are shown stippled; their nuclei have been omitted for clarity. An axial cell (depicted in white) always has its own nucleus. The vermiform subtype that has more than one axial cell is the stem nematogen (*c*). The nematogen phase (*d*) contains axoblasts and vermiform embryos. The transitional phase (*e*) contains axoblasts, vermiform embryos, infusorigens, and infusoriform embryos. The rhombogen phase (*f*) contains only infusorigens and infusoriform embryos. Ab, axoblast; AC, axial cell; an, axial cell nucleus; C, cells of the calotte; CC, central cell of the infusorigen; Ig, infusorigen; IE, infusoriform embryo; pn, paranucleus of the infusorigen; T, trunk; V, an external ciliated cell in the verrucciform state (see text); VE, vermiform embryo. (*a, b*, Photographs by P. Horvath.)

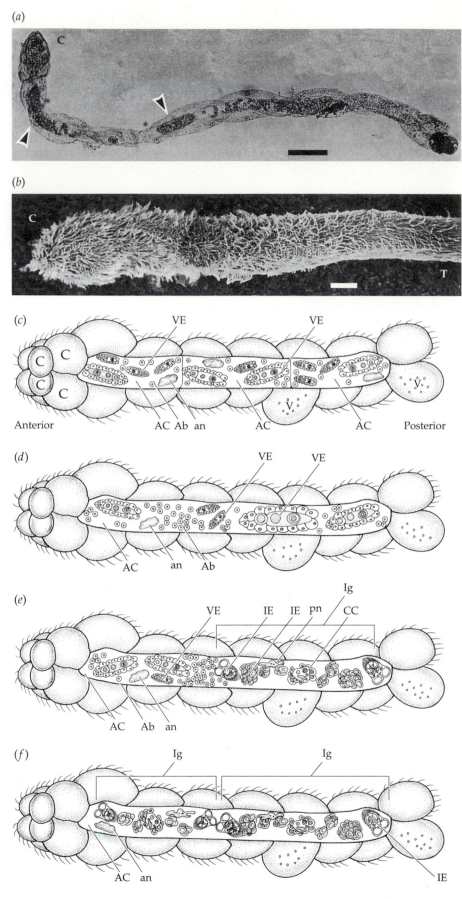

(a)

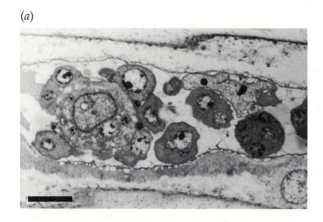

(c)

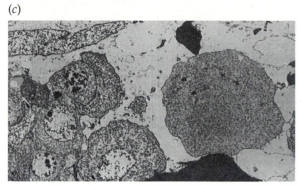

(b)

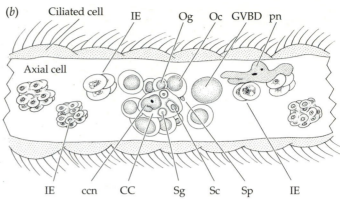

(d)

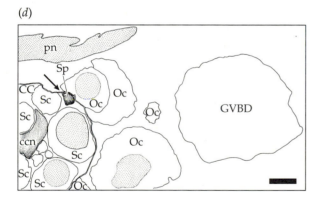

Figure 3.2 The infusorigen. (*a*) Light micrograph of a stained longitudinal 1-μm-thick section through a portion of an infusorigen from dicyemid species *Dicyemennea californica* from cephalopod host *Octopus bimaculoides*. Scale bar = 10 μm. (*b*) Schematic diagram of a longitudinal section through the center of an infusorigen. Note that an oocyte at the point of germinal vesicle breakdown is depicted in the diagram but is not present in the micrograph in (*a*). (*c*) Transmission electron micrograph of a portion of an infusorigen from dicyemid species *Dicyemennea californica* from cephalopod host *Octopus bimaculoides*. (*d*) Tracing of above transmission electron micrograph. Note the small tailless sperm with its dark nucleus that is about to erupt out of the central cell (arrow). Scale bar = 5 μm. CC, central cell; ccn, nucleus of the central cell; GVBD, oocyte at the point of germinal vesicle breakdown; IE, young infusoriform embryo; Oc, oocyte; Og, oogonium; pn, paranucleus; Sc, spermatocyte; Sg, spermatogonium; Sp, sperm. (*a*, photograph by P. Horvath; *c*, photograph by P. L. Dudley and P. Horvath.)

Figure 3.3 The infusoriform. (*a*) Differential interference contrast light micrograph of an unstained whole mount of an infusoriform embryo of a dicyemid species from cephalopod host *Octopus bimaculatus*. This fully developed infusoriform embryo is still inside of the axial cell of a vermiform dicyemid and is almost ready to escape. (*b*) Scanning electron micrograph of an infusoriform of a dicyemid species from cephalopod host *Octopus bimaculoides*. The posterior half of the infusoriform is heavily ciliated, with the cilia tapering to a point. Scale bar = 4 μm. (*c*) Schematic diagram of a sagittal section through an infusoriform. Only a few of the total 39 cells in this infusoriform's body are depicted here. ApC, apical cell; CaC, capsule cell; ci, cilia; EnC, enveloping cell; GC, germinal cell (inside of urn cell); rf, refringent body (inside of apical cell); UC, urn cell; UCn, nucleus of the urn cell. Most dicyemid species have two urn cell nuclei per urn cell. (*a*, photograph by P. Horvath; *b*, photograph by P. L. Dudley.)

(a)

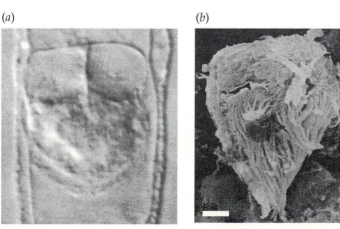

(b)

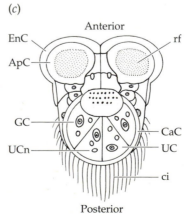

(c)

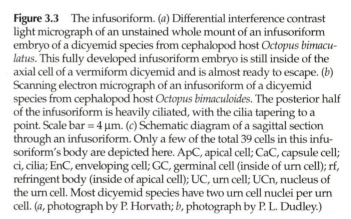

(a)

(b)

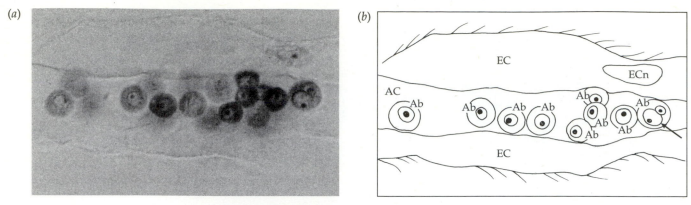

Figure 3.4   The axoblasts of the vermiform. (*a*) Light micrograph of stained axoblasts inside of the axial cell of a vermiform (whole mount). The dicyemid species is from cephalopod host *Octopus bimaculatus*. (Photo by P. Horvath.) (*b*) Tracing of above micrograph. One of the axoblasts is in the process of dividing mitotically (arrow). Ab, axoblast; AC, axial cell; EC, external ciliated cell; ECn, nucleus of an external ciliated cell.

Figure 3.5   The vermiform's environment: the cephalopod kidney surface and renal sac. The cephalopod kidney consists of a segment of the vena cava that passes through a special enclosed area called the renal sac. The portion of the vena cava inside this renal sac has many protuberances arranged in grapelike clusters called renal appendages. Urine and infusoriform dicyemids (not pictured) are expelled through the renal pore, an opening in the renal sac that leads to the mantle cavity of the cephalopod. See text for details. (Renal sac and renal appendages after Schipp and von Boletzky 1975.)

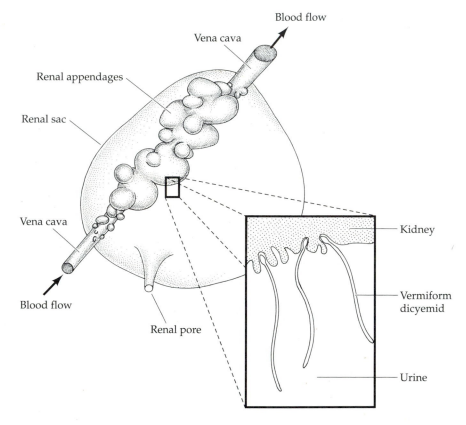

dergoes a series of mitotic divisions (see Figure 3.6*a*1 and Figure 3.7*a*) that give rise to a vermiform embryo with only one axial cell. The cell lineage for the development of the vermiform embryo has been worked out in detail by Furuya et al. (1994, 1996). A newly released young vermiform (Figure 3.6*a*2) swims around in the urine of the renal sac while its single axial cell and external ciliated cells grow in size and its axoblasts multiply (Figure 3.6*a*3). After this period of growth, the vermiform attaches itself to the host cephalopod's kidney, as described above. Within the single axial cell of this newly attached vermiform (Figure 3.6*a*4), the production of vermiform embryos by the axoblasts continues (see Figure 3.6*a*5 and Figure 3.7*a*). This cycle repeats (Figure 3.6*a*6) until the vermiform population within the cephalopod host's renal sac grows very large.

Lapan and Morowitz (1972) demonstrated that when the vermiform population gets very large, an unknown

(*a*) Asexual cycle (*b*) Sexual cycle

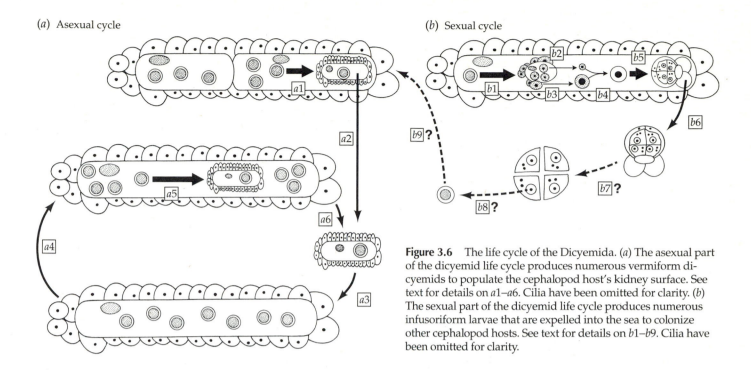

Figure 3.6 The life cycle of the Dicyemida. (*a*) The asexual part of the dicyemid life cycle produces numerous vermiform dicyemids to populate the cephalopod host's kidney surface. See text for details on *a*1–*a*6. Cilia have been omitted for clarity. (*b*) The sexual part of the dicyemid life cycle produces numerous infusoriform larvae that are expelled into the sea to colonize other cephalopod hosts. See text for details on *b*1–*b*9. Cilia have been omitted for clarity.

signal secreted by the vermiform dicyemids into their environment reaches a critical concentration, which induces some axoblasts to follow a developmental path that produces infusorigens rather than vermiform embryos (see Figure 3.6*b*1 and Figure 3.7*b*). Lapan and Morowitz (1975) developed a medium in which vermiforms could survive for up to 3 months *in vitro*. They placed previously uncrowded vermiforms whose axoblasts were developing into vermiform embryos into a test tube so that they became densely crowded at the bottom of the tube. Within 24 hours, the axoblasts of these crowded vermiforms produced infusorigens. These vermiforms were removed from their fluid. This fluid alone was then added to a dish containing uncrowded vermiforms whose axoblasts were only developing into vermiform embryos. Within 24 hours, the added fluid induced axoblasts in these uncrowded vermiform dicyemids to produce infusorigens.

After some axoblasts inside the axial cell of a given vermiform begin the developmental pathway leading to the formation of the infusorigen, the numerous remaining axoblasts pass out of the axial cell. These degenerating axoblasts are absorbed by certain specialized external ciliated cells that have gone into a verruciform state (Lameere 1919; Nouvel 1933; McConnaughey 1951). Verruciform cells (see Figure 3.1*c–f*) are swollen, have lost most of their cilia, and are filled with many types of vesicles, granules, lysosomes, and digestive vacuoles (Matsubara and Dudley 1976).

Infusorigen Stage

The infusorigen is a hermaphroditic reproductive complex that never leaves the vermiform axial cell in which it develops. Infusorigen structure is not strongly symmetri-

cal, the number of cells varies, and there are few cell types. Both oogenesis and spermatogenesis occur in spurts throughout the life of the infusorigen (Austin 1964). During spermatogenesis, male germ cells develop inside the cytoplasm of a large central cell, erupting from the central cell as tailless sperm when mature. In oogenesis, an oogonium mitotically divides to form an oogonium and a primary oocyte, both of which adhere to the central cell's surface. The primary oocyte then rapidly grows very large (McConnaughey 1983). A sperm (Figure 3.6*b*2) attaches to a growing oocyte's surface (Figure 3.6*b*3) and the oocyte detaches itself from the central cell (McConnaughey 1963).

The primary oocyte is only penetrated by the sperm when it has fully grown. Sperm penetration then triggers germinal vesicle breakdown in the oocyte. This is followed by the completion of the meiotic divisions in the oocyte (Austin 1965). Polar bodies are expelled and the two pronucleus stage follows (Furuya et al. 1992). The infusoriform zygotes formed via self-fertilization (Figure 3.6*b*4), and the young 2-cell and 4-cell stage infusoriform embryos are considered part of the infusorigen stage. In addition, the infusorigen stage includes a single giant free nucleus called the **paranucleus** (Whitman 1883), which floats nearby and touches some of the detached oocytes and young infusoriform embryos (see Figure 3.2 and Figure 3.7*b*).

Infusoriform Stage

Furuya et al. (1992) examined the cell lineage of the infusoriform embryo of *Dicyema japonica*. Differentiation of cells in the infusoriform embryo is complete after four to eight rounds of cell division and results in an infusoriform embryo composed of 37 cells. Early cleavages leading to the

Figure 3.7   Cell commitment and the axoblast. An axoblast can follow different developmental pathways, depending on certain conditions (see text for details). (*a*) This axoblast is shown following the developmental pathway leading to the formation of a vermiform embryo of *Dicyema japonica* via a series of mitotic divisions. Dashed lines indicate a single mitotic division. The two black arrows indicate two series of mitotic divisions contributing to the formation of the external ciliated cell layer of the vermiform embryo. The large gray arrow indicates a series of mitotic divisions accompanied by apoptosis that results in the formation of the axial cell and the first axoblast. Cilia have been omitted for clarity. Note that the axial cell of the developing vermiform embryo engulfs the first axoblast. Ab, axoblast; AC, axial cell of the vermiform embryo; acn, nucleus of the axial cell of the vermiform embryo; ECC, external ciliated cell; FAB, first axoblast of the vermiform embryo. (*b*) This axoblast is shown following the developmental pathway leading to the formation of a young hermaphroditic infusorigen. Dashed lines indicate a single mitotic division. Note that the central cell of the developing infusorigen engulfs the progenitor of male germ cells. The arrows represent two series of mitotic divisions, one leading to the production of the male germ cells, the other to the production of the female germ cells. The cytoplasm and cell membrane of the progenitor cell of the paranucleus is shown to disintegrate. The remaining nucleus grows in size without dividing to become the paranucleus. Ab, axoblast; cc, central cell; ccn, central cell nucleus; mci, mother cell of the infusorigen; o, female germ cells; og, oogonium; ogp, progenitor of female germ cells; pn, paranucleus; pnp, progenitor cell of the paranucleus; s, male germ cell; sgp, progenitor of male germ cells. (After Furuya et al. 1996.)

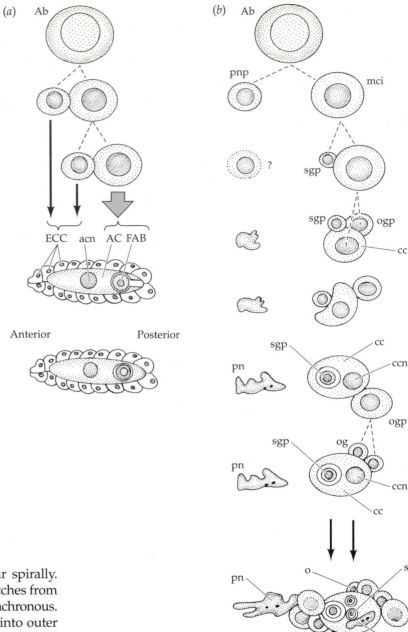

20-cell stage of the infusoriform embryo occur spirally. After the 20-cell stage, the cleavage pattern switches from spiral to bilateral, and the cleavages become asynchronous. The infusoriform embryo is roughly arranged into outer cells and inner cells. The inner cells are derived from the blastomeres of the vegetal hemisphere and differentiate late, while the outer cells are derived from the animal hemisphere and differentiate early. The innermost engulfed germinal cells are an exception in that they are derived from the vegetal pole but differentiate early. Their engulfment by the urn cells is one of the last cellular rearrangements in the development of the infusoriform embryo.

When fully formed, the infusoriform embryo (Figure 3.6*b*5) has external ciliated cells that enclose the posterior portion of its body and two huge anterior apical cells that contain large quantities of magnesium inositol hexaphosphate (Lapan 1975). These heavy cells create a negative buoyancy that aids the infusoriform stage to swim downward after being expelled with the host cephalopod's urine through the host's funnel (Figure 3.6*b*6). Once in the sea, the fate of an infusoriform is unknown. It is possible that

an infusoriform may directly colonize the next benthic cephalopod host. Female octopi are known to aerate their developing young by squirting them with their funnels (Cousteau and Diolé 1973). This would expose the future octopi to freshly expelled infusoriforms.

Each host cephalopod species usually harbors a specific dicyemid species or complex of dicyemid species (Pickford and McConnaughey 1949). This host specificity supports the idea that species of Dicyemida may be directly passed on to a new host by the host's mother. Exactly how the colonization of a new host kidney might be accomplished is not understood. The actual dicyemid vector might be an entire infusoriform dicyemid, urn cells expelled from the infusoriform dicyemid (Figure 3.6b7), or expelled germinal cells (Figure 3.6b8). The germinal cell (Figure 3.8) has been proposed as the direct precursor to the vermiform stem nematogen of the next cephalopod host (Figure 3.6b9), but evidence is scant (Lapan and Morowitz 1975; McConnaughey 1951; McConnaughey 1963; Nouvel 1948; Short and Damian 1966).

Development of Multicellurity

The Dicyemida may be useful in studying the evolution of compartmentation and the evolution of gonads in multicellular organisms. However, where the Dicyemida fit in the phylogenetic scheme is not clear. In the past, parasitologists regarded the Dicyemida as a group of degenerate flatworms (Stunkard 1937). More recently, one phylogenetic tree based on 5S rRNA sequences suggests instead that dicyemids might be "the most ancient group of multicellular animals" (Hori and Osawa 1987). Another phylogenetic tree based on 5S rRNA sequences has suggested that the Dicyemida developed multicellularity independently of the Metazoa, and, furthermore, that the Dicyemida arose from the same stock that gave rise to the green algae and the red algae (Krishnan et al. 1990). Yet another phylogenetic tree, this time based on 18S rDNA sequences (Katayama et al. 1995), places the dicyemids within a monophyletic unit of triploblastic animals (Bilateria). The life cycle of the Dicyemida does include the development of a sexual infusorigen stage that exists exclusively within an asexual vermiform. This situation is similar to that seen in higher plants, where a reduced gametophyte exists within a large sporophyte (McConnaughey 1963). It has also been suggested that the Dicyemida be included in the Protozoa because of the absence of collagenous connective tissue (Cavalier 1993). The Dicyemida may turn out to be allied with metazoans, protozoans, or even plants.

Summary

Whatever the Dicyemida is related to, the development of infusoriform embryos appears to be less complex than that of any multicellular animal (Furuya et al. 1992). One of the themes that emerges from dicyemid development is the sequestration of reproductive elements by compartmentalizing them inside of a single large cell. This may illustrate one strategy by which gonads in metazoans could have evolved.

Literature Cited

Austin, C. R. 1964. Gametogenesis and fertilization in the Mesozoan *Dicyema aegira*. *Parasitol.* 54: 597–600.

Austin, C. R. 1965. *Fertilization*. Prentice-Hall, Englewood Cliffs, NJ.

Bresciani, J. and T. Fenchel. 1965. Studies on dicyemid mesozoa. I. The fine structure of the adult (nematogen and rhombogen stage). *Vidensk. Medd. fra. Dansk naturh. Foren.* 128: 85–92.

Cavalier, S. T. 1993. Kingdom protozoa and its 18 phyla. *Microbiol. Rev.* 57: 953–94.

Cousteau, J.-Y. and P. Diolé. 1973. *Octopus and Squid*. Doubleday & Company, Garden City, New York.

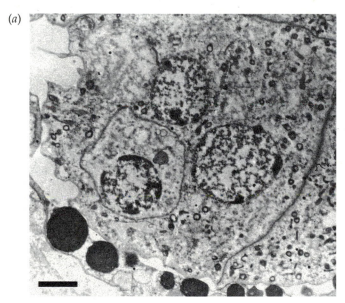

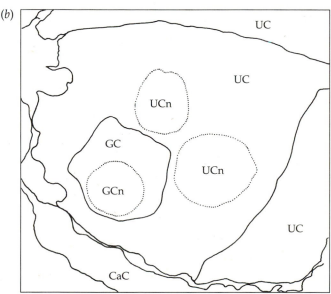

Figure 3.8   The germinal cell of the infusoriform. (*a*) Transmission electron micrograph of the germinal cell from dicyemid species *Dicyemennea californica* from cephalopod host *Octopus bimaculoides*. Scale bar = 5 μm. (*b*) Tracing of the micrograph. Note that the urn cell contains two urn cell nuclei and one engulfed germinal cell. CaC, capsule cell; GC, germinal cell; GCn, germinal cell nucleus; UC, urn cell; UCn, urn cell nucleus. (Photograph by P. L. Dudley and P. Horvath.)

Furuya, H., K. Tsuneki and Y. Koshida. 1992. Development of the infusoriform embryo of *Dicyema japonicum* (Mesozoa: Dicyemidae). *Biol. Bull.* 183: 248–257.

Furuya, H., K. Tsuneki and Y. Koshida. 1994. The development of the vermiform embryos of two mesozoans, *Dicyema acuticephalum* and *Dicyema japonicum*. *Zool. Sci.* 11: 235–246.

Furuya, H., K. Tsuneki and Y. Koshida. 1996. The cell lineages of two types of embryo and a hermaphroditic gonad in dicyemid mesozoans. *Dev. Growth Diff.* 38: 453–463.

Hochberg, F. G. 1982. The "kidneys" of cephalopods: a unique habitat for parasites. *Malacologia* 23: 121–134.

Hochberg, F. G. 1983. The parasites of cephalopods : A review. *Mem. Natl. Mus. Victoria* 44: 109–145.

Hori, H. and S. Osawa. 1987. Origin and evolution of organisms as deduced from 5S ribosomal RNA sequences. *Mol. Biol. Evol.* 4: 445–472.

Katayama, T., H. Wada, H. Furuya, N. Satoh and M. Yamamoto, 1995. Phylogenetic position of the dicyemid mesozoa inferred from 18S rDNA sequences. *Biol. Bull.* 189: 81–90.

Krishnan, S., S. Barnabas and J. Barnabas. 1990. Interrelationships among major protistan groups based on a parsimony network of 5S rRNA sequences. *BioSystems* 24: 135–144.

Lameere, A. 1916. Contributions a la connaissance des Dicyémides, premiére partie. *Bull. Biol. France-Belgique* 50: 1–35.

Lameere, A. 1919. Contributions a la connaissance des Dicyémides, troisiéme partie. *Bull. Biol. France-Belgique* 53: 233–275.

Lapan, E. A. 1975. Magnesium inositol hexaphosphate deposits in mesozoan dispersal larvae. *Exp. Cell. Res.* 94: 277–282.

Lapan, E. A. and H. J. Morowitz. 1972. The mesozoa. *Sci. Amer.* 227: 94–101.

Lapan, E. A. and H. J. Morowitz. 1975. The dicyemid mesozoa as an integrated system for morphogenetic studies. *J. Exp. Zool.* 193: 147–160.

Matsubara, J. A. and P. L. Dudley. 1976. Fine structural studies of the dicyemid mesozoan, *Dicyemennea californica* McConnaughey. I. Adult stages. *J. Parasitol.* 62: 377–389.

McConnaughey, B. H. 1951. The life cycle of the dicyemid mesozoa. *Univ. Calif. Publ. Zool.* 55: 295–335.

McConnaughey, B. H. 1963. The mesozoa. In *The Mesozoa*, E. C. Dougherty (ed.) University of California Press, Berkeley, pp. 151–165.

McConnaughey, B. H. 1983. Mesozoa. In *Reproductive Biology of Invertebrates*, Vol. 1: *Oviposition and Oosorption*, K. G. Adiyodi and R. G. Adiyodi (eds.), pp. 135–145. John Wiley & Sons, New York.

Noble, E. R., G. A. Noble, G. A. Schad and A. J. MacInnes. 1989. In *Parasitology: The Biology of Animal Parasites*, 6th Ed., pp. 432–434. Lea & Febiger, Philadelphia.

Nouvel, H. 1933. Recherches sur la cytologie, la physiologie et la biologie des Dicyémides. *Annales de l'Institute Océanographique, Monaco* 13: 165–255.

Nouvel, H. 1947. Les Dicyémides 1re partie: Systématique, génerations vermiformes, infusorigène et sexualité. *Arch. Biol.* 57: 59–219.

Nouvel, H. 1948. Les Dicyémides 2e partie: Infusoriforme, tératologie, spécificité du parasitisme, affinités. *Arch. Biol.* 59: 147–223.

Pickford, G. E., and B. H. McConnaughey. 1949. The *Octopus bimaculatus* problem: A study in sibling species. *Bull. Bingham Oceoanographic Collection* 12: 1–66.

Ridley, R. K. 1968. Electron microscopic studies on dicyemid Mesozoa. I. Vermiform stages. *J. Parasitol.* 54: 975–998.

Schipp, R. and S. von Boletzky. 1975. Morphology and function of the excretory organs in dibranchiate Cephalopods. *Fortschr. Zool.* 23: 89–110.

Short, R. B. and R. T. Damian. 1966. Morphology of the infusoriform larva of *Dicyema aegira* (Mesozoa: Dicyemidae). *J. Parasitol.* 52: 746–751.

Stunkard, H. W. 1937. The physiology, life-cycles, and phylogeny of the parasitic flatworms. *Amer. Mus. Novit.* 908: 1–27.

van Benéden, É. 1876. Recherches sur les Dicyemides, survivants actuels d'un embranchement des Mésozoaires (suite). II. Rhombogénes. *Bull. Acad. Roy. Belg. A. Sci. Ser. 2*, 42: 35–97.

Whitman, C. O. 1883. A contribution to embryology, life-history, and classification of the dicyemids. *Mittheil. aus der zool. St. zu Neapel* 4: 1–89.

CHAPTER 4

Poriferans, the Sponges

Paul E. Fell

SPONGES (PHYLUM PORIFERA) ARE AMONG THE SIMPLEST multicellular animals. They are solely aquatic and pump environmental water through their bodies. This current of water functions in feeding, respiratory gas exchange, elimination of wastes, and the transport of reproductive elements such as spermatozoa, eggs, embryos, or larvae (depending on the species).

Sponges exhibit different grades of complexity. The least complex sponges, certain calcareous forms, have an elementary tubular morphology (Figure 4.1). The outside of the sponge is covered by a simple squamous epithelium, the **pinacoderm**, interrupted by numerous **porocytes** with pores (**ostia**) that lead to the central cavity, the **spongocoel**. The spongocoel is lined by **choanoderm,** which is constituted by special flagellated cells called **choanocytes**. The beating of the flagella of the choanocytes pulls water into the spongocoel through the ostia and moves it out of the sponge through an opening at the top of the tube known as the **osculum**. Not only do the choanocytes create a flow of water through the sponge, but they also play a major role in removing suspended food particles from the water. The **mesohyl**, situated between the pinacoderm and choanoderm, contains skeletal spicules and various types of amoeboid cells, including the **archaeocytes**. These archaeocytes are called **totipotent** cells since they are capable of regenerating all other cell types of the sponge.

The demosponges are the most common and most highly developed members of the phylum. They possess an elaborate internal water-transport system (Figure 4.2). The **inhalant canals** branch extensively into successively smaller channels that terminate at numerous **choanocyte chambers**. More specifically, water enters the sponge through ostia of porocytes in the **dermal membrane**, empties into **subdermal spaces**, and then flows into the inhalant canals to the choanocyte chambers. **Exhalant canals** carry water from the choanocyte chambers back to the surface of the sponge. As they approach the surface these canals coalesce, forming progressively larger channels that ultimately open at chimney-like oscula. The presence of many choanocyte chambers and canals, together with the subdermal spaces, results in a large total surface area exposed to the medium which, as it moves through the sponge, comes in close proximity to most cells. The external surface of the sponge is delimited by pinacoderm, and pinacoderm lines the subdermal spaces and canals. The mesohyl, as in simpler sponges, contains skeletal elements, amoebocytes, and other cell types. Sponge morphology is highly dynamic. For example, in many sponges the positions of the canals and oscula are continually changing.

The simple tubular form of some calcareous sponges is known as **asconoid**. The much more complex organization of the demosponges and certain calcareous sponges, characterized by numerous small choanocyte chambers and an extensive canal system, is called **leuconoid**. Still other calcareous sponges exhibit an intermediate level of complexity with relatively few large choanocyte chambers. This condition is **syconoid**. Finally, the hexactinellids (glass sponges) possess a syncytial organization and large saccate choanochambers. The divisions of the sponge phylum and their characteristics are outlined in Table 4.1. Many aspects of sponge development remain unknown and represent a fertile area for investigation.

Sexual Reproduction

Gametogenesis

Sponges lack true gonads and special reproductive ducts. In most sponges the reproductive elements are distributed throughout much of the mesohyl (Figure 4.3). Typically oocytes, often surrounded by a flattened follicular epithelium, are individually scattered within the sponge. How-

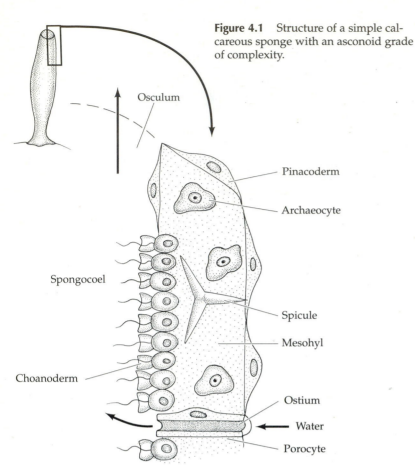

Figure 4.1   Structure of a simple calcareous sponge with an asconoid grade of complexity.

Osculum

Pinacoderm

Archaeocyte

Spongocoel

Spicule

Mesohyl

Choanoderm

Ostium

Water

Porocyte

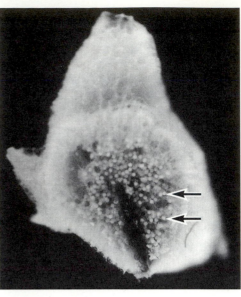

Figure 4.3   Sectioned specimen of *Haliclona loosanoffi* showing large oocytes and embryos (arrows) within the mesohyl.

Figure 4.2   Structure of a demosponge with a leuconoid grade of complexity. The disposition of reproductive elements is shown.

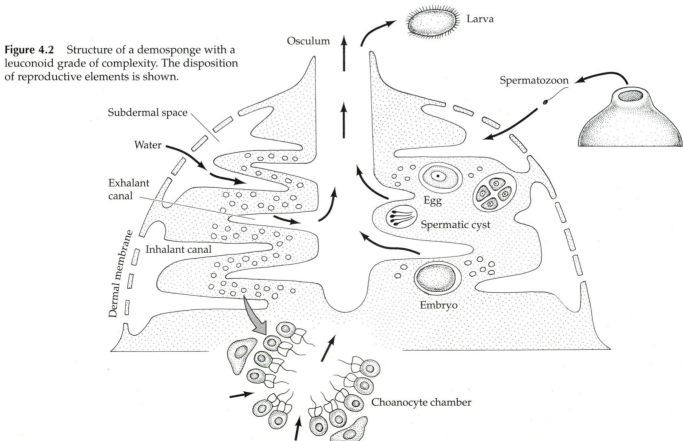

Osculum

Larva

Spermatozoon

Subdermal space

Water

Exhalant canal

Dermal membrane

Inhalant canal

Egg

Spermatic cyst

Embryo

Choanocyte chamber

TABLE 4.1   *Major Taxonomic Groups Within the Phylum Porifera*[a]

Subdivisions	Distinguishing features	Examples
Class Calcarea	Calcium carbonate spicules	
Subclass Calcinea	Coeloblastula larva	*Clathrina, Ascandra*
Subclass Calcaronea	Amphiblastula larva	*Sycon, Grantia*
Class Demospongiae	Skeleton typically formed by siliceous spicules and/or collagen	
Subclass Homoscleromorpha	Incubate embryos; blastular larva	*Octavella*
Subclass Tetractinomorpha	Oviparous; blastular larva in many cases	*Cliona, Suberites, Tetilla, Chondrosia*
Subclass Ceractinomorpha	Typically incubate embryos; parenchymella larva	*Microciona, Haliclona, Halichondria, Spongia, Spongilla, Ephydatia*
Class Hexactinellida	Siliceous spicules: tissues largely syncytial; trichimella larva	*Euplectella*

[a] Some small, poorly studied groups have been omitted.

ever, in a few species, including commercial bath sponges (*Spongia* and *Hippospongia*), oocytes and embryos occur in small, localized clusters. Spermatozoa develop within discrete spermatic cysts, which consist of a mass of spermatogenic cells usually enclosed by a simple squamous epithelium (Figure 4.4).

Sponges exhibit different patterns of sexual differentiation. Many sponges are **contemporaneous hermaphrodites,** simultaneously producing eggs and spermatozoa. In other sponges the sexes appear to be separate. However, in some cases individuals may change their sex over time, yielding first one type of gamete and then the other.

Figure 4.4 Mature spermatic cyst of *Spongilla lacustris*. Note the heads (Sp) and flagella (F) of the spermatozoa and the epithelial cells (CC) enclosing the cyst. (From Paulus 1989).

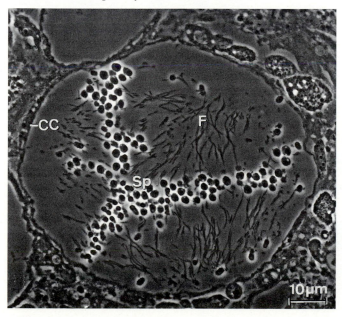

Such animals are **successive hermaphrodites.** In most cases it has not been possible to clearly distinguish between completely separate sexes (**gonochorism**) and successive hermaphroditism. Mechanisms of sex determination in sponges are completely unknown.

Moreover, the cellular origin of sponge gametes has not been firmly established through experimentation. Detailed morphological studies suggest that the gametes of certain sponges may arise from choanocytes, whereas those of others may develop from archaeocytes. In still other species, it appears that oocytes originate from archaeocytes and spermatogenic cells proceed directly from choanocytes.

The earliest recognizable spermatogenic cells, the spermatogonia, may divide mitotically one or more times or may transform directly into primary spermatocytes. In a number of species, the primary spermatocytes already possess a flagellum. These cells undergo the first meiotic division, forming secondary spermatocytes; the latter complete meiosis, producing spermatids. As the spermatocytes divide, they become progressively smaller. Sister-secondary spermatocytes and sister-spermatids frequently are connected by cytoplasmic bridges resulting from incomplete cytokinesis. Such connections are lost as the spermatids differentiate into spermatozoa (Reiswig 1983; Simpson 1984).

Mature spermatozoa possess a small nucleus containing condensed chromatin, one or more mitochondria, and a long flagellum that extends from a pair of centrioles situated near the nucleus. The spermatozoa of *Oscarella* exhibit a definitive **acrosome** (Figure 4.5*a*), and those of *Suberites* possess a few Golgi-derived vesicles positioned anterior to the nucleus (Figure 4.5*b*). However, the spermatozoa of other species completely lack any structure resembling an acrosome (Figure 4.5*c*). The function of the sperm acrosome in those sponges that possess one is unknown. The acrosome in the spermatozoa of many animals contains enzymes, which digest extracellular materials surrounding the egg during fertilization, and also egg-binding protein.

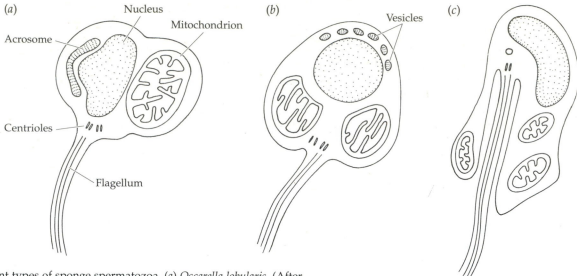

Figure 4.5 Different types of sponge spermatozoa. (*a*) *Oscarella lobularis.* (After Baccetti et al. 1986.) (*b*) *Suberites massa.* (After Diaz and Connes 1980.) (*c*) *Spongilla lacustris.*(After Paulus 1989.)

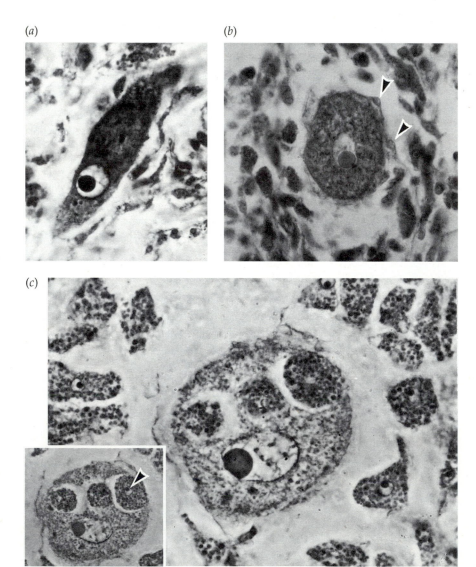

In certain sponge species some of the cytoplasm is discarded during the differentiation of spermatozoa.

In some sponges the process of oogenesis begins with the appearance of small primary oocytes scattered throughout the mesohyl. In other sponges, the primary oocytes are preceded by oogonia, which may undergo one or more mitotic divisions. The smallest oocytes are often amoeboid (Figure 4.6a) but as they enlarge, they typically become more or less spherical in shape. The initial growth of the oocytes usually occurs in the absence of yolk formation; but during the period of major growth, yolk granules of various types accumulate within the cytoplasm. The nucleus of the primary oocytes is large and contains a prominent nucleolus. The oocytes of many sponges become surrounded by a thin follicular epithelium at a relatively early stage (Figure 4.6b), but a delimiting epithelium is absent

Figure 4.6 Oogenesis in *Haliclona ecbasis*. (*a*) Small amoeboid oocyte. (*b*) Oocyte surrounded by follicle cells (arrow heads) and nurse cells. (*c*) Phagocytosis of nurse cells by oocyte. The inset shows an engulfed nurse cell with a persisting nucleus (arrow head) in the same oocyte. (From Fell 1969.)

from the oocytes of calcareous sponges (Fell 1983; Simpson 1984).

In most sponges, the oocytes develop asynchronously within individuals during the reproductive period. However, in many **oviparous** species (i.e., species in which the eggs or embryos are shed from the body), there is a single cycle of synchronized oocyte differentiation each year. Some calcareous sponges (e.g., *Sycon*), which incubate embryos, may initiate annually two or more synchronized cycles of oogenesis and embryonic development.

It appears that much of the yolk accumulated by the developing oocytes of most sponges is derived from **nurse cells**, or **trophocytes**. Such cells are defined here as cells that supply macromolecules and/or organelles to developing oocytes and in some cases to embryos. Unfortunately, for sponges there is little information concerning the nature of the contributions made by the nurse cells, the synthetic activities of the oocytes, or the chemical composition of the yolk. Archaeocyte-like amoebocytes function as nurse cells in many species of sponge; however, choanocyte-derived cells often serve in this capacity in the calcareous sponges. When the oocytes of many species reach a certain size, nurse cells begin to accumulate around them. This aggregation of nurse cells presumably is triggered by a substance(s) produced by the oocyte. Often a large number of nurse cells is associated with each oocyte. In many sponges, the nurse cells infiltrate among the follicular cells and are engulfed by the oocyte (Figure 4.6c). The macromolecules and organelles of the nurse cells are thereby transferred to the developing egg. In other sponges, at least during certain stages of oocyte growth, there appears to be little or no phagocytosis of nurse cells. In such cases the role of the nurse cells is uncertain. A number of observations suggest that the oocytes may take up molecules released by the adjacent nurse cells. This uptake may involve transport across membranes, pinocytosis, and/or possibly other mechanisms. Furthermore, in some sponges, materials may be transported from nurse cells to oocytes by way of cytoplasmic bridges. However, there is no conclusive evidence for these latter processes, which have been proposed solely on the basis of morphological studies (Fell 1983).

Although in many sponges most of the yolk apparently derives more or less directly from nurse cells, this is not the case for certain oviparous sponges with small eggs. There is no evidence of nurse cells in some of these sponges, and the oocytes appear to synthesize at least much of the yolk material. The oocytes possess well-developed rough endoplasmic reticulum and Golgi complexes, which are consistent with yolk synthesis, but such synthesis has not been unequivocally demonstrated by biochemical studies. As the oocytes ingest and/or synthesize yolk materials, yolk granules accumulate within their cytoplasm. Yolk granules often are dispersed throughout the cytoplasm of the larger oocytes. However, a region with little yolk or containing only small yolk granules may surround the nucleus (Figure 4.7).

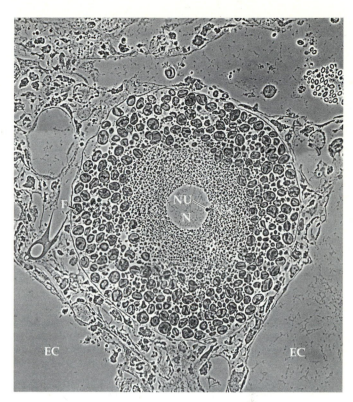

Figure 4.7 Fully developed oocyte of *Spongilla lacustris*. Note the central nucleus (N) with small nucleolus (Nu), the zone of fine granular yolk surrounding the nucleus, and the peripheral yolk spheres. The oocyte is enclosed by a follicular epithelium (F), which is separated by a short distance from the exhalant canals (EC). ×290 (From Saller and Weissenfels 1985.)

The meiotic divisions of the oocytes with the production of polar bodies have been observed in some calcareous sponges and only a few demosponges. The fact that these divisions have not been observed in more species may be due in part to the small size of the chromosomes and the abundance of yolk. Although in a number of sponges the meiotic divisions occur following fertilization, in *Sycon* they apparently precede it.

The ovulated eggs of certain oviparous sponges are surrounded by somatic cells. In some cases, such cells are brought into the interior of the developing embryos and apparently serve a nutritive function.

Spawning and Fertilization

Spermatozoa are released by way of the exhalant canal system and oscula into the surrounding water (Figure 4.8) where they may fertilize similarly shed eggs. However, for many oviparous species, it is not known whether fertilization is internal or external and thus whether eggs or zygotes are released from the sponge. The extruded eggs may be separate from one another or enclosed within a common gelatinous matrix that binds them to the parent sponge and the adjacent substratum. In a majority of sponges, fertilization and embryonic development occur

Figure 4.8 Rapid release of spermatozoa by a 1.5-m-tall specimen of *Verongia archeri* producing a plume of "smoke" 3 m high. (Photograph courtesy of H. M. Reiswig.)

body enlarge. Finally, the carrier cell sinks deeply into the side of the oocyte and transfers the spermiocyst to it. This fertilization sequence is based on an analysis of sections of fixed sponges and should be corroborated using other techniques. Even if the proposed sequence is shown to be essentially correct, the mechanisms by which various parts of the process are carried out need to be elucidated. Among the many unanswered questions are the following: How do the choanocytes distinguish between spermatozoa of the same and different species? Once a choanocyte has engulfed a spermatozoon, what brings about its transformation into a carrier cell? How is the spermiocyst transferred from the carrier cell to the oocyte?

It appears that fertilization usually takes place at the end of oocyte growth and before the meiotic divisions occur, although carrier cells containing spermiocysts may be associated with smaller oocytes. On the other hand, in some sponges fertilization evidently occurs when the oocytes are still relatively small, but the spermiocyst un-

within the egg follicles in the mesohyl. When fertilization is internal, spermatozoa are carried into the vicinity of eggs by the inhalant water current.

In order to accomplish internal fertilization, the spermatozoa must be able to locate the eggs within the mesohyl and penetrate barriers existing between the water transport system and the eggs. Stages of internal fertilization have been observed in only few, primarily calcareous sponges. In these sponges it appears that the spermatozoa are conveyed from the canal system to the eggs by modified choanocytes known as **carrier cells**.

Apparently a choanocyte near the egg engulfs a spermatozoon and, after losing its collar and flagellum, transports the spermatozoon to the egg (Figure 4.9). The spermatozoon, contained within a membrane-bound vesicle in the cytoplasm of the carrier cell, changes into a **spermiocyst**. Its flagellum is lost and its head and mitochondrial

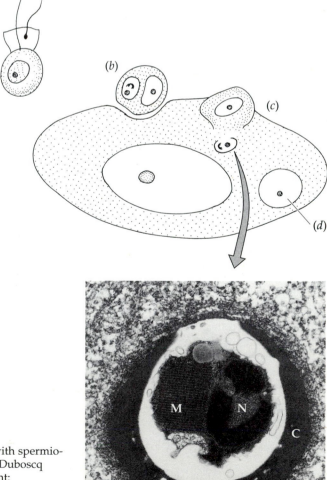

Figure 4.9 Internal fertilization in a calcareous sponge such as *Sycon*. (*a*) Spermatozoan making contact with a choanocyte. (*b*) Carrier cell with spermiocyst. (*c*) Transfer of spermiocyst to oocyte. (*d*) Male pronucleus. (After Duboscq and Tuzet 1937; Franzen 1988; Gallissian 1989, and others.) Enlargement: Spermiocyst enclosed by electron dense material (C) within the oocyte cytoplasm of *Sycon sycandra*. Note the sperm nucleus (N) and the modified mitochondrion (M). ×26,000 (From Gallissian 1989.)

(a)

(b)

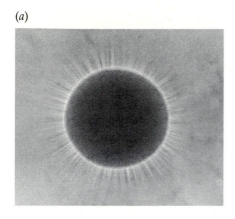
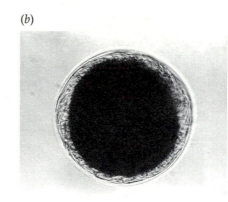

Figure 4.10   The spawned egg of *Tetilla japonica*. (*a*) Unfertilized egg with radiating collagen fiber bundles. (*b*) Egg 10 minutes following fertilization, with fiber bundles retracted into the perivitelline space. (From Watanabe and Masuda 1990.)

dergoes little change within the cytoplasm of the oocyte until the oocyte attains its full size and completes meiosis.

After the meiotic divisions have taken place, the nucleus of the fertilizing spermatozoon swells, becoming the male pronucleus. At the same time, the female pronucleus forms and increases in size. Then the two pronuclei meet near the center of the egg and fuse.

Egg surface changes at fertilization have been examined in only one oviparous sponge, *Tetilla*. The spawned eggs are surrounded by a clear **vitelline membrane** through which numerous radiating collagen fiber bundles extend (Figure 4.10*a*). Following fertilization, the vitelline membrane lifts off the surface of the egg, beginning at the point of sperm entry, and forms the **fertilization membrane**. Membrane elevation is complete after about one minute. At first, the long fiber bundles project through the fertilization membrane, but within 10 minutes they are completely retracted into the perivitelline space (Figure 4.10*b*). Then the fertilized egg adheres to the substratum and continues its development into a small sponge.

Embryonic and Larval Development

In most sponges, cleavage evidently begins soon after fertilization and the production of the zygote. However, the zygotes of commercial bath sponges (*Hippospongia* and *Spongia*) undergo substantial growth, from approximately 50 μm to about 300 μm in diameter, before the cleavage divisions are initiated. Each zygote is surrounded by two to four layers of nurse cells. Some of the nurse cells apparently transfer materials to the growing zygote by means of cytoplasmic bridges; others are phagocytized by the zygote.

Cleavage is total (**holoblastic**) in all sponges in which this process has been observed (Figure 4.11); however, the

pattern of cleavage differs among various species. In many sponges the blastomeres of the early embryo are equal or nearly equal in size, whereas in other sponges they are very unequal. During the early cleavage stages, blastomeres may possess more than one nucleus. In such cases cytoplasmic division lags behind nuclear division. The early developmental stages of some sponges are difficult to interpret. Oocytes, advanced embryos and larvae are readily distinguished, but a cleavage sequence is not apparent. This may be due to the presence of large quantities of yolk that obscure cytological details (Fell 1989).

In the calcareous sponges and a few demosponges cleavage results in the formation of a hollow ball of cells known as a **coeloblastula**. Calcareous sponges of the subclass Calcinea (e.g., *Clathrina* and *Ascandra*) produce simple, flagellated **coeloblastular larvae** that are transformed into solid **stereoblastulae** during their free-swimming existence. At first, a few cells at the posterior pole separate from the blastular wall and enter the spacious **blastocoel** (the cavity of the coeloblastula) (Figure 4.12; Top). These cells lose their flagellum. Later there is a much more extensive ingression of cells that obliterates the blastocoel.

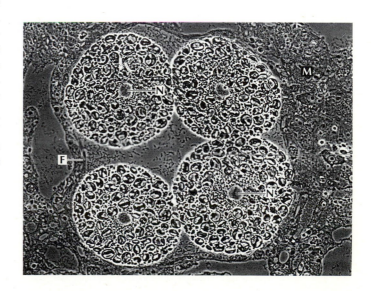

Figure 4.11   Early cleavage stage of *Spongilla lacustris*. Note the nuclei (N) of the blastomeres, the follicular epithelium (F), and the adjacent mesohyl (M). ×285 (From Saller and Weissenfels 1985.)

(a)

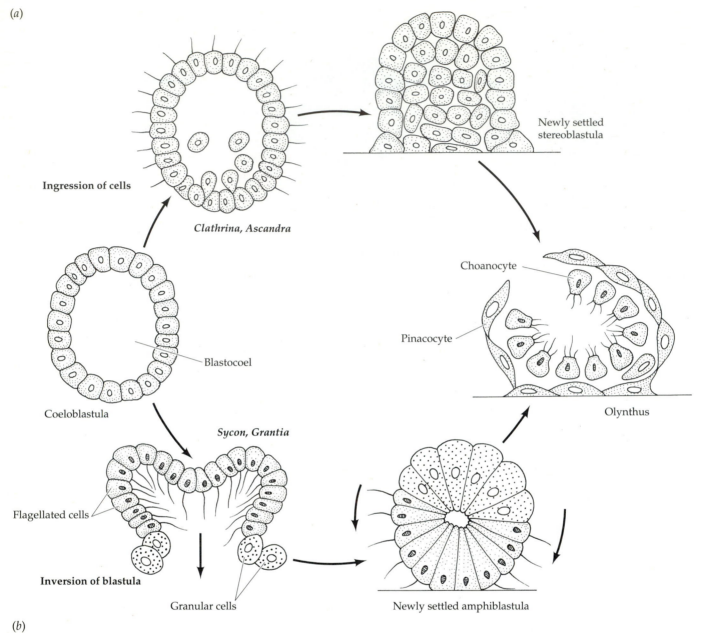

Ingression of cells

Clathrina, Ascandra

Newly settled stereoblastula

Blastocoel

Coeloblastula

Sycon, Grantia

Choanocyte

Pinacocyte

Olynthus

Flagellated cells

Inversion of blastula

Granular cells

Newly settled amphiblastula

(b)

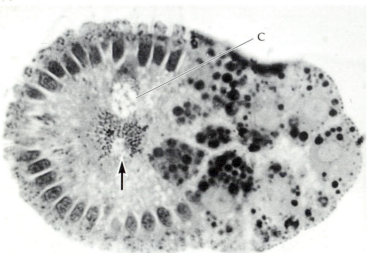

Figure 4.12 Larval development and metamorphosis in calcareous sponges. (a) Top left: Ingression of cells into the blastocoel of a coeloblastula (left center), forming a stereoblastula in sponges such as *Clathrina reticulum* and *Ascandra falcata* (After Borojević 1969). Bottom left: Inversion of the blastula and development of the amphiblastula larva in sponges such as *Grantia compressa, Sycon raphanus,* and *Scypha ciliata* (After Duboscq and Tuzet 1937; Franzen 1988; and others). Following settlement the larva develops into a small sponge, the olynthus (right center). (b) The amphiblastula larva of *Leucandra abratsbo.* The flagellated columnar cells are on the left and the large granular cells are on the right. Note the small central cavity (arrow) and a cruciform cell (C). (From Amano and Hori 1992.)

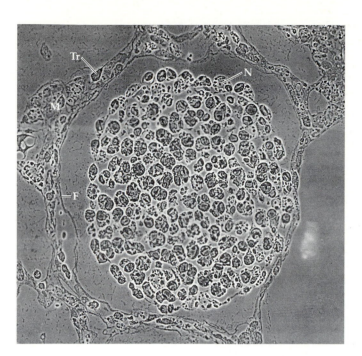

Figure 4.13 Late cleavage stage embryo of *Spongilla lacustris*. Note the nuclei (N) of the blastomeres, granular yolk within these cells, the follicular epithelium (F) and a trophocyte (Tr) within the mesohyl (M). ×260 (From Saller and Weissenfels 1985.)

This ingression is initiated at the equator of the larva and progresses first posteriorly and later anteriorly.

Calcareous sponges of the subclass Calcaronea (e.g., *Sycon* and *Grantia*) possess highly developed **amphiblastula larvae**. Two types of cells appear during the development of the blastula: flagellated columnar cells with small nuclei and rounded granular cells with large nuclei. The more numerous columnar cells constitute most of the blastular wall and their flagella are directed into the blastocoel. The larger granular cells occupy the pole of the embryo that is in contact with the parental choanoderm (Figure 4.12*a*; Bottom). As development continues, an opening appears among the granular cells, and the blastula turns inside out. This inversion results in the formation of the amphiblastula with external flagella and a small central cavity. During the course of inversion, the larva moves from the mesohyl into the adjacent choanocyte chamber.

The fully developed amphiblastula (Figure 4.12*b*) possesses two types of cells in addition to the flagellated columnar cells and the granular cells. Four cruciform cells, one in each quadrant near the equator of the larva, apparently differentiate from columnar cells in the blastula. Their function is unknown, but they may be secretory or perhaps photoreceptive. The central cavity of the larva contains a small number of yolk cells (nurse cells) of maternal origin. Symbiotic bacteria may also be present in the central cavity.

In most demosponges, cleavage produces a solid stereoblastula (Figure 4.13) that develops into a more or less

highly differentiated **parenchymella (parenchymula)** larva. During larval development, cells of different sizes and characteristics appear at stages that vary according to species. Typically, small cells accumulate at the surface of the developing larva and form a flagellated epithelium in which each cell possesses a single flagellum. In some species, the entire surface of the parenchymella is flagellated; in others, the posterior end or both the posterior and anterior ends of the larva are bare. The parenchymellas of certain sponges are characterized by a band of especially long flagella surrounding the bare or flagellated posterior end (Figure 4.14*a*). In some cases, vesicular cells are scattered within the flagellated epithelium and may be concentrated at the anterior pole of the larva. The function of these cells is unknown, but they possibly aid in larval attachment to the substratum at the time of metamorphosis.

The center of the parenchymella is occupied by various types of amoebocytes and frequently other kinds of cells that are characteristic of the adult sponge. Many parenchymellas carry a bundle of siliceous spicules that is often situated in a median to posterior position (Figure 4.14*b*). In the larvae of some species, the spicules are produced by scleroblasts within the central mass, whereas in others the spicules are first produced at the periphery of late cleavage-stage embryos and are only later brought into the interior of the developing larva. Fibrils of collagen may also be found within the central mass. The parenchymellas of a number of sponges possess choanocyte chambers and those of certain freshwater sponges (*Spongilla* and *Ephydatia*) also exhibit pinacocyte-lined cavities that are rudiments of the canal system. In these sponges, the larvae may possess most, if not all, of the cells types of the adult sponge, with the exception of reproductive elements. The parenchymella larvae of some species also contain symbiotic bacteria.

The parenchymellas of the freshwater sponges are characterized by the possession of a large anterior cavity delimited by pinacoderm. The cavity first appears as a small pinacocyte-lined space situated in the anterior hemisphere of the early larva. It then expands to half the volume of the larva.

Although most demosponges produce parenchymella larvae, some demosponges possess blastular larvae that develop from either a coeloblastula or a stereoblastula, and a few lack a free-living larval stage. In the amphiblastula larva of *Chondrosia reniformis*, the cavity is filled by a large number of maternal nurse cells. *Tetilla japonica* is an oviparous sponge with direct development. The zygote adheres to the substratum and develops into an adult sponge.

The hexactinellids (glass sponges) possess highly differentiated **trichimella larvae** (Figure 4.15). A broad median zone of the solid larva bears flagella, but the anterior and posterior poles do not. The peripheral cells of the median zone are multiflagellated and are covered by a thin pinacoderm through which the flagella project. Internally, well developed choanochambers are present in the poste-

(a)

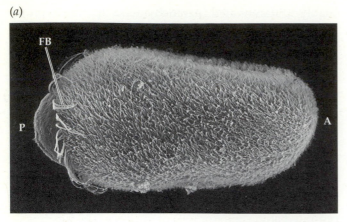

(b)

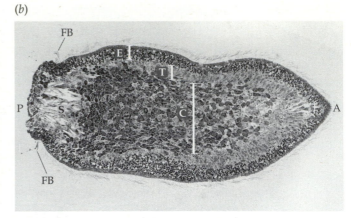

Figure 4.14   Parenchymella larva of *Haliclona tubifera*. (*a*) Scanning electron micrograph showing the pattern of flagellation. The posterior pole (P), which lacks flagella, is surrounded by a band of very long flagella (FB) ×295. (*b*) Longitudinal section of the larva showing the band of long flagella (FB) that surrounds the posterior pole (P), the posterior bundle of spicules (S), the flagellated pseudostratified columnar epithelium (E), the subepithelial cell layer (T) and the central cellular mass (C). ×210 (From Woollacott 1993.)

rior hemisphere and large lipid granules occupy the anterior pole. Siliceous spicules that are distinctly different from the spicules of the adult sponge constitute the larval skeleton. (Boury-Esnault and Vacelet 1994).

The development of many sponges has been described in varying detail from a morphological perspective, but very little is known about developmental mechanisms. For example, there is almost no information concerning the times/stages at which cells become committed to par-

Figure 4.15   Trichimella larva of the hexactinellid sponge, *Oopsacas minuta*. (After Boury-Esnault and Vacelet 1994.)

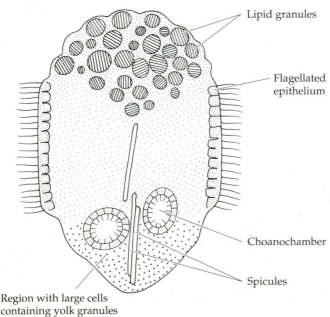

Lipid granules

Flagellated epithelium

Choanochamber

Spicules

Region with large cells containing yolk granules

ticular courses of differentiation or how the positioning of cells within the embryo may influence their development. Furthermore, the extent to which various types of differentiated cells, given appropriate conditions, are able to embark on new paths of differentiation has not been determined. Clearly, there is a need for experimental studies.

As noted previously, most sponges exhibit internal development. The eggs of many such sponges contain abundant nutrient stores and there appears to be little, if any, transfer of materials from the parent sponge to the developing embryos. For most sponges, incubation of embryos and larvae seems to serve primarily a protective function, reducing the probability of predation, physical damage, and other hazards. However, in some sponges with internal development, the embryos apparently ingest parental cells and/or cellular products as sources of nutrition. This phenomenon is particularly striking in many calcareous sponges. Similarly, a number of oviparous sponges release eggs or zygotes that are surrounded by somatic cells. These cells are subsequently taken into the embryo.

Free-Living Larvae and Metamorphosis

Sponge larvae are released from the parent sponge by way of the exhalant canal system and oscula. Larval release by a number of species takes place during the morning in response to light signals.

Sponge larvae are nonfeeding and generally short-lived. The larvae of most sponges are planktonic. However, the larvae of a few species are benthic and creep over the substratum. Planktonic larvae also move over the substratum immediately prior to settlement. During this exploratory phase, creeping and swimming may alternate. Larval attachment and the beginning of metamorphosis typically occur within a few hours to several days following the release of the larva, but in some cases these processes may be delayed for two or more weeks. The larvae of many species exhibit responses to gravity and/or light that appear to be important for dispersal and for determining where they will ultimately settle.

The various types of sponge larvae differ from one another not only morphologically but apparently also in the degree of differentiation of their component cells. Consequently, development into an adult sponge involves a different set of processes for each kind of larva.

The coeloblastular larva of certain calcareous sponges (subclass Calcinea) undergoes dramatic changes during its free-living existence. The cells of the blastular epithelium progressively lose their flagella, and some of them immigrate into the blastocoel which is ultimately obliterated. These processes convert the larva into a steroblastula consisting of morphologically identical cells (Figure 4.12a). Following attachment, the larva begins to spread over the substratum and lacunae appear among its cells. The peripheral cells, which become separated from the inner cell mass by lacunae, flatten and differentiate into pinacocytes. The merger of other lacunae gives rise to the spongocoel. Later choanocytes and scleroblasts appear. It is hypothesized that the cells of the newly settled larva are developmentally uncommitted and that they differentiate on the basis of their positions within the cellular mass. When the early larva of *Ascandra* is cut into anterior and posterior halves, each half develops into a small but otherwise normal sponge.

The amphiblastula larva of some calcareous sponges (subclass Calcaronea) usually attaches to the substratum by the anterior flagellated pole (Figure 4.12a); and within a few hours, the flagella are lost. The larva becomes a flattened disc, consisting of a mass of inner cells derived from the flagellated columnar cells of the swimming amphiblastula surrounded by larger pinacocytes that apparently develop from the larval granular cells. During the early stages of metamorphosis, the inner-mass cells exhibit residual flagellar rootlets, which provide evidence of their origin, but these structures later disappear. There is evidently little or no mitotic activity during this initial transformation. The peripheral cells of the inner mass apparently give rise to the cells of the mesohyl. Scleroblasts, spicules, and collagen fibrils appear within this layer. The central cells of the inner mass differentiate into choanocytes. A space appears among these cells and progressively expands. This is the spongocoel lined by choanoderm. Finally, inhalant pores and an osculum appear. The juvenile (called an **olynthus**) has a simple tubular structure; it later develops the more complex organization characteristic of the adult.

Although morphological studies suggest that the granular cells of the amphiblastula give rise to the pinacoderm and most of the rest of the young sponge is derived from the flagellated columnar cells, certain experimental results do not seem to be entirely consistent with this interpretation. When the amphiblastula of *Sycon* is cut into anterior and posterior halves, the flagellated cells of the anterior hemisphere form a simple blastula, whereas the granular cells of the posterior hemisphere give rise to a quasinormal sponge. The problem of cellular lineages could be resolved if a method can be found for specifically labeling the different types of larval cells and following their differentiation during metamorphosis.

As indicated earlier, the swimming parenchymella larvae of certain freshwater sponges are adultlike in their organization. Metamorphosis involves replacement of the locomotory flagellated epithelium by pinacoderm and continued development of the canal system. Following attachment of the larva, amoebocytes situated directly beneath the flagellated epithelium phagocytize the flagellated cells and take their place at the surface, where they transform into pinacocytes. This sequence of events begins at the anterior pole, where the larva attaches to the substratum, and progresses around to the opposite pole. The amoeboid cells in contact with the substratum secrete a basal lamella of collagen. The large larval cavity becomes subdivided into components of the exhalant canal system, and subdermal spaces form beneath the surface pinacoderm. Then pores appear in the dermal membrane and an osculum connects the exhalant canal system with the external environment, establishing a functional water-transport system.

The parenchymellas of most demosponges are less highly developed than those of the freshwater sponges, lacking choanocyte chambers, rudiments of the canal system and, in some cases, spicules. In some species, it appears that all of the components of the adult sponge are derived from the inner cell mass of the larva, whereas in other species transformation of the outer flagellated epithelial cells into other types of cells may occur during metamorphosis. The first situation evidently exists in sponges like *Microciona prolifera*. The surface of the larva of *Microciona* is completely covered by a flagellated epithelium that encloses a relatively large inner cell mass. When the flagellated epithelial cells of recently released larvae were selectively labeled with ^{125}I and then larvae were fixed at different stages of metamorphosis, autoradiograms of sectioned larvae showed that the labeled cells moved into the inner mass and were phagocytized by archaeocytes. Radioactive label accumulated within archaeocyte phagosomes and was later excreted. Developing choanocytes did not exhibit label exceeding background levels (Misevic et al. 1990).

On the other hand, the larva of *Halichondria melanadocia* is characterized by a thick flagellated epithelium, composed of tall columnar cells, and only a small diffuse mass of internal cells. In this case, as well as others, it seems likely that the surface epithelium is not a terminally differentiated larval structure but rather makes a contribution to the development of the postmetamorphic adult (Woollacott 1990).

The large larva of *Mycale contarenii* (1.5 mm long) is especially suitable for surgical experimentation. When the flagellated epithelium, together with a few amoebocytes, was separated from the inner cell mass and placed in culture, large numbers of typical choanocyte chambers developed from the epithelium within a few hours. The inner cell mass also gave rise to choanocyte chambers and eventually to a functional sponge when it was cultured in isolation from the flagellated epithelium. However, in such cultures the choanocyte chambers were initially smaller than normal and the choanocytes, which apparently developed from archaeocytes, were atypically large (Borojevic 1966). While these results are subject to various interpretations, they suggest the possibility that in *Mycale* the flagellated cells of the larva may give rise to the choanocytes of the young sponge. In the larva of *Haliclona permollis*, the flagellated epithelial cells possess a natural marker in the form of special granules. During metamorphosis such granules can be observed in a sequence of stages from larval flagellated cells, through amoeboid cells, to differentiating choanocytes. This fact suggests that the larval flagellated cells of *Haliclona* give rise to choanocytes; however, the differentiation of some choanocytes from other types of cells, including archaeocytes, can not be ruled out (Amano and Hori 1996). Studies using a variety of techniques may be required to demonstrate unequivocally the fates of various types of larval cells during metamorphosis. Furthermore, it appears highly probable that as studies are extended to more species, a number of developmental patterns will be found.

Asexual Reproduction

Sponges may reproduce asexually by a number of mechanisms (Figure 4.16). The simplest form of asexual propagation is fragmentation. Even small sponge fragments often possess all of the elements required to produce a new individual. In most cases, this process appears to involve little more than wound closure and some limited reorganization. Fragmentation of erect branching sponges may occur as a result of water turbulence during storms. Similarly, sponges attached to aquatic vegetation may be torn into two or more fragments as the plants are moved about by wind and currents. Fragmentation appears to be the primary method of propagation for some sponges, although they also reproduce by sexual means.

Some sponges produce external buds that eventually detach and take up an independent existence. This form of propagation has not been extensively studied. In certain sponges, buds are essentially protrusions of the body, which possess a functional organization from the beginning of their formation, and separation from the parent is analogous to fragmentation. In other sponges, the buds consist of archaeocyte-rich cellular masses with no evidence of choanocyte chambers or canals. Following excision, such buds attach to the substratum and develop into small sponges.

Many freshwater and some estuarine/marine sponges produce **gemmules**. Often a sponge produces hundreds to thousands of gemmules attached to the substratum at the base of the animal and/or dispersed throughout much of its body. Each gemmule consists of a mass of nutrient-laden cells enclosed by a collagenous capsule that in many cases contains spicules. The cells, which are all morphologically similar and analogous to blastomeres of the early embryo in many respects, are called **thesocytes**. In the gemmules of many freshwater sponges, these cells are binucleate. Gemmules enable the sponge to persist during particularly adverse conditions (Figure 4.17). While the parent sponge may perish, the gemmules survive in a dormant state, and when the adverse conditions are over, the gemmules develop into active sponges (Simpson 1984; Fell 1993).

The formation of a gemmule begins with the aggregation of mesohyl cells to form a dense spherical mass. In most freshwater sponges, such aggregates contain archaeocytes, nurse cells (trophocytes) and spongocytes. Within the interior of the gemmule rudiment, the archaeocytes phagocytize the nurse cells and transform into thesocytes. A transient squamous epithelium is formed by archaeocytes at the periphery of the gemmule; beyond this, the **spongocytes** constitute a columnar epithelium that secretes the gemmule capsule. The development of the ep-

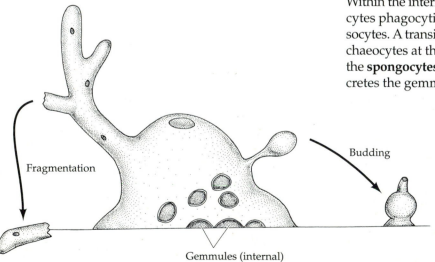

Fragmentation

Budding

Gemmules (internal)

Figure 4.16   Modes of asexual reproduction in sponges.

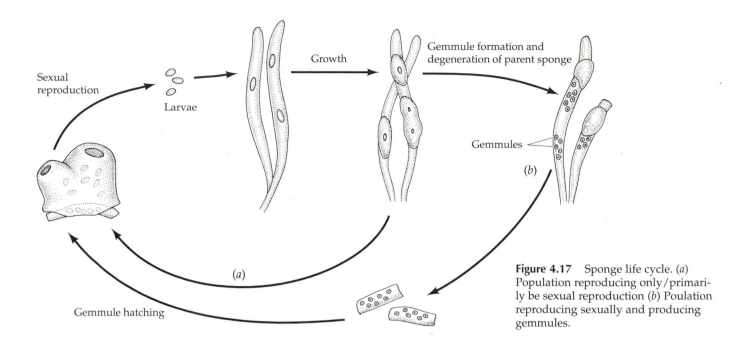

Figure 4.17 Sponge life cycle. (*a*) Population reproducing only/primarily be sexual reproduction (*b*) Poulation reproducing sexually and producing gemmules.

ithelia and the formation of the capsule start at one pole of the gemmule rudiment and progress around to the opposite pole where an aperture, the **micropyle**, is made. **Sclerocytes**, dispersed within the mesohyl, produce often-distinctive gemmular spicules that are carried to the developing gemmules, inserted between the spongocytes, and incorporated into the capsule (Figure 4.18).

Some time after they produce gemmules, sponges usually degenerate, leaving the gemmules exposed to the environment. In certain cases, the gemmules are firmly cemented to the substratum either individually or in a continuous pavement-like sheet. However, the gemmules of many species are held in place only by a meshwork of parental skeleton. Since gemmules are usually discrete entities, they may become separated from one another and dispersed. The likelihood of this occurring depends upon the strength of the skeletal framework in which they are contained and the frequency and intensity of physical disturbances.

At the time of gemmule hatching, the thesocytes begin to divide and, as they divide, their nutrient platelets are progressively broken down. This leads to the appearance of cells called **histoblasts** that are largely devoid of nutrient reserves. Such cells are especially numerous beneath the micropyle. The micropyle then opens, presumably as a result of enzymatic activity, and the histoblasts migrate

out of the gemmule capsule onto the substratum. These cells form a two-layered pinacoderm that radiates out from the micropyle and encloses the mass of cells that remains within the gemmule capsule (Figure 4.19). The layer of pinacoderm in contact with the substratum produces a collagenous basal lamella. As hatching continues, the thesocytes give rise to archaeocytes and scleroblasts, which populate the mesohyl that develops between the two layers of pinacoderm. Later, choanocyte chambers and rudiments of the canal system develop. Apparently. the choanocyte chambers derive from clusters of archaeocytes and the lining of the emerging canals, like the rest of the pinacoderm, is formed by histoblasts. Further devel-

Figure 4.18 Developing gemmule of *Ephydatia fluviatilis*. The spongocyte epithelium (SpB) secretes the inner layer (IS), pneumatic layer (VS), and outer layer (not yet formed) of the gemmule capsule which possesses an aperture, the micropyle (MM). A gemmular spicule, or gemmosclere (AD), may be seen within the capsule. (From Langenbruch 1981.)

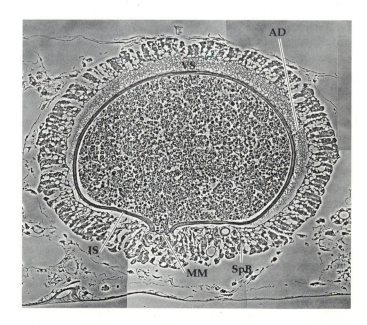

opment of the canal system and the appearance of one or more oscula result in the production of a functional sponge. Although a single gemmule is capable of developing into a new individual, frequently large numbers of gemmules hatch at about the same time and give rise to one relatively large sponge.

Gemmules typically undergo a period of dormancy that may take one of two general forms, **quiescence** or **diapause**. Quiescence is a dormant state imposed by unfavorable environmental conditions, such as low temperature and desiccation. This type of dormancy ends when favorable conditions are reestablished. Before the degeneration of the parent sponge, such nondiapausing gemmules are kept from hatching by a substance(s) produced by active sponge tissue (Rasmont, 1962).

In contrast, diapause is a dormant state that is maintained by an endogenous mechanism even in the presence of benign environmental conditions. The breaking of diapause often requires, or is at least accelerated by, exposure to low temperature and may be followed by a period of gemmule quiescence. For example, recently formed gemmules of some freshwater sponges will not hatch when kept in water at 20°C for more than 6 months. However, if such gemmules are first exposed to temperatures of about 3°C for 2 to 3 months, they readily hatch upon being transferred to 20°C (Fell 1987). Diapause tends to assure that gemmules produced during the summer or fall do not hatch until the following spring, even if the parent sponge degenerates long before the onset of unfavorable environmental conditions. The nature of the endogenous factor(s) that imposes dormancy in diapausing gemmules is unknown.

The gemmules of many sponges are able to survive severe environmental stresses. For example, the gemmules of certain sponges can tolerate being frozen in water at −20°C, and some gemmules are capable of surviving desiccation and exposure to extreme high temperatures. Finally, a few estuarine sponges produce gemmules that are tolerant of large fluctuations in environmental salinity.

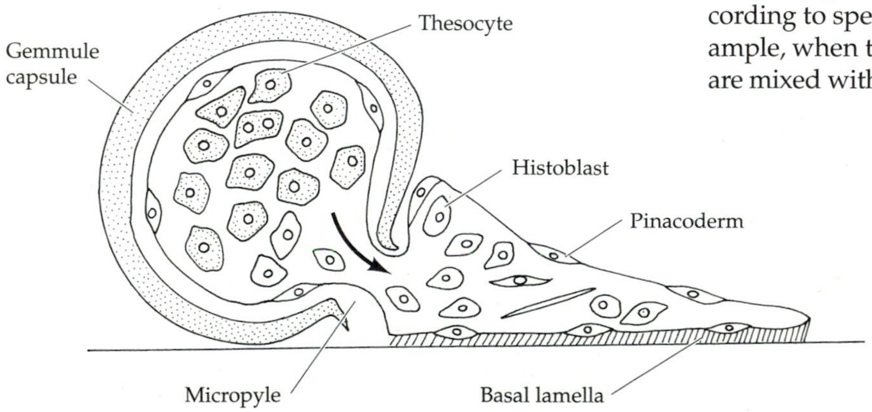

Tissue Regression and Regeneration

A number of sponges respond to adverse environmental conditions by tissue regression, in the course of which oscula, subdermal spaces, canals, and choanocyte chambers are lost. These sponges may persist as compact masses of cells for up to several months; then when conditions become favorable, they regenerate all of the structures that characterize the active sponge. Sponges, including *Microciona* and *Halichondria*, may undergo tissue regression during the winter in temperate regions. Low temperatures, possibly together with other factors, appear to initiate tissue regression and maintain the regressed condition. Tissue regression also is induced in *Microciona* by low environmental salinities. Although the occurrence of reversible tissue regression has been well documented, little is known concerning cellular activities during either the degenerative or regenerative phases.

Aggregation and Reconstitution by Dissociated Cells

The dissociation of the cells of an organism represents an extreme form of fragmentation. Wilson (1907) discovered that sponges can readily be reduced to a suspension of cells by squeezing pieces of sponge through fine silk cloth and that the dissociated cells of some species are able to reaggregate and then reconstitute a functional sponge. It appears that such reconstitution is brought about to a large extent by rearrangement of the various types of cells composing an aggregate. However, some cells, such as pinacocytes, which are sometimes poorly represented in the aggregates, may differentiate from archaeocytes. When different types of dissociated cells are separated from one another on a Ficoll density gradient, only archaeocytes exhibit the capacity to regenerate a complete sponge in the initial absence of other kinds of cells. This result provides strong evidence for the totipotency of archaeocytes (Buscema et al. 1980).

In certain cases the reaggregation of sponge cells exhibits species specificity. When the dissociated cells of two different species are mixed, they completely sort out according to species in the course of reaggregation. For example, when the cells of *Microciona prolifera* (red-orange) are mixed with those of *Haliclona occulata* (bluish purple),

Figure 4.19 Hatching of a sponge gemmule. (After Höhr 1977 and others.)

all of the aggregates are red-orange or bluish purple; there are none of intermediate color (Humphreys 1963).

Because of the ease with which sponge cell suspensions can be obtained and manipulated, they have been used for experimental studies of cellular cohesion. Upon placing marine sponges in Ca^{2+} and Mg^{2+}-free sea water, molecules involved in binding their cells together are solubilized and the cells become dissociated. The cell-binding molecules can then be separated from the cells by centrifugation. At 20°C cells dissociated in this way produce new cell binding molecules and reaggregate, but at 5°C they do not. However, when the solubilized cell binding molecules are added with Ca^{2+} to the dissociated cells at 5°C, the cells are able to form aggregates. The cell-binding molecules may exhibit species specificity, promoting only the aggregation of homologous cells (Humphreys 1963).

The cell-binding molecules of *Microciona* appear to be of a single type, a proteoglycan-like component called aggregation factor. It is a complex molecule consisting of a circular backbone from which 15–16 arms extend (Humpreys et al. 1977). This molecule binds Ca^{2+} and interacts with receptors on the cell surface. It appears that cellular cohesion results from Ca^{2+}-dependent binding between aggregation-factor molecules attached to the surface receptors of adjacent cells (Figure 4.20). Both attachment of aggregation factor to the cell surface and binding between aggregation-factor molecules apparently are mediated by multiple low-affinity carbohydrate interactions (Misevic and Burger 1990).

Life-History Strategies

The relative importance of sexual and asexual reproduction in the maintenance and spread of sponge populations is undetermined for most species, but seems to be highly variable. Furthermore, this may differ for different populations of a particular species, depending upon environmental conditions. It appears that all sponges reproduce sexually but, in some cases, only a small fraction of new individuals is produced in this way. As indicated earlier, some sponges are propagated to a large extent by detached fragments or buds. For gemmule-producing species, gemmule masses, as well as active sponges, may be fragmented, adding new individuals to the population. In the absence of such fragmentation, masses of gemmules serve to maintain individual sponges during unfavorable periods, but do not increase their number. Fragments and buds are often substantially larger than larvae and may experience a much lower rate of mortality. On the other hand, for many sponges the sexual production of large numbers of larvae and the relatively long-term survival of at least a few settled larvae provide the primary, if not only, means of propagation. In still other cases, both sexual and asexual reproduction appear to be of major importance in sponge life cycles. Sexual reproduction produces new individuals with unique genotypes, whereas asexual propagation multiplies the number of individuals carrying preexisting genotypes.

Some sponges are short-lived, normally surviving little more than a year; other sponges may live for 50 or more years. Although few studies have examined the age structure of sponge populations, it appears that many sponges probably have life spans of intermediate length. Certain sponges with relatively long life spans do not start to reproduce sexually until they are several years old. On the

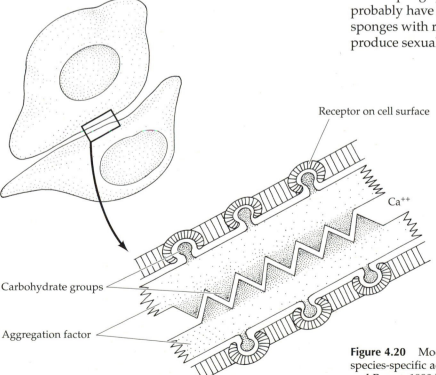

Receptor on cell surface

Ca^{++}

Carbohydrate groups

Aggregation factor

Figure 4.20 Model of how aggregation factor may mediate species-specific adhesion of sponge cells. (Modified from Misevic and Burger 1990.)

other hand, some sponges begin to produce gametes when they are about 2 to 3 weeks old and only a few millimeters in greatest dimension. Such rapid maturation may result in the production of multiple generations of offspring each year.

The timing of various events in the life cycle of a sponge may vary in different parts of its geographical range and even in different habitats within a restricted area. For example, in Connecticut *Haliclona loosanoffi* persists only as gemmules during the winter; however, in North Carolina this sponge is present in its active form throughout the winter but is absent except for gemmules during the middle of the summer. As would be expected, the timing of sexual reproduction by *Haliclona* is also different in these two localities. In Louisiana *Eunapius fragilis* living in temporary water bodies forms gemmules during the spring before the habitats dry up but in permanent streams of the same region, it produces gemmules during the fall.

Summary

Sponges present diverse paths of development—different types of sex determination, different ways of generating larvae, and different types and degrees of sexual and asexual reproduction. In addition, some elements of sponge development—the widespread use of phagocytosis for nutrition and sperm transport, and species-specific cell aggregation—appear to be characteristic of this phylum. Knowledge of sponge development may be critical to our understanding of the origins of the metazoa, yet we know very little about the molecular and cellular mechanisms of Poriferan differentiation and morphogenesis.

Literature Cited

Amano, S. and I. Hori. 1992. Metamorphosis of calcareous sponges. I. Ultrastructure of free-swimming larvae. *Inv. Reprod. Dev.* 21: 81–90

Amano, S. and I. Hori. 1996. Transdifferentiation of larval flagellated cells to choanocytes in the metamorphosis of the demisponge *Haliclona permollis*. *Biol. Bull.* 190: 161–172.

Baccetti, B., E. Gaino and M. Sará. 1986. A sponge with acrosome: *Oscarella lobularis*. *J. Ultrastruct. Molec. Struct. Res.* 94: 195–198.

Borojević, R. 1966. Étude experimentale de la différenciation des cellules de l'éponge au cours de son développement. *Dev. Biol.* 14: 130–151.

Borojević, R. 1969. Étude de développement et de la différenciation cellulaire d'éponges Calcaires Calcinéennes (Genre *Clathrina et Ascandra*). *Ann. Embryol. Morph. Fr.* 2: 15–36.

Boury-Esnault, N. and J. Vacelet. 1994. Preliminary studies on the organization and development of a hexactinellid sponge from a Mediterranean Cave, *Oopsacas minuta*. In *Sponges in Time and Space*, R. W. M. van Soest, van Kempen, Th. M. G. and J. C. Braekman, (eds.), pp. 407–415. A.A. Balkema, Rotterdam.

Buscema, M., D. De Sutte and G Van de Vyver. 1980. Ultrastructural study of differentiation processes during aggregation of purified sponge archaeocytes. *Roux's Arch. Dev. Biol.* 188: 45–53.

Diaz, J.-P. and R. Connes. 1980. Étude ultrastructurale de la spermatogenèse d'une Démosponge. *Biol. Cell.* 38: 225–230.

Duboscq, O. and O. Tuzet. 1937. L' ovogénèse, la fécondation, et les premiers stades du développement des éponges calcaires. *Arch. Zool. Exp. Gén.* 79: 157–316.

Fell, P. E. 1969. The involvement of nurse cells in oogenesis and embryonic development in the marine sponge, *Haliclona ecbasis*. *J. Morphal.* 127: 133–150.

Fell, P. E. Porifera. In *Reproductive Biology of Invertebrates I.Oogenesis, Oviposition, and Oosorption, 1983*, pp. 1–29; *IVA. Fertilization, Development, and Parental Care, 1989*, pp. 1–41, and *VIA. Asexual Propagation and Reproductive Strategies, 1993*, pp. 1–44. K. G. Adiyodi and R. G. Adiyodi (eds.) John Wiley & Sons, New York and Oxford & IBH Pub. Co., New Delhi.

Fell, P. E. 1987. Influences of temperature and desiccation on breaking diapause in the gemmules of *Eunapius fragilis* (Leidy). *Int. J. Inv. Reprod. Dev.* 11: 305–316.

Franzen, W. 1988. Oogenesis and larval development of *Scypha ciliata* (Porifera, Calcarea). *Zoomorphology* 107: 349–357.

Gallissian, M.-F. 1989. Le spermiokyste de *Sycon sycandra* (Porifera, Calcarea): étude ultrastructurale. *C.R. Acad. Sci. Paris*, Sér. III 309: 251–258.

Höhr, D. 1977. Differenzierungsvorgänge in der keimenden Gemmula von *Ephydatia fluviatilis*. *Roux's Arch. Dev. Biol.* 182: 329–346.

Humphreys, S., T. Humphreys and J. Sano. 1977. Organization and polysaccharides of sponge aggregation factor. *J. Supramolec. Struct.* 7: 339–351.

Humphreys, T. 1963. Chemical dissolution and *in vitro* reconstitution of sponge cell adhesions. I. isolation and functional demonstration of the components involved. *Dev.Biol.* 8: 27–47.

Langenbruch, P.-F. 1981. Zur Entstehung der Gemmulae bei *Ephydatia fluviatilis* L. (Porifera). *Zoomorphology* 97: 263–284.

Misevic, G. N. and M. M. Burger. 1990. Multiple low-affinity carbohydrates as the basis of cell-cell recognition in *Microciona prolifera*. In *New Perspectives in Sponge Biology*, K. Rützler (ed.), pp. 81–90. Smithsonian Institution Press, Washington, D.C.

Misevic, G.N., V. Schlup and M. M. Burger. 1990. Larval metamorphosis of *Microciona prolifera*: evidence against the reversal of layers. in *New Perspectives in Sponge Biology*, K. Rützler (ed.), pp. 182–187. Smithsonian Institution Press, Washington, D.C.

Paulus, W. 1989. Ultrastructural investigation of spermatogenesis in *Spongilla lacustris* and *Ephydatia fluviatilis* (Porifera, Spongillidae). *Zoomorphology* 109: 123–130.

Rasmont, R. 1962. The physiology of gemmulation in freshwater sponges. In *Regeneration* (Twentieth Symposium, Society for the Study of Development and Growth) D. Rudnick (ed.), pp. 3–25. Ronald Press, New York.

Reiswig, H. M. 1983. Porifera. in *Reproductive Biology of Invertebrates*. II. *Spermatogenesis and Sperm Function*, K. G. Adiyodi and R. G. Adiyodi (eds.). pp. 1–21. John Wiley & Sons, New York.

Saller, U. and N. Weissenfels. 1985. The development of *Spongilla lacustris* from the oocyte to the free larva (Porifera, Spongillidae). *Zoomorphology* 105: 367–374.

Simpson, T. L. 1984. *The Cell Biology of Sponges*, Springer-Verlag, New York.

Watanabe, Y. and Y. Masuda. 1990. Structure of fiber bundles in the egg of *Tetilla japonica* and their possible function in development. In *New Perspectives in Sponge Biology*, K. Rützler (ed.), pp. 193–199. Smithsonian Institution Press, Washington, D.C.

Wilson, H. V. 1907. On some phenomena of coalescence and regeneration in sponges. *J. Exp. Zool.* 5: 245–258.

Woollacott, R. M. 1990. Structure and swimming behavior of the larva of *Halichondria melanadocia* (Porifera: Demospongiae). *J. Morphol.* 205: 135–145.

Woollacott, R. M. 1993. Structure and swimming behavior of the larva of *Haliclona tubifera* (Porifera: Demospongiae). *J. Morphol.* 218: 301–32.

Radiate Animal Phyla

CHAPTER 5

Cnidarians, the Jellyfish and Hydras

Vicki J. Martin

THE PHYLUM CNIDARIA IS A DIVERSE GROUP CONTAINING the hydras, jellyfish, sea anemones, and corals, as well as the less familiar hydroids, sea fans, siphonophores, and zoanthids (Table 5.1). There are about 9000 living species, the majority of which are marine, except for the hydras and some additional freshwater hydrozoans. Most cnidarians live in shallow water and are either sessile, sedentary, or pelagic. Sessile forms are common on pier pilings, rocky coasts, and coral reefs.

Cnidarians possess radial symmetry, often modified as biradial or quadriradial. There is one main axis of symmetry, the oral-aboral axis, that extends from mouth to base. All parts are arranged concentrically around this axis. Most notable are tentacles: short, slender evaginations of the body wall that encircle the oral end and function in defense and the capture of food.

The basic cnidarian body plan is simple, consisting of a solid body wall enclosing a central digestive cavity (Figure 5.1). This space is the only body cavity found in cnidarians. The body wall is composed of an outer epithelium, the **epidermis**, and an inner epithelium, the **gastrodermis**, that surrounds the gastrovascular cavity. The epidermis consists of epitheliomuscular cells, glandular cells, sensory receptor cells, ganglionic neurons, interstitial cells, and stinging cells called **cnidocytes** or **nematocytes** (Figure 5.2). The gastrodermis has nutritive muscle cells, glandular cells, some neurons, and interstitial cells. Between the two epithelia is an extracellular layer, the **mesoglea**. The mesoglea varies in thickness depending on the organism, ranging from a thin, noncellular layer to a thick, fibrous, jellylike layer with or without wandering cells. Biochemical and immunohistochemical evidence shows that the mesoglea is composed of Type IV collagen, fibronectin, heparan sulfate proteoglycan, and laminin (Sarras et al. 1991). During embryogenesis two distinct germ layers form, the ectoderm and the endoderm, which give rise to

the two adult epithelia, the epidermis and gastrodermis, respectively. Thus cnidarians are **diploblastic**.

The gastrovascular cavity is essentially a simple endoderm-lined tube with a single opening, the mouth, at the oral end. This opening functions as both mouth and anus, serving for food intake and ejection of indigestible material. The cavity is often referred to as a **coelenteron**, as it functions both for digestion of food and for circulation. Throughout the phylum there is a tendency toward complication of the cavity by the extension of branches and pockets; in some groups, such as the scyphozoans, cubozoans, and anthozoans, the cavity is divided into compartments by gastrodermal-mesogleal projections called **septa**.

Cnidarians exhibit a tissue grade of construction with only a limited degree of organ development. They possess digestive, muscular, nervous, sensory, and often skeletal systems in a primitive stage of development. For example, the nervous system is a simple nerve net composed of naked and largely nonpolar neurons. The muscle system is formed by epitheliomuscular cells, cells that combine features of epithelial cells and muscular cells. These are the most primitive muscle cells found in the metazoans. There are no distinct respiratory, excretory, circulatory, or reproductive systems. Gonads consist merely of aggregations of sex cells (Figure 5.1).

Many cnidarians are dimorphic, meaning they exhibit two different adult morphologies during their life histories (Figure 5.3, Table 5.1). One form, the **polyp**, is usually sessile and has the shape of an elongated cylinder. The aboral end, the basal disc, is attached to the substrate, while the opposite free end, the oral end, bears the mouth and tentacles. The other morph, the **medusa**, is generally free-swimming and is a bell-, bowl-, or saucer-shaped animal with marginal tentacles. Compared with the polyp, the medusa has shortened along the oral-aboral axis, has expanded radially, and the mesoglea has increased in thick-

TABLE 5.1 *General Characteristics of Cnidarians*

	Body forms	Growth patterns	Sexual reproduction	Larval forms	Asexual reproduction
Class Hydrozoa					
Order Hydroida (Hydroids)	Polyp Medusa	Colonial Solitary	Dioecious Hermaphroditic	Planula Actinula	Budding Transverse fission Longitudinal fission Frustules
Order Milleporina (Fire corals)	Polyp Medusa	Colonial	Dioecious Hermaphroditic	Planula	Budding
Order Stylasterina (Hydrocorals)	Polyp	Colonial	Dioecious	Planula	Budding
Order Trachylina	Medusa	Solitary	Dioecious Hermaphroditic	Planula Actinula	Budding Frustules
Order Siphonophora	Polyp Medusa	Colonial	Hermaphroditic	Planula Calyconula Siphonula Conaria Rataria	Budding Cormidial buds
Order Actinulida	Polyp	Solitary	None recorded	None	Budding
Class Scyphozoa					
Order Stauromedusae	Sessile polyp-like medusa	Solitary	Dioecious	Planula	Budding
Order Coronatae	Medusa Scyphistoma	Solitary Colonial scyphistomae	Dioecious	Planula	Budding Strobilation
Order Semaeostomeae	Medusa Scyphistoma	Solitary	Dioecious Hermaphroditic	Planula	Budding Strobilation Podocysts Pedal stolons
Order Rhizostomeae	Medusa Scyphistoma	Solitary	Dioecious	Planula	Budding Strobilation Ciliated buds Podocysts
Class Cubozoa					
Order Cubomedusae	Medusa Small scyphistoma	Solitary	Dioecious	Planula	Budding
Class Anthozoa, Subclass Alcyonaria (Octocorallia)					
Order Stolonifera (Organ pipe coral)	Polyp	Colonial	Dioecious Hermaphroditic	Planula	Budding
Order Telestacea	Polyp	Colonial	Dioecious Hermaphroditic	Planula	Budding
Order Alcyonacea (Soft corals)	Polyp	Colonial	Dioecious Hermaphroditic	Planula	Budding
Order Coenothecalia (Helioporacea) (Blue corals)	Polyp	Colonial	Dioecious Hermaphroditic	Planula	Budding
Order Gorgonacea (Sea fans, sea whips)	Polyp	Colonial	Dioecious Hermaphroditic	Planula	Budding
Order Pennatulacea (Sea pens, sea pansies)	Polyp	Colonial	Dioecious	Planula	Budding

TABLE 5.1   (*continued*)

	Body forms	Growth patterns	Sexual reproduction	Larval forms	Asexual reproduction
Class Anthozoa, Subclass Zoantharia (Hexacorallia)					
Order Actiniaria (Sea anemones)	Polyp	Solitary	Dioecious Hermaphroditic	Planula Edwardsia Halcampoides	Budding Transverse fission Longitudinal fission Pedal laceration
Order Madreporaria (Scleractinia) (Stony corals)	Polyp	Colonial Solitary	Dioecious Hermaphroditic	Planula Edwardsia Halcampoides	Budding
Order Zoanthidea	Polyp	Colonial Solitary	Dioecious Hermaphroditic	Zoanthina Zoanthella	Budding
Order Antipatharia (Black corals)	Polyp	Colonial	Dioecious Hermaphroditic	Planula	Budding
Order Ceriantharia (Tube anemones)	Polyp	Solitary	Hermaphroditic	Cerinula	Budding
Order Corallimorpharia (Coral-like anemones)	Polyp	Solitary Colonial			Budding

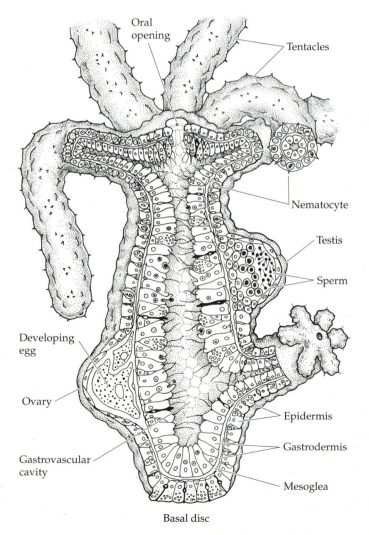

ness. Because of the expanded mesoglea, these forms are commonly referred to as *jellyfish*. Each jellyfish has a convex side, the **exumbrella**, directed upward and a concave underside, the **subumbrella**. The mouth is located in the center of the subumbrella and opens into the gastrovascular cavity. Tentacles dangle from the margin of the bell. Some cnidarians have only the **polypoid** (polyp-like) form, some only the **medusoid** (medusa-like) form, and others have both in their life cycles. In cases in which both polypoid and medusoid forms occur, the polyps reproduce by asexual methods, buding off medusae (Figure 5.4). The medusae, in turn, bear the eggs and sperm used for sexual reproduction. The life cycle thus consists of an alternation of the asexual polypoid form with the sexual medusoid form. This condition is referred to as **alternation of generations**, also called **metagenesis**. Both polyp and medusa generations are diploid; only the eggs and sperm are haploid.

Figure 5.1   Solitary polyp of the freshwater hydra. The anterior oral end bears the mouth and tentacles while the posterior aboral end, the basal disc, attaches to a substrate. An outer epidermis, a middle mesoglea, and an inner gastrodermis make up the body wall, which surrounds a gastrovascular cavity. The hydra reproduces asexually by budding and sexually by producing eggs or sperm. The gonads are simple aggregations of sex cells in the epidermis.

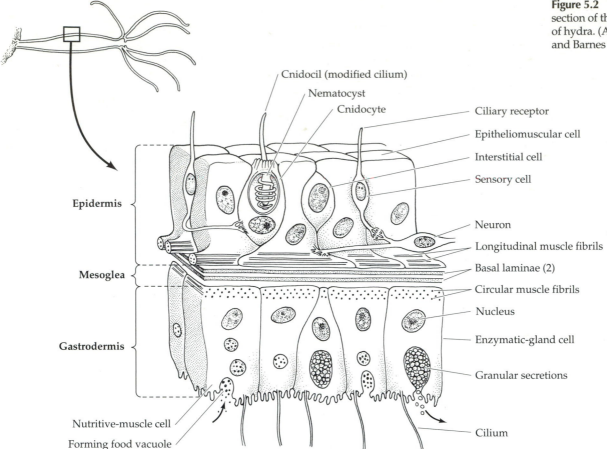

Figure 5.2 Longitudinal section of the body wall of hydra. (After Ruppert and Barnes 1994.)

Complications of this basic alternation of generations scheme are abundant throughout the phylum (Figure 5.5). Furthermore, asexual reproduction often occurs at stages other than medusoid budding. Most polyp stages reproduce themselves by budding; if the buds remain connected, huge colonies form. Some medusae also bud medusae. Certain groups tend to accentuate the polyp or medusoid generation with a concomitant simplification of the other.

Both gonochoristic (dioecious) and hermaphroditic individuals are found throughout the phylum (Table 5.1). Asexual reproduction is extremely important for the propagation of the phylum, and numerous types exist throughout the classes.

Many hydrozoans and anthozoans form extensive colonies of polyps. The description of a simple hydroid colony will serve to illustrate this common growth form

Figure 5.3 Dimorphic body forms. (*a*) Polyp. (*b*) Medusa. (After Ruppert and Barnes 1994.)

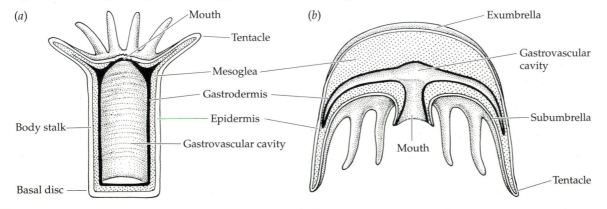

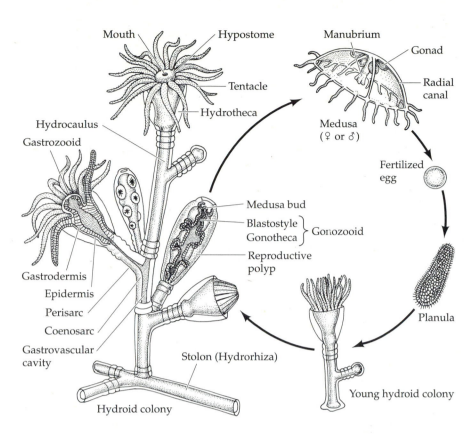

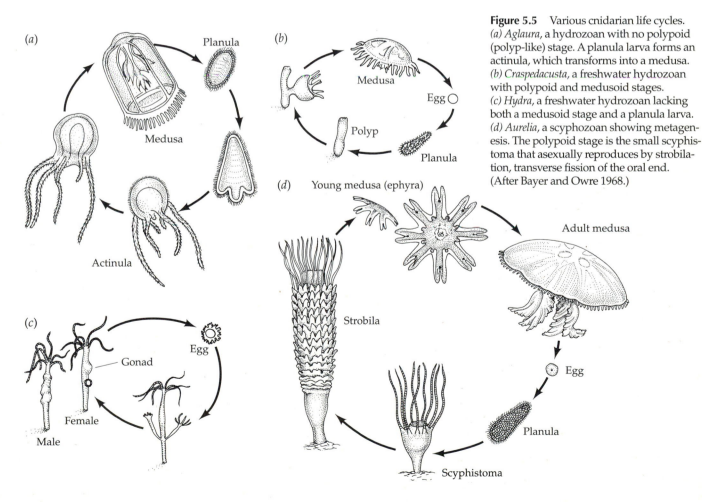

Figure 5.4 Life cycle of the colonial hydroid *Obelia* showing alternation of generations and formation of a planula, a common cnidarian larval form. (After Ruppert and Barnes 1994.)

Figure 5.5 Various cnidarian life cycles. (*a*) *Aglaura*, a hydrozoan with no polypoid (polyp-like) stage. A planula larva forms an actinula, which transforms into a medusa. (*b*) *Craspedacusta*, a freshwater hydrozoan with polypoid and medusoid stages. (*c*) *Hydra*, a freshwater hydrozoan lacking both a medusoid stage and a planula larva. (*d*) *Aurelia*, a scyphozoan showing metagenesis. The polypoid stage is the small scyphistoma that asexually reproduces by strobilation, transverse fission of the oral end. (After Bayer and Owre 1968.)

(Figure 5.4). A hydroid colony consists of several interconnected polyps, all sharing a common gastrovascular cavity. The body layers of the colony members are all continuous. The oral end of each polyp with its mouth and tentacles is called the *hydranth*. Below the hydranth is the stalk, or hydrocaulus, of the polyp. The stalk connects the hydranth to the base of the polyp, the basal disc, which may extend rootlike stolons, or hydrorhiza, over the substrate. Single polyps or branches of polyps arise from these stolon regions, producing different growth and budding patterns. Most colonies grow to large sizes; however, the individual polyps of the colony are quite small.

Many colonial hydroids are surrounded in part by a nonliving supportive cuticle, the perisarc. The living tissue to the inside of the perisarc is the coenosarc. In some forms the perisarc covers just the stolon region and hydrocaulus, whereas in others the perisarc extends upward to encase the hydranth as well. The perisarc surrounding the hydranth, called the *hydrotheca*, may be open or closed by a lid.

Most hydroid colonies exhibit polymorphism in that they have structurally and functionally distinct types of polyps (zooids). **Gastrozooids**, or feeding polyps, are the most abundant forms and are specialized for capture and ingestion of food. These polyps possess a mouth surrounded by a whorl of tentacles armed with cnidocytes. In many species the gastrozooids are also defensive. Some colonies have specialized defensive polyps, loaded with cnidocytes, named **dactylozooids**, and/or specialized reproductive polyps called **gonozooids**.

All hydroid colonies produce medusoid buds via asexual means; these buds, in turn, produce the gametes. Depending on the species, the medusoids may break free from the colony as free-swimming medusae or be retained by the colony. If free-swimming, they are usually small and transparent. The medusoid buds are commonly referred to as **gonophores**, meaning bearers of gonads. Gonophores may arise from different regions of the gastrozooids, and in some hydroids only the gonozooids produce gonophores. Such gonozooids lack mouths and tentacles and show little resemblance to polyps.

Overview of Sexual Reproductive Patterns of Cnidarians

Cnidarians represent an early phase of metazoan evolution. In the simpler cnidarians embryogenesis may appear **anarchic** (lacking order, regularity, or definiteness), whereas in the more advanced forms one sees complex mosaic patterns of embryogenesis.

Diploblastic cnidarians occupy a strategic position for examining evolution. This chapter presents a general overview of the sexual reproductive patterns of the Cnidaria. Also included are brief summaries on regeneration and asexual reproduction. The development of a freshwater hydra (*Hydra vulgaris*), a marine hydroid (*Pennaria tiarella*), and an anthozoan (*Nematostella vectensis*) is presented in some detail to enable the reader to fully appreciate the beauty of simplicity.

Gametogenesis

ORIGIN OF GERM CELLS.   Germ cells arise from **interstitial cells** (Chapman 1974; Tardent 1985). Interstitial cells are migratory stem cells that divide to replenish the population and/or differentiate into several somatic cells (nerve cells, nematocytes, glandular and mucous cells) and the gametes. Thus interstitial cells are multipotent cells. The interstitial cell system of the freshwater hydra is a heterogeneous population of interstitial cells composed of multipotent stem cells, unipotent stem cells, progenitor cells, and cells about to enter a differentiation pathway (Figure 5.6) (Teragawa and Bode 1990; Bode 1996). Littlefield (1985) has provided strong evi-

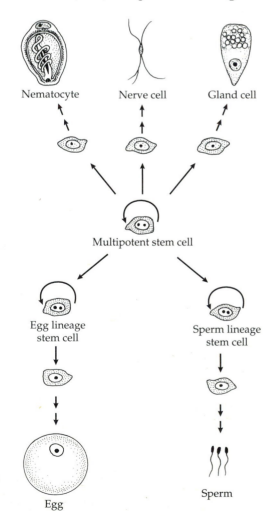

Figure 5.6   Interstitial cell lineages in hydra. Stem cells divide to replenish themselves and send committed interstitial cells down a differentiation pathway. (After Littlefield 1994.)

dence that there are subpopulations of interstitial cells specifically committed to the germ line (Figure 5.6). In her scheme, multipotent interstitial stem cells divide to maintain the population; some of these cells become committed and enter either the nerve cell, nematocyte, or glandular/mucous cell pathways and complete differentiation. With respect to the germ cells, some of the multipotent interstitial stem cells give rise to unipotent stem cells, which undergo both self-renewal and gamete differentiation. There is a unipotent stem cell for the egg lineage and a unipotent stem cell for the sperm lineage. Using a fluorescent monoclonal antibody to label the cells, Littlefield (1985) clearly showed that there is a distinct subpopulation of interstitial cells developmentally restricted to sperm production. This antibody stains the entire sperm lineage but does not stain egg or somatic lineages. In fact, stem cells committed to the sperm or egg lineages have been isolated from males and females of three gonochoristic hydra species. Littlefield further demonstrated that the germ line cells, and not the somatic cells, are responsible for determining the sexual phenotype in hydra (Littlefield 1994).

GONAD FORMATION. Cnidarian gonads are not distinct organs, but are merely massive assemblages of gametes in the interstitial spaces of the body tissue (Figure 5.1). Interstitial cells committed to egg or sperm differentiation accumulate in specific areas and form the gametes; these gamete-rich areas of tissue are considered primitive "gonads." However, no accessory cells of somatic origin participate in formation of "testes" or "ovaries." Accumulation and growth of the sex cells often cause the local epithelia to bulge, stretch, and eventually rupture. This rupturing results in gamete release. In hydrozoans such as hydra, the germ cells assemble in the ectoderm between the epithelial cells. In hydrozoans that bud medusae, the interstitial cells of the medusae accumulate in the ectoderm along the manubrium or radial canals. The "gonads" of scyphozoans, cubozoans, and anthozoans are gastrodermal; the germ cells accumulate within either the endoderm or the mesoglea of the gastric septa. In scyphozoans that lack septa, the gonads form on the floor of the stomach. Hydrozoans release their gametes directly to the outside during spawning, whereas scyphozoans, cubozoans, and anthozoans shed their gametes into the gastric cavity and later expel them to the exterior through the mouth.

SPERMATOGENESIS. During spermatogenesis, interstitial cells committed to the sperm lineage proliferate to produce large clusters or nests of cells (Figure 5.1). Within each nest the cells form a syncytium; they are held together by intercellular bridges resulting from incomplete cytokinesis following mitosis. The nests of cells subsequently undergo morphological differentiation, forming sperm intermediates that enter meiosis to produce spermatids. Spermatids are then modified to form functional sperm. Due to synchrony of development within the nests, large numbers of sperm are produced and released during the breeding season. This process is similar to spermatogenesis in higher metazoans.

OOGENESIS. Oogenesis is best described for the hydrozoans (Tardent 1985). Interstitial cells committed to the egg lineage divide to produce large groups of oogonial cells. These oogonial cells stream together and fuse, forming enormous syncytial masses (Figure 5.1). Within each mass one of the cells remains as the oocyte, while the rest, often referred to as **nurse cells, shrunken cells**, or **pseudocells**, disintegrate and become incorporated into the cytoplasm of the developing oocyte. The nuclei of these nurse cells often persist throughout embryogenesis (Martin et al. 1994). Final maturation of oocytes into eggs, i.e, meiosis, may occur prior to or after spawning. Oogonial fusion has also been seen in some anthozoans. Alternatively, representatives from all four classes are seen in which the oogonia do not fuse, but simply grow, produce yolk, and enlarge to form oocytes. In some scyphozoans and anthozoans, developing oocytes receive nutritive materials from the coelenteron; specialized endodermal cells apparently transfer "food" to the oocytes. Depending on the species, a single egg or hundreds of eggs are produced at any one time.

SPERM STRUCTURE. The functional cnidarian sperm is considered primitive; its morphology resembles that of other spermatozoa (Figure 5.7) (Franzén 1966). A small head houses a conical or cylindrical nucleus. The midpiece contains two to five small, round mitochondria and a pair of centrioles axially aligned at right angles. The distal centriole serves as the flagellar basal body and often has accessory structures, centriolar satellites and pericentriolar processes, associated with it (for a review see Thomas and Edwards 1991). Sperm swim by means of a long (30–90 μm) flagellum that shows the typical 9+2 pattern of microtubules. Cnidarian sperm, with one exception, lack an acrosome (Stagni and Lucchi 1970; Carré 1984). In many marine hydrozoans, Golgi-derived vesicles are found in the anterior tip of the sperm between the nucleus and the plasma membrane (Summers 1970). Flocculent material has also been seen in this region in other species (West 1978). Such vesicles and materials may represent the forerunner of the acrosome (Hinsch and Clark 1973).

Carré (1984) described a fully differentiated acrosomal complex in the siphonophore *Muggiaea kochi*. The complex contains a large Golgi-derived vesicle and a separate system of Golgi-derived saccules. These sperm undergo an acrosomal reaction characterized by exocytosis of the acrosomal vesicle, formation of an acrosomal process, and fusion of the tip of the acrosomal process with the oolemma. This is the only example of an acrosomal com-

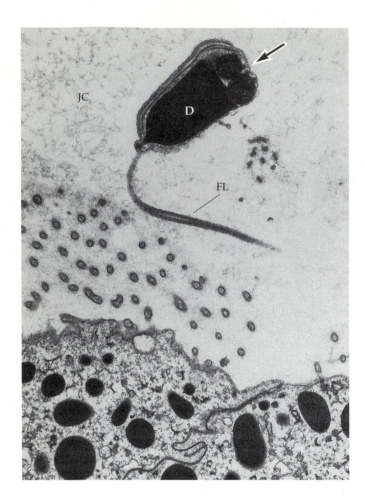

Figure 5.7   Sperm at the surface of a hydrozoan egg. The head (D), midpiece (arrow) containing mitochondria, and flagellum (FL) of the sperm are seen. The egg is covered with a jelly coat (JC) and microvilli project into this coat. ×12,000.

maternal RNA stored in the egg direct development through gastrulation.

Release of Gametes: Spawning and Brooding

Most cnidarians reproduce at specific times during the year. The length of the breeding season and the time of year when reproduction occurs vary from species to species. Fertilization is predominantly external in this phylum, and the majority of its members simply release their gametes into the water where they live. Such broadcast spawning is usually synchronized so that males and females of the same species spawn at the same time, thus enhancing the chance of egg and sperm encounter. In certain cases, various forms of behavior enhance reproductive success. For example, individuals often aggregate just prior to reproduction. Large swarms of medusae appear periodically in the water during breeding. Some sea anemones aggregate just prior to spawning (Uchida and Yamada 1968).

Several environmental factors may be involved in the synchronization and coordination of spawning. Numerous species respond to changes in intensity of illumination, either lunar or solar. Some individuals spawn shortly after dark, after a period of darkness, or after a dark exposure followed by light (Ballard 1942; Fadlallah 1983). Many activities are stimulated by light or dark; these include oocyte maturation, gonad rupture, gonophore swelling, and gonophore detachment and pulsation (Campbell 1974).

During spawning the gonadal tissue ruptures, releasing the gametes into the water for hydrozoans or into the gastrovascular cavity of scyphozoans, cubozoans, and anthozoans. Gametes released into the gastric cavity are shed to the outside via the mouth when the body wall contracts. In some species the gametes are shed to the outside through tentacle tips, aboral pores, or genital pores.

Brooding of young has also evolved in all four classes of cnidarians. Some hydroids lacking a free-swimming medusa stage undergo internal fertilization and development in the female gonophore, which later releases a planula or an actinula larva. Many cubozoan, scyphozoan, and anthozoan eggs are fertilized while in the gonad or following release into the gastrovascular cavity. Such fertilized eggs may be retained in the gastric cavity until some later stage of development, usually the planula larva stage, and then released to the outside. Many scyphozoans brood their young within their oral arms, releasing planulae just prior to their settlement. Some sea anemones and coral brood externally; their young adhere to the exterior of the parent or the colony (Fadlallah 1983).

plex and an acrosomal reaction known to occur in the cnidarians.

EGG STRUCTURE.   Cnidarian eggs are morphologically diverse. Some eggs have a smooth surface, while others are covered with microvilli and a jelly coat (Figure 5.7) (Ballard 1942; Clark and Dewel 1974). Some sea anemone eggs display beautiful cytospines, spirally arranged clusters of 50–70 large microvilli with actin cores (Schroeder 1982). The eggs of the hydroid B*ougainvillia multitentaculata* are covered with thousands of nematocytes, which form a solid layer (Szollosi 1969). These maternally derived nematocytes are present throughout early development and are eventually incorporated into the planula larva. Eggs contain varying amounts of yolk distributed throughout the cytoplasm. Some eggs have a relatively yolk-free ectoplasm surrounding a more yolky endoplasm. In some cases cortical granules are found; in others they are absent (Clark and Dewel 1974). Some eggs contain remnants of engulfed nurse cells; others do not (Martin et al. 1994; Boelsterli 1977). Eggs of siphonophores and trachyline medusae have large granules that are packed into the endoplasm (Freeman 1983b). High levels of RNA and proteins are found in many eggs, and Edwards (1975) showed that in *Hydractinia echinata*, high concentrations of

Cnidarian Embryogenesis

Fertilization

Species-specific sperm chemoattractants are released by some female gonophores and unfertilized eggs (Miller 1979, 1981; Honegger 1981). These attractants may be localized to the fertilization site or distributed over the entire surface of the egg (Carré and Sardet 1981; Freeman and Miller 1982). Such attractants affect the direction and speed of sperm migration.

Fertilization is predominantly external in water, although internal fertilization has evolved in all four classes. Internal fertilization occurs when sperm are shed into the water and are carried, drawn, or attracted into an egg-bearing polyp or medusa. Internal fertilization may occur in the gonophore lumen of hydrozoans and in the gonad or gastrovascular cavities of scyphozoans, cubozoans, and anthozoans. The young zygotes are either expelled immediately or retained for brooding and later released as planulae. At least one known example of copulation occurs in the sea anemone *Sagartia troglodytes* (Brusca and Brusca 1990). In this species the pedal discs (feet) of a receptive male and female are pressed together, creating a chamber into which gametes are released and fertilization occurs. The copulatory position is maintained for several days, until the planula larvae have formed. In the cubozoan *Tripedalia cystophora* the male transfers a **spermatophore**, a packet of sperm in a protective casing, directly to the female using its tentacles (Werner 1973).

Eggs are generally fertilized after oocyte meiosis is completed, that is, after both polar bodies have been discharged (Freeman 1987). The site of polar body release on the egg surface is specialized for fertilization; this site is functionally differentiated during egg maturation (Freeman 1987). In many species a localized depression in the surface of the egg marks the fertilization site; in others no morphological markers form, but molecular specificity is seen in the area of polar body discharge (Miller 1981). Directly beneath the region of polar body extrusion lies the female pronucleus; this area is the animal pole of the egg.

The cytological events associated with fertilization in cnidarians are poorly understood. Although cnidarian sperm lack a well-developed acrosome, in some species projections from sperm interact directly with the egg surface (Clark and Dewel 1974). Some anthozoan eggs release their cortical granules during fertilization. The cortical material may dissipate rapidly, or in some cases create a layer around the fertilized egg (Clark and Dewel 1974; Dewel and Clark 1974).

Cleavage and Blastulation

At least four types of cleavage patterns are seen in the cnidarians: (1) regular, holoblastic, radial; (2) irregular holoblastic; (3) unequal, holoblastic, anarchic; and (4) superficial. Variations, combinations, and modifications of these patterns exist throughout the classes. In regular, holoblastic, radial cleavage the divisions are total, usually equal, more or less regularly timed, and produce a radial pattern. The first two divisions are perpendicular to each other and pass through the animal-vegetal axis of the egg; the third cleavage is equatorial and may or may not be shifted toward the animal pole. Such a pattern produces regular tiers of blastomeres, even by the 64-cell stage. Examples include many hydrozoans, such as hydra (Honegger 1981; Martin et al. 1994), most scyphozoans (Berrill 1949), and several anthozoans: some sea anemones (Metschnikoff 1886), several stony corals (Fadlallah 1983), and various octocorals (Wilson 1884). Many hydrozoans, scyphozoans, and anthozoans, however, tend to be much less regular in their divisions; their blastomeres adhere rather loosely and have a tendency to roll into new positions between cleavage divisions. Such divisions may be synchronous or asynchronous. Some hydrozoans, such as *Pennaria tiarella* (Martin and Archer 1986b), produce blastomeres of unequal size, and no one embryo cleaves in exactly the same pattern. Such anarchic divisions may be synchronous or asynchronous. Generally, however, anarchically cleaving embryos produce a spherical blastula of more or less equal-sized cells. Hydrozoans and anthozoans with very yolky eggs undergo a superficial (epiblastic) kind of cleavage. In these eggs nuclear divisions occur under conditions in which the initial cytokineses are suppressed and cellularization begins after a number of mitoses have taken place. Such a pattern produces an embryo with a single layer of epithelial cells surrounding a central uncleaved mass of yolky vacuoles (Hyman 1940; Fadlallah 1983). Little is known about cubozoan development except that cleavage is total (Yamaguchi and Hartwick 1980).

In many cnidarian embryos the cleavage furrowing is asymmetric or unilateral (Campbell 1974; Martin et al. 1994). Here the cleavage furrow appears on one side of the blastomere, usually at the animal pole, and extends inward across the cell. As the furrow bisects the blastomere, the divided regions behind the furrow adhere together. Unilateral furrowing produces heart-shaped cleavage stages (see Figure 5.22) and may result in rotation of blastomeres. Thus cleavage is completed at the animal pole minutes or hours before the furrow reaches the vegetal pole; in fact, the next division often begins before the previous cleavage is done. Such unilateral furrowing is due to the eccentric placement of the nuclei toward the animal pole (Rappaport 1963).

Two common types of blastulae are formed (Figure 5.8): a solid stereoblastula filled with cells and lacking a blastocoel, and a hollow coeloblastula that contains a single layer of cells surrounding a large blastocoel. Both forms are found in hydrozoans, scyphozoans, and anthozoans. In superficially cleaving embryos a **periblastula** is produced; such a blastula consists of an outer layer of cells surrounding a central mass of yolk.

Figure 5.8   Types of gastrulation found in the cnidarians. *(a)* Invagination. *(b)* Unipolar ingression. *(c)* Invagination plus unipolar ingression. *(d)* Invagination plus multipolar ingression. *(e)* Multipolar ingression. *(f)* Delamination of a coeloblastula. *(g)* Delamination of a stereoblastula. (After Tardent 1978.)

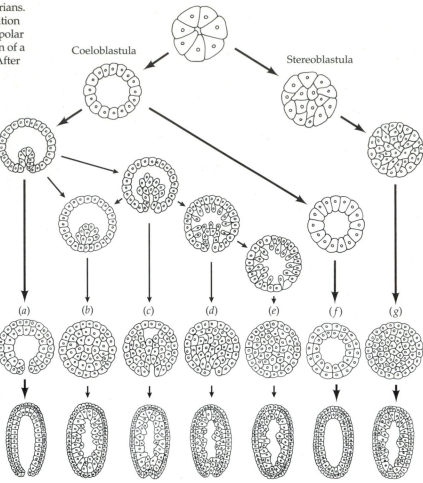

Gastrulation

During gastrulation the two embryonic tissue layers, the ectoderm and endoderm, are segregated and the mesoglea is synthesized. Gastrulation in cnidarians is diverse, and multiple patterns are found (Figure 5.8). They include unipolar ingression, multipolar ingression, invagination, delamination, epiboly, and mixed patterns (Hyman 1940; Campbell 1974). During **unipolar ingression** the endoderm forms by inward migration of individual cells at the vegetal pole. In **multipolar ingression** individual cells detach themselves from all points of the blastula wall and move to the interior as endoderm. During **invagination** the posterior region of the blastula bends inward, an indentation, the **blastopore**, is formed, and a sheet of cells migrates to the interior. **Delamination** may be primary or secondary. In **primary delamination** the endoderm arises by radial division of blastomeres such that one daughter cell remains at the surface (ectoderm) and one daughter cell moves to the interior (endoderm). **Secondary delamination** is characteristic of some stereoblastulae, in which tissue segregation occurs by a simple histological rearrangement producing an outer ectoderm and an inner endoderm. A variation of secondary delamination, termed

syncytial delamination, is found in yolky eggs that produce periblastulae. Here a less uniform layer of endodermal cells, produced by delamination of the outer cells, forms to the interior of the outer cells. The outer cells now represent the ectoderm (McMurrich 1890). Uchida and Yamada (1968) reported that certain hydrozoans gastrulate by **epiboly**, the migration of apical cells down over the rest of the embryo to envelop it with what then become ectodermal cells. In mixed patterns of gastrulation (a frequent occurrence in hydrozoans), embryos employ some combination of ingression, delamination, or invagination.

Berrill (1949) concluded that scyphozoans with small eggs (30–150 μm in diameter) gastrulate via unipolar ingression; those with intermediate-sized eggs (150–230 μm in diameter) gastrulate by ingression, invagination, or a combination of the two; whereas those with the largest eggs (300 μm in diameter) gastrulate by invagination. While many scyphozoans do adhere to Berrill's scheme, exceptions are also found. Invagination, ingression, and delamination are common among the anthozoans (Hyman 1940; Chia and Spaulding 1972).

The end product of gastrulation is a bilayered ciliated **planula larva** (Figure 5.9). The planula is cylindrical, has

(a)

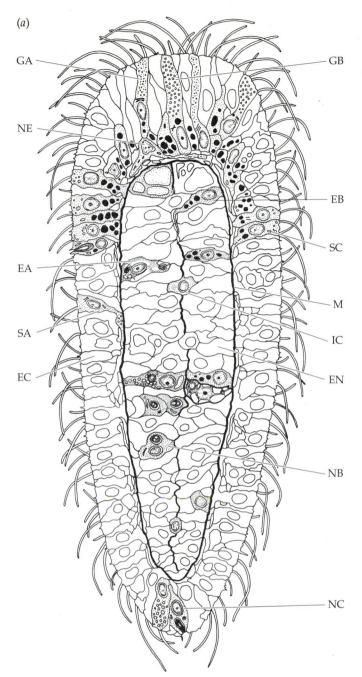

(b)

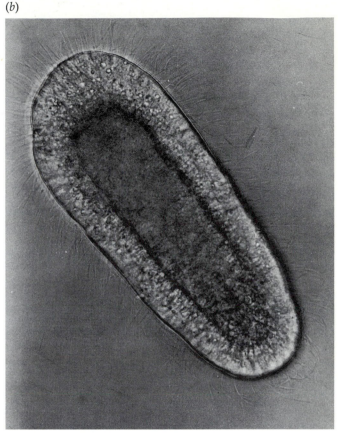

Figure 5.9 *(a)* Presettled hydrozoan planula. The ectoderm (EC) consists of supportive epitheliomuscular cells (SC), glandular cells (GA, GB), nerve cells (NE), sensory cells (SA), and nematocytes (NC). A mesoglea (M) separates the ectoderm and endoderm (EN). Supportive endodermal cells (EA, EB), nematoblasts (NB), and interstitial cells (IC) make up the endoderm. (After Martin et al. 1983.) *(b)* Typical ciliated hydrozoan planula of *Mitrocomella* showing an enlarged anterior end and a narrower posterior end. ×275.

distinct anterior and posterior ends, and is composed of an ectoderm and an endoderm separated by a thin mesoglea. Planulae creep or swim in the water, anterior end directed forward, for several hours to several days, during which time they elongate and undergo cell differentiation. The degree of histological and morphological differentiation of planulae varies depending on the species. A gastrovascular cavity may or may not be present. The ectoderm of mature planulae consists of epithelial (epitheliomuscular, glandular, and sensory) and interstitial cells, their derivatives (ganglionic cells, nematocytes), and differentiative intermediates (ganglionic neuroblasts and nematoblasts) (Martin and Thomas 1977, 1980, 1981a,b; Martin et al. 1983; Martin and

Archer 1986a,b; Martin 1988a,b, 1990, 1991). The endoderm is composed of epithelial cells (digestive, glandular), interstitial cells, nematoblasts, and neuroblasts (Martin 1988a).

The embryonic interstitial stem cell system has been extensively characterized by our laboratory (Martin and Archer 1986a; Martin 1991). In the planula the interstitial cells either divide to replenish the population or differentiate into two classes of somatic products: nematocytes and ganglionic neurons. Interstitial cells arise during gastrulation in the central core of the endoderm along the entire length of the embryo; throughout embryogenesis and larval development these cells emigrate to and populate the ectoderm (Figure 5.10). Many of the interstitial cells

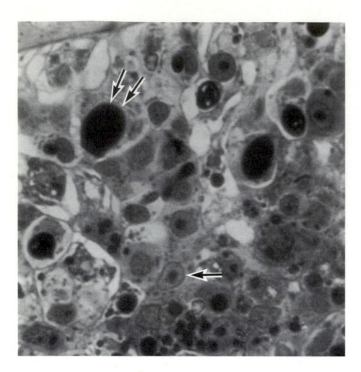

Figure 5.10   Endodermal interstitial cells (single arrow) of the mature planula of *Pennaria tiarella*. The cells are characterized by a large nucleus with a prominent nucleolus. Nematoblasts (differentiating nematocytes; double arrows) are derivatives of the interstitial cells. ×500.

Sexual Larval Forms

Throughout the phylum, both dioecious and hermaphroditic individuals are found, and most cnidarians produce sexual larval forms during development (Table 5.1). The larvae are generally planktonic and serve as a means for dispersal of the group. The typical cnidarian larva is the planula, an ovoid to fusiform, radially symmetrical, ciliated entity (Figure 5.9). Planulae are of two types: planktotrophic or lecithotrophic. **Planktotrophic planulae** feed on plankton in the water or detritus on the bottom. These feeding planulae are potentially long-lived and have a well-developed gastrovascular cavity. **Lecithotrophic larvae** do not feed, deriving their nutrition from stored yolk reserves; these planulae usually lack a gastrovascular cavity and are short-lived in the plankton. Some planulae even possess endosymbiotic algae that provide nutrition (Muscatine and McAuley 1982).

Hydrozoan planulae range from 100 to 1000 μm in length depending on the species, lack a mouth, and are lecithotrophic. They also lack a gastrovascular cavity, but in some species one may begin to form before metamorphosis. The hydrozoan planula develops into either a primary polyp or an actinula larva; in many siphonophores the planula gives rise to several intermediate larval forms: calyconula, siphonula, conaria, or rataria (Hyman 1940; Carré 1967, 1969). These intermediates eventually transform into the various adult siphonophores.

enter either the nematocyte or the neural differentiation pathway in the endoderm before migrating to the ectoderm; however, migration to a position in the ectoderm is required before differentiation is completed (Martin 1991). As planulae undergo metamorphosis, this larval stem cell population undergoes dramatic changes: (1) certain larval derivatives disappear (neurons), (2) new types of derivatives differentiate (neurons and nematocytes), (3) the distribution patterns change along the body axis, and (4) migration patterns become more complex (Martin and Archer 1997). Both the larval epithelial cells and the interstitial cell system contribute to the formation of the adult polyp. The larval epithelial cells undergo changes in their shape, thus producing the epithelial cells of the adult. Larval interstitial cells and nematoblasts are carried over into the adult; these cells constitute the adult stem cell system.

Figure 5.11   Life cycle of the colonial hydroid *Tubularia* showing formation of an actinula larva. *(a)* Medusae bud from the head region of the polyps and remain attached. Eggs are fertilized inside the attached medusae (gonophores) by free-swimming sperm. *(b)* A planula larva develops within each attached gonophore. *(c,d)* Each planula forms an actinula larva that is released from the gonophore. *(e,f)* The actinula attaches to a substrate and produces a new hydroid colony. (After Ruppert and Barnes 1994.)

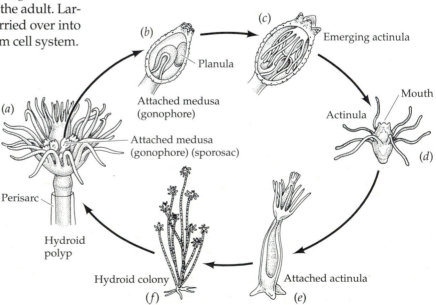

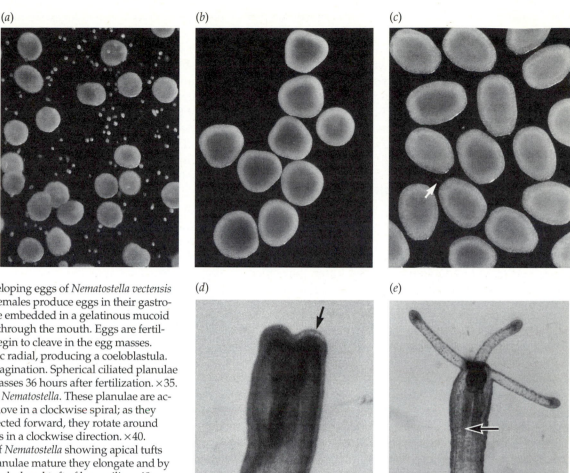

Figure 5.12 (a) Developing eggs of *Nematostella vectensis* within a jelly mass. Females produce eggs in their gastrodermis; these eggs are embedded in a gelatinous mucoid mass that is released through the mouth. Eggs are fertilized externally and begin to cleave in the egg masses. Cleavage is holoblastic radial, producing a coeloblastula. Gastrulation is by invagination. Spherical ciliated planulae hatch from the egg masses 36 hours after fertilization. ×35. (b) Young planulae of *Nematostella*. These planulae are active swimmers that move in a clockwise spiral; as they swim, aboral end directed forward, they rotate around their longitudinal axes in a clockwise direction. ×40. (c) Mature planulae of *Nematostella* showing apical tufts (arrow) of cilia. As planulae mature they elongate and by day 3 produce an apical, aboral tuft of large cilia. ×40. (d) Swimming planula of *Nematostella* growing tentacle buds (arrow) at the oral end. By 5 days many of the planulae produce at their oral ends four tentacle buds surrounding a mouth. At this time the first septa appear within the gastrovascular cavity. ×70. (e) Juvenile sea anemone. By day 7 most of the planulae stop swimming, attach to the bottom via their aboral poles, and metamorphose into juvenile sea anemones. These baby anemones possess three to four tentacles and septa (arrow) within the gastrovascular cavity. By 2–3 weeks the young anemones have eight tentacles and eight septa; by 4 weeks they have twelve tentacles and additional septa; and by 2 months possess sixteen tentacles and are approaching sexual maturity. ×95.

In some hydroids and trachylines the planula transforms into an **actinula larva** (Figure 5.11). The actinula resembles a stubby hydra; it is a motile, ciliated, pear-shaped, polyp-like larva bearing eight tentacles and an adhesive basal disc (Van de Vyver 1968; Bouillon 1975). In the Tubulariidae the planula stage is passed within the gonophore of the female polyp, and the motile form that emerges is the actinula (Pyefinch and Downing 1949). Actinula larvae give rise to polyps or medusae.

Scyphozoan and cubozoan planulae are small, ranging from 100 to 400 µm in length. With one exception (*Haliclystus*), they are uniformly ciliated and swim with an enlarged anterior end directed forward. These larvae lack

mouths and are nonfeeding; a gastrovascular cavity may appear just prior to metamorphosis (Martin and Chia 1982). At metamorphosis most planulae of scyphozoans and cubozoans transform into a polyp-like larva, the **scyphistoma** (Figure 5.5). The scyphistoma exhibits tetramerous symmetry, and its gastrovascular cavity is divided by four longitudinal septa. Scyphistomae may produce new scyphistomae by asexual budding or, in some species, produce medusae by transverse fission. Some pelagic scyphomedusae have eliminated the scyphistoma, and the planula transforms directly into a young medusa (e.g., *Pelagia*).

Anthozoan planulae are ciliated, swim with their aboral ends directed forward, and vary in length from less than 100 µm to over 5 mm. Many anthozoan planulae are planktotrophic, and many possess a well-developed aboral sensory structure consisting of cells bearing a tuft of elongated cilia (Figure 5.12) (Widersten 1968, 1973). Anthozoan planulae transform into two larval types, the **edwardsia** and the **halcampoides** forms, which are actually intermediate stages in the metamorphosis of the planula. These intermediate larvae resemble mature anemones and are classified by their number of septa. The edwardsia larva has eight septa, while the halcampoides larva possesses twelve septa. In addition, some anthozoans have distinctive larval stages (Tardent 1978). Cerianthids have an elongate four-tentacled **cerinula larva** that is a bilateral form of the edwardsia (Van Beneden 1898). The order Zoanthidea has both **zoanthella** and **zoanthina larvae** (Hyman 1940).

Postembryonic Development

Metamorphosis

Many free-swimming planulae change from active swimmers to slow swimmers or even crawlers as they near the time of settlement and metamorphosis. Often they creep along a substrate, changing their direction as if testing the environment. The planula must choose the site and conditions in which its polyp will survive.

Settlement is not a random process, nor is it determined by an internal clock (Chia and Bickell 1978). It is believed that planulae receive an external stimulus that signals the presence of an appropriate habitat. If they do not receive an appropriate signal, they will not undergo metamorphosis, and eventually will die. For the planulae of *Hydractinia echinata*, a colonial marine hydroid, the stimulus is provided by bacteria growing on hermit crab shells (Müller 1973a,b; Leitz and Wagner 1993). How such bacteria stimulate settlement and metamorphosis of planulae is not completely known. Recently, Leitz (1993) proposed that the biochemical mechanism for metamorphic activation of planulae involved a signal transduction pathway using the phosphatidylinositol cycle. He suggested that bacteria somehow activate this signal pathway in a small group of neurosensory cells in the anterior region of the planula. In fact, several other investigators have previously hypothesized that neurons are involved in transmitting the metamorphic stimulus throughout the larva (Martin and Thomas 1981b; Thomas et al. 1987; Brumwell and Martin 1994). Martin (1992) showed that hydrozoan planulae produce the neuropeptide RFamide; this peptide is formed by neurosensory cells in the anterior region of the planula and by ganglionic neurons scattered throughout the larva. She proposed that the neurosensory cells, upon appropriate environmental stimulation, release RFamide, which in turn stimulates the ganglionic neurons to expel their neuropeptides. These neuropeptides then function to relay the metamorphic stimulus to all the larval cells, which then respond accordingly.

When planulae undergo metamorphosis into polyps, several things occur: (1) the anterior (aboral) end of the planula transforms into the base and stolon of the polyp; (2) the posterior (oral) end of the planula forms the head and tentacles of the polyp; (3) certain larval derivatives disappear; (4) new types of cells differentiate; (5) distribution patterns of cells change; and (6) cellular migration patterns become more complex. Hydrozoan metamorphosis is simple (Figure 5.13; Martin and Archer 1997). The planula ceases swimming, loses its cilia, and attaches to the substrate by its anterior (aboral) pole. Both glandular secretions and nematocytes may be used for securing planulae to a substrate. Shortly after attachment, the posterior (oral) end of the larva contracts toward the aboral pole until it eventually disappears into the attached pole (Figures 5.13b,c). A tiny circular disc is formed. Next, a tiny bleb appears in the center of the disc and begins to elongate in an upright direction, forming a pawn shape (Figure 5.13d). Three distinct regions of the pawn are evident: an anterior head, a mid-stalk region, and a posterior base. The pawn undergoes growth and reshaping to produce a crown stage (Figure 5.13e). During the crown stage, general features of the adult polyp—head, stalk, and base— begin to take shape; a clear separation between the head and the stalk is evident. Tentacles evaginate from the head region, and a mouth breaks through at the tip of the head, producing an immature polyp (Figure 5.13f). A primary polyp is produced when a row of long filiform tentacles and a row of short capitate tentacles adorn the head (Figure 5.13g). A mouth is present at the very tip of the head, just above the whorl of capitate tentacles, and a perisarc covers the stalk and basal region of the polyp. If the polyp is to form a colony, it extends stolons and asexually buds additional polyps, all of which remain connected together.

Metamorphosis of scyphozoan planulae resembles that of hydrozoans, with one addition: formation of septa in the gastric cavity. As the scyphozoan planula develops into a scyphistoma, four septa form in the gastric cavity through outfoldings of the endoderm and mesoglea from the body wall. This process begins at the oral pole and progresses aborally; thus the gastric region of the developing polyp becomes subdivided. Transformation of anthozoan planulae into polyps resembles scyphozoan metamorphosis, but has a few novel features (Figure 5.12). Metamorphosis of anthozoan larvae usually begins before planulae settle; features of the polyp actually appear in the swimming planula. A mouth and pharynx form from the blastopore region if gastrulation occurred by invagination; if not, then cells in the oral region of the ectoderm rearrange to create the mouth and pharynx. Gastric septa arise from outpocketings of the endoderm and mesoglea. When eight septa have formed, the metamorphosing planula is an edwardsia larva; when twelve have formed, it is a halcampoides larva. These larvae may attach via their aboral poles at this time; tentacles may or may not have formed by settlement. If not, then shortly after settlement the tentacles arise as outpocketings of the developing oral region. Thus the juvenile polyp is formed. Additional tentacles and septa appear as the polyp ages.

Formation of Body Axes: Head, Foot

Classic experiments by Teissier (1931) and Freeman (1980, 1983a,b) showed that the uncleaved hydrozoan egg has a built-in polarity that specifies the anterior-posterior axis of

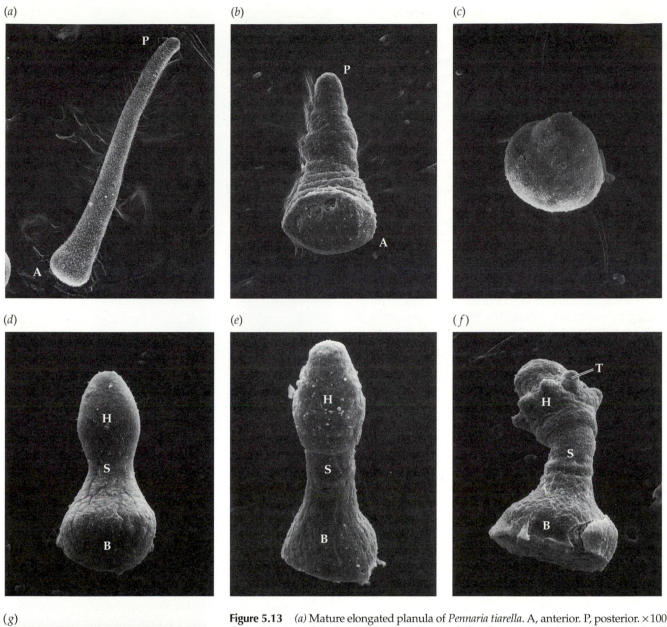

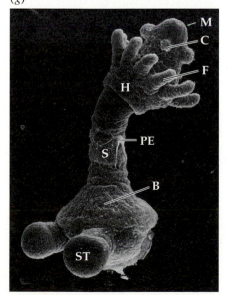

Figure 5.13 (a) Mature elongated planula of *Pennaria tiarella*. A, anterior. P, posterior. ×100. (From Martin and Archer 1986b.) (b–g) Stages of planular metamorphosis of *Pennaria tiarella*. (Photographs courtesy of W. Archer; all magnifications ×150.) (b) Attaching planula. The anterior attachment pole spreads over the substrate while the posterior end of the planula contracts down toward the attachment pole. Thus the planula becomes short and fat. A, anterior; P, posterior. (c) Disc. (d) Pawn. H, head; S, stalk; B, base. (e) Crown. H, head; S, stalk; B, base. (f) Immature polyp. H, head; T, tentacles; S, stalk; B, base. (g) Primary polyp. H, head; F, filiform tentacles; C, capitate tentacles; S, stalk; B, base; ST, stolon; PE, perisarc; M, mouth.

the planula larva. This polarity is set up while the egg is in the gonophore. During late stages of oocyte growth the germinal vesicle comes to lie on one side of the oocyte (Freeman 1990). This region of the egg is the animal pole, where the polar bodies will be discharged and where first cleavage will begin. The point on the egg surface where first cleavage is initiated becomes the posterior pole of the planula larva (oral end of the adult). Thus, the anterior-posterior axis of the planula is set up in the embryo as a consequence of the first cleavage. Freeman (1980) proposed that this generalization probably applies to all cnidarian eggs. At present, the basis for this polarity is not known (Freeman 1990).

When the planula undergoes metamorphosis, the primary polyp that forms has one head at the apical end and one foot at the basal end. There is evidence that at the planula stage of development, the posterior region has already been determined to form the head of the polyp and the anterior region has been determined to form the foot and stolon region of the polyp (Müller et al. 1977; Schwoerer-Böhning et al. 1990). Thus, the pattern of the polyp is generated under the influence of a concealed anterior-posterior prepattern found in the larva. How this prepattern is established is not known.

In adult polyps there is evidence that multiple factors play a role in the patterning processes that set up and maintain the head and the foot. Such factors include morphogens, short-range interactions based on local cell properties, and *HOM/HOX* genes (Schaller et al. 1979; Bode and Bode 1984; Bode et al. 1988; Shenk et al. 1993a,b). These factors may also be operating in the planula and thus become the basis for organismal polarity after the development of the planula (Müller et al. 1977). At least one morphogen has been identified in the planula of *Hydractinia echinata* (Müller et al. 1977), and one *HOM/HOX* gene product has been seen in embryonic hydra (Martin, personal observation).

Asexual Reproduction

Asexual reproduction is extensive in the Cnidaria and is an important vehicle for propagation of the phylum (Table 5.1). Within the hydrozoans, numerous asexual processes result in the creation of solitary individuals or the formation, growth, and specialization of colonial forms. Most polyp stages reproduce themselves by budding, whereas medusae bud from polyps and also from other medusae (Hyman 1940). All buds are genetically identical to their parent. In the freshwater hydra, budding is the chief means of reproduction. A bud develops as a simple evagination of the body wall from a region of the body column designated the budding zone (Figure 5.14). This zone is located at the lower end of the gastric region above the foot. Each bud contains an extension of the parent gastrovascular cavity and both epithelia. The bud rapidly elongates and forms a mouth and tentacles at its distal end. Eventu-

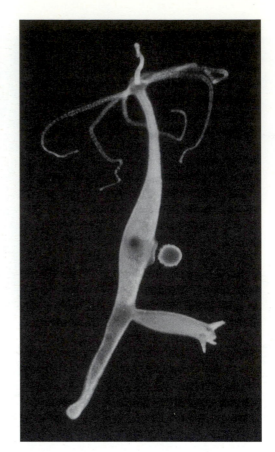

Figure 5.14   Budding hydra with an egg. ×25.

ally the bud detaches from the parent and becomes an independent hydra. The entire process requires 2 days, and well-fed hydra bud every 3–4 days. At times multiple buds may arise simultaneously from the budding zone.

The vast majority of the hydrozoans are colonial, and in these forms budding usually occurs from stems and stolons. Unlike hydra, however, in the colonial forms the buds remain attached to each other; therefore, each polyp is connected to every other polyp. Thus the whole colony shares a single ectoderm, endoderm, mesoglea, and gastrovascular cavity. Such colonies may be sessile (**hydroids**) or floating (**siphonophores**). Although budding is the primary mode of asexual propagation in the hydrozoans, other minor forms exist: transverse fission (e.g., *Protohydra*), longitudinal fission (hydra), and formation of frustules (e.g., *Craspedacusta*) (Hyman 1940). Frustules are nonciliated planula-like buds that creep or lie on the bottom and form polyps; they constrict from stolons and polyps (Hyman 1940). Some siphonophores produce chains of individuals called cormidia, which may break free to begin a new colony; two types of siphonophore larvae, the calyconula and siphonula, bud (Carré 1967, 1969).

Scyphozoans reproduce asexually by budding from the polyp or stolon, by production of ciliated planuloid buds, or by formation of **podocysts** (encysted polyp rudiments) (Berrill 1949). Scyphistomae may bud new polyps from the side of the stalk, like hydra, or extend hollow stolons from the stalk or base that bud off new scyphistomae (e.g., *Aurelia* and *Chrysaora*) (Gilchrist 1937; Berrill 1949). In some scyphistomae, such as *Cassiopea*, the head region pinches off free-swimming ciliated buds (Figure 5.15*a*). These buds resemble planulae and are called **planuloid buds.** Each bud, containing ectoderm, mesoglea, and endoderm, swims with its distal/aboral end directed anteriorly, rotating from right to left upon its longitudinal axis. After receiving an appropriate stimulus for metamorphosis, the planuloid bud attaches via its distal/aboral end and transforms into a scyphistoma (Figure 5.15*b*) (Van-Lieshout and Martin 1992). The distal end of the bud forms the stalk of the scyphistoma, while the proximal region forms the head of the scyphistoma. In well-fed scyphistomae, production and release of planuloid buds require 24 hours, and metamorphosis can occur in less than 24 hours if an adequate substrate is available (Van-Lieshout and Martin 1992). Semaeostome and rhizostome scyphistomae, under adverse environmental conditions, produce podocysts. These encapsulated structures, formed from the foot region of the scyphistoma, consist of epidermal and mesenchymal cells encased within a chitinous cyst. After a period of dormancy the cyst breaks open, yielding either a small polyp or a planula-like structure that settles to form a polyp (Chapman 1968).

One of the most spectacular examples of transverse fission in cnidarians occurs in the semaeostome and rhizostome scyphozoans. During certain times of the year, depending on temperature, light, and nutrition, the scyphistomae of these scyphozoans undergo **strobilation**, the production of young medusae (**ephyrae**) by transverse fission of their oral ends (Spangenberg 1965). In some species strobilation is monodisc, with a single ephyra forming at one time (e.g., *Cassiopea*) (Figure 5.16). In other groups it is polydisc; i.e., the entire scyphistoma undergoes a series of successive transverse constrictions such that the ephyrae are stacked like saucers at the oral end of the polyp (see Figure 5.5). A large scyphistoma such as *Chrysaora* may form 13 to 15 ephyrae at one time. As ephyrae are formed, they break away from the oral end of the polyp one by one. After strobilation, scyphistomae remain as polyps and usually repeat the process the following year. The average life span for the scyphistoma is 1–3 years. The young microscopic ephyra, characterized by a deeply incised bell margin, will grow and develop into a sexually mature adult scyphomedusa. Depending on the species this may take a few months to a few years.

(*a*)

(*b*)

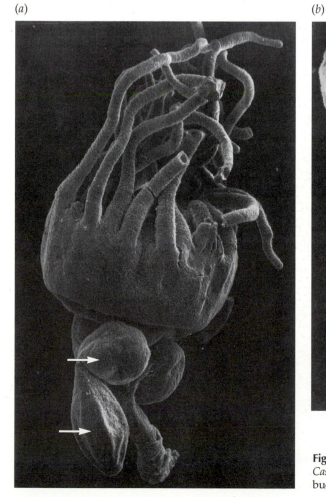

Figure 5.15 (*a*) Planuloid buds (arrows) attached to the scyphistoma of *Cassiopea xamachana*. ×60. (*b*) Young scyphistomae developing from planuloid buds of *Cassiopea xamachana*. ×115. (From VanLieshout and Martin 1992.)

Figure 5.16 Monodisc strobilation of *Cassiopea* scyphistoma. ×30. (Photograph courtesy of J. VanLieshout.)

disc extend lobes that fragment, or pieces of the disc adhere to a substrate and break free as the animal moves about. In both cases the detached pieces of tissue form tiny anemones (Stephenson 1928)

Regeneration

The potential for regeneration is extensive in all classes of the Cnidaria. Some species can be totally reconstituted from isolated groups of cells (Achermann 1980), whereas in other individuals some structures are incapable of regeneration or show limited potential for regrowth. The regenerative abilities of the freshwater hydra have been thoroughly studied (Bode et al. 1988; Javois 1992; Shenk et al. 1993b). When the head or foot of a hydra is removed, the missing structures are re-formed by the remaining tissue through a process of cell rearrangement (**morphallaxis**) (Figure 5.17). If a piece of tissue is removed from the gastric region, it will regenerate missing structures, maintaining the original polarity. Thus the apical end of the excised piece will regenerate head structures, while the basal end will regenerate foot structures. A polarity or gradient of dominance exists from the oral to aboral end. This gradient is reflected in the rate of regeneration, as an aboral piece will regenerate more slowly than a more oral piece

Many anthozoans reproduce asexually by fission, budding, or pedal laceration. Both longitudinal fission (e.g., *Sagartia*) and transverse fission (e.g., *Nematostella*) have been described for various sea anemones and often result in large groups, or clones, of genetically identical individuals (Hyman 1940; Hand and Uhlinger 1995). Budding from the column or pedal disc has been reported for some sea anemones, and in one rare example, little anemones actually bud from the tentacles. Budding is extremely important for the growth and development of colonies of scleractinian (stony) corals and alycyonarians. Such colonies originate from a single sexually produced polyp that buds new polyps that then remain connected. Colonies may be formed by polyps budding from other polyps, from basal stolons between polyps, or from a shared fleshy layer, the coenchyme. Pedal laceration is a common form of multiplication for many sea anemones (e.g., *Metridium, Sagartia, Aiptasia, Phellia, Bunodes,* and *Heliactis*) (Hyman 1940). In this process, portions of the pedal

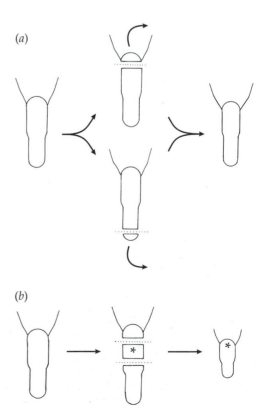

Figure 5.17 Hydra regeneration. (*a*) Removal of the head or the foot results in regeneration of the missing structure through cell rearrangements. (*b*) A piece of tissue excised from the gastric region regenerates the missing structures, maintaining the original polarity. Apical tissue (asterisk) produces head structures while basal tissue forms a new foot. (From Javois 1992.)

(Campbell 1974). In all cases the regenerated animal is complete, correctly proportioned, yet smaller than the original hydra.

The marine hydroids exhibit great variability in regenerative behaviors. Some species, such as *Podocoryne carnea*, like hydra, regenerate both distal head and proximal foot or stolon structures (Müller et al. 1986). Others, such as *Tubularia* and *Corymorpha*, regenerate almost exclusively distal head parts; for example, if a piece of their body column is excised, both ends usually regenerate heads (Child 1941; Berrill 1961; Tardent 1963). In many hydroid colonies any cut surface or free end where hydranths have been removed can produce stolons (Hyman 1940). Pieces of gastrozooids, dactylozooids, and gonozooids from *Hydractinia* are capable of regenerating their own kind of hydranth. Isolated gonophores and tentacles of hydroids are incapable of regeneration and eventually die. If gonophores are removed from colonies, the colonies do not regenerate these structures at the sites from which they were excised.

Hydromedusae may re-form lost parts such as tentacles, manubria, and a certain amount of umbrella tissue. However, if large chunks are excised from their bells, they usually close along the wound and do not regenerate the lost portion (Hyman 1940). Hydrozoan planulae such as *Pennaria tiarella* can replace oral and aboral ends if bisected. These regenerated larvae metamorphose to produce normal, yet small, primary polyps.

Regeneration of scyphozoan medusae is similar to that of hydromedusae. Bisected bells or bells in which quadrants have been removed close together at the cut surfaces without replacing the lost parts. However, in most medusae, marginal structures, rhopalia, oral arms, some umbrellar tissue, and gastric pouches usually regenerate perfectly (Lesh-Laurie and Suchy 1991). Apical and basal fragments of some scyphistomae regenerate both distal (head) and proximal (foot) structures when bisected; however, some species lack proximal regeneration altogether (Lesh-Laurie and Suchy 1991). In *Aurelia*, an entire scyphistoma can form from the distal one-third of a tentacle (Lesh-Laurie and Corriel 1973). Aboral fragments of scyphozoan planulae or planuloid buds produce complete polyps during metamorphosis. Oral fragments, however, produce polyps with head structures only; the polyps lack stalks and feet (Curtis and Cowden 1971; Neumann 1977).

In anthozoans regeneration is variable. In some forms, *Metridium*, *Sagartia*, and *Harenactis*, when the column is bisected, the aboral portion regenerates a new oral disc, while the oral piece fails to re-form a pedal disc. Instead, it may regenerate tentacles on the aboral surface (Hyman 1940). Isolated tentacles and pieces of pharynx, oral disc, or septa usually show no regenerative abilities. An exception is seen in the primitive anemone *Boloceroides*, in which cast-off tentacles regenerate into complete animals (Hyman 1940). And finally, in various anemones, pieces of the pedal disc of any size have been shown to regenerate, producing whole anemones.

Class Hydrozoa: Sexual Reproduction of *Hydra vulgaris*

Of all the cnidarian species, hydra have been relatively well studied. In *H. vulgaris* and *H. utahensis*, gametogenesis occurs spontaneously, and no specific environmental cues have been identified (Tardent 1968; Campbell 1989). Gametogenesis can be induced in *H. oligactis* and *H. hymanae* by lowering the temperature (Park et al. 1965; Littlefield et al. 1985) and in *H. magnipapillata* by decreasing food intake (Sugiyama and Fujisawa 1977).

Gametogenesis

In hydra, gametes arise from a population of interstitial cells (I-cells); these I-cells are dispersed within intercellular spaces of the epitheliomuscular cells. During gametogenesis I-cells accumulate in specific ectodermal regions of the body column to form gonads. Each gonad consists of clusters of I-cells, which cause the overlying ectodermal epithelium to bulge; these bulges are supported by the overlying stretched ectodermal cells (see Figure 5.1).

In *H. vulgaris*, a dioecious species, testes are conical protrusions provided with an opening, the nipple, through which sperm exit; these testes are found along the entire upper body column from beneath the head to the budding region (Figure 5.18*a*). Ovaries appear as large ectodermal bulges in the lower half of the body column near the budding zone. Each ovary contains one to several developing eggs covered by a thin layer of unspecialized flattened epitheliomuscular cells (Figure 5.18*b*).

Detailed accounts of spermatogenesis have been described for several species of hydra, and the events are similar to those in higher metazoans, including humans

Figure 5.18 (*a*) Male hydra with testes (arrow). ×20. (*b*) Female hydra forming an egg. Oogonial cells stream together and fuse; during this time the hydra exhibits a distinctive splattered-egg morphology. ×5.

(*a*)

(*b*)

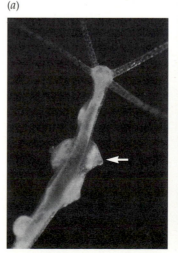
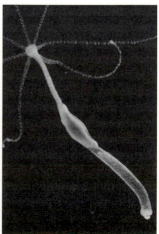

(Fawcett and Slautterback 1959; Stagni and Lucchi 1970; Littlefield et al. 1985; Littlefield 1994). During spermatogenesis, I-cells committed to the sperm lineage proliferate throughout the entire body column, forming nests containing 4–32 cells (Figure 5.19). Each nest consists of a syncytium of I-cells connected by intercellular bridges (Fawcett and Slautterback 1959). Once formed, these nests of I-cells migrate to testes-forming regions, where they undergo morphological differentiation into sperm intermediates; these intermediates undergo meiosis to form the spermatids and sperm. In *H. oligactis* the entire sperm lineage from I-cell to sperm can be followed using labeled antibodies. Using these antibodies, Littlefield reported that the time required to complete spermatogenesis is less than 3 days, and that proliferation of I-cells proceeds for 2 weeks prior to the development of morphologically identifiable testes (Littlefield et al. 1985). Within the testes distinct regions of maturation are found, with the youngest cells, the spermatocytes, lying closest to the mesoglea and the mature sperm located just beneath the ectodermal epithelium of the testis. In *H. vulgaris*, as long as testes persist, sperm are continuously released.

The sperm of hydra is the typical cnidarian type; it consists of a conical head with no distinct acrosome, a midpiece with five concentrically arranged mitochondria of equal size, and a flagellar tail 30 μm in length. Hydra sperm lack the proacrosomal vesicles that have been found in the spermatozoa of several marine cnidarians (Hinsch and Clark 1973). Once released from the testes, hydra sperm remain active for about 3 hours.

Figure 5.19   Diagram of the entire sperm and egg differentiation pathways in *Hydra oligactis*. In response to lowered temperatures, males and females produce sperm and eggs. (After Littlefield 1994.)

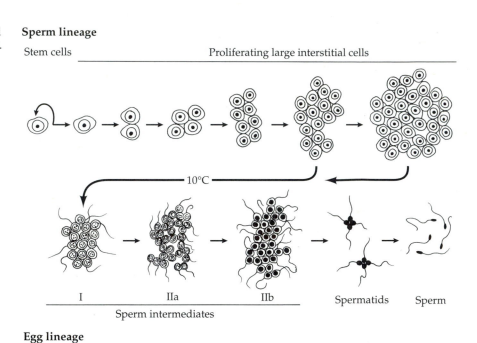

Sperm lineage

Stem cells — Proliferating large interstitial cells

10°C

I   IIa   IIb   Spermatids   Sperm

Sperm intermediates

Egg lineage

Stem cells — Proliferating large interstitial cells

10°C

Fusion of nurse cells   Developing oocyte   Mature egg

In females of *H. vulgaris*, gamete differentiation begins with the proliferation of I-cells, producing multiple clusters containing 4 -32 cells (Figure 5.19); these clusters are located in the interstitial spaces of the ectoderm between the mesoglea and the epitheliomuscular cells. During egg formation one of these cells becomes the oocyte, while the others begin to synthesize yolk. These yolk-forming eggs are called nurse cells. It is not known how the decision is made to become oocyte or nurse cell. The oocyte increases in mass by extending fingerlike processes in between the epithleliomuscular cells; these processes fuse with, and/or engulf, large numbers of nurse cells. Females at this stage show a distinctive "splattered-egg design" along the body column, much like the appearance of a dropped breakfast egg in a frying pan (Figure 5.18*b*). As the oocyte grows in mass it assumes a shieldlike appearance and finally becomes spherical (Figure 5.20*a*), distending the overlying ectoderm. Egg formation, starting with the proliferation of I-cells, requires 3–4 days for completion. Oogenesis is carried out solely by cells of the oocyte lineage; once they are segregated from the I-cells, no somatic cells are involved in egg production.

As soon as the egg becomes spherical, polar body formation occurs. The egg nucleus migrates to the apical end of the cell, the point farthest away from the mesoglea, where it undergoes meiosis, resulting in the extrusion of two polar bodies. The site of polar body release marks the future head end of the adult. The overlying ectoderm in the region of polar body discharge ruptures, exposing the egg (Figure 5.20*b*). The ruptured ectoderm recedes around the edge of the egg, forming a raised ring of tissue at the base of the egg referred to as the *egg cup*. Each egg measures 100–500 µm in diameter and is loosely attached to

the egg cup by thin strands of tissue. Eggs are now fertilizable for 2 hours; if they are not fertilized during this time, they will swell and degenerate within the day.

Each month a single female can produce from one to six eggs at one time. If a single egg is formed, it is large, measuring 400–500 µm in diameter; whereas if multiple eggs are formed, they are smaller and measure 100–300 µm in diameter. When multiple eggs arise at the same time, usually there are two to three egg cups produced on opposite sides of the lower body column. As many as three eggs per cup have been observed, the average being two eggs per cup.

The unfertilized egg of hydra is surrounded by a jelly coat, and numerous short microvilli project from the surface (Figure 5.20*c*). Directly beneath the egg membrane in the region of polar body discharge lies the egg pronucleus. Scattered throughout the cytoplasm of the egg are thousands of spherical, refractile pycnotic nuclei (dying nuclei); these nuclei represent the remnants of the nurse cells. Pycnotic nuclei persist in the developing embryo until hatching.

Fertilization

Fertilization in hydra occurs at the site of emission of the polar bodies. In this region of the egg the surface is slightly indented to form a cuplike depression, the **fertilization pit** (Figure 5.21). The jelly coat is present in this region. Directly beneath the pit is the egg pronucleus; sperm penetration is restricted to this particular region of the egg surface (Honegger 1983). Honegger (1983) reported that as many as 50 sperm can enter the fertilization pit, but only a single sperm fuses with the plasma membrane of the egg. Fusion is initiated by contact between the membrane of

(*a*)   (*b*)   (*c*)

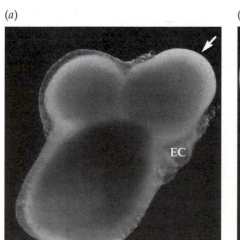
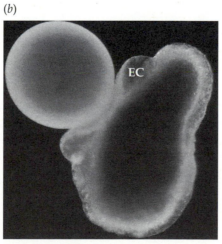

Figure 5.20   (*a*) Female hydra producing eggs. One egg has ruptured through the overlying ectoderm (arrow) while the other has not. EC, egg cup. ×60. (*b*) Emerging hydra egg. The ectoderm has ruptured and the egg is exposed to the surrounding environment. EC, egg cup. ×60. (*c*) Cross section of a hydra egg filled with dark-staining nurse cell nuclei. The functional nucleus of the egg is located directly beneath the surface at the animal pole (arrow). This is the fertilization site. EC, egg cup. ×70.

Figure 5.21 Early development of hydra. (*a*) Uncleaved egg, showing the fertilization pit, indentation, at the future oral end of the adult. (*b*) Two-cell stage. (*c*) Four-cell stage. (*d*) Eight-cell stage, showing two tiers of cells. (*e*) Late cleavage stage. (*f*) Blastula.

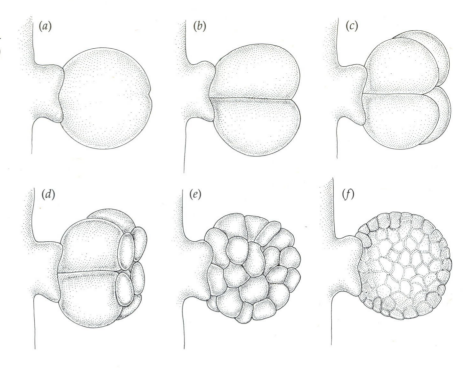

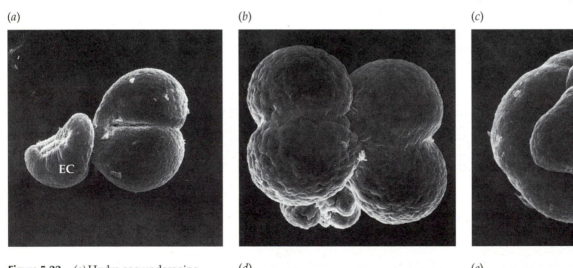

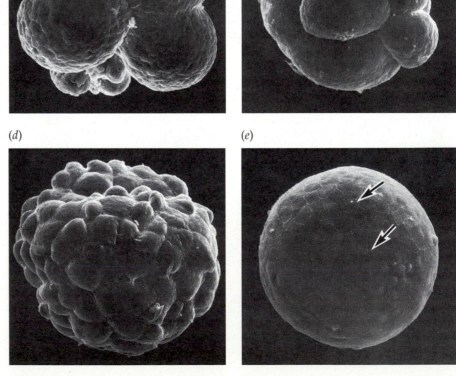

Figure 5.22 (*a*) Hydra egg undergoing first cleavage. Note the distinctive heart shape created by unilateral furrowing. EC, egg cup. ×85. (*b*) Four-cell embryo of hydra. ×225. (*c*) Forming 8-cell embryo of hydra, showing two tiers of cells. ×170. (*d*) Late cleavage stage of hydra. ×160. (*e*) Coeloblastula of hydra. Note the microvilli (arrows) that delineate the margins of the cells. ×225.

the lateral part of the sperm head and the egg surface at the bottom of the pit region. The time required for a sperm to penetrate the jelly coat and fuse with the egg membrane is extremely short, requiring no more than 30–40 seconds. Neither microvilli nor a fertilization cone form at the site of gamete fusion. In *H. carnea*, electron-dense material is released by the egg upon insemination, but cortical granules are not discharged and a fertilization envelope is not formed (Honegger 1983). During fusion both the sperm head and the midpiece are incorporated into the egg. Fusion of male and female nuclei occurs 8 to 10 minutes after insemination, and the fertilization pit disappears about 15 minutes after insemination.

Events of Embryogenesis

CLEAVAGE AND BLASTULATION. If sperm are present in the surrounding water, first cleavage begins 2 hours after the egg breaks through the mother's body column (Figures 5.21 and 5.22*a*). First cleavage and all subsequent divisions are unipolar; that is, the cleavage furrow progresses inward from one side of each cell. First cleavage is equal holoblastic; the cleavage furrow starts at the apical end of the egg (the side away from the egg cup, the site of polar body discharge) and progresses inward to the basal pole of the egg (i.e., perpendicular to the mother's body column), forming two blastomeres. Second cleavage, which begins shortly before the first division is complete, also occurs in an apical-basal direction and is perpendicular to the first (Figure 5.22*b*). It is equal holoblastic, resulting in the formation of four equal-sized blastomeres. The third division is equatorial (parallel to the adult body column) and starts at the end of the embryo closest to the head of the mother

hydra (Figure 5.22*c*). This division occurs much closer to the apical end of the embryo, resulting in two tiers of unequal-sized blastomeres. Because the furrow starts at one end, the division of the four blastomeres is asynchronous. Ideally, at the end of the third cleavage, the embryo consists of four large cells (closest to the egg cup) and four small cells (farthest from the egg cup). Thereafter, cleavage planes become oblique and cell divisions asynchronous, forming unequal-sized blastomeres, with those near the apical end generally smaller than those at the basal end (Figure 5.22*d*). Up to the 32-cell stage, cleavage divisions occur roughly every 60 minutes. Thereafter, until the formation of the blastula, the cell division rate slows down, and the cells tend to become more uniform in size.

By 6–8 hours after fertilization a hollow spherical blastula (coeloblastula) has formed (Figure 5.22*e*). The wall of the blastula is composed of approximately 76 rectangular cells arranged in a single layer around the fluid-filled blastocoel. The outer surface of the blastula resembles a patchwork quilt, as the outlines of the cells are clearly delineated by rows of short microvilli.

GASTRULATION. Shortly after formation of the blastula, gastrulation begins via multipolar ingression (Figure 5.23*a*). Individual cells detach from their neighbors and migrate into the blastocoel in a manner similar to that of primary mesenchyme cells in sea urchin embryos.

Figure 5.23 (*a*) Cross section of a hydra blastula, showing cells (arrows) ingressing into the blastocoel. ×180. (*b*) Cross section of a hydra gastrula, revealing two distinct layers. The outer layer (E) will give rise to the epidermis of the adult polyp, while some of the cells of the inner layer (EN) will contribute to the gastrodermis of the adult. Nurse cell nuclei (arrows) are abundant in the inner layer of cells. ×240.

(*a*)

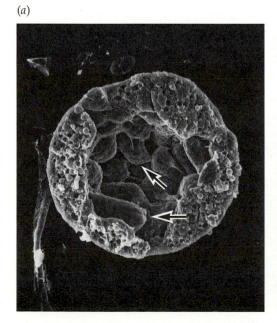

(*b*)

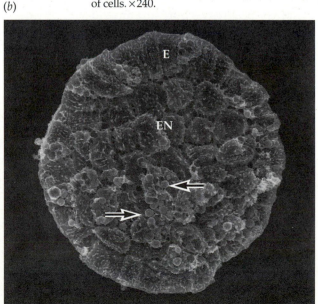

In the filling blastula the outer layer of cells changes shape to become more columnar; the nurse cell nuclei are rearranged and are found in the basal tips of the columnar cells. Cells start ingression by elongating and extending their basal surfaces into the blastocoel. Simultaneously, the apical and lateral contacts with neighboring cells are reduced and eventually severed, resulting in these roughly spherical blastomeres moving into the blastocoel. These central blastomeres are especially rich in nurse cell nuclei. Ingression begins at the apical end of the embryo, opposite the egg cup, and progresses outward from that point all around the embryo. There is an apparent wave of contraction as the wave of ingression travels over the embryo surface.

The process of ingression occurs in about 4 hours, and when complete, the embryo consists of an outer layer of columnar cells and a central mass of unorganized spherical cells (Figure 5.23b). By the completion of gastrulation the majority of the nurse cell nuclei are packaged in the central blastomeres; only a few of these pycnotic nuclei remain in the outer layer. In addition, no cavity remains, and the embryo has compacted, becoming somewhat smaller. During gastrulation the number of cells increases roughly fourfold, from about 76 cells to 315 cells.

Gastrulation is completed by 10–12 hours postfertilization. There is little visible activity for the next 10–12 hours.

CUTICLE DEPOSITION.   The next stage involves the formation of the cuticle, a thick protective layer that is also referred to as the **embryotheca** (Figures 5.24a,b). About 20–24 hours postfertilization, the cells of the outer layer begin to extend filopodia into the surrounding medium. Cuticular material is deposited in layers around the filopodia and on the apical surfaces of the cells. The process is complete 40–48 hours postfertilization. By this time the outer layer of cells has flattened against the cuticle. Cuticle deposition starts on the basal side in the egg cup region and slowly progresses outward around the embryo in an apical direction. The final distribution of material is somewhat asymmetric, with the cuticle being thickest (~60 μm) at the basal end and thinnest (~45 μm) at the apical end. After deposition is complete, a second very thin layer is deposited by the outer cells beneath the cuticle. Once both cuticle layers are formed, the embryo detaches from the parent, although detachment may occasionally occur at any point from the 2-cell stage on. When detachment occurs has no bearing on the progress of embryogenesis.

The length of time spent in the cuticle is variable, ranging from 3 to 18 weeks in the laboratory. This corresponds to a dormant overwintering period of the embryo in nature. However, two important events occur during this time. The outer layer of cells continues to flatten against the cuticle, and the cells become very narrow along their apical-basal axes but expand their lateral axes (Figure 5.24c). These cells are devoid of nurse cell nuclei. The inner region of the embryo is solid and is filled with spherical cells rich in nurse cell nuclei. The interstitial cells arise during this time, appearing among the interstitial spaces of the outer flattened layer (Figure 5.24c). The I-cells are single, contain a prominent nucleus with nucleolus, and are found throughout the outer layer. During this dormancy period the cell division rate is low, and cuticle morphology remains unchanged.

BILAYER FORMATION.   During the last stage of embryogenesis, two epithelial layers that closely resemble the layers of the adult animal and a mesoglea are formed. Because the cuticle stage is of variable length, the timing of the onset of bilayer formation is not easy to predict. However, an embryo in the final stage can be recognized by a clear morphological change 2 days before hatching. Before this time, the embryo is opaque due to the refractility of the several thousand nurse cell nuclei, but 2 days before hatching, the outer layer of the embryo becomes translucent due to the expansion of the outer layer of cells that are devoid of nurse cell nuclei (Figure 5.25). This layer now consists of low columnar epithelial cells, large numbers of interstitial cells, a few nematoblasts, and some neurons producing the neuropeptide RFamide. This layer closely resembles the ectoderm of the adult.

The formation of the ectoderm precedes the formation of the endoderm. Once the ectoderm has formed, many of the cells of the unorganized interior mass line up along the ectoderm and change their shape from spherical to columnar. This alignment occurs in different parts of the embryo simultaneously, thus producing a complete endodermal layer. Cells that are not incorporated into the forming endodermal layer degenerate, producing an internal cavity, the future gastrovascular cavity of the adult. Interstitial cells appear in the apical interstitial spaces of the endodermal layer. Once the two layers are formed, the mesoglea is synthesized, and the cuticle of the embryo begins to loosen.

By the bilayer stage, the rate of cell division has accelerated, and the cells of the I-cell lineage constitute about 40% of the total cells. The remainder are the epithelial cells.

HATCHING OF EMBRYO.   Once the bilayer has formed, the embryo begins to pulsate in a rhythmic contraction and expansion, which continues until hatching is complete. As the embryo pulsates, perforations and channels appear in the cuticle, indicating that it is beginning to break down. About 48 hours after formation of the two epithelial layers, the cuticle cracks open on its thinnest side, the original apical side of the embryo, where the head forms. The spherical embryo emerges from the cuticle through this crack (Figures 5.26a,b). For the next 2.5 hours the periodic elongations and contractions continue, and the shape of the animal changes from a compact sphere to an elongate cylinder. During this time the hatchling embryo is enveloped in an acellular accessory layer, which eventually ruptures as the hatchling in-

(a)

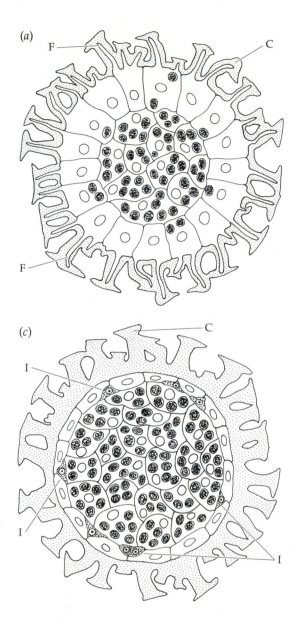

(b)

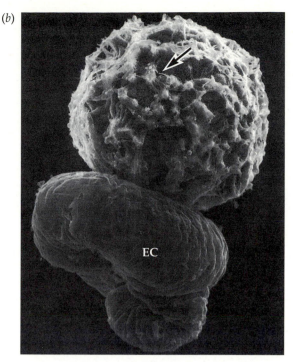

Figure 5.24 *(a)* Early cuticle stage of hydra. The cells of the outer layer extend tiny filopodia (F) upon which cuticular material (C) is deposited. Note that the darkly colored nurse cell nuclei have accumulated in the central cells. *(b)* Early cuticule of hydra. Note the cuticular material (arrow) being deposited at the surface. EC, egg cup. ×180. *(c)* Mid-cuticle stage of hydra. Interstitial cells (I) arise in the interstitial spaces of the outer ectodermal layer. Note the flattened morphology of the cells of the ectoderm. The inner cells are filled with nurse cell nuclei; no nurse cell nuclei remain in the ectoderm. C, cuticle. (Drawings by J. VanLieshout.)

creases in length. Upon rupture of the acellular layer, the hatchling increases its activity, gyrating and wiggling until it is completely free of the cuticle.

During the final stages of hatching the apical end of the hatchling undergoes dramatic morphological changes. Before the rupture of the accessory layer, the apical end has the shape of a smooth dome (Figure 5.26*b*). Within 15 minutes after rupture, the apical end narrows into a more conical shape, and one to five tentacle buds evaginate in a ring below the tip (Figure 5.26*c*). At this point the head of the hatchling is morphologically fairly complete. The foot

Figure 5.25 Bilayer stage. The two germ layers, the ectoderm (E) and the endoderm (EN), are distinct and are separated by a mesoglea (M). A forming gastrovascular cavity (G) is present. Interstitial cells (I) are found in both germ layers. C, cuticle. (Drawing by J. VanLieshout.)

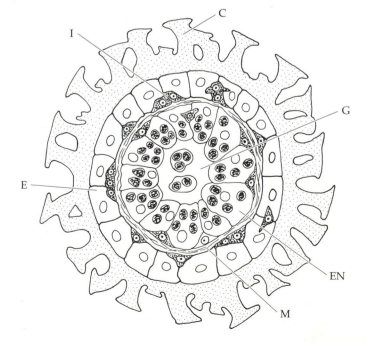

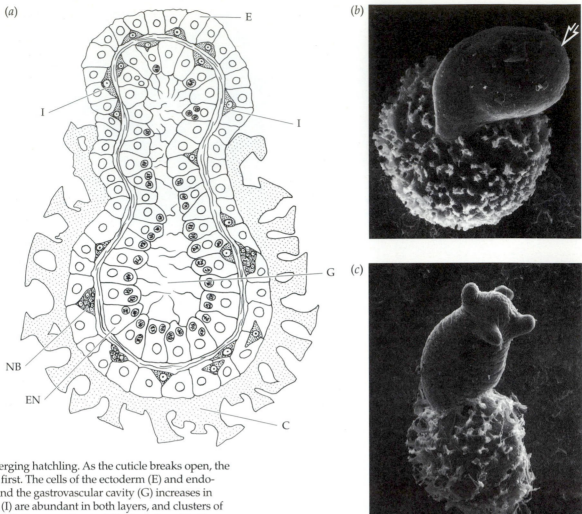

Figure 5.26 *(a)* Emerging hatchling. As the cuticle breaks open, the hatchling exits head first. The cells of the ectoderm (E) and endoderm (EN) expand and the gastrovascular cavity (G) increases in size. Interstitial cells (I) are abundant in both layers, and clusters of nematoblasts (differentiating nematocytes) (NB) are found in the ectoderm. C, cuticle. (Drawing by J. VanLieshout.) *(b)* Hydra hatchling emerging from the cuticle. Note the dome-shaped anterior head (arrow). ×95. *(c)* Hydra hatchling with tentacles. ×95.

is also quite complete since, upon removing itself from the cuticle, the hatchling is able to stick to the surface of the petri dish—a sign of a functional foot.

The hatchling has a full complement of the somatic cell types that an adult normally has. The number of cells doubles from about 5000 to 10,000 cells as the embryo progresses from the bilayer stage to the hatchling stage. This increase is entirely due to increases in the cell populations of the interstitial cell lineage. The epithelial cell numbers do not change during the last two stages.

The hatchling has the developmental capabilities of a normal adult. It regenerates as does a normal adult. Bisection of a hatchling results in the regeneration of a head at the apical end of the lower half and a foot at the basal end of the upper half. Young hydra bud within 10 days of hatching, and become sexual 4 to 8 weeks after hatching.

PATTERNING AND MORPHOGENESIS. Not only does a hatchling consist of two epithelial layers and interspersed cells of the interstitial cell lineage, but it also has a distinct form. The animal has an apical-basal axis that is polarized, with a head at the apical end and a foot at the basal end. Further, the head is composed of two parts: the apical dome is the mouth region, and below that is the tentacle zone, from which emerge the evenly spaced tentacles.

Two aspects of patterning occur very early. The apical-basal axis, or the axis of the body column, as well as the specification of the apical end, is set up at the time first cleavage is complete. By immobilizing embryos in agar in a known orientation, we have shown that the head always forms at the apical end where the first cleavage is initiated, and that the first cleavage plane is parallel to the future body axis of the adult.

Although the location at which a head will form is established very early in development, the details of the patterning of the head and the formation of the structures of the head are very late events. Since both epithelial layers are involved, these events most likely are initiated after these layers are established at the bilayer stage. Within a 15-minute period shortly after hatching, several small tentacles appear, evenly spaced in a ring in the tentacle zone of the head, directly below the hypostome, or mouth region. Therefore, the location of the tentacles, and the rearrangement of cells necessary for the emergence of the tentacles, must have been complete by hatching. Also, the mouth has formed by the time of hatching, as it is capable of opening and ingesting food shortly after hatching. Thus, the apical ends of both the ectoderm and endoderm must have been designated to form the mouth, and the endoderm must have undergone the appropriate modifications to allow the mouth to open. Finally, as soon as the hatched animal leaves its cuticle, it is capable of secreting sticky materials that allow it to adhere to substrates, indicating that the foot is fully formed.

Members of the *HOM/HOX* homeobox genes have recently been identified in hydra (Schummer et al. 1992; Shenk et al. 1993a,b). One of these, *Cnox-2*, resembles the *Deformed* gene of *Drosophila*. The expression pattern of *Cnox-2* suggests that it may be involved in axial pattering of hydra, as it is strongly expressed in the epithelial cells of the body column and foot (Figure 5.27), but only weakly expressed in the head (Shenk et al. 1993a). Experimental evidence also supports the role of *Cnox-2* in hydra pattern formation. When body column tissue is converted into head tissue by regeneration, transplantation, or during budding, *Cnox-2* expression is reduced, indicating that expression of this gene is suppressed during head formation

(Shenk et al 1993b). In young hatchlings the adult expression pattern is observed immediately after hatching. Furthermore, *Cnox-2* expression in hydra embryos has been shown to begin as early as the gastrula stage in a few cells (V. Martin, personal observation). Thus, *HOM/HOX* genes may be functioning early in hydra embryos to set up and maintain the head and foot.

PATTERNING AND DEVELOPMENTAL GRADIENTS. The ability of hydra to form a head or a foot is due to developmental gradients that exist in the tissue (for a review, see Javois 1992). Head gradients originate from the head and control head formation; foot gradients emanate from the foot and control foot formation. In the head end of hydra a gradient of head-forming potential, called **head activation**, is maximal, resulting in the formation of an anterior head. Head activation is graded along the body column, decreasing down the body column toward the foot. A second gradient, **gradient of head inhibition**, also originates from the head; this gradient is maximal in the head and reduced down the body column. Head inhibition prevents body column tissue from forming additional heads. Similar gradients of **foot activation** and **foot inhibition** arise from the foot end of hydra; they are maximal in the foot and decrease up the body column. Foot activation promotes foot development at the basal end of the animal, while foot inhibition prevents body column tissue from forming multiple feet.

The head and foot gradients work together to maintain the overall morphology of hydra and to prevent the formation of spontaneous structures (heads or feet) along the body column. Activation gradients stimulate head or foot formation; activation is a stable tissue property. Inhibition gradients, however, suppress head or foot formation; inhibition is a labile tissue property. It is the relative levels of activation and inhibition along the body axis that determine whether a structure will form.

Gradient interactions can explain the polarity of hydra regeneration (Figure 5.17). When a piece of tissue is removed from the hydra body column, the apicalmost end of the tissue has the highest levels of head activation, while the basalmost end has the highest levels of foot activation. As inhibition is labile, this gradient drops throughout the tissue. Thus the end with highest head activation regenerates a head, and the end with highest foot activation regenerates a foot.

Four factors, called **morphogens**, have been isolated from hydra and shown to play roles in establishing and maintaining the developmental gradients of activation and inhibition. A morphogen is a substance that exists in a con-

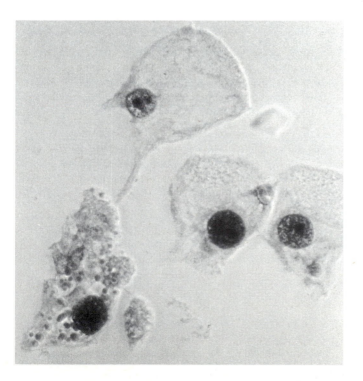

Figure 5.27 Hydra adult epithelial cells expressing *Cnox-2*, a *HOM/HOX* homeobox gene. Note the intense nuclear stain. (Photograph courtesy of A. Shenk, R. Steele, and H. Bode.)

centration gradient along the body axis; morphogens specify position along a body axis. Two activators, **head activator** (HA) and **foot activator** (FA), and two inhibitors, **head inhibitor** (HI) and **foot inhibitor** (FI), have been identified. The activators (HA and FA) are small peptides with molecular weights near 1000; they are present in hydra tissue in a graded fashion, with the highest levels of HA in the head and the highest levels of FA in the foot. Of the two activators, HA is best characterized, and its amino acid sequence is known: pGlu-Pro-Pro-Gly-Gly-Ser-Lys-Val-Ile-Leu-Phe. Both HA and FA are found in nerve cells, and both stimulate the proliferation of epithelial cells, interstitial cells, and nematocyte precursors. HA promotes differentiation of head-specific nerve cells, and FA enhances differentiation of foot-specific nerve cells. The inhibitors (HI and FI) are nonpeptide, hydrophilic molecules with molecular weights less than 500. They exist in a graded fashion along the body axis, with HI being highest in the head and FI highest in the foot. Both inhibitors are produced by nerve cells.

It is believed that the activators and inhibitors are released by nerve cells into the intercellular spaces of hydra, where they act on target cells. As there are high concentrations of nerves in the head and the foot, the highest concentrations of the morphogens are found within these two regions. Both activators and inhibitors diffuse away from their highest concentration points, creating gradients along the body axis. It is known that the activators diffuse only a short distance from their sites of release and tend to function locally, whereas the inhibitors diffuse over longer distances and act more globally. Target cells at any given location along the body axis assess their local parameters, namely concentrations of activators and inhibitors, to determine their position in the organism. Such positional information is used by cells to ensure that they differentiate appropriately.

We do not know when, during hydra embryogenesis, the activators and inhibitors are first expressed. However, by the bilayer stage a distinct head region and a distinct foot region are identifiable. Two distinct clusters of RFamide-positive neurons, denoting the future head and foot areas, are seen on opposite sides of the spherical embryo. Thus it is probable that the activators and inhibitors are present by the bilayer stage.

Acknowledgments

This chapter is dedicated to my mother, Margaret S. Martin, for her constant support and love. I acknowledge with thanks and sincere appreciation the skill and patience of William E. Archer. I am grateful for the assistance of Joan Smith, Jason VanLieshout, and Doolittle Martin in the preparation of the manuscript. I am indebted to Hans Bode and Lynne Littlefield for their continuing support, advice, and fruitful discussions. This work was supported by National Science Foundation Grants DCB-8702212, DCB-8942149, DCB-9046094, DUE-9552116, Career Advancement Award DCB-8711245, and grants from Sigma Xi and the Indiana Academy of Science.

Literature Cited

Achermann, J. 1980. The fate and regeneration capacity of isolated ecto- and endoderm in polyps of *Podocoryne carnea* M. Sars. In *Developmental and Cellular Biology of Coelenterates*, P. Tardent and R. Tardent (eds.). Elsevier/North Holland, Amsterdam, pp. 273–279.

Ballard, W. 1942. The mechanism for synchronous spawning in *Hydractinia* and *Pennaria*. *Biol. Bull.* 82:329–339.

Bayer, F. and H. Owre. 1968. *The Free-Living Lower Invertebrates*. The Macmillan Company, New York.

Berrill, N. 1949. Developmental analysis of scyphomedusae. *Biol. Rev.* 24:393–410.

Berrill, N. 1961. *Growth, Development and Pattern*. Freeman, San Francisco.

Bode, H. 1996. The interstitial cell lineage of hydra: A stem cell system that arose early in evolution. *J. Cell Sci.* 109:1155–1164.

Bode, P. and H. Bode. 1984. Patterning in Hydra. In *Pattern Formation, A Primer in Developmental Biology*, G. Malacinski and S. Bryant (eds.). Macmillan, New York, pp. 213–241.

Bode, P., T. Awad, O. Koizumi, Y. Nakashima, C. Grimmelikhuijzen and H. Bode. 1988. Development of the two-part pattern during regeneration of the head in hydra. *Development* 102:223–235.

Boelsterli, U. 1977. An electron microscopic study of early developmental stages, myogenesis, oogenesis and cnidogenesis in the anthomedusa, *Podocoryne carnea* M. Sars. *J. Morphol.* 154:259–290.

Bouillon, J. 1975. Sur la reproduction et l'écologie de *Paracoryne huvei* Picard (Tubularoidea-Athecata-Hydrozoa-Cnidaria). *Archiv. Biol.* (*Brux.*) 86:45–96.

Brumwell, G. and V. Martin. 1994. Localization of RFamide in a cnidarian planula. *Dev. Biol.* 163:548.

Brusca, R. and G. Brusca. 1990. *Invertebrates*. Sinauer Associates, Sunderland, MA.

Campbell, R. 1974. Cnidaria. In *Reproduction of Marine Invertebrates*, A. Giese and J. Pearse (eds.). Academic Press, New York, pp. 133–199.

Campbell, R. 1989. Taxonomy of the European *Hydra* (Cnidaria: Hydrozoa): A re-examination of its history with emphasis on the species *H. vulgaris* Pallas, *H. attenuata* Pallas and *H. circumcincta* Schulze. *Zool. J. Lin. Soc.* 95:219–244.

Carré, D. 1967. Etude du développement larvaire de deux Siphonophores: *Lensia conoidea* (Calycophore) et *Forskalia edwardsi* (Physonecte). *Cah. Biol. Mar.* 8:233–251.

Carré, D. 1969. Etude du développement larvaire de *Sphaeronectes gracilis* (Claus, 1873) et de *Sphaeronectes irregularis* (Claus, 1873), Siphonophores Calycophores. *Cah. Biol. Mar.* 10:31–34.

Carré, D. 1984. Existence d'un complexe acrosomal chez les spermatozoides du cnidaire *Muggiaea kochi* (Siphonophore Calycophore): Différenciation et réaction acrosomale. *Int. J. Inv. Reprod. Dev.* 7:95–103.

Carré, D. and C. Sardet. 1981. Sperm chemotaxis in siphonophores. *Biol. Cell* 40:119–128.

Chapman, D. 1968. Structure, histochemistry and formation of podocyst and cuticle of *Aurelia aurita*. *J. Mar. Biol. Assoc., U.K.* 48:187–208.

Chapman, D. 1974. Cnidarian histology. In *Coelenterate Biology*, L. Muscatine and H. Lenhoff (eds.). Academic Press, New York. pp. 1–92.

Chia, F. and L. Bickell. 1978. Mechanisms of larval attachment and the induction of settlement and metamorphosis in coelenterates: A review. In *Settlement and Metamorphosis of Marine Invertebrate Larvae*, F. Chia and M. Rice (eds.). Elsevier, New York, pp. 1–12.

Chia, F. and J. Spaulding. 1972. Development and juvenile growth of the sea anemone, *Tealia crassicornis. Biol. Bull.* 142:206–218.

Child, C. 1941. *Patterns and Problems of Development.* University of Chicago Press, Chicago.

Clark, W. and W. Dewel. 1974. The structure of the gonads, gametogenesis, and sperm-egg interactions in the Anthozoa. *Am. Zool.* 14:495–510.

Curtis, S. and R. Cowden. 1971. Normal and experimentally modified development of buds in *Cassiopea* (phylum Coelenterata; class Scyphozoa). *Acta Embryol. Exp.* 3:239–259.

Dewel, W. and W. Clark. 1974. A fine structural investigation of surface specializations and the cortical reaction in eggs of the cnidarian *Bunodosoma cavernata. J. Cell Biol.* 60:78–91.

Edwards, N. 1975. Patterns of macromolecular synthesis in a developing hydroid. *Acta Embryol. Exp.* 1975:177–200.

Fadlallah, Y. 1983. Sexual reproduction, development and larval biology in scleractinian corals: A review. *Coral Reefs* 2:129–150.

Fawcett, D. and I. Slautterback. 1959. The occurrence of intercellular bridges in groups of cells exhibiting synchronous differentiation. *J. Biophys. Biochem. Cytol.* 5:453–460.

Franzén, Å. 1966. Remarks on spermiogenesis and morphology of the spermatozoon among the lower metazoa. *Ark. Zool.* 19:335–342.

Freeman, G. 1980. The role of cleavage in the establishment of the anterior-posterior axis of the hydrozoan embryo. In *Developmental and Cellular Biology of the Coelenterates*, P. Tardent and R. Tardent (eds.). Elsevier/North Holland, Amsterdam, pp. 97–108.

Freeman, G. 1983a. The role of egg organization in the generation of cleavage patterns. In *Time, Space and Pattern in Embryonic Development*, W. Jeffrey and R. Raff (eds.). Alan R. Liss, New York, pp. 171–196.

Freeman, G. 1983b. Experimental studies on embryogenesis in hydrozoans (Trachylina and Siphonophora) with direct development. *Biol. Bull.* 165:591–618.

Freeman, G. 1987. The role of oocyte maturation in the ontogeny of the fertilization site in the hydrozoan *Hydractinia echinata. Roux's Arch. Dev. Biol.* 196:83–92.

Freeman, G. 1990. The establishment and role of polarity during embryogenesis. In *The Cellular and Molecular Biology of Pattern Formation*, D. Stocum (ed.). Oxford University Press, pp. 3–30.

Freeman, G. and R. Miller. 1982. Hydrozoan eggs can only be fertilized at the site of polar body formation. *Dev. Biol.* 94:145–152.

Gilchrist, F. 1937. Budding and locomotion in the scyphistomas of *Aurelia. Biol. Bull.* 72:99–124.

Hand, C. and K. Uhlinger. 1995. Asexual reproduction by transverse fission and some anomalies in the sea anemone *Nematostella vectensis. Invert. Biol.* 114:9–18.

Hinsch, G. and W. Clark. 1973. Comparative fine structure of Cnidaria spermatozoa. *Biol. Reprod.* 8:62–73.

Honegger, T. 1981. Light and scanning electron microscopic investigations of sexual reproduction in *Hydra carnea. Int. J. Inv. Reprod. Dev.* 3:245–255.

Honegger, T. 1983. Ultrastructural and experimental investigations of sperm-egg interactions in fertilization of *Hydra carnea. Roux's Arch. Dev. Biol.* 192:13–20.

Hyman, L. 1940. *The Invertebrates: Protozoa through Ctenophora.* McGraw-Hill, New York.

Javois, L. 1992. Biological features and morphogenesis of Hydra. In *Morphogenesis, An Analysis of the Development of Biological Form*, E. Rossomando and S. Alexander (eds.). Marcel Dekker Inc., New York, pp. 93–127.

Leitz, T. 1993. Biochemical and cytological bases of metamorphosis in *Hydractinia echinata. Mar. Biol.* 116:559–564.

Leitz, T. and T. Wagner. 1993. The marine bacterium *Alteromonas espejiana* induces metamorphosis in the hydroid *Hydractinia echinata. Mar. Biol.* 115:173–178.

Lesh-Laurie, G. and R. Corriel. 1973. Scyphistoma regeneration from isolated tentacles in *Aurelia aurita. J. Mar. Biol. Assoc., U.K.* 53:885–894.

Lesh-Laurie, G. and P. Suchy. 1991. Cnidaria: Scyphozoa and Cubozoa. In *Microscopic Anatomy of Invertebrates*, Vol. 2: *Placozoa, Porifera, Cnidaria and Ctenophora*, F. Harrison and J. Westfall (eds.). Wiley-Liss, New York, pp. 185–266.

Littlefield, L. 1985. Germ cells of *Hydra oligactis* males. I. Isolation of a subpopulation of interstitial cells that is developmentally restricted to sperm production. *Dev. Biol.* 112:185–193.

Littlefield, L. 1994. Cell-cell interactions and the control of sex determination in hydra. *Seminars in Developmental Biology* 5:13–20.

Littlefield, L., J. Dunne and H. Bode. 1985. Spermatogenesis in *Hydra oligactis.* I. Morphological description and characterization using a monoclonal antibody specific for cells of the spermatogenic pathway. *Dev. Biol.* 110:308–320.

Martin, V. 1988a. Development of nerve cells in hydrozoan planulae. I. Differentiation of ganglionic cells. *Biol. Bull.* 174:319–329.

Martin, V. 1988b. Development of nerve cells in hydrozoan planulae. II. Examination of sensory cell differentiation using electron microscopy and immunocytochemistry. *Biol. Bull.* 175: 65–78.

Martin, V. 1990. Development of nerve cells in hydrozoan planulae. III. Some interstitial cells traverse the ganglionic pathway in the endoderm. *Biol. Bull.* 178:10–20.

Martin, V. 1991. Differentiation of the interstitial cell line in hydrozoan planulae. I. Repopulation of epithelial planulae. *Hydrobiologia* 216/217:75–82.

Martin, V. 1992. Characterization of a RFamide-positive subset of ganglionic cells in the hydrozoan planular nerve net. *Cell Tissue Res.* 269:431–438.

Martin, V. and W. Archer. 1986a. Migration of interstitial cells and their derivatives in a hydrozoan planula. *Dev. Biol.* 116:486–496.

Martin, V. and W. Archer. 1986b. A scanning electron microscopic study of embryonic development of a marine hydrozoan. *Biol. Bull.* 171:116–125.

Martin, V. and W. Archer. 1997. Stages of larval development and stem cell population changes during metamorphosis of a hydrozoan planula. *Biol. Bull.* 192:41–52.

Martin, V. and F. Chia. 1982. Ultrastructure of a scyphozoan planula, *Cassiopeia xamachana. Biol. Bull.* 163:320–328.

Martin, V. and M. Thomas. 1977. A fine-structural study of embryonic and larval development in the gymnoblastic hydroid *Pennaria tiarella. Biol. Bull.* 153:198–218.

Martin, V. and M. Thomas. 1980. Ultrastructure of the nervous system in the planula larva of *Pennaria tiarella. J. Morphol.* 166:27–36.

Martin, V. and M. Thomas. 1981a. The origin of the nervous system in *Pennaria tiarella* as revealed by treatment with colchicine. *Biol. Bull.* 160:303–310.

Martin, V. and M. Thomas. 1981b. Elimination of the interstitial cells in the planula larva of the marine hydrozoan *Pennaria tiarella. J. Exp. Zool.* 217:303–323.

Martin, V., F. Chia and R. Koss. 1983. A fine-structural study of metamorphosis of the hydrozoan *Mitrocomella polydiademata. J. Morphol.* 176:261–287.

Martin, V., L. Littlefield and H. Bode. 1994. Hydra embryogenesis. *Dev. Biol.* 163:549.

McMurrich, J. 1890. Contributions on the morphology of the Actinozoa. II. On the development of the Hexactiniae. *J. Morphol.* 4:303–330.

Metschnikoff, E. 1886. *Embryologische Studien an Medusen. Ein Beitrag zur Genealogie der Primitiv-Organe.* Alfred Hölder, Vienna.

Miller, R. 1979. Sperm chemotaxis in the hydromedusae. I. Species-specificity and sperm behavior. *Mar. Biol.* (Berl.) 53:99–114.

Miller, R. 1981. Sperm-egg interactions in hydromedusae. In *Recent Advances in Invertebrate Reproduction*, T. Adams and W. Clark (eds.). Elsevier/North Holland, Amsterdam, pp. 289–317.

Müller, W. 1973a. Metamorphose-induktion bei planulalarven. I. Der bakterielle induktor. *Roux's Arch. Dev. Biol.* 173:107–121.

Müller, W. 1973b. Induction of metamorphosis by bacteria and ions in the planulae of *Hydractinia echinata*; an approach to the mode of action. *Publ. Seto. Mar. Biol. Lab.* 20:195–208.

Müller, W., G. Plickert and S. Berking. 1986. Regeneration in hydrozoa: Distal versus proximal transformation in *Hydractinia*. *Roux's Arch. Dev. Biol.* 195:513–518.

Müller, W., A. Mitze, J. Wickhorst and H. Meier-Menge. 1977. Polar morphogenesis in early hydroid development. Action of caesium, of neurotransmitters and of an intrinsic head activator on pattern formation. *Roux's Arch. Dev. Biol.* 182:311–328.

Muscatine, L. and P. McAuley. 1982. Transmission of symbiotic algae to eggs of green hydra. *Cytobios.* 33: 111- 124.

Neumann, R. 1977. Polyp morphogenesis in a scyphozoan: Evidence for a head inhibitor from the presumptive foot end in vegetative buds of *Cassiopea andromeda*. *Roux's Arch Dev. Biol.* 183:79–83.

Nishimiya-Fujisawa, C. and T. Sugiyama. 1993. Genetic analysis of developmental mechanisms in hydra. XX. Cloning of interstitial stem cells restricted to the sperm differentiation pathway in *Hydra magnipapillata*. *Dev. Biol.* 157:1–9

Park, H., N. Sharpless and A. Ortmeyer. 1965. Growth and differentiation in Hydra. I. The effect of temperature on sexual differentiation in *Hydra littoralia*. *J. Exp. Zool.* 160:247–254.

Pyefinch, K. and F. Downing. 1949. Notes on the general biology of *Tubularia larynx* Ellis & Solander. *J. Mar. Biol. Assoc., U.K.* 28:21–43.

Rappaport, R. 1963. Unilateral furrowing in *Hydractinia* and *Echinorachnius*. *Anat. Rec.* 145:273–274.

Ruppert, E. and R. Barnes. 1994. *Invertebrate Zoology*. Saunders College Publishing, New York.

Sarras, M., M. Madden, X. Zhang, S. Gunwar, J. Huff and B. Hudson. 1991. Extracellular Matrix (Mesoglea) of *Hydra vulgaris*. I. Isolation and characterization. *Dev. Biol.* 148:481–494.

Schaller, H., T. Schmidt and C. Grimmelikhuijzen. 1979. Separation and specificity of action of four morphogens from hydra. *Roux's Arch. Dev. Biol.* 186: 139–149.

Schroeder, T. 1982. Novel surface specialization on a sea anemone egg: "Spires" of actin-filled microvilli. *J. Morphol.* 174:207–216.

Schummer, M., I. Scheurlen, C. Schaller and B. Galliot. 1992. HOM/HOX homeobox genes are present in hydra (*Chlorohydra viridissima*) and are differentially expressed during regeneration. *EMBO J.* 11:1815–1823.

Schwoerer-Böhning, B., M. Kroiher and W. Müller. 1990. Signal transmission and covert prepattern in the metamorphosis of *Hydractinia echinata* (Hydrozoa). *Roux's Arch. Dev. Biol.* 198:245–251.

Shenk, M., H. Bode and R. Steele. 1993a. Expression of *Cnox-2*, a HOM/HOX homeobox gene in hydra, is correlated with axial pattern formation. *Development* 117:657–667.

Shenk, M., L. Gee, R. Steele and H. Bode. 1993b. Expression of *Cnox-2*, a HOM/HOX gene, is suppressed during head formation in hydra. *Dev. Biol.* 160:108–118.

Spangenberg, D. 1965. A study of strobilation in *Aurelia aurita* under controlled conditions. *J. Exp. Zool.* 160:1–10.

Stagni, A. and M. Lucchi. 1970. Ultrastructural observations on the spermatogenesis in *Hydra attenuata*. In *Comparative Spermatology*, B. Baccetti (ed.). Academic Press, New York, pp. 357–362.

Stephenson, T. 1928. *The British Sea Anemones, Vol. 1*. Roy. Society London, London.

Sugiyama, T. and T. Fujisawa. 1977. Genetic analysis of developmental mechanisms in hydra. 1. Sexual reproduction of *Hydra magnipapillata* and isolation of mutants. *Growth Dev. Diff.* 19:187–200.

Summers, R. 1970. The fine-structure of the spermatozoon of *Pennaria tiarella* (Coelenterata). *J. Morphol.* 131:117–129.

Szollosi, D. 1969. Unique envelope of a jellyfish ovum: The armed egg. *Science* 163:586–587.

Tardent, P. 1963. Regeneration in the hydrozoa. *Biol. Rev.* 38:293–333.

Tardent, P. 1968. Experiments about sex determination in *Hydra attenuata* (Pall). *Dev. Biol.* 17:483–511.

Tardent, P. 1978. Coelenterata, Cnidaria. In *Einleitung zum Gesamtwerk Morphogenetische Arbeitsmethoden und Begriffssysteme*, Reihe 1, Lieferung l:A-I, F. Seidel (ed.). Gustav Fischer, Stuttgart, pp. 69–415.

Tardent, P. 1985. The differentiation of germ cells in Cnidaria. In *The Origin and Evolution of Sex*, H. Halvorson and A. Monroy (eds.). Alan R. Liss, New York, pp. 163–197.

Teissier, G. 1931. Etude expérimentale du développement de quelques hydraires. *Ann. Sci. Nat. Ser. X* 14:5–60.

Teragawa, C. and H. Bode. 1990. Spatial and temporal patterns of interstitial cell migration in *Hydra vulgaris*. *Dev. Biol.* 138:63–81.

Thomas, M. and N. Edwards. 1991. Cnidaria: Hydrozoa. In *Microscopic Anatomy of Invertebrates*, Vol. 2: *Placozoa, Porifera, Cnidaria, and Ctenophora*, F. Harrison and J. Westfall (eds.). Wiley-Liss, New York, pp. 91–183.

Thomas, M., G. Freeman and V. Martin. 1987. The embryonic origin of neurosensory cells and the role of nerve cells in metamorphosis in *Phialidium gregarium* (Cnidaria, Hydrozoa). *Int. J. Invert. Reprod. Dev.* 11:265–287.

Uchida, T. and M. Yamada. 1968. Cnidaria. In *Invertebrate Embryology*, M. Kumé and K. Dan (eds.). Nolit Publ. House, Belgrade, pp. 86–116.

Van Beneden, E. 1898. Die anthozoen der plankton-expedition. *Ergebn. Plankton Expedit.* 2(K,3).

Van de Vyver, G. 1968. Etude du développement embryonnaire des hydraires athécates (gymnoblastiques) à gonophores. III. Discussion et conclusions générales. *Arch Biol.* (*Liège*) 79:365–379.

VanLieshout, J. and V. Martin. 1992. Development of planuloid buds of *Cassiopea xamachana* (Cnidaria: Scyphozoa). *Trans. Am. Microsc. Soc.* 111:89–110.

Werner, B. 1973. Spermatozeugmen and paarungsverhalten bei *Tripedalia cystophora* (Cubomedusae). *Mar. Biol.* 18:212–217.

West, D. 1978. Ultrastructural and cytochemical aspects of spermiogenesis in *Hydra hymanae*, with reference to factors involved in sperm head shaping. *Dev. Biol.* 65:139–154.

Widersten, B. 1968. On the morphology and development in some cnidarian larvae. *Zool. Bidrag. Uppsala* 37:139–182.

Widersten, B. 1973. On the morphology of actiniarian larvae. *Zool. Scripta.* 2:119–124.

Wilson, E. 1884. The development of *Renilla*. *Phil. Trans. Roy. Soc. London* 174:723–815.

Yamaguchi, M. and R. Hartwick. 1980. Early life history of the sea wasp, *Chironex fleckeri* (Class Cubozoa). In *Developmental and Cellular Biology of Coelenterates*, P. Tardent and R. Tardent (eds.). Elsevier/North-Holland, Amsterdam. pp. 11–16.

CHAPTER 6

Ctenophorans, the Comb Jellies

Mark Q. Martindale and Jonathan Henry

THE CTENOPHORA, DERIVED FROM THE GREEK "COMB-bearers" (sometimes called *comb jellies, moon jellies, sea walnuts,* or *sea gooseberries*) is a very well-defined and successful phylum of marine predators. Some species of ctenophores can be very abundant in surface waters and representative species are found in all of the world's oceans. Only about 200 species of ctenophores have been described, although estimates have suggested that up to 75% of all ctenophores are as yet undescribed, and represent predominantly deepwater forms. Ctenophores vary widely in size and shape (Figure 6.1). Some deepwater forms are known to exceed 1.5 m in length; however, the more common forms range in size from 1–20 cm. Most ctenophores are virtually transparent. For this and a number of other reasons, including their fragile nature that interferes with their collection and preservation, the evolutionary success of the ctenophore body plan has been generally underappreciated.

While the prominence of adult ctenophores may not be apparent, ctenophore embryos have played a central role in the development of our understanding of embryonic development within the Metazoa. Any good student of embryology knows that one of the most important moments in experimental embryology occurred when Hans Driesch separated the first two blastomeres of an indirect-developing sea urchin embryo (1892). Driesch found that these half-embryos "regulated" to give rise to two normal, miniature plutei larvae. These results led Driesch towards his view of embryos as "totally harmonic equipotential systems." In the same year, however, C. Chun reported the results of experiments he had started in 1877 in which he followed the development of separated blastomeres of two-cell ctenophore embryos, and observed the exact opposite result. Half-embryos gave rise to adults possessing exactly half of the normal set of adult structures. These results provided the best data for a view of embryos as being composed of a "mosaic" of individual parts. These two contradictory observations in different organisms helped shape the vocabulary and paradigms for developmental biologists for the next century. It might be noted here that Driesch, a powerful and arrogant individual, ridiculed Chun's results and made personal attacks on Chun's integrity. Chun was vindicated, however, when T. H. Morgan collaborated with Driesch and repeated Chun's results (1895). Although Driesch and Morgan observed essentially the same result as Chun, they attempted to interpret their results in such a way as to rob Chun of his important place as one of the first experimental embryologists. The many favorable attributes of ctenophore embryos have been utilized throughout this century to make fundamental progress in the way we think about development and regeneration.

Body Plan

Although ctenophores bear a superficial resemblance to medusa "jellyfish" from the phylum Cnidaria, they are quite distinct in a number of ways. Ctenophores have been described as **biradially symmetrical** animals consisting of an outer epidermis separated from an inner **gastrodermis** by a thick and largely acellular extracellular matrix called the **mesoglea**. Although individual muscle (longitudinal, radial, circumpharyngeal, and tentacular) and mesenchymal cells reside in the mesoglea and may be in close association with epithelial tissue, they are not organized into a defined layer of mesodermal tissue. Hence, ctenophores are often referred to as being **diploblastic**. Some authors, however, argue that the presence of definitive muscle cells make them true **triploblasts**. Regardless of this semantic problem, the ctenophore body plan, or **bauplan**, is quite distinct from other animal phyla.

Figure 6.1 Adult ctenophores of various shapes and sizes. (*a*) A lobate stage *Mnemiopsis* seen from the side. The aboral end is up and the oral end is facing down. Note the small copepods in the distal portion of the pharynx. This animal is 10 cm along its longest axis. (*b*) A *Pleurobrachia* with the oral end facing up showing the extensive tentacle apparatus this animal uses to capture crustaceans and other small prey items. The body is approximately 2 cm in diameter. (*c*) An adult lobate *Ocyropsis* several centimeters in length. (*d*) An atentaculate *Beroë* that feeds on other "jello plankton," primarily other ctenophores.

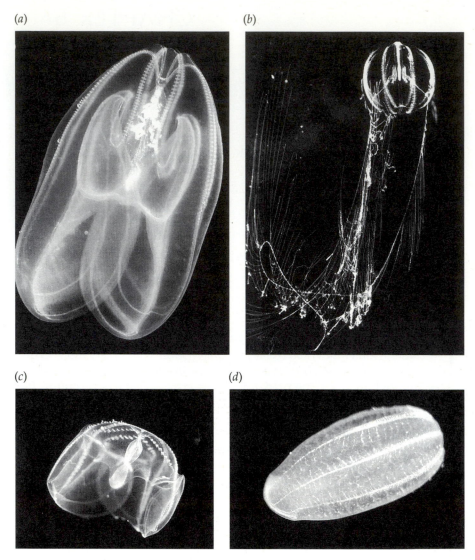

(*a*)   (*b*)

(*c*)   (*d*)

Ctenophores are characterized by eight longitudinal rows of **comb plates** located on the outer epidermis that the animal uses to locomote through the water column (Figure 6.2). Each comb plate is composed of thousands of aligned cilia that beat in **metachronous** waves to propel the animal mouth first though the water, albeit at a modest pace. The direction and beat frequency of each ctene row is under nervous control, thus giving the animal the ability to adjust its body posture and direction of travel. This mode of locomotion is distinct from the locomotion seen in the Cnidaria, in which the medusa moves via a "pumping action" through the isotropic contraction of its swimming bell. Most ctenophores possess two tentacles with which they capture prey. These tentacles posses a particular cell type, called the **colloblast**, which is used for the adhesion of prey to the tentacles. Colloblasts are similar to, but distinct from, nematocysts found in the Cnidaria, and along with the ciliated comb plates and the apical sensory organ, represent distinct **synapomorphies** defining this phylum.

The major body axis in ctenophores is the **oral-aboral axis** (Figure 6.2). Historically, ctenophores have been described as biradially symmetrical, with two planes of symmetry occurring around the oral-aboral axis. The sagittal, or esophageal, plane passes through the flattened plane of the esophagus in the undistended condition. The second plane is orthogonal to the sagittal plane and passes through the two tentacles (termed the tentacular plane). In addition to these two planes of mirror symmetry, a third oblique plane can be identified that passes through the two **anal pores** (Figures 6.2, 6.3). The anal pores are located in two diagonal quadrants at the aboral end of most ctenophores. They connect to the stomach by the endodermally derived anal canals. The line that connects the two pores defines the **anal axis** and together with the oral-aboral axis, defines a plane of two-fold rotational (but not mirror) symmetry referred to as the **anal plane.**

Internally, the mouth is connected to an ectodermally derived **pharynx** (or stomodeum) that leads distally to a region called the *esophagus*. In the lobate ctenophore *Mne-*

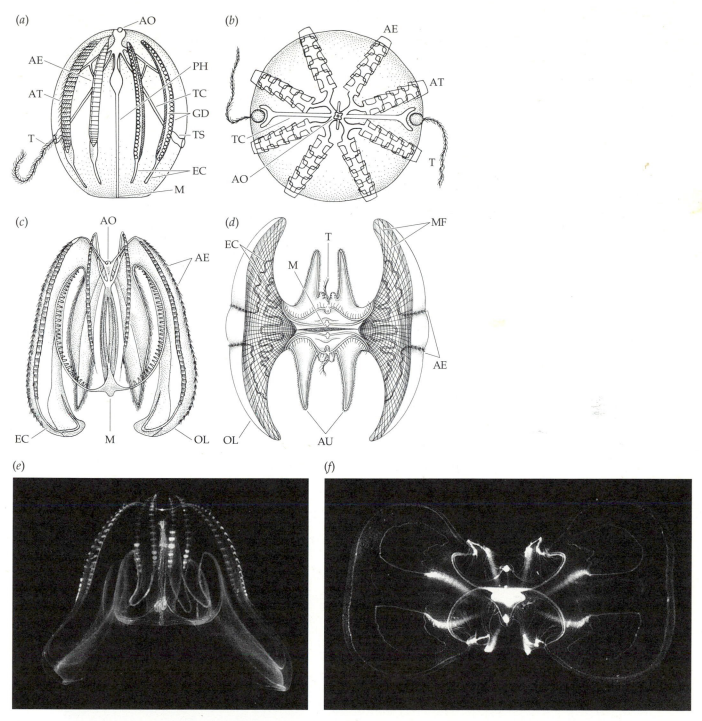

Figure 6.2 The ctenophore body plan. (*a*) Diagram of an adult member of the Cydippida seen from the side. The aboral pole contains the apical sense organ and is seen at the top. Four of the eight ctene rows are seen (although the pair of ctene rows on the right side are not shown so that the gonads associated with the endodermal meridional canals can be seen). The two tentacles define the tentacular axis (in the plane of the paper) while the sagittal or esophageal axis runs perpendicularly to the tentacular axis. (*b*) Aboral view of a cydippid showing all eight ctene rows and both the plane of symmetry that passes through the two tentacles (tentacular plane) and one running in the plane of the esophagus (esophageal plane). These two planes generate four nearly identical quadrants. Note the endodermal canals that run under each ctene row and to each tentacle. (*c*) Side view of an adult lobate ctenophore showing the oral lobes and auricles. The apical organ becomes recessed at the aboral pole. (*d*) Oral view of a lobate ctenophore. Note the position of the tentacles and the two large oral lobes and auricles that are used to capture prey items. A series of muscle fibers can be seen in the inside surface of the oral lobes; the endodermal canals can be seen coursing their way through the mesoglea. (*e, f*) Two photographs showing a lateral (*e*) and an oral view (*f*) of an adult lobate, *Mnemiopsis*. Note that *e* and *f* are the same views as the diagrams seen in *c* and *d*. AE, adesophageal ctene row; AO, apical organ; AT, adtentacular ctene row; AU, auricles; EC, endodermal canals; GD, gonad; M, mouth; MF, muscle fibers; OL, oral lobes; PH, pharynx; S, stomach; T, tentacle; TC, tentacular canal; TS, tentacle sheath.

miopsis the esophagus contains specialized cilia used for the processing of prey items. The esophagus is flattened in a plane perpendicular to the tentacles and connects to the endodermally derived stomach (or **infundibulum**), which is continuous with a complex system of **endodermal meridional canals**. These canals run subjacent to the ctene rows, in association with the tentacle sheathes, the apical organ, and to varying degrees through other regions of the mesoglea (Figures 6.1, 6.2). The **gonads** and light producing **photocytes** are also associated with the endodermal canals as they run under the ctene rows. The endodermal canals serve as a simple "circulatory system," transporting nutrients to the periphery of the animal. No other vascular system is present. Virtually all undigested food is expelled through the mouth, although material from the stomach can be seen to pass through the two anal pores, but no other excretory system is present (Hyman 1940).

At the end opposite the mouth (the aboral pole) is an **apical sensory organ** that contains a **statocyst** (Figure 6.3). The statocyst consists of cells containing mineral deposits called **lithocytes**, which are supported upon four groups of **balancing cilia** or **polster cells**. New lithocytes are generated at a regular rate from the floor of the apical organ throughout the life of the animal and become positioned on top of the balancing cilia to replace lost ones. The lithocytes and the balancing cilia are housed under a group of nonmotile cilia called **dome cilia** or the **cupula**. The statocyst serves as a "gravity amplifier" and regulates beat direction and synchrony within the locomotory comb rows via tracts of ciliated cells called the **ciliated grooves** (Figure 6.3). Thus, an intact apical organ is required for ctene row coordination and normal locomotory behavior. In most species (except the Beroida), the apical organ is connected to the endodermal canal system by the infundibular (or aboral) canal. Two epidermal specializations, called

Figure 6.3 Different views of the apical sense organ. (*a*) The nonmotile dome cilia house a statocyst consisting of mineral containing lithocytes perched atop four groups of balancing cilia. The balancing cilia are connected to the ciliated grooves, which lead to the eight ctene rows. The two polar fields—epidermal specializations of unknown function—project from the apical organ in opposite directions along the esophageal or sagittal axis. (*b*) A longitudinal view of the apical organ, indicating the relationship of the lithocytes to the balancing cilia. New lithocytes are released from the floor of the apical organ throughout the life of the animal. (*c*) Aboral view of the apical organ, indicating the position of the two anal pores. The anal pores are the openings of the endodermal anal canals, which occur in diametrically opposed quadrants. The anal pores define a plane of rotational symmetry called the *anal axis*. (*d*) Micrograph of a cydippid larvae of the lobate ctenophore *Mnemiopsis*, indicating the relationship of the apical organ to the ctene rows. This larvae is approximately 300 μm in diameter. BC, balancing cilia; CG, ciliated grooves; CR, ctene rows; DC, dome cilia; L, lithocytes; T, tentacular axis.

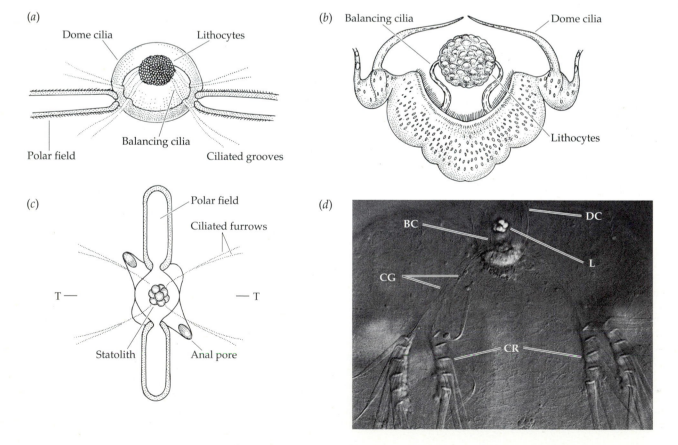

(*a*) Dome cilia — Lithocytes — Balancing cilia — Polar field — Ciliated grooves

(*b*) Balancing cilia — Dome cilia — Lithocytes

(*c*) Polar field — Ciliated furrows — T — T — Statolith — Anal pore

(*d*) BC — DC — L — CG — CR

polar fields, are present on the aboral pole, although their function is currently unknown. Excellent reviews of the ctenophore body plan (Hyman 1940; Hernandez-Nicaise 1991; Brusca and Brusca 1990) and the functioning of the apical organ (Tamm 1982) and its relationship to ctene row beating have been published, and readers are encouraged to consult these works for further details.

The nervous system in ctenophores has been difficult to characterize and little direct physiological information is known about it. Previous workers used classical nerve cell stains such as silver nitrate and methylene blue to visualize polygonal lattices or **nerve nets** of isopolar neurons similar to those seen in some cnidarians (Hertwig 1880; Bethe 1895; Hernandez-Nicaise 1973a). Further support for the existence of a definitive nervous system comes from electron microscopical (EM) investigations that reveal numerous synapses in the floor of the apical organ, underlying the ctene rows, associated with the tentacles, and various sensory structures (see Horridge 1974; Hernandez-Nicaise 1991; and Tamm 1982 for reviews). Unfortunately, the visualization of nerve cells is difficult to perform and reproduce. Physiological evidence for discrete nervous elements have been difficult to obtain. There is good evidence that inhibition of comb plate activity and the coordination of tentacle contraction are both mediated by nervous elements (Tamm 1982). These data indicate that a nervous system regulates both ciliary and muscular effectors; however, much more work remains to be done to characterize the elements of the ctenophore nervous system and their control of ctenophore behavior.

Evolutionary Relationships and Systematics

The most thorough treatment of the systematics of the Ctenophora has been published by Harbison (1985). Although ctenophores bear a superficial resemblance to the cnidarians, their relationship with other metazoan phyla is still open for discussion. Previous workers have postulated associations with virtually all other groups of basal metazoans, including the poriferans (Schneider 1902), cnidarians (Chun 1892; Haeckel 1896; Hyman 1940), flatworms (Lang 1884; Mortensen 1912a), trochophore larvae (Hatschek 1911), and even the deuterostomes (Nielsen 1995). The statocyst–ctene row complex, mesenchymal muscle cells, colloblasts, life history strategies, and the stereotyped pattern of embryogenesis (to be described below) all set the ctenophores apart from the cnidarians. There is little reason, however, to align them with any other phylum. Figure 6.4 indicates four scenarios for the relationship of the ctenophores to other major metazoan phyla. The traditional view proclaims that the ctenophores and cnidarians were grouped into a single group, the **coelenterata** (Figure 6.4*a*). More recent analyses have recognized the ctenophores and cnidarians as separate phyla;

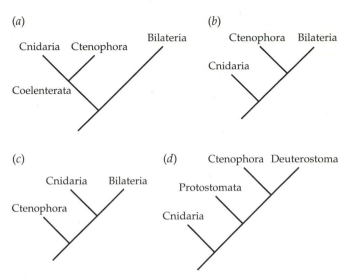

Figure 6.4 Possible scenarios of the evolutionary relationships of the ctenophores with other extant animal groups. (*a*) Traditional view of the cnidarians and the ctenophores grouped together in the phylum Coelenterata. (*b*) Separation of the cnidarians and ctenophores as distinct phyla, with the ctenophores being the sister group to the entire group of bilaterally symmetrical animals. (*c*) A similar situation in which the ctenophores diverged prior to the appearance of the cnidarians. (*d*) An alternate view, which places the ctenophores as a sister group to the deuterostomes.

however, no agreement on the relative position of the two groups has been achieved. One thought is that the ctenophores arose after the Cnidaria (Figure 6.4*b*), and would thus represent the sister group to the rest of the bilaterally symmetrical animals (the Bilateria). Molecular analyses may provide some clues as to the evolutionary position of the ctenophores. Initial studies using 18S rRNA sequences (Wainright et al 1993) suggest that the ctenophores diverged from other major metazoan phyla prior to the divergence of the cnidarians (Figure 6.4*c*). The virtual absence of ctenophores in the fossil record has made it difficult to resolve the relationships both within the phylum and with other groups. Stanley and Stürmer (1983) have reported a specimen from the Devonian that shows essentially the same body plan as modern forms. A possible ctenophore-like animal (*Fasiculus*) has been reported in Burgess Shale sites from the Mid-Cambrian (Collins et al. 1983), although this assignment is still tentative.

The Ctenophora have historically been split into two classes, the Tentaculata (six orders) and Atentaculata (or Nuda) (one order) (Harbison and Madin 1982; Table 6.1), although good arguments for as many as 11 monophyletic groups have been proposed (Harbison 1985). Relationships within the phylum are as perplexing as those relating the ctenophores to other extant metazoan phyla (Harbison 1985). For example, it is not at all clear whether the first ctenophores had tentacles. One scenario would have the Beroids losing their tentacles and assuming other

TABLE 6.1 *Two Classes of Ctenophores*

Class	Genera
Nuda	*Beroida*
Tentaculata	*Cydippida, Lobata, Platyctenida, Cestida, Ganeshida, Thalassocalycida*

specializations necessary for an active predatory existence. These specializations would include special chemosensory cells surrounding the lips of Beroids and macrocilia that are used to close the lips during active searching for prey items, and to hold and rip apart soft-bodied prey items (Tamm and Tamm 1991). Another scenario proposes that the ancestral ctenophores were atentaculate, and thus tentacles appeared secondarily. The independent invention of tentacles by ctenophores may not be too far-fetched due to the fact that they possess colloblasts, a cell-capture design that significantly differs from the nematocysts found in the Cnidaria (Hernandez-Nicaise 1991). Metschnikoff (1885) reported the transient existence of tentacle sheath anlaga during the development of the atentaculate *Beroë* (a finding that appears to never have been substantiated), which would make the argument for the existence of tentacles in ancestral ctenophores much more compelling.

Life History

Embryos in all but one order of ctenophore (the Beroids) pass through a stage referred to as a **cydippid larva** that displays all of the defining features of the phylum (Figures 6.1, 6.2). The cydippid does not appear to represent a true larval form, due to the lack of a radical metamorphosis following this stage. The postembryonic phase can be followed by differential growth of discrete regions or structures such as the **oral lobes** and **auricles** of the Lobates, or loss of elements such as the ctene rows in the creeping ctenophores (Platyctenids), which assume a benthic existence. Therefore, even though it is convenient to refer to the cydippid stage as a "larval stage," ctenophores are direct developers. There have been no reports of colonial ctenophores, or alteration of generations, which represent complex life-history strategies seen in the Cnidaria.

All ctenophores are carnivores. Those with tentacles generally utilize these structures to capture small crustaceans or other plankton. Following collision with the tentacles, the colloblasts discharge and the muscular core of the tentacles contract, bringing the prey item closer to the oral opening. Many ctenophores then initiate a coordinated burst of ctene row activity that results in reorienting the mouth so that captured prey item may be swallowed (referred to as *spin ingestion*). In some orders, the role of the tentacles in feeding is reduced. For example, in the Lo-

bates the oral ends of the esophageal axis are modified to serve as funnels and small ciliated fingerlike extensions called **auricles** are used to flick small prey items towards the mouth. One group of ctenophores, the **platyctenida**, or the "creeping" ctenophores, lose their ctene rows shortly after embryogenesis is completed and assume a benthic existence. These animals retain their apical organ and tentacles, however, and feed on benthic invertebrates.

The most dramatic variation in life history seen in ctenophores is displayed by members of the Beroida (class, Nuda), which retain their eight ctene rows and biradial symmetry but do not form tentacles. These animals are not passive feeders, but rather are active predators consuming tentaculate ctenophores or other forms of "jello plankton." Beroids show unique features that are indicative of their different life styles. They are in general larger, more active, and have specialized cilia that are used to keep their mouth closed during forward swimming (Tamm and Tamm 1991) and for the maceration of soft-bodied prey. Beroids also have special chemoreceptors on their lips which they use to discriminate edible prey. For example, some species of *Beroë* found in the Puget Sound off Washington will eat only members of the class Cydippida, while other species will eat only species of the class Lobata.

One interesting feature of some ctenophore species is their ability to generate progeny at very early stages of "larval" life. This process was termed **dissogeny** by Chun (1880, 1892) and has been reported for members of the Lobata and Cydippida. Not all individuals of a single spawn become precociously reproductive, and those that are reproductive at an early age will also become reproductive at later adult stages (Martindale 1987). The ability to generate rather limited numbers of progeny at very young ages appears to be an adaptation to ephemeral ecological conditions and/or high predation rates (Stanlaw et al. 1981; Reeve and Walter 1978). It is also of interest to note that only four of the eight possible gonads, the four associated with the ctene rows adjacent to the esophagus, become precociously reproductive. The idea that these adesophageal gonads differ from the adtentacular gonads with respect to larval reproduction is supported by additional findings. "Half-animals" generated by bisecting early ctenophore embryos become reproductive at a higher rate than normal animals, and in these partial embryos only the adesophageal gonads generate functional gametes (Chun 1880, 1892; Martindale 1987).

Gametogenesis

Most, but not all (Harbison and Miller 1986), ctenophores are functional hermaphrodites, possessing both male and female gonads. The gonads are present in pairs associated with the endodermal canals located underneath each of the eight ctene rows (Figure 6.5a-c). Within a pair of ctene rows of the same quadrant, the testes face one another, and the ovaries are located on the "outside" of the mirror

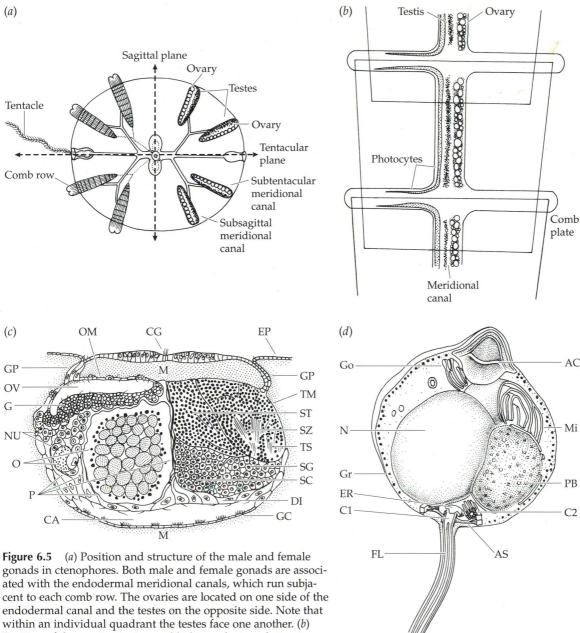

Figure 6.5 (*a*) Position and structure of the male and female gonads in ctenophores. Both male and female gonads are associated with the endodermal meridional canals, which run subjacent to each comb row. The ovaries are located on one side of the endodermal canal and the testes on the opposite side. Note that within an individual quadrant the testes face one another. (*b*) Position of the ovaries, testes, and light-producing photocytes subjacent to a subtentacular comb row. The aboral pole is towards the top of the page. (From Freeman and Reynolds 1973.) (*c*) Cross-section through the comb-row/endodermal canal complex showing both male and female gonads and their relationship to their gonopores. (After Dunlap-Pianka 1974; from Hernandez-Nicaise 1991.) (*d*) Ultrastructural reconstruction of a *Beroë* sperm. (After Franc 1973, in Hernandez-Nicaise, 1991.) Ac, acrosome; AS, anchoring star-shaped structure; C1, distal centriole; C2, proximal centriole; CA, lumen of meridional canal; CG, ciliated groove; DI, digestive cells; EP, epidermis; ER, endoplasmic reticulum; FL, flagellum; G, glandular epithelium; GC, ciliated gastrodermal cell; Go, Golgi complex; GP, gonopore; Gr, M, mesoglea; Mi, mitochondrion; N, nucleus; NU, nurse cells; O, oocytes; OM, ovarial membrane; OV, oviduct; P, interstitial processes of digestive cells; PB, paranuclear body; SC, spermatocytes; SG, spermatagonia; ST, spermatids; SZ, mature spermatozoids; TM, testicular membrane; TS, testicular sinus.

plane of symmetry (Figure 6.5*a*). Gametes are generally shed into a cavity (the testicular sinus and oviduct) before being shed into the water column through gonoducts that exit between individual comb plates (Figure 6.5*c*). Some workers have described gametes being shed through the mouth; however, these reports are likely to have been the result of mishandling of the adults. This handling might have caused gametes to enter the meridional canals associated with gonadal tissue, which are continuous with the stomodeum.

Ctenophore oocytes and mature eggs have a **centrolecithal** organization. This means that there is a thin layer of fluid cytoplasm, consisting primarily of endoplasmic reticulum and mitochondria, known as the **ectoplasm** that surrounds a larger central core of dense yolk spheres, the **endoplasm**. The oocyte nucleus resides at the boundary of these two plasms until the oocyte is shed from the gonad. The endoplasmic yolk spheres in the lobate and cydippid animals that have been studied appear to be composed largely of protein and carbohydrate, with very little lipid (Dunlap-Pianka 1974). Electron microscopic investigations of some ctenophore eggs have identified a third region, the *subcortical plasm*, which lies underneath the ectoplasmic layer that contains small PAS-positive (Periodic Acid-Shiff polysaccharide staining) particles dispersed in a more fluid cytoplasm (Dunlap-Pianka 1974). It would be interesting to examine members of the order Beroida, since these eggs float and this low-density phenomenon is often associated with the presence of lipid reserves.

Ctenophore eggs and embryos are some of the most optically clear specimens available for study. The eggs are relatively large, ranging from about 120 μm in diameter in the cydippid *Pleurobrachia bachei* to over a millimeter for *Beroë ovata*. The eggs of *Mnemiopsis leidyi* are intermediate in size, approximately 220 μm. An acellular membrane, called the **vitelline membrane**, is generated by the oocyte and swells during the ovulation process. Oocytes are also covered by a jelly coat, which is added by secretory cells of the oviduct wall during spawning (Dunlap-Pianka 1974).

Embryonic development is quite rapid in ctenophores (see below) and the generally ephemeral existence of these soft-bodied animals is reflected in the speed and mechanism of gamete formation. Eggs are formed in association with **nurse cell complexes** that transfer contents to the developing gamete. This situation is similar to that seen in rapidly developing insect embryos such as *Drosophila*, in which germ line sister cells assume a large role in transferring nutrients and other maternal substances to the developing oocyte (see Chapter 14 in this volume). Greve (1970) has estimated that oogenesis takes as little as 2 days in the North Sea forms *Pleurobrachia pileus* and *Beroë gracilus*, and might be even more rapid in warm water forms.

The best description of gametogenesis in ctenophores has been performed on several species of ctenophore found in Puget Sound, including the lobate ctenophore *Bolinopsis microptera* (Dunlap 1966; Dunlap-Pianka 1974). Each developing oocyte develops as a syncytium formed by the incomplete cytoplasmic divisions of a germ cell precursor. **Cytoplasmic bridges** remain during approximately eight rounds of division generating a complex of approximately 100 interconnected cells (Figure 6.6). One cell becomes enlarged (the presumptive oocyte) while retaining a connection to three groups of mitotic sister cells (the nurse cell complex). The nurse cell connections bear a consistent relationship to the site where the oocyte nucleus lies. It has been proposed that this relationship might play a role in axial determination in ctenophore embryos (see below). Nutrients and other substances from the enlarged nurse cells enter the developing oocyte throughout all stages of oogenesis (Dunlap 1966; Dunlap-Pianka 1974). Cteno-

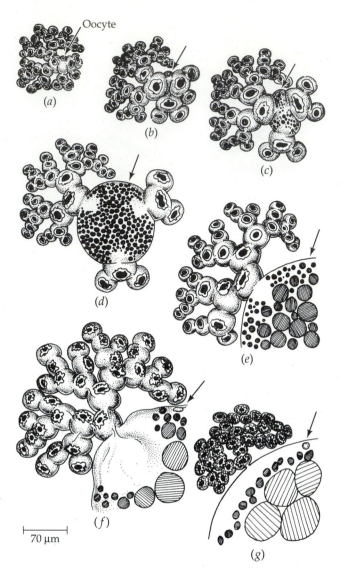

Figure 6.6 Formation of the oocyte nurse cell complex in the lobate *Bolinopsis*. The presumptive oocyte is indicated with the arrow. In *a–d* one entire nurse cell complex is shown while the other two complexes are represented by the three most proximal nurse cells. Only a single nurse cell complex is shown in *f* and *g*. Stages *a–g* indicate different stages of vitellogenesis, while *f* and *g* indicate the formation of the ectoplasmic cortical regions at the final stages of oogenesis. (After Dunlap-Pianka 1974.)

phores are rather unique in this sense because the oocyte nucleus itself remains small and presumably inactive, while the oocyte nucleus in other animals (insects being the other major exception) makes a contribution to its cytoplasmic constituents (Dunlap 1966; Dunlap-Pianka 1974).

Detailed EM investigations reveal that the centrolecithal organization of the ctenophore egg is directly related to the properties of the nurse cell complex (Dunlap 1966; Dunlap-Pianka 1974). Proximal nurse cells (i.e., those that retain the closest connections to the developing oocyte) are enlarged and metabolically active early during oogenesis. These cells generate the yolk spheres that make up the central endoplasm. During the latter stages of oogenesis, the proximal nurse cells shrink and the distal nurse cells contribute the outer ectoplasm seen in the periphery of mature eggs and embryos. The nurse cells then detach from the oocyte and presumably degenerate. A detailed molecular analysis of the maternal contributions to oogenesis has yet to be performed. Of particular interest would be the nature of the materials contributed by the distal nurse cells. The ectoplasm is segregated to distinct lineages of micromeres during the early stages of embryogenesis and probably plays a role in the precocious determination of cell fates in ctenophore embryos (see below).

Spermatogenesis in ctenophores operates in much the same way as in other animals. Within the testicular stroma large cohorts of sperm develop in synchrony, undoubtedly sister cells of germ cell divisions. EM investigations have shown that individual spermatocytes within a cohort are connected by cytoplasmic bridges (Dunlap-Pianka 1974). Sperm morphology is considered to be primitive in the sense that sperm heads are usually spherical to pyriform and contain the haploid nucleus, a Golgi-derived acrosomal vesicle, two centrioles, and a small number (one to a few) of mitochondria (Figure 6.5*d*). Sperm heads are connected to the distal centriole, giving rise to a normal flagellum composed of the standard 9 + 2 microtubular array. A definitive midpiece is not present, however, with the distal centriole being attached to a star-shaped structure at the base of the sperm head (Franc 1973; Hernandez-Nicaise 1991). Motile sperm are shed into a testicular sinus before being released into the open ocean through the gonopores.

Fertilization

Spawning in many ctenophores is controlled by photoperiod. For example, adult specimens of the lobate ctenophore *Mnemiopsis* will spawn approximately 8 hours after the lights are turned off, (although the timing of spawning varies considerably for different species). Some species spawn in response to the onset of daylight. Sperm are generally shed first and can be seen as white lines in the testicular sinuses underneath the ctene rows. Oocytes can be seen in the gonads soon after, and these are then released into the ovarian sinus before being shed through the gonopores. The comb plates overlying the gonads generally stop beating during the sporadic shedding of both sperm and eggs, although they quickly resume to help disperse gametes. Oocytes complete their first and often the second maturation division within the confines of the ovarian sinus. Both maturation divisions proceed regardless of the presence of sperm. In some species the polar bodies remain on the surface and mark the site of the underlying oocyte nucleus, but in others (e.g., *Mnemiopsis*) they drift away from the egg and thus are not visible during subsequent development.

Because most ctenophores are functional hermaphrodites, fertilization has been difficult to study. Under normal circumstances eggs are fertilized as they are shed from the gonopores, or very soon thereafter. The best studies of fertilization have been performed in *Beroë ovata* by Sardet and his colleagues. In this species, isolated individuals are able to produce viable offspring in only approximately 20% of the cases (Carré et al. 1990, 1991). This level of self-sterility is regulated at the level of the vitelline membrane. If this membrane is removed, then virtually 100% of the eggs of isolated individuals are self-fertile. By obtaining unfertilized eggs, it has been possible to study the fertilization process using sophisticated video, electron microscopic, and confocal microscopic techniques.

While some aspects of fertilization are similar to those found in other organisms, other aspects are rather unique in ctenophores. Motile sperm can attach to the outer egg membrane anywhere around its circumference and undergo a standard acrosome reaction. The flagellum becomes rigid as the sperm and egg membranes fuse and the egg cortex contracts around the sperm head. A fertilization cone forms around the sperm head as it enters the egg cortex, a process that takes up to 20 minutes from the time of sperm-egg contact. A local exocytosis of some vesicular material takes place around the site of sperm entry. This is not a global event, as seen in other animals, but only occurs in the local vicinity of the fertilizing sperm. Once the sperm head and its contents have entered the egg cytoplasm, the male aster forms. The aster is surrounded by a unique region of the cytoplasm containing mitochondria and other organelles. This causes a local thickening of the cortical region and the formation of large microvilli at the site of sperm entry (Carré et al. 1990, 1991).

Some unique features of ctenophore fertilization have been observed. One interesting phenomenon is the normal occurrence of polyspermy in fertilized ctenophore eggs. For example, it is not unusual to have as many as 20 sperm pronuclei in one *Beroë* egg (Carré et al. 1991). There does not appear to be an electrical "fast block" to polyspermy, which is found in most marine animals, and there is no substantial change in the extracellular membranes or investments that prevents the fusion of additional sperm. The lack of an effective block to polyspermy is not due simply to the large size of the *Beroë* egg (over 1 mm), because the modestly sized (220 µm) *Mnemiopsis* egg also becomes polyspermic. In addition, the male pronuclei remain close to the site of entry and the female pronucleus travels to various male pronuclei, eventually "selecting" one to fuse with (**syngamy**). This occurrence stands in contrast to what has been observed in many animal embryos, where the male pronucleus migrates toward the female pronucleus. *In vitro* observations of precleavage nuclear movements indicate that the female pronucleus may undergo a complicated migratory pathway before "choosing" an appropriate male pronucleus with which to fuse. Shortly thereafter the other male pronuclei degenerate.

The process by which a single male pronucleus is chosen is a complete mystery at this time, although it is clearly an important feature of fertilization and subsequent development in ctenophores.

Cleavage

Cleavage in ctenophores (Figure 6.7) displays none of the features of either spiral or radial cleavage so commonly found in other metazoan phyla. First cleavage normally occurs very close to the site of syngamy. All cleavages are **unipolar**, that is, the cleavage furrow begins at one pole of the cell instead of around the entire circumference (Figure 6.8). The first two cleavages run through what will become the future oral-aboral axis and give rise to four equal-sized cells (Figures 6.7, 6.8). Subsequent cell divisions occur synchronously within each of the four cell quadrants throughout the remaining course of development. The plane of first cleavage passes through the future sagittal plane and the second through the future tentacular plane (Figure 6.7). Third cleavage begins in a plane parallel to the first cleavage plane, but shifts obliquely such that it generates four slightly smaller **blastomeres** located on the

"ends" of the embryo (the so-called **E blastomeres**) and four larger cells which remain in the "middle" of the embryo (called the **M blastomeres;** see Figure 6.7). Immediately before fourth cleavage a major portion of the ectoplasm converges toward the site opposite the initiation of first cleavage (aboral pole). At the vegetal (aboral) pole each one of the eight cells generates a series of smaller cells, called **micromeres** that are enriched in ectoplasm. In some species this ectoplasm is slightly pigmented and can be seen to segregate into the aboral micromeres.

Figure 6.7 Diagram of early development from the zygote through the 60-cell stage. The plane of first cleavage corresponds to the sagittal (esophageal) plane (S) while the second cleavage plane passes through the tentacular plane (T). Note that at the 4-cell stage, two nonequivalent, diametrically opposed pairs of cells are generated (EM/ and EM\). The four cells divide synchronously. At the 8-cell stage the E and M blastomeres are born. The / and \ designations are included only in blastomere names for M macromere derivatives because only differences in the fates of adjacent M macromeres are known to date. At the 16-cell stage, the first round of micromeres is born at the aboral pole. Each cell divides to give rise to the 32-cell stage. At the end of the 32-cell stage each cell divides, except for the 2M macromeres. This division results in the formation of the 60-cell stage.

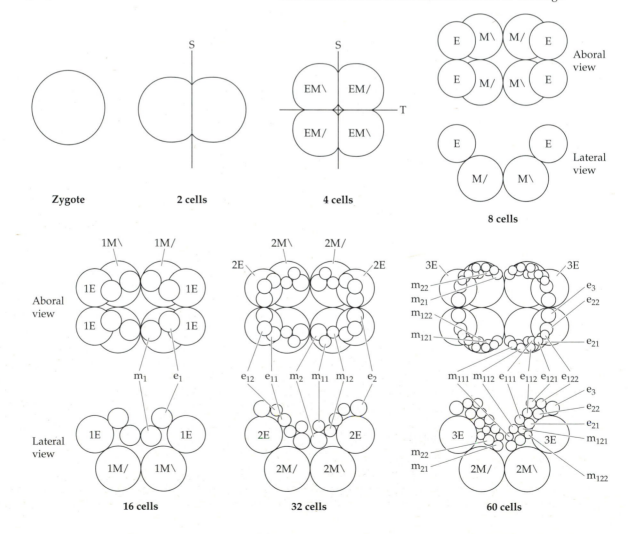

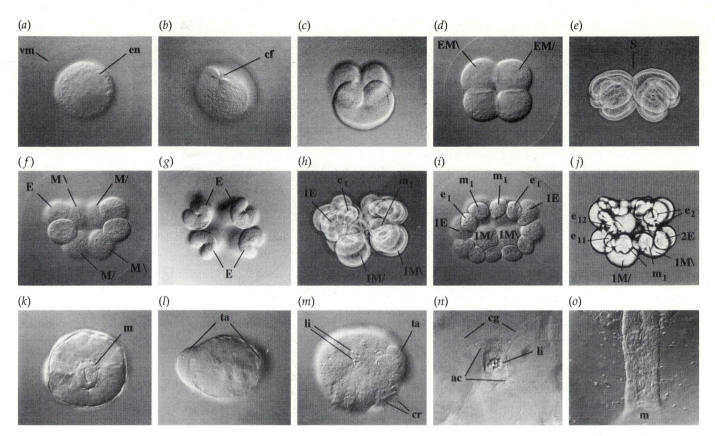

Figure 6.8 Differential interference contrast micrographs of *Mnemiopsis leidyi* development. (*a*) A fertilized but uncleaved embryo. Note outer vitelline membrane (vm) and the boundary between the ectoplasm and the centrally located endoplasm (en). (*b*) The initiation of the unipolar first cleavage furrow (cf). The plane of first cleavage defines the sagittal plane (S). (*c*) Unipolar cleavages of a 2-cell-stage embryo. The plane of second cleavage corresponds to the tentacular plane. (*d*) A 4-cell-stage embryo. Note the two pairs of diagonally opposed cells (EM\ and EM/). (*e*) Third cleavage. (*f*) Oblique lateral view of an 8-cell-stage embryo. Note the four E and four M blastomeres. (*g*) Aboral view of an 8-cell stage-embryo in which the four E blastomeres are beginning to divide. (*h*) Lateral view of an 16-cell-stage embryo. Each blastomere generates a smaller micromere toward the aboral pole. (*i*) Aboral view of a 16-cell-stage embryo. (*j*) Oblique lateral

view of an embryo at the end of the 16-cell stage. The e_1 micromeres divide a few minutes ahead of the m_1 micromeres. (*k*). Oral view of the mouth (m) of a late gastrula-stage embryo. (*l*) A slightly older embryo focused on the two tentacle sheaths (ts) prior to their invagination. (*m*) Oblique aboral view of the developing apical organ. Note the lithocytes (li) from each of the four quadrants converging toward the aboral pole. Four ctene rows (cr) and a tentacle sheath can also be seen. (*n*) Aboral view of the apical organ in a newly hatched cydippid larva. The ciliated grooves (cg) connect the apical organ to the eight ctene rows. Note the two endodermal anal canals (ac) connect to the outside via the anal pores. (*o*) Lateral view of the stomodeum of a newly hatched cydippid. The mouth (m) is located at the bottom and the aboral pole toward the top. Note the laterally radiating mesenchymal cells which are migrating through the mesoglea.

The aboral micromeres are labeled by a lowercase letter corresponding to the macromere of origin and a subscript that indicates the rank order of their production. Capital letters are used to denote the macromeres, and a numerical prefix is used to indicate how many rounds of micromeres have been born. Thus, at the 16-cell stage there are four e_1 and four m_1 micromeres and four 1E and four 1M **macromeres**. The E macromeres generate three rounds of **aboral micromeres** (e_1, e_2, e_3) and the M macromeres generate two rounds of aboral micromeres (m_1 and m_2). The micromeres begin to divide in a stereotypical manner soon after they are formed to ultimately generate a large number of cells. The daughter cells are labeled with successive subscripts. For instance, the two daughter cells of m_1 are called m_{11} and m_{12} and the daughters of m_2 are called m_{21} and

m_{22}. The aboral micromeres continue to proliferate to give rise to a cap of small cells located on the aboral pole (Figure 6.9). The nomenclature introduced above differs slightly from that previously used, but it has some advantages over the older system. The new system makes it easier to unambiguously distinguish these cells when discussing them. For example, it is easier to discern the identity of a cell named 1M ("one M") from m_1 ("m one") than it was using the older system, where one must also distinguish between uppercase and lowercase letters (M_1 as "capital M1" versus m_1 as "small m1" (Reverberi 1971).

Yatsu (1911) describes a minor variation in the cleavage program of the Mediterranean ctenophore *Callianira bialata*. In this form the E macromeres are supposed to give rise to an extra (fourth) set of aboral micromeres. This is the only

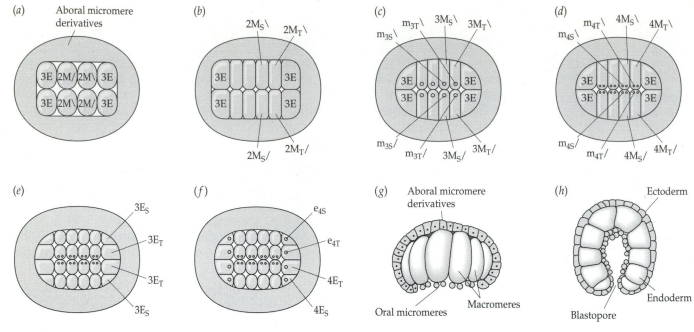

Figure 6.9   Diagram of gastrulation and the birth of oral micromeres in ctenophores. (*a–f*). Oral pole views of ctenophore embryos. Descendants of the aboral micromeres move by epiboly towards the oral pole. (*b*) The four 2M macromeres divide in the sagittal plane to give rise to eight macromeres. Each macromere gives rise to two small micromeres (*c–d*). The 3E macromeres then divide along the tentacular plane (*e*) before giving rise to a single round of oral micromeres (*f*). (*g*) Lateral view of gastrulation in ctenophore embryos. The aboral micromeres have moved down to the oral pole (the site of oral micromere formation). (*h*) The embryo elongates along the oral-aboral axis. The blastopore forms at the oral pole and will later become the mouth.

variation in the cleavage program in ctenophores that we are aware of, and deserves further investigation.

Until recently the four quadrants of the adult, each derived from one of the first four blastomeres, were treated as though they were equivalent to one another. Recent cell lineage experiments have revealed that the fate of one diagonal pair of cells at the 4-cell stage differs fundamentally from that of the other diagonal pair (Martindale and Henry 1995). Thus, it has become necessary to distinguish these two pairs of cells. It is not possible to discriminate the different diagonal pairs at the 4-cell stage unless the orientation of the first two cleavage planes is carefully followed. One can unambiguously identify derivatives of both sets of diagonal cells at the next cleavage division (8-cell stage), however, when the E cells are generated at the tentacular poles. By convention, when one looks at the aboral pole of the embryo, the pair consisting of the top-right and bottom-left cell is called the "/" (slash) pair and the top-left and bottom-right pair is referred to as the "\" (backslash) pair. For many situations, it is not necessary to discriminate the differences between / and \ pairs because differences in cell fates have not been detected, so that this nomenclature is only invoked when necessary.

Gastrulation and Organogenesis

Gastrulation proceeds as the embryo flattens in the oral-aboral dimension. The cap of small aboral micromere descendants proliferates and moves by epiboly towards the oral pole. Following the production of the aboral micromeres, the 3E and 2M macromeres continue to divide. The four 2M macromeres divide first in the sagittal plane to give rise to $2M_S$ (closest to the sagittal plane) and $2M_T$ (closest to the tentacular pole, T) (Figure 6.9). Each of the four $2M_S$ and four $2M_T$ cells subsequently give rise to two rounds of much smaller cells at the oral pole called **oral micromeres**, m_3 and m_4. According to Farfaglio (1963) and Revereri and Ortolani (1963) the four $2M_S$, four $2M_T$ cells, and the four 3E cells then divide equally in the tentacular plane. The eight 4E macromeres then each give rise to a small oral micromere (e_4), but the 5M macromeres do not give rise to additional micromeres. The embryo elongates in the oral-aboral axis as the macromeres and the oral micromeres buckle up into the middle of the embryo (Figure 6.9*h*). The opening at the animal pole becomes the mouth. Unfortunately, no mechanistic studies of gastrulation have been reported so that it is not yet possible to know how or where the forces that shape the embryo arise. For example, it would be interesting to know what the role of differential cell adhesion is, or whether the leading edge of micromeres "pulls" the rest of the animal cap down to the oral pole, or whether mitotic activity is responsible for "pushing" the animal cap cells, or whether the large macromeres themselves are playing an active role in the movement of the animal cap cells.

The various organ primordia begin to form soon after gastrulation. The tentacle sheaths appear as ectodermal thickenings (in tentaculate ctenophores) and the asynchronously beating precursors of the ctene rows appear in the aboral hemisphere. At first the ctene plate cilia are not arranged in discreet rows. As additional comb plate cilia appear, each quadrant of the embryo will generate a pair of ctene row precursors. The sensory organ forms on the aboral pole. The precursors of the nonmotile dome cilia first appear some distance away from the pole (Figure 6.8*m*). These cilia converge at the aboral pole as the lithocytes develop. The lithocytes are released from the floor of the apical organ and retained under the dome cilia until they are trapped by the newly formed balancing cilia. The tentacle sheathes form by invagination, and as the ectodermal structures are elaborated, the mesoglea hydrates and the embryo swells to a nearly spherical shape. The ctene rows begin to beat in a coordinated fashion at this time and the young cydippid begins to swim. The tentacles are generated by stem cells in the tentacle bulb and usually do not extend out from the tentacle sheathes for a day or so. Tentacles are generated throughout the life of the animal as they are prone to damage during the feeding process. The entire process of embryogenesis is rapid, taking as little as 16 hours in some species.

Cell Lineage

The regularity of development seen in ctenophore embryos suggests that the ultimate fates of identified cells in the embryo may become determined at early stages of development. An accurate **fate map** is essential for experimental investigations of the mechanisms of cell fate determination. The early workers attempted to follow the fates of individual cells by examining living and fixed embryos at different stages of development. An obvious improvement in these types of observational studies is to label individual cells and follow them throughout the course of development. The first attempt to generate an accurate ctenophore fate map utilized chalk particles (Ortolani 1963; Reverberi and Ortolani 1963). While these studies provided a great deal of information, chalk particles are not accurate cell lineage markers. A finite number of chalk particles may be placed on the outside surface of specific cells but they may not be distributed to each and every daughter cell. Furthermore, they are often so large that it is impossible to label small cells, and the particles can be displaced or lost during subsequent development.

In order to generate a higher resolution fate map, cell lineage studies were recently repeated on the lobate ctenophore, *Mnemiopsis leidyi* with the intracellular injection of the fluorescent dye DiI. (Martindale and Henry 1995, 1996, 1997). DiI is a lipophilic fluorescent dye. When injected into living cells it labels all of the membranous organelles in the cytoplasm, including those that give rise to new cell surface membrane. The dye remains confined to the plane of the lipid bilayer and therefore will label only the descendants of the injected cell. By systematically injecting single cells during each stage of embryogenesis and then determining which cells are labeled in the cydippid larvae, it is possible to identify the fates of all of the cells derived from the injected blastomere. The first capital fact revealed by these experiments is that the first cleavage plane (identified by injecting one cell at the 2-cell stage) always corresponds to the sagittal plane. One cell will make one complete tentacle apparatus and the other cell the opposite one. The second cleavage plane is perpendicular to the first and corresponds to the tentacular plane. These results confirm earlier studies (Ortolani 1963; Reverberi and Ortolani 1963; Freeman 1976a). These injection experiments indicated that the ectodermal derivatives of each of the first four cells remain confined roughly to one quadrant of the cydippid, but that the mesodermal and endodermal derivatives mix freely with descendants from other quadrants. The results of intracellular labeling techniques have been particularly useful for identifying the fates of cells born at later stages of development. A summary fate map is shown in Figure 6.10.

The E and e Lineages

Each of the e_1 micromeres of the 16-cell-stage embryo contributes to a pair of comb rows, part of the tentacle, tentacle bulb, tentacle sheath, portions of the floor of the apical organ, and epidermis located around the tentacle sheath that extends to the mouth. Each of the e_2 micromeres forms dome cilia, part of the floor of the apical organ, part of the pharynx, and a strip of epidermis located around the tentacle sheath that extends from the apical organ to the mouth. Each of the e_3 micromeres gives rise to the pharynx and a narrow string of beadlike epidermal cells extending between the tentacle sheath and the mouth.

The 3E macromeres (which eventually give rise to a set of small micromeres and the oral pole) give rise to longitudinal muscles located beneath the epidermis, stellate mesenchymal cells scattered throughout the mesoglia, radial muscle fibers that span from the pharynx to the epidermis, and a portion of the endoderm. The fate of the oral micromeres has not been followed due to their extremely small size, but their fate can be inferred by studies in which the macromeres were injected after the birth of the oral micromeres. If macromeres are injected following the birth of the oral micromeres, they generate endodermal derivatives exclusively; however, the exact endodermal lineages are difficult to define. No two embryos give rise to exactly the same pattern of endodermal staining. Endodermal cells appear to mix extensively, and labeled progeny from one injected cell can be found in all four quadrants. These results indicate that the oral micromeres from the E lineage generate a wide variety of muscle and mesenchymal cells as well as possibly contributing to endodermal derivatives.

Figure 6.10 Cell lineage diagram indicating the fates of the early blastomeres of the lobate ctenophore *Mnemiopsis leidyi*. There are definitive differences in the fates of derivatives of the 2M blastomeres derived from the / and \ quadrants. The / and \ nomenclature is not invoked for cases in which there is no known difference. AO, apical organ.

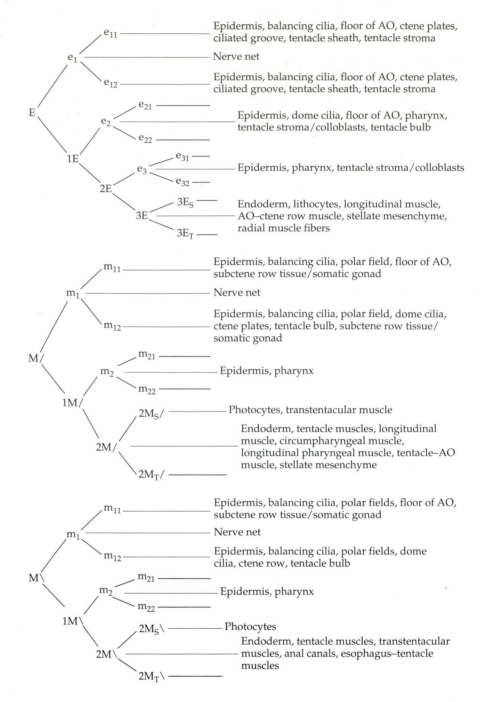

The M and m Lineages

The m_1 micromeres each form dome cilia, the polar fields, balancing cilia, part of the floor of the apical organ, epidermis located between the two ctene rows that extends from the apical organ to the mouth, and a portion of the ctene rows and subctene row tissue, likely to be the somatic gonad. The m_1 contribution to the ctene rows is somewhat variable, but usually includes the **polster cells** from the adesophageal regions of ctene rows in the quadrant derived from the injected cell (Martindale and Henry 1997). The m_2 micromere all contribute to the formation of the

pharynx, part of the tentacle bulb, and epidermis located between the ctene rows that extends from the apical organ to the mouth.

We have found that the fates of the four M macromeres and their progeny are distinctly different. The four 2M macromeres of the 32-cell-stage embryo (which eventually give rise to two rounds of oral micromeres and macromeres) are comprised of two different pairs situated diagonally opposite from one another that give rise to distinctly different cell fates. When the embryo is viewed from the aboral pole, with the tentacular plane running horizontally, one pair of cells consists of the top-right and bottom-left 2M cells, called the 2M/ (2M slash) pair. The other pair consisting of the top-left and bottom-right 2M cells, referred to as the 2m\ (2M backslash) pair. The 2M\ pair of macromeres generates the muscular core of the tentacle, the transtentacular muscle fibers, the esophageal-tentacular muscle fiber, endoderm including the tentacular canal, and both anal canals. The 2M/ pair of macromeres gives rise to longitudinal, circumpharyngeal, and longitudinal pharyngeal muscle cells, tentacle core muscles, transtentacular fibers, tentacle-apical organ fibers, endoderm associated with the gut, and multipolar stellate mesenchymal cells. Both 2M/ and 2M\ macromeres contribute to the formation of endodermal tissue after they have given rise to the oral micromeres. Endodermal contributions from the M lineages behave in a manner similar to that described for the E lineages, i.e., no definitive regions can be said to be generated by individual lineages. It is known that the $2M_S$ macromere (formally called the M3), but not the $2M_T$ macromere (formerly called the M4), generates light producing photocytes (Freeman and Reynolds 1973), probably from the oral micromeres: but this point has not yet been verified.

Summary of Cell Lineage Studies

The DiI lineage tracing experiments have revealed a number of important differences from the previous fate map (Reverberi and Ortolani 1963; Ortolani 1989). One of the most important findings is the fact that the m_1 micromeres take part in the formation of the ctene rows, whereas it was previously thought that only the e_1 micromeres gave rise to these structures. Both e_1 and m_1 micromeres also give rise to epidermis, balancing cilia, and part of the floor of the apical organ. Only e_1 micromeres appear to give rise to the ciliated grooves that connect the ctene rows to the apical organ. The lineage analyses also found that the lithocytes, including the reserve lithocytes made in the floor of the apical organ, are generated from the E macromeres following the formation of the three rounds of aboral micromeres (the 3E macromeres). The ctenophore nervous system is, at least in part, generated from the e_1 and m_1 lineages; however, additional experiments are needed to identify the total complement of micromere and/or macromere blastomere precursors from which the nervous system is derived.

These results also indicate that the aboral micromeres generate all of the ectoderm, the macromeres generate endoderm, and the oral micromeres give rise to the mesoderm. This result indicates that the mesodermal lineage is more closely related to that of the endoderm and not the ectoderm. This confirms earlier studies which suggested that ctenophores possess endomesoderm (Ortolani, 1963; Reverberi and Ortolani, 1963) and not ectomesoderm. This is an important issue from an evolutionary standpoint because it allies the ctenophores with the higher protostomous Bilaterians (virtually all of which posses

some mesoderm derived from endodermal precursors). Cnidarians, on the other hand, are thought to have only ectomesoderm (Hyman 1940).

Experimental Embryology

Spatial and Temporal Localization of Developmental Potential in Ctenophore Embryos

The regularity of the cleavage program in ctenophore embryos led many embryologists to believe that a great deal of spatial information (referred to as *germinal localization*) existed in the early embryo. Blastomere deletion and isolation experiments tended to support this view of ctenophore development. For example, if blastomeres are separated at the 2-cell stage and grown to cydippid larvae (Figure 6.11*d*), each one will possess four ctene rows, one tentacle, and a "half" of an apical organ (Chun 1892; Driesch and Morgan 1895; Martindale 1986). When the first four blastomeres are isolated, each gives rise to a pair of ctene rows (Figure 6.11*e*). The ctene rows are formed primarily (but not exclusively) from the e_1 micromeres generated at the aboral pole during the fourth cleavage. No ctene rows form if all four of these micromeres are deleted. Furthermore, two ctene rows are missing for each e_1 micromere that is removed (Farfaglio 1963; Martindale 1986; Martindale and Henry 1996, 1997). These experimental results could indicate that the proper organ-forming determinants are prelocalized to the correct locations. These results also suggest that ctenophore embryos are highly "mosaic" and unable to "regulate" to form the missing structures normally derived from the deleted cells.

There is, however, ample experimental evidence to argue that organ-forming regions are not prelocalized in ctenophore eggs. If a ctenophore egg is cut into two fragments and then fertilized, each half develops into a normal cydippid (Driesch and Morgan 1895; Yatsu 1912; Freeman 1977). Normal cydippid larvae, complete with apical organs, develop from fertilized eggs that had been bisected equatorially (i.e., perpendicular to the aboral pole) at the start of first cleavage (Yatsu 1912; Freeman 1977). These results indicate that factors responsible for giving rise to the apical organ and ctene rows are not localized to their presumptive locations until sometime after the onset of first cleavage (Figure 6.11*a–c*)

The localization of developmental potential for the formation of ctene rows has been examined in several species, including the atentaculate *Beröe*, (Zeigler 1898; Fischel 1903; Yatsu 1912; Houliston et al. 1993), the lobate *Mnemiopsis* (Freeman and Reynolds 1973; Freeman 1976a, b) and two species of the cydippid *Pleurobrachia* (Freeman 1977). In these studies of these ctenophore embryos, various regions were removed with fine glass needles at various stages of development, to see whether there was a reduction in the ability of the operated embryos to generate

Figure 6.11 Results of various operative procedures in ctenophore embryos. (*a*) Both halves of an oocyte bisected along its equator develop in to normal cydippid following fertilization. (*b*) Both halves of an oocyte bisected along the future oral-aboral axis also develop into normal larvae. (*c*) If fertilized eggs are bisected through their equator at the time of first cleavage only one of the two fragments continues to develop. The fragment containing the nucleus at the oral pole continues to develop and gives rise to a normal larvae. (*d*) The result of separating the first two cells in the ctenophore embryo. Each blastomere gives rise to a "half-larva" consisting of four ctene rows, a small tentacle apparatus, and a half of an apical organ. (*e*) Results of separating and rearing each cell at the 4-cell-stage. Each blastomere gives rise to a small larvae containing two ctene rows, a rudimentary apical organ, and tentacle apparatus. (After Freeman 1976a, 1977.)

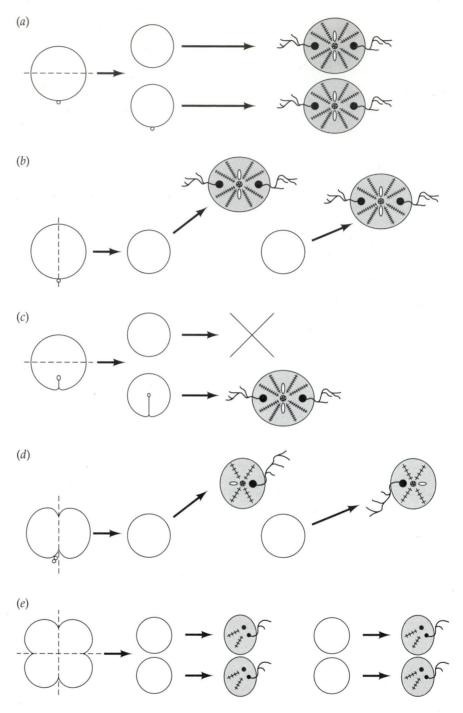

comb rows (Freeman 1976a, 1977; Freeman and Reynolds 1973). The results indicate that the factors responsible for generating comb rows are localized homogeneously throughout the egg cytoplasm, but they become transiently localized to the site of first cleavage (Houliston et al. 1993) and then progressively toward the aboral end of the embryo by the 4-cell stage (Figure 6.12). Freeman showed that the developmental potential for ctene row production remained in the aboral end of 4-cell embryos, but that by the end of the 4-cell stage, this potential had shifted away from the middle of the embryo toward the ends represent-

ing the presumptive tentacular region of each of the four cells (Figure 6.12). This represents the site where the E macromeres will be generated at third cleavage. Ctene row potential is shunted asymmetrically again at the next division (16-cell stage) so that the much smaller e_1 micromeres, but not the 1E macromeres, inherit ctene row determinants.

Freeman (1976a; Freeman and Reynolds 1973) also investigated the localization of developmental potential necessary for the production of the light-producing photocytes (Figure 6.5*b*). By using similar techniques of cell

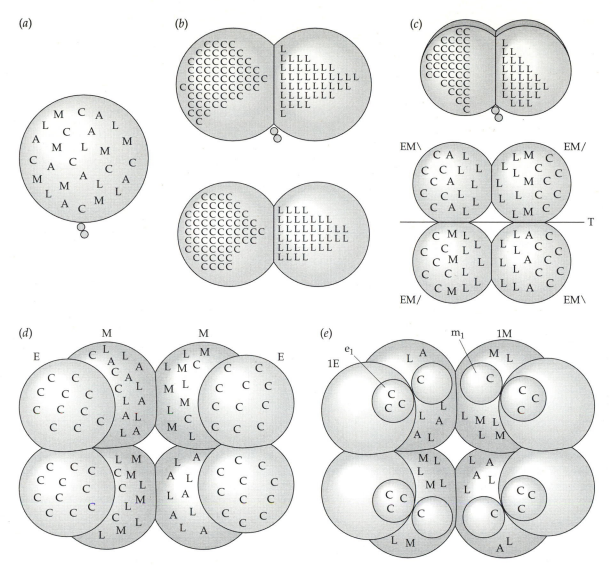

Figure 6.12 Diagram of the distribution of developmental potential for comb plates (C), light-producing photocytes (L), anal canals (A), and circumpharyngeal muscle (M). (*a*) All of these factors are evenly distributed in the uncleaved egg. (*b*) Two views (lateral view, top; aboral view , bottom) of a 2-cell-stage embryo. Comb plate determinants begin to move towards the aboral pole while photocyte potential remains confined to the central region of the embryo. For clarity, comb plate and photocyte potential are indicated in only one cell, but both cells contain an identical pattern of factors for both cell types. Nothing is known about the distribution of anal canal and circumpharyngeal muscle determinants at the 2-cell stage, so they are not shown here. (*c*) Same views of a 4-cell embryo as seen in *b*. Comb plate potential is moving towards the outer edge of each blastomere while photocyte potential is moving towards the oral pole along the sagittal plane of each cell. At this stage, factors for anal canals and circumpharyngeal muscle are localized to diametrically opposed blastomeres. T = tentacular plane. (*d*) Aboral view of an 8-cell-stage embryo indicating that photocyte potential is now exclusively localized to the M blastomeres while comb plate potential is localized largely, but not exclusively, to the E blastomeres. (*e*) Aboral view of a 16-cell-stage embryo, indicating that comb plate potential is localized to the e_1 and m_1 micromeres, while photocyte, anal canal, and circumpharyngeal muscle potential remains localized to the 1M macromeres. (*b* and *c* after Freeman 1976a.)

dissection and isolation, it was found that the localization of photocyte potential displayed a somewhat complimentary pattern of segregation. The ability to generate photocytes is present ubiqitously in the uncleaved egg, but becomes segregated to the four M macromeres by the 8-cell stage. Cell isolation experiments have indicated that the potential to form photocytes is segregated at each of the subsequent cell divisions of the M lineage to the vegetal macromeres, and not to the aboral m micromeres. Only the 3M macromeres possess the ability to produce photocytes (Freeman and Reynolds 1973). It is likely that the small micromeres given off at the oral pole of these macromeres generate these cells.

Thus, at each cell division there appears to be an unequal segregation of developmental fate. Visible movements of cytoplasmic materials take place during these cell divisions. The ectoplasm can be seen to concentrate in the aboral micromeres at the 8-cell stage, while the macromeres retain most of the yolky endoplasm. Centrifugation experiments have shown that the cortical ectoplasm is associated with the ability to form ctene rows (La Spina 1963) even when inherited by M cells (Freeman and Reynolds 1973), while the ability to form photocytes is associated with the yolky endoplasm (Freeman and Reynolds 1973). Thus, it is apparent that the segregation of ectoplasm into aboral micromeres is an important step in the localization of comb plate potential during the early cleavage divisions.

A recent investigation of the early development of the lobate ctenophore *Mnemiopsis* has shown that in addition to the epigenetic localizations of ctene row determinants and photocyte potential, an early segregation of developmental potential takes place along an axis oblique to both the sagittal and tentacular axes (Martindale and Henry 1995). Two diagonally opposing cells at the 4-cell stage (EM\) give rise to the endodermal anal canals while the other two cells (EM/) give rise to circumpharyngeal and longitudinal muscle cells. Cell deletion experiments have shown that if one diagonal pair of cells is isolated from the other, they will give rise to either the anal canals or circumpharyngeal muscles, but not to both. These data indicate that each of the first two cells gives rise to two daughter cells that are not equivalent. This segregation is significant because previously the first asymmetry of cell fates was thought to occur with the formation of the E and M lineages at the 8-cell stage (see the results of cell isolation experiments described above). While E and M macromeres can be identified based on cell size and position at the 8-cell stage, the cells at the 4-cell stage are identical in size and can not be distinguished from one another unless the site of first and second cleavage is fol-

lowed. Additional studies are required before anything can be said about the position of these determinants prior to the four-cell stage. This unique form of **diagonal determination** has not been described in any other metazoan embryo.

The Origins of Axial Information in Ctenophore Embryos

The evidence described above indicates that there is little preexisting spatial information in the uncleaved ctenophore egg and that developmental potential is segregated over time into the appropriate cell lineages. When, and by what mechanism does axial information arise in the ctenophore embryo? In most embryos the only overt axis present in the unfertilized egg is the **animal-vegetal** axis. The animal pole is generally defined by the location of the female pronucleus and the position where the polar bodies are given off. There are often other manifestations of this axis, such as a gradient in the distribution of yolk or pigmentation. In virtually all embryos the first two cleavage planes intersect at the animal and vegetal poles so that the animal pole ends up in the head of the larva or adult organism, while the vegetal pole becomes the initiation site of gastrulation and the vicinity of either the mouth (as in protostome development) or anus (as in deuterostome development).

In ctenophores, the site of first cleavage often does not correspond to the site of polar body formation. This result may be due to the fact that the female pronucleus often travels long distances to "visit" different male pronuclei before pronuclear fusion occurs. This phenomenon begs the question, Do the first two cleavages intersect along a preexisting animal-vegetal axis or can the cleavage program be initiated at any location? A series of detailed marking experiments performed by Freeman (1977)

Figure 6.13 Labeling experiments that indicate that the site of first cleavage is causally involved with the establishment of the oral-aboral axis. (*a*) If the site of polar body formation is labeled and it corresponds to the site of first cleavage, the oral pole of the resulting cydippid larvae will be labeled. (*b*) If the site of first cleavage is labeled but does not correspond to the site of polar body formation, the oral pole contains the label. (*c*) If the site of polar body formation is labeled but the site of first cleavage is distant from the labeled site, the oral pole is not labeled. This last experiment indicates that the act of labeling does not entrain the oral-aboral axis.

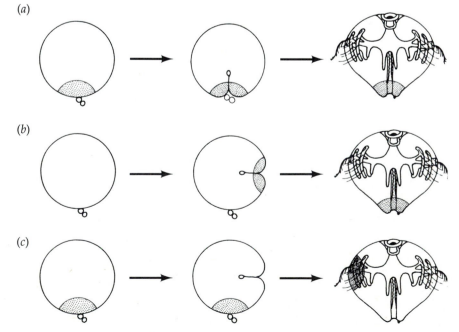

demonstrated that in normal embryos the position of the polar bodies may have a variable location with respect to the larval axes but that the site of first cleavage closely corresponds to the position of the mouth in the cydippid larvae (Figure 6.13). Freeman marked the site of polar body formation in uncleaved ctenophore embryos and then centrifuged the embryos such that the site of first cleavage occurred at a site distant from the site of polar body formation. The site of first cleavage was marked by a different label and the older cydippid larva examined. The results showed that the site where first cleavage is initiated is associated with the establishment of the oral-aboral axis, and that there is no meaningful axial information (e.g., an animal-vegetal axis) inherent in the uncleaved egg that directs subsequent development.

Additional work by Freeman (1976*a,b*) indicates that the mechanism responsible for the asymmetric localization of developmental potential to the E and M cells is intimately tied to a so-called **cleavage clock**. Cleavage in ctenophore embryos can be reversibly inhibited with either a metabolic poison (2,4- dinitrophenol) or the microfilament inhibitor, cytochalsin-B. When these drugs are used to inhibit one cleavage division, and then washed out, the subsequent cell division will generate either the blastomere configuration it would normally have undergone before treatment, or it will generate the advanced configuration of blastomeres present

in the untreated controls (Figure 6.14). If first cleavage is reversibly inhibited, then the embryo invariably generates a normal 2-cell embryo and continues to develop according to the normal sequence of cell divisions (Figure 6.14*b*). If the first cleavage is allowed to occur and the second (or a subsequent) cleavage is inhibited, then when the drug is washed out and the next division is allowed to proceed, it most often assumes the character of the division of an untreated control embryo (Figure 6.14*c,d*). These results indicate that the timing of the cleavage clock is set by some event related to the first cleavage division and that the normal size and orientation of subsequent cleavages are not related to the previous division, but are related to the ticking of an intrinsic cleavage clock that has not been disturbed as a result of these treatments.

Cell isolation experiments performed with the reversible inhibition of cleavage have also shown that the appropriate segregation of developmental potential is linked to the character of the cell division (Freeman 1976a). For example, when the second cleavage division is inhibited and the blastomere configuration seen in Figure 6.14*c* is obtained, it is possible to separate the E cells from the M cells and assay their developmental potential. E cells make comb cilia but do not produce light, and M cells produce light but do not produce comb cilia. These are precisely the results expected for a "normal" third cleavage regardless of the fact that these embryos did not pass through a normal 4-cell stage blastomere configuration.

The cleavage program does not just passively carve up certain cytoplasmic domains. If 2-cell embryos are compressed such that the orientation of the mitotic spindles is altered, they can make a 4-cell embryo that has the blastomere configuration of half of an 8-cell embryo (Figure 6.14*f*). These embryos have two E and two M macromeres. This configuration of blastomeres is identical in appearance to one-half of a normal 8-cell embryo, or to an em-

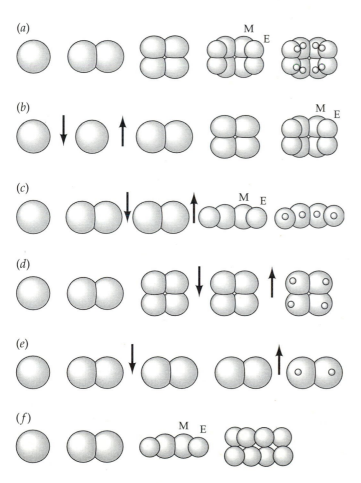

Figure 6.14 Experiments indicating the features of the ctenophore "cleavage clock." (*a*) Normal pattern of cell cleavages in undisturbed ctenophore embryos. (*b*) If uncleaved embryos are treated with cytochalasin B for one cleavage period and then washed out, the embryo begins cleaving one cleavage cycle behind untreated controls. The arrows indicate the time of addition and withdrawal of treatment. (*c*) If 2-cell-stage embryos are treated the same way, most cases skip a cleavage cycle. They cleave to give rise to E and M blastomeres as if they were half of a normal 8-cell-stage embryo. (*d*) The same phenomenon occurs if the cleavage is blocked following the second cleavage. (*e*) More than one cleavage can be blocked and the resulting embryo will cleave in way similar to the untreated controls. (*f*) If a 2-cell-stage embryo is compressed between first and second cleavage, an arrangement of blastomeres similar to half of an 8-cell-stage embryo can be produced. The developmental potential of the E and M blastomeres produced one cleavage early is fundamentally different from that obtained in *c* however. This last experiment indicates that factors specifying E and M cell fates have not become localized to their correct position until the onset of the eight-cell stage. See text for details. (After Freeman 1976a,b.)

bryo in which the second cleavage was reversibly inhibited (Figure 6.14c). However, if these E and M blastomeres derived from the altered cleavage spindle experiment are isolated, both types of blastomeres will give rise to comb plates. This result indicates that the premature formation of E cells at the tentacular poles does not result in the asymmetric localization of comb plate determinants. There is a coordination of the intracellular localization of ctene row determinants with the early cleavage program that ensures that developmental potential is segregated to the appropriate cell lineages, at the appropriate times, during early embryogenesis.

Freeman's experiments in which different regions of the developing embryo were removed at different times during the early cleavage program indicate that although there may be some degree of localization of comb row and photocyte potential occurring throughout the cell cycle, most of the localization occurs in association with cell division (Freeman 1976a). These results indicate that developmental potential is segregated by a complex process involving alterations in the cytoskeleton. A likely source of this organizing information is the spindle and the associated microtubules. The site of the formation of the first mitotic spindle, and hence of the first cleavage furrow, becomes the oral pole. Furthermore, first cleavage normally restricts the developmental potency of each of the first two blastomeres. Before first cleavage, each half of the zygote will give rise to a normal cydippid, but after first cleavage, each blastomere only gives rise to a "half-animal." Houliston and colleagues (1993) waited until the onset of first cleavage in *Beroë* and then quickly bisected the embryo along the presumptive first cleavage furrow. When these fragments were raised, each gave rise to a "half-animal," indicating that the completion of a normal cleavage furrow is not necessary for the change in developmental potential. While somewhat indirect, these results argue that changes in the localization of organ forming determinants may be mediated by changes in microtubule or microtubule associated factors.

The Role of Cell-Cell Interactions in Ctenophore Development

Many of the experiments described above suggest that cell fates are solely determined by virtue of cytoplasmic determinants inherited by different cell lineages. This view suggests that the embryo consists of a mosaic of individual parts with no interaction between them. The first reported case of inductive interactions in ctenophore embryos involved the deletion of e_1 micromeres (Martindale 1986). Under normal circumstances each ctene row is composed of the overlying ctene plates and an endodermal canal complex and associated gonadal tissue. The endodermal canals appear to be derived from the macromeres following oral micromere production. If an e_1 micromere is destroyed shortly after its birth then the resulting larva is deficient in a pair of ctene rows (Farfaglio 1963) *and* the underlying endodermal canal. More recent cell lineage ex-

periments have shown that the ctene rows are also generated from the m_1 micromeres (Martindale and Henry 1997). The failure of the ctene rows to appear following deletion of the e_1 micromeres indicates that m_1 micromere derivatives also require the presence of e_1 micromere derivatives in order to generate ctene rows. The relationship of the e_1 and m_1 micromeres is reminiscent of the situation seen with regard to the production of muscle cells in ascidian embryos, in which there are some lineages that will make muscle cells autonomously and some lineages that require inductive interactions for their formation (see Chapter 17, "Tunicates," by Jeffery and Swalla).

Regeneration

It is perhaps not too surprising to learn that many adult ctenophores have the ability to regenerate missing portions of their bodies (Mortensen 1912; Coonfield 1936; Franc 1970). Due to its fragile nature, the ctenophore body is prone to damage and without this ability it is doubtful whether many would survive to adulthood. This regenerative ability has been carried to an extreme in the benthic "creeping" ctenophores (order Platyctenea), which are known to undergo asexual fission (Tanaka 1931; Dawydoff 1938; Freeman 1967) in addition to normal patterns of sexual reproduction. These animals routinely shed small fragments of tissue at their periphery, which subsequently go on to reconstitute the complete complement of adult pattern elements (an apical organ and two tentacles, in as much as these animals do not possess ctene rows as adults).

Interestingly, individuals of *Beroë* collected from the eastern coast of the United States, the Gulf of Mexico (Florida and Texas), and Puget Sound do not appear to have an extensive capacity to regenerate (Martindale, personal observation). It is not clear why this is so. Perhaps tentaculate animals possess stem cells throughout their lives to replace damaged tentacles, and these or related lineages of stem cells are capable of replacing other missing structures. Another possibility is that the mesoglea is more rigid in beroids than in most other ctenophores and that they do not injure as easily. If one assumes that the atentaculate condition is derived, beroids may have lost the ability to replace missing parts along with the loss of their tentacle renewal system.

When cuts are made parallel or perpendicular to the aboral axis in the lobate *Mnemiopsis*, each piece faithfully replaces the missing structures. For example, when oral pieces are separated from aboral regions, they regenerate an apical organ (and the complimentary piece regenerates a mouth). When cuts are made parallel to the oral-aboral axis, each piece normally regenerates the normal complement of adult tissues (eight ctene rows, two tentacles, and a complete apical organ). For example, if an animal is cut along the oral-aboral axis such that one piece possesses one tentacle and four ctene rows, while the complementary piece possesses a tentacle, four ctene rows, and a complete apical organ, each piece only replaces the pattern elements necessary to form the normal complement of adult structures.

Under normal circumstances, when an animal is cut, the free edges fuse. Cells migrate to the cut edge and condense into cords. When there are similar cells in the vicinity, existing cells will participate in the formation of the new structures. For example, if a section of a ctene row is extirpated, the endodermal canals and overlying ctenes associated with the cut ends of the ctenes will stretch and make contact with each other. Initially, the endodermal canals are solid, but as they grow, a lumen forms that is continuous with the canals on either end. Additional ctene plates gradually appear in the epidermis above the endodermal cords. In situations in which there are not identical cell types or structures in the vicinity of the missing structures, the new structures are generated from a population of cells that migrate to the wound site.

In the majority of cases in which animals are cut along the oral-aboral axis, the anlage that give rise to the new structures are initially formed at the cut edge, in a pattern consistent with standard **intercalary regeneration** (Mar-

tindale 1986). In some cases, however, the wound heals but the new structures are not generated immediately. In these cases one of the intact structures, such as a tentacle apparatus, splits into two separate structures, and cells migrating into the intervening space replace the missing structures (Martindale 1986).

In some situations, when cuts are made parallel to the oral-aboral axis, one of the two pieces will fail to regenerate. These animals heal the wound site, but do not replace any of the missing structures. These deficient adults possess one-half of the normal complement of adult structures (one tentacle and four ctene rows). Close examination of the apical organ in these cases shows that there are only half the normal number of balancing cilia and polar furrows. For some reason these animals remain as "half-animals" and fail to revert back to the normal condition for the rest of adulthood (Chun 1880; Freeman 1967; Martindale 1986). "Half-animals" that form during adulthood as a result of the failure to regenerate completely are indistinguishable from "half-animals" that result from bisecting ctenophore embryos (Figure 6.11d) prior to ctene row coordination (Figure 6.15). For ctenophores, the "half-animal"

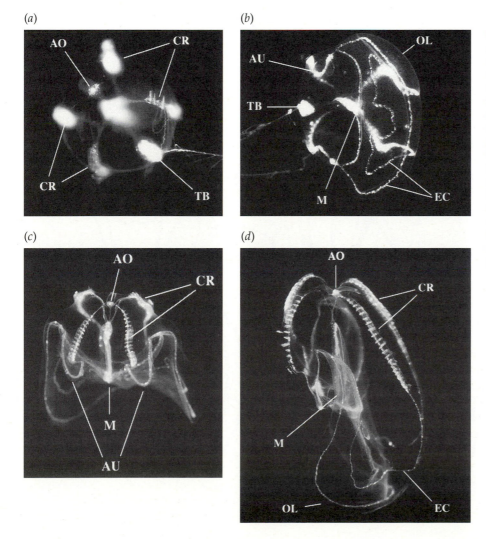

Figure 6.15 "Half-animals" of the lobate *Mnemiopsis*, derived from the failure of adults to regenerate following bisection along the oral-aboral axis. These animals are indistinguishable from "half-animals" that are generated by bisection during embryogenesis prior to ctene row coordination. (*a*) Aboral view of a young cydippid showing four ctene rows (CR) and one tentacle bulb (TB). The apical organ (AO) has only one polar field. (*b–d*) Oral, oblique, and lateral views of fully mature lobate stage adult "half-animals." These animals consist of a single oral lobe (OL), with a mouth (M) and endodermal canals (EC), but possess only half the normal number of auricles (AU), tentacle bulbs (TB) and ctene rows.

phenotype is a **metastable developmental state** (Freeman 1967). The study of "half-animals" has been very useful for understanding various aspects of ctenophore biology. For example, dissogeny (precocious reproduction during early larval stages) occurs more frequently in "half-animals" (Chun 1880; Martindale 1987) and "half-animals" have been exploited for experiments investigating the control and onset of regenerative ability (see below).

What controls the expression of pattern elements in regenerating ctenophores? There is good evidence that individual pattern elements possess an intrinsic sense of polarity. For example, if a ctene row is cut out of an adult lobate ctenophore and rotated 180 degrees, it heals with the outer epidermis. However, the ctene row complex does not connect to the closest end of the intact ctene row. Rather, the new aboral end of the grafted piece will turn around and grow aborally until it connects to the oral end of the cut row and the new oral end of the rotated piece will grow back to the aboral end of the intact segment (Figure 6.16). Over time the entire grafted piece will rotate so that it assumes its normal orientation (Coonfield 1937). These results indicate that the rotated piece retains a sense of its original polarity and maintains this axial information in the presence of its altered surroundings.

The structure that appears to play the greatest role in controlling the process of regeneration is the apical organ. Under normal circumstances, if adult ctenophores are bisected along the oral-aboral axis, each piece regenerates a normal apical organ. When metastable "half-animals" that possess a half apical organ are bisected in the same way, the piece that contains the intact half apical organ forms a "half-animal," while the piece that does not contain any apical organ often regenerates into a normal whole animal (Freeman 1967; Martindale 1986). If a whole apical organ is transplanted onto a "half-animal" in which the apical organ has been removed, it regenerates as a whole animal (Freeman 1967). These results indicate that the apical organ controls the extent of regeneration. On the other

Figure 6.16 Experiment indicating the stability of axial information in fragments of adult ctenophores. *a–c.* A piece of a ctene row is extirpated, rotated 180 degrees, and replaced in an adult *Mnemiopsis.* (*d*) The ctene row heals, but the endodermal canals from the ends of the ctene row do not make attachments to the nearest cut ends of the existing ctene rows. Rather, they grow back and attach to the ends to which they were originally attached. (*e*) Over time the entire grafted piece will rotate and reintegrate in its former configuration.

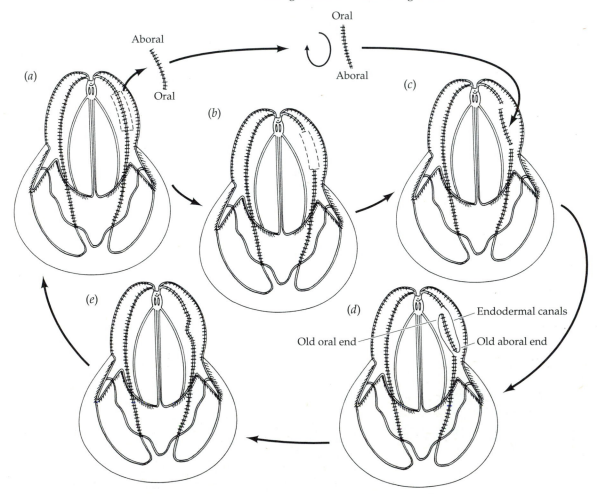

hand, there is definitely some feedback between the peripheral elements of the ctenophore body and the apical organ. If the half apical organ is removed from a "half-animal," then only a half-apical organ is reformed.

One should note the apparent paradox between the regulative/regenerative properties displayed by ctenophore embryos versus adults. Ctenophore *embryos* are not capable of replacing portions of their body when their precursors are experimentally removed, whereas *adults* of the same species can regenerate readily. Two fundamental questions may be asked. (1) When does the change in regulative ability occur? and (2) Does the **adult action system** merely reconstitute the pattern established during embryogenesis, or does the adult possess an innate "sense" of wholeness associated with the adult body plan? The first question has been addressed by bisecting ctenophore embryos at progressively later stages of development. Embryos bisected prior to the time the apical organ becomes functional (detected by the coordination of the ctene row beat) develop and remain as "half animals," while those bisected after this time will give rise to "whole animals" (Martindale 1986). It is not known whether the change in regenerative ability is directly correlated to features related to the apical organ (e.g., synaptic conductivity), or to some other event (such as formation of some lineage of regenerative cell population). It is clear, however, that regenerative potential changes over a relatively short period of time (about 15 minutes) rather than gradually in a stepwise fashion (Martindale 1986).

The second question has been addressed by challenging metastable "half-animals" to undergo further regeneration. Embryos bisected prior to the time of ctene row coordina-

tion grow up to form "half-animals." If these "half-animals" are later challenged to regenerate by cutting them parallel to the oral-aboral axis, many of them regenerate "whole animals." This result indicates that the tissues retain some innate sense of wholeness even though the animal did not develop as a "whole animal." Furthermore, adult animals do not just have a sense of the correct *number* of pattern elements, they also have the ability to replace missing cell types. If particular-organ forming regions, such as those which will give rise to the apical organ or to the ctene rows, are removed from the embryo before the time of ctene row coordination, they do not form during the embryonic period. If these deficient animals are cultured for a few extra days, however, the normal set of structures and cell types are spontaneously replaced (Martindale 1986; Martindale and Henry 1996). These results indicate that the adult action system is capable of replacing pattern elements and cell types that were never present during embryogenesis. Furthermore, it means that cell types did not have to differentiate to impart neighboring cells with their memory in order for these to be "regenerated" during adulthood. Therefore, the notion that regenerating adults merely reconstitute the body plan established during embryogenesis is too simplistic. It would be particularly interesting to know whether the same molecular controls of cell fate specification occur during embryogenesis and regeneration, because most of the cellular aspects of pattern formation during these times do not appear to be shared. For example, nothing comparable to the early cleavage program or gastrulation occurs during the regeneration response, yet both programs generate identical structures.

How do we explain the stability of the "half-animal" phenotype in light of the extensive ability of adult ctenophores to regenerate? One approach invokes the idea of positional information in the form of the **polar coordinate model** (French et al. 1976). In this model, the circumference of the ctenophore body is assigned a sequence of numbers (like those on a clock face) that reflect the biradial symmetry properties of the adult (Figure 6.17). If a region of the body is removed, there is a positional discontinuity and the missing structures are faithfully replaced by inter-

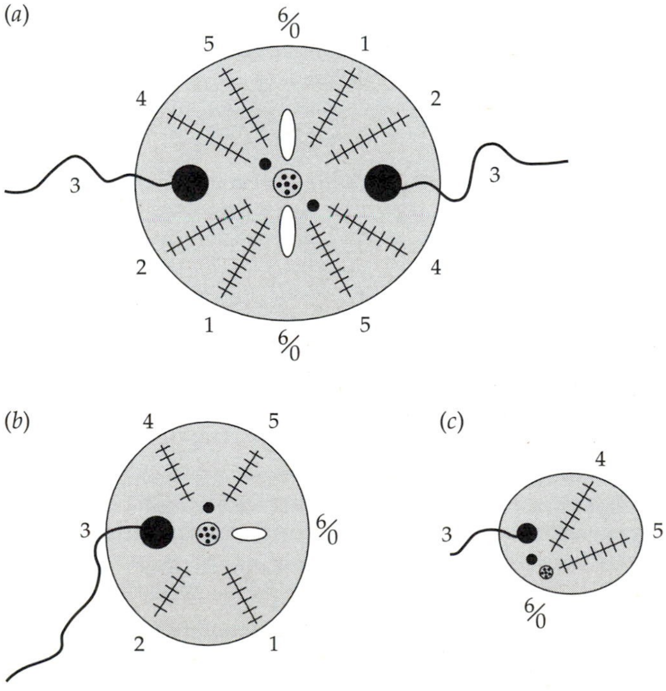

Figure 6.17 Modified polar coordinate model invoked to explain the stability of the "half-animal" phenotype. (*a*) Positional values from 0–6 are assigned along the circumference of the ctenophore body plan like a clockface. These positional values reflect the symmetry properties of adult ctenophores with diagonally opposite quadrants, e.g., the anal pores (black dots between positions "4" and "5"). Intercalary regeneration ensues when there is a gap or discontinuity in positional values. For example, if a tentacle apparatus is removed, positions "2" and "4" would be juxtaposed, thus initiating a regeneration event to replace the missing region. (*b*) This model predicts that "half-animals" are metastable, but that "quarter-animals" (*c*) or "three-quarter-animals" (not shown) are not.

calary regeneration. If the adult is cut in half, then the animal may heal and not detect an interruption in positional values. The result of this kind of experiment would be the production of a metastable "half-animal." Only "half-animals," and not other types of fragments, are predicted to be "stable" by this model.

Summary

Ctenophore embryos have had a long and storied history in experimental embryology. Their rapid development and crystal-clear cells have made them attractive material for observing many aspects of development. Furthermore, these embryos have played a central role in our understanding of how developmental potential becomes localized during embryogenesis. In particular, we have seen how the segregation of developmental potential in ctenophores is an active consequence of the cleavage program, rather than a function of a set of prelocalized factors. The localization process begins as the site of first cleavage organizes the major body axis, the oral-aboral axis, and continues as asymmetries in developmental potential arise at successive cleavage divisions. The current evidence indicates that the second cleavage asymmetrically distributes the developmental potential for anal canals and circumpharyngeal muscles, while the third cleavage partitions comb plate progenitors from photocyte progenitors. The cleavage plan is directly responsible for these events and displays many features of a "cleavage clock."

In addition to the role of the cytoskeleton in the segregation of developmental potential during the early cleavage program, the existence of inductive interactions is also clearly seen. The absence of endodermal canals and ctene rows following the destruction of e_1 micromeres can now be interpreted clearly, as the fate map shows that these structures are not derived solely from the deleted blastomeres. The ability to regenerate during the adult stage of the life history of ctenophores also indicates an ability to sense deficiencies and indicates that the adult and embryo have significantly different, and largely independent, mechanisms for generating the normal adult body plan.

Embryological studies can be informative, although not definitive, in establishing relationships between extant animal phyla. The site of gastrulation in ctenophores (the oral pole) corresponds to the site where first cleavage is initiated, a situation not seen in any other protostome or deuterostome phyla. Furthermore, the occurrence of diagonal determination is more reminiscent of an origin from a radially symmetrical ancestor. On the other hand, the mesoderm is generated by endodermal precursors not ectodermal ones. This result tends to align these animals with the higher Bilateria and not the Radiata. Additional work on ctenophore embryos will reveal further details of the mechanistic basis for the segregation of factors responsible for cell fate determination and the evolution of metazoan development.

Literature Cited

Bethe, A. 1895. Der subepitheliale Nervenplexus der Ctenophoren. *Biol. Zbl.* 15: 140–145.

Brusca, R. C. and G. J. Brusca. 1990. *The Invertebrates.* Sinauer Associates, Sunderland, MA.

Carré, D., C. Sardet. and C. Rouvière. 1990. Fertilization in Ctenophores. In *Mechanism of Fertilization.*, B. Dale (ed.). NATO ASI series V. H 45. Springer-Verlag, Berlin, pp. 626–636.

Carré, D., C. Rouvière and C. Sardet. 1991. *In vitro* fertilization in ctenophores: Sperm entry, mitosis, and the establishment of bilateral symmetry in *Beroe ovata. Dev. Biol.* 147: 381–391.

Chun, C. 1880. Die Ctenophoren des Golfes von Neapel. *Fauna Flora Golfes Neapel* 1: 1–311.

Chun, C. 1892. Die Dissogonie, eine neue Form der geschlechtlichen Zeugung. *Festsch. Zum siehenzigsten Geburtstage Rudorf Leuckarts* 77–108.

Collins, D., D. Briggs and S. C. Morris. 1983. New Burgess Shale fossil sites reveal middle Cambrian faunal complex. *Science* 222: 163–167.

Coonfield, B. 1936. Regeneration in *Mnemiopsis leidyi* Agassiz. *Biol. Bull.* 71: 421–428.

Coonfield, B. 1937. The regeneration of plate rows in *Mnemiopsis leidyi* Agassiz. *Proc. Nat. Acad. Sci.* 23: 152–158.

Dawydoff, C. 1938. Multiplication asexuëe, par lacération, chez les Ctenoplana. *C. R. Acad. Sci. Paris* 206: 127–128.

Driesch, H. 1892. The potency of the first two cleavage cells in echinoderm development. Experimental production of partial and double formations. In *Foundations of Experimental Embryology*, B. H. Willier and J. M. Oppenheimer (eds.). Hafner, New York.

Driesch, H. and T. H. Morgan. 1895. Zur Analysis der ersten Entwickelungsstadien des Ctenophoreneies. *Arch. Entwicklungsmech. Organ.* 2: 204–224.

Dunlap, H. L. 1966. Oogenesis in the Ctenophora. Ph.D. Thesis, University of Washington, Seattle, Washington.

Dunlap-Pinaka, H. 1974. Ctenophora. In *Reproduction of Marine Invertebrates, Vol. 1, Acoelomate and Pseudocoelomate Metazoans*, A. C. Giese and J. S. Pearse (eds.). Academic Press, New York. pp. 201–265.

Farfaglio, G. 1963. Experiments on the formation of the ciliated plates in Ctenophores. *Acta Embryol. Morphol. Exp.* 6: 191–203.

Fischel, A. 1903. Entwicklung und Organdifferenzierung. *Arch. Entwicklungsmech. Organ.* 15: 679–750.

Franc, J.-M. 1970. Évolution et interactions tissulaires au cours de la régénération des lèvres de Beroe ovata (Chamisso et Eysenhardt), Cténaire Nudicténide. *Cahiers Biol. Marine* 11: 57.

Franc, J.-M. 1973. Etude ultrastructurale de la spermatogénèse du Cténaire Beroe ovata. *J. Ultrastruct. Res.* 42: 255–267.

Freeman, G. 1967. Studies on regeneration in the creeping ctenophore *Vallicula multiformis. J. Morphol.* 123: 71–84.

Freeman, G. 1976a. The role of cleavage in the localization of developmental potential in the ctenophore *Mnemiopsis leidyi. Dev. Biol.* 49: 143–177.

Freeman, G. 1976b. The effects of altering the position of cleavage planes on the process of localization of developmental potential in ctenophores. *Dev. Biol.* 51: 332–337.

Freeman, G. 1977. The establishment of the oral-aboral axis in the ctenophore embryo. *J. Embryol. Exp. Morphol.* 42: 237–260.

Freeman, G. and G. T. Reynolds. 1973. The development of bioluminescence in the ctenophore *Mnemiopsis leidyi. Dev. Biol.* 31: 61–100.

French, V., P. J. Bryant and S. V. Bryant. 1976. Pattern regulation in epimorphic fields. *Science* 193: 969–981.

Greve, W. 1970. Cultivation experiments on North Sea ctenophores. *Helgol. wiss. Meer.* 20: 304–317.

Haeckel, E. 1896. *Systematische Phylogenie der wirbellosen Thiere (Invertebrata).* G. Reimer, Berlin.

Harbison, G. R. 1985. On the classification and evolution of the Ctenophora. In *The Origins and Relationships of Lower Invertebrates*, S. C. Morris, J. D. George, R. Gibson and H. M. Platt (eds.). Clarendon Press, Oxford.

Harbison, G. R. and Madin, L. P., 1982. Ctenophora. In *Synopsis and classification of living organisms*, Vol. 1., S. P. Parker (ed.) McGraw Hill, New York. pp. 707–715.

Harbison, G. R. and R. L. Miller. 1986. Not all ctenophores are hermaphrodites. Studies on the stematics, distribution, sexuality and development of two species of *Ocyropsis*. *Mar. Biol.* 90: 413–424.

Hatschek, B. 1911. *Das neue zoologische System.* W. Engelmann, Leipzig.

Hernandez-Nicaise, M.-L. 1973. Le système nerveux des Cténaires. I. Structure et ultrastructure des réseaux épithéliaux. *Z. Zellforsch.* 137: 223–250.

Hernandez-Nicaise, M.-L. 1991. Ctenophora. In *Microscopic Anatomy of Invertebrates, Vol. 2: Placozoa, Porifera, Cnidaria, and Ctenophora*. Wiley-Liss, New York pp. 359–418.

Hertwig, R. 1880. Ueber den Bau der Ctenophoren. *Jenaische Z. Naturwiss.* 14: 313–457.

Horridge, G. A. 1974. Recent studies on the Ctenophora. In *Coelenterate Biology: Reviews and New Perspectives*. L. Muscatine and H. M. Lenhoff (eds.). Academic Press, New York. pp. 439–468.

Houliston, E., D. Carré, J. A. Johnston and C. Sardet. 1993. Axis establishment and microtubule-mediated waves prior to first cleavage in *Beroe ovata*. *Development* 117: 75–87.

Hyman, L. H. 1940. *The Invertebrates. Protozoa through Ctenophora.* McGraw Hill, New York, pp. 662–695.

Lang, A. 1884. Die Polycladen (Seeplanarien) des Golfes von Neapel und der angrenzenden Meerabschniite. *Fauna Flora Golf. Neapel* 11: 1–688.

La Spina, R. 1963. Development of fragments of the fertilized egg of Ctenophores and their ability to form ciliated plates. *Acta Embryol. Morphol. Exp.* 6: 204–211.

Martindale, M. Q. 1986. The expression and maintenance of adult symmetry properties in the ctenophore, *Mnemiopsis mccradyi* . *Dev. Biol.* 118: 556–576.

Martindale, M. Q. 1987. Larval reproduction in the ctenophore, *Mnemiopsis myccradyi* (order Lobata). *Mar. Biol.* 94: 409–414.

Martindale, M. Q. and J. Q. Henry. 1995. Diagonal development: establishment of the anal axis in the ctenophore *Mnemiopsis leidyi*. *Biol. Bull.* 189: 190–192.

Martindale, M. Q. and J. Q. Henry. 1996. Development and regeneration of comb plates in the ctenophore *Mnemiopsis leidyi*. *Biol. Bull.* 191: 290–292.

Martindale, M. Q. and J. Q. Henry. 1997. Reassessing embryogenesis in the Ctenophora: The inductive role of e$_1$ micromeres in organizing ctene row formation in the "mosaic" embryo, *Mnemiopsis leidyi*. *Development*, In Press.

Metschnikoff, E. 1885. Vergleichend Embryologische Studien. IV. Ueber die gastrulation und mesodermbildung der Ctenophoren. *Z. Wiss. Zool.* 42: 648–656.

Mortensen, T. 1912a. Ctenophora. *Danish Ingolf-Exped,* Vol. 5(2). H. Hagerup, Copenhagen.

Mortensen, T. 1912b. On regeneration in Ctenophores. *Dunsk Naturhist. For. Kopenhaven Vidensk. Meddel.* 66: 45–51.

Nielsen, C. 1995. *Animal Evolution: Interrelationships of the Living Phyla.* Oxford University Press, London.

Ortolani, G. 1963. Origine dell'organo apicale e dei derivati mesodermici nello sviluppo embrionale di Ctenofori. *Acta Embryol. Morphol. Exp.* 7: 191–200.

Ortolani, G. 1989. The ctenophores: A review. *Acta Embryol. Morphol. Exp.* 10: 13–31.

Reeve, M. R., and M. A. Walter. 1978. Nutritional ecology of ctenophores—a review of recent research. *Adv. Mar. Biol.* 15: 249–287.

Reverberi, G. 1971. Ctenophores. In *Experimental Embryology of Marine and Fresh-water Invertebrates*, G. Reverberi (ed.). North-Holland Publ. Co., Amsterdam, pp. 85–103.

Reverberi, G. and G. Ortolani. 1963. On the origin of the ciliated plates and mesoderm in the Ctenophore. *Acta Embryol. Morphol. Exp.* 6: 175–199.

Schneider, K.C. 1902. *Lehrbuch der vergleichenden Histologie der Tiere.* G. Fischer, Jena.

Stanlaw, K. A., M. R. Reeve and M. A. Walter. 1981. Growth, food, and vulnerability to damage of the ctenophore *Mnemiopsis mccradyi* in its early life history stages. *Limnol. Oceanogr.* 26: 224–234.

Stanley, G. D. and W. Stürmer. 1983. The first fossil ctenophore from the Lower Devonian of West Germany. *Nature* 303: 518–520.

Tamm, S. L. 1982. Ctenophora. In *Electrical Conduction and Behavior in 'Simple' Invertebrates*. G. G. Shelton (ed.). Clarendon Press, Oxford. pp. 266–358.

Tamm, S. L. and S. Tamm. 1991. Reversible epithelial adhesion closes the mouth of *Beroë*, a carniverous marine jelly. *Biol. Bull.* 181: 463–473.

Tanaka, H. 1931. Reorganization in regenerating pieces of Coeloplana. *Kyoto Imp. Univ. Coll. Sci. Ser. B,* 7: 223–246.

Wainright, P. O., G. Hinkle, M. L. Sogin and S. K. Stickel. 1993. Monophyletic origins of the metazoa: An evolutionary link with fungi. *Science* 260: 340–342.

Yatsu, N. 1911. Observation and experiments on the Ctenophore egg. II. Notes on the early cleavage stages and experiments on cleavage. *Ann. Zool. Japan* 7: 333–346.

Yatsu, N. 1912. Observation and experiments on the Ctenophore egg. III. Experiments on germinal localization of the egg *Beroe ovata*. *Ann. Zool. Japan* 8: 5–13.

Ziegler, H. E. 1898. Experimentelle Studien ueber die Zellteilung. III. Die Furchungszellen von *Beroe ovata*. *Arch. Entwicklungsmech. Organ.* 7: 34–64.

SECTION IV

Bilateral Animal Phyla: Acoelomates and Pseudocoelomates

CHAPTER 7

Platyhelminthes, the Flatworms

Charles H. Ellis, Jr. and Anne Fausto-Sterling

T HE MEMBERS OF THE PHYLUM PLATYHELMINTHES SHARE A common adult body plan, triploblastic, acoelomate, bilateral, and flattened. They possess a well-organized epidermis and a simple nervous system. The mesodermal structures include well differentiated longitudinal and circular muscle layers (Figure 7.1, inset), simple excretory cells, and the parenchyma, whose organization only now is becoming understood (Rieger et al. 1991). There is no body cavity, although mesoderm precursor cells arise in much the same way as those of the coelomate annelids and molluscs. Most platyhelminthes have a well-defined gut with a single opening, and possess a muscular pharynx. Even with their anatomical similarities, these varied worms present extreme diversity in life cycle strategies, gonad organization, and developmental patterns. Figure 7.1 shows the relationships of this phylum's key orders, with illustrations of those whose development is discussed here.

The class **Turbellaria** comprises the free-living flatworms and a few similarly organized commensal or symbiotic forms. Planarians (triclads), their close relatives the proseriates, the polyclads, and a wide variety of smaller marine or aquatic worms are members of this diverse group. Acoels are commonly thought to be the most primitive, lacking a pharynx and a definitive gut, with digestion performed by phagocytic endodermal cells within the parenchyma. Superficially, the organization seems much like the cnidarian planula larva, suggesting that acoels represent the link between the radiata and the bilateral metazoans. However, their parenchymal ultrastructure, and the unique acoel cleavage pattern strongly suggest that they may be an offshoot and not the stem group (Smith and Tyler 1985).

The classes **Monogenea**, **Trematoda** (flukes), and **Cestoda** (tapeworms) are entirely parasitic and show life cycles ranging from simple to quite complex. This chapter will focus on turbellarian development, with only a brief summary of fluke and tapeworm life cycles.

Asexual reproduction is rare among free-living flatworms; it exists in a few groups, some that reproduce by fission and others with parthenogenic eggs that develop without being fertilized. Asexual reproduction is more common in the larval stages of the parasitic forms, providing an efficient method of propagation in the intermediate host. Some flatworm types have broad powers of regeneration. In transected planarians, both anterior and posterior fragments can regenerate fully, but other kinds of turbellarians can form new parts only at the cut surface of an anterior half. One could question whether the ability to form two individuals in response to injury is truly an asexual reproductive strategy. Yet planarians' enormous regenerative capacities have fascinated experimenters for many years, leading to many important insights about cell differentiation.

Sexual Reproduction

A look at flatworm sexual reproductive strategies shows a few common threads. Virtually all species are **hermaphroditic,** some showing **protandry** (the testes mature before the ovaries), but many have simultaneous male and female fertility. Fertilization is internal and zygotes leave the parent's body encapsulated in protective capsules. Self-fertilization is unlikely, given the organization of male and female copulatory structures and turbellarians' common mode of hypodermic insemination. Embryonic development takes place within the capsule. Many turbellarians show direct development, with a fully formed worm emerging from the capsule. Larval development is seen in polyclad flatworms and in the monogenes, trematodes, and cestodes, parasites whose larvae infect their hosts.

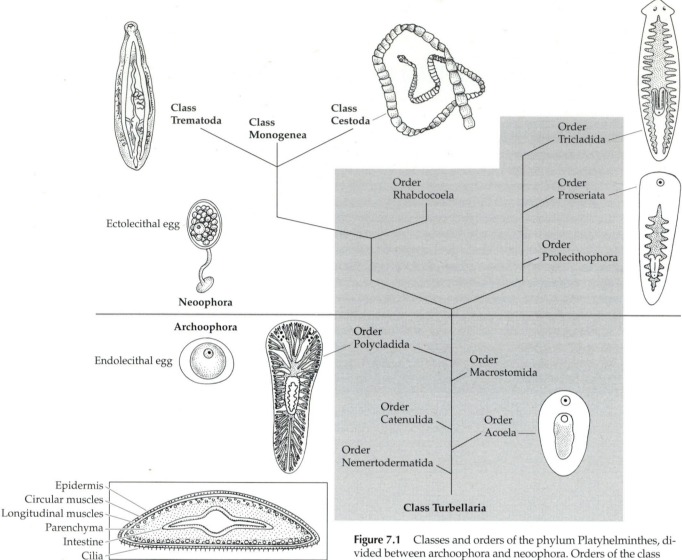

Figure 7.1 Classes and orders of the phylum Platyhelminthes, divided between archoophora and neoophora. Orders of the class Turbellaria are shown in gray.

Endolecithal and **ectolecithal** yolk deposition patterns provide the major difference by which flatworms can be placed into two distinct categories with very different motifs of early embryogenesis. Acoel and polyclad turbellarians and a few other orders have typical endolecithal eggs with yolk in the oocyte. Other turbellarian orders, and all the parasitic classes, are unique. They produce **yolk-free oocytes**, and provide separate **yolk cells** to support development in the capsule (hence *ecto*, outside; *lecithos*, yolk). Systematists thus divide the phylum into two groups based on yolk distribution. The **archoophora**, with endolecithal eggs, are regarded as the primitive type from which the ectolecithal **neoophora** evolved.

All flatworm embryos develop within some sort of egg capsule. In neoophorans this packaging keeps embryonic and yolk cells together until the yolk is internalized to the gut at gastrulation. Because these contrasting yolk distribution patterns form the basis for quite different developmental programs, we must take a comparative approach to flatworm development. It would be a mistake to generalize and claim a single model pattern for this phylum.

It is relatively easy to culture archoophoran embryos, observe living developmental stages, and carry out experiments with them. This ease of observation has led many to accept archoophoran embryos as the platyhelminth model. Conversely, early neoophoran embryos are far more difficult subjects, as each lies buried within a dense yolk cell mass. They are best examined by reconstructing serial sections of capsules fixed at various times after deposition. Culture of live neoophoran embryos has not been particularly successful. It is no wonder there are so few descriptive or experimental studies on early embryology of ectolecithal species.

Reproductive Systems and Gametogenesis

In most turbellarians the male and female gonads are separate organs found within the parenchyma and delimited by a thin epithelium. Even though primordial germ cell origins have been studied in only a few groups, the common view is that they derive from neoblasts, the totipotent, undifferentiated cells that are so important in planarian regeneration (Hendelberg 1983; Gremigni 1983; Baguñà and Boyer 1990).

Testes consist of one or more pairs of follicles, each surrounded by a thin epithelium and containing germ cells in all stages of spermatogenesis. The typically thin, elongate spermatozoa range from 40–100 μm in length (Hendelberg 1983). Flatworm spermatozoa generally do not possess an acrosome or similar body. Most possess two flagellae (Figure 7.2a) and are motile, even while in the male reproductive tract. The organization of the flagellar axonemal complex varies among turbellarian orders, the 9+1 pattern being most common (Hendelberg 1983). However, 9+2 and 9+0 patterns are found in some orders, and a few examples show flagellae modified to form an undulating membrane or are filiform, "aflagellar" sperm, but with an internal axonemal complex. In the parasitic classes, spermatozoa with a single flagellum are common, as are filiform sperm.

From a testicular follicle, spermatozoa enter a **sperm duct** (vas deferens) which carries them to the male copulatory apparatus (Figure 7.3). Mature sperm are often stored in a **seminal vesicle** before they pass into the penis by way of a prostatelike gland. Flatworm penises are generally complex muscular structures with a hard,

hollow central stylet surrounded by a number of accessory barbs. While some cases of male–female genital copulation are known, it seems that most flatworms pursue **hypodermic insemination**, injecting spermatozoa into the recipient's parenchyma through the body wall (Galleni and Gremigni 1989). An injected sperm wanders among the parenchymal cells until it reaches an egg in the female tract. This process can occur even in species whose female genitalia appear to be structured for standard copulation.

While the male reproductive systems of turbellaria share a common architectural theme, female systems can be organized in two different ways. In the archoophora (Figure 7.3a), the **ovary** is a single organ producing oocytes with copious intracellular yolk, even though the eggs remain relatively small (100–200 μm; Baguñà and Boyer 1990). Yolk distribution in acoel and polyclad eggs seems to be relatively uniform and **oligolecithal**. The mature oocyte passes through an **oviduct** to the female genital structures, coming to lie in a female **bursa**, where just after fertilization the capsule (Figure 7.2b) forms around the zygote before it exits to the outside.

Neoophoran worms (Figure 7.3) have two distinct female organs, the **germarium** (or ovary), producing **yolk-free oocytes** (60–100 μm diameter) and **vitelline glands** specialized to produce the smaller **yolk cells** (up to 40 μm; Gremigni 1983). These yolk glands are usually arranged as bilateral pairs in the parenchyma near the adult's gut diverticula. Yolk cells are transported toward the female bursa through **vitelline ducts**, which often join the oviduct as a common **ovovitelline duct**. In many smaller forms, such as proseriates, each germarium will have 4–8 oocytes, releasing one at a time. The oocyte arrives in the female bursa together with a large number of yolk cells and is fertilized by a sperm. Bursal glands secrete a capsule around the whole group of cells, and the capsule leaves the body through a female pore to be attached to the substratum by a short stalk (Figure 7.2c).

Developmental History

Fertilization

Very little is known of the details of fertilization in the flatworms. Together, the unusual hypodermic insemination scheme and intracapsular development make it impossible directly to observe egg-sperm interaction. However, it has become widely accepted that fertilization occurs at metaphase of the primary oocyte stage (Henley 1974). This process has been well described in many turbellarians, particularly in some polyclads and one rhabdocoel species.

Cleavage and Gastrulation

Even with the dichotomy between endolecithal and ectolecithal development, all platyhelminths exhibit **spiral,**

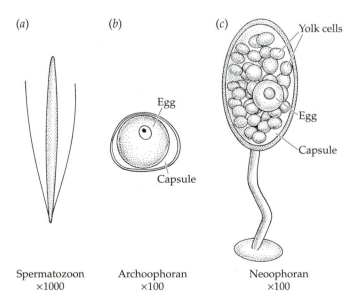

| (a) | (b) | (c) | Yolk cells |

Spermatozoon ×1000   Archoophoran ×100   Neoophoran ×100

Egg
Capsule
Egg
Capsule

Figure 7.2 Flatworm gametes. (a) Spermatozoon. (b) Archoophoran egg capsule (polyclad). (c) Neoophoran egg capsule (proseriate).

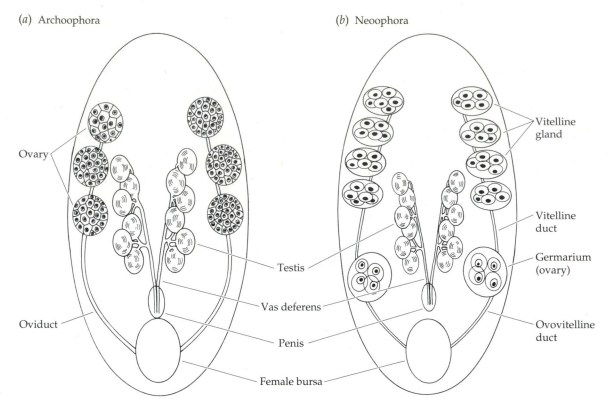

Figure 7.3 Flatworm reproductive systems. (*a*) Archoophoran (polyclad) reproductive system. (*b*) Neoophoran (proseriate; triclad) reproductive system.

holoblastic cleavage. The archoophorans (polyclads and acoels) proceed with developmental patterns, key cell lineages, and epibolic gastrulation much like the other spiralian phyla. But even these orders diverge at the second cleavage (Figure 7.4). Polyclad worms form the usual blastomere quartets based on four macromeres, but acoels show a unique duet spiral cleavage (Gardiner 1895) starting at the 4-cell stage, with a pair of smaller blastomeres lying offset above two macromeres. This leads to a duet-based pattern in the later divisions (Figure 7.4*a–d*) where cell lineages are based on symmetric A and B families.

Among neoophoran orders there are two patterns of blastomere behavior after the 8-cell stage. The embryos of the proseriate *Monocelis fusca* maintain a coherent mass of embryonic blastomeres throughout cleavage, even forming a kind of coeloblastula (Giesa 1966). On the other hand, triclads such as *Dendrocoelum lacteum* (Hallez 1887; Koscielski 1966) exhibit "**blastomere anarchy**," in which blastomeres separate from each other within the yolk mass. Once blastomere clusters have formed certain organ and tissue rudiments, they finally grow together to form the definitive worm. In both of these ectolecithal

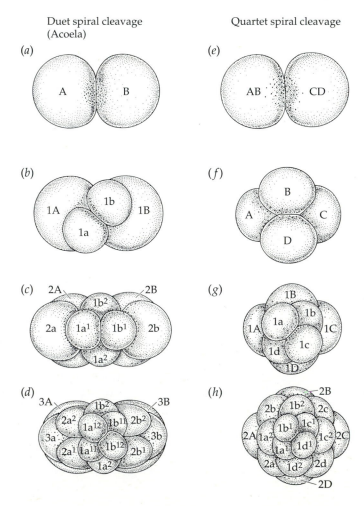

Figure 7.4 Duet versus quartet patterns of archoophoran cleavage. (*a–d*) Duet spiral cleavage (acoela). (*e–h*) Quartet spiral cleavage. (*a–d* after Gardiner 1895.)

forms, the "gastrulation" process is unlike that of other animals, in more ways than in the method of internalizing yolk cells within the developing gut.

These variations on the spiralian developmental theme make it impossible to present a single general picture of platyhelminth development. We cannot even reach a simple conclusion about determinate or regulative development. Blastomere deletion experiments (Boyer 1986, 1987, 1989) clearly show determinate development in a polyclad. However, experiments with acoels suggest their early stages are regulative (Boyer 1971). Some would claim that a planarian's prodigious regeneration capacity suggests regulative development, but there is no firm experimental evidence on blastomere potencies in early neoophoran embryos. Rather than generalize, we must look at three key examples to understand how turbellarians develop.

HOPLOPLANA INQUILINA, A POLYCLAD WITH ENDOLECITHAL DEVELOPMENT.

Hoploplana inquilina is a polyclad that lives as a commensal in the mantle cavity of a marine snail. The details of its development (Figure 7.5) were first described by Surface (1907), whose observations about cell lineages have been confirmed by blastomere deletion experiments and cell marking (Boyer 1986, 1987, 1989; Henry et al. 1995; Boyer et al. 1996).

These embryos are relatively easy to study because adults cultured in sea water will still deposit eggs. For experimentation, eggs and sperm can be obtained from the male or female reproductive tracts for fertilization in vitro. These embryos develop without the capsule that would have normally have formed within the female bursa (uterus) after fertilization. In this way, Surface (1907) could describe the appearance of polar bodies after eggs were laid, and could observe living embryos at all cleavage stages. *H. inquilina* develops relatively rapidly for a turbellarian. The eight-lobed, ciliated **Müller's larva** is complete in five to seven days after fertilization. Some polyclad species have a similar, but four-lobed Götte's larva.

Figure 7.5   Cleavage and development in the polyclad *Hoploplana inquilina*. (*a*) Zygote. (*b*) 2-cell stage, animal view. (*c*) 4-cell stage, animal view. (*d*) 8-cell stage, animal view. (*e*) 16-cell stage, animal view. (*f*) 45-cell stage, lateral view. (*g*) Gastrulation; epiboly. (*h*) Gastrula. (*a–f, h–i* after Surface 1907; *g* and *k* after Korschelt and Heider 1895; *j* after Ruppert, 1978.)

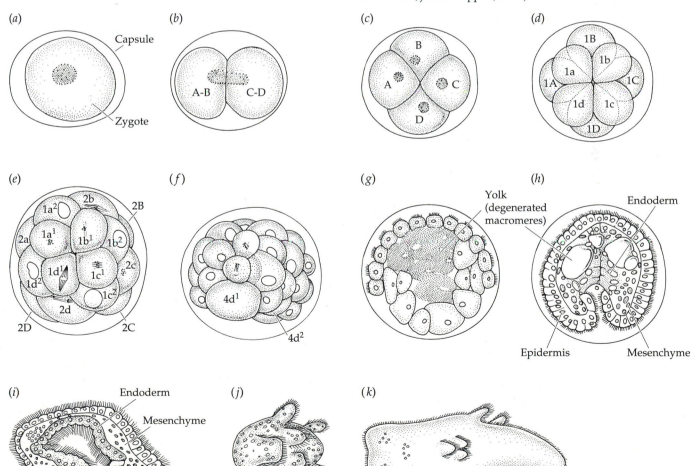

The first cleavage proceeds from the animal to the vegetal pole and is nearly equal, producing the AB and CD blastomeres.[*] Both blastomeres divide unequally to form a four-cell stage with B and D blastomeres noticeably larger than A and C. Cleavage to the eight-cell stage is unequal and to the right (dextral), producing a quartet of slightly smaller micromeres 1a–1d arrayed in the typical spiral manner. As further spirally arranged quartets form at the 16- and 32-cell stages, the micromeres of the d line tend to be a bit larger than the others. These cleavages show the typical alternation of dextral (right-handed) and sinistral (left-handed) divisions. After the 64-cell stage, gastrulation by epiboly begins as the smaller micromeres derived from the first three quartets spread over the surface of the four macromeres (4A–4D) and over the fourth micromere quartet (4a–4d). As gastrulation proceeds, 4d divides to form $4d^1$ and $4d^2$, while 4a–4c and all four macromeres degenerate to become a yolk mass within the embryo. The blastopore becomes the single opening of the gut system, clearly defining the polyclad as a **protostome**.

The current view of polyclad cell lineages (Figure 7.6) is little different from Surface's description. As with other spiralian phyla, 4d is a mesentoblast, $4d^1$ produces the mesoderm, and $4d^2$ the endoderm. The remaining micromeres form the ectodermal structures and some mesectoderm.

Henry et al. (1995) followed lineages in *H. inquilina* by labeling individual blastomere surfaces with DiI at the 2- or 4-cell stage. Detection of DiI at the larval stage showed each larval quadrant to derive from the respective A, B, C, or D quadrant of the embryo (Figure 7.7). The results are consistent with Boyer's (1987) findings showing corresponding larval lobes or eyespot missing when a specific blastomere has been eliminated. After these initial experiments, Boyer, Martindale and Henry (1996) went on to inject DiI into individual macromeres and micromeres up to the 64-cell stage. Dye localization in Müller's larva confirmed that the 4d blastomere is the true mesentoblast and that 2b is a "mesectoblast." These blastomere fates are identical to those seen in the other spiralia, particularly annelids and molluscs.

After gastrulation, the embryo flattens slightly, and develops the ciliation and lobes, characteristic external features of Müller's larva. While Surface did not bring *Hoploplana* larvae through metamorphosis, the process has been seen in other polyclads (as early as Lang 1884). In a gradual process, the larva flattens and elongates, the adult pharynx forms at the stomodeum, and the larval lobes disappear.

***MONOCELIS FUSCA*, A PROSERIATE WITH ECTOLECITHAL DEVELOPMENT.** Proseriates are small, free-living turbellarians, abundant in the fauna of the marine interstitial environment, living and reproducing in the wet spaces between sand grains. Because *Monocelis* is a neoophoran, it pro-

Figure 7.6 Polyclad cell lineages. (After Baguñà and Boyer 1990.)

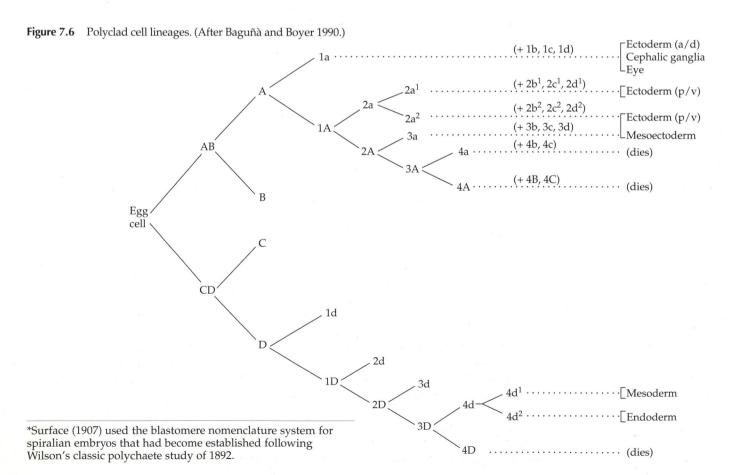

---

[*]Surface (1907) used the blastomere nomenclature system for spiralian embryos that had become established following Wilson's classic polychaete study of 1892.

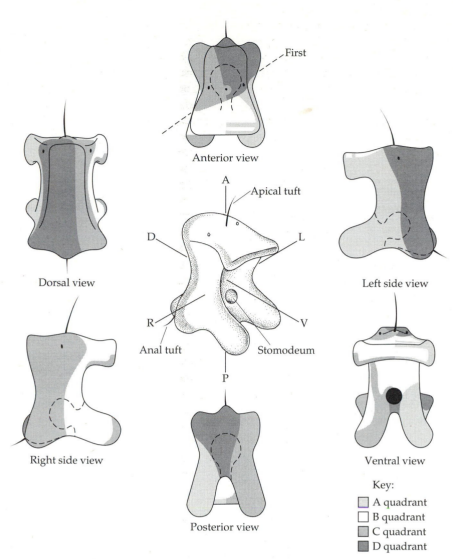

Labels on figure: First; Anterior view; A; Apical tuft; D; L; R; V; Anal tuft; Stomodeum; P; Dorsal view; Right side view; Posterior view; Left side view; Ventral view

Key:
- ☐ A quadrant
- ☐ B quadrant
- ☐ C quadrant
- ■ D quadrant

Figure 7.7 Axial specification in *Hoploplana* larva. Shadings show larval areas containing DiI stain derived from the respective injected A, B, C, or D blastomere (quadrant) of the 4-cell embryo. (After Henry et al. 1995.)

duces but a few egg capsules, each containing a single zygote among a mass of yolk cells. As each capsule is deposited, it is affixed to a sand grain.

It is very difficult to observe a single embryo's early development when it is completely surrounded by yolk cells. Thus there are no direct cell lineage observations or experiments reported on neoophoran embryos. However, cleavage patterns and early developmental history can be reconstructed from histological sections of capsules fixed at different times (Figure 7.8). In this way, Giesa (1966) described *Monocelis fusca* development at a level of detail that stands as a paragon of how these studies can be pursued.

Monocelis development takes about 30 days from egg deposition until a juvenile worm hatches. This timing is typical for many temperate marine neoophorans. Sperm transfer is probably by the hypodermic route. While *Monocelis'* genital structures could permit ordinary copulation, there is no evidence of a sperm storage organ or of large sperm populations within the female bursa.

Giesa's (1966) reconstruction of sectioned embryos shows the first cleavage to be somewhat unequal, produc-

ing an AB blastomere and a slightly larger CD cell. Orientation of mitotic spindles at the second division suggests that it is a sinistral cleavage, with smaller A and C cells budding from the larger B and D blastomeres. The third cleavage is unequal and dextral, producing micromeres 1a–1d lying offset above the 1A–1D macromeres in a perfect spiralian 8-cell pattern. Traditional cell lineage studies have always required an investigator to trace blastomere offspring as they divide or to mark the cells with vital dyes. As this tracing could not be done with *Monocelis*, it is not possible to discuss specific cell fates beyond this point. It is reasonable to assume that the first micromere quartet produces the usual ectodermal derivatives, but we cannot ascertain whether a 4d cell is a true mesentoblast as in *Hoploplana*.

By the time *Monocelis* has 16–24 cells, it is a fairly tight cell mass with a small central cavity. Giesa calls this a *blastocoel* and, with the support of sections from a few other proseriate species, concludes that this order forms a coeloblastula. However, from here gastrulation proceeds by a pattern peculiar to the ectolecithal situation.

The blastula flattens, and several cells, the mesenchyme

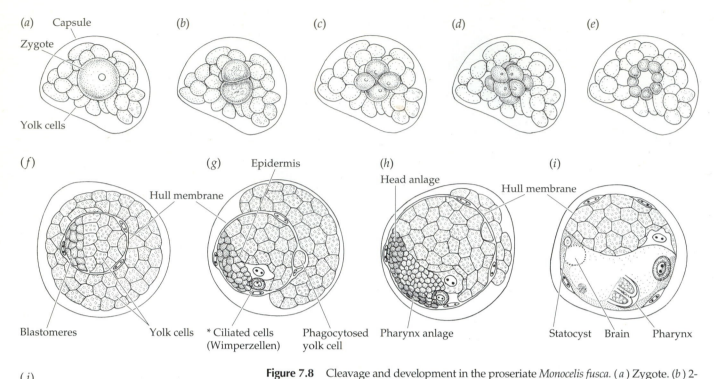

(a) Capsule
Zygote
Yolk cells

(b)

(c)

(d)

(e)

(f)
Hull membrane
Blastomeres   Yolk cells

(g) Epidermis
* Ciliated cells
(Wimperzellen)
Phagocytosed
yolk cell

(h) Head anlage
Hull membrane
Pharynx anlage

(i) Hull membrane
Statocyst   Brain   Pharynx

(j)

* Not referenced in text

Figure 7.8   Cleavage and development in the proseriate *Monocelis fusca*. (a) Zygote. (b) 2-cell stage. (c) 4-cell stage. (d) 8-cell stage. (e) Blastula. (f) Early gastrulation. (g) Mid gastrulation. (h) Late gastrulation. (i) Later development. (j) Juvenile worm in capsule. (After Giesa 1966.)

precursors, sink in to lie between the animal and vegetal epithelia. Eight small cells at the surface spread by epiboly around the embryo's ventral surface and then upward away from the embryonic mass as the thin **hull membrane** (*Embryonalhülle*, illustrated for a triclad by Hallez 1887) around some of the yolk cells. This sphere of thin extraembryonic epithelium expands through the yolk mass until all yolk cells are contained within it. The individual yolk cells appear to cross the membrane intact, phagocytized by specialized vitellophage cells within the hull membrane. Later the embryonic epidermis spreads epibolically up and around most of the remaining yolk cells, which thus reside in embryo's gut. There they break down to provide the nutrition for the rest of development and early posthatching life.

Giesa draws an analogy to classical gastrulation, identifying four of the hull membrane cells as *Urmundzellen* (blastopore cells), and the hull opening as a blastopore.

As the hull membrane expands, the embryonic cells remain together, proliferate, and begin to differentiate, while a number of small abortive blastomeres degenerate within the yolk mass. The embryo elongates and, as the epidermis expands, rudiments of key internal organs appear:

statocyst, brain (cerebral ganglion), pharynx, and gut endothelium. Later developmental stages can be observed alive through the tanned, but transparent, capsule wall. The worm now contains yolk instead of being buried within it. Proseriates hatch as fully formed juveniles, never having gone through a larval stage.

Giesa's detailed description of *Monocelis* development gives a clear picture of how embryos can develop in the ectolecithal mode, incorporating external yolk during gastrulation. This development is also obvious in the planarian species that have been studied. However, the triclads' unusual anarchic blastomere behavior during cleavage makes it difficult to comprehend how cell clusters, separated in the yolk mass, can form rudiments, communicate, and arrange into a coherent organism.

DENDROCOELUM LACTEUM, A PLANARIAN SHOWING BLASTOMERE ANARCHY.   A freshwater turbellarian found in streams, *Dendrocoelum lacteum* has stood as a key example of planarian development since the extensive description published by Hallez in 1887 (Figure 7.9). While the later developmental stages resemble those of *Monocelis*, the early stages are quite different. After fertilization, several zygotes, perhaps three or four, are encapsulated with one mass of yolk cells. All can develop into worms within the same capsule, each incorporating its share of the yolk mass. *Dendrocoelum* and other triclads show direct development, without larva, taking several weeks to complete development.

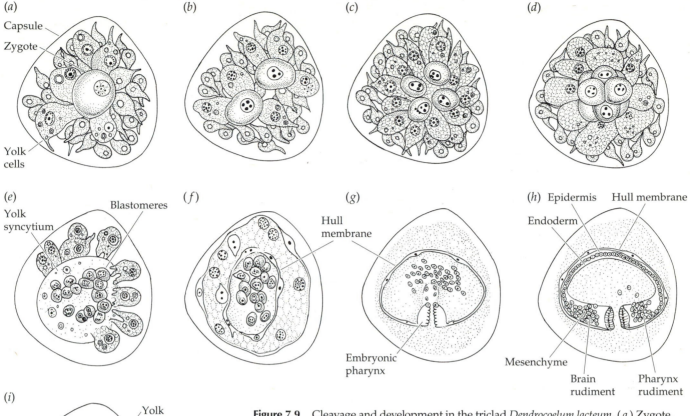

(a)
Capsule
Zygote

Yolk
cells

(b)

(c)

(d)

(e)
Yolk
syncytium
Blastomeres

(f)

(g)
Hull
membrane

Embryonic
pharynx

(h) Epidermis   Hull membrane
Endoderm

Mesenchyme
Brain
rudiment
Pharynx
rudiment

(i)
Brain
Yolk
Pharynx

Epidermis
and mesenchyme
Endoderm

Figure 7.9 Cleavage and development in the triclad *Dendrocoelum lacteum.* (*a*) Zygote. (*b*) 2-cell stage. (*c*) 4-cell stage. (*d*) 8-cell stage. (*e*) 16-cell stage (blastomere anarchy). (*f*) Hull membrane formation. (*g*) Embryonic pharynx stage. (*h*) Epidermis completed. (*i*) Juvenile worm. (*a–g* after Hallez 1887; *h* after Hallez 1887 and Korschelt and Heider 1895; *i* after Korschelt and Heider 1895.)

Hallez made his observations with fixed, stained embryos obtained by opening capsules surgically under the microscope. While this raises the possibility that disruption of the embryo might produce an apparent anarchic blastomere separation, modern photographs of *Dendrocoelum* sections (Koscielski 1966) confirm this anarchy.

The first cleavage produces two equal cells that only barely contact each other. Each of these blastomeres again divides equally and the 4-cell embryo then appears to cleave once, spirally, to make a fairly typical 8-cell stage. The cells then migrate a short distance away from each other within the yolk cell mass.

As development progresses from 16 to approximately 64 embryonic cells, the yolk mass undergoes a major change. Near each embryo the yolk cells fuse to form a multinucleate **yolk syncytium**, continuing in later development until the entire yolk is syncytial. Within the syncytium, at the periphery of each loose group of blas-

tomeres, several cells unite to form a hull membrane, and another group of blastomeres comes together as an **embryonic pharynx**. This organ provides a path for further yolk material to move into area bounded by the hull membrane and later into the embryonic gut. It will later be replaced by the definitive **pharynx**.

As gastrulation continues, epidermal rudiments form ventrally and spread over the inner hull-membrane surface by epiboly to form a complete epidermis. Simultaneously a group of endodermal cells produces the embryonic gut, enclosing the internal yolk mass. Cells remaining between the epidermis and endoderm proliferate to produce the mesenchyme, forming transitory mesodermal bands just beneath the epidermis. The longitudinal and circular muscle layers develop in this region, and remaining mesodermal cells proliferate to form the parenchyma. A brain rudiment forms in the ventral region anterior to the embryonic pharynx, while the precursor of the true pharynx develops just posterior. As each embryo reaches final stages, the embryonic pharynx disappears, now replaced by the definitive pharynx. Newly hatched worms do not begin to feed immediately; they continue to use the yolk remaining in the gut.

The Parasitic Classes

The platyhelminths of classes Monogenea, Trematoda, and Cestoda are parasitic forms whose larvae are important for invasion of hosts. Some show asexual reproduction at larval stages, but sexual reproduction is the primary mode of propagation. The adults can be prodigious gamete producers (each tapeworm continually forms multiple reproductive segments, or **proglottides**), producing thousands of eggs each day, for months or years. Fertilization is internal, and many of these forms appear to engage in true copulation. They are neoophoran, and show intracapsular developmental patterns similar to those described above. Two classic life cycles can illustrate the kinds of postembryonic larval strategies found in these classes.

THE LIFE CYCLES OF *SCHISTOSOMA*, THE BLOOD FLUKE OF HUMANS, AND *TAENIA*, THE TAPEWORM.

A digenean trematode, whose two host species are humans and snails, *Schistosoma mansoni* lives as a small adult in the veins of the intestine or bladder walls (Figure 7.10). Schistosomiasis is a very debilitating condition, and is most severe when embryonated eggs deviate from their normal path, invading liver and lung and developing there. Normally, the capsules break through the blood vessel walls into the intestine and leave the body in feces. Embryonic development in the aquatic environment ends as the miracidium larva hatches from the capsule. The miracidia then infect the intermediate host, a certain snail species, within which

they transform into sporocysts, which propagate asexually within the snail and then transform into free-swimming cercaria larvae. When a person swims or wades in water where cercaria are present, the microscopic larvae will attach to the skin and burrow into it, losing their tails in the process. They move in the blood to lungs and liver and then to the intestine or bladder, gradually metamorphosing to adults. The cycle begins again.

Taenia saginatus, the beef tapeworm, is a **cestode** parasitic in humans with cattle as its intermediate hosts (Figure 7.11). In a human, the adult attaches to the host's intestinal wall by a complex structure with suckers, the scolex. A zone of cells at the posterior end of this segment continually forms numerous small reproductive segments, or proglottides, creating a tapeworm perhaps a meter or more in length. Each mature proglottid contains a complete reproductive system. When fertilization and egg encapsulation are complete, the uterus is filled with embryonated eggs that are going through early developmental stages. These **gravid proglottides** break from the end of the worm and are passed in the feces.

In unsanitary habitations, it is possible for cattle to come in contact with human feces. If a cow ingests feces with encapsulated *Taenia* larvae, the capsules will be digested, releasing the **oncosphere** stage. These migrate via the circulation to invade muscle tissue, where they form cysts. Within the **cysticercus** (or bladder worm), major developmental transformations occur as a new scolex begins to differenti-

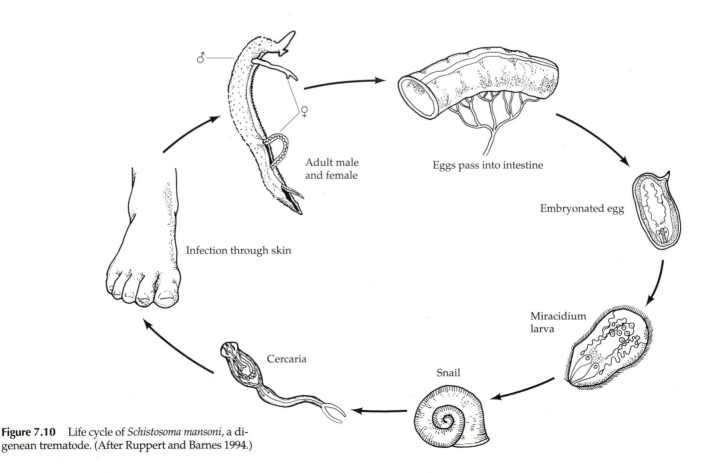

Figure 7.10 Life cycle of *Schistosoma mansoni*, a digenean trematode. (After Ruppert and Barnes 1994.)

Adult male and female

Eggs pass into intestine

Embryonated egg

Infection through skin

Miracidium larva

Cercaria

Snail

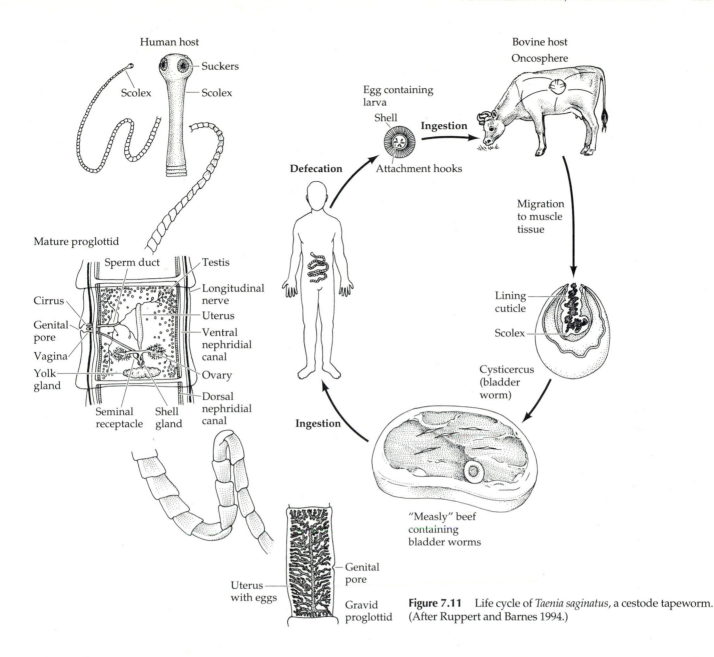

Figure 7.11 Life cycle of *Taenia saginatus*, a cestode tapeworm. (After Ruppert and Barnes 1994.)

ate. When a person eats infected beef, the cysticercal wall is digested so that the scolex can evert and attach to the intestinal wall to begin again the production of proglottides.

Comparative and Evolutionary Issues

The apparent diversity of platyhelminth developmental patterns raises a number of unanswered questions and provides a few insights as to where the flatworms fit into the phylogenetic picture. The latter question is important because adult anatomy has long been used to support theories that this phylum is ancestral to the bilateria. These issues are considered in depth by Baguñà and Boyer (1990) and in modern invertebrate biology textbooks (e.g., Brusca and Brusca 1990; Ruppert and Barnes 1994).

The platyhelminths nearly universally exhibit quartet-based spiral cleavage, and the archoophoran state is clearly more primitive than the neoophoran. This shows the phylum's close relationship to higher spiralian forms (annelids, molluscs). However, traditional arguments have placed the acoels, with their simplified anatomy, as the stem type, like an advanced cnidarian "planula" and a link to the radiata. However, the acoel's duet-based spiral cleavage is unlike any other spiral pattern and may truly be derived from the quartet pattern. If this is the case, the polyclads may be closer to the earliest line of platyhelminthes, with the acoel's anatomical simplicity secondarily derived. Continuing to the neoophoran clade, it would appear that the parasitic classes diverged from an ectolecithal, probably turbellarian, ancestral line.

Developmental adaptations to the ectolecithal situation make it difficult to compare gastrulation among the archoophora and neoophora. Epibolic gastrulation and the mesoderm-endoderm fates of the 4d blastomere line further cement the relationship between the polyclads and the coelomate spiralians. But in neoophorans the epibolic formation of the hull membrane should not be treated as true gastrulation. The establishment of mesodermal bands, a key step in any type of gastrulation, does not occur until the dorsal epidermal migration, long after the hull membrane and embryonic pharynx are formed. These cell movements are probably the gastrulation events. Hull membrane epiboly can be compared with gastrulation associated events in other phyla, such as annelids (the formation of provisional epithelium in leech development) and vertebrates (yolk sac epiboly in fishes).

Another key issue is the fate of the mesoderm and the relationship of platyhelminthes to the coelomates. No member of the phylum possesses a coelomic cavity. Except for the formation of longitudinal and circular muscle layers, the mesodermal bands are transitory in development, resolving into a parenchyma, which traditionally has been considered to be little differentiated and poorly organized. Recent ultrastructural studies on a wide variety of turbellaria (Rieger et al. 1991) reveal the parenchyma to be rich in differentiated cell types. The epidermal organization also proves more like that of some coelomates than was previously believed. Together, the polyclad developmental history and these anatomical findings can be used to support a hypothesis that the platyhelminthes derive from a coelomate or precoelomate ancestor and that their acoelomate habit is derived by loss of the cavity and lining. The alternative theory, proposed in the late nineteenth century by both Metschnikoff and von Graff, continues to stand fairly well, that the phylum comes directly from the radiata by way of a truly acoelomate, planula-like progenitor, represented by the acoels. However, the acoelan duet-based, spiral cleavage pattern is seen nowhere else, and it is easier to envision a transition from cnidarian cleavage patterns to a quartet-based spiral pattern. The controversy remains to be resolved.

To a modern developmental biologist, the triclad pattern of blastomere anarchy raises a major question. How is positional information conferred upon and interpreted by blastomeres separated from each other within a random yolk mass? It is relatively easy to conceive means to establish anterior-posterior and dorsal-ventral polarity in the archoophorans, or in *Monocelis*, where the embryonic cells remain in direct contact. But with vitelline cells coming from multiple vitellaria and packaged with apparent randomness in the capsule, preestablished gradients of morphogenetic factors seem unlikely. Yet individual blastomere groups are able to differentiate into organ anlagen, which somehow come together in the precise pattern that forms the juvenile triclad. If the apparent anarchy is but an artifact of tissue preparation pushing blastomeres apart, this process is not an issue. However, the several examples now suggest it to be a real phenomenon. New experimental techniques are needed to study neoophoran cell fates and potencies directly and then to see how these cells interact across the distances in the yolk mass.

Embryonic development of the flatworms is ripe for further investigation. Few species in even fewer families have been studied at all, severely limiting our knowledge of even the most basic developmental events. This lack might be a surprise to those who think of the long history of planarian regeneration research.

Flatworm Regeneration

In contrast to humans, whose lost appendages never regenerate, the planarian epitomizes the idea of regeneration. Cut off its head and it grows a new one. Divide its head laterally into three, four, or more parts and a three, four, or multiheaded worm will result. Cut it into ten parts and ten complete, albeit smaller, planaria grow, one from each fragment. For a general overview of planarian regeneration, see Goss 1969 and Brøndsted 1969.

Yet not all free-living platyhelminthes regenerate. Taxonomists divide nonparasitic platyhelminthes into two major groups, those with the ancestral type of egg (archoophora) and those with a more recently evolved egg type in which the yolk is produced and packaged separately from the ovary and developing oocyte (neoophora). As noted in the previous section, neoophoran development appears to be more flexible or regulatory than archooporan embryogenesis. This difference in plasticity extends as well to the adults belonging to these two taxonomic subdivisions.

There are five orders of free-living Platyhelminthes in the archoophora taxon (Figure 7.12). Extensive studies on regeneration exist for only one of these—the marine Polyclads, which have only limited regenerative abilities. Among the five neoophoran orders, detailed information on regeneration exists primarily for the Tricladida, and among the triclads, only for those living in fresh water. (The vast group of large terrestrial triclads is fascinating but virtually unstudied—a terrain wide open for the budding biologist—while the salt- or brackish-water triclads are few in number and more difficult to raise in the laboratory.)

Reproductive Modes of Freshwater Planaria

Even among the freshwater triclads, however, regeneration is not a universal trait. To understand its appearance we must consider it in relationship to the organism's overall life style. The freshwater planaria are often divided into three groups—the Dugesiidae, the Planariidae, and the Dendrocoelidae (Figure 7.13). The Dugesiidae and Planariidae have similar life styles. First, adults survive and reproduce for several years; that is, they are iteroparous. Second, although most strains reproduce sexually, some become parthenogenetic while others become fissiparous, reproducing exclusively via binary fission (Figure 7.14). In samplings from the southern New England area, 10% to 20% of the populations of the most regionally common flatworm, *Dugesia tigrina*, are exclusively fissiparous (Fausto-Sterling, personal observations). The others are either exclusively oviparous (but may exhibit asexual reproduction via parthenogenesis) or show an occasional ability to un-

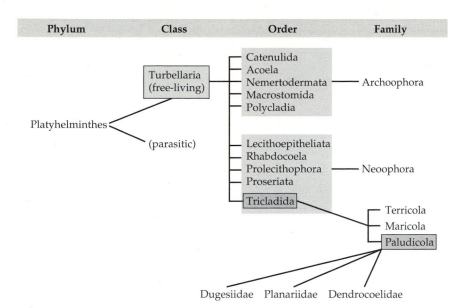

Figure 7.12 A tentative scheme of platyhelminthes classification. The current literature contains continued debate about the taxonomic relationships of the variously named groups.

dergo binary fission. The famous regenerating flatworms all belong to these two taxonomic groups. In contrast to the Dugesiidae and Planariidae, Dendrocoelidae tend to be semelparous; they live and reproduce for only a single season. Furthermore. they are exclusively sexual, never undergoing binary fission. And they are poor regenerators. If one cuts a Dendrocoelid in half, the head piece can regenerate the missing tail, but the tail cannot make a new head.

Calow and Read (1986) offer a working hypothesis to explain the differences in reproductive modes exhibited by the three main groups of freshwater triclads. The Dendrocoelidae have evolved a unique structure—the anterior adhesive organ, which they use to attach themselves to some surface. There, they lie in wait, attacking living prey that happens within their reach. In contrast, their sister groups glide along the the river or pond bottoms or on submerged rocks and plants, taking live prey as they find it and scavenging on fresh kills as well. Calow and Read note that earlier work had suggested that the "sit and wait" behavior enabled by the adhesive organ was metabolically more efficient and may have led to the evolution of increased juvenile survival and higher growth rates among the Dendrocoelidae. These changes might, in turn, have favored the evolution of a semelparous life style, which requires larger animals that can produce more young with better chances of survival than those from iteroparous organisms.

Calow and Read also connect the semelparous life style to the lack of fission and a limited regenerative capacity. A worm that reproduces only once in its lifetime must pour most of its metabolic resources into the reproductive effort, thus producing the greatest number of large, healthy juveniles possible. They suggest that such reproductive effort comes at the expense of other metabolic activities such as mitotic rate and the maintenance of embryonic stem cells utilized during regeneration or the growth of new parts following fission. If this were true, the evolution

of semelparity could have led to a loss of regenerative ability. Similarly, binary fission would be maladaptive, since fissioning worms have less body mass and thus fewer resources to put into sexual reproduction. Although Calow and Read offer some limited tests of their hypotheses, more work is still needed. Life cycle variation between planarian groups provides rich material for further explorations of life cycle evolution and the maintence and loss of the ability to regenerate.

Mechanisms of Regeneration

Although biologists have studied planarian regeneration for about a century, the answers to most questions concerning the mechanisms of regeneration remain elusive. Three research areas will be highlighted in this chapter—the question of which cells participate in regeneration, the question of how flatworms maintain their knowledge of

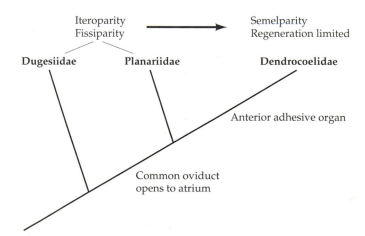

Figure 7.13 Correlations between taxonomic groups, life cycles, and the ability to regenerate and undergo binary fission. As noted by Calow and Read, the arrow represents a strong tendency, not an absolute claim. (Adapted from Calow and Read 1986.)

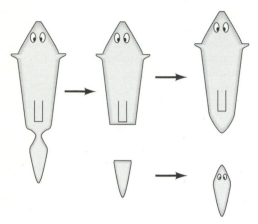

Figure 7.14 Spontaneous binary fission in planaria. At 24°C, the small tail piece eventually differentiates a pharynx and will become a feeding worm within two weeks.

whether to regenerate a head or a tail (maintenance of anterior-posterior polarity), and the effort to understand the molecular mechanisms governing the regenerative process. As work on the mechanics of regeneration progresses, we should be able to use it to better understand the evolution of diverse life styles and developmental capacities of the several families of freshwater planaria.

After years of controversy, scientists now agree that those planaria capable of regeneration maintain and utilize a reservoir of embryonic stem cells called **neoblasts**. Depending upon the species, neoblasts can comprise up to 30% of the total number of cells in an adult planarian. Neoblasts perform two roles—to replace cells that die in the course of normal physiological turnover and to provide the cellular raw material for regeneration. A lingering question remain as to whether neoblasts are the *only* cellular resource utilized during regeneration. It is possible that in addition to neoblasts, cells arising from the differentiation of differentiated cells at or near the wound surface contribute at least in a minor way to the population of cells participating in regeneration. (For a review of this question see Baguñà et al. 1994.)

Figure 7.15 The time course (at 25°C) of head regeneration in *Dugesia tigrina* following amputation. Stage 1 involves wound healing, which occurs within 15 minutes after cutting. Stage 2, in which blastema formation (unpigmented region) has begun, is reached after about 16 hours. Stages 3 and 4 (extensive blastema formation) are reached within 1.5 and 2 days, respectively. In stage 5, reached at day 3, eye spot differentiation becomes visible under a dissecting scope. Eye differentiation continues to stage 6 (3.5 days), while head morphology changes distinctively in stage 7 (4.4 days). Stages 8 and 9 (5.5 and 7 days after amputation) illustrate the appearance of auricles and the completion of head regeneration.

Regeneration is a rapid event. Within as little as 15 minutes after amputation, epithelial cells at the edge of the wound close over the lesion. Within a day, groups of undifferentiated cells accumulate beneath the wound epithelium, and as the days progress the aggregate of regenerating cells, called a **blastema**, grows exponentially. A combination of cell migration and cell division probably contributes to the increase in blastema size. The blastema cells themselves don't divide, but cells quite close (less than 500 μm) to the blastemal border provide the raw recruits for blastemal growth. Within 4–6 days, newly differentiated structures become evident within the growing blastema and under optimal conditions regeneration is complete within 9–10 days after amputation (Figure 7.15; Baguñà et al. 1994). This pattern differentiates planarian blastema formation from the pattern of dedifferentiation and blastemal cell multiplication and redifferentiation followed by organisms such as insects, amphibians, and annelids (Baguñà and Saló 1984).

Planaria have a strong sense of head and tail (anterior-posterior polarity). When cut, the anterior cut surface almost always regenerates head structures and the posterior cut surface regenerates tail structures. Two hypotheses exist to explain the maintenance of antero-posterior polarity. One calls for an inductive interaction between the newly formed epithelium, which covers the wound during the initial healing process, and the underlying blastema. The other suggests some sort of molecular gradient of antero-posterior determining factor(s). Despite a large body of experimental literature, no clear evidence supports either point of view.

A classic observation, first made by Thomas Hunt Morgan (1904), illustrates the quandary. If one makes extremely thin transverse cuts (Figure 7.16a), normal antero-posterior polarity is sometimes disrupted, resulting in a double-headed worm. If, however, one first makes a single transverse cut, and allows a short time to elapse before making a second cut (Figure 7.16b), the thin transverse fragment regenerates with normal polarity. Chandebois (1976) has championed an interpretation of these results, which utilizes the concepts of inductive interaction and positional information. She argues that under normal circumstances confrontation of the dorsal and ventral epidermis during wound healing governs subsequent regeneration. Wound-healing, however, is asymmetrical: at the anterior wound site the dorsal epithelium covers the ex-

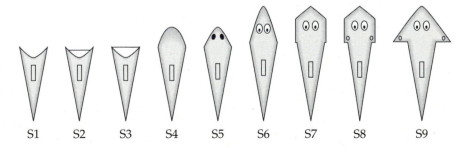

S1   S2   S3   S4   S5   S6   S7   S8   S9

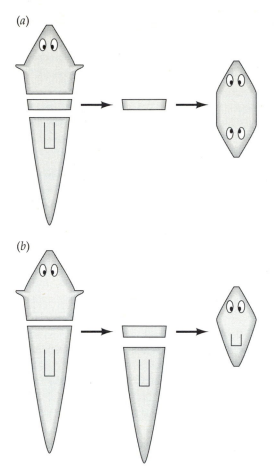

Figure 7.16 (*a*) A narrow, transverse anterior piece often leads to the production of a double-headed Janus-like organism. (*b*) If the anterior wound border is allowed to heal briefly before the posterior cut is made, regeneration proceeds normally.

posed stump, while posteriorly the ventral epitelium covers the wound. Chandebois believes that such asymmetrical healing brings together cells with different positional information and induces the regeneration of the missing positions. She suggests that when the initial fragment is very small, wound-healing is often abnormal, resulting in a notable number of double-headers. In contrast, those arguing for a morphogenetic antero-posterior gradient, hold that very thin transverse cuts are too narrow for the establishment of a gradient, but that if one waits before making a second cut, the larger organism establishes a gradient, the effects of which are maintained even after a second cut creates a narrow transverse fragment (Goss 1969). When we finally unravel the events of regeneration at the levels of gene activation and tissue specification, the differences between these two theories may well turn out to be moot. For the time being, however, an understanding of the mechanisms that maintain and determine antero-posterior polarity remains beyond our grasp.

This is not to say that nothing is known about head and tail regeneration. Head determination—the time when cells in the regenerate first commit themselves to their ultimate fate as head structures—occurs within the first 12 hours of regeneration (at 24°C; events move more slowly at cooler temperatures). Determination of more posterior structures such as the pharynx occurs another 12–24 hours later. Once the various events of determination have occurred, cell differentiation begins within two to four days after cutting. Head structures such as the neural ganglia appear within the blastema itself or at the stump/blastemal border. The pharynx, on the other hand, develops entirely within the original animal. This difference in location has led Baguñà et al. (1994) to raise the possibility that pattern formation in planaria involves a mixture of two classically identified processes: morphallaxis (a term coined by T. H. Morgan to designate the dedifferentiation of old tissues within the amputation stump and their remodeling for new functions) and epimorphosis (the differentiation of new cells and the construction of new structures using the newly divided cells accumulated in the blastema). Since neoblasts exist throughout the organism, however, they could still be the cells giving rise to a new pharynx. Resolution of the matter awaits the development of more sophisticated cell marking techniques.

Molecular Mechanisms

One of the most remarkable features of contemporary developmental biology is the degree to which genes and gene complexes involved with basic aspects of embryogenesis and pattern formation are conserved throughout the animal world. Although work on the existence and function of such genes in planaria is in its infancy, there is every reason to expect a rapid evolution of knowledge in this field. Already we know that planaria have 13 homeobox-containing genes distributed among gene families known to be of developmental importance in other organisms. Probably the most interesting of these uncovered to date are those homologous with the *Antennapedia* and *eyeless* genes made famous by work in the fruitfly *Drosophila* (Baguñà et al. 1994). To date, however, no evidence has emerged to implicate these genes in the early stages of pattern formation during regeneration. Preliminary work suggests that the POU III class of genes may be involved in nerve cell differentiation, while the NK class may be expressed late in regeneration, during the period of cell differentiation (Saló et al. 1995). In the coming years much interesting information will emerge concerning the genomic organization and developmental functions of these genes in planaria and perhaps, finally, planaria will reveal to us their secret of regeneration.

Summary

Planarians' prodigious abilities to regenerate have fascinated scientists for at least the past one hundred years. At the descriptive and phenomenological levels, we have learned a great deal about their abilities to replace lost body parts. Only recently, however, has it become possible to hope for deeper mechanistic understandings of the processes. Nevertheless, years of work lie ahead before clearer understandings emerge.

As with regeneration, we have more than a century of insights from descriptive flatworm embryology, yet we possess detailed knowledge of only a few species. Current experimental approaches to archoophorans such as *Hoploplana* are securing platyhelminthes' place at the base of the spiralian clade and open up vistas for further research. However, the difficulties of separating neoophoran embryos from their yolk cells leave us wondering about cell lineages, gastrulation, and the mechanism of yolk cell movement across the hull membrane. The question of embryonic positional information in separated anarchic cell masses of a triclad embryo is particularly vexing—if blastomere anarchy is as it has been described.

There is much material here for the aspiring developmental biologist! Furthermore, it is becoming evident that the diversity of planarian species and life styles provide ample material for investigating questions in the evolution of development. To understand more about the ecological, historical and genetic factors involved in the evolution of the varied planarian ways of life provides a fascinating challenge for scientists of the future.

Literature Cited

Baguñà, J. and B. C. Boyer. 1990. Descriptive and experimental embryology of the turbellaria: Present knowledge, open questions, and future trends. In *Experimental Embryology in Aquatic Plants and Animals*, H. J. Marthy (ed.). Plenum, New York, pp. 95–128.

Baguñà, J. and E. Saló. 1984. Regeneration and pattern formation in planarians. I. The pattern of mitosis in anterior and posterior regeneration in *Dugesia (G.) tigrina*, and a new proposal for blastema formation. *J. Embryol. Exper. Morphol.* 83: 63–80.

Baguñà, J., E. Saló, R. Romero, J. Garcia-Fernández, D. Bueno, A. M. Muñoz-Marmol, J. R. Byascas-Ramirez and A. Casali. 1994. Regeneration and pattern formation in planarians: Cells, molecules and genes. *Zool. Sci.* 11: 781–795.

Boyer, B. C. 1971. Regulative development in a spiralian embryo as shown by cell deletion experiments on the acoel *Childia*. *J. Exp. Zool.* 176: 97–106.

Boyer, B. C. 1986. Determinative development in the polyclad turbellarian *Hoploplana Inquilina*. *Int. J. Invert. Repro. Dev.* 9: 243–251.

Boyer, B. C. 1987. Development of *in vitro* fertilized embryos of the polyclad flatworm *Hoploplana inquilina* following blastomere separation and deletion. *Roux's Arch. Dev. Biol.* 196: 158–164.

Boyer, B. C. 1989. The role of the first quartet micromeres in the development of the polyclad *Hoploplana inquilina*. *Biol. Bull.* 177: 338–343.

Boyer, B. C., J. Q. Henry and M. Q. Martindale. 1996. Dual origins of mesoderm in a basal spiralian: Cell lineage analyses in the polyclad turbellarian *Hoploplana inquilina*. *Dev. Biol.* 179: 329–338.

Brøndsted, H. V. 1969. *Planarian Regeneration*. Pergamon, New York.

Brusca, R. C. and G. J. Brusca. 1990. *Invertebrates*. Sinauer Associates, Sunderland, MA.

Calow, P. and D.A. Read. 1986. Ontogenetic patterns and phylogenetic trends in freshwater flatworms (Tricladida): Constraint or selection? *Hydrobiologia* 132: 263–273.

Chandebois, R. 1976. Histogenesis and Morphogenesis in Planarian Regeneration. *Monographs in Developmental Biology 11*. S. Karger, Basel.

Galleni, L. and V. Gremigni. 1989. Platyhelminthes-Turbellaria. In *Reproductive Biology of Invertebrates IV, Part A: Fertilization, Development, and Parental Care*, K. G. Adiyodi and R. G. Adiyodi (eds.). Wiley-Interscience, New York, pp. 63–89.

Gardiner, E. G. 1895. Early development of *Polychaerus caudatus* Mark. *J. Morphol.* 11: 155–176.

Giesa, S. 1966. Die Embryonalentwicklung von *Monocelis fusca* Oersted (Turbellaria, Proseriata). *Z. Morphol. Ökol. Tiere* 57: 137–230.

Goss, R. 1969. *Principles of Regeneration*. Academic Press, New York.

Gremigni, V. 1983. Platyhelminthes-Turbellaria. In *Reproductive Biology of Invertebrates I: Oogenesis, Oviposition, and Oosorption*, K. G. Adiyodi and R. G. Adiyodi (eds.). Wiley-Interscience, New York, pp. 67–107.

Hallez, P. 1887. Embryogénie des dendrocoeles d'eau douce. Paris.

Hendelberg, J. 1983. Platyhelminthes-Turbellaria. In *Reproductive Biology of Invertebrates II: Spermatogenesis and Sperm Function*, K. G. Adiyodi and R. G. Adiyodi (eds.). Wiley-Interscience, New York, pp. 75–104.

Henley, C. 1974. Platyhelminthes (Turbellaria). In *Reproduction of Marine Invertebrates*, Vol. I, A. S. Giese and J. S. Pearse (eds.). Academic Press, New York, pp. 267–343.

Henry, J., M. Q. Martindale and B. C. Boyer. 1995. Axial specification in a basal member of the spiralian clade: Lineage relationships of the first four cells to the larval body plan in the polyclad turbellarian *Hoploplana inquilina*. *Biol. Bull.* 189: 194–195.

Kato, K. 1968. Platyhelminthes. In *Invertebrate Embryology*, M. Kume and K. Dan (eds.). NOLIT, Belgrade, pp. 125–143.

Korschelt, E. and K. Heider. 1895. *Textbook of the Embryology of Invertebrates*, Part I. Swan Sonnenschein, London.

Koscielski, B. 1966. Cytological and cytochemical investigations on the embryonic development of *Dendrocoelum lacteum* O.F. Müller. *Zool. Pol.* 83–102.

Morgan, T. H. 1904. The control of heteromorphosis in *Planaria maculata*. *Roux's Archiv. Entwicklungsmech. Organism* 17: 693–695.

Rieger, R. M., S. Tyler, J. P. S. Smith and G. E. Rieger. 1991. Playthelminthes: Turbellaria. In *Microscopic Anatomy of Invertebrates, Vol. 3: Platyhelminthes and Nemertinea*, F. W. Harrison and B. J. Bogitsch (eds.). Wiley-Liss, New York, pp. 7–140.

Ruppert, E. E. 1978. A review of metamorphosis of turbellarian larvae. In *Settlement and Metamorphosis of Marine Invertebrate Larvae*, F. S. Chia and M. E. Rice (eds.). Elsevier, New York, pp. 65–82.

Ruppert, E. E. and R. D. Barnes. 1994. *Invertebrate Zoology*, 6th Ed. Saunders College Publishing, Fort Worth, TX.

Saló, E., A. M. Muñoz-Marmol, J. R. Byascas-Ramirez, J. Garcia-Fernàndez, A. Mirales, A. Casali, M. Cormominas and J. Baguñà. 1995. The freshwater planarian *Dugesia (G.) tigrina* contains a great diversity of homeobox genes. *Hydrobiologia* 305: 269–275.

Smith, J. P. S. III and S. Tyler. 1985. The acoel turbellarians: Kingpins of metazoan evolution or a specialized offshoot? In *The Origins and Relationships of Lower Invertebrates*, S. C. Morris, J. D. George, R. Gibson and H. M. Platt (eds.). Oxford University Press, Oxford, pp. 123–142.

Surface, F. M. 1907. The early development of a polyclad, *Planocera inquilina*. *Proc. Acad. Nat. Sci. Philadelphia* 59: 514–559.

Thomas, M. B. 1986. Embryology of the turbellaria and its phylogenetic significance. *Hydrobiologia* 132: 105–115.

CHAPTER 8

Nematodes, the Roundworms

Einhard Schierenberg

WHY STUDY NEMATODES? AT FIRST GLANCE NEMATODES do not look like particularly attractive organisms. In terms of conventional esthetics their outer appearance is of limited appeal. A typical roundworm is shaped like an elongated, thin cigar (Figure 8.1) and can range in size from less than a millimeter (many free-living soil nematodes) to several meters (a parasite in the whale). It has neither legs, wings, appendages, eyes, segments, colored structures nor other features that might catch the eye of an uninitiated scholar. Exceptions to this rule exist in that some species carry pigmented eye spots, bizarre appendages, huge lips, a strikingly structured body surface, or can become transformed into a ball-shaped bag filled with eggs, but except for their role as parasites, nematodes have gained little attention from the general public.

Nematodes have adapted to nearly all habitats on earth, from high mountains to deep sea, from the Antarctic to the desert, from the protonephridia of an earthworm to the heart of a dog. Free-living species play an important role in the food chain and comprise a considerable fraction of the world's biomass. The many parasitic species cause tremendous damage in plants and animals and endanger the health and even the life of hundreds of millions of people, particularly in tropical regions.

That nematodes look insignificant does not mean that they are unattractive animals to study scientifically. Nevertheless, most of the important contributions to our understanding of nematode development probably came from researchers who were not primarily interested in nematodes per se but selected them as suitable systems because their simplicity favored the study of questions of more general relevance.

Structure and Function

Nematodes make use to an extreme degree of a physical principle widely used in nature: hydrostatic pressure. The body of nematodes can be considered a tube filled with liquid (Figures 8.1 and 8.2). The liquid-filled body cavity exerts a strong hydrostatic pressure and in this way keeps the body inflated (for modifications, see Bird 1991). A rigid **cuticle** consisting predominantly of collagen-like proteins excreted by the underlying layer of **hypodermis** (skin) cells (Johnstone 1994) prevents explosion of the animal. This construction is called a **hydroskeleton**. However, as it consists of only a single chamber (like a cheap air matress), a leak anywhere in the system will be disastrous. If the cuticle is seriously damaged, the pressure is released, the body collapses, and the animal shrivels and dies. It would be interesting to know whether the fact that nematodes did not develop the potential for regeneration is correlated to this feature. In most areas surrounding the body the hypodermis is just a thin layer, but in four places (dorsal, ventral, lateral) the cell bodies with nuclei are also located. These nucleated regions form hypodermal ridges that extend away from the body wall (Figure 8.2).

For locomotion, nematodes contain only longitudinal **body muscle**. The spindle shaped cells are organized in four quadrants, separated by the hypodermal ridges. The body muscle cell in nematodes is special in that it can be subdivided into a contractile part adjacent to the hypodermis and a noncontractile part projecting into the body cavity. From this latter part long axon-like innervation processes run to the ventral or dorsal nerve cord to receive synaptic input from the motoneurons (Figure 8.2). Nema-

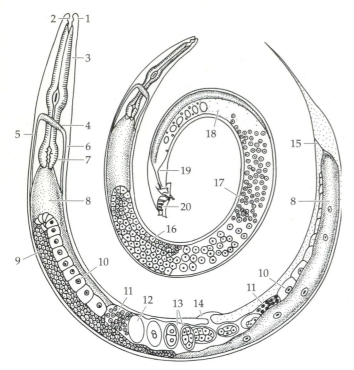

Figure 8.1   General anatomy of a nematode. Larger animal outside, female; smaller animal inside, male. 1, one of six lips; 2, mouth; 3, pharynx ; 4, nerve ring; 5, dorsal nerve cord; 6, ventral nerve cord; 7, posterior pharynx bulb; 8, intestine; 9, oogonia; 10, oocytes; 11, sperm; 12, freshly fertilized egg; 13, early cleavage stages; 14, vulva; 15, anus; 16, spermatogonia; 17, sperm; 18, vas deferens; 19, cloaca; 20, copulatory bursa. (After Meglitsch and Schram 1991.)

todes move in a snakelike way, lying either on their right or left side, with two phase-shifted contractile waves running along their body axis through the dorsal and ventral muscle-quadrants, respectively.

The **alimentary tract** starts with a mouth typically surrounded by six lips, then a **pharynx** with one or two bulbs, and a **gut** running through most of the body ending in an anus (Figures 8.1 and 8.3a). Because of the high body pressure, feeding poses a problem. On the one hand, the gut must generate an even higher hydrostatic pressure to form a lumen through which food can be transported. On the other hand, this type of construction risks the lethal condition wherin the nematode ejects its gut contents through the opened mouth to the outside while trying to ingest food. This problem is solved by the pharynx forming a lock-chamber system.

The most prominent part of the nervous system is the **nerve ring,** a neuropile (tight network of nerve cell extensions forming many synapses) surrounding the thinner part (**isthmus**) of the pharynx (Figure 8.2). It receives synaptic input from nerve bundles coming from **cephalic sensilla** in the head region. From the nerve ring to the posterior run a **ventral** and **dorsal nerve cord** (Figure 8.1) and less prominent lateral ones, which are all intercon-

nected by neuronal extensions, forming **commissures**. Most neurons are located in the head region (Figure 8.3b).

Nematodes do not possess a well developed **excretory system** as found in plathelminths or annelids. A single large H-shaped **excretory cell** (H cell) forms two canals running through the lateral hypodermal ridges through nearly the entire length of the body. It exits to the outside adjacent to the pharynx. The main function of this cell seems to be the control of internal osmotic value by pumping water out of the body cavity.

Most nematodes appear in two sexes, either female and male or hermaphroditic and male. While the males are generally smaller, they differ particularly in the kind of germ cells found in their gonad and in the end of their tail, which functions as a copulatory organ (Figure 8.1). Sex ratios differ widely, from 1:1 to the complete absence of males. In a later section on developmental questions we will take a closer look at the mechanism controlling sex determination. More detailed descriptions of nematode anatomy and function can be found elsewhere (e.g., Wood 1988; Bird 1991; Malakhov 1994; Riddle et al. 1997).

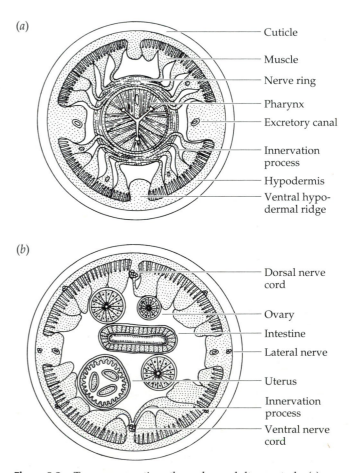

Figure 8.2   Transverse sections through an adult nematode. (a) Pharyngeal region. (b) Middle region. (After Lee 1965.)

(*a*)

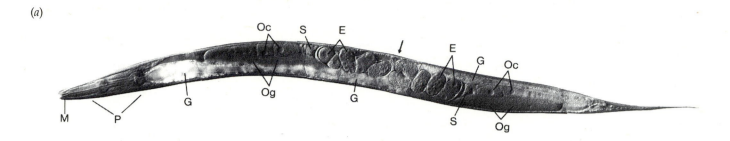

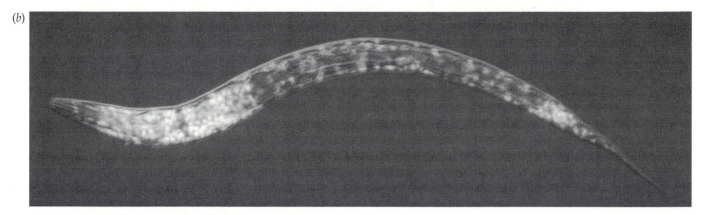

(*b*)

Figure 8.3 *Caenorhabditis elegans*. (*a*) Living, gravid adult hermaphrodite, by Nomarski optics. E, embryos; G, intestine; M, mouth; Oc, oocytes; Og, oogonia; P, pharynx; S, sperm; arrow, position of vulva (out of focus). (*b*) First juvenile stage (L$_1$), fixed and stained to visualize nuclei. Double exposure, by Nomarski optics and epifluorescence. Most accumulated nuclei in the head region belong to neuronal cells whose extension forms the nerve ring. Those nuclei in the tail region reflect the presence of posterior ganglia.

Life Cycles

After hatching from the egg, nematodes typically pass through four juvenile stages ("larvae," L$_1$–L$_4$) separated by molts in which the old cuticle is shed and replaced by a new one formed underneath (Johnstone 1994). During juvenile stages and as an adult, the animal grows to a smaller or larger extent, and the gonad develops from a small primordium to a very prominent organ. For instance, the parasite *Ascaris* (see the section on "Classic Studies") and free-living *Caenorhabditis elegans* develop from eggs of very similar sizes but differ as adults in body length by several hundredfold and with respect to egg production by even several thousandfold.

In the adult stages, various nematode species reproduce either after copulation of different sexes (**amphimixis**; Figure 8.4 *b,c*), self-fertilization of animals that are able to produce both types of gametes by themselves (**hermaphroditism**, Figure 8.4*a*), or by activation of the egg without sperm (**parthenogenesis**). Several variations and combinations of these basic schemes have been described. For example, rare males can occur in hermaphroditic cultures (either spontaneously or induced by environmental cues),

sperm may penetrate the oocyte but not contribute its genetic material to the developing egg (**pseudogamy**), or two modes of reproduction may alternate between subsequent generations (**heterogony**; Figure 8.4*d*).

Embryogenesis in nematodes can completely or partially take place inside the mother, or cleavage may start only after eggs have been laid. The more or less transparent, rigid eggshell (mechanical protection) and an underlying thin lipid layer ("vitelline membrane," a chemical protection) make the early embryo independent of a supporting maternal environment. The duration of embryogenesis is species specific and temperature dependent. Under optimal conditions it may be as short as 12 hours in one species or as long as several weeks in another. Four different examples of a nematode life cycle are shown in Figure 8.4.

Taxonomy and Phylogeny

About 20,000 nematode species have been described so far, but the estimate of how many species exist in total varies dramatically between 80,000 and 1 million (Malakhov 1994). Independently of which number comes closer to reality, nematodes are the animals to look for if you want to find a so-far undescribed species (and become immortal by giving it your name).

Nematode taxonomy generally relies essentially on morphological characteristics. There are only a small number of experts who know where to look for those tiny anal papillae (which may even require scanning electron

Figure 8.4 Nematode life cycles. (*a*) *Caenorhabditis elegans*, free-living . (*b*) *Ascaris lumbricoides*, parasitic except during embryogenesis, when eggs are passed out of the host with feces. (*c*) *Wuchereria bancrofti*, microfilaria blocking lymph glands may cause elephantiasis; parasite switching between two hosts. (*d*) *Rhabdias bufonis*, alternating between amphimictic, free-living and hermaphroditic, parasitic generations. L_1–L_4, four juvenile stages.

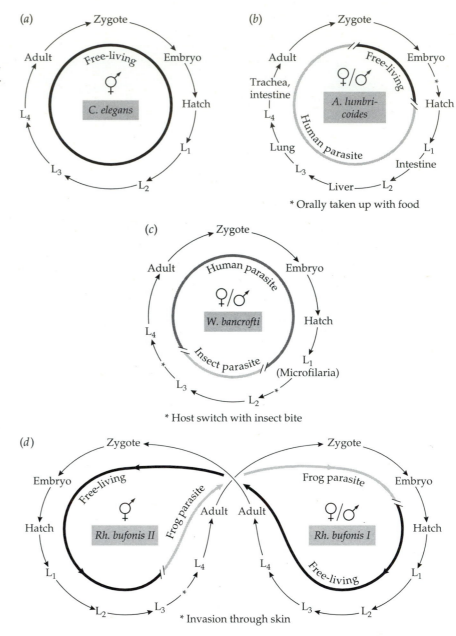

microscopy) or for minor differences in the mouth structure necessary to identify a rare species. Much work is still needed to clarify how many of the described "species" are real and which, despite certain morphological peculiarities, are just variants of the same species. Breeding tests between different isolates have answered the question in some cases but many of the "species" described decades ago are difficult to find again and therefore cannot readily be taken under scrutiny.

Even for a broad classification within nematodes, no generally accepted scheme exists, a lack which again is due to their morphological uniformity, the huge number of species, and an insufficient study of the group in general. The question of whether nematodes should be considered a separate phylum or would better be placed together with other inconspicuous animals like gastrotrichs, nematomorphs, rotifers, and so on into a phylum (Aschelminthes or Nemathelminthes) is treated variously in the literature (for a summary, see Malakhov 1994).

Since it has become clear that purely morphological features are probably not sufficient to identify all nematodes unequivocally on a species level and that the present taxonomy does not properly reflect the pathway of evolution, attempts have been initiated recently to come closer to a "phylogenetic taxonomy" by looking at structural variations of molecules with assumed constant rates of modification over evolutionary time scales (Vanfleteren et al. 1994). However, as long as additional data remain scarce, we must live with only preliminary ideas of how nematodes are related to each other.

A particularly important question is, Who are the closest relatives of nematodes? This issue has been diversely discussed in the past, as nematodes appear to represent quite an isolated group with distinct peculiarities. The suspicion, based on morphological data, that they are most closely related to gastrotrichs (tiny aquatic invertebrates) has recently obtained additional support. Some marine nematode species, interpreted as being the most primitive ones, do not express the characteristic early pattern of asymmetric cleavages and can therefore be more easily related to other invertebrates (Malakhov 1994).

Classic Studies

The most eminent pioneer in the analysis of nematode development was Theodor Boveri (1862–1915), whose first graduate student (doing his Ph.D. thesis on nematodes), Hans Spemann, won the Nobel Prize in 1935 for his work on the "organizer" in the amphibian embryo. In the 1880s Boveri had started to study embryogenesis of the nematode *Ascaris megalocephala* (today, *Parascaris equorum*), a parasite found in the intestine of horses. The females produce millions of eggs, releasing more than 100,000 per day; after fertilization the eggs start to cleave only after being released from the anaerobic environment. In addition, development can be temporarily interrupted, even for weeks, by placing the embryos into the refrigerator. Despite these favorable features, but in recognition of the very limited possibilities for experimental interference, Boveri wrote in 1910, "... I must admit that, if Würzburg [Southern Germany] were located at the sea, other objects and problems would have fascinated me before what I have worked on here...". The objects Boveri considered were most likely sea urchin embryos, with which he experimented during the summer months at the Marine Biological Station in Naples (Italy). However, at least some of Boveri's countrymen appreciated the importance of worms. The influential geneticist Richard Goldschmidt wrote a popular book, *Introduction to the Science of Life, or Ascaris*, wherein he used this nematode to teach biology to the public.

Normal Development and Theoretical Considerations

In fixed and stained early cleavage stages Boveri found a peculiar but reproducible division pattern accompanied by a loss of chromatin in most cells. The first division generates two cells, one of which at the next division passes through a process called **chromatin diminution**. Here, during metaphase, major parts of the chromatin break off and are released into the cytoplasm, where they remain visible for some time before degenerating. In contrast, the daughters of the other cell preserve their complete chromatin content when they divide. However, with the next division of these two daughter cells, one goes through chromatin diminution and the other does not. This pattern is repeated for several additional cleav-

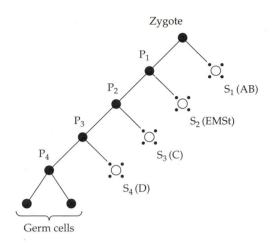

Figure 8.5 Scheme of early cleavage in *Ascaris*. Filled circles, germline cells; open circles with dots, somatic sister cells having passed through chromatin diminution. (From Boveri 1910.)

ages. In the end, all cells except a single one contain the reduced amount of chromatin. Boveri interpreted his observations (Figure 8.5) as an early separation of "soma" (S_1–S_4, those cells which form the body, differentiating into the various tissues and perishing with the death of the individual) from "germline" (zygote–P_4, blastomeres with the complete genetic information leading to potentially immortal germ cells that differentiate into gametes). Lineage studies on *Ascaris* and decades later on the nematode *Caenorhabditis elegans* (which like most nematodes does not perform chromatin diminution) proved his interpretation correct.

About the same time Boveri made his observations on the *Ascaris* embryo, another German zoologist, August Weismann, proposed a model of inheritance that also included a strict separation in the early embryo of a germline from the soma. Besides other implications, such a mechanism should exclude the inheritance of acquired traits (an assumption then favored by many biologists including Darwin). Today, since we know that an early soma/germline separation is not necessary to prevent genetic fixation of personal experience, a different role must be attributed to this process in the nematode embryo (see the later discussion on soma/germline separation).

Analysis of Embryos with Altered Early Cleavage

Boveri made additional crucial observations in *Ascaris* towards a better understanding of how embryonic cells are determined to enter the alternative pathways of soma or germline (Boveri 1910). By chance he observed cases in which the uncleaved egg performed a double mitosis, with two cleavage spindles crossing each other at right angles, apparently the result of dispermic fertilization (entrance of two sperm, each contributing a centriole; see the section "Development: An Overview"). This way four daughter cells emerged from one zygote. Boveri discerned three different patterns of further development of such *Simultan-Vierer* ("simultaneous-four"; Figure 8.6*a–c*). Depending on the spatial orientation of blastomeres, some eggs contained two somatic cells (performing chromatin diminution) and two germline cells (retaining the full chromatin complement); others featured three somatic cells and one germline cell; and, finally, some had one somatic cell and three germline cells.

In a second assay Boveri centrifuged *Ascaris* zygotes prior to and during first cleavage in order to cause an abnormal distribution of cytoplasmic components to both daughter cells. He found that, despite the formation of visible layers of cytoplasmic particles (e.g., yolk, lipids), further development was usually not affected. However, a few of the embryos divided equally and both emerging blastomeres behaved like germline cells (Figure 8.6*d–f*). According to Boveri, the cleavage spindle was exactly at right angles to its normal orientation in these cases (centrifugation had prevented the typical spindle rotation) with the result that both blastomeres had obtained an identical composition of cytoplasmic material. From his observations he developed the following conception: (1) even under experimental conditions only blastomeres with the "value" of the normally formed cells can be generated, and (2) even very small differences in cytoplasmic composition between two cells are sufficient and necessary to shunt them into one of two possible differentiation pathways.

In summary, Boveri's findings indicated that it is the cytoplasmic environment that controls the behavior of a nucleus and consequently that it is the cytoplasmic composition of a cell and not the nucleus that determines its fate. In the controversial debate of those days his results were a strong argument against the contention that the generation of cell diversity involves differential segregation of chromosomes to both daughter cells. After Boveri's untimely death, interest in nematode embryology declined for several decades.

Caenorhabditis elegans: A Model for Developmental Biologists

How to Become a Model Organism

Nearly half a century after Boveri's studies another nematode, this time a small free-living species, started a promising career as a model organism. Like many of his contemporaries, Sydney Brenner, a molecular biologist working in Cambridge, England, was seeking new challenges when he wrote to his colleague Max Perutz in 1963, "It is now widely realized that nearly all the 'classical' problems of molecular biology have either been solved or will be solved in the next decade.... Our success with bacteria has suggested to me that we could use the same approach to study specification and control of more complex processes in cells of higher organisms."

Brenner looked for a metazoan organism with a simple structure, short life cycle, and amenable genetics, in which molecular genetics analogous to that of bacteria could be combined with cellular analysis, especially of behavior. A few months later he stated in a proposal to the Medical Research Council, "We think we have a good candidate in the form of a small nematode worm…" (S. Brenner, quoted in Wood 1988).

During the following years Brenner established methods for culturing his nematode of choice on agar plates with a thin layer of bacteria as a food source and for isolating and characterizing mutants; he localized about a hundred genes on the six linkage groups and then published

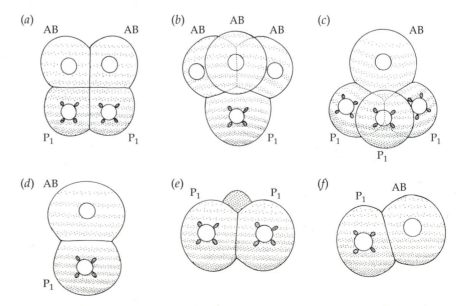

Figure 8.6 Abnormal development of *Ascaris* embryos. (*a–c*) Tetrapolar cleavage after dispermic fertilization results in 4-cell stages with three different combinations of cell types. (*d–f*) Development after centrifugation. (*d*) Normal first cleavage with differential segregation of cytoplasmic components indicated by gradually changing shading. (*e*) Rare cases with equal distribution of cytoplasmic components leading to two germline cells, often going along with ball-like extrusion of cytoplasm. (*f*) Despite reduced cytoplasmic differences, normal cell specification has taken place. (After Boveri 1910.)

his pioneering paper "The genetics of *Caenorhabditis elegans*" (Brenner 1974). Scientists who had previously worked in Brenner's laboratory spread the news and started their own laboratories, predominantly in the United States. In a kind of self-amplifying process, the accumulating new results lured more and more scientists to join the "*C. elegans* community." Besides the attractiveness of the system itself, additional factors contributed to the rapid establishment of *C. elegans* as a model system. Introductory courses were offered to newcomers, a *C. elegans* newsletter was launched, a *C. elegans* Genetics Center was established to distribute worm stocks (which can be frozen away in liquid nitrogen) to interested scientists all over the world, and every other year an International Meeting on *C. elegans* has been organized since 1977 (participants: about 60 in 1977, about 800 in 1995). Last, but not least, all scientists profited (and still do so) from a spirit of openness and cooperation present from the very beginning of this venture. Today *C. elegans* is probably the most completely described metazoan organism in terms of its anatomy, development, and behavior and its genetic control.

Development: An Overview

C. elegans is a 1-mm small, diploid, self-fertilizing hermaphrodite (Figure 8.3*a*), more precisely a female with a hermaphroditic gonad (Figure 8.7). However, rare males (about 1:700) which can mate with hermaphrodites occur occasionally because of an infidelity of sex chromosome segregation during meiosis (**nondisjunction**). In the late hermaphroditic juvenile, first sperm are produced from the germ cells in both arms of the gonadal tube and stored in a spermatheca. Sperm in nematodes generally differ from those in other organisms in that they bear no flagellum but form pseudopodia for migration (Figure 8.8; Roberts and Stewart 1995). Despite the lack of flagella, sperm can migrate considerable distances, for example, from the region of the vulva to the spermatheca after copulation from males; in hermaphrodites, sperm motility is not critical, as the eggs are pushed by the contracting gonadal tube through the sperm on their way to the vulva. Sperm motil-

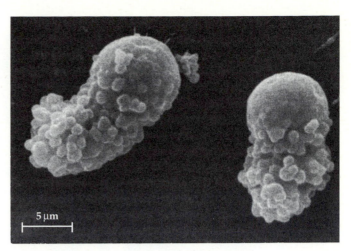

Figure 8.8　Scanning electron micrograph of *C. elegans* spermatozoa. Note the absence of flagella and the presence of pseudopodia used for locomotion. (Courtesy of S. Ward, University of Arizona.)

ity is not based on an actin-myosin contractile system, but seems to function in an analogous way (Roberts and Stewart 1995). How sperm find their way is not known.

After a limited number of sperm have been produced (about 150/gonadal arm), the differentiation program of germ cells in the hermaphrodites switches and oocytes develop (see the discussion on sex determination). These are much larger than sperm, partly because they massively incorporate yolk, which is produced by the gut, exported into the body cavity, and from there taken up by the gonad. In the distal part of the gonadal arm, meiosis starts in the germ cells (Figure 8.7), but is only completed after fertilization with the formation of the second polar body. Through muscle-induced contractions of the gonadal tube, mature oocytes are squeezed into the spermatheca where they are fertilized by a single sperm, a process indicating the presence of a reliable block against polyspermy. The sperm introduces a centriole, which organizes the cleavage spindle while the maternal centriole deteriorates. About 45 minutes after fertilization two pronuclei form at opposite poles of the egg. The oocyte pronucleus (positioned anteriorly) migrates towards the sperm pronucleus, then they migrate back into the center and fuse (Figure 8.9*a–d*). The pattern of early development in *C. elegans* (small, free-living, bacteria-feeding, and short-lived [3 weeks]) is very similar to that described for *Ascaris* (up to 40 cm long, parasitic, feeding on the intestinal contents of its host, living for many months) by Boveri and others. First cleavage generates a larger somatic cell AB and a smaller germline cell P_1. In a series of further unequal cleavages, five somatic founder cells (AB, MS, E, C, D) and a primordial germ cell P_4 are generated (Figure 8.9*e–i*). More details are given in later sections.

In general, gastrulation starts very early in nematodes (Skiba and Schierenberg 1992). In *C. elegans*, after the generation of P_4 in the 24-cell stage, the two daughters of the

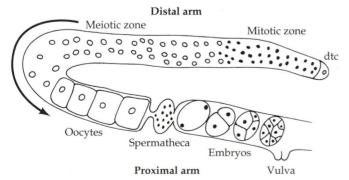

Figure 8.7　Mature gonad of adult *C. elegans*. Filled circles, nuclei in mitotic phase; open circles, nuclei in meiotic phase; dtc, distal tip cell controlling meiosis onset.

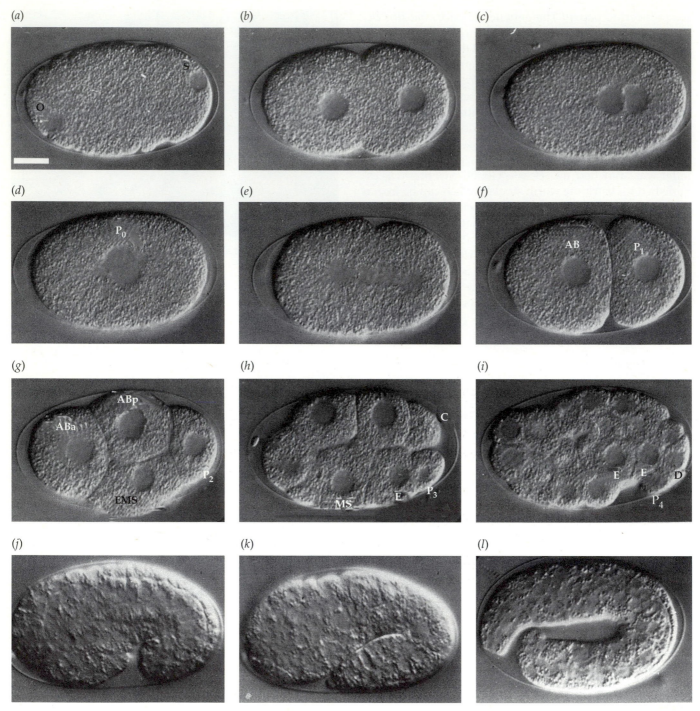

Figure 8.9 Early development of *C. elegans* (*a–d*). Pronuclei from oocyte (o) and sperm (s) form at opposite poles, meet and fuse to form the zygote. (*f–i*) Unequal cleavages in the germline lead to the sequential formation of five somatic founder cells (AB, MS, E, C, D) and the primordial germ cell P_4. For daughter cells a lowercase letter is added to the name of the mother cell, indicating relative positions to each other, e.g., ABa (anterior) and ABp (posterior). (*i*) After the 24-cell stage, gastrulation starts with the immigration of the two gut precursor (E) cells. (*j–l*) In the second half of embryogenesis, a ventral indentation forms and the embryo stretches into a worm. Left lateral view. Bar = 10 μm. (From Schierenberg 1986.)

E cell migrate from the ventral side into the center of the embryo (Figure 8.9*i*). A prominent blastocoel is absent and only small intercellular spaces are present. During embryogenesis, the E cells will form a gut consisting of only 20 cells (Figure 8.10). The next cells to migrate through the same ventral opening (**blastopore**) are the two daughters of P$_4$, at around the 100-cell stage, taking position below the gut primordium (Figure 8.11*b*). At the same time, descendants of the MS cell move inward from a more anteri-

or-ventral position. While the small blastopore widens to form a ventral cleft , C- and D-derived myoblasts follow in the posterior half. They are positioned on the left and right sides of the gut primodium (Schierenberg and Strome 1992), resulting in a three-layered (**triblastic**) embryo. Finally, AB cells contributing to the pharynx are translocated interiorly, while hypodermal cells from both sides move downwards, closing the ventral cleft. About 6 hours after fertilization, gastrulation is completed (Figure 8.11*c,d*).

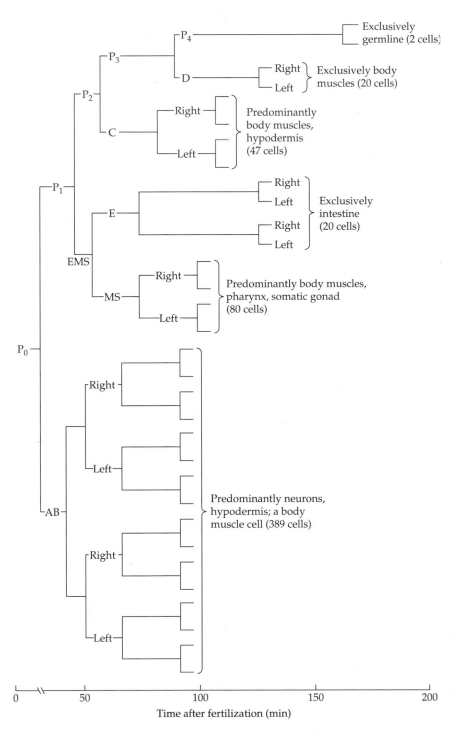

Figure 8.10 Early lineage tree of *C. elegans*. Within the first hour after onset of cleavage, five somatic founder cells (AB, MS, E, C, D) and the primordial germ cell (P$_4$) are generated. Each founder cell constitutes a cell lineage in which a fixed number of cells (given in parentheses) with invariant fates is produced. Left-right decisions are made early within individual lineages, as indicated. Note that among the descendants of the anterior AB daughter, cells become bilateral partners that do not occupy corresponding positions in left and right lineage branches.

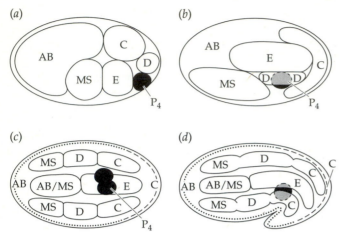

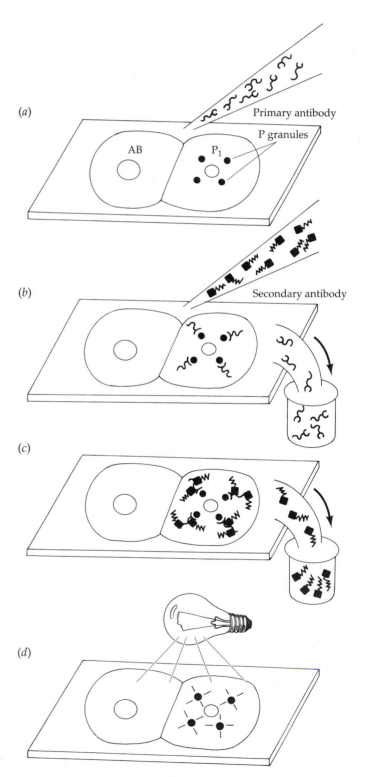

Figure 8.11   Schematized positions of founder cells and their descendants in various stages of *C. elegans* embryogenesis. (*a*) 26-cell stage, prior to the onset of gastrulation. 16 AB, 4 MS, 2E, 2C, D, and P_4 are present. (*b*) 102-cell stage after immigration of E, P_4, and D descendants. (*c*) end of proliferation phase, more than 500 cells present. (*d*) early morphogenesis phase, more than 500 cells present; only one of two P_4 descendants is visible. In (*c, d*) AB and MS cells have formed a pharynx primordium and descendants of MS, D, and C have formed four body muscle quadrants. Dotted and stippled lines mark the hypodermis contributed by AB and C, respectively. Left lateral views (*a, b, d*), ventral view (*c*). (From Schierenberg and Strome 1992.)

Methods of Analysis

To study gross morphology and behavior, to count progeny, or to transfer specimens from one culture plate to another, a dissecting microscope with illumination from below is most suitable. Anatomy and development are usually studied under a compound microscope at magnifications between 100× and 1000×. The classic studies on nematode embryogenesis by Boveri and others were restrict-

This roughly coincides with the end of the first half of embryogenesis, called the **proliferation phase** because most cleavages take place in that phase. For a more detailed description of gastrulation in *C. elegans*, see Bucher and Seydoux, 1994. In the second half of embryogenesis, called the **morphogenesis phase**, only few further divisions occur and the ball-shaped embryo stretches into a worm (Figure 8.9*j–l*). After only 12 hours of embryogenesis (at 25°C), a 0.2-mm small worm hatches which then passes through four juvenile stages ($L_1–L_4$) separated by molts before reaching adulthood in 2–3 days (Figure 8.4*a*).

Due to the rapid development, the transparent eggshell and cuticle of the hatched animal, the invariant cleavage and differentiation program of cells, and (last, but not least) years of intensive work by the scientists involved, the development of *C. elegans* has been described completely from the zygote to the adult on a cell-by-cell basis (see the section egg to organism). This description has not been possible for any other organism so far.

Figure 8.12   Indirect immunofluorescence. The object to study (here an early nematode embryo) is attached to a microscope slide and chemically fixed. (*a*) A primary antibody (here one against P granules is added). (*b*) After the primary antibody has bound to its target, the secondary antibody with a fluorescent tag is applied. (*c*) After that has bound to the primary antibody and unbound antibody molecules have been washed off (*d*), illumination with light of appropriate wavelength (e.g., $\lambda = 450–490$ nm) lights up the P granules (see Figure 8.13). Antibody molecules are not drawn to scale.

ed by the fact that only simple brightfield illumination was available. A major breakthrough in optical design called *Differential Interference Contrast* optics (DIC or Nomarski optics, after its inventor) made it technically possible to trace all cell divisions in *C. elegans* (as will be discussed later). Particularly suitable for small and transparent objects like a soil nematode and its embryos, it generates an excellent image of high resolution and contrast but a small depth of focus (Figures 8.3 and 8.9). In this way, many optical sections can be laid through the specimen, subdividing the complex overall information into separate portions.

A method widely used and particularly suitable to visualize specific structures in whole mounts of the nematode embryo is indirect immunofluorescence. An antibody generated against a certain protein (primary antibody) is added to fixed specimens (Figure 8.12*a*) , where some of it binds to the corresponding antigen and the rest is washed off (Figure 8.12*b*). In a second step, an antibody mixture (polyclonal secondary antibody) is added. This mixture carries a fluorescent tag and is directed against proteins of the organism in which the primary antibody was raised (usually mouse or rabbit). Some of it binds to the primary antibody and the rest is washed off (Figure 8.12*c*). Inspection under the fluorescence microscope indicates the distribution of the protein of interest indirectly via its fluorescing binding partner (Figure 8.12*d*; Figure 8.13). The two-step procedure saves the individual marking of each primary antibody and increases the intensity of the (often weak) signal as several secondary antibody molecules can bind to one molecule of primary antibody.

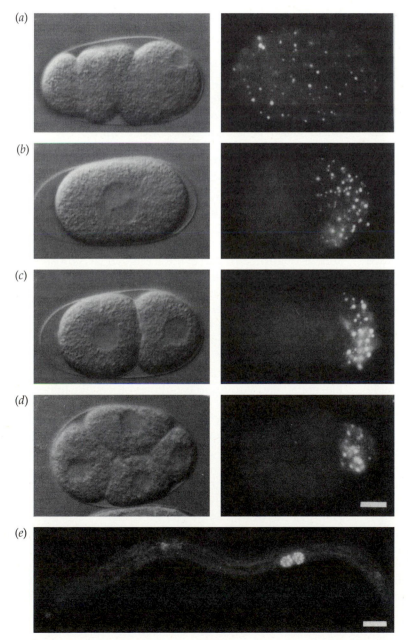

(a)

(b)

(c)

(d)

(e)

Figure 8.13 Segregation of germline-specific P granules in *C. elegans*. (*a–d*) Nomarski optics of the 1- to 4-cell stage (left), and the corresponding stages of fixed embryos stained with an antibody against P granules (right). (*e*) Young juvenile with two germline cells (daughters of the primordial germ cell P_4). Left lateral view. Bars = 10 μm. (From Strome 1988.)

To understand developmental strategies and underlying mechanisms, a detailed description of normal development is required, but is not sufficient. Experimental interference and analysis of induced deviations from the normal pattern of development can help to define individual steps and dissect out individual components playing a role in each of them. A few examples are given in the sections that follow on soma/germline separation and on sex determination.

Most frequently, manipulations are performed on the level of cells or on the level of genes. Cells can, for example, be ablated with a laser microbeam coupled to a microscope, and they can be isolated and recombined in abnormal configurations. In recent years, experiments of this kind in combination with the above-mentioned immunofluorescence have been tremendously useful in revealing specific cell–cell interactions in the nematode that could not have been readily assessed by any other available methods (Lambie 1994; Sommer et al. 1994; Tax and Thomas 1994).

While methods for cell manipulations (particularly with a laser microbeam) are well suited for application in nematodes, where important decisions are made on the level of individual cells, the genetic approach (so far essentially restricted to C. elegans) is conceptually similar to that in other amenable systems and has been partially adopted from methods used in bacteria and Drosophila. Here only a brief outline of the standard experimental procedure is given. Details, useful modifications, and alternative methods are described elsewhere (Wood 1988; Epstein and Shakes 1995; Riddle et al. 1997).

In searching for interesting genes, animals are treated with a mutagenizing agent or are irradiated to induce mutations, which—if occuring in the germline—are transmitted to the progeny. As mutations are usually only generated in one of the two copies of a given gene present in a diploid organism and in addition are recessive, one has to wait for the F_2 generation (self-fertilization in C. elegans automatically leads to homozygosity) to identify mutants that are altered in their morphology, development, or behavior. It is obviously easier to look for viable mutants, but lethal ones may be of particular interest because they are defective in essential genes. Once a mutant has been isolated, co-segregation tests can be done by mating with males carrying visible markers, representing each of the six chromosomes (resulting, for instance, in short and fat animals ["dumpies"] or those with twisted cuticle rotating around their long axis ["rollers"]). Such tests can be used to assign the new gene to a "linkage group" (chromosome). The exact location on a chromosome can then be determined by two- and three-factor crosses. To take advantage of the normally rare males needed for these crosses, males can simply be mated to hermaphrodites to establish a male stock with 50% males in the cross-progeny.

With procedures standard to molecular biologists, a gene can be cloned and sequenced, thus giving hints as to its function (e.g., it may belong to a known gene family or have a DNA-binding motif) and to phylogenetic relationships (e.g., it may show sequence homologies to genes found in other organisms). The genetic assay has proven to be a very powerful means to dissect development into single steps controlled by individual genes, to understand how they work on a molecular level, and how they interact with others. Drosophila, C. elegans, and mouse are the most prominent examples for the enormous potential of such a "genetic dissection of development" (Wilkins 1993). An example showing how far the genetic approach can go is given in the section on sex determination.

Selected Developmental Questions Studied in Nematodes

As C. elegans is by far the best-understood representative of nematodes, most of the data reported in this section come from the analysis of this species. To what extent the conclusions drawn from these studies are applicable to nematodes in general, needs to be determined.

From Egg to Organism: Cell Lineage Analysis

Cellular development of C. elegans is invariant from individual to individual and can be described in the form of a unique lineage tree. It turned out that, as in Ascaris (Figure 8.5), a handful of somatic founder cells and a primordial germ cell are generated during the early phase of embryogenesis (Figure 8.9). Each of these founder cells constitutes a cell lineage in which members divide (at least at first) synchronously, with cell cycle rhythms different from those in other lineages. In each lineage an invariant number of cells with fixed fates is produced (Figure 8.10). At the end of embryogenesis exactly 558 cells are present (in the male, 560). About 10% of these continue with additional postembryonic divisions, generating new structures like gonad and vulva and extending others like hypodermis and musculature. The adult is built from 959 somatic cells (the male with its complex tail, 1031) plus a variable number of germ cells (up to about 2000) dependent on environmental conditions (Wood 1988).

Several developmental principles have been deduced from the description of division patterns and cell fates by observation alone, without any experimental interference. Here just two examples are given.

Nematodes contain many of the tissues like skin, muscle, neurons, and gut that are found in all higher organisms. From a designer's point of view one might expect that each of the somatic founder cells would produce one cell type (clonal origin of tissues). However, this is not the case. Most of the founder cells contribute to several tissues and germ layers. For instance, body muscle cells (identical in their appearance and function) are derived from MS, C, and D, plus a single one from AB (Figure 8.10), thus being polyclonal in their origin.

During development of the *C. elegans* hermaphrodite, exactly 131 cells die at distinct times and positions in the lineage, often soon after their birth, without having fulfilled any obvious function. One explanation for these many deaths (nonrescue by laser ablations of neighboring cells indicate that they are suicides) is the idea that the lineage tree is constructed in a modular fashion from a limited number of repeatedly used "cleavage motifs" (called *sublineages*), reducing the complexity of the required genetic program (Wood 1988). Fine tuning is then made according to local requirements, and unnecessary cells are eliminated. The genetic control of programmed cell death (**apoptosis**) involves homologous genes in other organisms including vertebrates, suggesting a conserved mechanism (Hengartner and Horvitz 1994; Yuan 1995) .

Establishing the Embryonic Axes

Unlike many other organisms, the unfertilized egg of *C. elegans* and other studied nematodes does not express obvious signs of an inherent polarity. However, prior to first cleavage the egg develops a clear anterior-posterior (a-p) polarity. Polar bodies are extruded at the anterior pole while the oocyte pronucleus (Figure 8.9) and germline-specific granules (Figure 8.13; see also the next section) are translocated to the posterior region as microfilaments accumulate anteriorly. Recent observations in manipulated *C. elegans* embryos and in other species indicate that it is the entry point of the sperm that determines the posterior pole and thus the orientation of the a-p axis in the oval-shaped egg (Goldstein and Hird 1996). Only the germline continues to express an a-p polarity, as visualized by the asymmetric localization of germline granules here and the a-p orientation of its cleavage spindle (Schierenberg and Strome 1992).

The dorsal-ventral (d-v) axis of the developing embryo can be defined in the 4-cell stage because the side where the EMS cell is located will become ventral and the side of the ABp cell will become dorsal (Figure 8.9).

With the definition of two axes, the left-right (l-r) axis is formally fixed as well. For specification of cell fate in the *C. elegans* embryo, the position of a cell along the d-v or the l-r axis appears not to be critical, but differential segregation of cytoplasmic components and specific cell–cell interactions are essential (see the next section). The hatched animal shows a high degree of bilateral symmetry, like most animals. Lineage analysis revealed that basic l-r decisions are made very early within individual lineages. The first division in the MS, C, and D lineages and the second division in the AB and E lineages generate two daughter cells, one contributing to structures on the left side of the embryo and the other to corresponding structures on the right side (Figure 8.10; Wood 1988, Appendix 3; Schierenberg and Strome 1992). No bilateral symmetry is established in the germ lineage.

Soma/Germline Separation and Specification of Cell Fate

The most prominent feature of early embryogenesis in nematodes is the presence of a germ line and its stepwise separation from the soma via a series of unequal cleavages (Figures 8.9 and 8.10). This characteristic cleavage pattern goes along with at least two important processes (a) the specific segregation of cytoplasmic components into individual blastomeres, and (b) inductive interactions between germline and somatic cells.

Removal of even major portions of cytoplasm from the anterior part of the uncleaved zygote of *C. elegans* does not interfere with the typical series of unequal cleavages. However, if cytoplasm is removed from the *posterior* pole, the egg loses its potential to make any unequal division and to generate the full complement of differentiated cell types. This suggests that cytoplasmic components are present in the posterior region of the fertilized egg, and these components are necessary for the establishment of a germline (Schierenberg and Strome 1992). Looking for components being unequally distributed in the zygote and later germ cells, investigators have detected cytoplasmic granules (called **P granules**) with the help of antibodies. These granules are already present in the oocyte and after fertilization become segregated into the successive generations of the germline (P_1–P_4; Figures 8.10 and 8.13; Strome et al. 1994). The function of these granules is still obscure. They are apparently not responsible for the unequal, germline-specific cleavage but may turn out to be needed for the differentiation of definite germ cells. Some indication has been found that they are involved in the survival of certain RNAs in the germline cells which are degraded in the soma (Seydoux and Fire 1994). This may be achieved by the association of RNA-binding proteins with P granules (Draper et al. 1996).

Making use of antibodies that visualize the specific differentiation of cells, the classical view of autonomous cell determination in nematodes has been radically overturned during recent years. Experiments involving the ablation or removal of blastomeres in *C. elegans* revealed the importance of inductive cell-cell interactions for proper cell specification, particularly during early embryogenesis and then again during postembryonic development. When certain cells are eliminated, the developmental program of others is altered to a greater or lesser degree (Tax and Thomas 1994). However, above all, cells seem to carry an internal differentiation program that may be modulated or even overruled by external inducing or repressing signals.

Particularly obvious are interactions between early germline cells expressing an anterior-posterior polarity and their somatic sisters. Here inductive interactions can be interpreted as a polarization of the somatic cells, resulting in two daughter cells that then differ, for instance, in their abilities to translate the equally distributed maternal mRNAs (Priess 1994). When the interaction is prevented, the descendants of the affected cell do not become different from each other. An example for this is the interaction between P_2 and EMS. If the germline cell is eliminated in time, its somatic sister, instead of dividing with longitudinal spindle orientation into an anterior muscle/pharynx precursor (MS) and a posterior gut precursor (E), will cleave with transverse spindle orientation and generate two cells that both differentiate like an MS cell (Goldstein 1995).

All inductive interactions in the early embryo so far revealed seem to take place between two adjacent cells and affect decisions between two alternative fates. For other interactions studied during postembryonic development, this is not necessarily true. The choice can be between more than two fates, and in addition, several cells may respond even over some distance to the same signal.

For most cases of cell-cell interaction, the underlying mechanism is not known. However, the best-studied example has revealed that a cell surface receptor and ligand are involved there. This case is particularly interesting because the same receptor is needed twice during development in two different induction processes: (a) during early embryogenesis, to make the daughter cells of AB different from each other (Curtis 1994; Tax and Thomas 1994); and (b) during postembryonic development, to control the decision of germline cells (Clifford et al. 1994; Ellis and Kimble 1994) to either perform mitosis (i.e., increase their number) or to enter meiosis (i.e., to become haploid germ cells ready for fertilization). Here we will look only at the second event.

In the *C. elegans* hermaphrodite, at the end of each gonadal tube (made from somatic tissue) adjacent to the most immature oogonia sits a cell called the **distal tip cell** (dtc). If the dtc is eliminated, the mitotic germ cells will prematurely start meiosis (Figure 8.7). Thus, here a single cell acts on many others (which form a syncytium); this interaction is an inhibitory one, preventing germ cells from entering meiosis. It appears that a diffusible substance is released from the dtc, affecting cells nearby, while for those further away the signal has become diluted below a critical threshold.

In a search for mutants with defects in germ cell production one was found that showed exactly the same defect as after dtc ablation. The responsible gene (called *glp-1*, for *g*erm*l*ine *p*roliferation defective) was identified. An antibody was made against the protein for which it codes, and this antibody binds to the surface of oogonia. Together with additional data, this outcome indicates that *glp-1* encodes the receptor for a signal released from the dtc. It was very satisfying to find upon cloning of the gene a strong homology to genes in other organisms that also function as signal receptors (Artavanis-Tsakonas et al.

1995). Recently, a gene has been identified that seems to function by producing the complementary partner in this process, i.e., coding for the respective ligand (Henderson et al. 1994).

Not all cases of interactions affecting the determination of cells involve germline participation. An example, where information transfer takes place only between somatic cells, is the induction of the vulva, a slitlike structure of the hypodermis in the middle of the hermaphrodite body through which eggs are laid. Laser ablation experiments and the analysis of mutants suggested that induction takes place in two distinct steps. In the first of these, two equipotent cells of the gonadal tube compete with each other for becoming the so-called **anchor cell** whose function is to attach the gonad to the hypodermis, the "loser" becoming a uterine precursor. Remarkably, the molecules involved in this interaction are very closely related to those found in the mitosis/meiosis decision described above (Stern and deVore 1994).

In a second step, a signal released from the anchor cell can induce six neighboring hypodermis cells to become different types of vulva precursors. The signal acts in a position-dependent manner, comparable to that of the dtc, but here the decision is made between three different fates. The hypodermis cell closest to the anchor cell performs a specific cleavage and differentiation program called the *primary fate*, the two on both sides of it take *secondary fates*, and the other three potential vulva precursor cells positioned more distantly are not affected by the anchor cell under normal conditions (expressing the *tertiary fate*), probably because the signal has become too weak. Laser ablation experiments demonstrated that in fact the six hypodermis cells are equipotent, each of them being able to enter the primary fate pathway if positioned close enough to the anchor cell. They are therefore called an **equivalence group**. Further analysis revealed that additional signals are required for the induction and stabilization of vulva cell fates, for example, that the equivalent cells influence each others' choices (Sternberg 1993).

Sex Determination

In nematodes there are different modes of reproduction (see the section on life cycles), requiring different sexes. The developmental program leading to an individual with the one or the other sex probably involves the most complex individual regulatory network revealed so far.

In *C. elegans* the basic mechanism of sex determination is well understood and gives an excellent example of the power of genetic analysis. In the framework of this chapter only a cursory description and some basic conclusions can be presented. Readers interested in further details are referred to recent reviews (Hodgkin 1992; Ellis and Kimble 1994; Clifford et al. 1994; Wilkins 1995).

In the previous paragraphs several examples were given in which the manipulation of single cells revealed a specific interaction for the first time and the result then prompted a search for mutants in which this interaction was defective. In contrast, the understanding of the sex determination mechanism comes exclusively from genetic and molecular analysis. Unlike the typical scheme found in most metazoan species, the two sexes in *C. elegans* are male and hermaphrodite, which is believed to be a secondary specialization (see the section "Development: An Overview").

Hermaphrodites contain two sets of five autosomes plus two X chromosomes. Males contain the same number of autosomes but only one X. As quite a few genes that have nothing to do with sex determination are located on the X chromosome, gene dosage needs to be regulated. The failure to establish a correct dosage compensation is lethal. In *C. elegans* a down-regulation of X chromosome transcriptional activity in the hermaphrodite takes place, but in other systems different solutions for the same problem have been developed (see the next section).

Every cell, be it germline or soma, has either a male or a female sex. This enables cells to generate secondary sexual traits like a vulva or a copulatory bursa, but also less conspicuous sex-specific differences, e.g., in the nervous system. While it is obvious that germ cells in both sexes have to pass through alternative differentiation pathways, in the hermaphrodite gonad a regulative switch must be activated, because germ cells first differentiate in a male-specific way and only afterwards female-specifically.

A central question has been whether the different aspects of sex-related cell differentiation are closely related to each other or are independent. The answer is that dosage compensation, somatic sex and germline sex are coordinately regulated in gene cascades that start as a single common pathway but then diverge into three independent branches (Figure 8.14). After genes that affect sex determination had been identified by their mutation, the analysis of epistatic effects in double mutants helped to unravel the order in which these genes act.

The primary signal for all of them is the ratio of X chromosomes to the set of autosomes (X:A ratio). That it is not the absolute number of X chromosomes can be demonstrated by constructing strains with a different ploidy, e.g., triploids with 2 X chromosomes (X:A ratio of 0.66); they are male. The threshold ratio must be somewhere above 0.7 because tetraploids with 3 X chromosomes (ratio 0.75) develop into hermaphrodites. This primary signal sets the state for a few sequentially ordered master regulator genes. These in turn affect dosage compensation and a group of intermediate regulator genes different for somatic and germline sex, which then act on terminal regulator genes that finally determine sexual differentiation of cells. The control of transient sperm production in the hermaphrodite appears to involve the temporary repressive function of a gene acting only on intermediate regulators in the hermaphroditic germline (Figure 8.14).

Development and Evolution

During the past century the search for homologies on the level of morphological structures has been central to the establishment of a unifying theory of evolution as laid out by Charles Darwin. The data collected became overwhelming, leaving no room for any serious alternative hypothesis based on plausible scientific grounds to explain the origin and diversity of species. Suggestions for phylogenetic trees based on morphological data have given a useful framework for ideas concerning the course of evolution and for classifying newly identified species. However, limitations are set by the amount and kind of

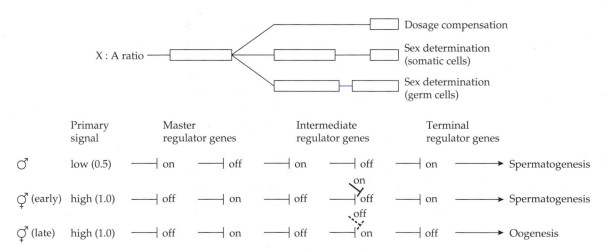

Figure 8.14 Sex determination in *C. elegans*. Upper part, general scheme for different pathways of dosage compensation and sex determination. Lower part, cascade of negatively regulated gene activities for sex determination in the germ cells of hermaphrodites and males. (After Hodgkin 1990.)

Acrobeloides maximus

$$P_0 - P_1 - P_2 - P_3 - AB - AB - EMS - AB - MS - E - C - AB$$

Caenorhabditis elegans

$$P_0 - AB - P_1 - AB - EMS - P_2 - AB - MS - E - C - P_3 - AB$$

Figure 8.15   Different sequence of early cleavages in two free-living soil nematodes. Note that only germline cells change their positions. See Figure 8.10 for cell names.

available data, complicating definitive interpretations. During the last decade, techniques of molecular biology have been developed that greatly simplify comparisons on the level of molecules. These new methods offer the chance to extend our understanding in cases where the classical approach has been unsatisfactory.

Besides structural elements, can differences and similarities in developmental events be found that are useful for phylogenetic considerations? Based on the description of *C. elegans* development, detailed comparative studies with other species have been initiated only recently. Here we will look at just two aspects, cell lineage and sex determination.

The complete cellular description of development in *C. elegans* (see Wood 1988) has raised many questions with respect to the economy of design. The lineage appears to consist of a patchwork of elements understandable only in the light of evolution. Why is the bilateral symmetry in the anterior AB cells achieved differently from the rest of the body (see the legend for Figure 8.10)? Why is the body muscle system derived from four different lineages? Why does just a single muscle precursor in the MS lineage divide postembryonically and quite extensively? Why are so many cells born just to die soon afterwards? In studying the lineage tree we get a frozen picture of ongoing evolution. Each division step, each cell fate is continuously exposed to selective pressure in an ever-changing environment. Therefore, cell lineages in different nematode species may be a good area to look for evolutionary modifications.

The sequence of cleavages, like the rest of development, is invariant in *C. elegans*. This goes along with a reproducible spatial positioning of cells. Comparative studies in early embryos of over a dozen other free-living nematodes revealed no difference in the division order of somatic cells. However, the timing of germline divisions varied

considerably between species. The most prominent difference from *C. elegans* is seen in the species *Acrobeloides maximus*, where all germline divisions occur first one after another before any somatic cells divide (Figure 8.15). The comparative studies revealed that the slower early development proceeds, the relatively faster the germline cell divisions are completed, supporting a model that germline cells must run through their typical division program in time before they prematurely lose their potential for asymmetric cleavage (Skiba and Schierenberg 1992).

The altered division order results in different spatial arrangements of cells. In addition, some other species—in contrast to *C. elegans*—show obvious variances in their arrangement of blastomeres and thus in cell contacts (Figure 8.16; M. Kutzowitz and E. Schierenberg, unpublished results). This strongly suggests that certain cell–cell inductive events as found in *C. elegans* (see the section "Establishing the Embryonic Axes") cannot—or at least not in the same way—take place in these nematodes.

Comparing postembryonic gonadal cell lineages of *Panagrellus redivivus* ("sour-paste nematode") with those of *C. elegans*, the observed differences suggest that the evolution of cell lineages involves just four types of modification (Sternberg and Horvitz 1981): (1) adoption of a cell fate normally associated with another cell; (2) altered segregation of developmental potential between two sister cells, resulting in a reciprocal switch in fate; (3) reversal in the polarity of a whole lineage branch; (4) alteration in the number of cell division rounds.

Another example, which illuminates the correlation between spatial pattern formation, lineage, and intercellular communication, is the comparative study of vulva development.

While many species—like *C. elegans*—have two gonadal arms and form their vulva in the center of the body, others have only one gonadal arm with their vulva positioned more toward the tail. The vulva in *C. elegans* is generated from three hypodermal cells that belong to a group of 12 cells equally spaced in a row along the body (see the section on soma/germline separation). As laser ablation experiments had shown that other members of this group are also competent to contribute to the vulva if close enough to the inducing anchor cell, it was to be expected that in those species with a posteriorly located vulva it was generated from more posterior members of these hypodermal cells. However, this turned out to be not the case. Instead, the same cells that make the vulva

Figure 8.16   Four different arrangements of blastomeres observed in the 6-cell stage of the free-living soil nematode *Cephalobus* sp., which expresses a sequence of cleavages similar to that of *Acrobeloides maximus* (Figure 8.15). See Figure 8.10 for cell names.

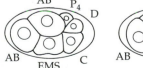 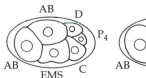 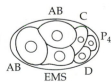 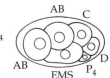

in *C. elegans* migrate to the posterior and execute there the typical vulva differentiation program (Sommer et al. 1994). One possible explanation for this unexpected finding could be that establishing hypodermis cells as a vulval equivalence group requires the action of at least one other gene; then a coincident shift in the expression domain of this gene would have induced additional, detrimental effects on other cells in the new area of action (Lambie 1994). The observed cellular rearrangement may indicate that the central position of the vulva (going along with a two-armed gonad) reflects a more primitive, original status.

Another surprising observation was made in conjunction with vulva formation in other nematodes. While some of the species behaved like *C. elegans* with the anchor cell being required for vulva induction, in others the ablation of the anchor cell and others had no effect on vulva formation (Sommer et al. 1994), suggesting that there the vulva develops without an external inducing signal. If this in fact is the case, it would be interesting to find out whether it represents a more original pattern or a later acquired modification.

As described in the section on sex determination, the sex determination pathway of *C. elegans* has been elucidated in its main components. Unfortunately, the lack of genetic data from other nematodes with different or varying distribution of sexes hardly allows any comparison with *C. elegans*. However, it is quite fascinating to compare the other well-understood system, *Drosophila*, in which gene cascades involved in sex determination have been revealed as they have in *C. elegans*.

The general organization of the sex determination pathways of the worm and the fly look suprisingly similar, both with the X:A ratio as the primary signal and then in having a hierarchical order of master, intermediate, and terminal regulator genes (Figure 8.14) controlling three different aspects: dosage compensation, sex in somatic cells, and sex in germline cells. However, the underlying molecular mechanisms differ fundamentally in several aspects (Hodgkin 1990). While in *C. elegans* gene dosage of the X chromosome is compensated by down-regulation in the hermaphrodite, there is up-regulation in the male of *Drosophila*. In *C. elegans* the interactions between genes in the sex determination cascade are negatively regulated, while in *Drosophila* these interactions are positively regulated. Therefore, the two pathways must have been evolved independently. Because of the striking similarity in general organization arrived at, this phenomenon has been termed "cybernetic convergence" (Hodgkin 1992). It seems likely that more examples of this kind will become apparent in the future.

A fundamental question is how complex developmental pathways like that for sex determination were established during evolution. Recently, a model has been suggested according to which this genetic pathway evolved in reverse order from the final step in the hierarchy to the first (Wilkins 1995).

Closing Remarks

In this chapter on nematodes only selected examples could be discussed. The study of nematodes has made many other important contributions in our understanding of development and will surely continue to do so. For instance, consider how the timing of various processes must be coordinated to give rise to a functional organism in the end. It is thought that temporal shifts play a central role in evolutionary progression. The analysis of heterochronic mutants in *C. elegans*, in which certain steps occur relatively too early or too late (Ambros and Moss 1994) offers the chance to understand how important the ordered sequence of events is, how it is controlled, and which variations were successfully established during evolution (see Figures 8.15 and 8.16).

Because of their relative simplicity and the fixed and fast developmental program, nematodes appear to be particularly suited to study in situ how organs are assembled and nerve cells find their targets. Many free-living nematodes live only for a number of weeks. What causes them to die? The study of long-lived mutants (Kenyon et al. 1993; Lakowski and Hekimi 1996) and changes in environmental parameters may help to better rationalize what factors influence the duration of life span.

Various aspects of nematode development are presently only studied in *C. elegans*, where appropriate methods are established and newly obtained results can be readily interpreted on a solid basis of already existing data. However, as it is not clear to what extent *C. elegans* can serve as a paradigm for nematodes as a whole, and to what extent those results have a more general relevance, it is necessary to extend the studies to other species as well. Comparisons with parasitic nematodes, for example, indicate that besides basic similarities, major differences exist, apparently due to adaptations to different environmental conditions. Comparative studies in free-living species could reveal why nematodes with different reproductive strategies (as described in the section on life cycles) live in close proximity in the soil without outcompeting each other.

In an international effort the genome of *C. elegans* is being sequenced; this project is expected to be completed in the near future. Sequencing will be a remarkable milestone, allowing quick physical availability and access to the structure of any newly identified gene. It will then be within the reach of imagination that all essential genes (estimated to be a few thousand) necessary for normal development of this nematode will be identified. The most intriguing question is: How much will the complete genetic information help us to understand what an organism is and does? Studies of gene function and interaction in *C. el-*

egans will likely be pursued as a standard. An extensive search for molecular homologies among nematodes and between these and other animal groups will broaden, perhaps even revolutionize, our perspective of developmental principles and phylogenetic relationships.

Finally, it may become understandable to the reader why nematodes, which may look rather unattractive at first glance, turn out to be the most fascinating organisms for many of those who became better acquainted with them.

Acknowledgments

I thank Randy Cassada for his helpful comments on the manuscript, Sam Ward for Figure 8.8, and Susan Strome for Figure 8.13.

Literature Cited

Ambros, V. and E. G. Moss. 1994. Heterochronic genes and the temporal control of *C. elegans* development. *Trends Genet.* 10: 123–127.

Artavanis-Tsakonas, S., K. Matsuno and M. E. Fortini. 1995. Notch signaling. *Science* 268: 225-232.

Bird, A. F and J. Bird. 1991. *The Structure of Nematodes.* Academic Press, San Diego.

Boveri, T. 1910. *Die Potenzen der Ascaris-Blastomeren bei abgeänderter Furchung.* Gustav Fischer Verlag, Jena.

Brenner, S. 1974. The genetics of *Caenorhabditis elegans. Genetics* 77: 71–94.

Bucher, E. A. and G. Seydoux. 1994. Gastrulation in the nematode *Caenorhabditis elegans. Sem. Dev. Biol.* 5: 121–130.

Clifford, R., R. Francis and T. Schedl. 1994. Somatic control of germ cell development in *Caenorhabditis elegans. Sem. Dev. Biol.* 5: 21–30.

Curtis, D. 1994. Translational repression as a conserved mechanism for regulation of embryonic polarity. *BioEssays* 16: 709–711.

Draper, B. W., C. C. Mello, B. Bowerman, J. Hardin and J. R. Priess. 1996. MEX-3 is a KH domain protein that regulates blastomere identity early in *C. elegans* embryos. *Cell* 87: 205–216.

Ellis, R. E. and J. Kimble. 1994. Control of germ cell differentiation in *Caenorhabditis elegans. Ciba Found. Symp.* 182: 179–192.

Epstein, H. and D. C. Shakes. 1995. *Methods in Cell Biology,* Vol. 48. Caenorhabditis elegans—*Modern Biological Analysis of an Organism.* Academic Press, San Diego.

Goldstein, B. 1995. An analysis of the response to gut induction in the *C. elegans* embryo. *Development* 121: 1227–1236.

Goldstein, B. and S. N. Hird. 1996. Specification of the anteroposterior axis in *Caenorhabditis elegans. Development* 122: 1467–1474.

Henderson, S. T., D. Gao, E. J. Lambie and J. Kimble. 1994. *lag-2* may encode a signaling ligand for the GLP-1 and LIN-12 receptors of *C. elegans. Development* 120: 2913–2924.

Hengartner, M. O. and H. R. Horvitz. 1994. The ins and outs of programmed cell death during *C. elegans* development. *Phil. Trans. Roy. Soc. London B.* 345: 243–246.

Hodgkin, J. 1990. Sex determination compared in *Drosophila* and *Caenorhabditis. Nature* 344: 721–728.

Hodgkin, J. 1992. Sex determination in the nematode *Caenorhabditis elegans. Sem. Dev. Biol.* 3: 307–317.

Johnstone, I. L. 1994. The cuticle of the nematode *Caenorhabditis elegans:* A complex collagen structure. *BioEssays* 16: 171–177.

Kenyon, C., J. Chang, E. Gensch, A. Rudner and R. Tabtiang. 1993. A *C. elegans* mutant that lives twice as long as wild type. *Nature* 366: 461–464.

Lakowski, B. and S. Hekimi. 1996. Determination of life-span by four clock genes. *Science* 272: 1010–1013.

Lambie, E. J. 1994. Variations on a vulval theme. *Curr. Biol.* 4: 1128–1130.

Lee, D. L. 1965. *The Physiology of Nematodes.* W. H. Freeman, San Francisco.

Malakhov, V. V. 1994. *Nematodes.* Smithsonian Institution Press, Washington, D.C.

Meglitsch, P. A. and T. R. Schram. 1991. *Invertebrate Zoology.* Oxford University Press, Oxford.

Priess, J. R. 1994. Establishment of initial asymmetry in early *Caenorhabditis elegans* embryos. *Curr. Opinion Genet. Dev.* 4: 563–568.

Riddle, D. L., T. Blumenthal, B. J. Meyer and J. R. Priess (eds.). 1997. *C. elegans II.* Cold Spring Harbor Laboratory Press, Cold Spring Harbor, NY.

Roberts, T. M. and M. Stewart. 1995. Nematode sperm locomotion. *Curr. Opinion Cell Biol.* 7: 13–17.

Schierenberg, E. 1986. Developmental strategies during early embryogenesis of *Caenorhabditis elegans. J. Embryol. Exp. Morphol.* 97 (suppl.): 31–44.

Schierenberg, E. and S. Strome. 1992. The establishment of embryonic axes and determination of cell fates in embryos of the nematode *Caenorhabditis elegans. Sem. Dev. Biol.* 3: 25–33.

Seydoux, G. and A. Fire. 1994. Soma–germline asymmetry in the distributions of embryonic RNAs in *Caenorhabditis elegans. Development* 120: 2823–2834.

Skiba, F. and E. Schierenberg. 1992. Cell lineages, developmental timing, and spatial pattern formation in embryos of free-living soil nematodes. *Dev. Biol.* 151: 597–610.

Sommer, R. J., L. K. Carta and P. W. Sternberg. 1994. The evolution of cell lineage in nematodes. *Development* (suppl.) 85–95.

Stern, M. J. and D. L. deVore. 1994. Extending and connecting signaling pathways in *C. elegans. Dev. Biol.* 166: 443–458.

Sternberg, P. W. 1993. Intercellular signaling and signal transduction in *C. elegans. Annu. Rev. Genet.* 17: 497–521.

Sternberg, P. W. and H. R. Horwitz. 1981. Gonadal cell lineages of the nematode *Panagrellus redivivus* and implications for evolution by the modification of cell lineage. *Dev. Biol.* 88: 147–166.

Strome, S. 1988. Generation of cell diversity during early embryogenesis in the nematode *Caenorhabditis elegans. Int. Rev. Cytol.* 114: 81–123.

Strome, S., C. Garvin, J. Paulsen, E. Capowski, P. Martin and M. Beanan. 1994. Specification and development of the germline in *Caenorhabditis elegans. Ciba Found. Symp.* 182: 31–51.

Tax, F. E. and J. H. Thomas. 1994. Receiving signals in the nematode embryo. *Curr. Biol.* 4: 914–916.

Vanfleteren, J. R., Y. van de Peer, M. Blaxter, A.A.R. Tweedie, C. Trotman, L. Lu, M.-L. van Hauwaert and L. Moens .1994. Molecular genealogy of some nematode taxa as based on cytochrome *c* and globin amino acid sequences. *Mol. Phylogen. Evol.* 3: 92-101.

Wilkins, A. S. 1993. *Genetic Analysis of Animal Development,* 2nd Ed. Wiley-Liss. New York.

Wilkins, A. S. 1995. Moving up the hierarchy: A hypothesis on the evolution of a genetic sex determination pathway. BioEssays 17: 71–77.

Wood, W. B. (ed.) 1988. *The Nematode* Caenorhabditis elegans. Cold Spring Harbor Laboratory. Cold Spring Harbor, NY.

Yuan, J. 1995. Molecular control of life and death. *Curr. Opinion Cell Biol.* 7: 211–214.

SECTION V

Bilateral Animal Phyla
Protostome Coelomates

CHAPTER 9

Nemerteans, the Ribbon Worms

Jonathan Henry and Mark Q. Martindale

MEMBERS OF THE PHYLUM NEMERTEA (ALSO KNOWN AS Nemertini, Nemertinae, or Rhynchocoela) are commonly referred to as the **ribbon worms** or **proboscis worms**, due to their characteristic shape, and to the presence of an eversible proboscis that is used in feeding. This phylum is comprised mainly of marine species, although there is one genus of freshwater forms (*Prostoma*), and a single genus of subtropical terrestrial species (*Geonemertes*).

The Basic Adult Nemertean Body Plan

Nemerteans are bilaterally symmetrical, unsegmented worms that possess a circulatory system and a complete digestive tract with a separate mouth and anus. The basic nemertean body plan is illustrated in Figure 9.1. The proboscis is located within a fluid-filled cavity called the **rhynchocoel**. This cavity opens anteriorly via a thin canal referred to as the **rhynchodaeum**, through which the proboscis is thrust. Muscular contraction and the resulting hydrostatic pressure serve to evert the proboscis. In many species the rhynchocoel opens separately from the mouth. In others (like the hoplonemerteans), the esophagus opens into the rhyncodeum. Some species (order: Hoplonemertea) have an armed proboscis that carries a pointed calcareous **stylet**. Venomous secretions help to subdue the captured prey and protect these worms from predation. Nemerteans move via ciliary gliding and rhythmic muscular contraction of the body. In many species it is clear that they travel along or through a secreted mucous trail. Excellent discussions of nemertean biology can be found by Gibson (1972), Brusca and Brusca (1990), and Turbeville (1996).

Previously it was thought that the nemerteans were **acoelomates**, lacking a true body cavity. Thus, some inves-tigators argued that the nemerteans are closely related to the flatworms (turbellarians). Nemerteans also possess **protonephridia** and in some species, rod-shaped secretory cells called **pseudocnids** or **rhabdoids**, which were thought to be homologous to **rhabdites** found in turbellarian flatworms. It is now clear that the nemerteans are, in fact, true **coelomates**. For instance, careful observations have revealed that the lateral blood vessels present in these organisms form within the mesoderm through a process of **schizocoely**. The rhyncocoel also forms in a similar manner. Thus, these structures represent true coelomic cavities (Turbeville and Ruppert 1985; Turbeville 1986). In addition, comparisons of 18S rRNA sequences indicate that the nemerteans are more closely related to the annelids and molluscs than the turbellarians (Turbeville et al. 1992).

Nemertean Taxonomy and Phylogeny

The phylum contains over 900 species (Gibson 1972), belonging to two major classes: the **Anopla** and the **Enopla** (Figure 9.2). Characteristics such as the location of the mouth, the organization of the body wall musculature and the nervous system, as well as the presence or absence of an armed proboscis, are used to subdivide this group into its various classes and orders. For instance, in the Anopla the mouth is located posterior to the brain, while the reverse is seen in the Enopla. The Anopla contains two principal orders: the **Paleonemertea** and the **Heteronemertea**. The Paleonemertea contain what have been argued by some investigators to be the more "primitive" species. This argument is based on certain aspects of their body plan, which share certain "primitive" features characteristic of other animal groups. These features re-

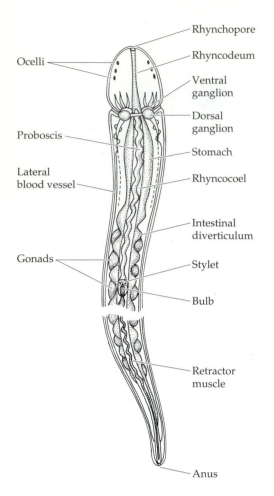

Ocelli

Proboscis

Lateral
blood vessel

Gonads

Rhynchopore

Rhyncodeum

Ventral
ganglion

Dorsal
ganglion

Stomach

Rhyncocoel

Intestinal
diverticulum

Stylet

Bulb

Retractor
muscle

Anus

Figure 9.1   Basic adult body plan of a typical hoplonemertean. The anterior end (head) is located upwards. While all nemerteans possess an eversible proboscis, only the hoplonemerteans have an armed proboscis bearing a stylet. Various structures are as labeled.

Hoplonemertea or "armed" nemerteans, which includes the freshwater and terrestrial species, and the **Bdellonemertea**, which includes a relatively small number of commensal species that live in the mantle cavity of certain molluscs or inside tunicates. The "armed" Hoplonemertea possess a calcareous, dagger-like sylet that is located at the end of the everted proboscis and is used to stab their prey. Basic nemertean taxonomy is illustrated in Figure 9.2 along with some typical distinguishing characteristics of the four orders.

Life History Strategies and Modes of Development

Development in this phylum has been described by a number of previous investigators (including: Hyman 1951; Gibson 1972; Friedrich 1979; Cantell 1989). Most nemerteans display direct development with the formation of a nonfeeding larva that resembles the adult. Some within the order Heteronemertea display indirect development with the formation of a planktonic, feeding larva (the **pilidium**) that undergoes a radical metamorphosis to give rise to the juvenile worm. A few species display a modified form of direct development that, in some respects, resembles the pilidium larva.

   Adult nemerteans have a tremendous capacity to regenerate missing parts. Generally, only anterior fragments containing the brain are able to regenerate the missing posterior portions. Some species, however, are able to reproduce by asexual fission. In these species any fragment that contains a portion of the brain or lateral nerve cord is able to replace the missing parts (Coe 1929, 1930).

late to the organization of the body wall musculature and the nervous system. Members of the Heteronemertea display indirect development involving the formation of a feeding, planktonic larval stage. This mode of development is considered by many to represent the typical ancestral life history for many metazoan phyla (Jägersten 1972; Willmer 1990).

   The Enopla class contains two principal orders: the

Figure 9.2   Current taxonomic classification within the Nemertea. Some special characteristics and the major mode of development are listed for each of the four nemertean orders.

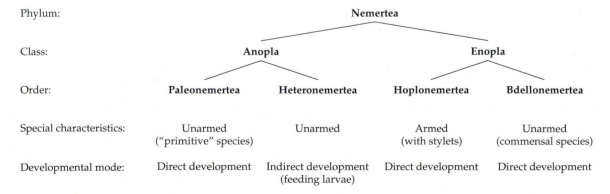

	Phylum:	**Nemertea**		
Class:		**Anopla**		**Enopla**
Order:	**Paleonemertea**	**Heteronemertea**	**Hoplonemertea**	**Bdellonemertea**
Special characteristics:	Unarmed ("primitive" species)	Unarmed	Armed (with stylets)	Unarmed (commensal species)
Developmental mode:	Direct development	Indirect development (feeding larvae)	Direct development	Direct development

Gametogenesis

Oogenesis

Most nemerteans are **dioecious** (having separate sexes); however, there are a few **hermaphroditic** species (particularly among the terrestrial and freshwater forms, such as *Prostoma*, and in some commensal species that occupy unusual ecological niches). Generally, the gonads are positioned laterally and lie between the diverticula of the digestive tract. Most nemerteans are **oviparous**, though some are **ovoviviparous** or **viviparous**. Many details regarding oogenesis have been described by previous investigators, including Riser (1974) and Bierne (1983).

The eggs may be shed directly into the surrounding water or deposited on various substrates in gelatinous masses. As the eggs are formed within the ovaries, they are endowed with an overt axial polarity, referred to as the **animal-vegetal axis**. Many nemertean species possess a small **peduncle** or stalk located at the vegetal pole, which apparently represents the point where the oocytes were connected in the ovaries (Wilson 1900). The opposite, animal pole represents the site where the polar bodies will form during the meiotic maturation divisions. The animal-vegetal axis bears a specific relationship to the early cleavage planes and the subsequent fates of the cells, as described below. In many species the eggs are surrounded by a extracellular **chorion** and a surrounding external **jelly coat**. Animal-vegetal polarity may also be evident in the shape of the chorion (see Figure 9.3). The jelly coat is very sticky in some species and functions, along with other secretions, to attach the eggs and embryos to the substrate.

Eggs range in size from 50 μm in *Procarinina remanei* (Nawitzki 1931) to 2.5 mm in *Dinonemertes investigatorus*, a pelagic nemertean (Coe 1926). Some calculations suggest that a large female of *Cerebratulus lacteus* may produce as many as one million oocytes in a single season (Coe 1943). There does not appear to be an absolute connection between egg size and the mode of development (see Table 9.1), although the larger egg sizes are associated with direct development and forms with indirect development appear to have smaller eggs. On the other hand, there are some direct-developing forms with very small eggs. A sufficient quantity of nutrients must be available to support the embryo through development. In *Lineus ruber*, groups of eggs are contained in a common sac (cocoon) and some of the eggs serve as **nurse eggs**, which are consumed by other developing individuals (Schmidt 1934). Interestingly, the congeneric species *L. viridis* does not display this "egg-eating" phenomenon; however, it forms larger eggs (Bartolomaeus 1984).

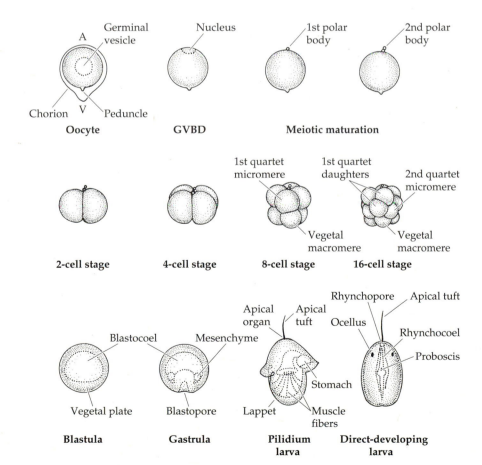

Figure 9.3 Embryo development, from the egg to the formation of various larval stages. Embryos are shown with animal poles directed upwards. Two different larvae, the pilidium and a direct-developing form, are shown with their apical ends directed upwards. Species forming a pilidium larva will undergo a radical metamorphosis to give rise to the juvenile worm (see Figure 9.8). On the other hand, the direct-developing larva is basically a miniature version of the juvenile worm (see Figures 9.6 and 9.7). A, animal; V, vegetal. Various structures are as labeled.

TABLE 8.1   *Comparison of Egg Size and Developmental Mode in Various Nemerteans*

Species[a]	Egg size (in μm)[b]	Developmental mode
Class Anopla, Order Paleonemertea		
Carinoma mutabilis	140	Direct
Cephalothrix rufifrons	105	Direct
Hubrechtella dubia	80–110	Pilidium
Procarinina remanei	50	Direct
Procephalothrix simulus	190	Direct
Tubulanus annulatus	120	Direct
T. polymorphus	350	Direct
T. punctatus	100	Direct
Class Anopla, Order Heteronemertea		
Cerebratulus lacteus	120	Pilidium
C. leydii	80	Pilidium
C. marginalis	140–160	Pilidium
Euborlasia elisabethea	140–150	Pilidium
Lineus lacteus	75	Pilidium
L. socialis	93–100	Pilidium + Asexual fission
L. torquatus	100	Pilidium
L. ruber	250	Schmidt's
L. viridis	300–400	Desor's
Micrura akkeshiensis	170	Iwata's
M. alaskensis	75	Pilidium
M. purpurea	100	Pilidium
Class Enopla, Order Hoplonemertea		
Amphiphorus formidabilis	250–350	Direct
A. incubator	1 mm	Direct
A. lactifloreus	150	Direct
A. ochraceus	200	Direct
Carcinonemertes epialti	70	Direct
C. carcinophila	58	Direct
Campbellonemertes sp.	200	Direct
Dinonemertes investigatorius	2.5 mm	Direct
Emplectonema gracile	80–170	Direct
Geonemertes agricola[a]	350–450	Direct
Gonomertes australiensis	180–200	Direct
G. parasita	250	Direct
Nemertopsis bivittata	100	Direct
Nipponnemertes pulcher	280–340	Direct
Paranemertes peregrina	300	Direct
Potamonemertes percivali	220	Direct
Prostoma graecense (rubrum)[a]	270	Direct
Tetrastemma caecum	300	Direct
T. grandis	200	Direct
T. phyllospadicola	400–500	Direct
Class Enopla, Order Bdellonemertea		
Malacobdella grossa	125–200	Direct

[a] All species listed are marine except for *Geonemertes agricola* (terrestrial) and *Prostoma graecense* (*rubrum*) (freshwater).

[b] Egg size given in μm except where mm specified.

Spermatogenesis

A number of investigators have examined the sperm of various nemerteans. Franzén (1956, 1983) has identified two main types of sperm. Those species that shed their sperm and eggs freely into the surrounding water display the more typical morphology, with small sperm heads and midbodies attached to an elongated flagella (i.e., *C. lacteus*). Together the head and midpiece may reach 8 μm in length in these species. Other species produce a modified sperm that possesses an elongated head and midpiece. For instance, the length of the head and midpiece totals 25 μm in the sperm of *L. ruber* (Franzen 1956). This "modified" type is found in species which do not release free-swimming sperm into the surrounding water, but instead fertilize their eggs internally or in close proximity to the adults while encased in mucous sheaths. All sperm appear to display the typical 9 + 2 arrangement of microtubules in the axoneme (Franzén 1983).

Ultrastructural examinations of nemertean sperm reveal that they contain a typical acrosomal vesicle located at the tip of the sperm head (Turbeville and Ruppert 1985; Stricker and Cavey 1986). However, in one species, *C. lacteus*, a distinct acrosomal reaction does not appear to take place at fertilization (Clark and Hinsch 1969).

Fertilization

Modes of Fertilization

Fertilization generally occurs externally. In some species like *L. ruber* and *L. viridis* a form of external "pseudocopulation" occurs by close contact of two or more individuals, typically inside a secreted mucous mass or cocoon (Franzén 1956; Cantell 1989). In such cases, sperm may make their way into the ovaries where internal fertilization can take place. In some forms accessory organs are involved in copulation. For instance, the bathypelagic (deep ocean dwelling) *Nectonemertes* has a pair of cephalic clasping tentacles and eversible penes that are thought to be involved in facilitating internal fertilization (Gibson 1972). In another pelagic nemertean, *Plotonemertes adhaerens*, the males have a pair of external sucker-like adhesive organs that may function in maintaining contact with the female. Obviously, there are a number of species with viviparous or ovoviviparous development that display internal fertilization (see Table 9.1).

In the case of *C. lacteus*, unfertilized eggs are released directly into the seawater. Germinal vesicle breakdown (GVBD) is initiated as a result of contact with the seawater. Following GVBD, the meiotic maturation divisions begin but the eggs become arrested in metaphase of the first meiotic division. After fertilization, the maturation divisions are completed and the polar bodies are produced at the animal pole (Freeman 1978).

Egg Activation and the Fast Block to Polyspermy

Like many other organisms, the nemertean egg cell membrane becomes electrically polarized as a result of fertilization and egg activation (Kline et al. 1986; Jaffe et al. 1986). The egg cell membrane of *C. lacteus* shifts from a negative resting potential to a positive one that persists for approximately one hour following fertilization (Kline et al. 1985). Experiments have demonstrated that this electrical **fast block to polyspermy** prevents additional sperm from entering the zygote until erection of more permanent nonelectrical barriers involving changes in the

cell membrane or **glycocalyx**. Studies indicate that an artificially imposed positive membrane potential will inhibit oocytes from becoming fertilized. Likewise, if the positive membrane potential is reversed by placing zygotes in low Na$^+$ seawater, they can become **polyspermic** (Kline et al. 1985). Nemerteans do not appear to elevate a **vitelline envelope** as a result of fertilization, and the chorion that surrounds the oocyte (like that of *C. lacteus*) is freely penetrated by the sperm.

Cleavage Pattern and Cell Lineage

Cleavage Pattern

Cell division proceeds in a regular and stereotypic fashion. Cleavage is **spiral** as in many other protostome phyla including the annelids, molluscs, and some turbellarian platyhelminths. The first and second cleavage divisions are meridional and occur along the animal-vegetal axis (see Figure 9.3). The second cleavage plane is orthogonal to the first. As both the first and second cleavage divisions are typically equal, all blastomeres in the resulting 4-celled embryo are the same size and lie an a single plane. Unlike the annelids and molluscs, there is usually no **cross-furrow** separating opposite blastomeres at the animal and vegetal poles (see Figure 9.4). Instead, all four cells meet together along a single line conforming to the animal-vegetal axis (Henry and Martindale 1994a). These four cells form the basis upon which subsequent development is organized. There are reports of a few nemerteans that display unequal first and second cleavage divisions. In the case of unequal-cleaving annelid and mollusc embryos, investigators have referred to the four cell quadrants as A, B, C, and D (Nusbaum and Oxner 1913; Dawydoff 1928). Typically, however, one cannot distinguish the identities of the four cell quadrants until much later during development. The third cleavage division is oblique to the animal-vegetal axis, and generates four animal blastomeres, referred to collectively as the **first quartet**. By convention these cells are referred to as **"micromeres"**; however, unlike the case seen in the annelids and molluscs, the first quartet cells of most nemerteans are typically larger than the four vegetal **macromere** cells due to the subequatorial plane of third cleavage (see Figures 9.3 and 9.4). This situation means that over half of the mass of the embryo is contained in the progeny of the first quartet. The spiral nature of the cleavage pattern is clearly evident in the 8-celled embryo. The first quartet is situated in a clockwise (**dextral**) direction relative to the vegetal macromeres. Subsequent quartets (second, third, etc.) are given off in alternating counter-clockwise (**sinistral**) and clockwise directions. This shifting arrangement results from the alternating oblique orientation of the cleavage spindles, the basis of the spiral cleavage pattern. Together the cells of the four quadrants appear to be arranged in a spiral pattern when viewed along the animal-vegetal axis (refer to Figures 9.3 and 9.4).

Figure 9.4 The developmental comparison of nemerteans with annelids and molluscs. The diagram illustrates the relationship between the first and second cleavage planes, and the future dorsoventral axis in nemerteans as compared to that of annelids and molluscs. Shading labels the progeny of one cell of the 2-celled embryo through the 8-cell stage, which serves to demarcate the plane of the first cleavage division. Dashed lines indicate the relative positions of the dorsoventral axis (labeled D–V). (*a*) In the case of the nemerteans, the dorsoventral axis corresponds closely to the plane of the first cleavage division in half of the cases examined; it runs at right angles to the first cleavage plane in the rest, as indicated by the progeny of the very large first quartet micromeres shown in the 8-cell-stage embryo. All views are from the animal pole, which is marked by the location of the polar bodies (small open circles). (*b*) In the case of other spiralians, such as the molluscs, the dorsoventral axis always runs at a 45° angle oblique to the first cleavage plane. The position of the first quartet of micromeres relative to the four vegetal macromeres and the location of the cross-furrows in typical annelid and mollusc embryos are shown. cf, vegetal cross-furrow cell; ncf, vegetal non-cross-furrow cell; 1q, first quartet micromere; M, macromere. (After Henry and Martindale 1994a.)

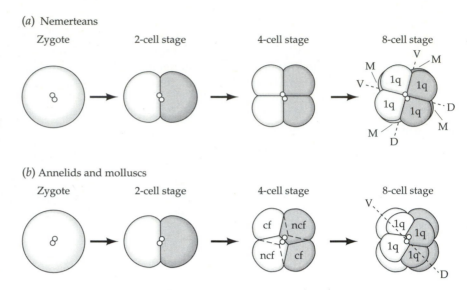

Cell Fate Map

The relationship between the first cleavage plane and future embryonic axes, such as the dorsoventral axis, has been examined in two species of nemerteans, the heteronemertean, *C. lacteus*, and the hoplonemertean, *Nemertopsis bivittata* (Henry and Martindale 1994a). This was accomplished by injecting one cell at the 2-cell stage with a "cell-autonomous" lineage tracer, a high-molecular-weight dye that is subsequently inherited by all of the progeny of the injected cell. As the boundary between labeled and unlabeled regions of the embryo and larva corresponds to the first cleavage plane, one can determine its relationship to the various

In equal-cleaving mollusc embryos, the cross-furrow separating the macromeres (see Figure 9.4) actually plays a key role in biasing the ultimate fates of these cells, one of which will ultimately be selected as the dorsal (D) macromere (van den Biggelaar and Guerrier 1983; also refer to the chapter on molluscan development). While all four of the vegetal macromeres can become the D quadrant, typically it is one of the two vegetal cross-furrow cells, because they occupy a more central position in the embryo. This position is favorable for establishing cell contacts with first quartet micromeres that determine the D macromere. This process of cell determination, however, does not appear to operate in nemertean embryos (see the discussion later in this section).

embryonic axes at later stages of development. When one examines the labeled larval ectoderm (mainly derived from the large first quartet micromeres) it was found in both *N. bivittata* and *C. lacteus* that the first cleavage plane corresponds to the plane of bilateral symmetry in roughly half of the cases, and bisects the dorsoventral axis in the rest (refer to Figure 9.4). As the second cleavage plane is orthogonal to the first, the resulting 4-celled embryos always contain the same discrete four cell quadrants, which have been referred to as the left and right ventral, and left and right dorsal cell quadrants. In the annelids and molluscs the first cleavage division bears a 45-degree, oblique orientation relative to the future dorsoventral axis (see Figure 9.4). Thus, one cell quadrant contributes to dorsal fates, one to ventral fates, and the others to left and right lateral fates. As explained below, the fates of the four cell quadrants in nemerteans are actually similar to, but not precisely the same as, those present in annelids and molluscs.

The ultimate fates of the four cell quadrants have been carefully examined in *C. lacteus*, where individual blastomeres have been traced through development to the formation of the pilidium larva (Hörstadius 1971; Henry and Martindale 1994a,b, 1996a, and in preparation). The contributions of each cell in the 8-celled embryo are illustrated in Figure 9.5. Most of the larval ectoderm is derived from the large first quartet micromeres. The ectodermal clones generated by these cells are organized primarily in a bilaterally symmetrical fashion. Thus, the four cell quadrants in *C. lacteus* have been referred to as the left and right ventral quadrants (LV, RV), and the right and left dorsal quadrants (RD, LD). Examination of the progeny of subsequent micromere quartets has revealed that these four cell quadrants can, in fact, be closely related to the A, B, C, and D cell quadrants, respectively, of other spiralians (Henry and Martindale 1996a and in preparation). All four cell quadrants contribute to the formation of the apical organ and

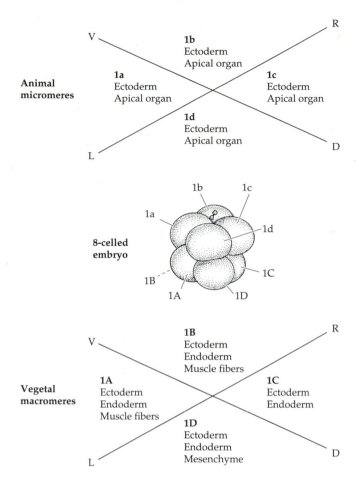

Figure 9.5 Simplified fate map relating the ultimate contributions of each blastomere in the 8-cell-stage embryo of *Cerebratulus lacteus* to various cell fates in the larva. To some extent cell fates are organized in a bilaterally symmetrical fashion, such as the larval ectoderm, and the larval musculature. The embryo contains four cell quadrants, A, B, C, and D, which correspond to the left and right ventral quadrants (LV, RV) and the right and left dorsal quadrants (RD, LD), respectively, of Henry and Martindale (1994a). Animal micromeres (the first quartet) are labeled in lowercase within each of the four-cell quadrants (i.e., 1a, 1b, 1c, 1d). Vegetal macromeres are labeled in uppercase letters (i.e., 1A, 1B, 1C, 1D). All four quadrants contribute equally to the formation of ectoderm, apical organ, and endoderm, while only the A and B quadrants generate larval muscle fiber cells (actually from their third quartet micromeres 3a, 3b). Notice, however, that the D quadrant generates a separate population of mesenchyme cells (actually from 4d), while the C quadant does not contribute to mesoderm formation.

imal-vegetal axis in the 16-celled embryo. He referred to these tiers of cells (arranged in order from the animal to the vegetal pole) as An1, An2, Veg1, and Veg2, respectively, following the nomenclature he used to label the cells present in sea urchin embryos. He observed that the animal derivatives of the first quartet of micromeres give rise to the apical tuft and most of the surface ectoderm. The vegetalmost derivatives of the first quartet give rise to a small portion of the surface ectoderm and most of the ciliated band located at the edges of the **lappets** (ciliated flaps extending posteriorly from the side of the mouth; see Figure 9.6*b*). The second quartet of micromeres gives rise to a small portion of the ciliated band on the inner surface, as well as ectoderm located on the inner surface of the lappets. In addition, these cells contribute to the formation of the esophagus. The vegetal macromeres contribute mainly to the formation of the endoderm and to a small portion of the esophagus.

The literature contains a great deal of discussion regarding the exact origins of the mesoderm in nemerteans (see discussions by Cantell 1989; Henry and Martindale 1994a). This point is of great significance in terms of understanding the relationship of this group to that of other spiralian phyla such as the annelids and molluscs. In the case of the annelids and molluscs, the mesoderm appears to arise from two sources, the **endomesoderm** and the **ectomesoderm** (Verdonk and van den Biggelaar, 1983). Most of the adult visceral mesoderm arises from the fourth quartet micromere of the D cell quadrant, the so-called **mesentoblast** cell (4d), which also gives rise to the intestine (thus the term *endomesoderm* or *mesentoblast*). In many species the fourth quartet cell of the D quadrant is distinctive. It may be of a greater size and arise before the fourth quartet micromeres of the other three cell quadrants. The mesentoblast subsequently divides in a bilateral fashion and gives rise to two (left and right) mesodermal, germinal bandlets. Ectomesoderm is generally derived from second and sometimes the third quartet micromeres, which also contribute to the formation of the ectoderm.

tuft. All four quadrants generate roughly one quarter of the surface ectoderm, and all four quadrants contribute to the formation of the gut. While the derivatives of the first and third quartets assume symmetrical positions relative to the median plane, the derivatives of the second and fourth quartets assume positions more typical of those generated by the A, B, C, and D cell quadrants of other spiralians (i.e., left, ventral, right and dorsal, respectively). A specific nomenclature has been developed to designate the progeny of the four cell quadrants, which follows that used for other spiralians. By convention, quartets of micromeres are labeled with lowercase letters, while the vegetal macromeres are labeled with uppercase letters, as in annelid and mollusc embryos. Thus, in nemerteans the first quartet cells are 1a, 1b, 1c, and 1d, while the corresponding macromeres are 1A, 1B, 1C, and 1D (see Figure 9.5). Subsequent quartets and their progeny are labeled using similar designations as in the annelids and molluscs. For example, as the first quartet of micromeres divide, they give rise to a set of four animal progeny designated with the superscript 1 ($1a^1$, $1b^1$, $1c^1$, $1d^1$), while the more vegetal progeny are designated with the superscript 2 ($1a^2$, $1b^2$, $1c^2$, $1d^2$).

Hörstadius (1937) examined the prospective contribution of each of the four tiers of cells arranged along the an-

Different reports would seem to suggest the the mesoderm in nemerteans may arise from different sources depending on the species examined. These reports include claims that the mesoderm is formed in a manner similar to that found in annelids and molluscs, as described earlier (Nusbaum and Oxner 1913) or that the mesoderm is generated strictly as ectomesoderm from two to all four cell quadrants (Lebedinsky 1897; Wilson 1900; Iwata 1957). Recent technical advances have now made it possible to carefully address this issue. Henry and Martindale (1994b, 1996a, and in preparation) have shown that the elaborate system of larval musculature in C. lacteus is, in fact, derived in a bilaterally symmetrical fashion from two of the four cell quadrants, specifically the third quartet micromeres of the A and B quadrants (3a and 3b, see Figure 9.5). This mesodermal population is generated as ectomesoderm. Additional loosely organized bandlets of mesenchyme cells are formed by the fourth quartet micromere of the D quadrant (4d), in a manner similar to that found in other spiralians. This mesoderm is generated as endomesoderm. The exact source of the adult mesoderm has not been clearly established, but may be derived from a dormant population of endomesoderm.

Blastula Formation

Following the early cleavage divisions, a hollow blastula-stage embryo (referred to as a **coeloblastula**) is generally reached (Figure 9.3). The central fluid-filled **blastocoel** begins its formation during early cleavage from the small cleavage cavities located between adjacent blastomeres. Some species form a solid **stereoblastula** instead. The blastula-stage embryos of nemerteans are generally covered with cilia. In addition, a tuft of elongated cilia (called the **apical tuft**) is usually present at the apical end (animal pole) extending from the apical organ, a small collection of elongated cells. Generally the cells at the opposite, vegetal end of the embryo are also elongated, and together they comprise what may be referred to as the **vegetal plate**.

In C. lacteus, mesodermal ingression begins during blastula stages and first appears within the blastocoel as paired populations of large ectomesodermal cells lying on opposite sides of the **blastopore**. These cells subsequently divide to give rise to a large population of wandering mesenchyme.

At this stage of development, some indirect-developing species such as Micrura alaskensis hatch from their surrounding chorion (Stricker 1987). Other species such as C. lacteus hatch during gastrula stages. As the embryos swim free in the water, they revolve around the animal-vegetal axis, with the animal pole (i.e., site of the apical organ and tuft) facing in the forward direction. Some direct-developing nemerteans such as N. bivittata do not hatch until much later in development after the larva has basically achieved the juvenile form (Henry and Martindale 1994a).

Gastrulation

Gastrulation generally occurs via invagination at the vegetal pole of the embryo to ultimately give rise to the endoderm (i.e., C. lacteus; Wilson 1900), although in some species gastrulation occurs via cellular ingression. For instance, gastrulation in C. marginatus occurs via a combination of invagination and unipolar ingression at the vegetal pole (Coe 1899). Invagination of the vegetal plate ultimately establishes the **archenteron** (gut/digestive tract), which is initially open via the **blastopore** at the vegetal pole (Figure 9.3). In many cases the blastopore subsequently closes and the mouth (**stomodeum**) forms as a secondary invagination on the ventral surface of the embryo at or close to the site of the blastopore. Thus, the blastopore effectively ends up forming the junction between the stomodeum and the midgut.

In direct-developing species, the ectoderm expands on the dorsal side; the original location of blastopore and ultimately the mouth are shifted anteriorly along the ventral side of the embryo. There is a tremendous proliferation and shifting of the ectoderm from the dorsal to the ventral surface in some embryos such as those of N. bivittata (Henry and Martindale 1994a). Such movements of the blastopore and development of the stomodeum and foregut are discussed by Iwata (1985) for a number of different species, and these observations have served as the basis of an alternative taxonomic classification (Iwata 1960a, 1985), which has not been widely accepted.

Larval Development and Metamorphosis

Direct vs. Indirect Development

Four main forms of development have been observed in nemerteans. Most species display direct development with the formation of a swimming stage that resembles the adult worm. Some of these species do not feed during this planktonic phase, while other species do appear to feed before settling to an adult, benthic existence (Iwata, 1960b; Jägerston 1972). One must be careful in considering the term *direct development*. Here the term is used to denote a path of development that leads to the establishment of the adult body plan (bauplan) *without* the formation of an intermediate feeding larval stage that possesses a distinct larval body plan.

A number of species (primarily within the order Heteronemertea) display indirect development with the formation of a feeding planktonic larva called the *pilidium*, within which the juvenile worm develops from a series of **imaginal discs**. Finally, there are a few species that display a form of development that appears to be closely related to the pilidium larva, and includes the formation of either **Desor's** or **Iwata's** larva (Desor 1848; Iwata 1958), each of which also possess imaginal discs. Some argue that these larvae represent intermediates between the direct- and indirect-developing forms while others believe these represent developmental modes derived from the pilidium larval condition.

Direct Development

Direct-developing species (Paleonemertea, Hoplonemertea, and Bdellonemertea) form a swimming stage ("larva") that basically resembles the adult (Figure 9.6a). The gastrula-stage embryo undergoes morphogenesis to give rise to an elongated, vermiform stage, which possesses a rhynchocoel, a proboscis, and a digestive tract, which may or may not be functional in the earlier stages of development (Figure 9.7). The rhynchodeum forms as a result of invagination of the ectoderm. The stomodeum (mouth) usually forms as a secondary invagination at or close to the site of the blastopore. Differential growth of the ectoderm typically moves the ultimate position of the mouth anteriorly along the ventral surface. The mouth may remain open externally, or the esophagus may subsequently fuse with the rhyncodeum, as in the case of *Malacobdella grossa* (Hammarsten 1918). The sequence and rate at which various structures appear during development differs depending on the species examined. The rhyncocoel and lateral blood vessels form within the mesoderm through a process of schizocoely, and represent components of the coelomic cavities (Turbeville and Ruppert 1985; Turbeville 1986). While direct-developing "larvae" are anatomically similar to the adult worms, they may also possess certain characteristics that are lost during settling. These features include an apical organ and an elon-

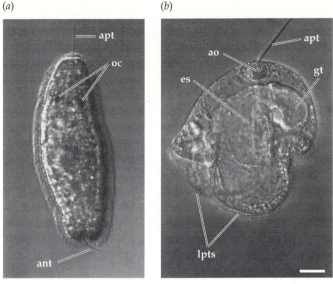

Figure 9.6 Photomicrographs showing the two main forms of development seen in nemerteans. (*a*) Free-swimming stage of development in the direct-developing hoplonemertean *N. bivittata*. This specimen is a five-day-old nonfeeding stage after hatching from the chorion. Note that there is an anterior apical ciliary tuft (apt), and a posteriorly located anal ciliary tuft (ant). Two pigmented ocelli (oc) can be seen in the anterior end. This example is viewed from the dorsal surface, with the anterior end located upward. (*b*) Free-swimming pilidium larva of the heteronemertean *C. lacteus*. This specimen is a 3-day-old feeding larva. Note the presence of the anterior, apical ciliary tuft (apt) extending from the apical organ (ao), and the two large posterior ciliated lappets (lpts). The esophagus (es) and blind gut (gt) are also clearly seen. This larva is viewed from the left lateral surface, with the anterior end located upward and the ventral side to the left. Scale bar = 25 μm.

Figure 9.7 Four stages of direct development. (*a*) Gastrula-stage embryo showing the invaginated archenteron (ar) that will ultimately give rise to the gut. The blastopore (bl), located at the vegetal pole of the animal-vegetal axis is still open at this stage. Numerous mesenchyme cells (ms) are located within the blastocoel. The thickened, apical organ (ao) is located within the ectoderm at the animal pole. An elongated apical tuft (at) of cilia emerges from the cells of the apical organ. (*b*) Somewhat older embryo after expansion of the dorsal ectoderm has begun to reposition the blastopore to the ventral side of the embryo. The endoderm (en) is seen enclosing the archenteron.

(*c*) Embryo after closure of the blastopore has occurred. The stomodeum (st) or mouth is seen as an invagination of the ectoderm on the ventral surface of the embryo. (*d*) Older swimming stage after the stomodeum (st) has formed a connection with the developing gut (gt). The proboscis rudiment (pr) is shown as an invagination of anterior ectoderm. The proboscis lies within the rhyncocoel (rc), which has formed within the mesoderm (ms). The apical tuft is no longer present at this stage. ec, ectoderm. The adult axes shown in (*d*) apply for all four stages illustrated; the axes are labeled A, anterior; P, posterior; V, ventral; and D, dorsal.

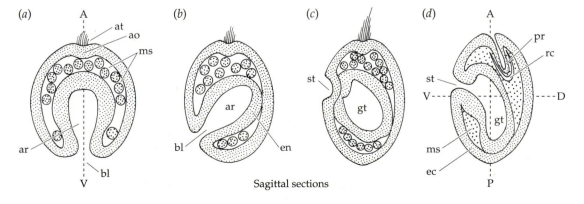

Sagittal sections

gated apical ciliary tuft at the anterior end, and in some species, there may also be a small posterior anal tuft of cilia (Figure 9.6a), lateral ciliated tufts, and a temporary "larval" cuticle. The settling and subtle metamorphosis of the juvenile worms are apparently under the control of certain biological cues that are poorly understood.

Pilidium Larva

Indirect-developing species (Heteronemertea) form a feeding larva referred to as the *pilidium* (first described by Müller 1847). This helmet-shaped larva contains a prominent elongated apical tuft of cilia that is formed by the apical organ (Figures 9.6b, 9.8). This latter structure consists of a small collection of elongated cells that appears to serve a sensory and locomotory functions. At this stage of development the gut is blind, as there is no anus. Two lateral ciliated flaps, called lappets, extend posteriorly from the sides of the mouth. A band of cilia that serves in loco-

motion and feeding extends around the edges of the lappets and the stomodeum. The larvae also possess a simple nervous system (Lacalli and West 1985). The larvae swim in the plankton where they feed for a prolonged period of time while the juvenile worm develops inside from a set of imaginal discs (Figure 9.8), which were first described by Metschnikoff (1869). These imaginal discs form primarily as invaginations of the larval ectoderm and give rise to the adult epidermis as well as various other ectodermal structures. The internal organs of the adult are derived from the mesoderm and endoderm of the larva. The first ectodermal invaginations to form are the paired cephalic discs. A pair of trunk discs also form along with a small pair of cerebral invaginations that contribute to the formation of the nervous system. A single dorsal imaginal disc also develops, but this originates through a process of delamination. Thus, there are a total of seven imaginal discs that contribute to the formation of

Figure 9.8 Diagram illustrating the formation of the juvenile worm within the pilidium larva of *Cerebratulus lacteus*. Both apical as well as left lateral views are shown for each of three different stages (*a–c*). The apical organ (ao), apical tuft (at) and ciliated lappet (lp) of the pilidium larvae are labeled. The juvenile worm forms from seven ectodermally derived imaginal discs. These include the bilaterally paired cephalic discs (cd), the paired cerebral discs (bd), the paired trunk discs (td) and the unpaired dorsal disc (dd). These discs eventually fuse to give rise to the ectoderm of the juvenile worm, while the larval gut (gt) and mesoderm contribute to the formation of internal structures. As the imaginal discs fuse they also create an outer amnion layer (am) that encases the juvenile

worm within the so-called amniotic cavity (ac). The proboscis (pr) forms from the fused cephalic imaginal discs. A pair of pigmented ocelli (oc) can be seen in the juvenile worm shown in (*c*). Ultimately the worm will emerge from its larval case during the final stages of metamorphosis. The larval axes are indicated in *a* and the adult axes are indicated in *c*: LV and AV, larval ventral and adult ventral; LD and AD, larval dorsal and adult dorsal; LL and AL, larval left and adult left; LR and AR, larval right and adult right; LA and AA, larval anterior and adult anterior; LP and AP, larval posterior and adult posterior. Note that the anteroposterior axis of the adult is shifted by approximately 90 degrees relative to its position in the larva. [Compare *a* and *c*.]

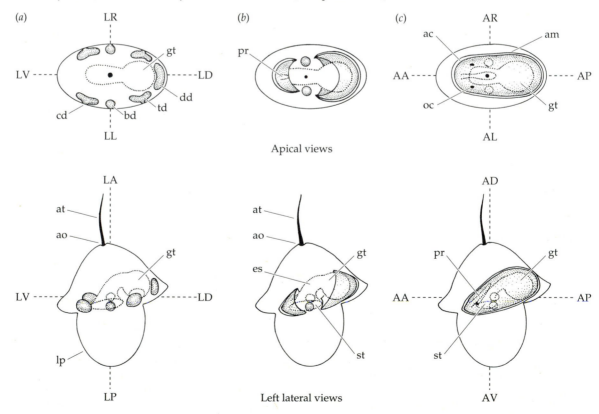

the juvenile worm in the pilidium larva (Salensky 1886, 1912). The two cephalic discs fuse to form the anterior end of the juvenile worm. The paired trunk discs fuse with the dorsal imaginal disc and the two cerebral discs. As these discs enlarge, they surround the digestive tract and eventually fuse. The inner surface of these imaginal discs gives rise to the larval ectoderm, while the outer surface (**amnion**) encloses the juvenile worm in a fluid-filled space referred to as the **amniotic cavity**. Most of the ectoderm is actually formed by the dorsal disc and the two trunk discs. The head ectoderm and the proboscis is derived from the fused cephalic discs. These discs also give rise to the cerebral ganglia. The larval gut is utilized in the development of the juvenile digestive tract. The mesoderm contributes to the formation of the muscula-ture, the blood vascular system, and the rhyncocoel. The cerebral organs are derived from the paired cerebral discs. This process is diagrammed in Figure 9.8 for *C. lacteus*. As the juvenile worm develops within the pilidium larva, the plane of bilateral symmetry is conserved; however, the anteroposterior and dorsoventral axes are shifted by 90° relative to those of the larva. After a prolonged period of time feeding in the plankton, the juvenile worm emerges via a dramatic metamorphic process. It has been reported that in many cases the juvenile worm devours the larval tissues during metamorphosis (Cantell 1966).

Desor's Larva

Some heteronemerteans such as *Lineus viridis*, *L. ruber*, and *Micrura akkeshiensis* produce larvae that appear to be close-ly related to the pilidium larva. These are the Desor's larva, Schmidt's larva, and Iwata's larva, respectively (Desor 1848; Iwata 1958). The Desor's larva is similar to the pilidi-um larva in that it, too, develops a series of imaginal discs that contribute to the formation of the juvenile worm (Fig-ure 9.9). On the other hand, the Desor's larva is a nonfeed-ing larva and it does not possess other features that charac-terize the pilidium larva, such as the ciliated lappets, an apical plate or apical tuft. The Desor's larva forms a total of eight imaginal discs, due to the addition of a small imagi-nal invagination that gives rise to the proboscis (Nusbaum and Oxner 1913; Gontcharoff 1960). The arrangement of these discs is the same as those in the pilidium larva (Fig-ure 9.9). As is seen in the pilidium larva, the axis of the adult worm appears to form at a right angle to that of the embryonic animal-vegetal axis. The definitive mouth of the juvenile worm ultimately forms as a secondary opening. The unpaired dorsal disc forms from a process of delami-nation, and after the formation of the imaginal discs takes place, an amnionic layer is not retained. The juvenile worm continues to develop within the larval integument, which is completely shed at metamorphosis.

While the Desor's larva of *L. viridis* is nonfeeding, *L.*

Figure 9.9 Diagram illustrating the formation of the Desor's larva. Both ventral as well as left lateral views are shown for each of three different stages (*a–c*). The juvenile worm forms from eight ectodermally derived imaginal discs. Similar to the case seen in the pilidium larva, these include the paired cephalic discs (cd), the paired cerebral discs (bd), the paired trunk dics (td), and an un-paired dorsal disc (dd). These discs eventually fuse and give rise to the ectoderm (ae) of the juvenile worm. Unlike the case of the pilid-ium larva, the proboscis (pr) forms from a separate imaginal rudi-ment or disc (pd). The gut (gt) and stomodeum (st) are also shown. An outer amnion layer is not established in the case of the Desor's larva. Ultimately, the worm will emerge from the larval integument (le) during the final stages of metamorphosis. The adult axes are indicated in *a* and apply to all three stages shown. A, anterior; P, posterior; L, left; R, right; D, dorsal; V, ventral. Note that the larval and adult axes are the same.

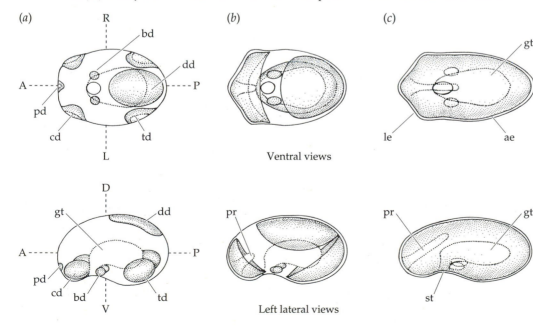

ruber forms a Desor's larva, which feasts on the undeveloped eggs ("nurse eggs") or on aborted embryos that share a common cocoon (Schmidt 1934). This unusual feeding larval stage is referred to as **Schmidt's larva**, after the investigator who first described this behavior. The eggs of *L. viridis* are much larger then those of *L. ruber* and the implication is that they possess ample reserves to support development without such a feeding requirement.

Iwata's Larva

The heteronemertean *M. akkeshiensis* forms a nonfeeding larva that is somewhat similar to the Desor's larva, but it forms only five ectodermal invaginations (Iwata 1958). These include two cephalic imaginal discs, two trunk discs and a single dorsal imaginal disc. This larva, the Iwata's larva (Figure 9.10), does possess some larval traits, including an elongated apical tuft and an apical organ. Unlike the Desor's larva, but similar to the pilidium, the Iwata's larva is free-swimming. While the juvenile worm maintains the same plane of bilateral symmetry and dorsoventral axis as that of the Iwata's larva, the worm's anterior-posterior axis is reversed by 180° (Figure 9.10). During metamorphosis the juvenile worm sheds the larval integument as it crawls out of the amniotic cavity.

Other members of the genus *Micrura* form pilidium lar-

vae. Thus, some investigators argue that the Iwata's larva is derived from that of the pilidium (Jägersten 1972). Iwata (1958) argues that the larva of *M. akkeshiensis* represents an intermediate between that of the pilidium and the Desor's larva. Investigators argue that the invention of lecithotrophic larvae such as the Desor's larva and the Iwata's larva represent an adaptation to harsh environmental conditions, such as life in the intertidal zone, where desiccation is a primary danger (Smith 1935; Jägersten 1972).

Evolution of Different Modes of Development

A reasonable question that one can ask is: Which condition is more closely representative of the ancestral mode of development in the nemerteans? The standard view holds that most metazoan phyla are derived from stocks that display a biphasic life history, with a feeding, free-swimming larval stage followed by some form of metamorphosis into the juvenile adult. As indicated above, indirect development, characterized mainly by the formation of a feeding pilidium larva, is restricted to the Heteronemertea. Most nemerteans display direct development, including those considered to be the most "primitive" nemerteans, the Paleonemertea. There is, however, one indirect-developing paleonemertean (*Hubrechtella dubia*; Cantell 1969) that forms a pilidium larva. Some suggest

Figure 9.10 Diagram illustrating the formation of the Iwata's larva. The juvenile worm forms from only five ectodermally derived imaginal discs. These include the paired cephalic discs (cd), the paired trunk dics (td) and an unpaired dorsal disc (dd). These discs eventually fuse and give rise to the adult ectoderm (ae) of the worm. The endoderm forms during gastrulation earlier in development. The location of the blastopore (bl) as well as the archenteron (ar) are shown in *a*. The differentiated gut (gt) is shown in *c*. The stomodeum or mouth (st) opens as a secondary invagination of the juvenile ectoderm. The proboscis (pr) forms from the fused cephalic imaginal discs. The Iwata's larva possesses an apical tuft of cilia (at). Ultimately, the worm will emerge from the larval integument (le) during the final stages of metamorphosis. (ac) amniotic cavity. Larval axes are indicated in *a*. LA, larval anterior; LP, larval posterior; LD, larval dorsal; LV, larval ventral. The adult axes are indicated in *c*. AA, adult anterior; AP, adult posterior; AL, adult left; AR, adult right; AD, adult dorsal; AV, adult ventral. Note that the adult anteroposterior axis is reversed relative to the larval anteroposterior axis.

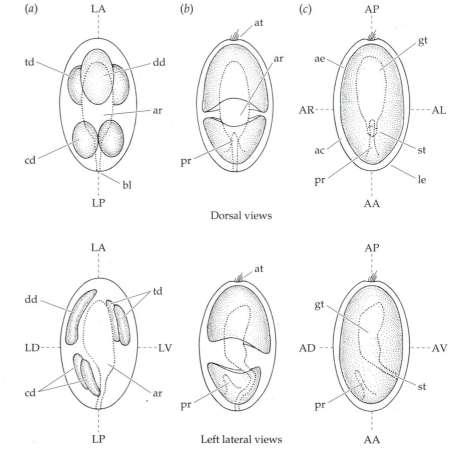

that this may indicate that the paleonemerteans were derived from an indirect-developing stock; however, *H. dubia* may actually belong to the Heteronemertea (Cantel 1969; Jägersten 1972). Thus, direct development, such as that displayed by the paleonemerteans and the hoplonemerteans, may actually represent the more primitive condition. The uncertainty of the systematic and phylogenetic relationships within the Nemertea makes an assessment of the ancestral life history condition problematic.

Experimental Embryology of Nemerteans: Mechanisms of Cell and Axis Determination

Localization of Developmental Potential along the Animal-Vegetal Axis

An important issue of development relates to the mechanisms that are utilized to establish particular cell fates and axial properties during development. Experimental studies reveal that a progressive restriction of developmental potential takes place along the animal-vegetal axis between the time of fertilization and the third cleavage division. A number of earlier investigators demonstrated that any fragment of the unfertilized *Cerebratulus* egg, if not too small, can be fertilized, and develops into a pilidium larva of normal appearance, complete with an apical tuft and a gut (Yatsu 1910; Hörstadius 1937). This includes both animal and vegetal fragments. On the other hand, isolated animal and vegetal half-embryos prepared at the 8- or 16-cell stage display a strictly mosaic pattern of development in accord with their normal embryonic fates (Zeleny 1904; Yatsu 1910; Hörstadius 1937). Animal halves form ciliated ectodermal larvae with a ciliated band, an apical organ and

tuft but do not form a digestive tract. Vegetal halves differentiate some ectoderm and a ciliated band; they form a gut, but contain no apical organ and tuft.

Freeman (1978) has demonstrated that factors responsible for the formation of the apical organ/tuft and the archenteron become localized along the animal-vegetal axis between the time of fertilization and the 8-cell stage. These factors are more widely distributed in the unfertilized egg and are subsequently localized to the animal and vegetal hemispheres after fertilization (see Figure 9.11). This localization was demonstrated by bisecting the eggs and embryos through the equatorial plane at various stages of development and following the development of the isolated animal and vegetal fragments. Those factors responsible for the formation of the gut are localized by the 4-cell stage, while those involved in the formation of the apical organ/tuft are localized by the 8-cell stage. In a control set of experiments it was found that embryos bisected along a meridional plane (along the animal-vegetal axis) are able to form larvae with both an apical organ/tuft and an archenteron at all stages examined. Freeman (1978) demonstrated that these localization events can be accelerated by treatments that stimulate aster formation (high salt treatment), or inhibited by treatments that retard aster formation (ethyl carbamate treatment). Thus, the factors that specify the apical organ and the gut are localized by a mechanism that involves asters that are formed during the meiotic maturation divisions and the early cleavage divisions. Presumably, the asters interact with other elements of the cytoskeleton that drive these localization events, and this interaction may involve microtubules, a principle component of the asters. These experiments demonstrate that mechanisms exist to ensure that essential morphogenetic factors become localized to the appropriate cells during the normal course of development.

Figure 9.11   Diagram showing the segregation of morphogenetic determinants. Small black dots indicate derminants responsible for the formation of the apical organ/tuft (shown in *a*), and the gut (shown in *b*). Determinants for the apical organ/tuft are ultimately segregated into the four large animal micromeres, while determinants responsible for the formation of the gut are segregated into the four vegetal macromeres. The gut determinants are localized more quickly than those for the apical tuft. All stages are illustrated with the animal pole facing upward. The small polar bodies (tiny open circles) formed during the meiotic maturation divisions mark the site of the animal pole. (After Freeman 1978.)

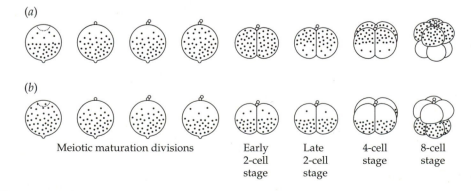

(a)

(b)

Meiotic maturation divisions   Early 2-cell stage   Late 2-cell stage   4-cell stage   8-cell stage

Experiments Investigating the Establishment of the Dorsoventral Axis

Other developmental axes, such as the dorsoventral axis, are not overtly apparent until later stages of development, when specific morphogenetic events associated with the formation of the larval or adult body plan take place. When are these axes first set up during development? As previously discussed, the first and second cleavage planes normally bear specific relationships relative to the dorsoventral axis and plane of bilateral symmetry in nemerteans (Henry and Martindale 1994a). One can ask whether the first and second cleavage divisions play a causal role in setting up the dorsoventral axis, or whether this axis is established prior to these events. By compressing the fertilized egg it is possible to entrain the orientation of the cleavage spindle, and thus reposition the plane of the first cleavage division. One can then follow the relationship between the altered first cleavage plane and future larval axes by injecting lineage tracer into one cell of the resulting 2-celled embryo. The results of such experiments indicate that one can, in fact, dissociate the early cleavage planes from an underlying set of axial properties to generate totally novel cell quadrant identities (Henry and Martindale 1995, 1996b). Thus, the plane of the first cleavage division is not itself involved in the establishment of the dorsoventral axis. These findings indicate that cell quadrant identity is established precociously relative to an underlying system of axial properties that reside within the fertilized egg. Under normal conditions some mechanism operates to orient the early cleavage divisions relative to these underlying axial properties (including the animal-vegetal and dorsoventral axes) to generate the four typical nemertean cell quadrants. Although the animal-vegetal axis appears to be established during oogenesis, it remains to be determined whether other axes, like the dorsoventral axis, also exist in the unfertilized egg, or whether they are set up as a consequence of fertilization.

Development of Isolated Blastomeres (Regulative vs. Mosaic Development in Nemerteans)

The ultimate fates of particular cells are determined at specific stages of development. Functionally, one can follow this process by examining the development of isolated blastomeres. Hörstadius (1937, 1971) observed that specific animal versus vegetal tiers of cells isolated from 8-, 16- and 32-cell embryos of C. lacteus develop in accord with their normal fates, as previously described. Unlike the case found in sea urchins (Chapter 16), Hörstadius concluded that inductive interactions do not control development with respect to the animal-vegetal axis in Cerebratulus.

On the other hand, blastomeres isolated at the 2-cell stage of Cerebratulus are able to give rise to complete miniature pilidium larvae with perfect bilateral symmetry. These larvae contain both a gut and an apical organ/tuft (reviewed by Hörstadius 1971). Thus, their development was interpreted to be completely regulative at this early stage. Blastomeres isolated at the 4-cell stage are also capable of some regulation; however, completely normal miniature pilidium larvae are not formed. Earlier investigators concentrated on the development of the gut as well as the apical organ and tuft in these partial embryos. These structures are not ideal for assessing regulative capacity, however, since all four cell quadrants normally contribute to their formation in the intact embryo (see the earlier section on the cell fate map). On the other hand, the larval musculature forms from only the A and B cell quadrants (Henry and Martindale 1994b, 1996a; Martindale and Henry 1995). Using a fluorescently labeled molecule (**phallacidin**) that binds to **filamentous actin**, Henry and Martindale examined the development of muscle fiber cells in quarter embryos derived from blastomeres isolated at the four-cell stage. In a number of cases, they observed muscle cell differentiation in all four quarter larvae derived from individual embryos of C. lacteus. Thus, the other two cell quadrants (C and D), which normally do not form larval muscle cells, are able to form muscles cells when cultured in isolation. These findings indicate that inhibitory cell-cell interactions (**inductive interactions**) originating from the A and B cell quadrants normally prevent progeny of the adjacent C and D cell quadrants from giving rise to muscle cells. The exact time at which the normal progenitors of the muscle cells are determined is unknown.

On the other hand, the isolated early blastomeres of the direct-developing Hoplonemertean N. bivittata do not display the same degree of regulation as those of C. lacteus (Martindale and Henry 1995). For instance, N. bivittata initially forms two pigmented ocelli; however, blastomeres isolated at the 2-cell stage give rise to half-animals that contain only a single eye spot. Furthermore, when one cell is deleted during the 4-cell stage, approximately 50% of the resulting three-quarter embryos will form two eye spots, while the other 50% generate only a single eye spot. These results indicate that the developmental potential to form a single eye spot is segregated to both of the cells in the 2-cell embryo and to only two of the cells in the 4-cell embryo.

Why should these two nemerteans exhibit such differences in regulative ability and the timing of cell fate commitment? It is conceivable that this may be related to differences in their mode of development. C. lacteus forms a feeding, planktonic larva, while N. bivittata undergoes a nonfeeding form of direct development. Direct development is generally considered to represent an accelerated embryonic condition (Raff 1987, 1992). Thus, one can argue that some decisions regarding the establishment of specific cell fates (such as those responsible for the formation of the ocelli) will be made precociously in direct-developing embryos. Although this correlation is by no means a universal one, a similar trend is found in the case of sea urchins. Blastomeres isolated at the 2-cell stage in indirect-developing sea urchins can form complete miniature pluteus larvae, while those of the direct-developing sea urchin *H. erythrogramma* fail to display complete regulation (Henry and Raff 1990).

The behavior of nemertean embryos differs from mollusc and annelid embryos. For instance, the development of blastomeres isolated from equal-cleaving mollusc embryos is generally very poor (Martindale and Henry 1995), such that cells isolated at the 2- and 4-cell stages typically do not display any regulation (Morrill et al. 1973; reviewed by Verdonk and Cather 1983). In the case of annelids and molluscs, the two eyes are generated by progeny of the lateral cell quadrants (A and C), but their formation requires an inductive interaction from the progeny of the dorsal cell quadrant (D) relatively early during development (Verdonk and Cather 1983). While the development of the ocelli in *N. bivittata* also appears to require inductive interactions, it appears that two adjacent cell quadrants are involved in regulating their formation (Martindale and Henry 1995).

Summary

The nemerteans represent a diverse group of organisms that display a wide range of life-history strategies based on both direct and indirect modes of development. Like other spiralians, nemerteans exhibit a highly conserved pattern of cell divisions (spiral cleavage); however, recent analyses indicate that nemertean cell lineages differ somewhat from those found in these other spiralians (Henry et al. 1995). Different mechanisms also appear to operate in the determination of embryonic axial properties and cell fates in nemertean embryos. In fact, the development of only a few nemerteans has been examined from an experimental standpoint. Further investigations with other species will undoubtably reveal additional discoveries for this fascinating group of organisms.

Literature Cited

Bartolomaeus, T. 1984. Zur Fortpflanzungsbiologie von *Lineus viridis* (Nemertini). *Helgoländer Meeresunters.* 38: 185–188.

Bierne, J. 1983. Nemertina, In *Reproductive Biology of Invertebrates, Oogenesis, Oviposition and Oosorption*, Vol. 1. K. G. Adiyodi and R. G. Adiyodi (eds.). John Wiley & Sons, Chichester, pp. 147–167.

Brusca, R. C. and G. J. Brusca. 1990. Nemertea, Chapter 11. In *Invertebrates*. Sinauer Associates, Inc., Sunderland, MA, pp. 315–334.

Cantell, C.-E. 1966. The devouring of the larval tissues during the metamorphosis of pilidium larvae (Nemertini). *Ark. Zool. Sr. 2*, 18: 489–492.

Cantell, C.-E. 1969. Morphology, development and biology of the pilidium larvae (Nemertini) from the Swedish west coast. *Zool. Bidrag. Fran Uppsala* 38: 61–111.

Cantell, C.-E. 1989. Nemertina. In *Reproductive Biology of Invertebrates, Fertilization, Development and Parental Care*, Vol. 4, Part A. K. G. Adiyodi and R. G. Adiyodi (eds.). John Wiley & Sons, Chichester, pp. 147–165.

Clark, W. H., Jr. and G. W. Hinsch. 1969. Some aspects of sperm-egg interaction in *Cerebratulus lacteus*. *Biol. Bull.* 137: 395.

Coe, W. R. 1899. On the development of the Pilidium of certain nemerteans. *Trans. Connecticut Acad.* 10: 235–262.

Coe, W. R. 1926. The pelagic nemerteans. *Mem. Mus. Comp. Zool. Harvard.* 49: 1–244.

Coe, W. R. 1929. Regeneration in nemerteans. *J. Exp. Zool.* 57: 109–144.

Coe, W. R. 1930. Asexual reproduction in nemerteans. *Physiol. Zoöl.* 3: 297–308.

Coe, W. R. 1943. Biology of the nemerteans of the Atlantic coast of North America. *Trans. Connecticut Acad.* 35: 129–328.

Dawydoff, C. 1928. Sur l'embryologie des *Protonemertes*. *C. R. Acad. Sci. Paris* 186: 531–533.

Desor, E. 1848. On the embryology of *Nemertes*. *Boston J. Nat. Hist.* 6: 1–18.

Franzén, Å. 1956. Remarks on spermiogenesis, morphology of the spermatozoan, and biology of fertilization among invertebrates. *Zool. Bidrag. Uppsala* 31: 355–482.

Franzén, Å. 1983. Nemertina. In *Reproductive Biology of Invertebrates. Spermatogenesis and Sperm Function*, Vol. 2. K. G. Adiyodi and R. G. Adiyodi (eds.). John Wiley & Sons, Chichester, pp. 159–170.

Freeman, G. 1978. The role of asters in the localization of the factors that specify the apical tuft and the gut of the nemertine, *Cerebratulus lacteus*. *J. Exp. Zool.* 206: 81–108.

Friedrich, H. 1979. Nemertini. In *Morphogenese der Tiere*, Lieferung 3: D5-I. F. Seidel (ed.). Gustav Fisher Verlag, Stuttgart.

Gibson, R. 1972. *Nemerteans*. Hutchinson University Library, London. 224 pp.

Gontcharoff, M. 1960. Le développement post-embryonnaire et la croissance chez *Lineus ruber* et *Lineus viridis* (Nemertes, Lineidae). *Ann Sci. Nat. (Zool.) Paris* (12) 2: 225–279.

Hammarsten, O. D. 1918. Beitrag zur Embryonalentwicklung der *Malacobdella grossa* (Müll.) Inaugural Diss. Stockholm. 95pp., Uppsala.

Henry, J. J. and R. A. Raff. 1990. Evolutionary change in the process of dorsoventral axis determination in the direct developing sea urchin, *Heliocidaris erythrogramma*. *Dev. Biol.* 141: 55–69.

Henry, J. Q. and M. Q. Martindale. 1994a. Establishment of the dorsoventral axis in nemertean embryos: Evolutionary considerations of spiralian development. *Dev. Genet.* 15: 64–78.

Henry, J. Q. and M. Q. Martindale. 1994b. Inhibitory cell-cell interactions control development along the dorsoventral axis in the embryos of *Cerebratulus lacteus*. *Biol. Bull.* 187: 238–239.

Henry, J. Q. and M. Q. Martindale. 1995. The experimental alteration of cell lineages in the nemertean, *Cerebratulus lacteus*: Implications for the precocious establishment of embryonic axial properties. *Biol. Bull.* 189: 192–193.

Henry, J. Q. and M. Q. Martindale. 1996a. The origins of mesoderm in the equal-cleaving nemertean worm, *Cerebratulus lacteus*. *Biol. Bull.* 191: 286-288.

Henry, J. Q. and M. Q. Martindale. 1996b. The establishment of embryonic axial properties in the nemertean worm, *Cerebratulus lacteus*. *Dev. Biol.* 180: 713–721.

Henry, J. Q., M. Q. Martindale and B. Q. Boyer. 1995. Axial specification in a basal member of the spiralian clade: Lineage relationships of the first four cells to the larval body plan in the polyclad turbellarian *Hoploplana inquilina*. *Biol. Bull.* 189: 194–195.

Hörstadius, S. 1937. Experiments on determination in the early development of *Cerebratulus lacteus*. *Biol. Bull.* 73: 317–342.

Hörstadius, S. 1971. Nemertinae In *Experimental Embryology of Marine and Fresh-Water Invertebrates*. G. Reverberi (ed.). North Holland Pub. Co., Amsterdam, pp. 164–174.

Iwata, F. 1957. On the early development of the nemertine, *Lineus torquatus* Coe. *J. Fac. Sci. Hokkaido Univ.* (6) (Zool.) 13: 54–58.

Iwata, F. 1958. On the development of *Micrura akkeshiensis. Embryologia* 4(2): 103–131.

Iwata, F. 1960a. Studies on the comparative embryology of the nemerteans with special reference to their inter-relationships. *Publ. Akkeshi Mar. Biol. Stat.* 10,

Iwata, F. 1960b. The life history of the Nemertea. *Bull. Mar. Biol. Stat. Asamuchi.* 10: 95–97.

Iwata, F. 1985. Foregut formation of the nemerteans and its role in nemertean systematics. *Amer. Zool.* 25: 23–36.

Jaffe, L. A., R. T. Kato and R. P. Tucker. 1986. A calcium-activated conductance produces a long-duration action potential in the egg of the nemertean, *Cerebratulus lacteus. J. Physiol.*

Jägersten, G. 1972. *Evolution of the Metazoan Life Cycle.* Academic Press, London and New York.

Kline, D., L. A. Jaffe and R. P. Tucher. 1985. Fertilization potential and polyspermy prevention in the egg of the nemertean, *Cerebratulus lacteus. J. Exp. Zool.* 236: 45–52.

Kline, D., L. A. Jaffe and R. T. Kado. 1986. A calcium-activated sodium conductance contributes to the fertilization potential in the egg of the nemertean worm, *Cerebratulus lacteus. Dev. Biol.* 117: 184–193.

Lacalli, T. C. and J. E. West. 1985. The nervous system of a pilidium larva: evidence from electron microscope reconstructions. *Can. J. Zool.* 63: 1901–1916.

Lebedinsky, J. 1897. Zur Entwicklungsgeschichte der Nemertinen. *Biol. Centralbl.* 17: 113–124.

Martindale, M. Q. and J. Q. Henry. 1995. Novel patterns of spiralian development: Alternate modes of cell fate specification in two species of equal-cleaving nemertean worms. *Development* 121: 3175–3185.

Metschinikoff, E. 1869. Studien über die Entwicklung der Echinodermen und Nemertinen. *Mem. de l'Acad de Sci. St. Petersbourg.* Ser 7, 14: 49–65.

Morrill, J. B., C. A. Blair and W. J. Larsen. 1973. Regulative development in the pulmonate gastropod *Lymnaea palustris* as determined by blastomere deletion experiments. *J. Exp. Zool.* 183: 47–56.

Müller, J. 1847. Über einige neue Tierformen der Nordsee. *Arch. Anat. Physiol.* 159–160.

Nawitizki, W. 1931. *Procarinina remanei.* Eine neue Paläonemertine aus der Kieler Förde. *Zool Jahrb. Anat.* 54: 159–234.

Nusbaum, J., and M. Oxner, M. 1913. Embryonalentwicklung des *Lineus ruber*, Müll. Ein Beitrag zur Entwicklungsgeschichte der Nemertinen. *Zeitchr wiss. Zoöl.* 107: 78–197.

Raff, R. A. 1987. Constraint, flexibility and phylogenetic history in the evolution of direct development in sea urchins. *Dev. Biol.* 119: 6–19.

Raff, R. A. 1992. Direct-developing sea urchins and the evolutionary reorganization of early development. *BioEssays* 14: 211–218.

Riser, N. W. 1974. Nemertinea. In *Reproduction of Marine Invertebrates*, Vol. 1. A. C. Giese and J. S. Pearse (eds.). Academic Press, New York and London, pp. 359–389.

Salensky, W. 1886. Bau und Metamorphose des *Pilidiums. Zeitschr. wiss. Zoöl.* 43: 481–511.

Salensky, W. 1912. Über die Morphogenese der Nemertinen. I. Entwicklungsgeschichte der Nemertine im *Pilidiums. Mem. Acad. Imp. St. Petersburg.* Ser 8, 30: No. 10, 1–74.

Schmidt, G. A. 1934. Ein zweiter Entwicklungstypus von *Lineus gesserensis-ruber* O. F. Müller. *Zoöl. J. Anat.* 58: 607–660.

Smith, J. E. 1935. Early development of *Cephalothrix rufifrons. Quart. J. Micr. Sci.* 77: 335–382

Stricker, S. A. 1987. Phylum Nemertea. In *Reproduction and Development of Marine Invertebrates of the Northern Pacific Coast.* M. F. Strathmann (ed.). University of Washington Press, Seattle, pp. 129–137.

Stricker, S. A. and M. J. Cavey. 1986. An ultrastructural study of spermatogenesis and the morphology of the testes in the nemertean worm *Tetrastemma phyllospadicola* (Nemertea, Hoplonemetrea). *Can. J. Zool.* 64: 2187–2202.

Turbeville, J. M. 1986. An ultrastructural analysis of coelomogenesis in the hoplonemertean *Prosorhochmus americanus* and the polychaete *Magelona* sp. *J. Morphol.* 187: 51–60.

Turbeville, J. M. 1996. Nemertini, Schnurwürmer. In *Spezielle Zoologie.* W. Westheide and R. M. Rieger (eds.). Teil I. pp. 265–275. Gustav Fischer, Stuttgart.

Turbeville, J. M. and E. E. Ruppert. 1985. Comparative ultrastructure and the evolution of nemertines. *Amer. Zool.* 25: 53–71.

Turbeville, J. M., K. C. Field and R. A. Raff. 1992. Phylogenetic position of Phylum Nemertini, inferred from 18s rRNA sequences: Molecular data as a test of morphological character homology. *Mol. Biol. Evol.* 9: 235–249.

van den Biggelaar, J. A. M., and P. Guerrier. 1983. Origin of spatial information. In *The Mollusca*, N. H. Verdonk, J. A. M. van den Biggelaar, and A. S. Tompa (eds.). Academic Press, New York, pp. 179–213.

Verdonk, N. H., and J. N. Cather. 1983. Morphogenetic determination and differentiation. In *The Mollusca*, N. H. Verdonk, J. A. M. van den Biggelaar, and A. S. Tompa (eds,). Academic Press, New York, pp. 215–252.

Verdonk, N. H., and J. A. M. van den Biggelaar. 1983. Early development and the formation of the germ layers. In *The Mollusca*, N. H. Verdonk, J. A. M. van den Biggelaar, and A. S. Tompa (eds.). Academic Press, New York, pp. 91–122.

Willmer, P. 1990. *Invertebrate Relationships. Patterns in Animal Evolution.* Cambridge: Cambridge University Press.

Wilson, C. B. 1900. The habitats and early development of *Cerebratulus lacteus. Quart. J. Micr. Sci.* (2) 43: 97–198.

Yatsu, N. 1910. Experiments on cleavage and germinal localization in the egg of *Cerebratulus. J. Coll. Sci. Imp. Univ. Tokyo.* 27: (Art. 17) 1–37.

Zeleny, C. 1904. Experiments on the localization of developmental factors in the nemertine egg. *J. Exp. Zool.* 1: 293–329.

CHAPTER 10

Sipunculans and Echiurans

John F. Pilger

THE INCLUSION OF THE SIPUNCULA AND ECHIURA PHYLA into a single chapter reflects the traditional phylogenetic interpretation that these organisms are, in some way, related. More recently, however, the view has gained strength that the sipunculans and echiurans may not be closely related to each other but that each is related to a different spiralian phylum. In order to obtain a more complete embryological understanding of the unity and diversity of life, one should consider both phyla, whether they are closely related or not. Both groups already provide rich material for embryological and developmental studies. Some species are used as model organisms for specific developmental phases, while others wait for their opportunity to show scientists the secrets they hold.

The Sipuncula

Anatomy

The sipunculan body is made up of the trunk from which extends an elongated introvert bearing the terminal head (Figure 10.1a). The **introvert** with the head is capable of being retracted into the trunk. In the retracted condition the body of smaller sipunculans looks like a small shelled peanut, thus their common name, peanut worms.

The trunk is oval to elongated and cylindrical. Many species bear numerous small **papillae**, wartlike bumps, on the surface of the body wall. Small wrinkles may circumscribe the body. A few species possess a hard **anterior shield** on the dorsal surface of the trunk and another at the posterior end (the **anal shield**). The anus is located mid-dorsally near the junction of the trunk and the introvert.

The mouth opens at the anterior end and is bordered dorsally or surrounded completely by variously formed tentacles. The mouth opens to a long esophagus that descends toward the posterior region of the trunk. There it makes a 180-degree turn to ascend to the anterior end (Figure 10.1b). The descending and ascending portions of the gut are twisted around each other, forming a helical mass. A thin spindle muscle extends from the posterior end of this loop to the posterior body wall and keeps the intestine from becoming hopelessly entangled in itself. The gut ends at the anus located on the mid-dorsal body wall.

Retraction of the introvert is accomplished by the action of retractor muscles that originate on the posterior body wall and insert on the head. Two to four of these muscles are present in most species. Extension of the head and introvert is accomplished by relaxing the retractor muscles and contracting the body wall to increase coelomic pressure.

Sipunculans have an annelid-like brain with a cerebral ganglion dorsal to the anterior esophagus. Circumesophageal connectives from the cerebral ganglion meet ventrally and connect to the ventral nerve cord. Although lateral branches innervate the body wall musculature, there is no evidence of metameric ganglia in the ventral nerve cord.

A pair of metanephridia are present on the ventral body wall. These organs have been modified to collect differentiated gametes from the coelomic fluid and store them until spawning. An excretory function has been suggested based on the microscopic anatomy of these organs, but the physiological data that would confirm this function is lacking.

The gonads arise from the coelomic peritoneum, usually at the base of the retractor muscles. Gametogenesis begins here, but soon the germ cells are shed into the coelomic fluid, where they continue their differentiation and maturation.

Illustrations of sipunculans date from the sixteenth century and in the first descriptions they were classified as

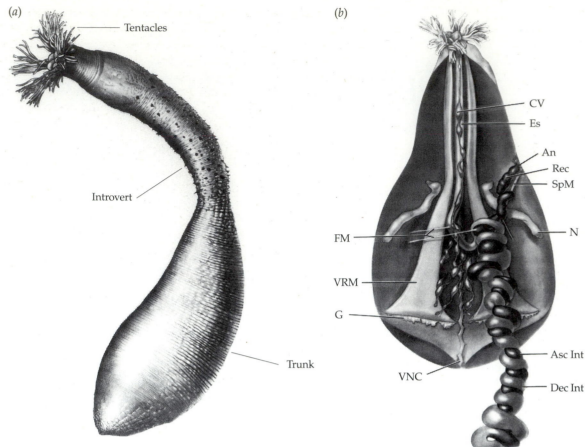

Figure 10.1   External and internal anatomy of a typical sipunculan, *Themiste alutacea*. (*a*) External anatomy. (*b*) Internal anatomy. An, anus; Asc Int, ascending intestine; CV, contractile vessel; Dec Int, descending intestine; Es, esophagus; FM, fixing muscle; G, gonad; N, nephridium; Rec, rectum; SpM, spindle muscle; VNC, ventral nerve cord; VRM, ventral retractor muscle. (From Rice 1993; illustration by C. B. Gast.)

members of the group Vermes (the catch-all category for wormlike creatures). By the nineteenth century some scientists considered them to be related to echinoderms (in particular the holothurians, sea cucumbers) and they were relegated with the Echiura and Priapula to a single phylum, the Gephyrea. This group (the name means "bridge") was constructed to bridge the phylogenetic gap between the annelids and the echinoderms, the group with which these organisms were thought to be allied. Now that the Gephyrea has been disbanded, former members of this catch-all group have risen to phylum status (Stephen and Edmonds 1972) but their phylogenetic relationships remain enigmatic.

Developmental Patterns and Larval Forms

Rice (1975a) defined four developmental pathways for sipunculans (Figure 10.2; Table 10.1). The first path includes those species whose egg develops directly to a small vermiform juvenile without the intervention of a lar-

val stage. Three species have been identified with this pattern. The second pattern has a single larval stage, a trochophore, that develops from the egg. This trochophore feeds on its own yolk reserves, i.e., is lecithotrophic, and then develops into a small crawling juvenile. Two species have been described with this pattern. Other sipunculans have two sequential larval stages in their life history, a trochophore and a **pelagosphera**. Both the trochophore and pelagosphera larvae of the third developmental category are lecithotrophic. Seven species are known with this type of development. Ten species follow the fourth developmental plan in their having a lecithotrophic trochophore followed by a plankton-feeding (that is, planktotrophic) pelagosphera.

Pelagosphera larva were originally thought to be small pelagic adult sipunculans of the genus *Sipunculus*. Now known to be a planktotrophic or lecithotrophic larval form intermediate between the trochophore and juvenile, they represent, for some species, a teleplanic larval stage of 2- to 6-month duration (in planktotrophic forms) that is capable of transoceanic passage in major currents (Hall and Scheltema 1975; Scheltema 1986; Scheltema and Rice 1990). As such, the pelagosphera contributes greatly to the biogeographic distribution of the phylum.

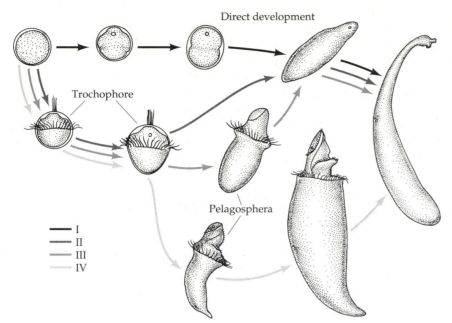

Direct development

Trochophore

Pelagosphera

— I
— II
— III
— IV

Figure 10.2 Diagrammatic summary of the chief developmental patterns in the Sipuncula. Category I: Direct development with no pelagic stages. Category II: Pelagic lecithotrophic trochophore transforms into vermiform stage. Category III: Pelagic lecithotrophic trochophore transforms into second larval stage, lecithotrophic pelagosphera, which then transforms into vermiform stage. Category IV: Pelagic lecithotrophic trochophore metamorphoses into second larval stage, planktotrophic pelagosphera. After prolonged period in the plankton the larva, having increased in size, undergoes a second metamorphosis into the juvenile form. (After Rice 1975b.)

Gametogenesis

The gonad is a small flap of tissue located at the base of the retractor muscles where they originate on the ventral body wall. Based on observations during the early development of *Golfingia* and *Phascolopsis*, germ cells are considered to be derived from the parietal peritoneum, but additional observations at the electron microscopic level of study are necessary to confirm this derivation. Only the earliest stages of gametogenesis occur in the gonad. The release of clusters of gametocytes into the coelomic fluid marks the end of the **gonadal phase** and the beginning of the **coelomic phase** of gametogenesis. After a period of differentiation within the coelomic fluid, the largest, most differentiated gametes become selectively removed from the fluid by the ciliated funnel (nephrostome) of the metanephridium. The gametes are stored here until spawning.

OOGENESIS. Studying the ovary of *Golfingia vulgaris*, Gonse (1956; Figure 10.3*a*) identified seven maturational stages progressing from the first stage, oogonia at the

Figure 10.3 (*a*) Diagram of a longitudinal section through the ovary of *Golfingia vulgaris*. The ovary is suspended by a thin mesentery at the origin of the retractor muscle. Seven stages of oogenesis are recognized on the basis of nuclear morphology. I, oogonia; II, transition from oogonia to the first meiotic prophase; III, leptonema stage; IV, zygonema stage; V, pachynema stage; VI, diplonema stage; VII, completion of the first meiotic prophase. (*b*) Summary of coelomic oocyte differentiation in *Golfingia vulgaris*. Six stages are recognized: O, oocytes recently detached from gonad, accessory cells are present; l, individual oocytes with follicle cells attached, growth begins; T, transition stage, follicle cells lost; 2, main growth phase; 3, slower growth period, RNA production accelerated, yolk platelets appear; M, fully differentiated, "mature" oocytes ready to be removed to the nephridium, germinal vesicle breaks down. (From Gonse 1956.)

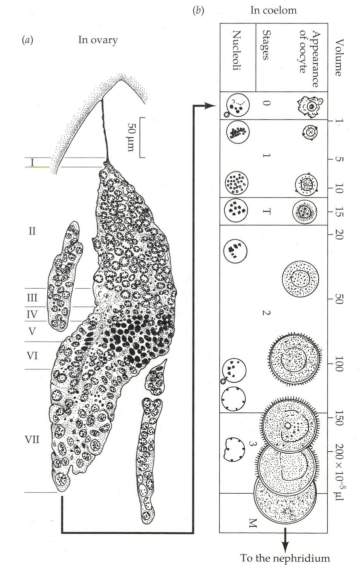

(*a*) In ovary

(*b*) In coelom

50 µm

Nucleoli	Stages	Appearance of oocyte	Volume
	0		1
	1		5
			10
	T		15
			20
	2		50
			100
	3		150
	M		200×10^{-5} µl

To the nephridium

TABLE 10.1. *Larval developmental types, distributed by taxa[a]*

	Larval Type			
	I	II	III	IV
Class Sipunculidea, Order Sipunculiformes				
Family Sipunculidae				
Sipunculus	—	—	—	X[b]
Siphonosoma	—	—	—	X
Phascolopsis	—	X	—	—
Class Sipunculidea, Order Golfingiiformes				
Family Golfingiidae				
Golfingia	—	—	XX	—
Nephasoma	X	—	—	X
Thysanocardia	—	—	X	—
Family Phascolionidae				
Phascolion	X	X	X	—
Family Themistidae				
Themiste	X	—	XXX	—
Class Phascolosomatidea, Order Phascolosomatiformes				
Family Phascolosomatidae				
Phascolosoma	—	—	—	XXX
Apionsoma	—	—	—	X
Antillesoma	—	—	—	X
Class Phascolosomatidea, Order Aspidosiphoniformes				
Family Aspidosiphonidae				
Aspidosiphon	—	—	—	XX

Source: From Cutler 1994
[a] Each letter X indicates one species. No data for five genera.
[b] Somewhat unique pattern.

proximal end of the gonad, where it attaches to the ventral body wall, to the last stage, cells near the completion of meiotic prophase I at the other end of the gonad. These stages have been confirmed in other sipunculans. At or near the end of prophase I, the gametes are released as small clumps into the coelomic fluid where maturation and differentiation continues. Gonse (1956; Figure 10.3b) further distinguished six stages of coelomic oocytes based on size, nuclear and cytoplasmic contents, and the loss of attached accessory cells. Among the most distinctive aspects of oocyte differentiation in the coelomic phase are growth and change in shape.

Sipunculan eggs are spherical, oval (tapered at both ends), or elongate (untapered). Rice (1989) presented egg-size data from 23 species. Spherical eggs range from 105–190 μm in diameter, while the oval ones are from 90–110 μm wide by 104–140 μm in length. Elongate eggs are described from *Nephasoma minutum* and a species of *Phascolion* and are about 260–280 μm and 124 μm long, respectively. Egg size appears to be positively correlated with the developmental pattern, the larger, more yolky eggs having greater propensity for direct development, the smaller eggs forming both trochophores and pelagosphera, and the intermediate-size eggs giving rise to trochophores without pelagosphera. To a lesser extent, the shape of the egg is related to the taxonomic position of the species.

Though vitellogenesis varies with the type of yolk produced, typically it begins when the eggs are shed into the coelom and continues throughout the coelomic phase. Glycogen, carbohydrate- and lipid-based granules constitute the bulk of the nutritional reserves produced through the middle of the coelomic oogenesis. Protein-based yolk platelets appear late in oogenesis, and lipid droplets may appear either early or late, depending on the species. The cytoplasmic distribution of these constituents varies considerably.

An egg envelope covers each egg in all sipunculans. Varying from 3–12 μm in total thickness, the envelope typically is made up of three layers of differing density (Figure 10.4). At the electron microscopic level of examination, one can see that the envelope is penetrated by microvilli that arise from the egg plasma membrane and branch as they emerge on the surface of the envelope. The tips of these microvilli may be important in species-specific sperm-egg recognition.

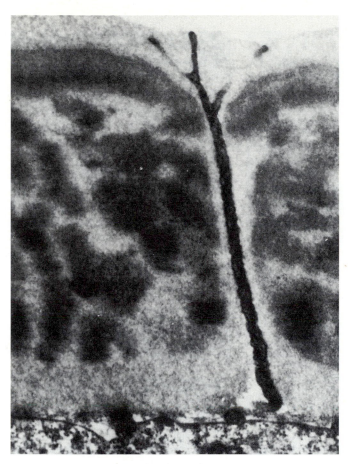

Figure 10.4 Transmission electron micrograph of the egg envelope of an unfertilized egg of *Aspidosiphon fischeri*. Note the branching microvillus and the patchy organization of the electron-dense fibrous material of the middle layer. ×25,380. (From Rice 1989.)

As coelomic eggs complete their differentiation, they are selectively removed from the coelomic fluid by the nephrostome and stored in the metanephridia until spawning. The mechanism whereby the most completely differentiated eggs are selected and incompletely differentiated eggs are left in the coelomic fluid is not known. The eggs of many species undergo germinal vesicle breakdown just prior to their being collected into the nephridium, but spawned eggs of *Sipunculus nudus* and *Nephasoma pellucidum* have been observed with the germinal vesicle intact.

SPERMATOGENESIS AND SPERMIOGENESIS. In contrast to oogenesis, much less is known about maturation and differentiation of sipunculan sperm. Spermatogenesis begins in the gonad but after a short period there, spermatocytes break free as small clumps and continue their differentiation in the coelomic fluid. Meiotic divisions, cytoplasmic reduction, and cytodifferentiation characterize this phase.

Mature spermatozoa of sipunculans are considered to have the primitive sperm morphology; a flattened acrosome and four or five mitochondrial spheres in the mid-

piece (Figure 10.5). Deviation from this morphology in a few species may be related to specific aspects of fertilization. For example, the modifications of the sperm of *Themiste pyroides*, an enlarged and pointed acrosome and modified mitochondria, may be important to enhance the ability of this sperm to pass through a thick jelly coat on the egg of that species (Rice 1975b).

GAMETOGENIC CYCLES. Cyclical gametogenic cycles have been reported for a few sipunculans. Studies of this phenomenon typically are conducted by monitoring the number and condition of coelomic gametes (usually oocytes) in specimens collected at intervals over a complete season or more. In most species, small oocytes are present in the coelomic fluid throughout the year (Rice 1975b; Pilger 1987; Amor 1993). Prior to the breeding season, the size-frequency distribution of oocytes becomes bimodal as oocytes move to larger size classes and then unimodal when they are collected into the nephridia.

Data from studies such as these have allowed scientists to infer relative rates of gametogenesis. If one has evidence for continuous oocyte production and knows that there is no differential selection against any size class, then one can infer that the frequency of oocytes in any particular size class is roughly proportional to the amount of time an oocyte spends in that size class. Oocyte size-frequency distributions, therefore, can be used to predict relative rates of oogenesis. *Themiste lageniformis* produces oocytes throughout the year. An oocyte size-frequency distribution (Figure 10.6) for this species shows a unimodal distribution during February and March and a bimodal distribution for other months. A substantial mode is never present in the range of 50–120 μm. The inference from this is that oocytes progress through the intermediate size range rather quickly and that oocyte growth in the smaller and larger size ranges is slower. The slow growth rate of the large and small oocytes may indicate that a biochemically complex process such as vitellogenesis may be taking place. From observations such as this, one can pose hypotheses and make predictions that can be tested experimentally.

Seasonal spawning activity has been reported in several species of Sipuncula. Most temperate species have breeding seasons that are well-defined, while tropical species seem to have broader and less well-defined breeding periodicity. Attempts to discover a correlation between spawning activity and tidal cycles or lunar phases have not produced positive evidence for such a relationship, although controlled experiments of these or other environmental parameters have not been conducted.

Asexual Reproduction

Three sipunculans have been reported to reproduce by asexual means. One species uses parthenogenesis (Pilger 1989) and the other two use transverse fission and budding (Rice and Pilger 1993).

Figure 10.5 Scanning electron micrographs of sipunculan sperm. (*a*) *Themiste pyroides.* × 12,500. (*b*) *Aspidosiphon fischeri.* × 7500. (*c*) Egg of *Nephasoma pellucida* with attached sperm. × 980. (*d*) Higher magnification of *c*. Note extended acrosomal filaments of sperm. × 85,000. (From Rice 1989.)

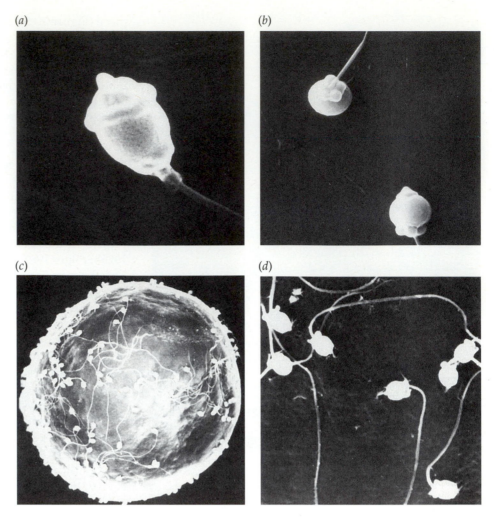

Themiste lageniformis, from Florida, has biased sex ratios where females outnumber males 24 to 1 (Pilger 1987). Where the species occurs in India, the ratio is 60 females to 1 male. A biased sex ratio such as this suggested that the reproductive mode was atypical and led to further observations, new questions, hypotheses, and experiments designed to test the hypotheses. It was determined that females isolated in dishes in the laboratory spawn eggs that undergo spontaneous activation and development; no evidence for sperm storage or hermaphroditism was found. The presence of males in the population, although few in number, suggested further that parthenogenesis was not obligatory, for why would males be present if they are not participating in fertilization at all? Are parthenoproduced offspring haploid or diploid? Are any offspring produced by fertilization (zygogenesis)? How are males and females produced? How is the biased sex ratio maintained?

To answer these questions and understand the cytogenetics of parthenogenesis in *Themiste*, nuclei from cells representing various stages of the life cycle of the organism were stained in a manner that allowed the amount of DNA to be quantified. It was discovered that the amount of DNA in cells of parthenogenetically produced individ-

uals was the same as that in cells of those *Themiste* produced by sexual means; that is, it was diploid. Thus, parthenogenetically produced individuals must have a mechanism to restore their cells to the diploid condition.

In examples of parthenogenesis such as this, the diploid condition may be restored by **apomixis** (meiosis is absent and diploid eggs are produced) or by **automixis** (the egg nucleus fuses with a polar body nucleus or two mitotic daughter cell nuclei fuse). It was reasoned from the data and observations on the early development of this species that parthenogenesis is the primary mode of reproduction and that the diploid condition is restored by automixis. Further, fertilization was considered to be facultative. For the sex ratio and sex determination, it was hypothesized that fertilization may produce both males and females but that parthenogenesis produces only females (**thelytoky**). Such a scheme, though not yet tested, would maintain the biased sex ratio present in populations of *Themiste lageniformis*.

Two species of sipunculans, *Sipunculus robustus* and *Aspidosiphon elegans*, are reported to undergo asexual reproduction by other means. Under stressed laboratory conditions *Sipunculus* was observed to undergo lateral budding from the posterior end and transverse fission in the poste-

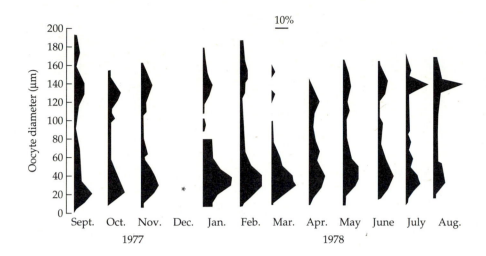

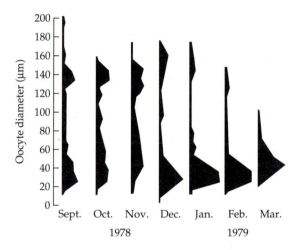

Figure 10.6 Size-percent frequency distributions of coelomic oocytes from *Themiste lageniformis*. Scale bar: 10%. An asterisk (*) indicates no measurements made. (From Pilger 1987.)

rior half of the trunk. These events were not observed in the field. *Aspidosiphon*, on the other hand, undergoes transverse fission regularly in the field. As many as 15% of specimens collected from its Caribbean habitats have a constriction at the posterior end of the trunk (Rice 1970). Portions of the internal organs within the posterior portion are used as anlage for the regeneration of new organs. The head and introvert arise from a small epidermal invagination immediately posterior to the constriction. The trunk regenerates a new posterior end.

Regeneration of various sipunculan tissues has also been reported. In several species, amputation of tentacles or the anterior introvert results in regeneration of the lost part. Regeneration and asexual reproduction by budding or transverse fission all demonstrate the plasticity of sipunculan tissues. These examples show that the fate of the cells is not determined even in adults, that some degree of metaplasia is possible. Virtually no formal research has been carried out on this phenomenon in these animals.

Sexual Reproduction

FERTILIZATION. The release of gametes may be in response to environmental conditions, but this relationship has not been systematically investigated. Early studies of fertilization of sipunculan eggs reported that

sperm penetrated existing pores in the egg envelope, but more recent work has shown that the sperm release the contents of their acrosomal vesicle, form an acrosomal filament, and produce a hole in the egg envelope that persists for several days.

Sipunculan eggs are triggered to complete maturation by the penetration of sperm (Figure 10.7). Once inside, the sperm pronucleus rotates approximately 180 degrees, migrates to the center of the egg, and swells slightly. Within nine minutes after the addition of sperm *in vitro*, the cytoplasm of the oval egg of *Apionsoma misakiana* retracts from the animal and vegetal poles of the surrounding egg envelope and the cell resumes meiosis. The first and second polar bodies are released at the animal pole by 29 and 52 minutes, respectively, while the male pronucleus waits in the central cytoplasm. A female pronucleus forms at the periphery of the ooplasm (60 minutes) and then migrates to the center (75 minutes), where it fuses with the male pronucleus to form the zygote nucleus (85 minutes).

CLEAVAGE. Sipunculan eggs undergo spiral, holoblastic, and unequal cleavage. Cleavage and cell lineage has been followed in *Golfingia vulgaris* and in *Phascolopsis gouldii* (Gerould 1906). The first cleavage is meridional and produces a small AB blastomere and large CD blastomere. In

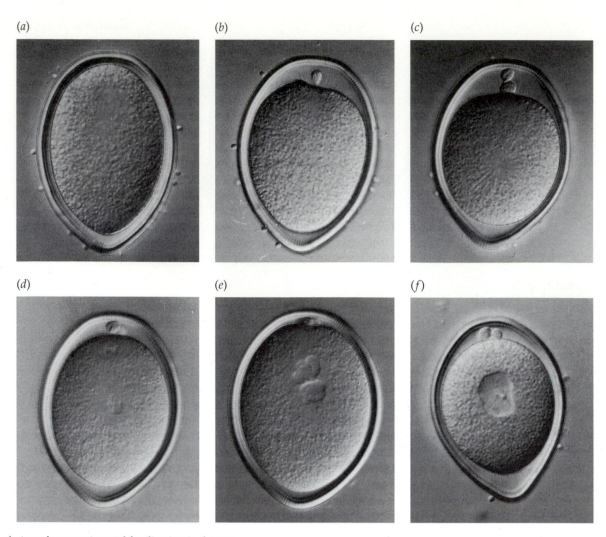

Figure 10.7 Completion of maturation and fertilization in the egg of *Apionsoma misakiana*. Light micrographs taken with Nomarski optics. Time measured is minutes after combination of eggs and sperm. (*a–e*) are the same egg. (*a*) At 9 minutes, cytoplasm has withdrawn from egg envelope at vegetal pole and first meiotic metaphase spindle has moved toward animal pole. (*b*) At 29 minutes, first polar body stage. (*c*) At 52 minutes, second polar body stage. Clear central area is site of formation of male pronucleus. (d) At 60 minutes, female pronucleus has formed at animal pole and male pronucleus in the central egg. (*e*) At 75 minutes, the vesicular female pronucleus and the male pronucleus have made contact. (*f*) At approximately 85 minutes, female and male pronuclei have united to form zygote nucleus. × 450. (From Rice 1989.)

the second cleavage, the mitotic apparatus in each cell aligns to produce an obliquely meridional, and unequal cleavage. The A, B, and C blastomeres are roughly equivalent in size, but the D blastomere contains approximately five times as much cytoplasm as each of the others. Just prior to the third cleavage, some of the cytoplasm and yolk in the cells of sipunculans that undergo lecithotrophic development becomes redistributed so that the mitotic apparatus in cells A, B, and C is displaced slightly toward the vegetal pole. In the D cell the cytoplasmic shift is absent or negligible. Thus, when the third cleavage occurs, it is parallel to the equator and, because of the

cytoplasmic redistribution, it establishes an upper quartet of blastomeres, i.e., micromeres, that are *larger* than the lower quartet of cells, typically referred to as macromeres. The relative absence of yolk in eggs of sipunculans that undergo planktotrophic development does not cause the displacement of the mitotic apparatus and, therefore, the micromeres are equal to or, as one would expect, smaller than the macromeres. Thus, the influence of yolk on early cleavage (Balfour's principle) is evident in sipunculans that produce lecithotrophic larvae.

When the embryo reaches the 48-cell stage (after approximately 10 hours of development), the cells in the apical plate display a typical molluscan cross (Figure 10.8). In this pattern the cells in the arms of the cross lie in the frontal and sagittal planes of the future embryo. The arms of the cross, thus, roughly define the dorsal and ventral regions of the future embryo. The C and D quadrant-cells lie in the dorsal hemisphere and the A and B lie in the ventral hemisphere. This arrangement of cells on the apical plate is typical of molluscan embryos, thus the term *molluscan cross*.

The fate of many of embryonic cells is known from the work of biologists around the turn of the century. The fate

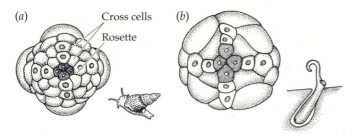

Figure 10.8 Embryos of a mollusk and a sipunculan, illustrating the rosette cells and typical molluskan cross. The arms of the cross are shown in the radii of the rosette in (*a*) mollusks and (*b*) sipunculans. (After Ruppert and Barnes 1994; *a* based on E. W. McBride 1914; *b* based on Gerould 1906.)

of most cells appears to be the same as those similar cells in other spiralian phyla. For instance, the first three quartets of micromeres that form in the embryo become ectoderm and ectomesoderm. The 4d cell, derived from 3D, produces two bands of mesoderm which, eventually, will become the trunk coelom, as it does in annelids. Other descendants of the 3D cell (4a, 4b, 4c, and the 4D cell, also referred to as 4Q in older literature) produce the endoderm.

BLASTULATION. During the next 14 hours (10–24 hours after fertilization) the embryo undergoes changes through which it becomes a blastula. In the apical plate, division leads to a group of four cells (the **definitive rosette**) that will give rise to an apical tuft of cilia. Cells at the margin of this rosette sink inward slightly, so that the apical plate appears somewhat elevated and surrounded by a depression, the apical groove. A blastocoel, the first or primary body cavity, is present in species with microlecithal eggs, but the position of the cavity in the blastula varies by species.

GASTRULATION. Gastrulation in sipunculans is accomplished by epiboly, invagination, or a combination of both processes. In species having eggs with a lot of yolk, i.e., macrolecithal eggs (developmental categories I–III), gastrulation takes place by epiboly (Rice 1988). In these embryos, cells of the somatic plate that lies on the dorsal side of the embryo divide bilaterally and grow later-

ally and ventrally over the vegetal endoderm cells in an epibolic fashion. The endoderm becomes internalized, except for a narrow region through which their cell apices maintain contact with the egg envelope (Figure 10.9*a*, *b*). This region represents the blastopore and is the spot where ectoderm will invaginate slightly to form the stomadeum, the ectodermal opening of the mouth.

Gastrulation in the planktotrophic species *Sipunculus nudus* occurs principally by invagination (Figure 10.9*c*, *d*). The endoderm cells move inward to form the primitive gut, the archenteron, and, in doing so, pull away from the egg envelope, creating a small posterior cavity. Cells from the **prototrochal** band extend anteriorly and posteriorly to cover this posterior cavity as well as the anterior cavity, the apical groove, that was formed around the apical plate.

In *Golfingia elongata* and two Phascolosomatidae species, gastrulation combines both epiboly and invagination (Cutler 1994; Rice 1988). Cells of the somatic plate grow over the macromeres and surround the embryo and a small stomodeal invagination has been formed.

Division of the 3D cell produces the 4D and 4d daughter cells. Descendants of the 4d cell, the **teloblasts**, move inward at the blastopore, and give rise to two mesodermal bands that, depending on the developmental category, will split into somatic and splanchnic mesoderm either while the larva is a trochophore or during metamorphosis of the trochophore (Figure 10.9*b*). In direct-developing sipunculans (developmental category I), the mesoderm is laid down when the ciliated prototroch band degenerates and before the embryo begins to elongate.

Development and Metamorphosis of the Trochophore

Sipunculan trochophores have three distinct regions: the pretrochal lobe with an apical plate, the prototroch, and

Figure 10.9 Gastrulation and mesoderm formation in the Sipuncula. Prototroch cells are marked by dashes; other ectoderm cells are clear; endoderm is dotted; mesoderm is barred. (*a*) Sagittal section of an embryo of *Golfingia vulgaris*, a lecithotrophic species, showing blastopore and endodermal cells attached at that point. (*b*) Cross section of an embryo of *Golfingia vulgaris*. Mesoderm has split into splanchnic and somatic layers. (*c*) Optical median section of embryo of *Sipunculus nudus*, a planktotrophic species, showing formation of ectodermal somatic plate and embolic gastrulation. (*d*) Optical median section of embryo of *Sipunculus nudus* at a later stage than *c*. Prototroch cells now surround the embryo. ag, apical groove; at, apical tuft; bp, blastopore; dc, dorsal cord; end, endoderm; mes, mesoderm; p, prototroch; pc, posterior cavity; r, rosette cells; sp, somatic plate; st, stomadeum. (After Rice 1975b.)

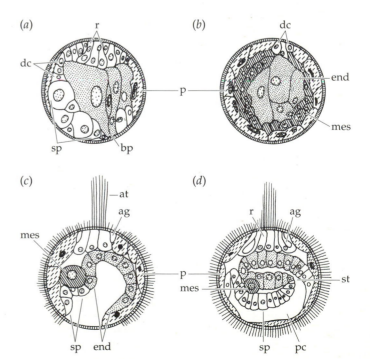

Figure 10.10 Trochophores of sipunculans. (*a*) Trochophore of *Golfingia elongata*, sagittal section; (*b*) trochophore of *Golfingia vulgaris*, ventral view, about 40 hours old; (*c*) late larva of *Golfingia*, sagittal view; (*d*) trochophore of *Sipunculus nudus*, lateral view. The process of shedding the egg envelope and underlying cell layer, serosa, has begun at the posterior end. a, anus; ag, apical groove; at, apical tuft; bo, buccal organ; cm, coelom; e, eye; end, endoderm; lg, lip gland; m, metatroch; p, prototroch; se, serosa; st, stomadeum. (*a* and *c* after Gerould 1906 and Rice 1975b; *b* after Åkesson 1961 and Rice 1975b; *d* after Hatschek 1883 and Rice 1975b.)

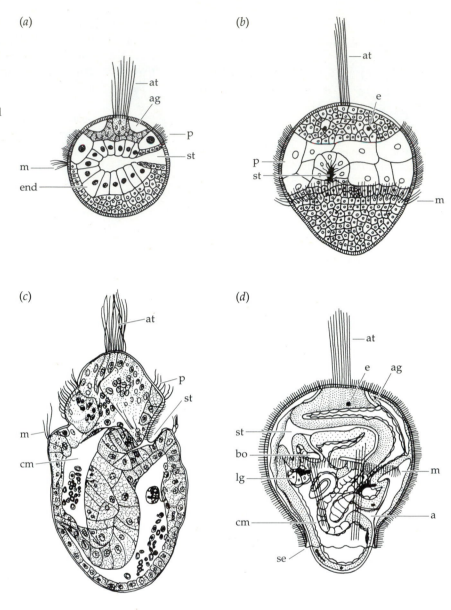

trunk or posttrochal lobe (Figure 10.10). In contrast to most protostomes, sipunculan trochophores lack protonephridia. The adult cuticle begins to form beneath the egg envelope, and preoral cilia develop from the rosette cells in the apical plate. Cells that surround the midregion of the embryo develop cilia at the larval equator, the **prototroch**. Posterior to the prototroch, a narrower band of large cells encircles the embryo and produces the longer cilia of the metatroch. Ocelli (eyes) form in the apical plate region during the second day and cause the larvae to be positively phototactic.

The ventral nerve cord begins as a few groups of longitudinally arranged cell clusters in the ventral ectoderm. The cells in these groups differentiate into a single longitudinal cord that, at its anterior end, splits, surrounds the esophagus, and joins dorsally with the supraesophageal ganglion, which has developed independently from the

apical plate. Retractor muscles arise from ectodermal cells along the sides of the apical plate by sinking below the ectoderm, elongating, and attaching to the posterior body wall. As the larval trunk elongates, these muscles are drawn into long, thin fibers. The circular muscles of the body wall first develop in the trochophore as a band just posterior to the prototroch and then spread in anterior and posterior directions. Longitudinal muscles of the body wall develop later.

By the end of the first day, the mouth is formed just posterior to the prototroch. Before the end of the second day of development, endoderm cells in the posterior trunk grow anteriorly under the dorsal body wall, where they attach to the ectoderm. Later, a small invagination forms in the ectoderm as the proctodeum and joins with the endoderm. Though contiguous from mouth to anus, the lumen of this primitive gut does not open until much later, after the yolk in its cells has been absorbed by the embryo. While enclosed in the egg envelope, sipunculan trochophores are incapable of feeding.

Depending on the developmental category of the species, metamorphosis of the trochophore may produce either a vermiform (wormlike) stage or a pelagosphera larva. In some species the egg envelope is shed, but in most it is transformed into the cuticle of the pelagosphera. When it is shed, the process is aided by the retraction and extension of the head made possible by the development of functional retractor muscles. Prototroch cells degenerate and slough into the coelom, where the yolk they contain becomes a nutritive source for the embryo. The prototroch region is replaced by ectodermal cells.

Development and Metamorphosis of the Pelagosphera

Rice (1967) defined the pelagosphera larva as "any sipunculid larva resulting from the metamorphosis of a trochophore that swims by means of a prominent ciliated metatroch, and in which the prototroch either has been lost or has undergone marked regression." Morphologically, the body of a pelagosphera is elongated, has a well-developed head and mouth, and has a broad metatroch (Figure 10.11). The posterior tip of the trunk may bear a terminal organ that performs both adhesive and sensory roles (Ruppert and Rice 1983).

Conversion of the trochophore to a pelagosphera is characterized by extensive development of the ciliary field on the ventral surface of the head in the region of the mouth (Figure 10.12). In some species a buccal organ forms in the mouth and glands develop in the lower lip. An enlarged stomach leads to the intestine and then the anus on the mid-dorsal body wall. The digestive tract is open and functional in planktotrophic forms but nonfunctional in lecithotrophic larvae.

Metamorphosis of the pelagosphera involves several changes and may take from 3 to 7 weeks to complete. Lecithotrophic forms generally metamorphose more slowly than planktotrophic pelagosphera. One important change is the loss of metatrochal cilia. With this primary locomotor organ gone, the larvae sink to the bottom of the sea. Metamorphosis may be augmented by settlement-inducing substances (Rice 1986). Qualitative characterization of a putative inducer has shown it to be a heat-labile, low-molecular-weight substance. The terminal organ may play a role in the discrimination of appropriate substrata for settlement.

During metamorphosis the head undergoes modifications that change the opening of the mouth from ventral to

(a)

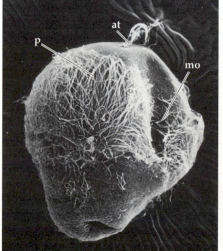

(b)

(c)

Figure 10.11 Scanning electron micrograph series of the metamorphosis to the pelagosphera in *Siphonosoma*. at, apical tuft; m, metatroch; mo, mouth region; l, lower ciliated lip; p, prototroch; s, stomadeum; to, terminal attachment organ. ×500. (Courtesy of M. Rice.)

(d)

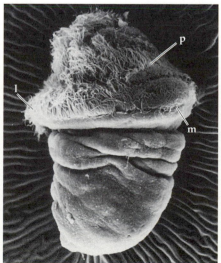

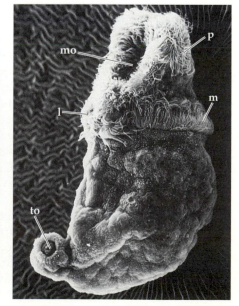

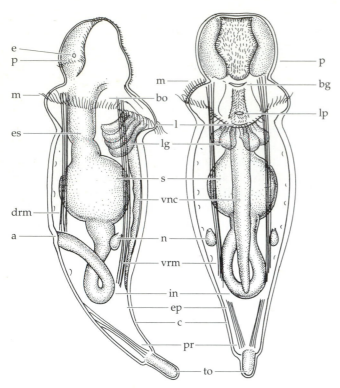

Figure 10.12   Pelagosphera larva of *Phascolosoma perlucens*. Diagram showing internal structures at 1 week of age. Left, lateral view; right, ventral view; a, anus; bg, buccal groove; bo, buccal organ; c, cuticle; drm, dorsal retractor muscle; e, eye; ep, epidermis; es, esophagus; in, intestine; l, lip; lg, lip gland; lp, lip pore; m, metatroch; n, nephridium; p, prototroch; pr, posterior retractors; s, stomach; to, terminal organ; vnc, ventral nerve cord; vrm, ventral retractor muscle. (After Rice 1975a, b.)

terminal. Williams (1977) reported that complete extension of the head of *Themiste lageniformis* is restricted by the action of a strong sphincter muscle in the introvert. In this restricted position, the head epidermis is held in contact with the epidermis of the introvert, and the cuticle between the two tissues dissolves. This allows the head and introvert epidermal tissues to fuse dorsally and laterally. Incomplete fusion in the ventral region provides a tube with a terminal opening, which serves as a connection between the larval mouth and the new juvenile mouth. In the trunk, metamorphosis consists of increase in size of the body.

In pelagosphera larvae, the brain is connected dorsally to the inside of the body wall and ventrally to the esophagus. Perhaps due to increased coelomic pressure during extension of the introvert, the body wall and the brain separate, leaving the latter attached only to the dorsal side of the esophagus.

Tentacles form as buds on the anterior region of the head. In some species, the tentacles surround the mouth, while in others, they may be limited to the region dorsal to the mouth. Tentacles may be digitate, palmate, or flap-like. The canals within the tentacles have been shown to be coelomic cavities (Pilger 1982).

Metamorphosis complete, a benthic vermiform stage is produced. A juvenile sipunculan will gradually grow from this small stage.

The Echiura

Although there are only about 150 species of Echiura, some have been model organisms for studies of fertilization and developmental processes. Echiurans in the family Bonellidae are well known for their extreme sexual dimorphism. Females are normal-sized individuals, but males are very small forms that reside in the female's reproductive tract and function only to fertilize her eggs as they pass out during spawning.

Anatomy

The echiuran body has two distinct regions, the trunk and the proboscis or **prostomium** (Figure 10.13*a*). Oval, spherical, or elongate, the trunk may be a few millimeters to several centimeters long. The mouth is located at the anterior end and the anus at the posterior end of the trunk. Longitudinal muscle bands are present in the body wall of some species; their number is a diagnostic characteristic of the species. One or more setae (usually a pair) project from the midventral body wall at the anterior end. In a few species, one or two rings of setae surround the anus. Nephridia that are modified to serve a reproductive function, **metanephridia**, open to the outside at **nephridiopores** located on the median ventral body wall.

The flat, elongate proboscis used for feeding provides the source of the common name of this phylum, the spoon worms. Fragile and often broken from specimens collected from the field, the proboscis may have a rounded, blunt, or bifid tip. The proboscis joins the trunk immediately dorsal to the mouth. While concealed in its mud or sand burrow, an echiuran will extend this retractable organ out onto the sediment surrounding the entrance in order to feed on organic matter deposited there. When *Bonellia viridis* is feeding, its proboscis is known to extend more than a meter from the trunk, which itself may be only 3 centimeters long. In echiurans of the genus *Urechis*, the proboscis is reduced to a stubby dorsal flap. These forms do not deposit feed but produce mucus nets in their burrows though which they pump water and capture suspended food material.

The internal anatomy of echiurans displays more variation than the external anatomy (Figure 10.13*b*). The mouth leads to a relatively short pharynx and esophagus before joining a very long, twisted intestine. Along much of its length, the intestine bears a collateral tube, the **intestinal siphon**. The digestive tract ends at the anus with the short rectum and intestinal caecum.

At the junction of the rectum with the anus a pair of anal sacs evaginate into the coelom. These sacs may be simple or extensively branched and may extend nearly

(a)   *(b)*

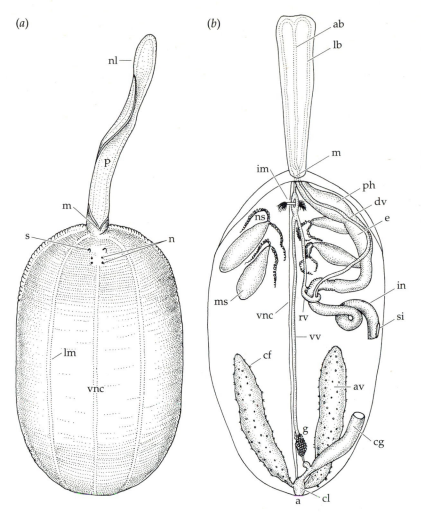

Figure 10.13 External and internal anatomy of an echiuran, *Listriolobus pelodes*. (*a*) Ventral view. (*b*) Generalized internal anatomy in dorsal view. Most of the coiled intestine has been omitted for clarity. The form of the nephrostome and its position on the metanephridium varies with the species. a, anus; ab, axial blood vessel; av, anal vesicle; cf, ciliated funnel; cg, ciliated groove; cl, cloaca; dv, dorsal blood vessel; e, esophagus; g, gonad; im, interbasal muscle spanning the proximal ends of setae; in, intestine; lb, lateral blood vessel; lm, longitudinal muscle band; m, mouth; ms, metanephridial sac; n, nephridiopores; nl, anterior nerve loop; ns, nephrostome; p, proboscis; ph, pharynx; rv, ring vessel; s, setae; si, siphon; vnc, ventral nerve cord; vv, ventral vessel. (*a* from Pilger 1980, 1993; *b* after Stephen and Edmonds 1972 and from Pilger 1993.)

half of the length of the trunk. Most anal sacs bear small ciliated funnels on their coelomic surface. These sacs are the functional excretory organs of the echiura.

Although most commonly there are two pair of nephridia, there may be a single nephridium or up to 40 pairs. One species, *Ikeda taenioides*, is reported to have between 200 and 400 unpaired nephridia. Because these nephridia function in the storage of differentiated gametes rather than in excretion, they are termed *metanephridia*. The nephrostome, the region of the nephridium that opens to the coelom, is usually elongated into ciliated and coiled lips that are specialized for the recovery of differentiated gametes from the coelomic fluid. Adult male bonellid echiurans are unique in that they possess a morphologically simpler and functional excretory organ, protonephridia. Their possession of this feature is correlated, as it is in other invertebrates (Ruppert and Smith 1988) with their small size and lack of vascular system.

Adult echiurans possess an unpaired ventral nerve cord that bifurcates at the anterior end to encircle the pharynx near the mouth. The confluence of these two branches on the dorsal side of the mouth extends into the

proboscis as the anterior nerve loop. There is no evidence of ganglia (a brain) at the confluence.

Most echiurans possess a closed circulatory system. The dorsal vessel extends from the gut into the proboscis, where it continues as the axial vessel (Figure 10.13*b*). At the tip it splits and returns to the base of the proboscis as the lateral vessels. Entering the trunk again, the vessels converge to form the ventral vessel, which extends the length of the trunk while attached to the ventral nerve cord. Blood in the posterior end of this vessel returns through intestinal sinuses and joins the dorsal vessel on the intestine. Blood flowing posteriorly in the ventral vessel may shunt to the dorsal vessel through the neurointestinal vessel. The circulatory system is absent in *Urechis caupo*.

The location of the gonad has not been determined in many echiurans. Where it is known, it has been described as being located on the ventral blood vessel near the posterior end of the body. Gametogenesis begins here but continues when the gametocytes are shed into the coelomic fluid. In one species, the testis was described as arising from the peritoneal tissue between the longitudinal muscle bands of the body wall. In spite of the popular use of *Ure-*

chis caupo as a model system by developmental biologists, the location of the gonad in that species is not known.

The first description of an echiuran was by Pallas (1766). The first taxonomic home of echiurans was the class Gephyrea in the phylum Annelida. They were grouped there with the Sipuncula and Priapula because these animals were thought to be a phylogenetic "bridge" between the annelids and holothurians. Later the sipunculans and priapulids were elevated to phylum status, but the echiurans remained in the Annelida. After the detailed developmental work of Newby (1940) showed the echiurans to be distinct from the annelids, he suggested the establishment of the phylum Echiuroidea. Later, the phylum name was changed to Echiura to conform to rules of taxonomic nomenclature.

Gametogenesis

Gametogenesis has three phases that are defined by the location of their events. The **gonadal phase**, when gametogenesis begins and the gametocytes (germ cells) are attached to the gonad; the **coelomic phase**, when gametes break free from the gonad and continue differentiation while floating free within the coelom; and the **storage phase**, when the most differentiated gametes are collected from the coelom and stored in the metanephridia until spawning. It has been suggested that the germinal cells are retroperitoneal, since follicle cells on oocytes of some species are of peritoneal origin. The gonad has not been examined beyond the light-microscopic level of study.

OOGENESIS. The gonadal phase of oogenesis has not been studied in detail. Some species produce oocytes that break free from the gonad as single cells, while in others they may break free as clusters of oocytes or with attached accessory or follicle cells (Figure 10.14). The origin or function of these cells has not been determined.

Based on oocyte size-frequency data and biochemical analysis, the coelomic phase of oogenesis is suspected to last about 4–6 months. Much of the focus of studies of oogenesis in the echiura have been on *Urechis caupo* and *Bonellia viridis*. As echiurans go, these species are atypical in that *Urechis* lacks a vascular system and, therefore, may undergo oogenesis (especially vitellogenesis) in a fashion that is not typical of other echiurans and *Bonellia* is characterized by prominent sexual dimorphism and internal fertilization. Information from many other echiuran species is necessary before broad generalizations can be made about oogenesis in this phylum.

Echiuran oocytes range from 60 μm to over 400 μm in diameter. In *Urechis caupo* it was reported that oocytes accumulate most of their constituents continuously during the coelomic phase, though at varying rates (Gould-Somero and Holland 1975).

At some point late in the coelomic phase of oogenesis, the most fully differentiated oocytes become selectively removed from the coelomic fluid by elongated nephros-

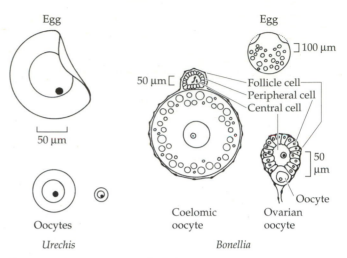

Figure 10.14 Oogenesis in the Echiura. *Urechis* oocytes differentiate without attached accessory or follicle cells. A prominent indentation is present in the differentiated egg. Oocytes of *Bonellia* have a cluster of cells attached that may enhance vitellogenesis in this relatively yolky egg. The egg, central cell, and peripheral cells are covered by a thin layer of follicle cells. (From Gould-Somero 1975; after Spengel 1879.)

tomal lips. The mechanism for this phenomenon is not known, but it is thought that, in *Urechis caupo*, the selection may be due to the formation of a prominent indentation on the oocyte as it nears the end of the coelomic phase. Other species do not produce eggs with indentations and must rely on other means such as a surface-recognition mechanism. Eggs sequestered in the elongated storage sac of the metanephridia probably do not undergo much further differentiation since the largest coelomic eggs are viable when tested directly after being removed from the coelom. *Urechis* eggs are removed from the coelom and stored in diakinesis of prophase I.

Very little is known about the stimuli that may be responsible for triggering spawning behavior of males or females. There is some evidence from *in vitro* observations to indicate that elevation of temperature may play a role, but the phenomenon and other potential stimuli have not been studied in a systematic fashion.

Bonellid males are reduced forms that reside in the metanephridium of females and fertilize eggs as they pass out of her body. Nothing is known of the mechanisms that trigger females to release eggs or males to release sperm as the eggs pass by.

SPERMATOGENESIS AND SPERMIOGENESIS Sperm separate from the gonad in syncytial clusters attached to a central mass of cytoplasm, the **cytophore**. On the basis of the size of the clusters and the presence or absence of a flagellum one can distinguish between primary and secondary spermatocytes, spermatids, and mature spermatozoa.

Echiuran sperm is considered to have a primitive morphology (Figure 10.15). The acrosome of echiuran sperm is

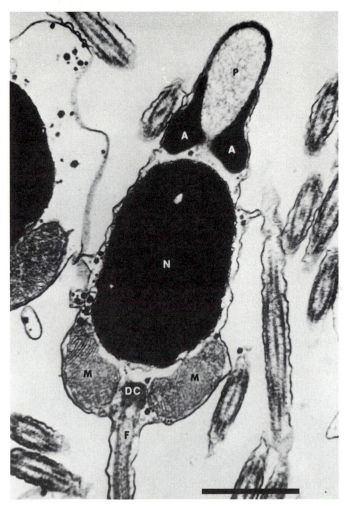

Figure 10.15 Sperm of *Listriolobus pelodes*. Transmission electron micrograph. A, acrosome; DC, distal centriole; F, flagellum; M, mitochondrion; N, nucleus; P, perforatorium. Scale bar: 1 μm. (From Pilger 1993.)

formed by the fusion of Golgi-produced vesicles. In bonellids the acrosome is cocked at a slight angle to the long axis of the nucleus, whereas in others it points directly anterior. A small space behind the acrosome is filled with homogenous, granular material. Condensation of the chromatin results in a round or oval nucleus. Mitochondria are present as a group of four spheres or as a single ring surrounding the distal centriole.

GAMETOGENIC CYCLES AND SPAWNING. Little information is available about gametogenic cycles in the Echiura. Pilger (1980) reported an annual cycle in *Listriolobus pelodes* off southern California. The gonad seems to be active in producing coelomic gametes throughout the year, but especially in late spring and summer. Coelomic oocytes reach maximum size and are collected into the storage organ during fall. Spawning occurs in winter and spring. Occasional notes about seasonal reproductive activity in other species indicate that egg production may be season-

al (e.g., winter in *Urechis unicinctus* and May–June in *Ikedosoma gogoshimense*) or continuous (as in *Urechis caupo*).

Fertilization

The events of fertilization are known best from *Urechis caupo*. Spawned in diakinesis, these eggs must complete maturation before cleavage. Sperm contact with the jelly coat of the egg triggers an acrosomal reaction wherein the acrosomal contents are released and acrosomal filaments are produced. Within a minute, the indentation in the egg rounds out. Shortly thereafter, a fertilization cone develops at the egg surface. Later, when the cone widens, the nucleus, mitochondria, and centriole pass through it and enter the ooplasm. Although the vitelline membrane elevates from the surface of the egg, the cortical granules are not released as is typical for many other invertebrates. Moreover, observations indicate that the cortical granules are actually retained until after the gastrula is formed.

An effective and rapid block to polyspermy is established within 10 seconds after sperm contact with the egg (Paul 1975a,b; 1976) and a secondary block is established a few minutes later. The rapid block is effective at the level of sperm-egg interaction. The late block is related to the formation of the fertilization membrane.

Parthenogenic egg activation has been reported in the Echiura. Exposure to dilute or ammoniacal seawater or seawater with trypsin is enough to trigger egg activation in *Urechis*, the latter treatment being most effective. In some cases, a few of these artificially activated eggs have developed to trochophore larvae. Recent work by Stephano and Gould (1995) has shown that a sperm acrosomal protein (P23) will induce parthenogenic cleavage in *Urechis caupo* under appropriate pH. This system is proving to be a valuable model for the investigation of the control mechanisms of cell division.

Cleavage

Cleavage of the zygote is spiral, holoblastic, and nearly equal. The first cleavage is meridional and produces the AB and CD blastomeres. The second cleavage plane is perpendicular to the first, but still meridional, and results in A, B, C, and D blastomeres. The third cleavage is equatorial and produces an upper quartet of micromeres that are rotated relative to the lower quartet of macromeres, as is typical of spiralians (Figure 10.16). In species with yolky eggs such as *Bonellia*, the yolk droplets concentrate at the vegetal pole so that the macromeres are slightly larger and more yolky than the micromeres.

By the 40-cell stage the apical cells have given rise to an apical rosette and to a group of cells that form an annelid cross (Figure 10.17). The arms of this cross project at a 45-degree angle to the sagittal and frontal planes of the future embryo. This contrasts to the molluscan cross that is present in the sipunculans (Figure 10.8). Blastulae develop 4–7 hours after fertilization in *Urechis caupo* and *Thalassema mellita*. Prototroch and apical tuft cilia form and project

Figure 10.16 Cleavage and early development in some echiurans. (From Gould-Somero 1975; *Bonellia*, after Spengel 1879.)

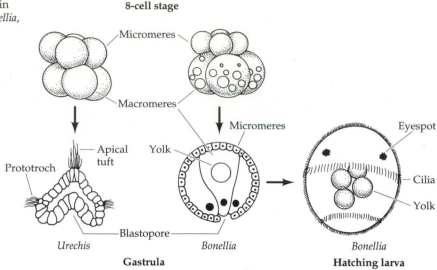

8-cell stage

Micromeres

Macromeres

Micromeres

Apical tuft

Prototroch

Yolk

Eyespot

Cilia

Yolk

Blastopore

Urechis **Gastrula** *Bonellia*

Bonellia **Hatching larva**

through the fertilization membrane. In most species the blastula is hollow, a coeloblastula, but in the bonellids, the blastula is a solid and unciliated, a stereoblastula.

Gastrulation

In *Bonellia*, division of the animal pole micromeres continues in such a fashion that they grow down over the vegetal hemisphere, producing a gastrula by epiboly (Figure 10.16). The macromeres that have been internalized by this process form endoderm. Cells at the lips of the blastopore migrate inward and represent the mesodermal precursors.

Gastrulation by invagination and ingrowth of the vegetal plate occurs in *Thalassema* (= *Lissomyema*) and *Urechis* (Figure 10.18a). The blastopore forms in both species in the same general fashion. The stomodeum forms from ectoderm that rolls inward at the blastopore.

Mesoderm arises from the 4d cell. Descendants of this cell produce two mesodermal bands in the blastocoel. The bands thicken and split into visceral and parietal peritoneum.

The Trochophore Larva

Trochophore larvae are formed in all echiuran species; a number of ciliary fields are present. The prototroch surrounds the trochophore anterior to the mouth and is the primary swimming organ (Figure 10.18b). In some species,

additional bands are present just posterior to the mouth. These auxiliary bands are termed the *mesotroch* (in *Urechis*) and the *metatroch* (in *Echiurus*). At the posterior end of the larva the telotroch surrounds the anus. The preoral lobe of most trochophores is covered by short cilia. A thin band of cilia, the neurotroch, may extend along the ventral midline from the mouth to the anus. In those species that possess it, bonellin pigment is usually present at this stage. Many trochophores seem to be negatively geotactic, swimming to the surface water even though their parents are benthic.

Trochophore larvae of *Echiurus* and *Listriolobus* possess protonephridia, but *Urechis* and *Bonellia* do not. The mouth forms at the blastopore and the anus forms very near to it and slightly posterior. A proctodeum, an invagination of ectoderm at the anus, is absent. The ventral nerve cord is derived from two parallel rows of cells that delaminate from the post-trochal ectoderm. Shortly after formation, the two rods fuse and form the single, ventral nerve cord.

Bonellid eggs, having been fertilized internally by dwarf males in the female reproductive tract, are yolky and lead to a lecithotrophic larva that hatches within two days and begins crawling over the substratum. A small, lecithotrophic, green-pigmented trochophore is produced by the third or fourth day.

Protonephridia are not present in the trochophore of *Bonellia viridis*. At metamorphosis, females develop meta-

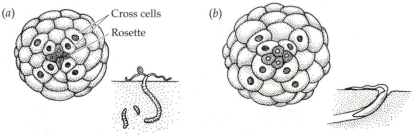

(a) Cross cells / Rosette (b)

Figure 10.17 Embryos of an annelid and an echiuran, illustrating the rosette cells and typical annelid cross. The arms of the cross are shown in the interradii of the rosette (a) in annelids and (b) in echiurans. (After Ruppert and Barnes 1994; a based on Wilson 1892; b based on W. W. Newby 1932.)

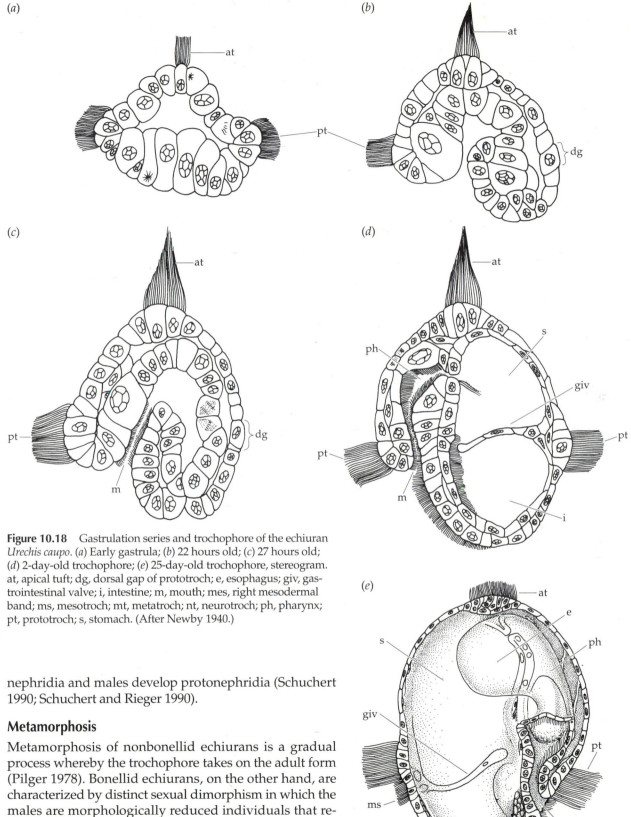

Figure 10.18 Gastrulation series and trochophore of the echiuran *Urechis caupo*. (*a*) Early gastrula; (*b*) 22 hours old; (*c*) 27 hours old; (*d*) 2-day-old trochophore; (*e*) 25-day-old trochophore, stereogram. at, apical tuft; dg, dorsal gap of prototroch; e, esophagus; giv, gastrointestinal valve; i, intestine; m, mouth; mes, right mesodermal band; ms, mesotroch; mt, metatroch; nt, neurotroch; ph, pharynx; pt, prototroch; s, stomach. (After Newby 1940.)

nephridia and males develop protonephridia (Schuchert 1990; Schuchert and Rieger 1990).

Metamorphosis

Metamorphosis of nonbonellid echiurans is a gradual process whereby the trochophore takes on the adult form (Pilger 1978). Bonellid echiurans, on the other hand, are characterized by distinct sexual dimorphism in which the males are morphologically reduced individuals that reside in the nephridia of the females. Bonellid larvae are considered to be sexually indifferent until settlement, at which time they will become male if they settle on or near a female and female if they settle away from the influence of an adult female. The trochophore can attach to the female proboscis by means of a special attachment organ. In *Bonellia viridis* this organ is a ventral field of secretory cells that produce adhesive material so that the larva can remain in contact with the female. In *Bonellia fuliginosa* the attachment organ is a muscular sucker. As such, bonellid metamorphosis differs from that of nonbonellid echiurans.

The change from a pelagic to a benthic existence is effected by the loss of the prototrochal cilia. This process seems to occur over a 4- to 5-day period in *Urechis* and *Lissomyema* during which time these larvae alternate between benthic and semipelagic existence. There is some evidence that echiuran larvae may select substrates that have a high organic content. Sensory structures that would be used to select these substrates have not been described.

In the trochophore of *Urechis caupo*, the stomach is separated from the intestine by a shelflike **gastrointestinal valve** (Figure 10.18*d, e*; 10.19*a*). Food passes from the stomach to the intestine through a small hole in the valve near the ventral midline. In the intestine, a ciliated groove runs from the stomach–intestine opening along the ventral midline to the anus. During metamorphosis, the stomach and larval tissue anterior to the gastrointestinal valve enlarges relative to the intestine. Thus, the gastrointestinal valve becomes distorted so that it forms a dorsal cover over the ventral ciliated groove and a caecum where it bulges into the intestine (Figure 10.19*b*). The posterior wall of the caecum ruptures, forming a new passage from the stomach directly to the intestine (Figure 10.19*c*). The gastrointestinal valve, thus, covers the ciliated groove and, according to Newby (1940), forms the midgut siphon of the adult (Figure 10.19*d*). Others have argued that the ciliated groove in the larva and the adult siphon are not related.

Externally, metamorphosis is characterized by the loss of the prototroch, neurotroch, and telotroch, the enlargement of the trunk, and elongation of the proboscis (Figure 10.19*e–g*). Regardless of its final adult shape, the proboscis is formed in essentially the same fashion in all species. The process can best be understood by imagining the preoral lobe being stretched out anteriorly. The dorsal portion of the circumesophageal nerve ring and blood vessels extend out with it and are present in the adult proboscis as the anterior nerve loop and the lateral and median blood vessels. Short cilia that covered the preoral lobe become restricted to the ventral region of the proboscis, and the musculature of the organ increases substantially.

The postoral region becomes the trunk by expansion, loss of the protonephridia, if present, and the formation of metanephridia. There have been several reports in the literature of transient metamerism in echiuran larvae. This metamerism is reported to be present in mesoderm and mesodermal derivatives, in ciliary bands, and in epidermal glands. Early reports took this to be evidence of vestigial segmentation and of a relationship between the Echiura and Annelida. However, these phenomena appear to be unrelated to segmentation in that they do not

have a significant spatial relationship with ventral nerve cord ganglia, as would be expected if they were remnants of true segmentation.

The larva of *Bonellia viridis* is sexually indifferent until settlement. Because of the extreme sexual dimorphism, males and females metamorphose differently, although adult organs are present in rudimentary form in the larva. Metamorphosis of the female, because it has the typical echiuran body form, is basically the same as in nonbonellid echiurans. The preoral lobe elongates to form the bifid proboscis, and the cilia in this region become restricted to the ventral surface. Photoreceptors present in the preoral lobe of the larva are carried to the tip of the proboscis, where they remain functional and allow the proboscis to respond to changes in illumination. The trunk basically forms by expansion and increased muscularization. Invaginations of the stomodeum and proctodeum open to the gut, making it functional and complete. Protonephridia regress and metanephridia are produced.

Male bonellids metamorphose into diminutive forms that produce sperm while inside the female body. At the time of metamorphosis they loose their prototrochal and metatrochal cilia but retain their uniform covering of short cilia over the body (Figure 10.20). The glandular region used by male *Bonellia viridis* to attach to the female expands posteriorly and is retained, but the muscular sucker used by male *Bonellia fuliginosa* is lost.

In direct contrast to bonellid female and nonbonellid metamorphosis, the male bonellid pretrochal lobe regresses, deflecting the anterior nerve ring posteriorly. Perhaps as a consequence of pretrochal lobe regression, the eyes are lost and the posttrochal region elongates but does not expand. Green bonellin pigment, characteristic of this group of echiurans, is lost in males. Reduced to a small, vermiform individual with an incomplete gut, simplified nervous system, protonephridia, and testis, the male bonellid enters the female nephridiopore and takes up residence in the **androecium**, an expanded region of the metanephridium past which eggs must move during spawning.

The suppression of adult structures during metamorphosis of males and the mechanism for sex determination is a very interesting problem for developmental biologists and is not understood although it has been studied for more than 80 years. Early studies suggested that when a larva settles and associates with an adult female, female differentiation is inhibited and a male is produced. In the absence of this association and stimulus at the time of settlement, a female develops. Aqueous extracts of the female proboscis or intestine were able to induce a significant number of sexually indifferent larvae to become males, whereas the remainder died or became "intersexes," individuals in which the sex could not be determined. Association with the sex-determining substance has a temporal component whereby anterior structures differentiate after a short exposure but progressively posterior structures re-

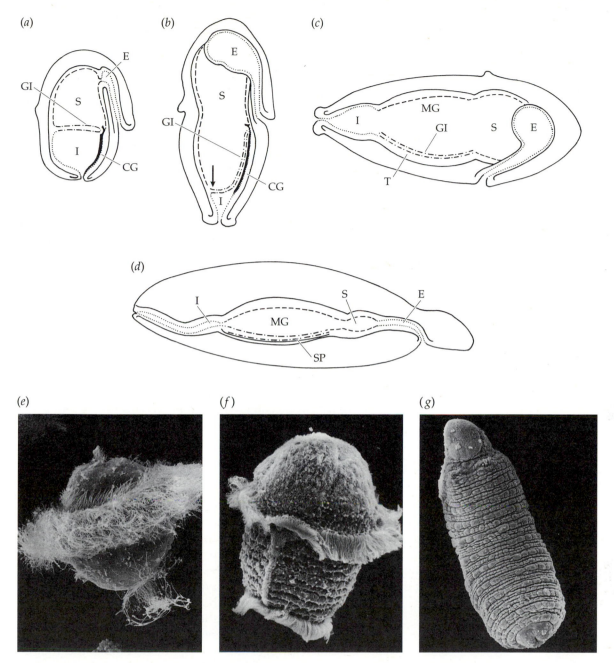

Figure 10.19 Metamorphosis of the *Urechis caupo* trochophore into a juvenile. The larval gastrointestinal valve transforms into the primordium of the adult midgut siphon. (*a*) 4-day-old larva; the gastrointestinal valve separates the larval stomach from the intestine. (*b*) 35-day-old larva; the dorsal and lateral margins of the gastrointestinal valve have shifted posteriorly. (*c*) 50-day-old larva; the gastrointestinal valve has ruptured dorsally, making a linear pathway from the midgut to the intestine. The valve lies on the floor of the midgut and transforms the ciliated groove into a ciliated tube. (*d*) 60-day-old larva; the siphon primordium lies along the ventral side of the midgut. (*e*) SEM of young *Urechis* trochophore larva. Note the prototroch, telotroch, and apical tuft. (*f*) 15-day-old *Urechis* larva showing well-defined neurotroch and expanded telotroch. (*g*) Juvenile *Urechis* after settling. E, esophagus; I, intestine; MG, midgut; S, larval stomach; SP, siphon primordium; T, ciliated tube. Arrow indicates the region of the caecum where the gastrointestinal valve will rupture. (*a–d* after Newby 1940, from Pilger 1978; photographs *e–g* © M. Apley.)

quire a longer exposure to be induced. Modification of the ionic concentration of various seawater elements also is able to modify the gender of 70-90% of the offspring. Environment may not be the sole factor in *Bonellia viridis* sex determination. Leutert (1974) estimates that about half of the individuals already have their sex determined before they settle. In these individuals, the environment does not appear to play a major role.

Experimental work in sex determination in *Bonellia viridis* seems to be following a path reminiscent of that of

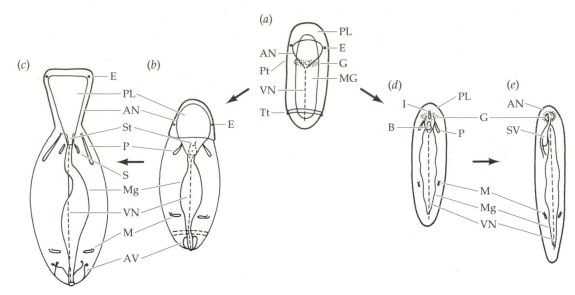

Figure 10.20   Metamorphosis of male and female *Bonellia viridis* from a sexually indifferent trochophore. (*a*) A sexually indifferent trochophore. (*b*) An intermediate stage in male metamorphosis; note the loss of trochal bands and post-trochal body cilia. (*c*) A juvenile female; the pretrochal lobe has elongated to form the proboscis; the trunk has enlarged greatly. (*d*) An intermediate stage in male metamorphosis; the trochal bands are lost and the pretrochal lobe has shortened considerably. (*e*) A juvenile male; the pretrochal lobe is reduced to a mere cap and the short body cilia are retained. AN, anterior nerve loop; AV, anal vesicle; B, blastopore; E, eye; G, gland complex; I, invagination; M, metanephridium; Mg, midgut; P, protonephridium; PL, pretrochal lobe; Pt, prototroch; S, setae; St, stomadeum; SV, seminal vesicle; Tt, telotroch; VN, ventral nerve. (After Baltzer 1925; from Pilger 1978.)

experimental studies of induction in vertebrate embryology. Therefore, invertebrate embryologists and developmental biologists interested in *Bonellia viridis* would be well advised to study the design of vertebrate experiments when approaching their problem.

Synthesis: Phylogeny and the Contribution of Embryology

A review of the taxonomic history of the Sipuncula and Echiura demonstrates the difficulty biologists have had finding a phylogenetic home for these invertebrates. Based on their adult morphology alone, they have been linked to parasitic worms, the Holothuria, Annelida, Priapula, Bryozoa, and the Phoronida. Clearly, morphological data from adult specimens alone has not produced a satisfactory understanding of the phylogenetic relationship between sipunculans and echiurans and other phyla.

Late in the nineteenth century, biologists turned to embryology to provide new information about the relationships among animal groups. This information provided biologists with embryological examples of unity and diversity among animals and new phylogenetic inferences

were drawn. Biologists had available to them information about fertilization, cleavage, cell lineage, gastrulation, fate maps, organogenesis, and metamorphosis. Although many of the initial studies dealt with common invertebrates such as echinoderms, molluscs, and tunicates, it was only a matter of time until this effort became directed toward smaller phyla whose evolutionary position had been enigmatic. Today, biologists have more detailed embryological data on many more species as well as molecular information from developmental biology. This body of information, therefore, has given invertebrate and evolutionary biologists new insight into the relationships among many animal phyla. Our understanding of the phylogenetic position of the Sipuncula and Echiura has certainly benefited from this information.

Although there are historical taxonomic linkages of the Sipuncula and Echiura to each other and to other phyla, the clearest affinities, though still ambiguous, seem to be with the annelids and the molluscs. These four phyla are united in their possession of a number of basic characters: spiral cleavage, the blastopore becoming the mouth, schizocoely, and the formation of a trochophore larva. Further, they are the only ones that share the production of a rosette and apical cross early in their development. Other important characters vary with the species and phylum.

Segmentation is perhaps the most significant trait of the annelids. If either the Sipuncula or Echiura were related to annelids, one would predict that some evidence of vestigial segmentation would be present during embryogenesis. Attempts to demonstrate transient segmentation or metamerism in sipunculan and echiuran larvae have not produced significant evidence to indicate that the trait may have been present in the ancestral forms of the phyla. For instance, in the Echiura, segmental ganglia are present in the ventral nerve cord, but the periodicity in this segmentation is not related to mesodermal segmentation, which is absent. Instead, the periodicity is correlated with rings of glands in the trunk epidermis. This may be the only connection to a segmental character, but it is considered to represent pseudometamerism. It constitutes extremely weak evidence on which to build a phylogenetic relationship to a phylum whose hallmark is segmentation. There is no evidence of segmentation during sipunculan embryogenesis. The recognition that true segmentation is absent from these phyla is sufficient to keep them from being included in the Annelida and from being considered as being evolutionary derivatives from the annelids.

Sipunculans seem to be more closely related to the molluscs than the annelids because their development includes the molluscan cross, and the buccal organ and lip of pelagosphera bear some resemblance to the oral structures and foot of molluscs. Many scientists consider the latter resemblance of questionable phylogenetic significance and not a homology. If sipunculans are related to the molluscs, they probably diverged from that line very early, before definitive molluscan characters evolved.

The Echiura share many more important features with the annelids than with the molluscs but, at the same time, they retain their place as a phylum distinct from the Annelida because they lack true segmentation. The ventral nerve cord of echiurans arises as paired cellular rows that eventually fuse into a single cord, similar to the way it forms in annelids. Further, they also produce annelid cross in the early embryo and an annelid-like trochophore. Adults echiurans produce setae (epidermal bristles) similar to the annelids. It seems most likely, then, that the echiurans and annelids evolved from a common ancestor but that the Echiura diverged from the annelid line before segmentation evolved. Clark (1969), Jägersten (1972), and Rice (1985) have reviewed the phylogenetic relationships of the Sipuncula and Echiura. Interested readers are referred to these papers for greater detail about embryological contributions to our understanding of these fascinating animals.

Literature Cited

Åkesson, B. 1961. The development of *Golfingia elongata* Keferstein (Sipunculidea) with some remarks on the development of neuroscretory cells in sipunculids. *Ark. Zool.* 13: 511–531.

Amor, A. 1993. Reproductive cycle of *Golfingia margaritacea*, a bipolar sipunculan, in subantarctic water. *Mar. Biol.* 117: 409–414.

Baltzer, F. 1925. Untersuchungen über die Entwicklung und Geschlechtsbestimmung der *Bonellia*. *Publ. Staz. Zool. Napoli* 6: 223–286.

Clark, R. B. 1969. Systematics and phylogeny: Annelida, Echiura, Sipuncula. In *Chemical Zoology*, Vol. 4. Academic Press, New York, pp. 1–68.

Cutler, E. B. 1994. *The Sipuncula: Their Systematics, Biology and Evolution.* Cornell University Press, Ithaca, NY.

Gerould, J. H. 1906. The development of *Phascolosoma*. (Studies on the embryology of the Sipunculidae II.) *Zool. Jahrbücher* 23: 77–162.

Gonse, P. 1956. L'ovogènese chez *P. vulgare*. II. Recherchés biométriques sur les ovocytes. *Acta Zool.* 37: 225–233.

Gould-Somero, M. C. and L. Holland. 1975. Oocyte differentiation in *Urechis caupo* (Echiura): A fine structural study. *J. Morphol.* 147: 475–505.

Hall, J. R. and R. S. Scheltema. 1975. Comparative morphology of open ocean pelagosphaera. In *Proceedings of the International Symposium on the Biology of the Sipuncula and Echiura*, M. E. Rice and M. Todorovic (eds.). Naucno Delo Press, Belgrade, pp. 183–197.

Hatchek, B. 1883. Uber Entwicklung von *Sipunculus nudus. Arb. Zool. Inst. Univ. Wien Zool. Stat. Triest.* 5: 61–140.

Jägersten, G. 1972. *Evolution of the Metazoan Life Cycle.* Academic Press, London.

Leutert, R. 1974. Zur Geschlectsbestimmung und Gametogenese von *Bonellia viridis. R. J. Embryol. Exp. Morphol.* 32: 169–193.

McBride, E. W. 1914. *Textbook of Embryology. Vol. 1, Invertebrata.* Macmillan and Co., London.

Newby, W. W. 1932. The early embryology of *Urechis caupo. Biol. Bull.* 63: 387–399.

Newby, W. W. 1940. The embryology of the echiuroid worm *Urechis caupo. Mem. Amer. Phil. Soc.* 16: 1–219.

Pallas, P. S. 1766. *Lumbricus echiurus.* In *Miscellania Zoologica Hagae Comitum*, pp. 146–151.

Paul, M. 1975a. The polyspermy block in eggs of *Urechis caupo*: Evidence for a "rapid" block. *Exp. Cell Res.* 90: 137–143.

Paul, M. 1975b. Release of acid and changes in light scattering properties following fertilization of *Urechis caupo* eggs. *Dev. Biol.* 43: 299–312.

Paul, M. 1976. Evidence for a polyspermy block at the level of the sperm–egg plasma membrane fusion in *Urechis caupo. J. Exp. Zool.* 196: 105–112.

Pilger, J. F. 1978. Settlement and metamorphosis in the Echiura: A review. In *Settlement and Metamorphosis of Marine Invertebrate Larvae*, F. S. Chia and M. E. Rice (eds.). Elsevier/North-Holland Biomedical Press, New York, pp. 103–112.

Pilger, J. F. 1980. The annual cycle of oogenesis, spawning, and larval settlement of the echiuran *Listriolobus pelodes* off southern California. *Pac. Sci.* 34: 129–142.

Pilger, J. F. 1982. Ultrastructure of the tentacles of *Themiste lageniformis* (Sipuncula). *Zoomorphology* 100: 143–156.

Pilger, J. F. 1987. Reproductive biology and development of *Themiste lageniformis*, a parthenogenic sipunculan. *Bull. Mar. Sci.* 41: 59–67.

Pilger, J. F. 1989. A cytophotometric study of parthenogenesis in *Themiste lageniformis* (Sipuncula). *Bull. Mar. Sci.* 45(2): 415–424.

Pilger, J. F. 1993. Echiura. In *Microscopic Anatomy of Invertebrates*, F. W. Harrison and M. E. Rice (eds.). Wiley-Liss, New York, pp. 185–236.

Rice, M. E. 1967. A comparative study of the development of *Phascolosoma agassizii, Golfingia pugettensis*, and *Themiste pyroides* with a discussion of developmental patterns in the Sipuncula. *Ophelia* 4: 143–171.

Rice, M. E. 1970. Asexual reproduction in a sipunculan worm. *Science* 167: 1618–1620.

Rice, M. E. 1975a. Observations on the development of six species of Caribbean Sipuncula with a review of development in the phylum. In *Proceedings of the International Symposium on the Biology of the Sipuncula and Echiura*, M. E. Rice and M. Todorovic (eds.). Naucno Delo Press, Belgrade, pp. 141–160.

Rice, M. E. 1975b. Sipuncula. In *Reproduction of Marine Invertebrates*, A. C. Giese and J. S. Pearse (eds.). Academic Press, New York, pp. 67–127.

Rice, M. E. 1985. Sipuncula: Developmental evidence for phylogenetic inference. In *The Origins and Relationships of Lower Invertebrates*, S. C. Morris, J. D. George, R. Gibson and H. M. Platt (eds.). Oxford University Press, Oxford, pp. 274–296.

Rice, M. E. 1986. Factors influencing larval metamorphosis in *Golfingia misakiana* (Sipuncula). *Bull. Mar. Sci.* 39: 362–375.

Rice, M. E. 1988. Sipuncula. In *Reproductive Biology of Invertebrates*, Vol. IV, Part A, K. G. Adiyodi and R. G. Adiyodi (eds.). Oxford and IBH Pub. Co., New Delhi, pp. 263–280.

Rice, M. E. 1989. Comparative observations of gametes, fertilization, and maturation in sipunculans. In *Reproduction, Genetics, and Distributions of Marine Organisms*, J. S. Ryland and P. A. Tyler (eds.). Olsen and Olsen, Fredensborg, Denmark, pp. 167–182.

Rice, M. E. 1993. Sipuncula. In *Microscopic Anatomy of Invertebrates*, F. W. Harrison and M. E. Rice (eds.). Wiley-Liss, New York, pp. 237–325.

Rice, M. E. and J. F. Pilger 1993. Sipuncula. In *Reproductive Biology of Invertebrates*, Vol. VI, Part A, K. G. Adiyodi and R. G. Adiyodi (eds.). Oxford and IBH Pub. Co., New Delhi, pp. 297–310.

Ruppert, E. E. and R. D. Barnes. 1994. *Invertebrate Zoology*. Saunders College Publishing, New York.

Ruppert, E. E. and M. E. Rice. 1983. Structure, ultrastructure, and function of the terminal organ of a pelagosphera larva (Sipuncula). *Zoomorphology* 102: 143–163.

Ruppert, E. E. and P. R. Smith. 1988. The functional organization of filtration nephridia. *Biol. Rev.* 63: 231–258.

Scheltema, R. S. 1986. Long-distance dispersal by planktonic larvae of shoal-water benthic invertebrates among central Pacific islands. *Bull. Mar. Sci.* 39: 241–256.

Scheltema, R. S. and M. E. Rice. 1990. Occurrence of teleplanic pelagosphaera larvae of sipunculans in tropical regions of the Pacific and Indian oceans. *Bull. Mar. Sci.* 47: 159–181.

Schuchert, P. 1990. The nephridium of the *Bonellia viridis* male (Echiura). *Acta Zool.* 71: 1–4

Schuchert, P. and R. M. Rieger. 1990. Ultrastructural observations on the dwarf male of *Bonellia viridis* (Echiura). *Acta Zool.* 71: 5–16.

Spengel, J. W. 1879. Beiträge zur Kenntnis der Gephyren. Die Eibildung, die Entwicklung und das Männchen von *Bonellia*. *Mitt. Zool. Sta. (Neapel)* 1: 347–419.

Stephano, J. L. and M. C. Gould. 1995. Parthenogenesis in *Urechis caupo* (Echiura). I. Persistence of functional maternal asters following activation without meiosis. *Dev. Biol.* 167: 104–117.

Stephen, A. C. and S. J. Edmonds. 1972. *The Phyla Sipuncula and Echiura*. British Museum (Natural History).

Williams, J. A. 1977. Functional development in four species of the Sipuncula. Ph.D. dissertation, Univ. Hawaii.

Wilson, E. B. 1892. The cell lineage of *Nereis*. *J. Morphol.* 6: 361–466.

CHAPTER 11

Gastropods, the Snails

J. R. Collier

THE GASTROPODA ARE THE LARGEST AND MOST DIVERSE class of molluscs. The gastropods or snails are unsegmented and asymmetrical molluscs, with a well-developed head, a foot, and a visceral mass typically contained in a spirally coiled shell (Hyman 1967). They occupy a wide variety of habitats and are the most successful group of this phylum.

Gastropods are of special interest for developmental studies because they have a regular pattern of cleavage whose cell lineage has been thoroughly studied. The marked animal-vegetal polarity of the gastropod egg emphasizes the importance of ooplasmic segregation in development. In some forms, for example *Ilyanassa* and *Bithynia*, vegetal cytoplasm is segregated into a cytoplasmic protrusion, a **polar lobe**, that can be removed during early cleavage and thereby provide an excellent opportunity for studying the role of egg cytoplasm in determination and differentiation.

From an evolutionary view the gastropods show a great deal of diversity in form and habitat and are the most extensively speciated group among the molluscs, which are second only to the arthropods in species diversity. The diversification of the gastropods is reflected in their adaptive radiation, which has occurred many times in their evolutionary history. Two examples of their adaptive radiation are modification for water circulation and gas exchange. These evolutionary changes, along with extensive adaptive radiation in the reproductive system, account for the success of some gastropods in being one of the few animal groups to inhabit marine, freshwater, and terrestrial environments.

Gastropod Systematics

The systematics of the gastropods (Barnes 1987) is outlined in Table 11.1. The gastropods are divided into three subclasses. The most primitive are members of the subclass Prosobranchia (anterior gills), and from some prosobranch-like ancestor there arose the subclasses Opisthobranchia (posterior gills) and Pulmonata (a lung).

Prosobranchia

Prosobranchs are torted snails with a mantle cavity containing gills located anteriorly in front of the visceral mass. (Modern gastropods are characterized by the rotation of the visceral mass and overlying shell 180 degrees with respect to the head and foot. This **torsion** occurs during development and will be discussed later. Gastropods retaining this condition are said to be *torted*. Those gastropods that reverse this condition later are said to be *detorted*.) Prosobranchs are mostly marine, though there are terrestrial and freshwater species. The sexes are usually separate and development is indirect, with the production of a **veliger larva**. The prosobranchia are the most numerous, widely distributed, and diverse of the gastropods, and include the limpets, abalones, turban snails, periwinkles, moon snails, whelks, conchs, and volutes.

Opisthobranchia

The opisthobranchs, the most highly specialized group of the gastropods, are almost entirely marine and hermaphroditic. The Opisthobranchia also have a veliger larva as the hatching stage. The more highly evolved members of the opisthobranchs are detorted, have lost the spiral shell, and have a reduced (or absent) mantle cavity. This subclass includes the sea slugs and their relatives.

Pulmonata

The pulmonates are hermaphroditic, most have direct development, and live in freshwater or on land, though there are a few marine species. The pulmonata are conservative in their structure but show the most extensive physiologi-

TABLE 11.1 *Systematics of the Class Gastropoda*

Taxon	Characteristics	Examples
Subclass Prosobranchia	Marine, freshwater, and terrestrial forms. Mantle cavity located anteriorly. External shell, usually with an operculum. Mostly dioecious.	
Order Archaeogastropoda	Primitive forms with two gills	*Patella, Trochus*
Order Mesogastropoda	Mostly marine, but many freshwater and terrestrial forms. One gill, one auricle, and one nephridium.	*Crepidula, Littorina, Viviparus, Bithynia*
Order Neogastropoda	One gill, one auricle, and one nephridium, but have a radula and a complex osphradium.	*Ilyanassa, Buccinum, Busycon, Urosalpinx*
Subclass Opisthobranchia	Nine orders of marine gastropods that are detorted 90°. Simultaneous or protandric hermaprodites.	
Subclass Pulmonata	One auricle and one nephridium. Vascularized lungs. Hermaphroditic.	
Order Basommatophora	One pair of tentacles. Mostly freshwater.	*Lymnaea, Physa, Biomphalaua, Bulinus, Acroloxus, Ancylus*
Order Stylommatophora	Two pairs of tentacles. Terrestrial.	*Helix, Limax, Arion, Deroceras, Bradybaena*
Order Systellommatophora	Slugs with anus at posterior end of body.	*Onchidium, Rhodope*

cal adaptation to different environments. Most land snails and terresterial slugs are in this subclass.

Gastropod Body Plan

The morphology of the gastropods is reviewed by Hyman (1967) and Barnes (1987), the prosobranchs by Fretter and Graham (1962) and Voltzow (1994), and the opistho-branchs by Gosliner (1994). The gastropod body parts are a head, foot, mantle, and visceral mass. A schematic of the body plan of a primitive gastropod is in Figure 11.1.

The **head** has tentacles, eyes, mouth, cerebral ganglia, and cerebral connectives. The **foot** is a large muscular organ whose primary function is locomotion and adhesion but is also used in mating, capture of prey, shaping egg capsules, and escape from predators. The prosobranchs have on their foot a horny structure, the **operculum**, that functions as a door to close the opening of the shell when the foot is retracted. The operculum is proteinaceous and sometime calcified. The **mantle** is an epithelium that lines the shell and covers the viscera. At its anterior end it forms a mantle cavity that contains respiratory organs and genital, excretory, and anal openings. The mantle cavity also contains the retracted head and foot. Anteriorly, the mantle folds into a **siphon** that allows for the intake of water. At the base of the siphon is the **osphradium**, whose function is chemoreception. The siphon and osphradium play an

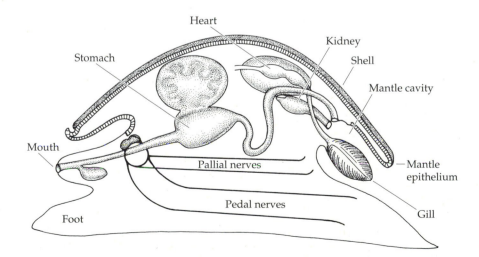

Figure 11.1 Gastropod body plan. (After van den Biggelaar et al. 1994.)

important role in detecting sources of food. The **visceral mass** contains a digestive tract, digestive gland (midgut gland, liver, or hepatopancreas), heart (one ventricle and two auricles), nephridium, and reproductive organs. The digestive gland, whose function is absorption of nutrients from partially digested food and the secretion of enzymes, is the largest organ in the visceral mass. The nephridium (often called metanepridium [Hyman 1967] or kidney) is the molluscan excretory organ. In most forms, except for the nudibranchs, the visceral mass is contained in and attached to an external **shell**, while the head and foot are protrusible from the shell. The shell is a conical spire composed of tubular whorls coiled about a central axis (the **columella**). The shell is produced by glandular cells in the mantle and is composed of calcium carbonate crystals embedded in a proteinaceous matrix.

Protostomian Development

Developmentally, the Gastropoda are protostomes and have spiral determinate cleavage. Their eggs have an animal-vegetal polarity that defines the anterior-posterior axis of the embryo. The animal pole is the anterior pole and the vegetal pole is the posterior pole. The cytoplasmic segregation of morphogenetic determinants, which are initially distributed along the animal-vegetal axis, is associated with the spiral cleavage of the gastropods. The localization of these determinants plays a major role in the early organization of the gastropod embryo.

Other protosomian characters are that the mouth is formed from a blastopore, and formation of mesoderm is from a **primary mesentoblast** (the 4d cell). Derivatives of the 4d cell (the Me^1 and Me^2 cells) are teloblasts that give rise to a bilateral set of mesodermal bands. A feature distinguishing gastropods from other protostomes is that formation of the coelom does not involve a schizocoelic split within the mesodermal bands. (Molluscs do not have a coelom between the digestive tract and the body wall; a true coelom is limited to the pericardial cavity, the lumen of the gonad, and the nephridia [Hyman 1967].) A free-swimming *trochophore* larva, characteristic of many protostomes, is not present in all gastropods, but is formed only in the most primitive prosobranchs (the archaeogastropods), which produce a free-swimming trochophore larva (Figure 11.2) that precedes the development of a veliger larva (Figure 11.3a). In all other gastropods the trochophore larva is suppressed or passed through quickly, and development is either direct, as in most pulmonates and nearly all freshwater opisthobranchs, or a free-swimming veliger larva is produced that metamorphoses into a young adult snail. Most pulmonates hatch as young snails that are anatomically complete except for the reproductive system.

Figure 11.2 Transformation of the *Patella* trochophore into a veliger larva. (*a*) View from the dorsal surface. (*b*) View from the side. ap tuft, apical tuft; an c, anal cell; ft, foot; prot, prototroch; shg, shell gland; vel, velum. (After Patten 1886.)

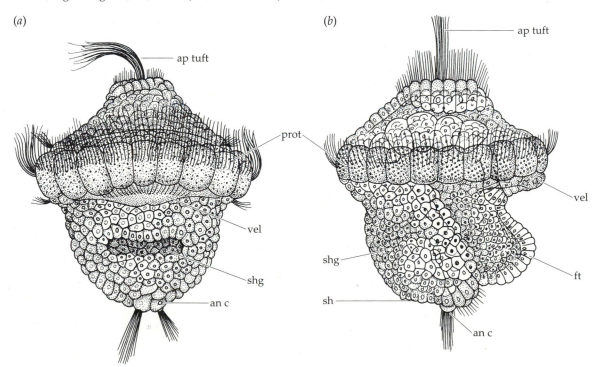

(*a*)

(*b*)

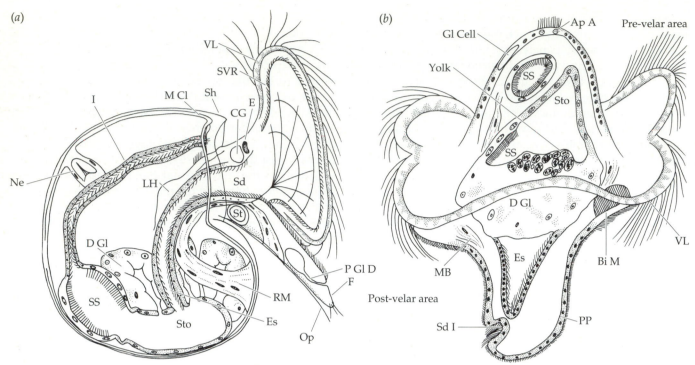

Figure 11.3 Normal and lobeless larvae of *Ilyanassa obsoleta*. (*a*) Normal 8- to 9-day veliger larva. The left velum is omitted. (*b*) Lobeless larva. Ap A, apical area; Bi M, birefringent mass; CG, cerebral ganglion; D Gl, digestive gland; E, eye; Es, esophagus; F, foot; Gl Cell, ectodermal gland cell; I, intestine; LH, larval heart; MB, muscle band; M Cl, mantle collar; Ne, nephridium; Op, operculum; PP, posterior protrusion; P Gl D, pedal gland duct; RM, retractor muscle; S S, style sac; Sd, stomodeum; Sd I, stomodeal-like invagination; Sh, shell; St, statocyst., Sto, stomach; S V R, secondary velar row; V L, velar lobe. (After Atkinson 1971.)

Reproductive Systems

Most gastropods reproduce sexually. The only form of asexual reproduction is parthenogenesis in three unrelated families of prosobranchs (Webber 1977; Aldridge 1983; Pointier et al. 1992). Many gastropods undergo metamorphosis before the adult is formed; however, in many pulmonates development is direct but the young adult that hatches has yet to develop a reproductive system (see Hyman 1967; Voltzow 1994; Gosliner 1994).

Gastropod reproductive systems are complex and varied. Jong-Brink et al. (1983) have pointed out that the development of a variety of reproductive systems such as hermaphroditism, sex reversal, and parthenogenesis has enhanced the possibility of survival and made it possible for gastropods to successfully establish themselves in a variety of environments.

The gastropod reproductive system consists of an ovary, a testis (in hermaphroditic forms a combined ovotestis), a sperm duct, an oviduct, and a variety of accessory glands or vesicles associated with the gonoducts. There are *seminal vesicles* that store endogenous sperm in hermaphroditic forms. Seminal vesicles have *sphincter*

muscles that relax to release sperm that are then moved by a ciliated epithelium along the *sperm duct* and often into a *vas deferens*. There are *seminal receptacles (spermatheca)* for storage of exogenous sperm (often for long-term storage), and in some forms a *copulatory bursa* for short-term storage of sperm received at copulation. In males there is a *prostate gland* composed of glandular cells and a ciliated epithelium. The prostate gland secretes a fluid into the sperm duct just before copulation, which dilutes the sperm and probably aids in sperm transport. At copulation, the penis transfers sperm into the mantle cavity of a female or into the female *gonopore.*

The female system contains, in addition to the sperm vessels described above, a number of cavities and glands. There are albumen, membrane, mucous, and capsule glands (sometimes replaced by a *"jelly gland"* in snails that lay their eggs in a gelatinous mass), which provide nutritive and protective elements for eggs.

Prosobranchs

Prosobranchs are dioecious. Primitive prosobranchs (archaeogastropods) shed their gametes and fertilization is external. In less primitive prosobranchs (neogastropods and some mesogastropods) fertilization is internal and eggs are deposited in an albumen-containing egg capsule that is attached to a substratum. In some prosobranchs, e.g., *Crepidula fornicata*, eggs are contained in a capsule that is attached to an internal brood chamber.

Some lower prosobranchs are protandric hermaphrodites; that is to say, individuals first function as males and then change to females for the rest of their lives. The prosobranch genus *Crepidula* (the slipper shell) is a classic example of protandry in a hermaphroditic genus. Individuals live stacked up on one another; generally an older individual living at the bottom of the stack is a female and younger individuals stacked onto an older female are initially males. The initial male phase is followed by a degeneration of the male reproductive tract with subsequent development into either a female or another male. The sex of each individual is influenced by the sex ratio of associated individuals, probably by the action of pheromones. The penis of an upper individual can reach the gonopore of a female individual below, thus allowing cross-fertilization. An older male will remain a male longer if it is attached to a female, but if isolated from its normal association with a female it will develop into a female (Barnes 1987).

Several species of neogastropoda, e.g., *Ilyanassa obsoleta*, are pseudohermaphroditic, a condition in which females possess male sexual traits such as a penis and, in some cases, have a sperm duct leading to the penis (Jenner 1979). The male sexual system is nonfunctional, and these snails are neither simultaneous nor protandric hermaphrodites.

In lower prosobranchs (archaeogastropods) the nephridal ducts serve as gonal ducts, while the neogastropoda gonad has a separate gonoduct that empties into the mantle cavity. As the oviduct reaches the posterior end of the mantle cavity, it is greatly expanded to form a uterus. The oviduct forms many branched sacs, the seminal receptacles; these store sperm received at copulation. Following the seminal receptacle(s) there is an albumen gland that secretes albumen around each egg or egg mass. Distal to the albumen gland is the capsule gland in those snails that enclose their eggs in an egg capsule. In some neogastropods there is in the female a foot gland that is used to mold the egg capsule.

Opisthobranchs

Opisthobranchs are almost exclusively simultaneous hermaphrodites, though some are protandric. Although hermaphroditic, they do not self-fertilize, but cross-copulate, with mutual insemination. Fertilization is internal and eggs are deposited in a gelatinous mass.

The hermaphroditic gonad (ovotestis) is a large organ located posteriorly and in contact with the digestive gland (Figure 11.4). Beeman (1970) cleared up earlier confusion over the combined roles of the ampulla and the seminal vesicle as storage sites for endogenous sperm in the sea hare (*Phyllaplysia taylori*). The sperm in the ampulla are inactive and do not fertilize eggs as they pass through the ampulla during spawning.

Distal to the ampulla there may be, in some species, a fork of the gonoduct into a ciliated sperm duct and a separate oviduct, or in other groups there may be a common spermoviduct. In the female reproductive system of most opisthobranchs there are two organs for reception of exogenous sperm, a seminal receptacle (sometimes there may be two), and a copulatory bursa. The seminal receptacle provides long-term storage for exogenous sperm while

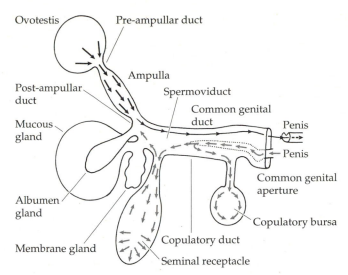

Figure 11.4 Reciprocal sperm transmission in the hermaphroditic sea hare *Phyllaplysia taylori*. Sperm movements are shown within only one individual. Heavy arrows show path of sperm emitted; light arrows, the path of sperm received. (After Barnes 1987; from Pennington and Chia 1985.)

the bursa is the initial site of sperm reception after copulation. In many groups the copulatory bursa has been lost and only seminal receptacles remain for sperm storage.

Pulmonates

The pulmonates are mostly simultaneous hermaphrodites, and reciprocal copulation with the exchange of sperm between individuals is the usual practice. Among aquatic forms eggs are deposited in a gelatinous mass, while terrestrial pulmonates deposit their eggs in an albumen-containing capsule whose wall has embedded in it calcite crystals.

In the ovotestis, spermatogenesis and oogenesis occur simultaneously and both eggs and sperm pass through a common hermaphroditic duct. The hermaphroditic duct has a number of diverticula that function as seminal vesicles for storage of endogenous sperm. As the sperm duct continues anteriorly it splits to form separate male and female ducts. The sperm duct continues anteriorly through a prostate gland and on to a penis.

After the female duct separates from the male duct it has a complex course until it ends at the female gonopore, where there are provisions for packaging eggs, receiving exogenous sperm, and for fertilization. At the point of separation of these two ducts there is an albumen gland, followed in many forms by a mucous gland, then an *oothecal gland* that secretes an egg case. The oothecal gland is followed by a vagina into which a seminal receptacle duct from the seminal receptacle enters and delivers exogenous sperm just before the female gonopore. The seminal receptacle stores exogenous sperm deposited during the reciprocal copulation that generally occurs in these hermaphroditic snails.

Differentiation of the Gonad

Jong-Brink et al. (1983) have reviewed the origin and differentiation of the molluscan gonad and have pointed out that the gonad and reproductive tract have both a mesodermal and an ectodermal origin. The gonad and part of the reproductive tract develop from pericardial mesoderm. Other parts of the reproductive tract and accessory gland cells develop from ectodermal invaginations. These origins have been established for two prosobranchs (*Littorina saxatilis*, Guyomarc'h-Cousin 1976; *Viviparus viviparus*, Griffond 1977). A similar dual origin of the gonad and reproductive system exists among some pulmonates (the Basommatophora, but not among the Stylommatophora). In the latter, the gonadal system develops from a single ectodermal source as shown for the slug *Arion circumscriptus* (Luchtel 1972a). Thus, there is variation among the gastropods, especially the pulmonates, as to a dualistic or monistic origin of the gonads and reproductive organs.

The Ovotestis

The gastropod ovotestis is organized into a number of small sacs or acini in which gametes and auxiliary cells develop. Among many fresh water snails, such as *Lymnaea stagnalis* (Joosse 1975), the acini are often embedded among the divertilculae of the digestive gland. In hermaphroditic snails male and female gametes may be produced in different or common acini. In basommatophoran pulmonates both gametes are produced in the same acinus, though in some cases there may be separate phases of sperm and egg production.

The acinar wall is composed, from the outward surface, of connective tissue with a basement membrane, a squamous epithelium composed of ciliated cells and germinal

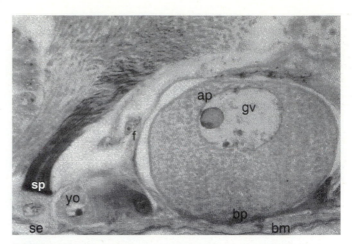

Figure 11.6 Eggs and sperm in ovotestis. An oocyte in the ovotestis of *Lymnaea stagnalis*. The oocyte has a large germinal vesicle facing the apical pole, while its basal pole rests on the wall of the ovotestis. In the lower-left corner, a group of spermatozoa can be distinguished. ap, apical pole; bm, basement membrane of the ovotestis; bp, basal pole; f, follicle cells; gv, germinal vesicle; se, Sertoli cell bearing spermatozoa; sp, spermatozoa; yo, young oocyte. (From van den Biggelaar et al. 1994.)

epithelia cells. Gametes of both sexes and auxiliary cells (Sertoli and follicle cells) develop from the germinal epithelium; in some cases, germ cells develop from *primordial germ cells* that are distinguishable in the epithelial layer. The arrangement and subsequent differentiation of cells within an acinus of a freshwater snail (*Biomphalaña glabrata*) is illustrated in Figure 11.5. Note the attachment of spermatogonia onto Sertoli cells, the attachment of oocytes to the basement lamina, and the follicle cells surrounding the oocytes. There are also depicted degenerating oocytes, which are probably phagocytosized by follicle cells. Geraerts and Joosse (1984) point out that the germinal epithelium is confined to the upper part of the acinus and that the developing oocytes migrate to the bottom of the acinus where they remain as sessile cells surrounded by follicle cells (Joosse and Reitz 1969). Developing spermatogonial cells attached to Sertoli cells pass over developing oocytes and migrate to the bottom of the acinus. Within an acinus the male gametes are oriented outwardly toward the lumen of the acinus while the female gametes are peripheral. The two gametes are separated by a layer of Sertoli cells. In Figure 11.6 a section of the ovotestis of *Lymnaea stagnalis* illustrates the detailed arrangement of oocytes and sperm in an ovotestis.

Gametogenesis

The cellular origin of the germ cells among the gastropods is not wholly resolved. Jong-Brink et al. (1983) review two major hypotheses for the origin of germ cells in mollusca: the origin from two cell lines (germinal and nongerminal

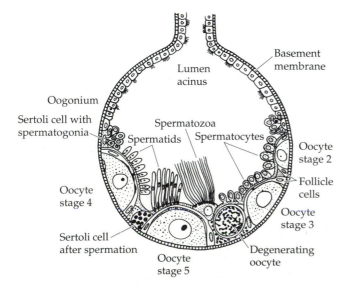

Figure 11.5 Scheme of a longitudinal section through an acinus of *Biomphalaria glabrata*. (After Jong-Brink and Joosse 1983.)

lines) established early in development and/or the origin from a single cell line, the **gonadal stem cell** (GSC). In the latter view, the GSCs divide to form both the gametes and the auxiliary cells (Sertoli cells, follicle cells, and cells of the mature spermoviduct). For example, in the mesogastropod *Viviparus viviparus* there is a single cell type that divides unequally to give rise to two cell types, one germline, one somatic (Griffond 1977). A similar pattern exists among the pulmonates, for example in the stylommatophoran *Helix aspersa* (Guyard 1971).

However, there are a number of cases in which **primordial germ cells** (PGCs) are distinguishable, for example, in several basommatophorans such as *Lymnaea stagnalis* and *Physa acuta* (Brisson and Regondaud 1971, 1977), stylommatophorans *Arion circumscriptus* and *Deroceras reticulatum* (Luchtel 1972a,b), and in the pulmonate slug *Bradybaena fructicum* (Moore 1977). In these cases, the PGCs represent a lineage separate from the gonads.

In *Crepidula fornicata* and *Buccinum undatum*, plasms similar to germ plasms in other organisms have been found in a polar lobe that is formed just before first cleavage (Dohmen and Lok 1975; Dohmen and Verdonk 1979a,b). There is some evidence that this plasm may be transmitted to the 4d cell in *Crepidula*, but its fate in *Buccinum* is unknown. Nieuwkoop and Sutasurya (1981) conclude, "...a true germ plasm has not been described so far (among the molluscs). In our opinion, germ cell formation in the mollusc is best classified tentatively as *intermediate* between the typically epigenetic and preformistic modes."

Spermatogenesis

A spermatogonium mother cell divides mitotically to produce 32 primary spermatogonia that grow and divide to form 64 secondary spermatogonia. These spermatogonia differentiate into primary spermatocytes that undergo a meiotic division to form 128 secondary spermatocytes, and then after a second meiotic division produce 256 haploid spermatids (Walker and MacGregor 1968; Webber 1977; Dohmen 1983). Developing male germ cells may be attached to Sertoli cells (nutritive and phagocytic cells), to the epithelium of the gonad, or they may be free in the lumen of the gonad. Sertoli cells may destroy abnormal sperm or all sperm during starvation.

Cytoplasmic bridges exist between the descendants of any one clone of spermatogonia. A clone is a set of 64 spermatogonia produced before the onset of meiosis. Incomplete cytokinesis during the mitotic division of the spermatogonium mother cell is responsible for the cytoplasmic bridges. The cytoplasmic bridges ensure synchronous division and differentiation within a clone of spermatocytes (Fawcett 1961; Hodgson, 1993).

After completion of meiotic divisions, spermatids begin to differentiate into mature sperm. This differentiation is called spermiogenesis and involves the disappearance of cytoplasmic bridges, loss of cytoplasm, formation of the sperm acrosome, differentiation of a flagellum, and condensation of chromatin. Dohmen (1983) describes these processes in some detail and reviews the literature on molluscan spermiogenesis. Dohmen recalls a number of forms, among them the opisthobranch *Aplysia*, whose sperm do not contain an acrosome. Spermatozoan structure and spermatogenesis in *Nassarius kraussianus* has been described by Hodgson (1993).

Many prosobranchs produce two kinds of sperm, a typical flagellated sperm that is capable of fertilizing eggs and a short vermiform atypical sperm that is not capable of fertilization. Typical or **eupyrene** sperm have a normal haploid complement of chromosomes. The **apyrene**—atypical sperm—may have less than or more than the haploid chromosome number, or they may have no chromosomes. The function of atypical sperm is not clear, but in some organisms they may have a nutritive function, i.e., upon degeneration they may provide an energy source for eupyrene sperm. Many apyrene sperm are larger than eupyrene sperm and store a yolk-like material. Eupyrene sperm have one flagellum; atypical sperm may have many flagella. It has been suggested that the atypical sperm provide transportation for attached eupyrene sperm. Apyrene sperm appear to originate from spermatogonia. Initially all spermatogonia are alike but follow divergent paths of differentiation at an early stage. Apyrene sperm show species-specific characteristics and the percentage of apyrene sperm is constant for a given species. Buckland-Nicks et al. (1982, 1983) have considered the origin of apyrene and eupyrene sperm.

Oogenesis

Oogenesis has been thoroughly studied among the pulmonates, less so among the prosobranchs, and even less for the opisthobranchs (for reviews, see Jong-Brink et al. 1983; Dohmen 1983; Geraerts and Joosse 1984; Tompa 1984; Webber 1977; and Voltzow 1994).

The general scheme for gastropod oogenesis is that primordial germ cells (PGCs) contained in a germinal epithelium form primary oogonia that divide mitotically to produce secondary oogonia. The secondary oogonia may enter into the first meiotic division and become primary oocytes, or they may form auxiliary cells. The meiotic division of the primary oocyte arrests during first prophase and is not completed until after ovulation. The second meiotic division and the formation of the second polar body occurs after fertilization. Thus, egg maturation is not completed until the egg is fertilized.

Putative PGCs have been identified in some pulmonates, but in no case has their determination by segregation of a "germ plasm" been unequivocally established. In many cases it appears that oogonia arise from a germinal epithelium that has its origin from gonadal stem cells (GSCs) derived from pericardial mesoderm.

PREMEIOSIS. After oogonial proliferation, the oogonia enter a premeiotic stage (Raven 1961) in which the oogonial nucleus enlarges to form a germinal vesicle. At this stage the chromosomes are in the first meiotic prophase and progress through the leptotene, zygotene, and pachytene stages, but arrest in the diplotene stage. Lampbrush chromosomes at this stage have been observed in *Ilyanassa* (Davidson 1976), *Bithynia, Planorbarius,* and *Lymnaea* (Bottke 1973). Presumably these chromosomes are actively transcribing a complex set of messenger RNAs, as has been described for *Xenopus* and other organisms (Anderson and Smith 1978).

PREVITELLOGENESIS. The premeiotic phase of oogenesis is followed by a previtellogenic phase during which the oocyte grows and is increasingly active in protein and RNA synthesis. In the prosobranch *Ilyanassa obsoleta* these RNAs are maternal messenger RNAs. Collier and McCarthy (1981) and Collier (1981) have shown that the isolated polar lobe of *Ilyanassa,* which contains only maternal mRNAs, synthesized the same set of proteins as does the fertilized egg. Ninety-five to one hundred nascent proteins, isolated by two-dimensional electrophoresis, were shown to be synthesized by the isolated polar lobe (Collier 1981). The contribution of mitochondrial protein synthesis was ruled out by the use of cycloheximide as an inhibitor of nonmitochondrial protein synthesis. Collier (1981) also showed that histones H1, H2A, H2B, H3 and H4, and high mobility group polypeptides 14 and 17 were synthesized in the isolated polar lobe and were therefore translated from maternal mRNAs. That actin mRNA is a maternal message in *Ilyanassa* was shown by hybridization of RNA from mature unfertilzed eggs with an actin gene probe (Collier 1989). Raff et al. (1976) have shown that microtubule proteins in *Ilyanassa* are translated from maternal mRNA. Thus, this prosobranch makes an extensive array of maternal messages during oogenesis.

Nucleoli are present in very young gastropod oocytes and greatly increase in volume during oogenesis. The previtellogenic nucleolus develops a protein portion in the prosobranchs *Bithynia* (Bottke 1973), *Patella* (Bolognari et al. 1976) and *Ilyanassa* (McCann-Collier 1979). In *Ilyanassa* this two-part nucleolus (amphinucleolus) has been shown by the autoradiographic studies of McCann-Collier (1979) to synthesize RNA (presumably rRNA) in its protein compartment. Kielbowna and Koscielski (1974) made similar observations in oocytes of the freshwater pulmonate *Lymnaea stagnalis.*

Follicle cells (variously called accessory cells, supportive cells, and in some cases nurse cells, which they definitely are not) are auxiliary cells of the ovary derived from the germinal epithelium. They are closely associated with and cover all or part of the developing oocyte that is not attached to the basement membrane of the acinus (see Figures 11.5 and 11.6). Their involvement in various aspects of oogenesis, vitellogenesis, formation of egg coats, and determination of egg polarity has been reviewed by Jong-

Brink et al. (1983). They may also participate in ovulation and hormone production.

Follicle cells are protein-synthesizing cells as indicated by their well-developed rough endoplasmic reticulum, seen, for example, in *Ilyanassa obsoleta* and *Lymnaea stagnalis* (Geraerts and Joosse 1984; Taylor and Anderson 1969). Even though follicle cells appear to be protein-producing cells, Geraerts and Joosse (1984) doubt that they contribute yolk proteins to the oocyte; they base their view in part on the absence of pinocytotic vesicles along the juncture of follicle cells and the oocyte. While pinocytosis is not universally reported in gastropod oocytes it does occur in some cases such as the uptake of ferritin by pinocytosis in *Planorbarius* and *Lymnaea* (Bottke and Sinha 1979; Bottke et al. 1982). Geraerts and Joosse (1984) also point out that in *Lymnaea stagnalis* cytoplasmic processes extend from follicle cells into the oocyte. Follicle cells of the prosobranchs *Lamellaria perspicua* (Webber 1977) and *Crepidula* (Silberzahn 1979) are completely resorbed into the developing oocyte. Thus, the nature of the proteins made by follicle cells and their relation to the oocyte is not clear and needs to be reconsidered.

VITELLOGENESIS. The vitellogenic phase of oogenesis is primarily concerned with the synthesis and accumulation of yolk. It is characterized by active synthesis of nonribosomal RNA (McCann-Collier 1979) and proteins, the accumulation of yolk platelets and yolk granules, and other nutritive materials such as lipid droplets and glycogen.

Yolk in gastropod oocytes most often occurs as lipid and proteinaceous granules. The protein yolk contains muco- or glycoproteins, phospholipids, ferritin, and basic proteins (Ubbels 1968) and is contained in a membrane-bound platelet or granule. In *Ilyanassa* the yolk platelet consists of a core of crystalline protein surrounded by granular cytoplasm and bounded by a unit membrane (McCann-Collier 1979). The lipid yolk granules are not bound by a membrane and the fatty yolk is mainly composed of neutral lipids.

In *Lymnaea stagnalis* (Ubbels 1968; van der Wal 1976a,b) and other freshwater snails such as *Biomphalaria* (Jong-Brink et al. 1976) there are two types of protein-containing yolk granules, **beta-** and **gamma-granules**. Only gamma-granules contain RNA (Ubbels 1968). In earlier literature it was suggested that gamma-granules developed from beta-granules; however, later observations on *Lymnaea stagnalis* (van der Wal 1976a,b) and other freshwater snails (de Jong-Brink et al. 1976) suggest that the two yolk granules arise independently.

Many pulmonates develop within a capsule containing an albuminous fluid that serves as the major nutrient until the larval snail hatches from the capsule (Morrill et al. 1976). That *Lymnaea* eggs deprived of the bulk of their yolk develop normally (Morrill 1964) but when removed from their egg capsule they do not develop normally beyond the gastrula stage show the importance of the capsular fluid and suggests that it is a major nutrient source. Lymnaeid capsule fluid contains polysaccharides, galatogen, and proteins. Morrill (1964) has shown that the proteins in Lymnaeid capsule fluid are species-specific.

It is not clear to what extent large- and intermediate-size yolk platelets are used during embryogenesis. In *Lymnaea* and *Ilyanassa*, the volume of yolk platelets does not materially decrease during embryogenesis. In Lymnaeid snails, degradation of yolk starts early but proceeds slowly, and intact yolk platelets are present in late embryos. Further, Morrill (1964) demonstrated that *Lymnaea* nucleated egg fragments lacking the bulk of yolk developed normally. In *Ilyanassa*, no significant disappearance of yolk platelets occurs until the veliger larva is fully differentiated. Thus, it appears that a major function of proteid yolk platelets may be the provision of food and an energy source during larval or postembryonic differentiation.

In addition to yolk platelets and yolk granules, other nutritive substances such as glycogen and **lipid droplets** accumulate during the vitellogenic phase. Lipid droplets (granules), which are generally very abundant in gastropod eggs, often appear in the oocyte cytoplasm before proteinaceous platelets. The source of lipid droplets is not always clear and when known does not follow a consistent pattern among all gastropods. In some cases their origin does not appear to be associated with any specific cellular organelle, in others their origin has been associated with Golgi apparatus, mitochondria, and Balbiani bodies (basophilic structures presumed to contain RNA and sometimes called a yolk nucleus). An exogenous source of oil droplets has been suggested for the pulmonate *Lymnaea* by Ubbels (1976) on the basis of his observation of large lipid droplets in follicle cells. In *Ilyanassa* and other species lipid droplets seem to come from Balbiani bodies and concentric lamellae of the ER (Gerin 1976). Fat droplets are used as a source of lipid for the synthesis of new membranes required during cleavage. They also contribute to egg polarity, as they are often localized in the animal hemisphere of the egg.

Glycogen is an important energy source in developing embryos and it occurs in a variety of forms in gastropod oocytes. In *Crepidula* it is localized as rosettes that occupy most of the space between cellular organelles; in *Ilyanassa*, it occurs as clusters. (Pucci-Minafra et al. 1969). Geuskens and d'Ardoye (1971) have suggested that glycogen in the *Ilyanassa* egg is localized as alpha rosettes, and in *Lymnaea* eggs glycogen is present in an amorphous form in the egg cytoplasm.

POLARITY. Gastropod eggs have a distinct animal-vegetal polarity, and in many eggs morphogenetic determinants synthesized during oogenesis are localized in the vegetal ooplasm. In a number of prosobranch eggs the vegetal cytoplasm containing determinants is partitioned into a mass of cytoplasm, the **polar lobe** (Figure 11.7), that is almost completely constricted from the egg just before first cleavage. In *Ilyanassa* (Clement 1952) and *Bithynia* (Cather and Verdonk 1974) polar lobes are isolatable and have been shown by deletion experiments to contain determinants that are essential for differentiation.

The animal-vegetal polarity of the gastropod egg appears to correspond to the apical-basal polarity of the oocyte as it is positioned in relation to the wall and lumen of the ovary. The basal point of attachment of the oocyte to the ovarian wall appears to become the vegetal pole of the egg, the free apical pole becoming the animal pole (Figure 11.6). Morphogenetic determinants are localized in the vegetal hemisphere of the egg. The germinal vesicle lies near the animal pole, and polar bodies are extruded both at the center of the animal pole and diametrically opposite the vegetal pole. The animal pole corresponds to the cephalic or anterior end of the future embryo and the vegetal pole to the caudal or posterior end.

Polarity is a stable attribute of oocytes and eggs. The original polarity of the oocyte is independent of elements that can be stratified or displaced by centrifugal forces that allow survival of the egg. Compression of the egg to alter the orientation of the mitotic spindle also fails to change the original polarity. The possible role of cytoskele-

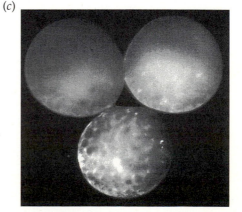

Figure 11. 7 (*a*) *Bithynia* polar lobe (pl). (*b*) 2-cell stage in which the vegetal body (vg bd) has been centrifugally dislocated into one blastomere; note that polar lobe components other than the vegetal body have not been translocated. (From van den Biggelaar et al. 1994.) (*c*) Trefoil stage of first cleavage of *Ilyanassa obsoleta* showing the large third polar lobe. The yolk-laden polar lobe is toward the bottom of the image. (From Koser and Collier 1976.)

tal elements such as microtubules and microfilaments in ooplasmic segregation and the maintenance of polarity is reviewed by van den Biggelaar and Guerrier (1983), but experimental evidence bearing on this point is meager for gastropod eggs.

Egg polarity is reflected by the distribution of cytoplasmic inclusions and molecular components along its animal-vegetal axis. In the animal hemisphere there is the germinal vesicle (and later the pronuclei) and an abundance of lipid droplets and mitochondria. Though lipid droplets and mitochondria are more abundant in the animal hemisphere, they extend into the equatorial region and vegetal hemisphere of the egg. Large- and intermediate-size protein yolk granules and platelets are concentrated in the vegetal hemisphere, though they do extend into the equatorial region and to a lesser extent into the animal hemisphere of the egg. Ribosomes and endoplasmic reticulum (both smooth and rough) are present in varying amounts. In yolky eggs they are displaced toward the animal hemisphere, though they are also interspersed in the yolk in the vegetal hemisphere.

A polar organization of the plasma membrane, based on an asymmetrical distribution of *intramembrane particles* (IMP), has been described in *Nassarius* eggs (Speksnijder et al. 1985). The IMP density was much higher at the vegetal pole than in other regions of the egg, raising the possibility that this characteristic of the polar membrane may be related to the localization of morphogenetic determinants, which are thought by some investigators to be localized in the cortex of the vegetal ooplasm. In *Nassarius, Buccinum,* and *Crepidula* eggs, a "vegetal pole patch" (VPP) has been described by Dohmen and van der Mey (1977). The VPP takes the forms of folded ridges of the plasma membrane or longer and larger microvilli that occur on the polar lobes of these eggs. These authors suggest that the VPP corresponds to the contact area of the oocyte with ovarian follicle cells and that the egg cortex may differentiate these patches in response to the ovarian environment. This special surface architecture is correlated with the localization of morphogenetic determinants in the vegetal pole of many eggs.

Fertilization

Copulation and internal fertilization occur in the opisthobranchs, pulmonates, and most prosobranchs except for some archeogastropods that spawn their gametes into the surrounding media. Fertilization occurs at the first metaphase of meiosis, when it induces the resumption of the meiotic divisions, or at the germinal vesicle stage of egg maturation (Longo 1983).

In those prosobranchs with internal fertilization, sperm are received into the mantle cave or into the female gonopore during copulation, stored in a seminal receptacle,

and finally passed by way of the sperm duct to the site of fertilization. The site of fertilization is often in the lumen of the albumen gland or at the entrance of the capsule gland. Indeed, in some cases the site is well into the lumen of the capsule gland, but in all cases, fertilization occurs before the eggs are encapsulated. After fertilization the eggs are surrounded by an egg capsule or embedded in a mucus coat. Prosobranch egg capsules are composed of several layers of proteins and glycoproteins and are quite resistant to enzymatic digestion and other chemical means employed for their removal. (For the composition of the *Ilyanassa* capsule, see Sullivan and Mangel 1984.) Egg capsules are either deposited onto a substratum such as eel grass, rocks, or shells, or they may be retained inside the mantle cavity in a brood chamber, or in other cases on the inner or outer surface of the shell.

Opisthobranchs are simultaneous hermaphrodites, and cross-fertilization occurs by reciprocal or nonreciprocal copulation (Hadfield and Switzer-Dunlap 1984). In reciprocal insemination, sperm are delivered by the insertion of the penis of each mating individual into the vaginal or common genital pore of its mate and there is a mutual exchange of sperm. Unilateral copulation may also occur, as in *Alplysia*. In some groups, hypodermic insemination is a common mode of sperm transfer wherein sperm are injected into a closed vaginal bursa or deposited randomly over the body surface. Before they are encapsulated, eggs are fertilized in a fertilization chamber of the female reproductive tract. In gastropods that can be ferilized artificially, an acrosome reaction has been observed (Dan 1956).

The block to polyspermy has not been extensively studied in the molluscs and has been studied even less in the gastropods. Indeed, in some pulmonates (*Lymnaea* and *Helix*) there appears to be no block to polyspermy, as more than one sperm enters the egg (but only one is successful in forming a pronucleus, while the supernumerary sperm degenerate). The mechanism for this selection is not understood.

In those eggs fertilized at the germinal vesicle stage, fertilization is followed by the breakdown of the germinal vesicle, the organization of a meiotic spindle at the periphery of the egg, and the formation and extrusion of polar bodies. Subsequent to polar body formation, the egg chromosomes are enclosed in a nuclear envelope to form a female pronucleus that migrates toward the center of the egg. Those eggs fertilized at metaphase I of meiosis proceed immediately to form polar bodies and a female pronucleus, as previously outlined. After sperm entry into the egg there is a breakdown of the sperm nuclear envelope, chromatin dispersion (decondensation), and the enclosure of the dispersed chromatin by a newly formed pronuclear envelope. These events result in the formation of a male pronucleus that lies near the cortex of the egg. After the two pronuclei are formed, they move toward the center of the egg and come to lie closely apposed to each other. Next, the two pronuclear envelopes break down and the chromosomes condense. Finally, both sets of chromosomes, one from the female pronucleus and the other from the male pronucleus, intermix and establish the first cleavage spindle.

Cleavage

Most gastropods have holoblastic spiral cleavage and bilateral cleavage. There are a few deviations from spiral cleavage during early development, and these occur in very yolky eggs, in which cleavage is meroblastic and excessive yolk obscures and distorts the cleavage plane.

Cleavage is of special importance in gastropod development because the developmental fate of blastomeres is established during early stages of cleavage. This highly determinant pattern of cleavage is important for the organism because the segregation of embryonic determinants and the rapid formation of larval structures are more effectively achieved by determinant cleavage than by indeterminant cleavage. For the student of development, cleavage provides an incisive experimental approach for studying the mechanisms of determination. Accordingly, the following discussion will consider in some detail the cleavage patterns of gastropods, from the nature of spiral cleavage, the dominant form of cell division in early gastropod development, to the establishment of cellular embryology by cell lineage studies, and the segregation of determinants for germ layers and embryonic structures.

Spiral Cleavage, a Derivative of Radial Cleavage

The word *spiral* was first used in relation to cleavage by Selenka (1881) in his description of the rotation of blastomeres in the cleavage of polyclad flatworms. Wilson (1892) proposed the recognition of a *spiral type* of cleavage as a derivative of radial cleavage. In radial cleavage, meridional cleavage furrows alternate with equatorial cleavage furrows. Meridional furrows are parallel to the polar axis of the egg and equatorial furrows are at right angles to this axis. The difference between radial and spiral cleavage is that in spiral cleavage the cleavage furrows are *oblique* to the egg axis.

The oblique inclination of the spindle in spiral cleavage is most evident at the third cleavage, when the first quartet of micromeres is formed; however, the obliquity of the spindles is present but less evident at the 2- and 4-cell stage. Thus, these early cleavages are *prospectively spiral*, as judged by the disposition of the spindles for future divisions (Conklin 1897). The first polar body of *Ilyanassa* is produced at an oblique angle to the egg axis (McCann-Collier 1984); she states that the "obliquity of the meiotic spindle and subsequent angle of extrusion of the first polar body is an early expression of the egg organization that later produces spiral cleavage."

Chirality of Spiral Cleavage

Spiral cleavage is achieved by the right to left (or left to right) alternation of cleavage planes that are oblique to the animal-vegetal axis of the egg. Thus, the essence of spiral cleavage is the *obliquity of the spindle orientation and the regular directional alternation of spindle planes*. The alternation from right-handed (dextrotropic or dextral) cleavage to left-hand (levotropic or sinistral) cleavage subscribes a spiral, thus *spiral cleavage*. Handedness or chirality of cleavage is judged to be right or left (clockwise or counterclockwise) as one looks down on the animal pole of the egg. If a micromere lies obliquely and to the right of its parent macromere, then cleavage is dextrotropic, if it lies to the left of its parent then cleavage is levotropic.

The third cleavage of gastropods is most commonly a right-handed (dexiotropic) division, but there are cases in which it is left-handed. The chirality of this division is a constant correlate with the left or right location of the D macromere, the orientation of mesodermal bands, the twisting of the visceral mass during development, and with the chirality of the shell (Figure 11.8 illustrates these relations).

In *Lymnaea peregra*, the handedness of cleavage and of the shell are determined by the maternal inheritance of a single locus with dextrality being dominant to sinistrality (Freeman and Lundelius 1982). These authors injected cytoplasm from dextral eggs into sinistral eggs and observed a reversal of the body asymmetry of animals from sinistral eggs. Injection of cytoplasm from sinistral eggs into dextral eggs did not affect the cleavage of dextral eggs. Thus, the inheritance of asymmetry in snails is truly a case of maternal inheritance in which the dextral gene is transcribed during oogenesis.

The Homology of Cell Lineages

Cell lineage describes the cell by cell origin of an embryonic or adult structure. Cell lineage studies during the latter half of the nineteenth century were the foundation of cellular embryology and its close relation to cytology and genetics, a relation that diminished for the last three-quarters of a century but has now reemerged with the development of molecular genetics. This field of cellular embryology was initiated by the studies of C. O. Whitman (1878) on the leech embryo *Clepsine*, of Rabl (1879) on the pond snail *Planorbis*, and Van Beneden and Julin (1884) on the ascidian *Corella*. "These researches demonstrated that the cleavage of the ovum, in some animals at least, is a perfectly ordered process, in which every individual cell in the early stages of development may possess a definite morphological value in the building of the body" (Wilson, 1925). Efforts to test these observations led to the experiments of Roux, Chabry, and Driesch and to the creation of experimental embryology.

The homology of cell lineages among the annelids, polyclad flatworms, and gastropods was emphasized by

Figure 11.8 Influence of cleavage chirality on the asymmetry of cleavage, mesodermal bands, adult body form, and the adult shell. (After Wilson 1925.)

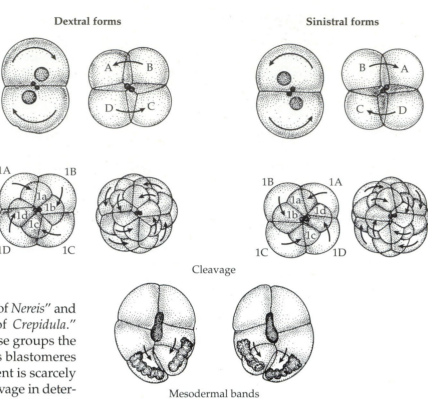

Dextral forms   Sinistral forms

Cleavage

Mesodermal bands

Adult body form

Adult shell

Wilson in his 1892 paper, "The Cell-Lineage of *Nereis*" and again by Conklin, in "The Embryology of *Crepidula*." Hyman (1951) remarked that, "Among these groups the cleavage pattern and the fate of the various blastomeres are so nearly identical that a common descent is scarcely to be doubted." So that the role of spiral cleavage in determinant development may be appreciated as a major body of information in developmental biology, it is worthwhile to repeat the resemblances between the early cleavages of these organisms. Quoting Conklin (1897, p. 193):

(1) the number and direction of cleavages is the same in all three up to the 28-cell stage; (2) in general the cells formed are similar in position and size, viz., there are four macromeres, three quartettes of micromeres, and the first quartette is surrounded by a belt composed of the second and third quartettes. The first quartette undergoes three spiral divisions in alternate directions, and the second quartette divides once. Here the resemblance with the polyclads ceases, though the annelid and gasterpod go one step further in these likenesses, viz., the three quartettes of micromeres are ectomeres in the annelid and gasterpod, and (4) in both these groups the mesoblast is formed from the cell 4d, which gives rise to paired mesoblastic bands.

Conformation and Terminology of Spiral Cleavage

In spiral cleavage the fate of each early blastomere has been established by cell lineage studies and in many cases affirmed by experimental analyses; accordingly, there is a system of nomenclature in which each cell has a name. The nomenclature is that of Wilson (1892) for "The Cell-Lineage of *Nereis*," as modified by Conklin (1897) in this classic cell lineage study of the "slipper shell" *Crepidula fornicata*. The following account of cleavage is based on the prosobranchs *Crepidula fornicata* and *Ilyanassa obsoleta*. The early cleavages of these two organisms are very similar except that the mesogastropod *Crepidula* has a small protrusion of vegetal cytoplasm at first cleavage (Conklin 1902), whereas the neogastropod *Ilyanassa* has a larger protru-

sion of vegetal cytoplasm at first cleavage. Consequently, the blastomeres of the two cleavage of *Crepidula* are equal in size in contrast to the unequal cleavage in *Ilyanassa*. This account of cleavage and cell lineage is adapted from Conklin (1897), Hyman (1951), Clement (1952), and Collier (1965). Morrill (1997) has given a detailed account of cleavage and morphogenesis in two freshwater pulmonate snails, *Lymnaea* and *Physa*.

Spiral cleavage is commonly holoblastic; the first and second cleavages are meridional to the egg axis and may be equal or unequal. They are prospectively spiral and nor-

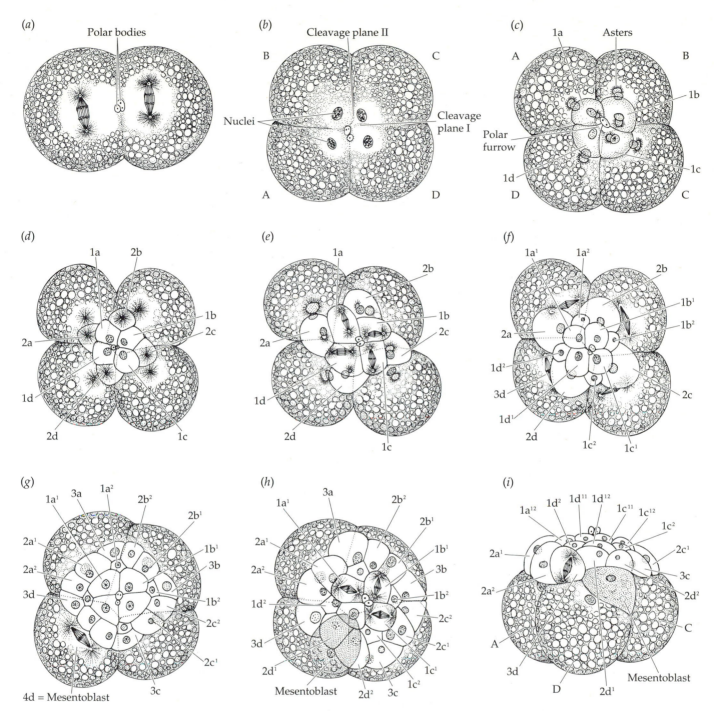

Figure 11.9 Conklin's drawings of *Crepidula* cleavage from the 2- to the 25-cell stage. (*a*) 2-cell stage. (*b*) 4-cell stage; *pb* are polar bodies laying in the cleavage furrow at the animal pole. (*c*) 8-cell stage; first quartet of micromeres shifted to the right of the cleavage furrows to yield a dextrotropic cleavage. *pf* is polar furrow. (*d*) 12-cell stage; second quartet of micromeres in a levotropic division. (*e*) 12-cell stage, with a spindle setup for the next cleavage as in (*f*), the 16-cell stage. (*g*) Spindle in D macromere is arrayed for the formation of the 4d cell. (*h*) Top view of 4d cell from the animal pole. (*i*) Side view of 4d. Designations of individual cells are as discussed in the text. (After Conklin 1897.)

mally dextrotropic. For the most part gastropod cleavage after the 4-cell stage proceeds by the division of sets of four cells. Each set of four cells, large cells (macromeres), or small cells (micromeres), is a quartet. Each quartet is designated by a coefficient and a letter (capital letter for macromeres and a lowercase letter for micromeres); the coefficient expresses the cell generation and the letter designates the quadrant of the egg from which a cell is derived.

In gastropods the first cleavage is meridional and oblique to the egg axis and produces two blastomeres (either equal or unequal), the AB and CD cells. The second cleavage, which is also meridional and dextrally or sinistrally oblique to the first cleavage plane, forms the A, B, C, and D cells. When the first cleavage is equal, these four cells will also be equal in size; if first cleavage is unequal, the D cell will be larger than the equal-sized A, B, and C cells. The C and D macromeres correspond to the left and right, respectively, posterior areas of the embryo; the A and B macromeres correspond to the left and right anterior areas of the embryo, respectively.

The third cleavage is universally horizontal to the egg axis and usually dextrotropic. Each of the four macromeres divides to form a small, yolk-free cell; thus, there is a set of four small cells, the *micromeres*. These cells are named 1a, 1b, 1c, and 1d, and they are the first quartet of micromeres, referred to collectively as (1a–1d). The micromeres do not lie exactly over their parent macromeres as they would in radial cleavage. Rather, their oblique cleavage furrow has resulted in their being shifted dextrally from the egg axis so they lie between the angles of their parent macromeres.

The coefficient of each micromere designates the quartet number (generally there are at least three and usually four quartets of micromeres formed during cleavage), and the letter of each micromere identifies its parent macromere. After the formation of the first quartet of micromeres, the parental macromeres are designated 1A, 1B, 1C, and 1D.

A second quartet of micromeres, 2a, 2b, 2c, and 2d (2a–2d) is formed by the division of macromeres 1A, 1B, 1C, and 1D. The macromeres now become 2A, 2B, 2C, and 2D; the coefficient of the macromere tells the number of micromeres it has formed, and the letter of each macromere shows its parental relation to its daughter micromere. At the next division the macromeres (2A–2D) divide to yield a third quartet of micromeres (3a–3d), the macromeres becoming 3A, 3B, 3C, and 3D.

In the interval between the formation of the second and third quartets of micromeres, the first and second quartets have cleaved to produce a total of 24 cells. When 1a divides, its daughter cells are $1a^1$ and $1a^2$. When $1a^1$ divides, its daughter cells are $1a^{11}$ and $1a^{12}$, and daughters of $1a^2$ are $1a^{21}$ and $1a^{22}$. The division of the first quartet of micromeres results from the fifth cleavage, and the division of the second quartet arises from the seventh cleavage. At this stage the *Crepidula* embryo consists of 24 cells. At the eighth and next cleavage, the 3D macromere divides precociously to produce the primary mesentoblast, the 4d cell. The division of other members of the fourth quartet are delayed for a few hours. The 4d cell is larger than other micromeres and contains some yolk. In some gastropods, for example *Ilyanassa obsoleta*, $1a^1$–$1c^1$ divides before the 3D macromere and the 4d cell is the twenty-eighth cell of the embryo.

Figure 11.9 is based on Conklin's drawings of the cleavage of *Crepidula* from the 2-cell to the 25-cell stage. Note the dextrotropic cleavage in (*e*) as the first quartet of micromeres are formed and note also the levotropic division of the second quartet of micromeres in (*d*). The mesentoblast (ME; 4d) is seen from the top in (*h*) and from the side in (*i*).

Figure 11.10 is a diagram of *Ilyanassa* polar lobe formation and cleavage beyond the 28-cell stage. Note that the formation of the first polar lobe is correlated with the first meiotic division in Figure 11.10(2), the formation of the third polar lobe in Figure 11.10(6) and (7), and its culmination in the trefoil stage in Figure 11.10(8). It is at this stage that the polar lobe can be isolated. *Ilyanassa* forms a total of five polar lobes. The fourth and fifth lobes shown in Figures 11.10(10) and (12) are only slight constrictions in the D macromere at its second and third division. In Figure 11.10(19–21) the division of the 4d cell to form left and right primary mesentoblasts (ME^1 and ME^2) and the later formation of the primary enteroblasts is also illustrated.

Lineage of the Germ Layers

In the 25-cell *Crepidula* embryo there has been a significant but incomplete separation of germ layers. Ectodermal lineages have been segregated from endodermal lines, ectomesoderm from ectodermal and endodermal cells, but the twenty-fifth cell (primary mesentoblast, 4d cell) has a mixed lineage of mesoderm and endoderm. The germ layer lineage of early and later stages of cleavage are summarized in Table 11.2.

The first three quartets of micromeres and their division products are ectodermal or ectomesodermal. Ectomesoderm, one of the two sources of mesoderm in *Crepidula* and *Ilyanassa*, is, according to Conklin (1897), derived from 2a, 2b, and 2c and forms the larval mesoblast. (Verdonk and van den Biggelaar [1983] question Conklin's view and point out that in gastropods ectomesoderm is from the third quartet, namely, derived from 3a, 3b, and 3c.) The larval mesoblasts are scattered mesenchymal cells that form unicellular muscle fibers found predominantly in the foot but also in the head vesicle and velum. The larval mesenchyme forms a large part of the larval musculature, which degenerates at metamorphosis.

The **primary mesentoblast** (ME, 4d) contains two germ layer components, an endomesoderm (that mesoderm derived from endodermal cells and exclusive of ectomesoderm derived from the second or third quartet of micromeres) and an endodermal component (primary and secondary entoblasts) that ultimately contributes to the formation of the posterior end of the intestine. Though the 4d blastomere is the twenty-fifth or twenty-eighth cell to be formed, it undergoes several divisions before the mesodermal and endodermal components are separated.

Specifically, the primary mesentoblast (4d) divides into right and left mesentoblasts (ME^1 and ME^2), each of which divides to form a right and left mesentoblast (Me^1 and Me^2) and a right and left primary enteroblast (E^1 and E^2). The latter are purely endoblastic and do not divide

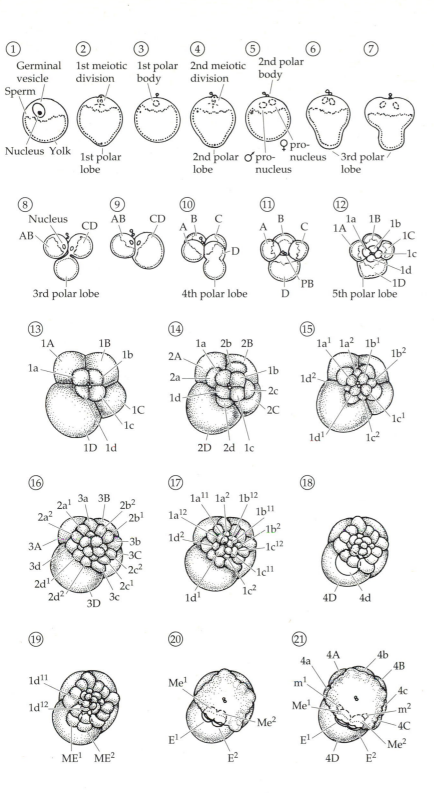

Figure 11.10 Diagram of *Ilyanassa* polar lobe formation and cleavage. Sketches (1–12) show the formation of the polar bodies and five polar lobes of *Ilyanassa*. Sketch 8 depicts the third polar lobe at its maximal constriction from the egg to form a trefoil stage, at which time the polar lobe is isolatable. Sketches 10 and 12 show polar lobes 4 and 5 as minor constrictions of the D blastomere. Pictures 13 through 18 depict the formation of the first three quartets of micromeres. Sketches 19 through 21 show the formation of the 4d cell and its division to produce the primary mesentoblasts (ME^1 and ME^2) and primary enteroblasts (E^1 and E^2). (After Collier 1965.)

again until much later; the mesentoblasts (Me^1 and Me^2) contain both mesoderm-forming and endoderm-forming components. The next division is of the mesentoblasts, to produce two purely mesoblastic cells (m^1 and m^2) and another pair of mesentoblasts (M^1e^1 and M^2e^2). The mesentoblasts divide to form a pair of mesoblastic teloblasts (M^1 and M^2) and a pair of secondary enteroblasts (e^1 and e^2).

Of these daughter cells of the 4d cell, the primary and secondary enteroblasts give rise to the posterior end of the intestine. The mesoblastic teloblasts (M^1 and M^2) are the pole cells of the mesoblast; they divide to form the mesodermal bands. Each teloblast on the right and left sides, respectively, of the embryo divide to form eight or nine cells that make up the **bilateral mesodermal bands**. The mesodermal bands grow in length and extend

TABLE 11.2 *Lineage of Germ Layers in* **Crepidula**

Endoderm	Ectoderm	Ectomesoderm	Endomesoderm
4d: Primary and secondary enteroblasts (E^1 and e^1) form posterior part of intestine.	**2d:** Primary somatoblast. Forms posterior "growing point," essential for shell formation.	**3a, 3b, 3c:** Ectomesoderm **Larval mesoblasts:** Forms myocytes of foot, head vesicle, and velum.	**4d:** Forms mesoblastic teloblasts that produce mesodermal bands. (See Endoderm for other components of 4d.)
4A–4C: Anterior floor of archenteron.			
5A–5D: Archenteron posterior to mouth.			
5C: Initiates asymmetry of embryo that results in torsion.			

around the periphery of the embryo. At first they are near the dorsal surface, but are later carried down over the yolk and come to lie near the ventral surface. The mesodermal bands form trunk or body mesoderm and do not extend beyond the anterior lip of the blastopore. In older embryos the cells of the anterior ends of the band become scattered mesodermal cells; the posterior ends remain distinct for a time but ultimately disappear. Relatively few of the endomesodermal cells contribute to the larval musculature.

In addition to the entoblasts derived from the 4d cell, other endodermal components are 4A–4C macromeres, which form the anterior floor of the archenteron, and derivatives of the fifth quartet of macromeres (5A–5D), which form the archenteron posterior to the mouth. Of these cells, 5D is of special interest because at a later stage it moves ventrally and begins a twist in the archenteric cavity and all parts of the embryo behind the mouth, thus marking the beginning of torsion.

The mixed lineage of the 4d cell found in *Crepidula* is not universal among gastropods. In many, the 4d cell contains only a mesodermal component; in others, the final segregation of endodermal and mesodermal components of the 4d cell is less tortuous. In the opisthobranchs there is no ectomesoderm and all mesoderm is derived from macromeres. *Viviparus* shows a most striking divergence in which the 4d cell is not a mesentoblast but an ordinary entoblast of the fourth quartet.

SPIRAL CLEAVAGE AND DETERMINANT DEVELOPMENT. An outstanding feature of spiralian development is the fixing of cell fates during early cleavage. The developmental fate of several lineages is summarized in Table 11.3. The specification of cell fates is most often accomplished by the action of cytoplasmic factors (**morphogenetic determinants**) that become apportioned into different cells. Determinants segregated into a specific cell regulate the expression of the zygotic genome whose gene products establish the fate and differentiation of the recipient cell and its descendants.

No determinant has yet been isolated and chemically characterized; however, from genetic analyses of *Drosophila* development (Driever et al. 1989; St. Johnston and Nusslein-Volhard 1992), determinants appear to be proteins acting as sequence-specific DNA binding proteins that regulate the expression of zygotic genes required for establishing the fate of cells in a developing embryo. A stabilized mRNA coding for such a protein could also function as a determinant. While we may anticipate that all or many determinants are transcriptional regulatory proteins, presently we are faced with situations in which, for most cases, a determinant can only be defined as a localized region of cytoplasm that has a demonstrable influence on determination and differentiation.

The animal-vegetal axis of the gastropod egg is organized during oogenesis, but the dorsal-ventral axis, specified by the appearance and position of the D quadrant, is set up epigenetically either by the differential segregation of egg cytoplasm into a specific blastomere or by cellular interactions between micromeres and macromeres. The D quadrant is identified by the appearance of the primary mesentoblast, the 4d cell, and, in equally cleaving eggs, the D quadrant cannot be recognized until the 4d cell appears.

Just before first cleavage, some eggs produce a protrusion of vegetal cytoplasm, a polar lobe, that plays an important role in establishing the dorsal-ventral axis and subsequent differentiation of gastropod embryos. Polar lobes contain a number of maternal determinants that were originally localized in the vegetal ooplasm, and it is the translocation of these determinants into the CD blastomere that creates asymmetrical cleavage and thus defines the D quadrant and dorsal-ventral polarity. Polar lobes are absent in the archaeogastropods in which cleavage of the macromeres are equal, small polar lobes are present in the mesogastropods, and large polar lobes occur in neogastropods. Polar lobes are absent among the equally cleaving opisthobranchs and pulmonates.

Bithynia (Figure 11.7a) and *Ilyanassa* (Figure 11.7c) are respective examples of eggs with small and large polar

---

TABLE 11.3 *Developmental Fates of Lineages in* Crepidula *and* Ilyanassa[a]

First Quartet of Micromeres (1a–1d)

Cells of the first quartet form head vesicle, apical sense organ, cerebral ganglion and eyes, cerebral commissures, cerebral-pedal connective, apical cell plate, and part of the preoral velum.

*1a and 1b are required for differentiation of the left and right eyes, respectively.

*Deletion of 1b causes synophthalmia and a slight reduction in the size of the velum.

Second Quartet of Micromeres (2a–2d)

2d is the primary somatoblast and forms a posterior growing point—a more or less regular row of cells that lies just ventral to the shell gland and immediately over the mesodermal bands. The cells of the posterior growing point contribute to the shell gland, foot, and ventral region of the embryo.

*2a is essential for development of left velar lobe, left eye, and left statolith.

*2b deletion shows no consistent pattern of defects.

*2c is required for normal development of the shell, development of the heart, right velar lobe, right statolith, and right eye.

*2d deletion results in the absence or reduction in size of the external shell.

Third Quartet of Micromeres (3a–3d)

3a, 3b, and 3c ectomesoderm, or "larval mesoblasts," give rise to myocytes in the foot, head vesicle, and velum.

Cells of the third quartet, along with those of the second quartet, contribute to the formation of the mouth, velum, shell gland, foot, and the external extretory cells.

*Deletion of 3a and 3b results in size reduction of the left and right velar lobs, respectively.

*Removal of 3c and 3d resluts in the absence of the right and left statocysts and right and left half of the foot, respectively.

Fourth Quartet of Micromeres (4a–4d)

*Deletion of the primary mesentoblast (4d) results in the absence of the intestine, heart, larval kidney, midgut, digestive gland, and shell.

---

[a] Items marked with an asterisk (*) are from deletion experiments with *Ilyanassa* (Clement 1967, 1986a,b); others are from the cell lineage of *Crepidula* (Conklin 1896).

---

lobes. Although the distribution of polar lobe cytoplasm determines the D quadrant in both *Bithynia* and *Ilyanassa*, there are some important differences between the polar lobes in these two eggs. *Bithynia* has a small polar lobe consisting of less than one percent of the egg volume, whereas *Ilyanassa* has a large polar lobe that is about one-third the egg volume.

Dohmen and Verdonk (1979a,b) have reviewed in detail the ultrastructure of the small polar lobes of *Bithynia*, *Crepidula*, and *Buccinum* and the large polar lobes of *Ilyanassa* and *Nassarius*. It is for the most part unclear where within the substructure of the cytoplasm morphogenetic determinants are localized nor is much revealed about the nature and operation of determinants. In the *Ilyanassa* egg, Crowell (1964) observed double membranous vesicles localized in the polar lobe, as did Schmekel and Fioroni (1975) in the polar lobe of *Nassarius reticulatus*. In *Ilyanassa*, Clement (1968) demonstrated that these vesicles can be displaced by centrifugation without displacing any morphogenetic factors.

Gerin (1971) described a "perinuclear corpuscle" in *Ilyanassa* oocytes and later (Gerin 1972) observed in the polar lobe a special organelle that he called a "double membranous vesicle." McCann-Collier (1977) found in *Ilyanassa* oocytes the presence of a multicomponent organelle, which she called a "polymersome." Some parts of the polymersome disappeared in late stages of oogenesis, leaving behind a head region that resembled the double membranous vesicles of Crowell (1964) and Gerin (1972), as well as the multimembranous vesicles observed by Pucci-Minafra et al. (1969) to be localized in the vegetal region of the *Ilyanassa* egg. It is not established that these membranous components found in the *Ilyanassa* egg are related to the cytoplasmic determinants localized in the polar lobe.

The clearest relation between structural components of the egg and the localization of determinants is in the *Bithynia* egg. During oogenesis a cup-shaped structure composed of electron-dense vesicles (Dohmen and Verdonk 1974), the **vegetal body** (see Figure 11.7a), is formed in the vegetal hemisphere of the *Bithynia* egg. At first cleavage the vegetal body is incorporated into the small polar lobe of the *Bithynia* egg (Figure 11.7a) . Removal of the polar lobe results in a partial larva that fails to form mesodermal bands, establish bilateral symmetry, or to differentiate eyes, foot, operculum, and shell (Cather and

Verdonk 1974). Centrifugal displacement of the vegetal body into one of the first two blastomeres (Figure 11.7*b*) with subsequent removal of the polar lobe results in normal development of 50 percent of the experimental embryos (van Dam et al. 1982). Thus the morphogenetic determinants contained in the *Bithynia* polar lobe appear to reside in the vegetal body. The normal development of lobeless eggs after displacement of the vegetal body into a blastomere gives positive evidence, as opposed to negative evidence obtained from deletion experiments, for the presence of determinants in the vegetal body.

In eggs with large polar lobes, the polar lobe cytoplasm is translocated asymmetrically into the first four blastomeres. The blastomere receiving the polar lobe cytoplasm at the 2-cell stage becomes the CD blastomere; the next division (which is also asymmetrical, and in some cases, such as in *Ilyanassa*, results from the formation of another polar lobe at second cleavage), produces C and D daughter cells. The larger D blastomere defines the D quadrant as a result of having received the determinants of the vegetal ooplasm contained in the polar lobe.

The importance of the vegetal cytoplasm contained in the large polar lobe of *Ilyanassa* eggs was first shown by Crampton (1896) who, in one of the earlier pieces of experimental work by American embryologists, saw that removal of the polar lobe resulted in failure of the embryo to form the primary mesentoblast cell. Crampton was unable to rear *Ilyanassa* embryos long enough to observe the effects of removing the polar lobe on organogenesis. This problem was reexamined by A. C. Clement in 1952,* at which time pasteurized seawater had entered the work of embryologists, and he was able to rear embryos outside of their egg capsules for many days and to see the influence of the polar lobe on organogenesis. Clement found that when the third polar lobe was removed at the trefoil stage the lobeless embryos failed to differentiate mesodermal bands, eyes, shell, foot, heart, and intestine. Later Atkinson (1971) made a histological study of *Ilyanassa* lobeless embryos and found that while many organs failed to differentiate (compare Figures 11.3*a* and *b* for the organogenesis of normal and lobeless *Ilyanassa* embryos) as reported earlier by Clement, a number of specific cell types did differentiate. For example, Atkinson saw that 97.5 percent of lobeless larvae differentiated velar, digestive gland, and style sac cell types, but none formed these organs. These findings suggest that, in addition to the possibility of containing specific determinants, the vegetal cytoplasm contains determinants that influence the organization of differentiated cells into tissues and organs.

Thus far, the localization of determinants has been related to eggs containing polar lobes and having unequal cleavage. What of equally cleaving gastropod eggs that

lack polar lobes such as *Limax, Lymnaea*, and *Patella*? That determinants are localized in the vegetal region of *Lymnaea* eggs is shown by abnormal development of centrifugally fragmented eggs that lack vegetal ooplasm (Morrill 1964).

In gastropods in which the first four blastomeres divide equally, dorsoventral polarity (as designated by the determination of the D quadrant) is established by cellular interactions between the first quartet of micromeres and the macromeres. This event has been extensively studied in the prosobranch *Patella vulgata* (van den Biggelaar 1977; van den biggelaar and Guerrier 1979), and the pulmonate *Lymnaea stagnalis* (Arnolds et al. 1983). The spatial organization in both unequally and equally cleaving eggs has been reviewed by van den Biggelaar and Guerrier (1983). Evolutionary aspects have been reviewed by van den Biggelaar and Haszprunar (1996).

In equally dividing gastropod eggs, the macromeres of the four quadrants are morphogenetically equivalent until between the fifth and sixth cleavage. At this time the 3D blastomere (the mother cell of the primary mesentoblast) divides precociously and forms the primary mesentoblast or 4d cell. If, as shown by van den Biggelaar and Guerrier (1979), contact between micromeres and macromeres is suppressed prior to this division, either by cell dissociation or by deletion of first quartet cells, normal differentiation of the macromeres does not occur.

Thus, in the absence of specific localized determinants as found in polar-lobe-containing eggs, in early cells of equally cleaving eggs determination is achieved by epigenetic interactions between cells of animal pole origin with cells of vegetal pole origin. van den Biggelaar (1977) postulated that "normally the median macromere that obtains the majority of contacts with 1st quartet micromeres becomes the mesentoblast mother cell." Further, the experimental results of van den Biggelaar (1977) and Arnolds et al. (1983) showed that this mode of determination is stochastic, as the contacts between animal-vegetal cells are haphazard. In other words, specific cells are not involved; rather, the median macromere to receive the largest number or most extensive accidental contact with and micromere or micromeres of the first quartet becomes the 3D macromere.

Gastrulation

In many gastropod blastulae, the blastocoel is either absent or very small, as in *Ilyanassa* and *Crepidula*. These are stereoblastula. In *Patella* and *Lymnaea*, the blastula develops a small lumen and becomes a coeloblastula.

Gastrulation is usually by epiboly, wherein a cap of micromeres at the animal pole divides and "overgrows" the vegetal macromeres. Epiboly continues until the macromeres become almost completely covered with micromeres except for a small slitlike opening, the blastopore, at the vegetal pole of the embryo.

In the pulmonates and some prosobranchs such as *Patella* and *Littorina*, gastrulation is by a form of invagina-

---

*See Atkinson 1986a,b for a biographical sketch and bibliography of A. C. Clement.

tion called emboly. Emboly is the extension of macromeres (or their daughter cells) into the blastocoel, where they form an archenteron (primitive gut). In *Patella*, the endoderm fills the blastocoel to form a stereogastrula that will later delaminate to form the archenteron.

Trochophore and Veliger Larvae

Indirect development in which larvae are produced as the end product of embryogenesis is widespread among the gastropods; therefore, larval forms will be discussed before organogenesis.

A larva is a postembryonic stage that is free-living and is capable of developing into an adult. Among the gastropods, larvae are mostly restricted to marine forms because larvae generally cannot withstand the wide range of environmental conditions found in freshwater.

Gastropods produce two kinds of larvae, a trochophore (Figures 11.2*a* and 11.11*b*) and a veliger (Figure 11.3*a*). In most snails the free-living larvae is a veliger. Only archaeogastropods have a free-living trochophore, and it soon develops into a veliger that metamorphoses into an adult. Though most gastropods with indirect development hatch as veligers, a trochophore-like stage occurs during embryogenesis; indeed, a trochophore-like stage also occurs early in the embryogenesis of pulmonates, which have direct development.

In addition to its role in gastropod development, the trochophore is of evolutionary significance because it is the strongest feature linking molluscs to annelids or other wormlike ancestors. Long-known as Loven's larva in honor of its discoverer, it was given its present name by Hatschek in 1878 when he developed his trochophore theory. Hatschek suggested that the trochophore is a larva of a form ancestral to all coelomate bilateral phyla. It is striking that the annelids and molluscs, which are structurally quite different, have nearly identical trochophore larvae. To what extent this commonality of larvae results from an ancestral relationship, or whether the molluscan trochophore is independently convergent with the annelid trochophore is debatable (Willmer 1990).

Hyman (1951) describes a trochophore larva as "a somewhat biconical creature with a protruding equator." It is a top-shaped creature with a band of cilia around its equator, the *prototroch*, and a tuft of cilia at its apical end, the *apical tuft*. The prototroch is a locomotor organelle, and the apical tuft is sensory organ that is made up of a group of cilia that extend from the apical plate of sensory cells. There may be additional bands of cilia, such as the *metatroch*, which passes below the mouth, and a circle of cilia around the anus, the *telatroch*. The external surface consists of a single cell layer of ectodermal epithelium. Internally, the trochophore contains a simple digestive tube consisting of a mouth, stomodeum, and a stomach that leads into an intestine, which exists by the anus. The en-

Figure 11.11 Sagittal sections of a *Patella* embryo (*a*) and a trochophore larva (*b*). (*a*) 1, apical sensory plate; 3, endoderm; 9, prototroch; 10, metatroch; 11, accessory ciliary girdles; 13, mesodermal band; 14, cells of mesodermal band becoming mesenchyme; 15, anal cells; 16, telotroch. (*b*) 1, apical tuft; 2, apical sensory plate; 3, trochoblasts; 4, stomodeum; 5, radular evagination; 6, stomach; 7, intestine; 8, anal cell; 9, telotroch; 10, shell gland. (After Hyman 1967.)

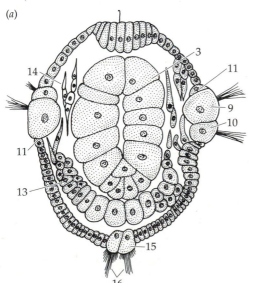

(*a*)

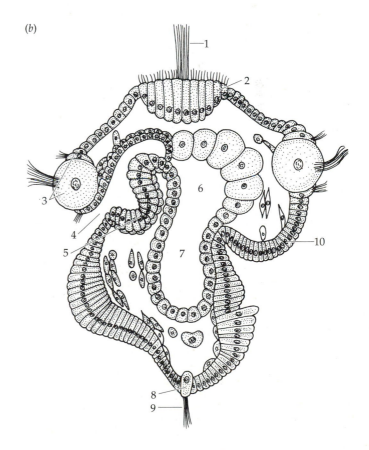

(*b*)

tire digestive tube is ciliated. There is a large blastocoel that extends between the external ectoderm and the gut. The blastocoel contains mesenchyme cells, muscle cells (from ectomesoderm), a pair of mesodermal bands (from endomesoderm), and a nephridium. The nervous system may be complex. Basically, it consists of a mass of ganglia beneath the apical sensory plate, longitudinal nerves that extend from these ganglia to radial nerve rings. The radial nerve rings encircle the horizontal axis of the trochophore and innervate its musculature and ciliary bands. There may be several radial nerve rings, but a major ring under-lies the prototroch. Sense organs such as eyes and stato-cysts are present. The eye detects only light and the stato-cysts are equilibrium organelles. The structure of the trochophore is illustrated in Figure 11.11. Hyman (1951) gives a detailed account of the musculature and nervous system of the trochophore.

A veliger is a molluscan larva that is characterized by the presence of a velum and a foot. (See Figure 11.3*a* for a detailed drawing of a veliger larva.) The velum, the pri-mary locomotive organ the veliger uses for swimming, consists of two large, ear-like lobes that bear many long cilia. The velar cilia also sweep food particles into a *food groove*, where they are trapped in mucus and carried to the mouth. The veliger larva is uniquely molluscan and it may develop from a free-living trochophore larva, as in some archaeogastropods, or it may hatch from an egg cap-sule or egg mass as the endpoint of embryogenesis. Dur-ing the course of embryogenesis most gastropod embryos go through a trochophore-like stage in which prototrochal cilia (derived from molluscan cross cells) and a primitive gut are present. In this situation the trochophore larva is said to suppressed. Such a suppressed trochophore stage exists even among pulmonates that have no larval stage but have a direct development in which baby snails hatch from the egg capsule. Thus, a connection with the tro-chophore is maintained in gastropod development.

The transformation of a trochophore into a veliger larva is well depicted for *Patella* by Patten (1886); some of his figures are reproduced herein and by Hyman (1967). When a trochophore transforms into a veliger (see Figure 11.2) the dorsal ectoderm between the prototroch and the anus differentiates into a shell gland, which forms a veliger shell. The edge of the shell gland is the edge of the mantle and the invagination of mantle cells in front of the shell gland forms the mantle cavity. The prototroch of the trochophore is transformed into the veliger velum.

The foot develops as a protrusion of columnar cells on the the ventral side of the trochophore. The dorsal surface of the foot secretes an operculum that will be cast off in those snails that do not have an operculum as an adult. Along the posterior-lateral surface of the foot, epithelial cells invaginate to form a pair of statocysts.

The embryogenic origin of the veliger and direct devel-opment are described in the following section on organo-genesis.

Organogenesis

Sources for gastropod organogenesis are Conklin (1897), Moritz (1939), Werner (1955), Fretter and Graham (1962), Wada (1968), Moor (1983), Barnes (1987), Gosliner (1994), and Voltzow (1994). Morrill's (1982, 1997) reviews of the natural history and development of the gastropod *Lym-naea* are valuable sources on pulmonate development. Craig and Morrill (1986) have made a scanning and trans-

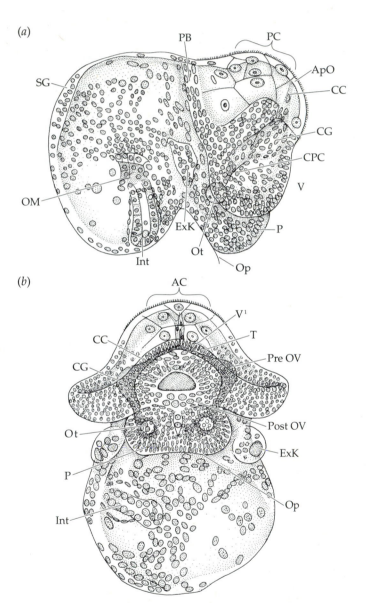

Figure 11.12   View of whole mounts of older embryos. (*a*) Side view of older embryo. (*b*) Ventral view of older embryo. Abbrevia-tions: AC, apical cell plate; ApO, apical cell organ; CC, cerebral com-missure; CG, cerebral ganglion; CPC, cerebro-pedal connective; ExK., external kidney; Int, intestine; OM, opening into mesenteron; Op, operculum; Ot, otocyst; P, foot; PB, posterior branch of velum; PC, posterior cell plate; Post OV, post-oral velum; Pre OV, pre-oral velum; SG, margin of shell gland; T, tentacle; V, velar ridge; V[1], first cell row of velum. (After Conklin 1897.)

(a)

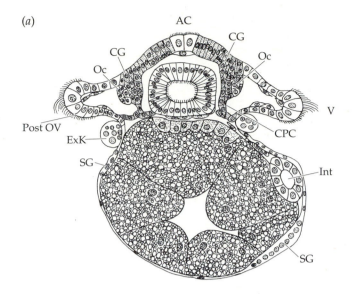

(b)

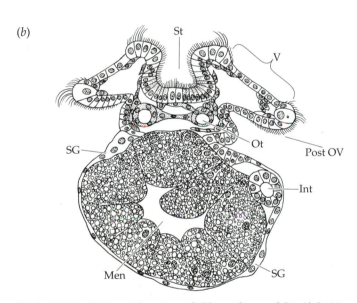

Figure 11.13 Horizontal sections of older embryos of *Crepidula*. (*a*) Section taken near the ventral side of the embryo. (*b*) Section taken nearer the ventral side than in (*a*). Abbreviations: AC, apical cell plate; CG, cerebral ganglion; CPC, cerebro-pedal connective; ExK., external kidney; Int, intestine; Men, mesenteron; Oc, ocellus; Ot, otocyst; P, foot; Post OV, post-oral velum; SG, margin of shell gland; St, stomodeum; V, velum. (After Conklin 1897.)

mission electron microscopic study of early development of *Ilyanassa obsoleta*, and Tomlinson (1987) has a detailed scanning electron microscope study of later stages of *Ilyanassa* development.

Figures 11.12 and 11.13 are whole mounts and sections of older *Crepidula* embryos (Conklin 1897) that illustrate early stages of organogenesis.

Nervous System

In comparison to other molluscs, the gastropods have a well-developed nervous system. Their nervous system is organized around a series of paired ganglia in which each member of a pair is either fused or connected to each other by a commissure; different ganglia—for example, the cerebral and pedal ganglian—are connected by a connective, the cerebralpedal connective.

The major ganglia common to many forms are the cerebral, buccal, pleural, pedal, intestinal, and visceral. All of these ganglia are paired, except for the visceral ganglion, which may be unpaired. Some snails have additional ganglia such as parietal ganglia located along the visceral nerve cord. Hyman (1967) discusses and illustrates the nervous system in a variety of gastropods.

Ganglia are formed from ectodermal cells derived from the first three quartets of micromeres. In many forms such as *Crepidula*, *Planorbis*, and *Lymnaea*, ectodermal cells in the region of a prospective ganglion become crowded and individual cells become columnar in shape. By continued cell division the prospective ganglion cells form a mass of cells on the inner side of the external epithelium (Figure 11.13*a*). This mass of cells separates from the epithelium and forms a ganglion. In other cases, as in *Patella* and *Limax*, ganglia are formed by invagination of ectodermal cells, which later separate from the external ectoderm. Commissures and connectives grow out from their related ganglia.

At the apical end of the *Crepidula* embryo there are four large cells located at the exact point where the polar bodies were extruded; these are the **apical plate cells** (Figure 11.14). The apical cells proliferate some cells into the head vesicle. These cells along with the apical cells form the apical sense organ. The apical sense organ is connected by a strand of cells to the cerebral ganglion. In *Crepidula*, the apical plate cells are covered with small cilia. In other forms, such as *Patella*, which has a trochophore stage, the apical plate bears long cilia, the apical tuft (Figures 11.2 and 11.11*b*).

Cells lateral to the apical plate divide rapidly to form the cephalic plate (Figure 11.14) from which the cerebral ganglia, tentacles, and eyes develop.

Sense Organs

THE EYE. The cellular origin of the gastropod eye has been established by deletion experiments in *Ilyanassa* (Clement 1971) and in *Bithynia* and *Lymnaea* (Cather 1976). In all cases, deletion of 1a results in the absence of the left eye, deletion of 1c the absence of the right eye. In *Ilyanassa* and *Bithynia*, eye development is polar lobe-dependent, because even when 1a and 1c cells are intact, eyes fail to appear if the polar lobe is removed.

Eyes develop just posterior to the cerebral ganglia from pockets of cells created by epithelial invaginations or by delaminations of external epithelial cells. These epithelial pockets separate from the overlying layer of external cells and differentiate into embryonic eyes (see C.P.C. in Figure 11.13*a*). Embryonic eyes contain presumptive photoreceptor cells, pigmented retina, and corneal cells (Gibson 1984). Embryonic eyes are retained during metamorphosis

Figure 11.14   Section through apical and cephalic plate of *Lymnaea stagnalis*. (After Raven 1958.)

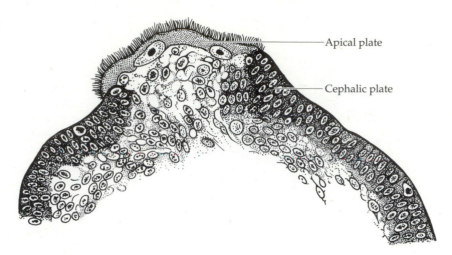

Apical plate

Cephalic plate

and later differentiate into adult eyes, which contain lens, cornea, retina, and a neuropile. Gibson (1984) has made a detailed study of eye development in *Ilyanassa obsoleta*.

STATOCYSTS. The statocysts are paired equilibrium organs that develop from epithelial invaginations along the posterior sides of the foot. The invaginated epithelial cells separate from the external epithelium, hollow out to form a large lumen, and secrete a statolith (calcium carbonate concretion). The epithelial cells of the statocyst differentiate into receptor cells that respond to contact with the statolith to provide orientation with respect to gravity. Statocysts are closely associated with the pedal ganglia but are innervated by connectives from the cerebral ganglia.

Shell

The shell is an important and distinguishing organ of the veliger larva and of the majority of adult snails. In all cases it is a "house of protection." In the veliger it gives a smooth external surface, protects the veliger organs, and in moments of stress the shell completely encloses the veliger when the retracted foot and operculum close the door to the shell. A shell develops early in all gastropods. If the phenotype of an adult snail is the absence of a shell, then the larval shell is cast off during metamorphosis.

The cellular origin of the veliger shell is usually from the 2d cell. In some cases, as in the neogastropod *Ilyanassa obsoleta*, in which shell differentiation has been extensively studied, the veliger shell has a dual orgin, from 2d and 2c. When Cather (1967) carbon-marked separately 2d and 2c, the carbon markings from 2d appeared in the shell gland and the carbon markings from 2c turned up in the mantle. After a single deletion of 2c or 2d, a partial shell was formed. After a dual deletion of 2c and 2d, a normal shell was absent (Cather 1967). Cather's observations have been confirmed and extended by McCain (1992). In her study of biomineralization in *Ilyanassa*, she did multiple deletions of 2a, 2c, 2d, 3c, and 3d and found that 2a modulated the formation of external shell by 2c and 2d. McCain also

showed that the calcium carbonate in *Ilyanassa* shell and statoliths is aragonite and calcite. Importantly, she showed by mineral analyses that the so-called internal shell masses, often seen in lobeless and -2c2d embryos, have calcite crystals of calcium carbonate but are lacking the aragonite crystals found in the external shell. McCain correctly suggests that these so-called internal shell masses be referred to as *internal birefringent masses*.

Initially, shell is formed by the shell gland and later by the mantle epidermis. Soon after gastrulation there appears on the dorsal-posterior surface of the embryo an oval field of columnar cells; these cells are the **shell field**. The extent of this field can be seen in Figure 11.15. Only the marginal cells of the shell field form the organic (proteinaceous) material that makes up the shell matrix. The shell field invaginates to form the shell gland. After this invagination brings the marginal cells of the shell field into close contact, they secrete an organic shell matrix (Eyster 1983, 1986; Kniprath 1981). This close apposition of the marginal cells makes the formation of a solid sheet of shell matrix possible. Otherwise a shell with a hole in its center would be formed. Figure 11.15 shows the shell gland, the pore of the invaginated shell gland, and a fragment of the organic shell matrix overlaying the shell gland pore. After the shell field invaginates, it evaginates to reconstruct an external shell field surrounded with a ring of marginal cells. These marginal cells, uniquely capable of making shell matrix, grow out over the mantle surface to form a "growing edge" of cells (Eyster 1983) that make shell matrix as they overgrow the mantle. Eyster (1986) suggests that mineralization of the shell may occur by the incorporation of calcium transported through cells lining the shell-field invagination, to the lumen of the shell gland, to the pore of the shell gland, and onto the surface of the overlaying shell matrix.

Foot and Operculum

The foot is a characteristic organ of the veliger larva; it aids in feeding by the beating of cilia on its surface, it develops and contains equilibrium organs (statocysts), secretes and

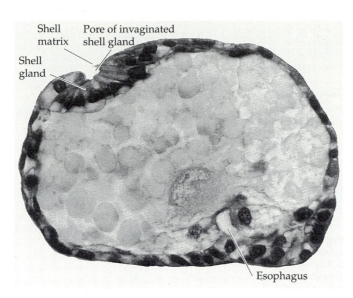

Shell matrix
Pore of invaginated shell gland
Shell gland
Esophagus

Figure 11.15   Sagittal section through shell gland of *Ilyanassa* embryo. (Collier, unpublished.)

contains the operculum (a door to the shell), and enables the metamorphosing veliger and adult snail to creep along a substratum. The foot also aids in circulation by its powerful contractions, which drive blood from the foot into the body cavity. This role is especially evident in many pulmonates that develop an enlarged vesicle, called a **podocyst**, whose contraction forces blood out of the foot.

The foot appears as a protrusion of cells on the ventral surface of the embryo (Figures 11.12*a,b*). The midline cells of the long foot have long cilia, which beat toward the mouth and aid in food collection.

The cellular origin of the foot needs to be clarified. Conklin (1897) contended that the foot originated from the 2d cell (the primary somatoblast) by way of the posterior growing center (ventral plate). Clement's (1971) deletion experiments showed that the foot has a dual lineage of 3c and 3d cells. When Clement deleted 3c and 3d micromeres from the *Ilyanassa* embryo, the respective right and left halves of the foot failed to develop. Thus, it is unlikely that the foot is from the lineage of the primary somatoblast (2d).

The dorsal surface of the foot secretes an operculum (Figure 11.12*a*). The operculum is made of a horny substance that is molded to fit the opening of the veliger shell. Thus, when the retractor muscle contracts the head, velum and foot are pulled into the shell and the operculum "closes the door." Whether the operculum is retained or cast off depends on the adult phenotype.

Stomodeum and Foregut

In some gastropods, such as *Patella* and *Lymnaea*, the **stomodeum** (the future mouth and esophagus) invaginates from the anterior edge of the blastopore. In *Crepidula* and *Ilyanassa*, the blastopore closes and a stomodeum invagi-

nates just forward of the anterior region of the closed blastopore. At first the stomodeum is closed at its inner end but soon opens and joins the mesenteron (midgut). Later, as the foot and posterior part of the embryo grow, the mouth is pushed forward. As the yolk-containing cells are shifted backward, the esophageal component of the stomodeum is directed posteriorly to establish its contact with the mesenteron.

The **radula** is a highly differentiated feeding organ located in the buccal cavity of snails. The radula is formed by and contained in a radula sac that develops as an evagination of the esophageal portion of the stomodeum. The origin of the radula in *Patella* is illustrated in Figure 11.11*b*.

The epithelium of the radula sac differentiates into a basal epithelium and odontoblasts (tooth-forming cells) that form the basement membrane and the radular teeth. In the adult snail, radular teeth are constantly worn away and replaced by odontoblasts at the rate of several rows of teeth per day. The buccal cavity and buccal glands develop as outpocketings of the stomodeum.

Mesenteron and Intestine

The archenteron is a primitive and undifferentiated cavity that will give rise to the gut. The roof of the archenteron is formed by the four yolky macromeres, its sides are formed by cells of the fifth quartet of micromeres, and its floor is composed of cells from the fourth quartet of micromeres, 4a–4c but not from the 4d cell. Some fourth quartet cells contribute to the anterior part of the mesenteron where it joins the stomodeum.

The archenteron will differentiate into the mesenteron, the anterior end of which will connect to the stomodeum, and the posterior end to the intestine. The anterior part of the mesenteron becomes ciliated and forms the stomach. Evaginations from the stomach give rise to the midgut gland or larval liver. In herbivorous gastropods there occurs an anterior evagination from the stomach that forms a style sac. The style sac secretes a crystalline style composed of amylase. The style sac is ciliated and rotates the crystalline style against a cuticularized gastric shield, wherein the amylase of the crystalline style is solublized and made available for digestion of carbohydrates.

The intestine develops from a posterior prolongation of the mesenteron. Part of its walls are made up of enteroblasts derived from the 4d cell. Derivatives of the enteroblasts are often called small-celled endoderm, in contrast to the large-celled endoderm descended from the yolky macromeres.

Pericardial Cavity: Heart, Nephridia, and Gonads

It is debatable (Willmer 1990) whether gastropods—or molluscs in general—have a true coelom. But it is known that they do have a pericardial cavity bounded by an epithelium of mesodermal origin. This pericardial cavity has an essential role in differentiation of the heart, nephridia, and gonads.

The anterior cells of the mesodermal bands that separate and become mesenchyme reaggregate into paired masses of mesenchymal cells. These cell aggregates hollow out to become the right and left **pericardial cavities**, which later fuse to form a single pericardial cavity (Figure 11.16a). The heart develops as an invagination of a dorsal thickening of the pericardial wall, which later constricts into an auricle and a ventricle of the heart (Figure 11.16a,b). The larval heart has anterior and posterior valves. Anteriorly, the heart pumps blood into the lacunar spaces of the velum and foot; posteriorly, it pumps into the visceral mass (Werner 1955).

Next, left and right primordia of the nephridia evaginate from the ventral pericardial wall (Figure 11.16b). Later, the gonad proliferates from the cells of the pericardial wall. Some prosobranch veliger larvae lack a protonephridium but have external excretory cells (external kidneys) that are heavily vacuolated and receive excretory products. In lower prosobranchs there are a pair of nephridia; in others there is a single nephridium. Nephridia empty into the mantle cavity by way of a nephridiopore, which often also serves as a gonoduct.

Torsion

Torsion is a 180-degree counterclockwise twisting of the visceral mass in relation to the fixed position of the head and foot. Torsion results in the translocation of the mantle cavity, gills, anus, and nephridiopores from the posterior position of the snail into the anterior-dorsal part of the body just behind the head. Internally, this creates a loop in the digestive tract and a twisting of the nervous system. These changes associated with torsion are illustrated in Figure 11.17.

Torsion begins during embryonic development and is completed in the veliger larva. Temporary torsion occurs in many veligers by differential and frequent contraction of the right retractor muscle. This muscle is attached to the shell and inserted into the velum of the larva or head of the adult. Permanent torsion is established by the differential growth of the right retractor muscle during development and metamorphosis.

There are snails that are untorted (i.e., have never been torted), snails that are torted, and snails that have been detorted. Detorted snails have been torted at some stage in their development but have undergone a complete or partial reversion of torsion. Many prosobranchs are torted; relatively few opisthobranchs are torted, most having undergone detorsion; and the pulmonates are largely detorted.

Much has been written about the function and evolution of torsion. Garstang (1928) suggested that torsion arose from a series of mutations that were advantageous to larvae but not adults. In the untorted primitive larva, the mantle is posterior to the larval head (Figure 11.17a). Complete torsion would then place the opening of the mantle cavity anteriorly (Figure 11.17d) whereby contraction of the retractile muscle would withdraw the delicate head and velum into the mantle cavity, thereby offering a protective housing. This housing of the velum would cause the larva to sink and thus escape from organisms that prey on plankton. Garstang (1928) expressed these views poetically in the first stanza of "The Ballad of the Veliger, or, How the Gastropod Got Its Twist":

> The Veliger's a lively tar, the liveliest afloat,
> A whirling wheel on either side propels his little boat;
> But when the danger signal warns his bustling submarine,
> He stops the engine, shuts the port, and drops below unseen.

Stasek (1972) has pointed out that this view has been virtually abandoned because, in primitive archaeogastropods, torsion is not completed until after the planktonic phase of the larval life cycle has ended; this argument appears valid for this group of prosobranchs, but it is not clear how extensive this time of torsion and planktonic existence is among other torted snails. Pennington and Chia

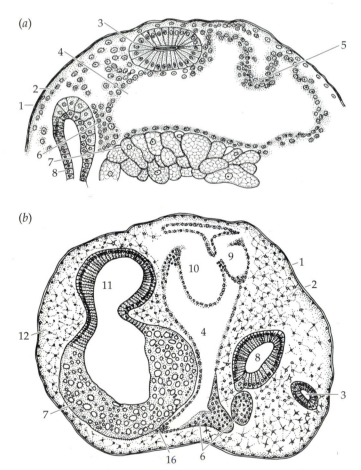

Figure 11.16 Transverse sections of *Viviparus* embryo showing derivatives of the pericardial cavity. (a) Early stage of heart development. (b) Later stage of heart development. 1, shell; 2, mantle epidermis; 3, intestine; 4, pericardial cavity; 5, primordium of heart; 6, primordium of nephridium; 7, midgut gland; 8, mantle invagination; 9, heart ventricle; 10, heart auricle; 11, stomach; 12, mesenchyme; 16, gonad. (After Hyman 1967.)

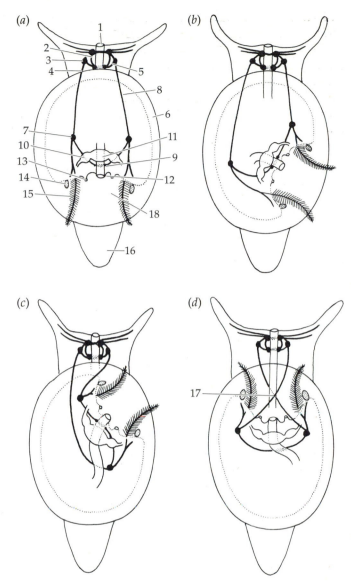

Figure 11.17 Torsion. (*a*) Untorted gastropod. (*b* and *c*) Intermediate stages of torsion. (*d*) Complete torsion. 1, mouth; 2, left cerebral ganglion; 3, left pleural ganglion; 4, left pedal ganglion; 5, circumenteric ring; 6, mantle edge; 7, left parietal ganglion; 8, pleurointestinal connective; 9, visceral ganglion; 10, left auricle; 11, ventricle; 12, anus; 13, left nephridiopore; 14, left osphradium; 15, left gill; 16, foot; 17, crossing pleurointestinal connectives; 18, mantle cavity. (After Hyman 1967.)

(1985) tested the putative protective function of torsion by subjecting torted and untorted larvae to planktonic prey and found that each group was equally preyed upon.

Stasek (1972) proposed that the likely function of torsion is to provide greater movement of the head, whose evolution parallels the evolution of torsion, and that the adaptive advantages of torsion were not restricted to either the larva or adult but were shared by the entire life cycle of some snails. The advantages of moving the opening of the mantle cavity anteriorly, as is achieved by tor-

sion, is to provide protection and greater movement of the enlarged gastropod head. In addition, osphradia (chemotactic sensing organs) and gills are placed anteriorly in the torted snail. The osphradia test the water into which the snail is moving and the gills are exposed to fresh "respiratory water." That torsion was neutral or disadvantageous to some gastropods is indicated by the extent that detorsion has occurred. It appears that torsion arose in many groups of snails and that when it had an adaptive advantage, at any stage in the life cycle, it was retained; otherwise, it was eliminated by detorsion.

Metamorphosis

The course of gastropod development varies greatly with respect to the direct and indirect development. In direct development, the embryo hatches from its egg capsule as a small snail. Indirect development results in the production of a free swimming larva that metamorphoses into an adult snail.

Prosobranch and opisthobranch snails usually have indirect development in which embryos hatch from their egg capsules as free-swimming veliger larvae. A trochophore or trochophore-like stage may occur during development before hatching. Most often the trochophore stage is suppressed and is passed before hatching occurs. Hatching from the egg capsule in many cases is associated with the digestion of the egg capsule plug. In the neogastropod *Ilyanassa obsoleta*, Sullivan (1984) has shown the presence of hatching activity in older embryos that is dependent on transcription and protein synthesis. Some archaeogastropds, such as *Patella, Trochus,* and *Acmaea,* which shed their gametes directly into the ocean, produce a free-living trochophore (Figure 11.2*a*) that soon develops into a veliger larva. The pattern of larval development shows great variability, even among closely related organisms. For example, embryos of some periwinkles (e.g., the genus *Littorina*) may hatch as veliger larvae, while others of this genus may brood their embryos in egg capsules until a baby snail develops.

Veliger larvae may live from days to weeks before they metamorphose into young snails. During this time their energy is derived from feeding on plankton and/or digesting stored yolk. During its planktonic life the veliger continues to grow, the differentiation of internal organs continues, the veliger shell enlarges, and the foot becomes longer. Most of the larval organs are retained as adult organs, with the notable exception of the velum and heart.

In order to metamorphose, the veliger must settle upon a suitable substrate. Most often there is a high level of selectivity for a particular settling site, and many forms can delay metamorphosis until a site acceptable for settling is found. Some larvae require specific organisms, such as hydroids, bryozoans, or ascidians, as a settling site. The onset of metamorphosis appears to be induced by the chemical nature of the substratum rather than by its physical properties.

As metamorphosis begins, the velum and the external kidneys are cast off. The veliger is no longer planktonic, but settles onto a substratum. At this point, the veliger foot is sufficiently developed to provide a creeping movement of the metamorphosing larva along a substratum. The operculum is lost in those species who do not have an operculum in the adult. The larval heart is repressed and replaced by an adult heart. The larval shell is frequently retained as the shell spire in those species that have an adult shell; in other cases, the larval shell is discarded. Following these abrupt events, the larva continues to grow and to develop the final features of the adult.

Freshwater and terrestrial pulmonates have direct development in which the trochophore and veliger stages are passed in the egg sacs and the embryos hatch as young snails. The hatched snails are completely differentiated except for the reproductive system. Free-living veliger larvae occur only in marine pulmonates.

Pulmonates with direct development often develop specialized and temporary structures such as an albumen sac, a podocyst, and a protonephridium that support development in an egg capsule. The albumen sac is a diverticulum from the archenteron which digests albumen contained in the egg capsule. The podocyst is an expansion of the foot and functions in gas exchange, excretion, and circulation. Cather and Tompa (1972) has shown that the podocytst has pinocytotic activity, indicating that it may uptake albumen by pinocytosis. The protonephridium is an excretory tubule that terminates in a flame cell and, in addition to having an excretory function, it serves as an osmoregulatory tubule.

Evolutionary Trends

The overall evolutionary trend among the gastropds has been the evolution of a large and mobile head; the evolution of a shell from a simple shield-like structure to a coiled protective house into which the head and foot can be withdrawn; and the evolution of torsion, which evolved in parallel with evolutionary changes in the head.

Prosobranch evolution is marked by the changes from the archaeogastropod to the derived forms of the meso- and neogastropods. The archaeogastropods contain paired organs such as gills, osphradia, auricles, and kidneys. As this group evolved into higher prosobranchs, the paired organs were replaced by only the left of each paired organ. Functionally, this involved evolution from a slow moving group into a more mobile group whose diet and feeding habits were extremely diverse. The derived prosobranchs developed elaborate inhalant siphons, and, in parallel, a more elaborate, often filamentous, osphradium. Among the carnivores and scavengers, the elaborate osphradium was a useful adaptation for the efficient location of food.

Among the higher gastropods, radular evolution paralleled changes in food sources and became most highly developed in the pulmonates. Most opisthobranchs and pulmonates also developed jaws, which in conjunction with the radula resulted in the efficient collection and maceration of food.

Adaptive radiation is shown in a number of systems among the gastropods. Striking examples include the reproductive system and the mechanisms for circulation of water and exchange of gases. Among the prosobranchs and opisthobranchs there have been many adaptive changes, such as the number of gills, location of gills as related to torsion and detorsion, role of the mantle in respiration, and finally in the culminating development of the mantle cavity into a lung in the pulmonates.

Acknowledgments

The preparation of this manuscript was supported by grants 665137 and 666156 from the City University of New York PSCBHE Research Award Program. I thank Sharon Moshel for her help in preparing the figures.

Literature Cited

Aldridge, E. 1983. Physiological ecology of freshwater prosobranchs. In *The Mollusca: Ecology*, W. D. R. Hunter (ed.). Academic Press, New York, pp. 329–358.

Anderson, D. M. and L. D. Smith. 1978. Patterns of synthesis and accumulation of heterogeneous RNA in lampbrush stage oocytes of *Xenopus laevis* (Daudin). *Dev. Biol.* 67: 274–285.

Arnolds, W. J. A., J. A. M. van den Biggelaar and N. H. Verdonk. 1983. Spatial aspects of cell interactions involved in the determination of dorsoventral polarity in equally cleaving gastropods and regulative abilities of their embryos, as studied by micromere deletions in *Lymnaea* and *Patella*. *Roux's Arch. Dev. Biol.* 192: 75–85.

Atkinson, J. W. 1971. Organogenesis in normal and lobeless embryos of the marine prosobranch gastropod *Ilyanassa obsoleta*. *J. Morphol.* 133: 339–352.

Atkinson, J. W. 1986a. Anthony Calhoun Clement (1909–1984). *Int. J. Inv. Reprod. Dev.* 9: 131–133.

Atkinson, J. W. 1986b. Publications of Anthony Calhoun Clement. *Int. J. Inv. Reprod. Dev.* 9: 135–138.

Atkinson, J. W. 1986c. An atlas of light micrographs of normal and lobeless larvae of the marine gastropod *Ilyanassa obsoleta*. *Int. J. Inv. Reprod. Dev.* 9: 169–178.

Barnes, R. D. 1987. *Invertebrate Zoology*. Saunders, Philadelphia.

Beeman, R. D. 1970. An autographic study of sperm exchange and storage in a sea hare, *Phyllaplysia taylori*, a hermaphroditic gastropod. *J. Exp. Zool.* 175: 125–132.

Bolognari, A., A. Licata and M. B. Ricca. 1976. Primary nucleolus and amphinucleoli in the oocytes of *Patella coerulea* (L.). (Moll. Gast.). *Experientia* 32: 870–871.

Bottke, W. 1973. Lampenburstenchromosomen und Amphinukleolen in Oocytenkernen de Schnekke *Bithynia tentaculata*. *Chromosoma* 42: 175–190.

Bottke, W. and J. Sinha. 1979. Ferritin as exogenous yolk protein in snails. *Roux's Arch. Entw. Org.* 186: 715–191.

Bottke, W., J. Sinha and I. Keil. 1982. Coated vesicle-mediated transport and deposition of vitellogenic ferritin in the rapid growth phase of snail oocytes. *J. Cell Sci.* 53: 173–191.

Brisson, P. and J. Regondaud. 1971. Observations relatives a l'origine dualiste de l'appareil genital chez quelques Gasteropodes Pulmones Basommatophores. *Comptes Rendus de l'Academis des Sciences de Paris.* 273: 2339–2341.

Brisson, P. and J. Regondaud. 1977. Origine et structure de l'Ebauche de la Gonade chez les Gasteropodes Pulmones Basommatophores. *Malacologia.* 16: 457–466.

Buckland-Nicks, J. A., D. Williams, F. S. Chia and A. Fontaine. 1982. Studies on the polymorphic spermatozoa of a marine snail. I. Genesis of the apyrene sperm. *Biol. Cell.* 44: 305–314.

Buckland-Nicks, J. A., D. Williams, F. S. Chia and A. Fontaine. 1983. Studies on the polymorphic spermatozoa of a marine snail. II. Genesis of the eupyrene sperm. *Gamete Res.* 7: 19–37.

Cather, J. N. 1967. Cellular interactions in the development of the shell gland of the gastropod, *Ilyanassa obsoleta. J. Exp. Zool.* 166: 205–244.

Cather, J. N. 1976. Cellular interactions in the early development of the gastropod eye, as determined by deletion experiments. *Malacological Rev.* 9: 77–84.

Cather, J. N. and A. S. Tompa. 1972. The podocyst in pulmonate evolution. *Malacological Rev.* 5: 1–3.

Cather, J. N. and N. H. Verdonk. 1974. The development of *Bithynia tentaculata* (Prosobranchia, Gastropoda) after removal of the polar lobe. *J. Embryol. Exp. Morphol.* 31: 415–422.

Clement, A. C. 1952. Experimental studies on germinal localization in *Ilyanassa* . I. The role of the polar lobe in determination of the cleavage pattern and its influence in later development. *J. Exp. Zool.* 121: 593–626.

Clement, A. C. 1967. The embryonic value of the micromeres in *Ilyanassa obsoleta* , as determined by deletion experiments. I. The first quartet cells. *J. Exp. Zool.* 166: 77–88.

Clement, A. C. 1968. Development of the vegetal half of the *Ilyanassa* egg after removal of most of the yolk by centrifugal force, compared with the development of animal halves of similar visible composition. *Dev. Biol.* 17: 165–186.

Clement, A. C. 1971. *Ilyanassa.* In *Experimental Embryology of Marine and Fresh-water Invertebrates.* G. Reverberi (ed.). North-Holland Company, Amsterdam and London, pp. 188–214.

Clement, A. C. 1986a. The embryonic value of the micromeres in *Ilyanassa obsoleta* , as determined by deletion experiments. II. The second quartet cells. *Int . J. Inv. Reprod. Dev.* 9: 139–153.

Clement, A. C. 1986b. The embryonic value of the micromeres in *Ilyanassa obsoleta,* as determined by deletion experiments. III. The third quartet cells and the mesentoblast cell. *Int. J. Inv. Reprod. Dev.* 9: 155–168.

Collier, J. R. 1965. Morphogenetic significance of biochemical patterns in mosaic embryos. In *The Biochemistry of Animal Development,* R. Weber (ed.). Academic Press, New York, pp. 203–241.

Collier, J. R. 1981. Protein synthesis in the polar lobe and lobeless egg of *Ilyanassa obsoleta. Biol. Bull.* 160: 366–375.

Collier, J. R. 1989. Cytoplasmic regulation of translation during *Ilyanassa* embryogenesis. *Development* 106: 263–269.

Collier, J. R. and M. E. McCarthy. 1981. Regulation of polypeptide synthesis during early embryogenesis of *Ilyanassa obsoleta. Differentiation* 19: 31–46.

Conklin, E. G. 1897. The embryology of *Crepidula. J. Morphol.* 13: 3–209.

Conklin, E. G. 1902. Karyokinesis and cytokinesis in the maturation, fertilization and cleavage of *Crepidula* and other gasterpoda. *J. Acad. Nat. Sci. Phil.* 12: 5–116.

Costello, D. P. and C. Henley. 1976. Spiralian development: A perspective. *Amer. Zool.* 16: 277–292.

Craig, M. and J. B. Morrill. 1986. Cellular arrangements and surface topography during early development in embryos of *Ilyanassa obsoleta. Int. J. Inv. Reprod. Dev.* 9: 209–228.

Crampton, H. E. 1896. Experimental studies on gastropod development. *Roux's Arch. Entwmech. Org.* 3: 1–19.

Crowell, J. 1964. The fine structure of the polar lobe of *Ilyanassa obsoleta. Acta Embryol. Morphol. Exp.* 7: 225–234.

Dan, J. C. 1956. The acrosome reaction. *Int. Rev. Cytol.* 5: 365–393.

Davidson, E. H. 1976. *Gene Activity in Early Development.* Academic Press, New York.

Dohmen, M. R. 1983. Gametogenesis. In *The Mollusca: Development,* N. H. Verdonk and J. A. M. van den Biggelaar (eds.). Academic Press, New York, pp. 1–48.

Dohmen, M. R. and D. Lok. 1975. The ultrastructure of the polar lobe of *Crepidula fornicata* (Gastropoda, Prosobranchia). *J. Embryol. Exp. Morphol.* 34: 419–428.

Dohmen, M. R. and J. C. A van der Mey. 1977. Local surface differentiations of the vegetal pole of the eggs of Nassarius reticulatus, *Buccinum undatum,* and *Crepidual fornicata* (Gastropoda, Prosobranchia). *Dev. Biol.* 61: 104–113.

Dohmen, M. R. and N. H. Verdonk. 1974. The structure of a morphogenetic cytoplasm present in the polalr lobe of *Bithynia tentaculata* (Gastropoda, Prosobranchia). *J. Embryol. Exp. Morphol.* 31: 423–433.

Dohmen, M. R. and N. H. Verdonk. 1979a. The ultrastructure and role of the polar lobe in development of molluscs. In *Determinants of Spatial Organization,* S. Subtelny and I. R. Konigsberg (eds.). Academic Press, New York, pp. 3–27.

Dohmen, M. R. and N. H. Verdonk. 1979b. Cytoplasmic localization in mosaic eggs. In *Maternal Effects in Development,* D. H. Newth (ed.). Cambridge University Press, Cambridge, pp. 127–145.

Driever, W., J. Ma, C. Nusslein-Volhard and M. Ptashne. 1989. Rescue of *bicoid* mutant *Drosophila* embryos by Bicoid fusion proteins containing heterologous activating sequences. *Nature* 342: 149–154.

Eyster, L. S. 1983. Ultrastructure of early embryonic shell formation in the opisthobranch gastropod *Aeolidia papillosa. Biol. Bull.* 165: 394–408.

Eyster, L. S. 1986. Shell inorganic composition and onset of shell mineralization during bivalve and gastropod embryogenesis. *Biol. Bull.* 170: 211–231.

Fawcett, D. W., S. Ito and S. Slautterback. 1959. The occurence of intercellular bridges in groups of cells exhibiting synchronous differentiation. *J. Biophys. Biochem. Cytol.* 5: 453–460.

Freeman, G. and J. W. Lundelius. 1982. The developmental genetics of dextrality and sinistrality in the gastropod *Lymnaea peregra. Roux's Arch. Dev. Biol.* 191: 69–83.

Fretter, V. and A. Graham. 1962. *British Prosobranch Molluscs: Their Functional Anatomy and Ecology.* Ray Society, London.

Garstang, W. 1928. Presidential address to the British Association for the Advancement of Science.

Geraerts, W. P. M. and J. Joosse. 1984. Fresh-water snails (Basommatophora). In *The Mollusca: Reproduction,* A. S. Tompa, N. H. Verdonk and J. A. M. van den Biggelaar (eds.). Academic Press, New York, pp. 141–199.

Gerin, Y. 1971. Etude par cytochimie ultrastructurale des corpuscules perinucleaires presents dans les jeunes oocytes de *Ilyanassa obsoleta* Say (Mollusca gastropoda). *J. Embryol. Exp. Morph.* 25: 423–438.

Gerin, Y. 1972. Morphogenese des vesicules a *double membrane* du lobe polaire D' *Ilyanassa obsoleta* Say, etude ultrastructurale. *J. Microscopie* 13: 57–66.

Geuskens, M. and V. de J. d'Ardoye. 1971. Metabolic patterns in *Ilyanassa* polar lobes. *Exp. Cell Res.* 67: 61–72.

Gibson, B. L. 1984. Cellular and ultrastructural features of the adult and the embryonic eye in the marine gastropod, *Ilyanassa obsoleta. J. Morphol.* 181: 205–220.

Gosliner, T. M. 1994. Gastropod: Opisthobranchia. In *Microscopic Anatomy of Invertebrates,* F. W. Harrison and A. J. Kohns (eds.). Wiley-Liss, New York, pp. 253–355.

Griffond, B, 1977. Individualisation et organogenesis de la gonade embryonnaire de *Viviparus viviparus* (L.) *Roux's Arch. Entw. Org.* 183: 131–147.

Guyard, A. 1971. Etude de la differenciation de l'ovotestis et des facteurs controlant l'orientation sexuelle des gonocytes de l'Escargot *Helix aspersa* Muller. Thesis Doct. Sci. Nat., Besancon.

Guyomarc'h-Cousin, C. 1976. Organogenese descriptive de l'appareil genital chez *Littorina saxatalis* Olivi (Gastropoda, Prosobranchia). *Bull. Soc. Zool. Fr.* 101: 465–476.

Hadfield, M. G. and M. Switzer-Dunlap. 1984. Opisthobranchs. In *The Mollusca: Reproduction*, A. S. Tompa, N. H. Verdonk and J. A. M. van den Biggelaar (eds.). Academic Press, New York, pp. 209–334.

Hodgson, A. N. 1993. Spermatozoan structure and spermiogenesis in *Nassarius kraussianus* (Gastropoda, Prosobranchia, Nassariinae). *Inv. Reprod. Dev.* 23: 115–121.

Hyman, L. H. 1951. *The Invertebrates: Platyhelminthes and Rhynchocoela.* McGraw-Hill, New York.

Hyman, L. H. 1967. *The Invertebrates: Mollusca I.* McGraw-Hill, New York.

Jenner, M. G. 1979. Pseudohermaphroditism in *Ilyanassa obsoleta* (Neogastropoda). *Science* 205: 1407–1409.

Jong-Brink, M. de, A. Witt, G. Kraal and H. H. Boer. 1976. A light and electron microscope study on oogenesis in the freshwater pulmonate snail *Biomphalaria glabrata. Cell Tissue Res.* 171: 195–219.

Jong-Brink, M. de, H. H. Boer, and J. Joosse. 1983. Mollusca. In *Reproductive Biology of Invertebrates*, K. G. Adiyodi and R. G. Adiyodi (eds.). John Wiley and Sons, New York, pp. 297–355.

Joosse, J. 1975. Structural and endocrinological aspects of hermaphroditism in pulmonate snails, with particular reference to *Lymnaea stagnalis* (L.). In *Intersexuality in the Animal Kingdom.* R. Reinboth (ed.). Springer-Verlag, Berlin and New York, pp. 158–169.

Joosse, J. and D. Reitz. 1969. Functional anatomical aspects of the ovotestis of *Lymnaea stagnalis. Malacologia* 9: 101–109.

Kielbowna, L. and B. Koscielski. 1974. A cytochemical and autoradiographic study on occyte nucleoli in *Lymnaea stagnalis* L. *Cell Tissue Res.* 152: 103–111.

Kniprath, E. 1981. Ontogeny of the molluscan shell field: A review. *Zool. Sci.* 10: 61–79.

Koser, R. B. and J. R. Collier. 1976. An electrophoretic analysis of RNA synthes in normal and lobeless *Ilyanassa* embryos. *Differentiation* 6: 47–52.

Longo, F. J. 1983. Meiotic maturation and fertilization. In *The Mollusca: Development*, N. H. Verdonk, J. A. M. van den Biggelaar and A. S. Tompa (eds.). Academic Press, New York, pp. 49–89.

Luchtel, D. 1972a. Gonadal development and sex determination in pulmonate molluscs I. *Z. Zellforsch.* 130: 279–301.

Luchtel, D. 1972b. Gonadal development and sex determination in pulmonate molluscs II. *Arion ater rufus* and *Deroceras reticulatum. Z. Zellforsch.* 130: 302–311.

McCain, E. R. 1992. Cell interactions influence the pattern of biomineralization in the *Ilyanassa obsoleta* (Mollusca) embryo. *Dev. Dynamics* 195: 188–200.

McCann-Collier, M. 1977. An unusual cytoplasmic organelle in oocytes of *Ilyanassa obsoleta. J. Morphol.* 153: 119–128.

McCann-Collier, M. 1979. RNA synthesis during *Ilyanassa* oogenesis: An autoradiographic study. *Dev. Growth Differ.* 21: 391–399.

McCann-Collier, M. 1984. Microscopic observations of cryptic polar body production and spiralian organization in the egg of *Ilyanassa obsoleta. Biol. Bull.* 167: 488–494.

Moor, B. 1983. Organogensis. In *The Mollusca: Development*, N. H. Verdonk and J. A. M. van den Biggelaar (eds.). Academic Press, New York, pp. 123–177.

Morrill, J. B. 1964. Protein content and dipeptidase activity of normal and cobalt treated embryos of *Lymnaea palustris. Acta Embryol. Morphol. Exp.* 7: 131–142.

Morrill, J. B. 1982. Development of pulmonate gastropod *Lymnaea.* In *Developmental Biology of Freshwater Invertebrates*, G. W. Harrison and R. W. Cowden (eds). Alan R. Liss, New York, pp. 399–483.

Morrill, J. B. 1997. Cellular patterns and morphogenesis in early development of freshwater pulmonate snails, *Lymnaea* and *Physa* (Gastropoda, Mollusca). In *Reproductive Biology of Invertebrates* VII: *Progress in Developmental Biology.* K. G. Adiyodi and R. G. Adiyodi (eds.). John Wiley and Sons, New York and Oxford and IBH Pub. Co, New Delhi. In Press.

Morrill, J. B., R. W. Rubin and M. Grandi. 1976. Protein synthesis and differentiation during pulmonate development. *Amer. Zool.* 16: 547–561.

Nieuwkoop, P. D., and L. A. Sutasurya. 1981. *Primordial Germ Cells in the Invertebrates: From Epigenesis to Preformation.* Cambridge University Press, Cambridge.

Patten, W. 1886. Embryology of *Patella. Arbeiten Zool. Inst. Univ. of Wien* 6: 1–25.

Pennington, J. T., and F. Chia. 1985. Gastropod torsion: A test of Garstang's hypothesis. *Biol. Bull.* 169: 391–396.

Pointier, J. P., B. Dellay, J. L. Toffart, M. Lefevere and A. R. Romero-Alvarez. 1992. Life history traits of three morphs of *Melanoides buberculatum* (Gastropoda, Thiaridae), an invading snail in the French West Indies. *J. Molluscan Studies.* 58: 415–423.

Pucci-Minafra, I., S. Minafra and J. R. Collier. 1969. Distribution of ribosomes in the egg of *Ilyanassa obsoleta. Exp. Cell Res.* 57: 167–168.

Rabl, C. 1879. Über die Entwicklun der Tellerschnecke. *Morphol. Jahrb.* 5: 562–655, pp. 32–38.

Raven, C. P. 1961. *Oogenesis: The Storage of Developmental Information.* Pergamon, Oxford.

Raven, C. P. 1963. The nature and origin of the cortical morphogenetic field in *Limnaea. Dev. Biol.* 7: 130–143.

Schmekel, L. A. and P. Fioroni. 1975. Cell differentiation during early development of *Nassarius reticulatus* (Gastropoda, Prosobranchia). *Cell Tissue Res.* 153: 503–522.

Selenka, E. 1881. Zur Entwicklungsgeschichte der Seeplanarien. *Zoolegische Studien* 2.

Silberzahn, N. 1979. Les cellules de la lignee femelle chez un hermaphrodite protandre *Crepidual fornicata*, (Mollusque, Prosobranche). *Ann. Soc. Fr. Biol. Dev. Paris* 17–18.

Speksnijder, J. E., M. M. Mulder, M. R. Dohmen, W. J. Hagen and J. G. Bluemink. 1985. Animal-vegetal polarity in the plasma membrane of a molluscan egg: A quantitative freeze-fracture study. *Dev. Biol.* 108: 38–48.

St. Johnston, D. and C. Nusslein-Volhard. 1992. The origin of pattern and polarity in the *Drosophila* embryo. *Cell* 68: 201–219.

Stasek, C. R. 1972. The molluscan framework. In *Chemical Zoology*, M. Florkin (ed.). Academic Press, New York.

Sullivan, C. H. 1984. Developmental alterations in the appearance of hatching activity in *Ilyanassa obsoleta* embryos. *Roux's Arch. Dev. Biol.* 193: 219–225.

Sullivan, C. H. and T. K. Mangel. 1984. Formation, organization, and composition of the egg capsule of the marine gastropod, *Ilyanassa obsoleta. Biol. Bull.* 167: 378–389.

Taylor, G. T. and E. Anderson. 1969. Cytochemical and fine structural analysis of oogenesis in the gastropod, *Ilyanassa obsoleta. J. Morphol.* 129: 211–248.

Tomlinson, S. G. 1987. Intermediate stages in the embryonic development of the gastropod *Ilyanassa obsoleta:* A scanning electron microscope study. *Int. J. Inv. Reprod. Dev.* 12: 253–280.

Tompa, A. S. 1984. Land Snails (Stylommatophora). In *The Mollusca: Reproduction.* A. S. Tompa, N. H. Verdonk and J. A. M. van den Biggelaar (eds.). Academic Press, New York, pp. 48–140.

Ubbels, G. A. 1968. A cytochemical study of oogenesis in the pond snail *Limnaea stagnalis.* Ph.D. Dissertation. University of Utrecht.

Van Beneden, E. and C. Julin. 1884. La Segmentation ches les Ascidians dans ses rapporte avec l'organization. *Arch. Biol.* 5.

van Dam, W. I., M. R. Dohmen and N. H. Verdonk. 1982. Localization of morphogenetic determinants in a special cytoplasm present in the polar lobe of *Bithynia tentaculata* (Gastropoda). *Roux's Arch. Dev. Biol.* 191: 371–377.

van den Biggelaar, J. A. M. 1977. Development of dorsoventral polarity and mesentoblast determination in *Patella vulgata*. *J. Morphol.* 154: 157–186.

van den Biggelaar, J. A. M. and P. Guerrier. 1979. Dorsoventral polarity and mesentoblast determination as concomitant results of cellular interaction in the mollusk *Patella vulgata*. *Dev. Biol.* 68: 462–471.

van den Biggelaar, J. A. M. and P. Guerrier. 1983. Origin of spatial organization. In *The Mollusca: Development*, N. H. Verdonk, J. A. M. van den Biggelaar and A. S. Tompa, (eds.). Academic Press, New York, pp. 179–208.

van den Biggelaar, J. A. M. and G. Haszprunar. 1996. Cleavage patterns and mesentoblast formation in the gastropoda: An evolutionary perspective. *Evolution* 50: 1520–1540.

van den Biggelaar, J. A. M., J. A. Wim, G. Dictus and F. Serras. 1994. Molluscs. In *Embryos, Color Atlas of Development*, B. Jonathans (ed.). Wolfe, Singapore, pp. 77–91.

van der Wal, U. P. 1976a. The mobilization of the yolk of *Lymnaea stagnalis* (Mollusca). I. A structural analysis of the differentiation of the yolk granules. *Proc. K. Ned. Akad. Wet., Ser. C* 79: 393–404.

van der Wal, U. P. 1976b. The mobilization of the yolk of *Lymnaea stagnalis* (Mollusca). II. The localization and function of the newly synthesized proteins in the yolk granules during early embryogenesis. *Proc. K. Ned. Akad. Wet., Ser. C* 79: 405–420.

Verdonk, N. H. and J. A. M. van den Biggelaar. 1983. Early development and the formation of germ layers. In *The Mollusca: Development*, N. H. Verdonk and J. A. M. van den Biggelaar (eds.). Academic Press, New York, pp. 91–122.

Voltzow, J. 1994. Gastropoda: Prosobranchia. In *Microscopic Anatomy of Invertebrates*, F. W. Harrison and A. J. Kohn (eds.). Wiley-Liss, New York, pp. 111–252.

Wada, S. K. 1968. Mollusca. In *Invertebrate Embryology,* M. Kume and K. Dan (eds.). National Library of Medicine, Washington, D.C., pp. 485–525.

Walker, M. and H. C. MacGregor. 1968. Spermatogenesis and the structure of the mature sperm in *Nucella lapillus* (L.). *J. Cell Sci.* 3: 95–104.

Webber, H. H. 1977. Gastropoda: Prosobranchia. In *Reproduction of Marine Inverebrates, Molluscs: Gastropods and Cephalopods*, A. C. Giese and J. S. Pearsres (eds.). Academic Press, New York, 1–77.

Werner, B. 1955. Über die Anatomie, die Entwicklung und Biologie des Veligers und der Veliconcha von *Crepidula fornicata* (L.). (Gastropoda: Prosobranchia). *Helgolander Wissenschaftliche Meeresuntersuchunger,* 5: 169–217.

Whitman, C. O. 1878. The embryology of clepsine. *Quart. J. Microsc. Sci.* 18: 215–314.

Willmer, P. 1990. *Invertebrate Relationships: Patterns in Animal Evolution.* Cambridge University Press, New York.

Wilson, E. B. 1892. The cell-lineage of *Nereis*. *J. Morphol.* 6: 362–442.

Wilson, E. B. 1925. *The Cell in Development and Heredity.* Macmillan Company, New York.

CHAPTER 12

Annelids, the Segmented Worms

Marty Shankland and Robert M. Savage

T HE ANNELIDS ARE A NUMEROUS AND WIDELY DISTRIBUTED phylum of segmented worms that includes such widely recognized animals as the earthworm and the leech. Annelids are traditionally allied with the molluscs and arthropods in the protostome branch of the phylogenetic tree. Like all protostomes, annelids have a bilaterally symmetric and triploblastic body plan composed of three discrete germ layers, with the internal body cavity arising as a coelom lined with derivatives of the middle or mesodermal layer.

In this chapter, we will focus on the development of the glossiphoniid leech as a model system for outlining the essentials of annelid embryology. The leeches represent one of the more highly derived annelid taxa, but they are also the best characterized with respect to the cellular and molecular events that underlie pattern formation (Stent et al. 1992; Shankland 1994). Readers desiring a broader discussion of annelid embryology are referred to the excellent overview of Okada (1988) and the comparative analysis of Anderson (1973).

The phylum Annelida is subdivided into three major groups: the polychaetes, the oligochaetes, and the leeches (Figure 12.1). The polychaetes are a diverse collection of predominately marine worms that show pronounced similarities to the molluscs in both their embryonic and larval development, and that are widely believed to represent the primitive stock from which the other two annelid groups arose. The oligochaetes (e.g., earthworms) and leeches display many similarities to the polychaetes in terms of adult body plan and early embryology, but also share a number of distinguishing features that have led to their being grouped together in the taxon Clitellata (Figure 12.1). The clitellate annelids deposit their eggs into an impervious cocoon secreted by a specialized glandular structure (the clitellum), and within that cocoon the eggs undergo a direct development that bypasses the free-swimming marine larval stage typical of most polychaetes. The majority of clitellate annelids lay their cocoons on land or in fresh water, and it seems likely that the developmental modifications characteristic of this group were a major factor in their successful invasion of the continents.

Leeches (class Hirudinida; Brusca and Brusca 1990) and oligochaetes are distinguished from one another by several anatomical and developmental characters. For instance, the leeches have lost the external bristles (chaetae) typical of the other two annelid groups,* and their terminal segments are modified to form discrete front and rear suckers for grasping onto the substrate. The formation of a specialized rear sucker has an important developmental corollary, namely that the leech embryo generates its full complement of body segments during a single, discrete phase of embryonic development. The other annelids do not display this sort of terminal specialization, and continue to elongate the body trunk during postembryonic development by the accretion of additional segments to the rear end of the body plan.

Adult Body Plan

Annelids are described as worms in the common usage because they have flexible, elongate bodies that lack skeletal support. In the leeches, the body wall is a muscular tube with discrete suckers at the front and rear ends (Figure 12.2a). The mouth is located ventrally at the center of the front or oral sucker, and the anus is situated dorsally at

---

*The one exception is the genus *Acanthobdella*, which is generally included with the leeches although it sports chaetae on the five anteriormost segments (Livanow 1906). Interestingly, the species *A. peledina* also resembles the other annelids in that it fails to form an oral sucker.

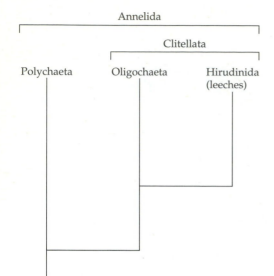

Figure 12.1 Taxonomic relationship of the major annelid groups. Derived groups are shown as side branches.

the junction of the rear sucker and the intervening mid-body region. There are generally no appendages, in contrast to the polychaetes which have bilaterally paired parapodia on many or all body segments. The integument is comprised of an epidermis, dermis, and three layers of subdermal musculature, and the central nervous system (CNS) is a ganglionated ventral nerve cord running the body's length (Figure 12.2*b*). The structure of the circulatory system varies greatly among leech species, consisting in some groups of a closed vascular system separate from the coelom, and in others of a combined haemocoel. Respiration occurs largely by passive diffusion across the integument, although a few species pump their blood through external gills (branchiae). Sawyer's (1986) text on leech biology goes into these and other topics in far greater detail than is possible here.

From an embryological standpoint, the leech body plan can be subdivided into two regions: an unsegmented **cephalic domain**; and a much larger segmented **trunk**

Figure 12.2 The glossiphoniid leech *Helobdella triserialis*, shown in dorsal view. Anterior is towards the top. (*a*) The leech body plan consists of a nonsegmental cephalic domain (hatched), and an elongate trunk composed of 32 serially homologous segments, whose boundaries are marked by transverse lines. The segments are enumerated according to the neuromeres of the CNS. The four most rostral segments (R1–R4) join with the non-segmental prostomium (P) to form the front sucker, a ventrally oriented concavity surrounding the mouth. The seven most caudal segments (C1–C7) form the disc-shaped rear sucker, the posterior edge of which extends out past the anus. There are also 21 intervening midbody segments (M1–M21). The digestive tract is subdivided into unsegmented foregut (hatched), and a segmented midgut (stippled) composed of crop, intestine, and rectum. (*b*) The leech CNS is a ganglionated nerve cord situated ventral to the digestive tract. The rostral and caudal neuromeres fuse into compound terminal ganglia, whereas the 21 midbody neuromeres (M1–M21) remain separate as discrete segmental ganglia. The supraesophageal ganglion is a nonsegmental derivative of the cephalic domain and encircles the gut tube (shown in profile) at the anterior end of the segmental nerve cord.

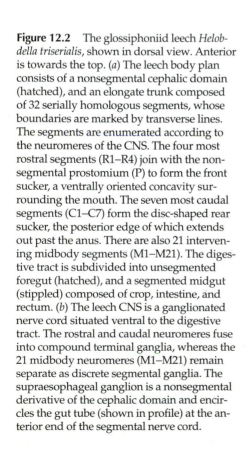

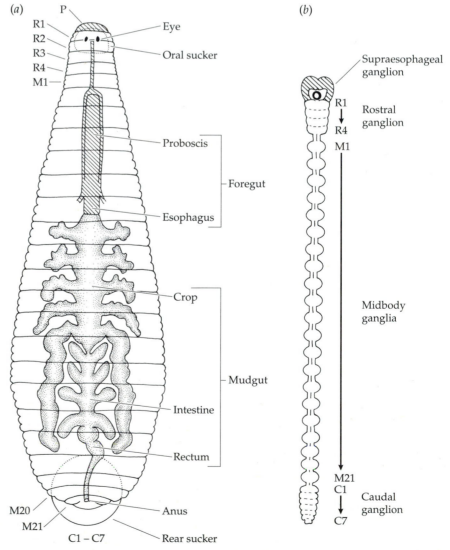

(Figure 12.2; see Figure 12.5*d*). Other annelids possess an unsegmented anal domain called the **pygidium**, but this latter structure is vestigial in the leech, probably as a consequence of evolutionary modifications associated with the formation of the rear sucker.

The cephalic domain arises from embryonic progenitor cells situated at the anterior end of the germinal plate. The external component of this domain is known as the **prostomium**,* and includes the anteriormost portion of the integument and the supraesophageal ganglion of the CNS (Figure 12.2). This ganglion arises anterior to the ventrally situated oral invagination of the embryo, and is joined to the more posterior segmental nerve cord by circumesophageal connectives. The supraesophageal ganglion separates from the prostomial body wall during later development (see Figure 12.5*d*), and becomes repositioned dorsal to the digestive tract in the mature body plan. The internal portion of the cephalic domain is the **foregut** (Figure 12.2*a*), which includes the feeding apparatus and anteriormost organs of the digestive tract. The glossiphoniid leeches feed by means of an extensible proboscis, which is initially formed on the outer surface of the embryo but is withdrawn into the body cavity as development proceeds (see Figure 12.5*d*).

The body trunk is much more extensive than the cephalic domain, and consists of a tandem array of segments that display a spatially iterated pattern of tissue organization (Figure 12.2). Each segment is externally subdivided into multiple annuli, one of which (the so-called neural annulus) bears a circumferential row of external sensilla. Internally, a typical midbody segment houses a bilateral pair of excretory organs (nephridia) and a single ganglion of the ventral cord. Each ganglion is comprised of approximately 400 neuronal cell bodies, and many of the individual neurons can be uniquely identified from one segment to the next by a shared constellation of differentiated properties (for reviews on the leech nervous system, see Drapeau et al. 1995). In the embryo the individual segments are at first separated internally by mesodermal **septa**, but the tissues comprising the septum lose coherence and dissipate as organogenesis proceeds. This is in contrast to other annelids, in which discrete intersegmental septa persist into the adult.

Cell lineage studies have shown that the repeating structure of the mature leech body trunk does not reflect a primary subdivision of the embryo into morphogenetically separable developmental compartments (Weisblat and Shankland 1985), as is the case in insects (Lawrence 1973). It would therefore be a mistake to treat the intersegmental boundary of the leech as an embryological concept. We

prefer instead to view the delineation of segments as a matter of anatomical convenience, and here employ the "neurocentric" scheme in which segments are defined according to the ganglia of the ventral nerve cord and the blocks of peripheral tissue innervated by those ganglia.

Despite their many overt similarities, the individual trunk segments manifest many precise differences in cellular composition that correlate with their positions along the anteroposterior (AP) body axis. Both the nephridia and reproductive system are restricted to only a subset of body segments, as are the gills in those species that possess them (Sawyer 1986). Segmental differences are especially prevalent in the CNS. In the midbody segments, the neuromeres (segmental units of the CNS) develop into discrete segmental ganglia joined by connective nerves. On the other hand, the four anteriormost and seven posteriormost neuromeres fuse into compound rostral and caudal ganglia that innervate multiple segments. There are also a number of well-characterized segmental differences in the formation and differentiation of individually identified neurons (for review, see Shankland and Martindale 1992). In accordance with the organization of the CNS, the leech body plan is subdivided into 4 rostral segments (R1–R4), 21 midbody segments (M1–M21) and 7 caudal segments (C1–C7), with the individual segments in each domain numbered in anterior-to-posterior order (Figure 12.2*b*).

The vast majority of leech species have 32 body segments, although certain of the more divergent groups have fewer. Unfortunately, there are many sources in the literature that number leech segments in excess of 32, an antiquated practice that derives from the mistaken belief that the supraesophageal ganglion should be counted as part of the segmental nerve cord. It is now well established that the supraesophageal ganglion receives no contribution from the iterated cell lineages that produce the trunk segments (Weisblat et al. 1984), nor does it exhibit any compelling similarity to a segmental ganglion in terms of neuronal composition or gene expression (Wedeen and Weisblat 1991; Nardelli-Haefliger and Shankland 1993). Pronounced morphogenetic differences have also observed between the prostomial and segmental mesoderm (Zackson 1982).

The portion of the digestive tract that develops in close association with the segmented body trunk is known as the **midgut** (Figures 12.2*a*,5*d*), and is subdivided in anteroposterior order into crop (or stomach), intestine, and rectum. During embryonic development, the organs of the midgut become connected to the foregut anteriorly and to the anus posteriorly. Due to the vestigial character of its pygidium, the leech has little or no invaginated hindgut. The crop and intestine have a segmental organization that is structurally integrated with the segmentation of the overlying mesoderm, and includes periodic patterns of both morphogenesis and gene expression (Nardelli-Haefliger and Shankland 1993; Wedeen and Shankland, in

*In polychaetes and oligochaetes the cephalic domain is subdivided into two discrete subunits known as the prostomium ("in front of the mouth") and peristomium ("surrounding the mouth"). In some species the peristomium is thought to arise by an integration of both segmental and nonsegmental tissues.

preparation). To date, this represents the only clearly documented example of endodermal segmentation for any animal species.

Reproduction and Gametogenesis

Annelids exhibit a number of diverse strategies for both sexual and asexual reproduction (Brusca and Brusca 1990). The polychaetes show the greatest diversity, but do manifest certain features that are likely to be primitive for the phylum as a whole. Most polychaete species have separate male and female genders that show only a limited degree of sexual dimorphism. They do not maintain permanent gonads, and gametogenesis tends to be dispersed throughout the coelomic lining in some or all body segments. The immature gametes or their precursors are shed directly into the coelomic cavity rather than into specialized gonoducts, and the mature gametes are eventually expelled into the surrounding water by a variety of mechanisms including passage through coelomoducts, nephridial excretory ducts, or transient ruptures in the body wall. In some polychaetes, a number of gamete-engorged segments separate from the remainder of the body to form a freely swimming **epitoke** that leaves the relative safety of the worm's benthic feeding habitat in order to spawn near the ocean surface.

The clitellate annelids (leeches and oligochaetes) share a number of reproductive modifications. All clitellate species have a single hermaphroditic gender, and some are capable of self-fertilization (Wedeen et al. 1990). The reproductive system is more complex than that of the polychaetes, with the gonads and their accessory structures being permanent organs located in a restricted number of segments at a fixed position along the body's length. The ovaries and testes are anatomically separate, and both sets of gametes are guided to the outside world by a system of gonoducts that connect to typically separate male and female gonopores situated on the ventral surface. In leeches, the oogonia divide mitotically to produce multiple cells connected by cytoplasmic bridges (Sawyer 1986). One or more of these cells differentiates as an oocyte, and the remainder function as nurse cells that directly transfer protein, RNA, and organelles into the enlarging oocyte(s).

Clitellate annelids employ contact mating and a direct exchange of sperm. In oligochaetes, the partner's sperm are transferred to storage organs (spermathecae) during a period of pseudocopulation, and fertilization occurs at a later time when the individual simultaneously expels its own eggs and the stored sperm. In leeches, fertilization is internal and can occur either by true copulation or by transdermal insemination—i.e., by injection of the sperm into the coelomic cavity, from which they migrate to the ovaries.

The mature oocyte is encased in a protein coat known as the vitelline envelope. This envelope lifts off the surface of the egg cell at the time of fertilization due to the discharge of the cortical granules. The intervening or perivitelline space is thereafter filled with a hygroscopic fluid, which maintains turgor pressure on the membrane and resists mechanical deformation of the otherwise fragile egg. Once fertilized, the leech egg proceeds to first meiotic metaphase, but is actively blocked at this step of the meiotic cycle so long as it is held within the ovary, and does not carry on with embryonic development until it has been released (Fernández et al. 1987). Following deposition, the fertilized egg renews its developmental progression, undergoing polar body formation and fusion of the male and female pronuclei at the so-called **animal pole**. The opposite end of the egg is defined as the **vegetal pole**, and the egg is morphologically symmetrical about this animal-vegetal axis.

In many spiralian embryos, the spatial patterning of embryonic development is dependent upon regulatory molecules that are heterogeneously distributed in the cytoplasm of the uncleaved zygote, and the segregation of developmentally potent cytoplasmic domains has been well documented in the glossiphoniid leech (Fernández et al. 1987; Astrow et al. 1987). Shortly after the formation of the polar bodies, the zygote begins to experience slowly moving waves of cortical contraction directed towards its two poles (Figure 12.3). The circumferential rings of contraction serve as foci for the segregation of a specialized cytoplasm called **teloplasm**, which excludes the numerous yolk platelets that fill nearly all of the remaining zygote cytoplasm, and is greatly enriched in organelles and RNA. Teloplasm is thought to play an important role in specifying the developmental fate of those blastomeres that inherit it during embryonic cleavage (Astrow et al. 1987). By the end of the first mitotic cycle, the rings of cortical contraction subside, with the teloplasm having been concentrated into a large pool at the animal pole and a smaller pool at the vegetal pole (Figure 12.3).

Regardless of the mechanism by which fertilization is achieved, all clitellate annelids encapsulate their newly deposited eggs in a durable and largely impermeable cocoon. The cocoon and its internal fluids are secreted by the clitellum, a glandular region of epidermis that can be readily associated with the female gonopore during egg deposition. In many clitellate species the cocoon fluid is rich in albumen and is consumed by the developing embryo as a major source of nutrients for its growth (Anderson 1973). However, the glossiphoniid leeches that are the focus of this chapter lay large yolk-rich eggs (0.4–2 mm diameter) and only begin to take external nourishment after hatching from the cocoon. As a consequence, glossiphoniid embryos will develop normally in a defined ionic medium if they are removed from the cocoon at any time after deposition, a feature which has greatly facilitated their utilization in experimental studies of early development.

Many glossiphoniid leech species exhibit complex brooding behaviors. In *Helobdella triserialis*, the hatched

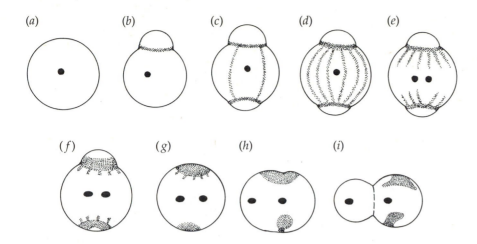

(a) (b) (c) (d) (e)

(f) (g) (h) (i)

Figure 12.3 Segregation of teloplasm during the first mitotic cycle of the glossiphoniid leech *Theromyzon rude*. Individual profiles depict half-hour intervals in the maturation of the zygote from the time of second polar body formation (*a*) to the time of first cleavage (*i*), with the animal pole oriented toward the top. Teloplasm is represented by stippling, and nuclei in black. Note that the teloplasm seems to flow along meriodonal rays (*c*–*e*), and accumulates at circumferential constrictions that slowly move towards the animal and vegetal poles of the egg (*b*–*f*). The first cleavage plane is eccentric, so that the teloplasm is predominantly segregated into only one of the two daughter blastomeres. (From Fernández et al. 1987.)

embryo remains attached to its maternal parent for a period of weeks, first by means of an adhesive gland, then through the muscular grip of its rear sucker. This species feeds exclusively on pond snails, and the juvenile leech takes its first several meals from snails that have been captured and dispatched by the brooding parent. Therefore, *Helobdella* embryos that have been removed from the parental cocoon for experimental purposes must be manually fed on snails wounded by the experimenter if they are to survive the transition into self-feeding juveniles.

In addition to sexual reproduction, some polychaetes and oligochaetes are also capable of undergoing asexual reproduction by either budding or fission (Okada 1988). Moreover, these worms have a robust regenerative capacity, and it is reported that certain polychaetes can reform the complete adult body plan from a single, isolated segment (Brusca and Brusca 1990). These topics lie outside the scope of this chapter, but they do emphasize the extraordinary plasticity of pattern formation and regulation in some annelid species. Leeches are distinct in that they are not capable of asexual reproduction, nor can they regenerate missing segments.

Cleavage

The annelid embryo develops via an invariant sequence of stereotyped cell divisions (Figures 12.4, 12.5). The early divisions are complete (holoblastic), and the first five to six rounds of division follow a stereotyped pattern known as **spiral cleavage**. Many invertebrate phyla employ this same cleavage pattern (e.g., molluscs; sipunculids; nemerteans), and are often lumped together under the rubric spiralians.

In spiral cleavage the first two cleavage planes are parallel to the animal-vegetal axis, and at right angles to one another (Figure 12.4*a*). These cleavages divide the zygote into four **macromeres**, which are designated A, B, C, and D in clockwise order when viewed from the animal pole.

The derivatives of these four macromeres are often described as embryonic quadrants.

The spiral character of the annelid cleavage pattern comes into evidence at the third round of division. Each of the four macromeres divides asymmetrically to produce a small animal pole daughter cell or **micromere**, and cleavage occurs at an oblique angle such that the quartet of micromeres is rotated 45 degrees with respect to the parent macromeres as seen from the animal pole (Figure 12.4*b*). The direction of this rotation alternates with successive rounds of division (Figure 12.4*b*). Odd-numbered cleavages—third and fifth—produce micromeres that are rotated clockwise (dextral cleavages) ; while even-numbered cleavages—fourth and, in some species, sixth—produce micromeres that are rotated counterclockwise (sinistral cleavages). The micromeres produced at the earlier cleavages themselves undergo a spiral pattern of subsidiary divisions, and thus the animal pole of the cleavage stage embryo becomes invested in a micromere cap.

One of the fundamental characteristics of spiralian development is the fact that the D quadrant becomes specialized during early development to produce a major portion of the adult body trunk. Spiralians are known to employ three distinct mechanisms for bringing about the unique specification of the D quadrant (Freeman and Lundelius 1992). In many annelid embryos, the first two cleavages are asymmetric and the largest macromere (D) inherits a cytoplasmic domain that confers a unique pattern of subsequent differentiation. For instance, both leech and oligochaete undergo asymmetric first and second cleavages in which the eccentric placement of the cleavage furrow ensures that the vast majority of the polar teloplasm is inherited by cell D (Figure 12.4*a*).

On the other hand, certain polychaete embryos undergo asymmetric cleavages that do not require eccentric placement of the cleavage furrow. In these latter species, the developmentally potent cytoplasm becomes sequestered into a vegetal protrusion known as the polar lobe, and this lobe remains attached to—and is ultimately

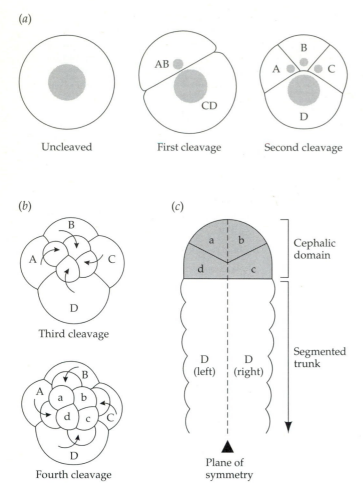

(a)

Uncleaved First cleavage Second cleavage

(b)

Third cleavage

Fourth cleavage

(c)

Cephalic domain

Segmented trunk

Plane of symmetry

Figure 12.4 Spiral cleavage of the annelid embryo as viewed from the animal pole. (*a*) The first two cleavage planes are parallel to the animal-vegetal axis, and separate the embryo into macromeres A, B, C, and D. In the leech, these cleavage planes are eccentric, and the polar teloplasm (gray) is distributed asymmetrically among the sibling blastomeres. The D macromere receives the vast majority of the teloplasm, and is thereby committed to a unique developmental fate. (*b*) Subsequent cleavages involve the formation of micromere quartets (lowercase letters) clustered around the animal pole. Note that both the first and second quartets are rotated 45 degrees with respect to their parent macromeres, and that rotation alternates from clockwise to counterclockwise at successive cleavages. (*c*) The lineal basis of bilateral symmetry differs in the cephalic domain and segmented body trunk, as shown here in a dorsal view of the anterior germinal plate. In the head, the a and d micromere clones on the left and the b and c micromere clones on the right are bilaterally homologous, such that the plane of symmetry corresponds to the second plane of cleavage. By way of contrast, the plane of symmetry bisects the D quadrant in the trunk, whose left and right sides arise from symmetric cleavage of the later D-quadrant blastomeres (see Figure 12.6).

resorbed by—only one of the two sibling blastomeres at each of the first two cleavages. Finally, there are some polychaete embryos that cleave symmetrically, and hence may specify their D macromere entirely by cell interactions in a manner akin to that seen in equal-cleaving molluscan embryos (Freeman and Lundelius 1992).

Many polychaetes retain the basal mode of spiral cleavage with three or more rounds of micromere production, and this pattern of cleavages is almost certainly ancestral to the annelids as a whole. However, some groups deviate from the full sequence of micromere-producing divisions. The D macromere of the glossiphoniid leech generates but a single micromere before undergoing symmetric cleavage to produce a large animal blastomere (cell DNOPQ) and a large vegetal blastomere (cell DM) (Figure 12.5*b*). The latter two cells inherit the majority of the teloplasm, and—following the production of additional micromeres—serve as the progenitors of the trunk ectoderm and mesoderm respectively. Experimental studies suggest that teloplasm contains one or more factor(s) that commit these latter two blastomeres to the production of trunk stem cells (Astrow et al. 1987). In addition, cell DNOPQ appears to inherit a second, non-teloplasmic domain of cytoarchitecture that commits it to an ectodermal fate, whereas the sibling blas-

tomere DM seems to become mesodermal by default (Nelson and Weisblat 1992).

In polychaete embryos that undergo the basal mode of spiral cleavage, the segmented trunk is likewise formed by a discrete pair of ectodermal and mesodermal progenitor cells (the primary and secondary somatoblasts), which are in many regards similar to those found in the leech. However, in the polychaetes these progenitors are not sibling blastomeres of a symmetrical cleavage, but arise instead as the second and fourth D-quadrant micromeres (Figure 12.5*a*). This observation has two important implications. First, it emphasizes that the designation of certain embryonic blastomeres as micromeres serves as a description of their formative cleavage pattern, and can vary between species despite precise homologies in developmental fate. Second, species differences in the cleavage pattern point out that cells of ostensibly identical developmental fate can arise via a distinct sequence of formative divisions, implying in turn that the exact sequence of formative divisions is not a strict requirement for the specification of cell fate. This latter idea receives further support from the experimental studies of Nelson and Weisblat (1992), who showed that perturbing cleavage orientation in leech embryos alters the segregation but not the formation of ectodermal and mesodermal cell lineages.

Embryonic Fate Map

Embryogenesis has been described by direct observation and histological reconstructions for a wide variety of annelid species. However, the analysis of cell lineage relationships took a major step forward with the introduction of intracellular lineage tracers (Weisblat et al. 1978; Gimlich and Braun 1985). By injecting individual blastomeres

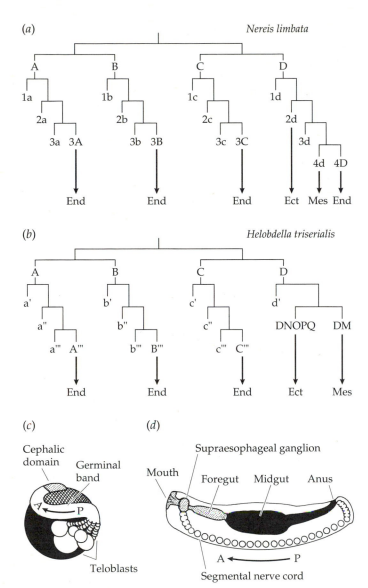

(a) *Nereis limbata*

(b) *Helobdella triserialis*

(c)

Cephalic domain

Germinal band

Teloblasts

(d)

Supraesophageal ganglion

Mouth   Foregut   Midgut   Anus

A ⟵ P

Segmental nerve cord

Figure 12.5 Fate map of the annelid embryo. *(a,b)* Comparison of the early embryonic cell lineage for the polychaete *Nereis* and the leech *Helobdella.* Individual cells are designated according to Wilson's standardized nomenclature for spiral cleavage in *(a)*, and according to Bissen and Weisblat (1989) in *(b)*. Micromere-forming divisions are shown as asymmetric crossbars, with the micromere displayed to the left. In both species, the A–C macromeres generate three micromeres apiece before giving rise to endoderm (End), and the D macromere produces the ectodermal and mesodermal progenitor cells for the segmented body trunk (labeled Ect and Mes, respectively). However, the exact sequence of D-quadrant divisions differs between the two species. *(c)* Fate map of a *Helobdella* embryo containing roughly 10^3 cells, shown in left-side view with the animal pole oriented towards the top. The macromeres (black) form the central bulk of the embryo. One population of micromere derivatives forms the primordium of the cephalic domain (stippled) at the animal pole, and is attached to the anterior end of the germinal band. This band is composed of segmental mesoderm and ectoderm (white), and derives from a set of teloblastic stem located in the D quadrant. Note that the future AP axis of the germinal band curves around the surface of the spherical embryo at this stage. A second population of micromere derivatives forms an expanding provisional integument (cross-hatched) that covers the animal hemisphere of the macromeres as well as the outer surface of the germinal band (not shown). *(d)* Left-side view of a *Helobdella* embryo near the end of organogenesis, with dorsal towards the top. Cellular domains are shaded as in *(c)*. The cephalic domain (gray) has separated into oral integument, supraesophageal ganglion, and foregut. The provisional integument has disappeared, and the midgut (black) has been internalized by the segmental mesoderm and ectoderm of the body trunk (white). The segmental ganglia of the ventral nerve cord are arrayed in a chain along the midline.

1. *Midgut.* Following the formation of the micromeres, the three remaining macromeres join together to create a midgut primordium. This primordium gives rise to the definitive endodermal layer of the gut wall (Nardelli-Haefliger and Shankland 1993), as well as a residual yolk mass that is digested during embryogenesis and the early stages of juvenile life.

2. *Cephalic domain.* As the micromeres proliferate, their descendants separate into two morphogenetically distinct cell populations (Smith and Weisblat 1994). One population remains at the animal pole of the early embryo, and forms the unsegmented cephalic domain of the adult leech. The cephalic ectoderm derives in large part from the first quartet of micromeres (Weisblat et al. 1984; Nardelli-Haefliger and Shankland 1993), while the foregut is produced by the secondary and tertiary micromeres of quadrants A–C.

3. *Provisional epidermis.* The second population of micromere derivatives separates from the cephalic domain to form a simple squamous epithelium that expands from the animal pole toward the vegetal pole during the process gastrulation (Smith and Weisblat 1994). This micromere-derived epithelium is defined as an extraembryonic membrane because it makes no contribution to the mature body plan—and hence is

with histologically detectable enzymes or labeled dextrans, it is possible to trace the fate of individual cell clones through both embryonic and in some cases postembryonic development, and to generate fate maps that are far more accurate and detailed than those put forward by the classical embryologists, who were largely forced to reconstruct cell genealogies by analysis of unlabeled tissues. Injectable lineage tracers have been employed with the greatest success in the large, readily accessible embryos of the glossiphoniid leeches *Helobdella* (Weisblat and Shankland 1985; Smith and Weisblat 1994), *Theromyzon* (Torrence and Stuart 1985) and *Haementeria* (Kramer and Weisblat 1985), but have also proven useful in the analysis of other annelid groups (Storey 1989).

The fate map of the glossiphoniid leech embryo (Figure 12.5*b–d*) can be subdivided into four principal cellular domains:

not part of the embryo proper—but rather forms the epidermal layer of a **provisional integument** that is later discarded in favor of a definitive integument composed of segmental tissues. The morphogenesis of the provisional integument is discussed in the section on gastrulation (see Figure 12.10).

4. *Segmental ectoderm/mesoderm.* In addtion to the micromere cap, the segmental ectoderm and mesoderm of the body trunk also arise from the D macromere. Cells DNOPQ and DM arise by symmetric cleavage of the D' macromere, and as described in the following section, they cleave to produce a set of large embryonic stem cells known as **teloblasts** (Figure 12.6) whose serially iterated descendant lineages generate the explicitly segmental organization of the trunk tissues.

Bilateral Symmetry

The symmetry properties of the developing annelid embryo are a source of much potential confusion due to the spiral character of the cleavage process. From a morphological standpoint, the uncleaved egg is radially symmetric about the animal-vegetal axis, but the early cleavages give rise to a bilaterally symmetric embryo in which there will be little or no subsequent mixing of right and left cell lineages across the median plane.

Wilson (1892) made a key insight into this problem in his careful study of the polychaete *Nereis*, namely that the symmetry properties of a given quartet of micromeres depends on the clockwise or counterclockwise nature of their formation. The progeny of the first and third micromere quartets (clockwise cleavages) become disposed such that the A- and D-quadrant derivatives are located on the left side of the embryo's median plane, whereas the B- and C-quadrant derivatives come to occupy symmetric positions on the right. The first micromere derivatives of the leech *Helobdella* show a similar pattern of symmetry (Figure 12.4c), which therefore dominates the disposition of epidermal and neural tissues in the prostomium (Weisblat et al. 1984; Nardelli-Haefliger and Shankland 1993).

In contrast, micromeres produced by counterclockwise cleavage of the *Nereis* embryo (cells 2a–2c, and the somatoblasts 2d and 4d) are rotated such that the D- and B-quadrant derivatives come to straddle the median plane on the dorsal and ventral sides of the embryo, respectively (Wilson 1892). Given that somatoblasts give rise to the segmented trunk tissues, this has an enormous consequence for the symmetry properties of the adult worm. A comparable situation holds true in the leech, in which the somatoblasts (cells DNOPQ and DM) arise from the D quadrant, and divide symmetrically across the median plane to produce bilaterally homologous right and left derivatives (Figure 12.6). Thus, the disposition of the embryonic quadrants with respect to the median plane is overtly dif-

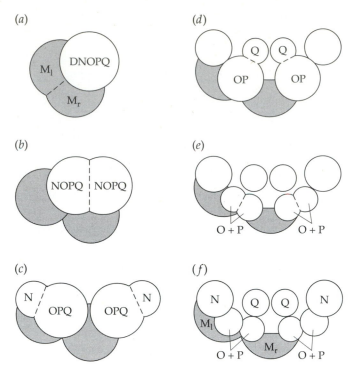

Figure 12.6 Cleavage divisions of the ectodermal (white) and mesodermal (shaded) progenitor cells in the glossiphoniid leech *Theromyzon rude*. Blastomeres are viewed from the animal pole, with the B quadrant oriented towards the top. (*a*) The vegetally situated mesodermal progenitor cleaves first to produce left and right M teloblasts (cells M_l and M_r). (*b*) The ectodermal progenitor (DNOPQ) cleaves symmetrically to produce right and left NOPQ cells. Note that bilateral symmetry is offset in the formative cleavages of the mesoderm and ectoderm. (*c*) Each NOPQ cell cleaves asymmetrically to produce an N teloblast and an OPQ cell. (*d*) Cell OPQ cleaves asymmetrically to produce a Q teloblast and an OP cell. (*e*) Cell OP cleaves to produce generative O and P teloblasts, which cannot be reliably distinguished on the basis of position (Weisblat and Blair 1984). (*f*) Final organization of the ten teloblastic stem cells that form the posterior growth zone for the leech's segmented body trunk. (From Fernández 1980.)

ferent in the cephalic and trunk portions of the mature annelid body plan (Figure 12.4c).

This distinction between quartets is not apparent in molluscan embryos, in which the first quartet micromeres manifest the same axial fate—dorsal, ventral, right or left—as their parent macromeres (Render 1991). This molluscan pattern of quadrant fates has also been reported for the polychaete *Chaetopterus* using an extracellular lineage tracer (Henry and Martindale 1987), raising the possibility that the morphogenetic establishment of bilateral symmetry may vary within the annelid phylum.

Segmentation

In the leech, the segmental character of the mature body trunk originates in the repetitive organization of the cell lineages derived from the embryonic teloblasts. First, we will describe the formation of the teloblasts from the D quadrant. The mesodermal progenitor (cell DM″)* cleaves equally to produce a single pair of bilaterally symmetric M teloblasts (Figure 6A). The ectodermal progenitor (cell DNOPQ‴) also cleaves symmetrically, and then undergoes additional cleavages to produce four bilaterally paired teloblasts designated as cells N, O, P, and Q (Figure 12.6*b–f*). (As described below, the generative O and P teloblasts can only be distinguished by the location or fate of their descendants, and are therefore often referred to without distinction as O/P teloblasts—Weisblat and Blair 1984.) Thus, each half of the segmented body trunk arises from a bilateral set of five teloblasts, with the four ectodermal teloblasts arranged in a crescent facing the micromere cap and the mesodermal teloblast lying beneath (Figure 12.6*f*). From this point forward the right and left sides of the trunk segments will develop in a largely independent manner, with essentially no mixing of cells across the midline.

The teloblasts of the leech embryo are large stem cells that undergo a regular sequence of highly asymmetric cell divisions to generate a linear column or **bandlet** of much

*Primes are used to denote the products of micromere formation; e.g., cell DM divided to produce two micromeres and cell DM″. For a detailed cell lineage, see Bissen and Weisblat (1989).

smaller daughters known as **primary blast cells** (Figure 12.7*a*). During subsequent development the blast cells will undergo several rounds of subsidiary mitosis, producing clones of approximately 10^2 differentiated descendants. The segmental organization of the teloblast's descendant lineage lies in the fact that its various blast cell progeny undergo the same sequence of subsidiary cell divisions (Zackson 1984), and thereby produce a tandem array of very similar descendant clones situated in different body segments (Shankland 1987a,b).

Primary blast cells produced by a given teloblast are designated by the same letter as the parent teloblast but in lowercase, with a subscripted number denoting the rank order of their birth (q_1, q_2, q_3, etc.). The first blast cell in each bandlet is attached to the prostomial micromeres (Sandig and Dohle 1988), and as the five ipsilateral bandlets lengthen, they begin to merge in parallel at their front ends. This merger produces a bilayered **germinal band**, in which the progeny of the mesodermal m blast cells lie deep to the ectodermal n, o, p, and q blast cells (Figure 12.7*c*). The germinal band elongates by the continual incorporation of progressively later-born blast cells at its trailing end (Figure 12.7*b*), until it contains the founder blast cells for all 32 body segments. Elongation displaces the band around the circumference of the embryo, with

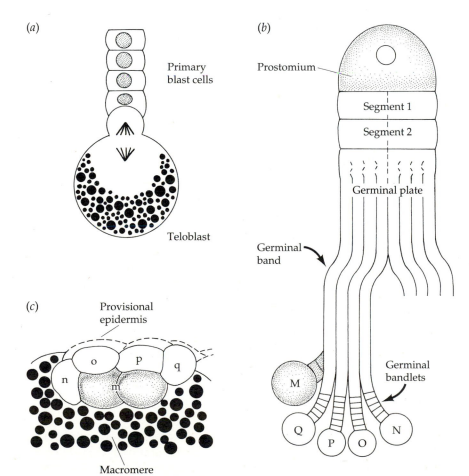

Figure 12.7 Formation of the segmented body trunk by the teloblastic stem cells. (*a*) Each teloblast undergoes a repeated sequence of highly asymmetric divisions to produce a linear array of much smaller primary blast cell daughters. Blast cells are generated at the yolk-free anterior pole of the teloblast. (*b*) The five ipsilateral blast cell chains (germinal bandlets) merge in parallel to form the germinal band. Right and left germinal bands then merge in parallel to form a bilaterally symmetric germinal plate, which undergoes a segmentally iterated pattern of organogenesis in anterior-to-posterior progression. (*c*) Cross section of the left germinal band, showing the stereotyped bilayered arrangement of ectodermal and mesodermal bandlets. Note that the primary m blast cell has already divided to produce two daughter blast cells. (*b* from Shankland 1994; *c* from Shankland and Weisblat 1984.)

the right and left bands eventually meeting at the far side and fusing to form a bilaterally symmetric **germinal plate**, which then undergoes an anterior-to-posterior wave of segmental differentiation (Figure 12.7b).

Within each teloblast lineage, the orderly birth of the primary blast cells reliably predicts the segmental location of their descendant clones in the segmental body plan of the adult. The front end of the band contains the first-born blast cells, and their descendants take part in forming the anteriormost body segment.* Progressively more posterior segments will be formed by the progeny of later-born blast cells in a precise and graded fashion (Figure 12.8). Thus, the length of the germinal band corresponds to the presumptive AP axis of the mature body trunk (Figure 12.5c).

One complicating factor in the assembly of segmental tissues along the AP axis of the leech germinal band is the fact that three teloblasts (M, O, and P) generate 1 primary blast cell/segment, whereas the other two teloblasts (N and Q) generate 2 primary blast cells/segment (Figure 12.8). All five teloblasts generate blast cells at essentially the same rate, and thus blast cells that are born and enter the germinal band at the same time are destined to take part in the formation of widely separate body segments according to their teloblast of origin (Weisblat and Shankland, 1985). The final organization of blast cell clones is not yet established when the blast cells first enter the germinal band, but rather requires that the blast cell clones in the n and q bandlets slide anteriorly past the m, o, and p bandlets by as much as half the body's length. It is only after the bandlets have slid past one another along the AP axis that the arrangement of blast cell clones within germinal plate accurately predicts the segmentally periodic organization of their descendant tissues in the mature leech.

Organogenesis does not begin until after blast cell clones fated to occupy the same segment have finally come into proximity with one another in the germinal plate. Lineage tracer injections have shown that each individual teloblast gives rise to a distinct and reproducible subset of the segmental tissues. The m blast cell clones give rise to all of the musculature, nephridia, connective tissue, and a small number of central neurons (Kramer and Weisblat 1985). The n, o, p, and q blast cell clones give rise to distinct subsets of the CNS, peripheral nervous system, and epidermis (Figure 12.9). The initial organization of the bandlets predicts the overall spatial relationship of their descendant tissues, as seen in the mediolateral organization of the N, O, P, and Q lineages. It should be noted, however, that neighboring blast cell clones intermingle extensively in both the AP and mediolateral axes (Figure 12.9). Readers interested in a detailed fate map of blast cell

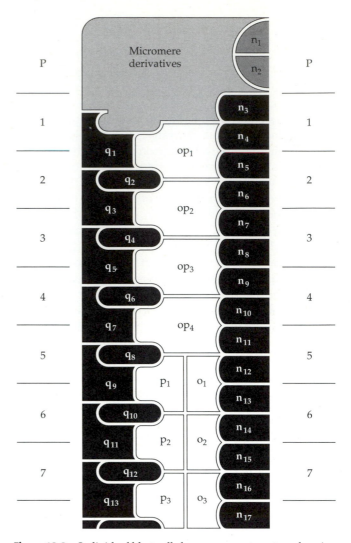

Figure 12.8   Individual blast cell clones occupy stereotyped positions in the segmental body plan of the adult leech, in accordance with their cell lineage history. This schematic shows the distribution of ectodermal blast cell clones on the left side of the anterior germinal plate in *Helobdella triserialis*, with segmental boundaries and numbers demarcated at the side. Individual blast cells are designated by lowercase letters corresponding to their teloblasts of origin and numbers representing birth rank in the stem cell lineage of the parent teloblast. Within each teloblast lineage the blast cell clones are aligned in order of their birth, with the first-born blast cells at the anterior end. Note that the N and Q teloblast lineages have two blast cell clones/segment, whereas the O and P lineages have one blast cell clone/segment. (The OP blastomere produces four op blast cells before cleaving to produce separate O and P teloblasts, resulting in a fork in the op bandlet.) The segmentally iterated pattern of blast cell clones continues posteriorly throughout the body's length. The first two n blast cells do not produce segmental tissues, but rather form an adhesive gland on the nonsegmental prostomium (P).

---

*The anteriormost body segment also receives some contribution from the nonsegmental micromeres, due to the fact that the constituent blast cell clones do not extend all of the way to the segment's anterior border (Figure 12.8). A similar situation occurs in certain other annelid species, in which the peristomium arises by an integration of unsegmented larval tissues with modified trunk somites.

clones are refered to papers by Weisblat and Shankland (1985) and Ramírez et al. (1995).

There are seven primary blast cells/hemisegment (1 m, 2 n, 1 o, 1 p, and 2 q), and in total approximately 440 blast cells that contribute to the segmented trunk of the adult

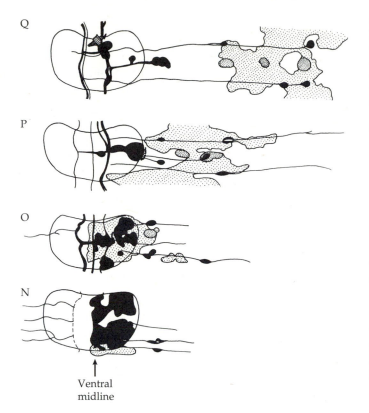

Ventral
midline

Figure 12.9   Camera lucida tracings of cellular contributions made by the N, O, P, and Q teloblast lineages in a typical midbody segment of *Theromyzon rude*. The ganglia of the ventral nerve cord (shown in outline) are located at the ventral midline, and the lateral edge of the germinal plate lies to the right. Neurons and their axons are shown in black, body wall epidermis is lightly stippled, and the epidermal specializations known as cell florets are heavily stippled. The teloblast lineages become intermingled, but nonetheless retain the same general mediolateral organization displayed by their precursor blast cells in the germinal band (see Figure 12.6c). (From Torrence and Stuart 1985.)

leech. A wide variety of experimental manipulations of the *Helobdella* embryo have been aimed at understanding the developmental events that specify the individual fates of these blast cells. It is clear that blast cells have an autonomous specification, which seems to be inherited from their teloblast of origin and accounts for bandlet-specific differences in their clonal fates (for review, see Shankland 1991). However, the o/p blast cells undergo an additional specification event in which they become committed to one of two alternative pathways of differentiation depending whether they are in the o bandlet or p bandlet position within the germinal band (Weisblat and Blair 1984). This conditional specification indicates that the differentiation of the o/p blast cell is influenced by cell interactions within the band, and several potential sources of positional information have been defined experimentally (Shankland and Weisblat 1984; Ho and Weisblat 1987; Huang and Weisblat, 1996).

The individual blast cells within a bandlet also exhibit unique specificities along the AP axis. For instance, the two alternating classes of blast cell produced by the N and Q teloblasts are distinct from one another in terms of generating the anterior and posterior halves of the segment, and appear to be differentially committed from the time of their birth (Bissen and Weisblat 1987). Moreover, homologous blast cell clones in different segments give rise to segment-specific descendant cell types in accordance with their birth ranks in the stem cell lineage of the parent

teloblast. To learn whether this segment identity is intrinsic to the blast cell as a result of its lineage or extrinsically determined by its environment, experiments were performed in which individual bandlets were shifted along the length of the germinal band so that some blast cell clones were forced to take part in the formation of segments inappropriate for their cell lineage history. Such experiments have shown that the segment identity of the blast cell is established at or shortly after the time of its birth from the parent teloblast, and is passed along to the descendant clone over several cell divisions in a lineage-dependent manner (Martindale and Shankland 1990; Gleizer and Stent 1993; Nardelli-Haefliger et al. 1994).

It is clear that patterns of cell lineage play a major role in the segmental patterning of the leech embryo. However, just as with the distinctions between bandlets, there are some segmental differences that require position-specific cell interactions. For example, certain ectodermally derived cells develop in a segment-specific manner because they receive trophic or inductive signals from mesodermal tissues that have already undergone a pattern of segmental diversification specified from a much earlier stage (Loer et al. 1987; Martindale and Shankland 1988).

In molecular terms, many of these segmental differences are likely to depend upon a differential expression of the *Hox* genes. This gene family is expressed in a regionalized manner along the AP axis of most higher animal species, and plays a major role in establishing segmental differences in other organisms (McGinnis and Krumlauf 1992). A number of *Hox* genes have been characterized from leeches (Aisemberg et al. 1993; Shankland 1994), and differing profiles of *Hox* gene expression would appear to be a determining factor in the segment-specific differentiation of segmentally homologous neurons (Wong et al. 1995). It is known, however, that the segment-specific expression of the leech *Hox* genes is not simply determined by the position of the blast cell clone along the AP axis of the germinal plate, but rather depends upon the birth rank that blast cell occupies within its particular teloblast lineage (Nardelli-Haefliger et al. 1994). Whether the primary blast cell obtains that information from its lineage history or from cell interactions occuring immediately after its birth remains unanswered, but it is clearly a question of central importance in understanding the evolution of mechanisms for *Hox* gene regulation.

Gastrulation

Gastrulation is a morphogenetic process in which the cell lineages of the early embryo become reorganized into three distinct germ layers—the ectoderm, mesoderm, and endoderm. In the leech, gastrulation is accomplished by a series of coordinated cellular rearrangements occurring over the surface of the large, yolk-rich macromeres that form the bulk of the early embryo.

1. An early event in the gastrulation of the leech embryo is the formation of a bilayered germinal band in which the mesodermal blast cells come to lie beneath the ectodermal blast cells (Figure 12.7*b*), anticipating the internal-to-external arrangement of their descendant tissues in the adult leech. The bilayered organization of the germinal band is apparent from the earliest stages of its formation, and most likely reflects a stable pattern of adhesive cell interactions. If the ectodermal teloblasts are ablated on one side of the embryo, the ipsilateral m bandlet often switches sides and joins the contralateral mesoderm in forming the deep layer of the other germinal band (Blair 1982).

2. A second major step in the gastrulation process is the internalization of the endoderm. In the leech, the endoderm is internalized by a process called **epiboly** in which both the germinal bands and the provisional integument spread from the animal pole to the vegetal pole over the surface of the endodermal macromeres (Figure 12.10). The epibolic movements of these two structures are tightly coordinated during normal development, although they can show a surprising degree of mechanistic independence in embryos in which one or the other structure has been ablated (Smith et al., 1996).

   Germinal band epiboly is easiest to appreciate if one first recognizes that the two ends of the band behave as fixed points. The anterior end of the germinal band remains attached to the prostomial micromeres at the animal pole (Sandig and Dohle 1988; Nardelli-Haefliger and Shankland 1993), while the posterior end of the germinal band remains fixed near the crescent of teloblasts at the equator of the D quadrant (Figure 12.10). (It is important to remember that the posterior end of the germinal band is not a particular group of cells, but rather reflects the dynamic point at which progressively younger blast cells are continually being incorporated into the band.) The elongating band loops outward between its two ends, such

that the apex of the loop moves over the surface of the macromeres in an oblique, circumferential direction (Figure 12.10*a*). The ultimate result of these movements is that the right and left bands meet and fuse at their outer edges (i.e., at the n bandlets) along a meridion that runs from animal-to-vegetal down the midline of the B quadrant, and then continues from vegetal-to-animal up to the equator of the D quadrant. Thus, the germinal plate is formed as a ribbon of tissue wrapped three-quarters of the way around the circumference of the macromeres, with the prostomium at its anterior end and the spent teloblasts at its posterior end (Figure 12.10*a*).

3. The provisional integument arises between the prostomial micromeres and the ectodermal teloblasts (Figure 12.5*c*), and the germinal bands form beneath the perimeter of its free edge (Figure 12.7*c*). The outer layer of this integument is a simple squamous epithelium of micromere derivation (Smith and Weisblat 1994), but it also has an inner muscular layer derived from the segmental mesoderm (Zackson 1982; Weisblat and Shankland 1985).

   Elongation of the germinal bands increases the perimeter of the area covered by the provisional integument, which continually expands to the fill the intervening space on the dorsal surface of the embryo (Figure 12.10*b,c*). As the right and left germinal bands fuse ventrally to form the germinal plate, the free edges of the provisional epidermis also fuse with one another to form a continuous layer. The completion of this process marks the end of gastrulation *per se*, in that the macromeres (endoderm) are now completely enveloped in a provisional epidermis (ectoderm) with a layer of provisional musculature (mesoderm) in between.

This mode of gastrulation is found in many annelid species, and is generally viewed as an evolutionary modification of a more primitive mechanism typified by the polychaete *Polygordius* (reviewed in Okada 1988). This species has a small embryo with a limited yolk store, and its spiral cleavages give rise to a hollow blastula with a well-defined central blastocoel. *Polygordius* gastrulates by an invagination of the vegetally situated macromeres into the blastocoel, coupled with an epibolic expansion of the micromere derivatives to seal over the site of macromere entry. In annelids such as the leech with larger, yolk-rich eggs, the macromeres inherit the bulk of the yolk and are capable of only limited morphogenetic movements, with the result that the blastocoel is reduced or absent, and the epibolic spread of the micromere cap over the surface of the macromeres becomes the sole mechanism for the internalization of the latter cells (Anderson 1973).

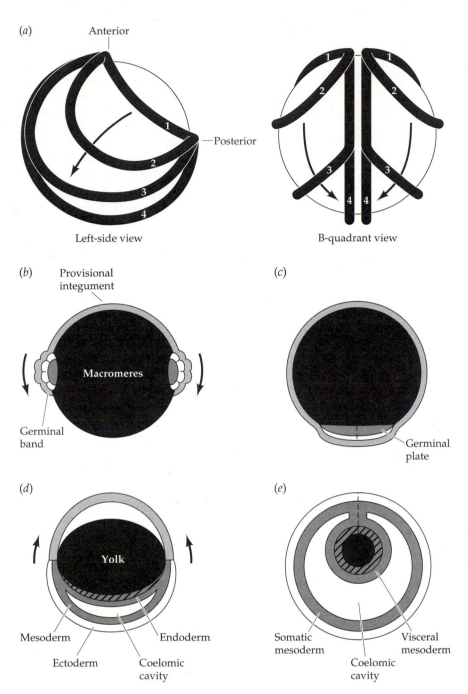

(a)

Anterior

Posterior

Left-side view

B-quadrant view

(b) Provisional integument

Macromeres

Germinal band

(c)

Germinal plate

(d)

Yolk

Mesoderm

Ectoderm

Endoderm

Coelomic cavity

(e)

Somatic mesoderm

Coelomic cavity

Visceral mesoderm

Figure 12.10 Gastrulation by epiboly. (*a*) Four steps in the epiboly of the germinal bands, shown in left-side and B-quadrant views of a *Helobdella* embryo, with the anterior pole oriented towards the top. During germinal band elongation, the anterior end remains fixed at the animal pole of the embryo, and the posterior end remains fixed at the equator of the D quadrant. The elongating band loops outward between these two fixed points, eventually contacting the contralateral band on the far side of the embryo where the two fuse to form the germinal plate. (*b*–*e*) Transverse sections taken at various stages of gastrulation. (*b*) Midway through epiboly (step 2 in part *a*), the germinal bands are lying on the surface of the macromeres. The rapidly expanding provisional integument covers the animal surface of the embryo and extends to the edges of the germinal bands. (*c*) At the end of epiboly (step 4 in part *a*), the germinal bands have fused at the ventral midline to form the germinal plate, and the provisional integument has sealed along its free edge to completely internalize the macromeres. (*d*) The germinal plate displaces the overlying provisional integument and begins to expand dorsally. Note the formation of a cavity (the coelom) within the germinal plate mesoderm. By this stage the macromeres have formed a layer of definitive endoderm on the inner surface of the germinal plate, as well as a residual yolk mass that will ultimately be digested. (*e*) The lateral edges of the expanding germinal plate fuse along the dorsal midline, and the last remnants of the provisional integument disappear. The gut is now internalized as a separate tube whose lumen contains the residual yolk.

4. Although not technically a part of gastrulation, the replacement of the provisional integument by the definitive integument represents another major morphogenetic event in establishing the final germlayer organization. The provisional epidermis initially covers the germinal plate (Figure 12.10*c*), but shortly after germinal band fusion it is displaced by definitive epidermal cells arising from the underlying segmental ectoderm. Over the course of several days the germinal plate then expands over the surface of the macromeres (Figure 12.10*d*,*e*), effectively retracing the route taken by the germinal band during epiboly.

This expansion occurs at the expense of the provisional integument, which is either lost or resorbed and makes little or no cellular contribution to the body wall of the adult leech.

Coincident with germinal plate expansion, the leech embryo extends its head through the vitelline membrane and begins the transformation from a sphere to a cylinder (Figure 12.5*c*,*d*). The right and left edges of the germinal plate eventually meet along the dorsal midline of the cylinder (i.e., the midline of the D quadrant), and there fuse to completely enclose the midgut in the definitive integument (Figure 12.10*d*,*e*).

Larval Development of Polychaetes

As described above, the glossiphoniid leech embryo develops directly from the zygote into a miniature version of the adult. The juvenile undergoes a substantial size increase and sexual maturation, but postembryonic development does not involve any significant reorganization of the body plan established during embryogenesis. This direct mode of development is characteristic of all clitellate annelids, but comparative studies indicate that it has evolved secondarily from a life cycle in which the ancestral annelid first developed into a free-swimming larva whose body plan had little resemblance to that of the adult worm, then underwent a period of postembryonic development terminating in metamorphosis. During metamorphosis, some larval tissues were discarded and others remodeled to take part in the formation of the adult worm. This latter indirect mode of development is still found in many species of polychaete, as typified by *Polygordius* (Figure 12.11). We must emphasize that the polychaetes are a highly diverse group (Anderson 1973) and that a detailed comparative analysis of their development is beyond the scope of this chapter. Still, a summary overview of certain key similarities and differences between the directly and indirectly developing annelids should provide the reader with a basic appreciation for the way in which morphogenetic mechanisms have evolved during the history of this phylum.

The *Polygordius* embryo undergoes spiral cleavage, and gastrulates by a combination of macromere invagination and epiboly to produce a microscropic larva that utilizes ciliary organs to effect its locomotion, sensory reception, and feeding (Figure 12.11*a*). Larvae of this particular anatomy are known as **trochophores**; they are found with a variable degree of modification in those phyla that manifest spiral cleavage. In light of this correlation, it is generally accepted that the formation of a trochophore-like larva is the vestige of an ancestral developmental program that preceded the radiation of the various spiralian phyla.

The *Polygordius* trochophore is subdivided into cellular domains (Figure 12.11) whose developmental fates can be reconciled in a relatively straightforward manner with the lineally homologous domains of the *Helobdella* embryo (Figure 12.5*c,d*). However, it is important to realize that homologous domains not only differ with respect to the exact structures they produce, but in some cases also behave quite differently with respect to the time-course of their differentiation, a phenomenon known as **heterochrony.**

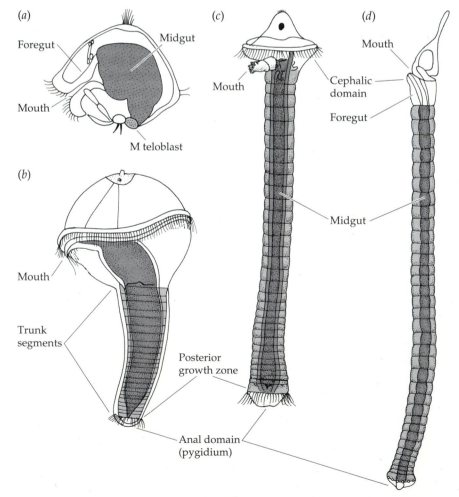

Figure 12.11 Indirect development in the polychaete *Polygordius*. (*a*) The embryo gastrulates to form a free-swimming trochophore larva, here shown in sagittal section. The body wall and foregut of the trochophore are derived from micromeres (white), while the midgut is derived from the macromeres (dark shading). The trochophore does not have a segmental organization, and the mitotically quiescent teloblasts are situated near the anus. (*b*) During the course of larval development, the ectodermal and mesodermal teloblasts generate a posterior growth zone, which forms serial trunk segments (light shading) in anterior-to-posterior order. Trunk segments form as rings around the elongating midgut, and separate the anal and cephalic domains. (*c*) End of larval development, and onset of metamorphosis. (*d*) Adult worm, in which the larval tissues have been remodeled to form the cephalic domain of the adult. (After Woltereck 1904 and Okada 1988.)

1. As occurs in the leech, the macromeres of the polychaete embryo become internalized during gastrulation and differentiate to form midgut. However, the polychaete midgut differs from that of the leech in that it first differentiates into a functional larval digestive tract, and is then remodeled to become the elongate midgut of the adult worm.

2. The polychaete micromeres (with the exception of 2d and 4d, see Figure 12.5a) differentiate at the end of embryonic gastrulation to form the larval integument and a number of internal organs. As with the midgut, some of these micromere derivatives also persist through metamorphosis and become part of the adult body plan, notably the foregut and the apical portion of the trochophore animal hemisphere, which gives rise to unsegmented head tissues (Figure 12.11). Thus, the cephalic domain of the adult polychaete consists of remodeled larval tissues.

In contrast, other micromere-derived larval tissues are discarded at metamorphosis and make no contribution to the adult polychaete. For example, this is typically the case for the ciliary organs situated around the equator and on vegetal hemisphere of the trochophore. Such structures may be directly homologous to the provisional epidermis of the leech embryo, since both arise from the same cell lineages (the epibolic descendants of the four first-quartet micromeres and the subsidiary D-quadrant micromeres; Smith and Weisblat 1994), and are eventually discarded in favor of definitive segmental tissues.

3. The indirectly developing polychaete embryo cleaves to produce progenitors for the segmental ectoderm and mesoderm, which are outwardly similar to the DNOPQ and DM cells of the leech embryo. However, the trochophore itself has no segmental organization, and the segmental trunk tissues are elaborated during postembryonic development in anterior-to-posterior sequence from a **posterior growth zone** situated in front of the anus (Figure 12.11). Some segments are added during the later development of the larva, but many polychaetes add a number of additional segments following metamorphosis. Each new segment is added as a ring around the larval digestive tract, and thus the body trunk elongates as a tube separating the future cephalic domain (mouth and animal hemisphere of the trochophore) at one end from the anal domain (vegetal hemisphere) at the other end (Figure 12.11).

Thus, one of the striking differences between polychaete and leech is a heterochrony involving the relative timing of gastrulation and segment formation (Figure 12.12). In polychaetes, the mesodermal progenitor cleaves to produce a bilateral pair of M teloblasts, which in many species divide repeatedly like their leech homologues to produce columns of segmented mesodermal tissue extending from the anus towards the mouth. However, the M teloblast of the polychaete initiates its stem cell divisions after gastrulation is complete, whereas the M teloblast of the *Helobdella* embryo completes the formation of its blast cell progeny midway through gastrulation (Figure 12.12). Thus, the M teloblast lineage of the glossiphoniid embryo has already begun to form a segmented coelomic mesoderm by the time gastrulation is finished (Zackson 1982).

Evolution of the Annelid Body Plan

By comparing the morphogenesis of homologous embryonic domains in different annelid species, it can be seen that the evolution of directly developing clitellate annelids such as the leech from their indirectly developing ancestors involved an ensemble of interrelated modifications in the developmental program. Clitellate annelids deposit their eggs within a protective cocoon, and provide them with a nutrient source (large stores of internal yolk or albuminous cocoon fluid, depending on the species) that frees the developing embryo from reliance upon external food sources. Without a selective pressure to maintain or-

Figure 12.12 The relative timing of gastrulation and segment formation is heterochronous when compared between the indirect development of the polychaete and the direct development of the glossiphoniid leech. In a typical polychaete, the M teloblasts begin dividing to produce segmental progeny around the end of gastrulation, and morphologically distinct segments are added by the posterior growth zone during larval and adult life. In the glossiphoniid leech, the M teloblasts divide to produce all of their segmental blast cells by the midpoint of gastrulation, and segmental morphogenesis is completed during embryonic development. The period of gastrulation is shown as a shaded bar, and is normalized for the two life cycles.

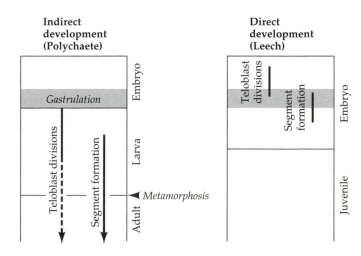

gans involved in larval feeding or locomotion, the micromere lineages involved in producing those organs evolved into a relatively undifferentiated provisional epidermis which—in the glossiphoniid leeches—is little more than a transient yolk sac for the gastrulating embryo. In concert with this reduction or loss of larval differentiation, the segmental mesoderm and ectoderm of the adult worm experience a precocious differentiation that begins much earlier in the gastrulation process (Figure 12.12). It is interesting to note that there are certain polychaete species that have also evolved life cycles in which they do not feed as trochophores, and that such species convergently evolved some of the same developmental modifications, including larger, yolk-rich eggs, failure to form larval ciliary organs, or precocious development of a segmentally organized mesoderm (Anderson 1973).

One of the major unanswered questions in annelid natural history is the means by which this phylum evolved its characteristically segmented body trunk. Annelids and arthropods are thought to have arisen from a common segmented ancestor (Brusca and Brusca 1990; Wedeen and Weisblat 1991), and—with the exception of a few groups such as the flies—body trunk formation in arthropods also involves a progressive, anterior-to-posterior addition of segments from a posterior growth zone situated near the anus. This mode of development is refered to as **teloblastic** growth (from Greek: *telos*, end; *blastos*, bud). In the leeches, oligochaetes (Storey 1989) and certain crustaceans (Dohle and Scholtz 1988), the posterior growth zone is composed of a fixed number of large, uniquely identifiable stem cells, the teloblasts, whose progeny undergo a stereotyped sequence of segmentally iterated cell divisions. In contrast, the posterior growth zone of some other arthropods, e.g., short germ band insects, is composed of a large number of outwardly indistinguishable cells whose progeny bear no clearly defined lineal relationships. In this latter group, segmentation apparently does not occur in close association with cell genealogies, but rather is thought to be dependent almost entirely on position-dependent cell interactions.

What then was the mode of segmentation in the posterior growth zone of the last common annelid/arthropod ancestor? If we assume that indirectly developing polychaetes are representative of the primitive stock from which direct developing annelids such as the leech have evolved, then it is clearly very important to discover to what degree the leech and the polychaete utilize the same or different cellular mechanisms of segment formation. Unfortunately, the nature of the cell lineage relationships within the posterior growth zone of the polychaete larva has not yet been examined with modern cell lineage-tracing techniques, due in large part to the small size of the cells in question. In some polychaetes the M teloblasts undergo iterated, asymmetric cell divisions similar to those seen in the leech, but this is not true of all species. Lineage

relationships are even more uncertain for the polychaete's ectoderm. Indeed, the so-called ectodermal teloblasts of the polychaete are not large, persistent stem cells like those seen in the leech, but rather undergo a combination of asymmetric and symmetric divisions (Wilson 1892) to produce a ectodermal growth zone that is also a functioning part of the larval epidermis (Anderson 1973).

If the cell lineage relationships in the posterior growth zone of the polychaete larva can be elucidated, this information will be of vital importance in shaping our understanding of how the segmentation process has evolved. One could envision that the use of stereotyped and iterated cell lineages to generate a segmentally organized body trunk is a primitive feature of annelids as a whole, in which case some or all of the extant polychaetes should also display this trait. On the other hand, if the polychaetes as a group generate segments without reliance upon stereotyped cell lineages, it would suggest that the lineal stereotypy of the teloblastic cell lineages in clitellate annelids is derived from some other, more primitive segmentation mechanism.

Literature Cited

Aisemberg, G. O., J. Wysocka-Diller, V. Y. Wong, and E. R. Macagno. 1993. *Antennapedia*-class homeobox genes define diverse neuronal sets in the embryonic CNS of the leech. *J. Neurobiol.* 24: 1423–1432.

Anderson, D. T. 1973. *Embryology and Phylogeny in Annelids and Arthropods.* Pergamon Press, New York.

Astrow, S. H., B. Holton and D. A. Weisblat. 1987. Centrifugation redistributes factors determining cleavage patterns in leech embryos. *Dev. Biol.* 120: 270–283.

Bissen, S. T. and D. A. Weisblat. 1987. Early differences between alternate n blast cells in leech embryo. *J. Neurobiol.* 18: 251–269.

Bissen, S. T. and D. A. Weisblat. 1989. The durations and compositions of cell cycles in embryos of the leech, *Helobdella triserialis. Development* 106: 105–118.

Blair, S. S. 1982. Interactions between mesoderm and ectoderm in segment formation in the embryo of a glossiphoniid leech. *Dev. Biol.* 89: 389–396.

Brusca, R. C. and G. J. Brusca. 1990. *Invertebrates.* Sinauer Associates, Sunderland, MA.

Dohle, W. and G. Scholtz. 1988. Clonal analysis of the crustacean segment: The discordance between genealogical and segmental borders. *Development Supplement* 104: 147–160.

Drapeau, P., W. B. Kristan and M. Shankland (eds.). 1995. *Leech Neurobiology and Development; J. Neurobiol.*, Vol. 27. John Wiley & Sons, New York.

Fernández, J. 1980. Embryonic development of the glossiphoniid leech *Theromyzon rude* characterization of developmental stages. *Dev. Biol.* 76: 245–262.

Fernández, J., N. Olea and C. Matte. 1987. Structure and development of the egg of the glossiphoniid leech *Theromyzon rude*: Characterization of developmental stages and structure of the early uncleaved egg. *Development* 100: 211–225.

Freeman, G. and J. W. Lundelius. 1992. Evolutionary implications of the mode of D quadrant specification in coelomates with spiral cleavage. *J. Evol. Biol.* 5: 205–247.

Gimlich, R. L. and J. Braun. 1985. Improved fluorescent compounds for tracing cell lineage. *Dev. Biol.* 109: 509–514.

Gleizer, L. and G. S. Stent. 1993. Developmental origin of segment identity in the leech mesoderm. *Development* 117: 177–189.

Henry, J. J. and M. Q. Martindale. 1987. The organizing role of the D quadrant as revealed through the phenomenon of twinning in the polychaete *Chaetopterus variopedatus*. *Roux's Arch. Dev. Biol.* 196:499–510.

Ho, R. K. and D. A. Weisblat. 1987. A provisional epithelium in leech embryo: Cellular origins and influence on a developmental equivalence group. *Dev. Biol.* 120: 520–534.

Huang, F. Z. and D. A. Weisblat. 1996. Cell fate determination in an annelid equivalence group. *Development* 122: 1839–1847.

Kramer, A. P. and D. A. Weisblat. 1985. Developmental neural kinship groups in the leech. *J. Neurosci.* 5: 388–407.

Lawrence, P. A. 1973. A clonal analysis of segment development in *Oncopeltus* (Hemiptera). *J. Embryol. Exp. Morphol.* 30: 681–699.

Livanow, N. 1906. *Acanthobdella peledina* Grube, 1851. *Zoöl. Jb. Anat.* 22: 637–866.

Loer, C. M., J. Jellies and W. B. Kristan. 1987. Segment-specific morphogenesis of leech Retzius neurons requires particular peripheral targets. *J. Neurosci.* 7: 2630–2638.

Martindale, M. Q. and M. Shankland. 1988. Developmental origin of segmental differences in the leech ectoderm: Survival and differentiation of the distal tubule cell is determined by the host segment. *Dev. Biol.* 125: 290–300.

Martindale, M. Q. and M. Shankland. 1990. Segmental founder cells of the leech embryo have intrinsic segmental identity. *Nature* 347: 672–674.

McGinnis, W. and R. Krumlauf. 1992. Homeobox genes and axial patterning. *Cell* 68: 283–302.

Nardelli-Haefliger, D. and M. Shankland. 1993. *Lox10*, a member of the *NK-2* homeobox gene class, is expressed in a segmental pattern in the endoderm and in the cephalic nervous system of the leech *Helobdella*. *Development* 118: 877–892.

Nardelli-Haefliger, D., A. E. E. Bruce and M. Shankland. 1994. An axial domain of HOM/Hox gene expression is formed by the morphogenetic alignment of independently specified cell lineages in the leech *Helobdella*. *Development* 120: 1839–1849.

Nelson, B. H. and D. A. Weisblat. 1992. Cytoplasmic and cortical determinants interact to specify ectoderm and mesoderm in the leech embryo. *Development* 115: 103–115.

Okada, K. 1988. Annelida. In *Invertebrate Embryology*, M. Kumé and K. Dan (eds.). Garland Publishing, New York, pp. 192–241. (English translation by J.C. Dan.)

Ramírez, F., C. J. Wedeen, D. K. Stuart, D. Lans and D. A. Weisblat. 1995. Identification of a neurogenic sublineage required for CNS segmentation in an annelid. *Development* 121: 2091–2097.

Render, J. 1991. Fate map of the first quartet micromeres in the gastropod *Ilyanassa obsoleta*. *Development* 113: 495–501.

Sandig, M. and W. Dohle. 1988. The cleavage pattern in the leech *Theromyzon tessulatum* (Hirudinea, Glossiphoniidae). *J. Morphol.* 196: 217–252.

Sawyer, R. T. 1986. *Leech Biology and Behaviour*. Clarendon Press, Oxford.

Shankland, M. 1987a. Differentiation of the O and P cell lines in the embryo of the leech. I. Sequential commitment of blast cell sublineages. *Dev. Biol.* 123: 85–96.

Shankland, M. 1987b. Differentiation of the O and P cell lines in the embryo of the leech. II. Genealogical relationship of descendant pattern elements in alternative developmental pathways. *Dev. Biol.* 123: 97–107.

Shankland, M. 1991. Leech segmentation: Cell lineage and the formation of complex body patterns. *Dev. Biol.* 144: 221–231.

Shankland, M. 1994. Leech segmentation: A molecular perspective. *BioEssays* 16: 801–808.

Shankland, M. and M. Q. Martindale. 1992. Segmental differentiation of lineally homologous neurons in the central nervous system of the leech. In *Determinants of Neuronal Identity*, M. Shankland and E. R. Macagno (eds.). Academic Press, New York, pp. 45–77.

Shankland, M. and D. A. Weisblat. 1984. Stepwise commitment of blast cell fates during the positional specification of the O and P cell lines in the leech embryo. *Dev. Biol.* 106: 326–342.

Smith, C. M. and D. A. Weisblat. 1994. Micromere fate maps in leech embryos: Lineage-specific differences in rates of cell proliferation. *Development* 120: 3427–3438.

Smith, C. M., D. Lans and D. A. Weisblat. 1996. Cellular mechanisms of epiboly in leech embryos. *Development* 122: 1885–1894.

Stent, G. S., W. B. Kristan, Jr., S. A. Torrence, K. A. French and D. A. Weisblat. 1992. Development of the leech nervous system. *Int. Rev. Neurobiol.* 33: 109–193.

Storey, K. G. 1989. Cell lineage and pattern formation in the earthworm embryo. *Development* 107: 519–532.

Torrence, S. A. and D. K. Stuart. 1985. Gangliogenesis in leech embryos: migration of neural precursor cells. *J. Neurosci.* 6: 2736–2746.

Wedeen, C. J. and D. A. Weisblat. 1991. Segmental expression of an *engrailed*-class gene during early development and neurogenesis in an annelid. *Development* 113: 805–814.

Wedeen, C. J., D. J. Price and D. A. Weisblat. 1990. Analysis of the life cycle, genome, and homeobox genes of the leech *Helobdella triserialis*. In *Cellular and Molecular Biology of Pattern Formation*, D. L. Stocum and T. Karr (eds.). Oxford University Press, New York, pp. 145–167.

Weisblat, D. A., R. T. Sawyer and G. S. Stent. 1978. Cell lineage analysis by intracellular injection of a tracer enzyme. *Science* 202: 1295–1298.

Weisblat, D. A. and S. S. Blair. 1984. Developmental indeterminancy in embryos of the leech *Helebodella triserialis*. *Dev. Biol.* 101: 326–335.

Weisblat, D. A. and M. Shankland. 1985. Cell lineage and segmentation in the leech. *Phil. Trans. Roy. Soc. Lond.* B 312: 39–56.

Weisblat, D. A., S. Y. Kim and G. S. Stent. 1984. Embryonic origins of cells in the leech *Helobdella triserialis*. *Dev. Biol.* 104: 65–85.

Wilson, E. B. 1892. The cell lineage of *Nereis*. *J. Morphol.* 6: 361–480.

Woltereck, R. 1904. Beiträge zur praktischen Analyse der *Polygordius* Entwicklung nach dem "Nordsee"- und dem "Mittelmeer"-Typus. I. Die für beide Typen gleichverlaufende entwicklungsabschnitt: Vorm Ei bis zum jungsten Trochophora-Stadium. *Arch. Entwick. Mech. Org.* 18: 377–403.

Wong, V. Y., G. O. Aisemberg, W. B. Gan and E. R. Macagno. 1995. The leech homeobox gene *Lox4* may determine segmental differentiation of identified neurons. *J. Neurosci.* 15: 5551–5559.

Zackson, S. L. 1982. Cell clones and segmentation in leech development. *Cell* 31: 761–770.

Zackson, S. L. 1984. Cell lineage, cell-cell interaction and segment formation in the ectoderm of a glossiphoniid leech embryo. *Dev. Biol.* 104: 43–60.

CHAPTER 13

Arthropods: The Crustaceans, Spiders, and Myriapods

Scott F. Gilbert

T HE PHYLUM ARTHROPODA COMPRISES THE LARGEST GROUP of animals on the planet. They range from tiny 1-mm-long mites to crabs extending over 4 m. The major extant arthropod groups are usually classified as follows:

Subphylum Cheliceriformes
Body divided into anterior prosoma (cephalothorax) and posterior abdomen (opisthosoma). Unbranched (uniramous) appendages; first prosomal appendages are chelicerae, second appendages usually pedipalps. Bipartite brain.
    Class Chelicerata (spiders, ticks, mites, scorpions, horseshoe crabs)
    Class Pycnogonida ("sea spiders")
Subphylum Uniramia
Body divided into a distinct cephalon (head) and either elongated or regionalized trunk. Four cephalic appendages, three pairs of postoral appendages (mandibles, maxillules, maxillae). Uniramous appendages. Tripartite brain.
    Class Myriapoda (centipedes, millipedes, etc.)
    Class Insecta (insects)
Subphylum Crustacea
Appendages biramous or uniramous, tripartite brain, distinct cephalon, five pairs of cephalic appendages, four pairs of postoral appendages.

This classification, however, is controversial and is in the process of being scrutinized. The principal difficulty involves deciding which morphological similarities uniting these groups are the result of close evolutionary relationships and which represent the convergence of traits due to two or more groups' sharing similar environments. There have been three traditional hypotheses of arthropod evolution that have been at odds with one another over the past decades. The first holds that there were three nonarthropod ancestors that each gave rise to a different group of arthropods (the Chelicerates, the Crustaceans, and the Unirames). This polyphyletic view has largely been rejected on the basis of morphological and molecular evidence of the past decade. The other views contend that there was a common ancestor for all three arthropod groups. One theory (called the TCC theory) holds that one group containing the trilobites (T), crustaceans (C), and chelicerates (C) separated from the unirames. The second hypothesis, the mandibulate hypothesis, suggests that the Unirames and Crustaceans are closely allied, and that the Chelicerates (and trilobites) diverged earlier (Figure 13.1a).

Recently, however, molecular evidence from two sources has converged to provide a fourth hypothesis for the arthropod lineage (Figure 13.1b). The first source involves relatively conserved ribosomal RNA sequences. These sequences should not be affected by natural selection and similarities should reveal common descent. The second source involves rare gene rearrangements in mitochondrial DNA. These rearrangements are so uncommon that the presence of such rearrangements in two groups is evidence for their close common ancestry. These new data suggest that the insects are not closely related to the myriapods, as has usually been assumed. Rather, the insects seem closely related to crustaceans (Boore et al. 1995; Friedrich and Tautz 1995).

If this relationship is confirmed, then features shared by insects and myriapods must be due to convergence. The features that have been used to link these two groups—uniramous appendages, a tracheal system, no appendage corresponding to the second antennae of crustaceans, and a mandible composed of a whole limb (where crustacean mandibles are formed only from a limb base)—have been called into question (Friedrich and Tautz 1995). The first four of these features may be accomplished by convergent evolution and are even found in common with some terrestrial arachnids. In addition, recent evidence from developmental biology and paleontology suggests that the insect mandible is also made from only its base (Panganiban

Figure 13.1 Phylogenetic trees of arthropod evolution. (*a*) The mandibulate hypothesis of arthropod evolution linking the myriapods with the insects. (*b*) An alternative scheme based on recent molecular and paleontological data, emphasizing the similarities between insects and crustaceans. The *darkly shaded* boxed region represents elements in common between insects and crustaceans and thought (by this hypothesis) to be derived from a close common ancestor. The *lightly shaded* boxed entities are thought to be shared by insects and myriapods due to their common terrestrial mode of life. (After Telford and Thomas 1995.)

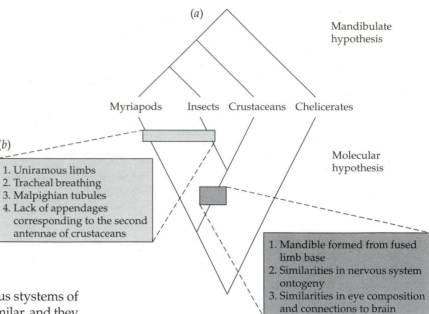

et al. 1994; Kukalová-Peck 1992). The nervous systems of insects and crustaceans are also extremely similar, and they are different from those of chelicerates or myriapods (Paulus 1979; Thomas et al. 1984; Whitington et al. 1993).

Molecular evidence corroborates the longstanding view (based on embryological and structural evidence) that the arthropods are related to the annelids (which can be grouped together as the Articulata). Not only do they form segments in a similar fashion, but they appear to be using a similar mechanism (especially the products of the *engrailed* gene) to define segmented boundaries and polarity (Patel et al. 1991; Wedeen and Weisblat 1991; Whitington et al. 1991).

There are numerous other groups that may or may not be related to arthropods—onychophorans, tardigrades, and pentastomids. The places of these phyla are not well established and will probably have to wait for evidence from the combined study of molecular and developmental biology (see Ballard et al. 1992). There is no way that we can do justice to the entire arthropod phyla in a book such as this, but we hope to introduce the arthropod phylum through two chapters. This chapter will outline the development of representative chelicerates, crustacea, and myriapods. The next chapter will outline the development of insects. What you may learn from this chapter is that there is a lot more work left to do.

Crustaceans

No science fiction writer could create life cycles stranger than those found among aquatic arthropods. However, while crustaceans are among the most widespread species on earth, we know very little about their embryology. There are very few studies following the development of any one species in detail.

The sexes in crustaceans are usually separate, although hermaphroditism is sometimes seen. Parthenogenesis is also common among branchiopods (shrimp such as *Daphnia*)

and some ostracods. The gonads of crustaceans are usually separate, derived from coelomic pouches in the trunk; but in barnacles (cirripedes), the gonads form in the head region. The gonads are connected to openings, the gonopores. The female crustacean may have **seminal receptacles** for storing sperm, and the ducts leading from the male gonad are often fused to form a single penis. The female lobster *Homarus* can store sperm for years in her receptacle, thereby allowing multiple broods to be born from the same mating event. The eggs of most crustaceans are protected by one of the parents until the larvae are free-swimming. The embryos are often "cradled" in sacs or adhered to the body wall with a glue secreted by the adult dermal glands.

For all their remarkable diversity, crustaceans share some characteristic developmental stages and processes. First, all crustaceans pass through a form of larval stage known as the **nauplius**. This "phylotypic stage" has been important in linking diverse animals into the crustacean class. For instance, for centuries, the taxonomic place of barnacles had been a mystery. However, when J. V. Thomson discovered its nauplius larva, it became clear the barnacles were a type of crustacean (see Winsor 1969). The nauplius larva can be free-living (as in the cases of crustaceans having little yolk in their eggs) or it can be passed through as part of embryonic development (as it is in species with yolky eggs).

The second characteristic of crustaceans is their cleavage. Most crustaceans divide holoblastically, and their cleavage is similar to that of flatworms, snails, and molluscs. In the malacostraca (crabs, shrimp, lobsters, krill), there is superficial cleavage that retains some of the characteristics of the spiral type (Anderson 1973). The third characteristic of crustacean development is the role that **teloblasts** play in forming the posterior portion of the body, posterior to the first maxillary segment. In most crustacea, these teloblasts are stem cells that produce the ectoderm and mesoderm for the posterior thorax and everything behind it, including the telson (posteriormost tail). (The anterior structures are made directly from the germinal disc.) There are, however, significant modifications of this pattern. In amphipods (sand fleas), there are no teloblasts and the posterior tissues form from the germinal disc along with the anterior structures. In some crustaceans such as *Artemia* (the relatively well-studied brine shrimp), the teloblasts have been replaced by a "growth zone" that undergoes no regular pattern of cleavage, but provides masses of cells for subsequent patterning.

Fertilization

The eggs of most crustacea are centrolecithal. In some species, such as the mysid opossum shrimp, *Hemimysis lamorn*, the yolk pervades the cytoplasm, leaving only a thin rim of perpheral cytoplasm around the embryo (Manton 1928). In other crustaceans, like the barnacle *Tetraclita rosea*, the yolk distribution changes during fertilization and becomes concentrated in one region of the egg (Anderson 1969). Crustacean spermatozoa are not "typical." Instead of being streamlined cells with little cytoplasm and large axonemic flagella, crustacean sperm are often large round or stellate cells that move by pseudopods or may even be nonmotile (Figure 13.2).

Figure 13.2 Spermatozoa from several crustaceans. (*a*) *Latona setifera*; (*b*)*Daphnella brachyura*; (*c*) *Moina paradoxa*; (*d*) *Polyphemus pediculus*; (*e*) *Squilla oratoria*; (*f*) *Leander aspersus*; (*g*) *Galathea squamifera*; (*h*) *Ethusa marascone* (lateral view); (*i*) *E. marascone* (upper side); (*j*)*Astacus fluviatilis*; (*k*) spermatophore of *Galathea*. (After Shiino 1968.)

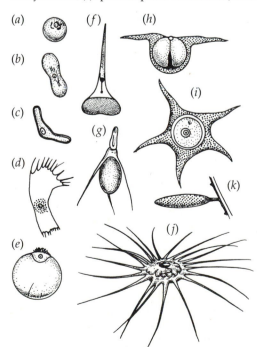

The spermatozoa of malacostrocans (crabs, lobsters, shrimp) and copepods are delivered to the female in packets called **spermatophores**. These oval or round packets are glued to the gonopore of the female. A secretion from the vas deferens is also packed inside the spermatophore. It expands when it is inside the female gonopore, thereby bursting the spermatophore sac and releasing the sperm. Knowledge of crustacean fertilization is limited by the opacity of the egg and the location of fertilization being within the female. Hudinaga (1942) has made an excellent study of fertilization in the shrimp *Penaeus japonicus*. Immediately after the egg is laid, sperm enter and the first polar body is formed. A number of sperm reach the egg and induce the egg to extend fertilization cones to each sperm. Originally, the cone appears to be transparent, but the cytoplasm of the egg soon rushes into it. The cone in which this ooplasm first reaches the attached sperm head begins to contract, and it draws the entire sperm into the interior of the cell (Figure 13.3). The other cones are retracted without engulfing their respective spermatozoa. A fertilization membrane forms around the egg and a second polar body is formed, indicating that the egg nucleus is haploid.

Cleavage

In the barnacle *Tetraclita rosea*, the egg has a definite polarity, much like that of a hen's egg. As the mitotic spindle of first cleavage begins to form, the creamy white yolk that had originally been spread uniformly throughout the egg becomes concentrated in the future posterior end (Figure 13.4*a*). This allows for a holoblastic division, similar to that of the telolecithal eggs of frogs. As division continues, the yolk-free cytoplasm streams anteriorly and the plane of division "rotates" from being oblique to transverse. Moreover, the mitotic spindle is displaced from its original oblique position to being stretched along the anterior-posterior axis. Thus, first division separates a relatively small yolk-free cell (AB) from a relatively large yolky one (CD; Figure 13.4*b*). The AB cell divides before the CD and splits into two cells, the left (A) and the right (B) blastomere (Figure 13.4*c*). Meanwhile, yolk-free cytoplasm accumulates around the CD nucleus, and this nucleus begins an oblique division, creating another relatively yolk-free cell, C, dorsal to the large yolky D blastomere (Figure 13.4*d*). The yolk-free cytoplasm remaining in the D blastomere accumulates around the nucleus. The third cleavage division (Figure 13.4*e*) in-

Figure 13.3 Sperm entry in *Penaeus japonicus*. (After Shiino 1968.)

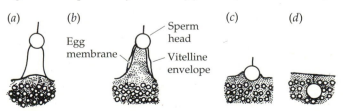

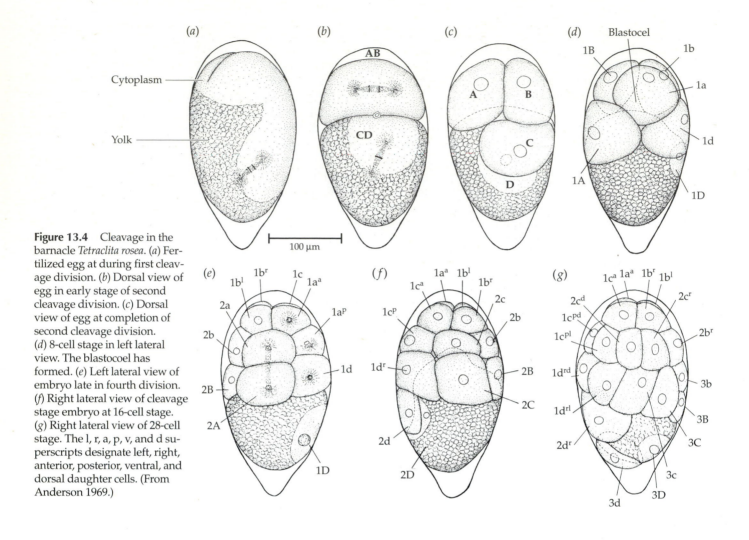

Figure 13.4   Cleavage in the barnacle *Tetraclita rosea*. (*a*) Fertilized egg at during first cleavage division. (*b*) Dorsal view of egg in early stage of second cleavage division. (*c*) Dorsal view of egg at completion of second cleavage division. (*d*) 8-cell stage in left lateral view. The blastocoel has formed. (*e*) Left lateral view of embryo late in fourth division. (*f*) Right lateral view of cleavage stage embryo at 16-cell stage. (*g*) Right lateral view of 28-cell stage. The l, r, a, p, v, and d superscripts designate left, right, anterior, posterior, ventral, and dorsal daughter cells. (From Anderson 1969.)

volves A dividing into 1A and 1a, et cetera. This creates a small blastocoel. As division continues (Figures 13.4 *f,g*), the yolk-free cells move more posteriorly. At the end of the fifth cleavage, the yolk-free blastomeres flatten and spread over the yolk cell, 3D, which remains exposed only at its most posterior ventral end. The nucleus of the 3D cell lies directly under this exposed region (Anderson 1969).

In the superficially cleaving egg of mysids (opossum shrimp), the central nucleus lies in a small stellate mass of relatively yolk-free cytoplasm, surrounded by a huge amount of yolk (Manton 1928). Strands from the yolk-free perinuclear cytoplasm extend to the rim of the egg where a thin layer of yolk-free cytoplasm resides. Early cleavage occurs within the yolk mass, forming nuclei surrounded by a small amount of yolk-free cytoplasm. Meanwhile, at the surface of the egg, cytoplasmic projections are made into the yolk, breaking up the yolk near the periphery. Some of the strands appear to connect to the stellate blastomeres in the center of the yolk. At the 16-cell stage, the blastomeres approach the surface of the egg, the majority aggregating at the animal pole of the egg where the cytoplasmic rim is thickest (Figure 13.5*a*). At about the 32-cell stage, the cells reach the cytoplasmic rim. This occurs first at the vegetal pole where the cells stretch out and become extremely thin. At the animal pole, the blastomeres become rounded while they are still within the yolk mass, and they become connected to one another by thin cytoplasmic bridges (Figure 13.5*b*). This animal pole cell layer gradually rises to the surface by the 128-cell stage, thus surrounding the yolk (Figure 13.5*c*). At this stage, the blastomeres change their positions over the yolk and form an accumulation at the posterior ventral edge of the animal pole. The cells at this region increase in thickness (Figure 13.5*d*).

Gastrulation and Naupliar Organogenesis

In *Tetraclita*, the 28-cell stage leaves only a small portion of the yolky cell uncovered. Three precocious sixth cleavage divisions occurring at the posterior end result in the near-complete coverage of the yolky cell and the penetration of three cells into the interior (Figure 13.6*a,b*). These interior cells (3A, 3B, and 3C) are the precursors of the mesoderm. Meanwhile, the yolky cell 3D divides to release a relatively nonyolky 4d from the large yolky 4D blastomere. The 4D blastomere is the endodermal precursor. The 4D cell divides in two, and each of these two daughter cells divides unequally. The two yolky cells that remain internal

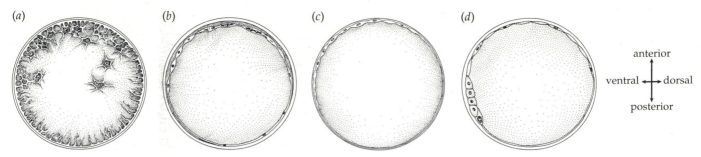

Figure 13.5   Cleavage in the mysid crustacean *Hemimysis lamornae*. (*a*) 12-cell embryo with blastomeres scattered through the yolk (dark). (The positions of the blastomeres have been projected onto a single plane in this picture.) Projections from the rim cytoplasm are seen entering the yolk. (*b*) Sagittal section through an embryo about the 64-cell stage, showing celll differences at the animal and vegetal poles. (*c*) At about the 128-cell stage, all blastomeres are external to the yolk. (*d*) The ventral thickening forms the germinal disc (the orientation is for (*d*) only). (After Manton 1928.)

Figure 13.6   Mesoderm formation in *Tetraclita rosea*. (*a*) Right lateral view of 31–33-cell embryo, with mesodermal migration advancing. (*b*) Transverse section of (*a*) at the point of the arrow in (*a*), showing the mesodermal 3A, 3B, and 3C blastomeres between the outer cells and the anterior midgut cell. (*c*) Transverse section through the caudal region of a slightly later-stage embryo. The anterior midgut cell has divided in half. Each of its descendants divides off a smaller posterior midgut cell to the outer surface. (*d*) Dorsal view of embryo with three anterior midgut cells and whose caudal papilla is filled with mesoderm. (*e*) Frontal section of (*d*) showing the relationships between the internal endoderm, external ectoderm, and the intermediately positioned mesoderm. (*f*) Presumptive fates of the *T. rosea* blastomeres mapped onto the 33-cell stage embryo. In this right lateral view, the anterior end is on the right. Dorsal side is upward. (From Anderson 1969.)

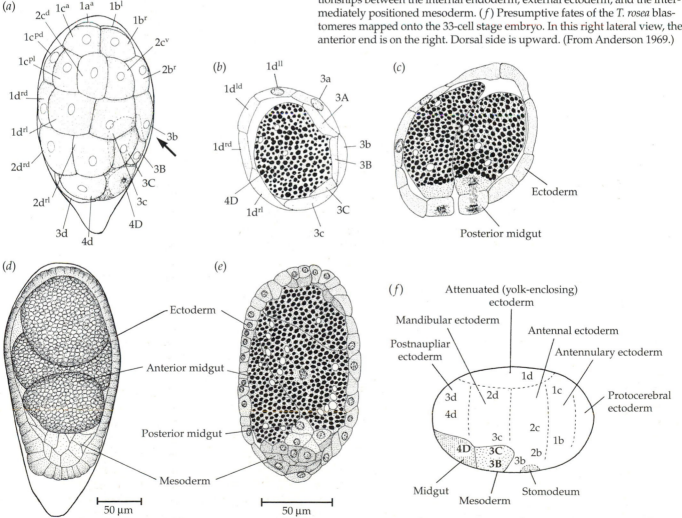

(a)   (b)

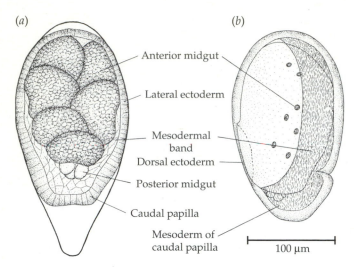

Figure 13.7 Mesodermal band formation in *Tetraclita rosea*. (a) Dorsal view of embryo with six anterior midgut cells. (b) Right lateral view of an embryo containing seven anterior midgut cells and whose mesodermal band extends from caudal papilla to the cephalic region. (From Anderson 1969.)

form the **anterior midgut rudiment**, while the smaller cells at the posterior ventral surface form the **posterior midgut rudiment** (Figure 13.6c). Shortly thereafter, the posterior midgut cells migrate into the embryo to lie immediately behind the mesoderm (Figure 13.6d,e).

The fates of the cells of the 33-cell embryo are diagrammed in Figure 13.6f. Organogenesis begins as the posterior mesodermal mass which has filled the **caudal papilla** proliferates cells that migrate anteriorly in bands, on either side of the anterior midgut (Anderson 1969). The bands eventually connect at the anterior of the embryo and grow ventrally until they meet at the ventral midline (Figure 13.7). Once this happens, cells divide and accumulate in three paired regions along the mesodermal bands anterior to the caudal papilla. Eventually, nearly all the mesodermal cells are either in one of these three thickenings or in the caudal papilla. The most anterior thickening gives rise to the **antennulary somite**, the middle thickening becomes the **antennal somite**, and the third thickening produces the **mandibular somite** (Figure 13.8). These somites generate the limb musculature of the nauplius larvae (Figure 13.9). At the posterior end of the mesoderm, two cells, one on each side, grow larger and become the **mesoteloblasts**. These cells will generate the mesodermal tissue for the trunk. The ectoderm of the trunk segments will be provided by the **ectoteloblasts** at the posterior end of the embryo.

These cells continue to divide while the other ectodermal cells differentiate and secrete their cuticle.

In crustaceans possessing ectoteloblasts, each round of asymmetric division of these precursor cells generates a single row of progeny. These single rows then divide to form one segment's worth of the body, a **genealogical unit**. Curiously, however, the boundaries of these segmental units do not correspond to the morphologically visible segments and are instead slightly out of register. Each genealogical unit comprises the posterior one-fourth of one segment and the anterior three-fourths of the next segment (Dohle 1976). Molecular studies have revealed that these genealogical units correspond to what are called parasegments in insects (Patel et al. 1989; Scholtz and Dohle 1996). In insects, these parasegments are not lineage units as in crustaceans, but instead have been identified on the basis of genetic and molecular criteria (Martinez-Arias and Lawrence 1985). This common parasegmental organization of the body plan during development is another feature that seems to argue for a close association of insects and crustaceans.

As the teloblasts make their geneological units, an eye develops on each side of the head. The anterior and posterior midgut cells collect into epithelial tubes and connect with each other to produce the gut which opens at both ends. However, there is very little yolk left in the anterior midgut cells, so the larvae will be released to acquire food.

In *Hemimysis*, the same nauplius condition is arrived at by a very different method (Manton 1928). The cells enclosing the yolk differ from each other, and certain regions are already committed to forming different portions of the larva (Figure 13.10). Most of gastrulation involves the epibolic streaming of cell layers between the germinal disc (the cells that cover the yolk at the end of cleavage) and the yolk. A blastoporal region forms in the **genital rudiment**. The cells migrating through the blastopore become the precursors of the endoderm and nearly all the mesoderm (including the mesoderm of the naupliar appendages). Although a small amount of extremely anterior mesoderm comes from the germinal disc, hardly any

(a)   (b)

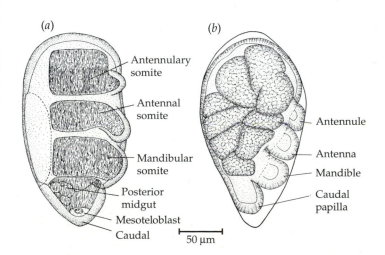

Figure 13.8 Formation of naupliar musculature. (a) Right lateral view of embryo in which mesodermal band condensations give rise to three somites. (b) Right lateral view of embryo in which the elongation of the naupliar limbs has begun. (From Anderson 1969.)

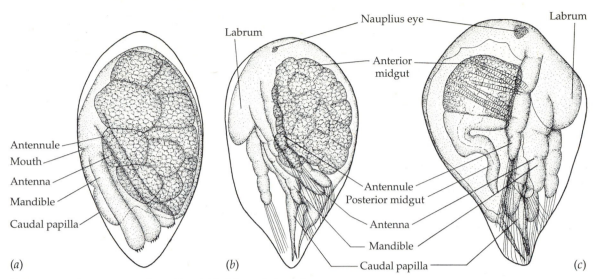

Figure 13.9 Construction of nauplius. (*a*) Left lateral view of embryo wherein naupliar arms have elongated in posterior dorsal direction and the secretion of cuticle has begun. The anterior midgut still contains numerous yolky deposits. (*b*) Right lateral view of embryo where secretion of cuticle is advanced, but the muscle movements have not begun. The anterior midgut is still yolky and the eye is poorly developed. (*c*) Embryonic development is complete and the nauplius is ready to hatch. The anterior midgut is nearly devoid of yolk and the eye is well developed. (From Anderson 1969.)

embryo, the endodermal cells surround the yolk, the mesodermal cells form their bands, caudal papilla, and associated mesoteloblasts, and the ectodermal teloblasts begin to form the thoracic and abdominal ectoderm. From here, a nauplius larva is formed.

other cells of the germinal disc form internal tissues. Most of the germinal disc forms the head and dorsal thoracic ectoderm. The remainder of the trunk ectoderm is formed from the ectodermal teloblasts, which migrate into the embryo through the blastopore (Figure 13.11). The ectoteloblasts form two rows of 15 cells each, and they divide synchronously to produce rows of ectoderm. The number of these ectodermal stem cells remains constant throughout ectoderm production. A row of eight mesoteloblasts moves forward under the ectoteloblasts, while head mesodermal cells continue to migrate into the embryo to form the head mesodermal bands. Once inside the

Figure 13.11 Gastrulation in *Hemimysis*. Cells migrate through the blastopore formed in the genital rudiment region. (*a*) Germ layers and blastopore region. (*b*) As gastrulation ensues, the genital rudiment has moved anteriorly and the head mesoderm cells, the mesodermal teloblasts (the precursors of the trunk mesoderm cells), and endoderm cells pass inwards. (*c*) Near the completion of gastrulation, only a few endoderm cells have yet to pass through the blastopore. The head mesoderm bands are forming, and the mesodermal teloblasts form the trunk mesoderm. Ectodermal teloblasts (which do not ingress into the embryo) form the trunk ectoderm and push the blastopore more anteriorly. en, endoderm; e1, e2, e3, new endoderm; E, ectodermal teloblast; G, genital rudiment; m, head mesoderm; m1, m2, new mesoderm; M, mesoderm teloblast. (After Manton 1928.)

Figure 13.10 Diagram showing the fates of various regions of the *Hemimysis* blastoderm in the area of the presumptive genital rudiment (where the blastopore will form). (After Manton 1928.)

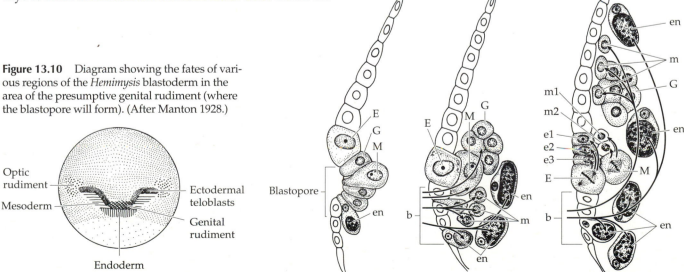

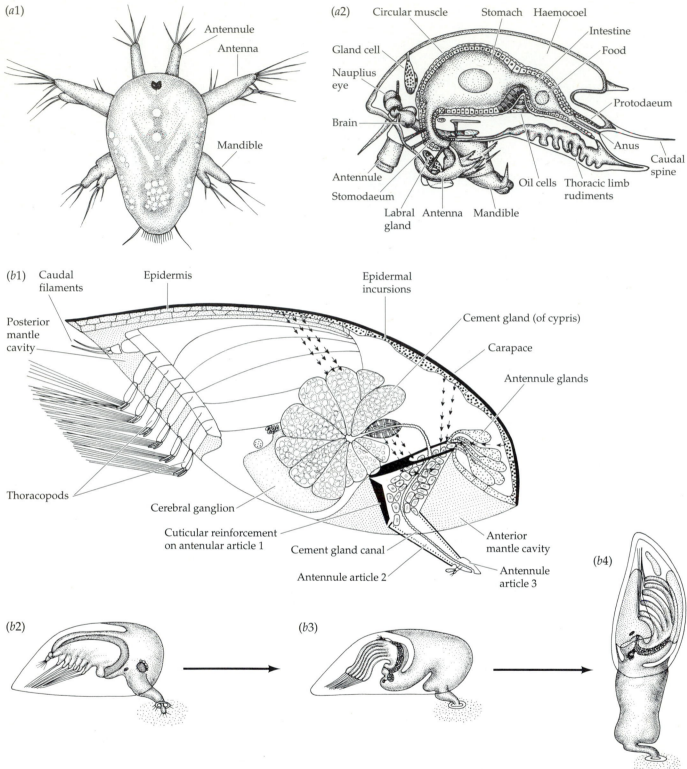

(a1)

Antennule
Antenna
Mandible

(a2)

Circular muscle   Stomach   Haemocoel
Gland cell
Nauplius eye
Brain
Antennule
Stomodaeum
Labral gland   Antenna   Mandible
Intestine
Food
Protodaeum
Anus
Caudal spine
Oil cells   Thoracic limb rudiments

(b1)

Caudal filaments   Epidermis
Posterior mantle cavity
Thoracopods
Cerebral ganglion
Cuticular reinforcement on antenular article 1
Cement gland canal
Antennule article 2
Epidermal incursions
Cement gland (of cypris)
Carapace
Antennule glands
Anterior mantle cavity
Antennule article 3

(b2)   (b3)   (b4)

Crustacean Larvae

The nauplius is the typical crustacean larva. It can be a feeding stage or can be passed over during a brood period. In certain rapidly developing species, the nauplius stage is skipped altogether. More often, the nauplius larva molts and is able to form a more specialized type of larva. There are many such types of larvae, and we will only

Figure 13.12   Crustacean larval types. (a1) Nauplius larvae; (a2) Internal anatomy of cirripede nauplius. (b1) Cyprid larva of rhizocephalan barnicle *Lernaeodiscus*. The arrows represent muscle fibers. (b2–b4) Cyprid larva of *Lepas* barnacle undergoing metamorphosis. (c) Zoea stage of the shrimp *Penaeus*. (d, e) Zoea and megalopa forms of the blue crab *Callinectes sapidus*. (After Brusca and Brusca 1990; a2 after Anderson 1973.)

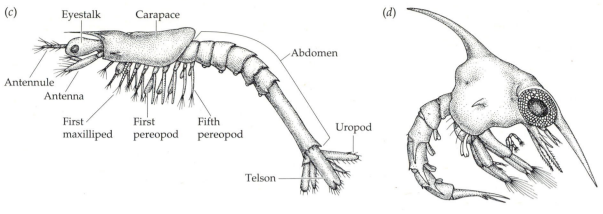

(c) Eyestalk Carapace Abdomen Antennule Antenna First maxilliped First pereopod Fifth pereopod Uropod Telson

(d)

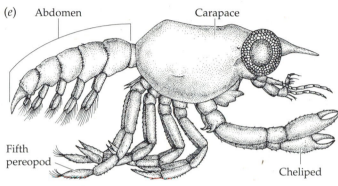

(e) Abdomen Carapace Fifth pereopod Cheliped

mention a few here. Some of these forms are outlined in Table 13.1. In most Maxillopoda (and in nearly all Branchiopods), the nauplius adds new segments with each molt, eventually forming a larva that no longer looks like a nauplius. While it still has only three pairs of appendages (which become the adult antennules, antennae, and mandibles), the **metanauplius larva** has a few more body somites and associated segments. In barnacles, a particular type of larva, the bivalved **cyprid larva**, emerges after the nauplius molts (Figure 13.12). In copepods, a small juvenile form, the **copepodite larva** comes from the nauplius. In certain shrimps, there is a cascade of larval types. Their nauplius larva molts to become a **protozoea** larva with its complete head appendages and its sessile eyes. After a few more molts, this form gives rise to the **zoea** larva, with its spiked helmet, stalked eyes, and three pairs of maxillipeds. The zoea generates a sexually immature juvenile form, the **postlarva** (sometimes called a **megalopa** or **glaucothöe**). There are obviously many ways to build a crustacean, and the multitude of blastula types, gastrulation processes, and larval forms has given rise to numerous debates over the phylogeny of these organisms.

Sex and Barnacle Life Cycles

Few humans have been as obsessed with crustacean life cycles as Charles Darwin. He spent so much time observing barnacles that his young children assumed that every adult man did the same. He noted that while barnacles were usually hermaphroditic, there were some fascinating exceptions. Several specimen of *Ibla cummingii* had tiny wormlike organisms in them. Darwin assumed these were parasites until he looked closer. The "barnacles" were all females. What he thought were "parasites" were the males. Darwin wrote to his mentor, Henslow, "the male, or sometimes two males, at the instant they cease being locomotive larvae become parasitic within the sac of the female, & thus fixed & half embedded in the flesh of their wives they pass their whole lives & can never move again." Darwin thought that such a domestic arrangement would certainly shock the pious naturalists who saw a moral in everything nature did, and he kidded Lyell about the pocket in the female shell in which "she kept a little husband." In a similar species, Darwin found the male to be merely a rudimentary head atop an "enormous coiled penis." Darwin also described transitions between the hermaphrodite and the "parasitic male" form, wherein the male was supplemental to the hermaphroditic adult. He wrote Hooker of his discovery of these males wherein the larva "fixes itself on the hermaphrodite [and] develops itself into a great testis." Indeed, they became little more than "mere bags of spermatozoa" (Darwin 1854; Desmond and Moore 1991). Other species of barnacles have evolved other elaborate ways of getting the sperm to the egg. In species where males and females may live apart from one another, the male can send out a "groping penis" which moves about the colony, eventually entering a female barnacle and depositing its sperm (Figure 13.13).

The rhizocephalan barnacles are parasitic upon other crustaceans, and their life cycles are certainly fantastic. The adult *Peltogaster paguri*, sitting atop a hermit crab, produces numerous broods of male and female nauplius larvae

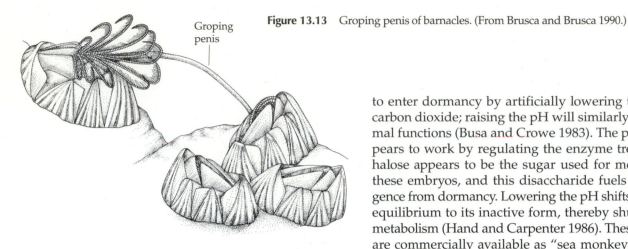

Groping penis

Figure 13.13 Groping penis of barnacles. (From Brusca and Brusca 1990.)

(Høeg and Lutzen 1985; Glenner and Høeg 1995; Figure 13.14). These become cyprid larvae that are competent to settle. Female cyprid larvae land on the limbs or thorax of host crabs and undergo a metamorphosis to the **kentrogon form** of the larva which develops a pair of antennules and an **insertion stylet**. The body of the larva goes *through* the stylet into the body of the crab. Eventually, the female emerges as an adult on the abdomen of its host. Here, it is called an **externa**. When the externa matures and acquires a mantle pore, it becomes attractive to male cyprid larvae, which settle within the aperture. Here, the male cyprid larvae develop into the **trichogon** form and implant part of their body contents into the female's receptacles. This deposit differentiates into spermatozoa, which fertilize the female's eggs.*

Environmental Regulation of Crustacean Development

There are numerous instances in the development of crustacean species where the alterations in development may be crucial to survival. In such instances, the animal has evolved a plasticity that enables the environment to modify its development to the more advantageous phenotype. One well-known example is that of the brine shrimp *Artemia salina*. The danger of being a brine shrimp is that one's salty environment is in danger of evaporating. *Artemia* has evolved mechanisms of encysting itself as a gastrula and remaining developmentally and metabolically dormant until it becomes rehydrated. This ability to suspend life functions is called **cryptobiosis**. The cue for entering dormancy appears to be the decrease of intracellular pH by more than a unit (from a pH around 7.9 to one around 6.3). *Artemia* gastrulae can be artificially induced

to enter dormancy by artificially lowering the pH with carbon dioxide; raising the pH will similarly restore normal functions (Busa and Crowe 1983). The pH signal appears to work by regulating the enzyme trehalase. Trehalose appears to be the sugar used for metabolism in these embryos, and this disaccharide fuels the reemergence from dormancy. Lowering the pH shifts the enzyme equilibrium to its inactive form, thereby shutting down metabolism (Hand and Carpenter 1986). These organisms are commercially available as "sea monkeys" or "space monkeys" and develop within a day into nauplius larvae.

The "water flea" *Daphnia* has evolved another developmental strategy to deal with unfavorable conditions. Here the unfavorable condition is the presence of predators. *Daphnia* are clonal populations. Each *Daphnia* female (and there are no other sexes in *Daphnia*) produces diploid eggs (releasing only one polar body) that are self-activated. But *Daphnia* juveniles are preyed upon by the large larvae of the phantom midge *Chaoborus* (Figure 13.15). *Daphnia* juveniles can respond to chemicals produced by *Chaoborus* larvae by developing spiked heads, neck teeth, and long tails. These changes in development appear to favor their surviving attempts by *Chaoborus* larvae to feed on them (Dodson 1989; Lampert et al. 1994). Such chemicals released by predators are termed *kairomones*. The environment is not merely a passive player in animal development. The human immune system and nervous system are other examples of this ability of development to be altered by environmental conditions (See Gilbert 1997; van der Weele 1995).

Chelicerata

The Chelicerates are a class of arthropods that includes horseshoe crabs, sea spiders, and arachnids. Spiders, along with scorpions, ticks, mites, and daddy long-legs comprise the *Arachnid* subclass of the Chelicerata. This section will focus on spider development. Spiders comprise the order Araneae and have two main body parts, the anterior **cephalothorax** (or **prosoma**) and the posterior abdomen (the **opisthosoma**). These are connected by the **pedicel**. Reproduction in spiders is a hazardous undertaking.[†] The male is usually much smaller than the female

---

* The kentrogon is one of very few larval forms whose functions have been sung in heroic couplets. Walter Garstang's poem "Kentrogon" concludes with the following (c. 1920, in Garstang 1985):

> So *Kentrogon*, like Charon, carries souls from light to dark,
> Himself at once both ferryman and passenger and bark,—
> A Phoenix all perverse who on his desperate day of doom
> Refanned his flame to start afresh within a living tomb.

[†] The black widow spider has a bad reputation, but usually the male mates and survives the experience. The males of lesser known species, such as *Araneus pallidus* or *Cyrtophora cicatrosa*, are not as lucky. The geometry of courtship makes it impossible for a male of these species to remain in position to mate, so the female eats his abdomen so that his pedipalp-containing head and thorax won't fall off her (Grasshoff 1964; Foelix 1982). In other species, the male spider must use its chelicera to prop open the female's jaw during mating.

TABLE 13.1   *Forms of development among crustaceans*

Taxon	Development type or larval type at time of hatching	Comments
Remipedia	?	Development not yet studied.
Cephalocarida	Metanauplius	Two eggs at a time fertilized and carried on genital processes.
Branchiopoda		
Anostraca	Nauplius or metanauplius	Embryos usuall shed from ovisac early in development; resistant (cryptobiotic) fertilized eggs accommodate unfavorable conditions.
Notostraca	Nauplius or metanauplius	Eggs brooded briefly, deposited on substrate; resistant (cryptobiotic) fertilized eggs accommodate unfavorable conditions.
Cladocera	Direct development, nauplius, metanauplius	All but *Leptodora* undergo direct development; *Leptodora* hatches as nauplius or metanauplii
Conchostraca	Nauplius or metanauplius	Embryos carried on thoracopods, then released.
Maxillopoda		
Ostracoda	Nauplius	Embryos usually deposited directly on substrate; many myodocopans and some podocopans brood embryos between valves until hatching as a reduced adult; usually 6 naupliar stages to juvenile.
Mystacocarida	Nauplius or metanauplius	Little is known about this group; eggs apparently laid free; 6 naupliar stages (?).
Copepoda	Nauplius	Usually 6 naupliar stages leading to a second series of 5 "larval" stages (copepodites).
Branchiura	Direct development; nauplius	Embryos deposited; only *Argulus* known to hatch as nauplii; others have direct development and hatch as juveniles.
Cirripedia	Nauplius	Six naupliar stages followed by unique larval form, the cyprid larva.
Tantulocarida	?	Development entails metamorphosis.
Malacostraca		
Phyllocarida	Direct development	All undergo direct development in female brood pouch, hatching as postlarval mancas (subjuveniles).
Eumalacostraca		
Hoplocarida	Antizoea or pseudozoea	Eggs brooded or deposited in burrow; hatch late as clawed pseudozoea larva, or earlier as an unclawed antizoea larva; both go through several molts before settling as juveniles.
Syncarida	Direct development	Free larval stages lost; eggs deposited on substrate.
Peracarida	Direct development	Embryos brooded in marsupium (brood pouch); usually released as mancas (subjuveniles with incompletely developed eighth thoracopods and differences in body proportions and pigmentation).
Eucarida		
Euphausiacea	Direct development	Embryos shed or briefly brooded; typically undergo nauplius → zoea → megalopa → juvenile → adult transition series.
Amphionidacea	Nauplius	Apparently brooded under thorax, but held by anterior pleopods; typically undergo nauplius → zoea → megalopa → juvenile → adult transition series.
Decapoda	Zoea; protozoea; nauplius	Dendrobranchiata shed embryos to hatch in water as nauplii or protozoea; all others brood embryos (on pleopods) that do not hatch until at least the zoea stage.

Source: Adapted from Brusca and Brusca 1990.

(a)

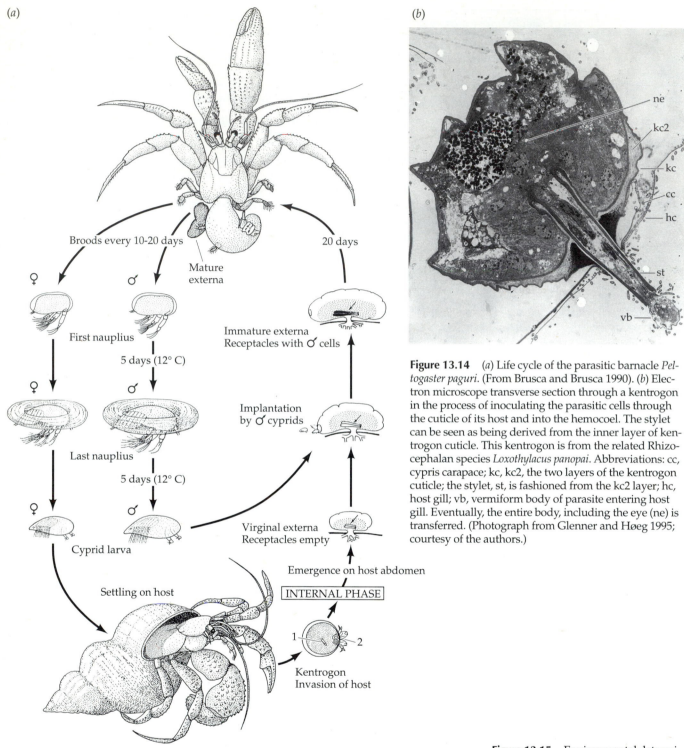

Broods every 10-20 days

20 days

Mature externa

♀ ♂

First nauplius

5 days (12° C)

♀ ♂

Last nauplius

5 days (12° C)

♀ ♂

Cyprid larva

Settling on host

INTERNAL PHASE

1 2

Kentrogon
Invasion of host

Emergence on host abdomen

Virginal externa
Receptacles empty

Implantation
by ♂ cyprids

Immature externa
Receptacles with ♂ cells

(b)

ne

kc2

kc

cc

hc

st

vb

Figure 13.14   (a) Life cycle of the parasitic barnacle *Peltogaster paguri*. (From Brusca and Brusca 1990). (b) Electron microscope transverse section through a kentrogon in the process of inoculating the parasitic cells through the cuticle of its host and into the hemocoel. The stylet can be seen as being derived from the inner layer of kentrogon cuticle. This kentrogon is from the related Rhizocephalan species *Loxothylacus panopai*. Abbreviations: cc, cypris carapace; kc, kc2, the two layers of the kentrogon cuticle; the stylet, st, is fashioned from the kc2 layer; hc, host gill; vb, vermiform body of parasite entering host gill. Eventually, the entire body, including the eye (ne) is transferred. (Photograph from Glenner and Høeg 1995; courtesy of the authors.)

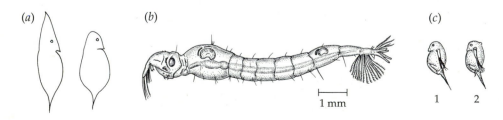

(a)   (b)   (c)

1 mm

1   2

Figure 13.15   Environmental determination of juvenile form in *Daphnia*. (a) Extremes of head and tail morphologies seen naturally in Daphnia populations by Woltereck in 1908. (b) The predatory larva of *Chaoborus americanus*. (c) Juveniles of *Daphnia pulex*, which respond to larval secretions by forming "neck teeth." (c1) Form found in control pond water; (c2) form found when pond water had contained *Chaoborus* larvae. (After Dodson, 1989).

and risks being prey. Thus, there is often a complex courtship wherein the male and female consent to reproduce rather than forage. The appendages of the mature male spider include a palp with a specialized tarsal segment. Palps are similar to legs although they are used for nonlocomotor functions. On the tarsus of the male palp is a pear-shaped bulb that has a duct that opens at the tip. A male spider will often make a **sperm web** and press his abdomen against it to expel a drop of sperm from his genital opening. He then collects the sperm in the tarsal bulbs (Figure 13.16). The bulb collects sperm into this duct and stores it while the male spider searches for a mate. (One of the distinguishing characteristics between spiders and daddy long legs is that the males of the latter group have

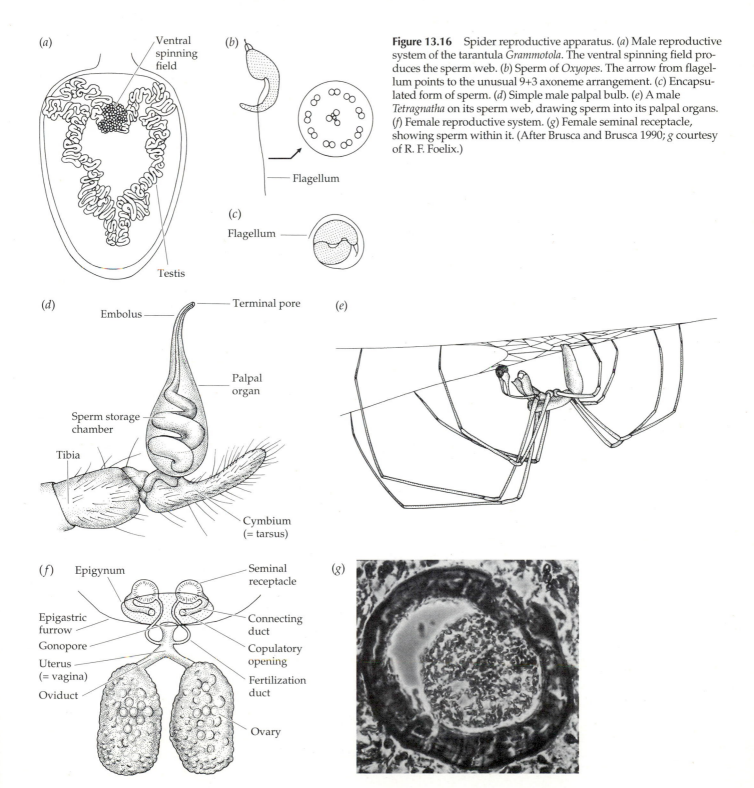

Figure 13.16 Spider reproductive apparatus. (*a*) Male reproductive system of the tarantula *Grammotola*. The ventral spinning field produces the sperm web. (*b*) Sperm of *Oxyopes*. The arrow from flagellum points to the unusual 9+3 axoneme arrangement. (*c*) Encapsulated form of sperm. (*d*) Simple male palpal bulb. (*e*) A male *Tetragnatha* on its sperm web, drawing sperm into its palpal organs. (*f*) Female reproductive system. (*g*) Female seminal receptacle, showing sperm within it. (After Brusca and Brusca 1990; *g* courtesy of R. F. Foelix.)

penes, as well as a divided prosoma.) During copulation, the male will put this palp into the **gonopore** of the female and expel the sperm into the female's **sperm receptacles**. Female spiders can store the sperm in receptacles and use it to fertilize several broods. (However, if the female molts, her receptacle molts, too, and she has to mate again.)

Females often don't lay their eggs until a week or so after copulation. Fertilization takes place just before the eggs are deposited, the sperm cells being released from the receptacle as the eggs pass through the uterus. The fertilized eggs are then squeezed out of the genital opening and become surrounded by a viscous liquid. This liquid will harden, cementing the eggs together. These eggs are protected by a silken cocoon, maintaining a moist environment for the developing embryos.

Spider development is divided into three periods (Vachon 1958; Table 13.2). The **embryonic period** encompasses development from fertilization until the spider obtains its morphological shape. The **larval period** encompasses the prelarval and larval stages. The **prelarva** that emerges has no claws or functional legs, and its reproductive organs are undeveloped. After a few more molts, however, the spider becomes a **larva** in which there is some mobili-

ty, an undifferentiated claw, and the beginnings of hair and spines. The larva still lacks reproductive organs. In the nympho-adult period, the **nymph** is a juvenile spider, complete in all but its reproductive organs. When these organs mature, the spider is considered adult (**imago**).

Fertilization and Cleavage

Spider eggs are encircled by two layers, the inner **vitelline envelope** and the outer **chorion**. Immediately beneath the vitelline envelope is a layer of relatively yolk-free cytoplasm, the **cortical layer**. This cortical cytoplasm is partitioned into patches, probably as a result of pressure from the oviduct (MacBride 1914). The large mass of deep cytoplasm is full of yolk, and processes from the cortical layer extend through it to the center of the egg wherein the female pronucleus resides. The fusion of egg and sperm pronuclei occurs within a few hours after the egg has been deposited (Montgomery 1908). The details of development differ widely between spider species, but we will use as an example the development of *Agelena labyrinthica* (Kishinouye 1891; Holm 1952). First nuclear division begins at about 6 hours after egg deposition, and the nuclei divide within a relatively yolk-free cytoplasm near the

TABLE 13.2 Developmental stages of spiders

Characteristic	Embryonic Not self-sufficient (yolk as food supply)	Larval Not self-sufficient (yolk as food supply) Prelarval	Larval	Nymph-Imaginal Self-sufficient (prey as food supply) Nymph	Adult
Mobility	None	None	Very little	Complete	Complete
Leg segmentation	Incipient	Incomplete	Complete	Complete	Complete
Hairs, spines	Neither	Neither	Both, but little differentiation	Large variety of both	Large variety of both
Claws	None	None	Simple	Differentiated	Differentiated
Cheliceral claw	None	Undifferentiated	Differentiated, no poison canal	Differentiated, poison canal	Differentiated, poison canal
Spinnerets	None	Barely differentiated	Differentiated, but without spigots	Functional	Functional
Reproductive organs	None	Undeveloped	Undeveloped	Developed but nonfunctional	Functional

Embryonic   Prelarval   Larval   Nymphal

Source: Adapted from Vachon 1958; Foelix 1982.

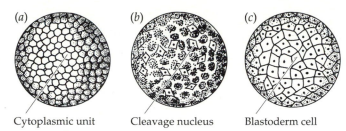

Cytoplasmic unit   Cleavage nucleus   Blastoderm cell

Figure 13.17   Early cleavage stages of *Agelena*. (*a*) External cytoplasm broken up into polygonal units. (*b*) Cells begin to migrate to periphery. (*c*) Cells form cellular blastoderm at periphery. (After Sekiguchi 1968.)

center of the yolky mass. The nuclei divide without forming cells. As they divide, however, the nuclei separate the yolky cytoplasm into cell-free columns. The nuclei migrate toward the periphery, and by the 64-cell stage, most all the nuclei can be found in the cortical cytoplasm (Figure 13.17). Complete cell membranes have not formed at this time, but after a few more cell divisions, each nucleus has coalesced with four to eight of the cytoplasmic patches to create large flattened cells (Holm 1952). Thus, by about 35 hours, a thin **blastoderm** has been created by superficial cleavage. This blastoderm encircles the yolk. At this point, the yolk sinks to the bottom portion of the blastula (probably defined by gravity), leaving a space, the **blastocoel**, which is filled with perivitelline fluid. Some of the nuclei enter into the yolky cytoplasm (either from the early blastoderm or possibly from being "left behind" during blastoderm formation) and initiate the formation of **yolk cells**.

Gastrulation and Inversion

Shortly after the formation of the large blastocoel, the blastoderm contracts. The upper portion of the blastoderm separates itself from the vitelline envelope and sinks down onto the yolk. (The vitelline fluid appears to pass between the cells and now becomes located between the flattened yolk-filled blastoderm and the vitelline envelope). The cells that had been at the upper portion of the embryo and that are now lying on the yolk begin to differentiate. These cells will become the ventral surface of the larva. Some of these

ventral cells aggregate to form a small diffuse **germinal disc**, which becomes full of small rounded cells. At about 85 hours after egg laying, a small indentation, the **primitive groove** (or **blastopore**), appears in the center of this disc. This is the space through which the mesodermal and endodermal precursors will migrate. As migration continues, the underlying cells spread over the upper surface of the yolk mass. The area containing this spreading mass of mesodermal and endodermal cells can be seen as a raised process and is called the **primitive plate**.

The primitive plate is originally flat and coherent, but one end of it buds off to become a secondary thickening, the **cumulus** (Figure 13.18). Cells in the cumulus are thought to be endodermal precursors and the precursors of the germ cells, but this interpretation remains controversial. The cumulus migrates to the margin of the germinal disc. By this time, cell division and migration have made the primitive plate several cell layers thick. The plate then expands to cover the region traversed by the cumulus, about 180 degrees of the developing embryo. Shortly thereafter, the mesoderm forms into a **germinal band**. This band develops aggregations of cells that form the **cephalic lobe**, the **caudal lobe**, and the five intervening pairs **somites**. These somites represent the precursors of the eight legs and two pedipalps. The cephalic lobe divides to form a cheliceral segment, and the caudal lobe splits to form the first abdominal segment (Figure 13.19).

Figure 13.19   Germinal band of the spider *Cupiennius salei* (about 130 hr) before inversion. (*a*) Prior to inversion, mesoderm has formed clear segmental somites. On the prosoma side, the extremeties have begun extension. (*b*) Longitudinal section of a slightly older embryo. Note the metameric coelomic cavities (dotted lines) that will form the musculature. (After Foelix 1982.)

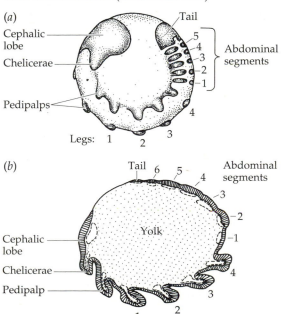

Figure 13.18   Surface views of *Agelena* embryo showing the primitive streak and the formation of the cumulus. (After MacBride 1914.)

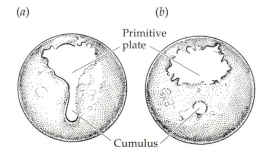

The abdominal segments are added sequentially to the posterior of the embryo. Internally, the coelomic cavities are formed by the separation of the mesoderm, and the walls of this cavity will eventually differentiate into muscles (Foelix 1982).

When the germinal band has expanded to the point that the caudal and cephalic lobes almost touch each other, a furrow forms that divides the embryo into right and left parts connected at the posterior point. The two sides move laterally until they finally meet in the dorsal midline (Figure 13.20). In this way, the originally convex germinal band has turned itself around (**inversion**). Meanwhile, in the cephalothorax, the anlagen of the pedipalps and legs divide into their seven segments. A slight constriction in the first abdominal segment delineates the border between the prosoma and the opisthosoma, and invaginations of the second and third abdominal segment begin differentiating into the book lungs and tracheae, respectively. Spinnerets begin to form on abdominal segments 4 and 5.

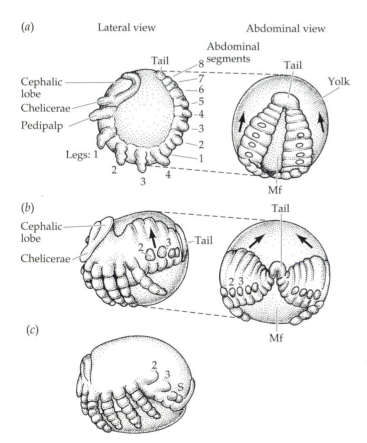

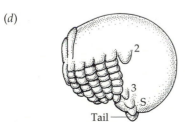

Organogenesis

Most organs develop from the three germ layers during the larval period. The ectoderm forms the integument, sensory organs, nervous system, silk glands, poison glands, book lungs, and trachea, as well as the foregut and hindgut. The endoderm cells form the midgut and the Malpighian tubules. The mesoderm cells differentiate into the reproductive organs, circulatory system, coxal glands, and muscles. Where the muscle walls reach the dorsal midline, they form the heart in the abdominal region and the dorsal aorta, continuous with the heart, in the cephalothorax. The nervous system originates from the invagination of the ectoderm, which forms the **cerebral grooves**. These grooves become deeper, and the walls thicken with neural cells to form the **supraesophageal ganglia** (colloquially called the brain). Other ectodermal thickenings appear as paired outgrowths on either side of the ventral median, and they separate from the ectodermal wall to join together to form the ventral nerve chain. As the segments fuse with one another shortly before hatching, the abdominal ganglia are pushed into the prosoma where they form the **subesophageal** ganglion. Spiders generally have a pair of central eyes and three pairs of lateral eyes. The central eyes emerge from invaginations of the ectoderm, which fold upon each other to become a lens and retina (Figure 13.21). The lateral eyes develop from ectodermal outpockets in the anterior dorsolateral ectoderm. The bottom of each outgrowth differentiates directly into a retina. At the base of the larval pedipalps, special structures called cuticular denticles or "egg teeth," help rip through the chorion sheath of the egg. At this time, the larval spiders molt for the first time (Holm 1940). Growth of spiders occurs only during each molt. While the cuticle is on, the spider is prevented from growing.

Myriapoda

Cleavage

In this section, we will concentrate on the development of a common European centipede, *Scolopendra cingulata*. The detailed anatomy of this myriapod was published in 1901 by Heymons, and little has been added since then (see Johannsen and Butt 1941). The eggs of this species are laid in June, about 5 cm into the soil, in clumps of around 20

Figure 13.20   Late embryonic development in *Agelena*. (*a*) Beginning of inversion. Germinal band splits along the midline (Mf) and each part moves laterally. (*b*) Anlagen of external extremities appear on abdominal segments 2 and 3. (*c*) Lateral body walls have merged at the dorsal midline. What had been lateral is now medial. The legs have added another segment. (*d*) At the completion of inversion (about 15 days after egg deposition), the embryo has covered the yolk mass, the prosoma and opisthosoma have flexed toward each other, and the abdominal appendages have moved apart and have differentiated into book lungs and tracheae (abdominal segments 2 and 3) and spinnerets (S, abdominal segments 4, 5). (After Foelix 1982.)

(a)

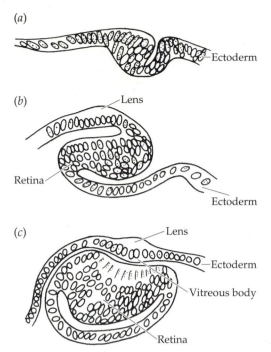

Ectoderm

(b)

Lens

Retina

Ectoderm

(c)

Lens

Ectoderm

Vitreous body

Retina

Figure 13.21 Development of the central eye in *Agelena*. The ectoderm involutes and the coils interact with one another to create the lens and the retina. (After Sekiguchi 1968.)

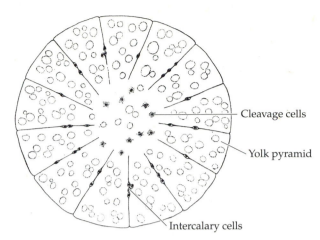

Cleavage cells

Yolk pyramid

Intercalary cells

Figure 13.22 Medial section through the yolky egg of the centipede *Scolopendra*. The "intercalary cells" are migrating towards the egg periphery. Some internal cleavage cells are also observed. Their function is not determined, but they may act to partition the yolk. (From Heymons 1901.)

ovoid eggs. The female curls about them until they are ready to hatch. The unfertilized egg is almost entirely full of yolk, with little or no peripheral periplasm. The zygote nucleus appears to be surrounded by a small volume of yolk-free cytoplasm, and each nuclear division appears to divide the yolk into discrete regions. However, these **"yolk pyramids"** are not membrane-bound and do not represent cells. Some of the nuclei appear to move through the channels caused by the partitioning of the yolk and migrate towards the periphery (Figure 13.22). Division

takes place both during this migration and at the cell periphery. The blastoderm eventually surrounds the entire egg, as single cells meander into intervening spaces on the periphery to multiply. Thus, the blastoderm of myriapod eggs seems to arise from isolated cells that have migrated from the central region of the cell to the edge of the oocyte cytoplasm.

Gastrulation and Germ Band Formation

The cell proliferation at one region on the presumptive ventral surface is more active than anywhere else, and this gives rise to a **germ disc**. Continuing cell proliferation makes the germ disc several cell layers thick and it is sometimes referred to as **embryonic rudiment.** It appears to be similar to the primitive streak in spiders, but it is sometimes referred to as the cumulus. (Figure 13.23). In

Figure 13.23 Formation of the embryonic rudiment in *Scolopendra*. (a) Cellularization of the peripheral layer of the cytoplasm. (b) Section through the early embryonic rudiment. (c) Initial expansion of the rudiment in a manner similar to that of the cheliceran primitive streak. (From Heymons 1901.)

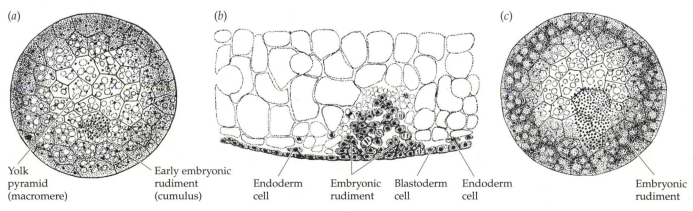

(a)

Yolk pyramid (macromere)

Early embryonic rudiment (cumulus)

(b)

Endoderm cell

Embryonic rudiment

Blastoderm cell

Endoderm cell

(c)

Embryonic rudiment

Figure 13.24   Mesodermal formation in *Scolopendra cingulata* at successive stages (*a*) and (*b*). Abbreviations: bc, blood cell; ec, endoderm cell; mes, mesoderm; md, membrana dorsalis; mv, membrana ventralis; y, yolk; yc, yolk cell. (From Heymons 1901.)

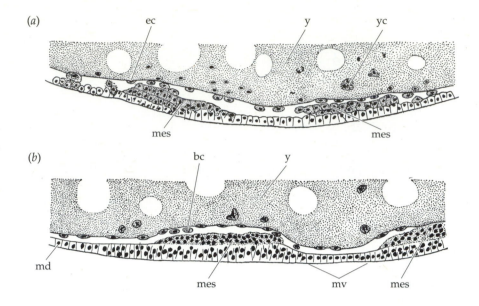

addition to this embryonic rudiment, other cells are being liberated individually into the yolk. These cells will form the "yolk cells" that enter into the central portion of the cytoplasm. The endoderm is formed in a similar manner, with cells being liberated from numerous places on the ventral surface of the egg. However, these cells do not enter into the yolky mass but stay on the outside, between the blastoderm and the yolk. They form a flattened layer along the yolk. The next cells liberated from the embryonic rudiment are the mesesenchyme cells. Meanwhile, the cells on the inner side of the embryonic rudiment migrate anteriorly as two lateral bands of mesodermal cells (Figure 13.24). The germ disc area now becomes the posterior region where both mesodermal bands meet.

Segmentation

The extension of the germ bands continue until they have reached the anterior of the egg, where they come together. This future anterior region will become differentiated into the mouth (at the medial line where mesoderm is lacking), the labrum, the preantennae, and antennae (Figure 13.25). The other segments form the jaw and legs. Near the posterior is a proliferation zone where further cell division cre-

Figure 13.25   Germ band extension and flexion in *Scolopendra*. (*a*) The germ band divided down the midline, connected only at the posterior and anterior ends. (*b*) A slightly later stage seen laterally. The anterior (acron, preantenna, antenna) segments and the posterior (telson) segment have come together. (*c*) The furrow is beginning to form in the ventral membrane. (*d*) Formation of the yolk furrow through the ventral membrane. Abbreviations: ant, antenna; cly, clypeus; lp, leg primordium; lr, labrum; mn, mandible; mpd, maxilliped; md, membrana dorsalis; mv, membrana ventralis; mx, maxillary segment; pren, preantenna; prz, proliferation zone; ster, sternite; tel, telson; ter, tergite. (From Heymons 1901.)

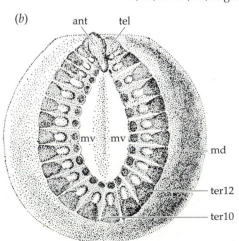

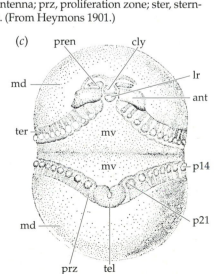

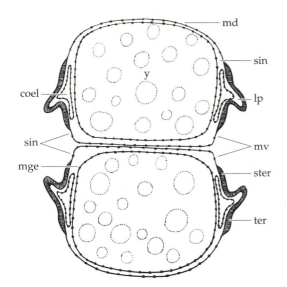

Figure 13.26 Cross section of the *Scolopendra* embryo shortly after flexure of the germ band. The embryo has "doubled over" on itself, so two segments are seen. Abbreviations: coel, coelom; lp, leg primordium; mge, midgut epithelium; md, membrana dorsalis; mv, membrana ventralis; sin, lateral blood sinus; ster, sternite; ter, tergite; y, yolk. (From Heymons 1901.)

the head and telson and by the two ectodermal bands (the ventral and dorsal membranes).

During this time, the ectodermal anlagen of the appendages begin to differentiate. The antennal segment develops preorally. The unpaired labrum develops as a lobe that overhangs the mouth. The preantennal segment (which is pushed anteriorly by the forward migration of the antennal segment) also elongates. The trunk appendages begin to form; the interior (mesal) portion becoming the **sternite**, the distal portion becoming the **tergite**, the part between them becoming the **leg bud** (Figure 13.26).

While the majority of mesodermal cells have become cuboidal and have attached themelves to the ectoderm, others, more flattened, have become associated with the endoderm. These latter cells will form the **visceral wall**, while the former cells will form the **somatic wall**. At first, these cells are in contact, but when these two layers separate, they create the **coelomic cavity**. A pair of these are formed in the maxilliped and in each trunk segment, six are formed in the head, and the telson and acron (the most anterior segment) lack them (Figure 13.27).

The initial coelom may be seen as being composed of three parts: a central part whose cavity extends into the developing appendage, a ventral portion beneath the sternite, and a dorsally extending portion beneath the tergite. Cells from the somatic mesodermal wall proliferate to

ates new cells for the mesodermal bands, and the posterior of the egg becomes the telson. The embryo curves such that the telson actually touches the anterior segments. The germ bands begin to separate, one on each lateral side, and the ectodermal strip between them, the **membrana ventralis**, becomes more pronounced. As the membrane ventralis is stretched, a transverse furrow forms in it, cutting through the yolk, and dividing the embryo into two parts, linked only by a thin dorsal bridge of blastoderm cells (the **membrana dorsalis**). The final embryonic form of this species contains (in addition to the head segments) two maxilliary segments, one maxipedal segment, 21 rump segments, and a large telson segment that bears the anus. The halves of the germ band are connected only at

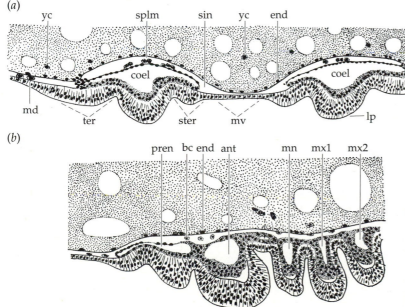

Figure 13.27 Development of the coelomic cavity in (*a*) the trunk segments and (*b*) in the head. Abbreviations: ant, antenna; bc, blood cell; coel, coelom; end, endoderm; lp, leg primordium; mn, mandible; md, membrana dorsalis; mv, membrana ventralis; mx, maxillary segment; pren, preantenna; sin, blood sinus; splm, splanchnic mesoderm; ster, sternite; ter, tergite; yc, yolk cell. (From Heymons 1901.)

form the longitudinal muscles and the musculature of the appendage. Cells of the visceral (splanchnic) mesoderm lining the gut endoderm will form the midgut muscular layer (Figure 13.28). At each lateroventral angle, the yolk sac becomes separated from the body wall, forming a **lateral blood sinus**. This is the anlagen of the definitive body cavity, the **schizocoel**. The blood cells are thought to come from splanchnic mesoderm. The schizocoelom expands dorsally and ventrally, while its central portion becomes limb muscle cells, fat cells, muscle strands, and peritoneum. At the dorsal junction of the visceral and somatic mesoderm, heart-forming **cardioblasts** appear. Germ cells are harbored in the dorsalmost region of the coelomic sacs.

The ectoderm differentiates in several ways. Along the hypodermis (skin), nine invaginations appear (in segments 3, 5, 8, 12, 14, 16, 18, and 20), which give rise to the primary tracheal trunks. The ventral nerve cord originates from a pair of longitudinal thickenings several cell layers deep, at the base of the sternite. Some of these ectodermal cells will differentiate into neural tissue. These paired nerves remain separate until after the first molt, when they are brought together at the ventral midline. In each segment, a ganglion arises from the neural masses, and lateral neurons project from them. Eventually the ventral nerve cord will consist of 24 pairs of ganglia, from the maxillipede segment to the pregenital and genital segments. The **subesophageal ganglion** of the head results from a fusion of three ganglia into one. The **supraesophageal ganglion** (the "brain") has three parts (forebrain, midbrain, and hindbrain) and is composed of four original ganglia. One of the major differences between the myriapods on one hand, and the insects and crustaceans on the other, is that the nervous system of myriapods does not form from neuroblasts (a dividing stem cell population of neural precursors). Rather, the earliest axonal pathways in myriapods appear to come from the posterior growth of axons originating in the brain (Whitington et al. 1991).

The alimentary canal is formed primarily by those endodermal cells that originally lined the ventral portion of the yolk. Before the coeloms form, these cells have spread over the entire yolk mass. The **stomodeum** (mouth) and **proctodeum** (anus) form from inpocketings of the ectoderm. Most of the yolk is used up shortly before the first molt, and the opening of the stomodeum and the proctodeum (the fusion of the ectodermal inpockets with the endoderm in these areas, creating a fuctional alimentary canal) occurs shortly after the first molt. The chorion ruptures at this point, and the organism must feed. After another molt, the limbs will be fully formed and the centipede will be able to move. After several more molts, the reproductive organs mature, and the centipede can reproduce a new generation.

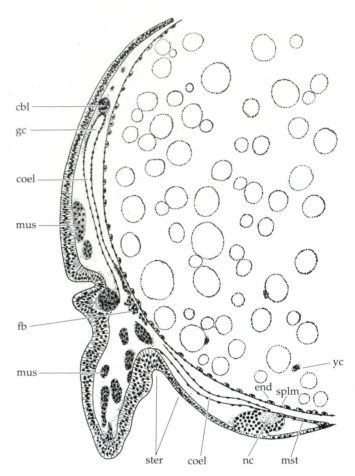

Figure 13.28 Cross section through the left half of a trunk segment of *Scolopendra cingulata*. Abbreviations: cbl, cardioblasts; coel, coelom; end, endoderm; fb, fat body; gc, genital coelom; mst, middle strand; mus, muscle; nc, nerve cord; splm, splanchnic mesoderm; ster, sternite; yc, yolk cell. (From Heymons 1901.)

Summary

The arthropods comprise the largest group of organisms on earth, yet we know very little of their embryology. Of the millions of arthropods on this planet, there is only one whose embryology is known in detail—the fruit fly *Drosophila melanogaster*. Realizing that our understanding of its development is like a pillar in a desert, an island in an ocean of relative ignorance, we turn our attention in the next chapter to the insects, particularly the fruit fly.

Acknowledgments

I would like to thank Nipam Patel, Sean Carroll, Tom Valente, Peter Smallwood, and Alan Harvey for their helpful discussions on arthropod lore.

Literature Cited

Anderson, D. T. 1969. On the embryology of the cirripede crustaceans *Tetraclita rosea* (Krauss), *Tetraclita purpurascens* (Wood), *Chthamalus antennatus* (Darwin) and *Chamaesipho columna* (Spengler) and some considerations of crustacean phylogenetic relationships. *Phil. Trans. Roy. Soc. Lond.* B 256: 183–235.

Anderson, D. T. 1973. *Embryology and Phylogeny in Annelids and Arthropods.* Pergamon Press, Oxford.

Anderson, D. T. 1994. *Barnacles: Structure, Function, Development, and Evolution.* Chapman and Hall, London.

Ballard, J. W. O., G. J. Olson, D. P. Faith, W. A. Odgers, D. M. Rowell, and P. W. Atkinson. 1992. Evidence from 12S ribosomal RNA sequences that onychophorans are modified arthropods. *Science* 258: 1345–1348.

Boore, J. L., T. M. Collins, D. Stanton, L. L. Daehler and W. M. Brown. 1995. Deducing the pattern of arthropod phylogeny from mitochondrial gene rearrangements. *Nature* 376: 163–165.

Brusca, R. C. and G. J. Brusca. 1990. *Invertebrates.* Sinauer Associates, Sunderland, MA.

Busa, W. B. and J. S. Crowe. 1983. Intracellular pH regulates transition between dormancy and development of brine shrimp (*Artemia salina*) embryos. *Science* 221: 366–368.

Darwin, C. 1854. *A monograph of the subclass Cirrripedia, with figures of all species. The Balanidae, the Verrucidae, etc.* Ray Society, London.

Desmond, A. and J. Moore. 1991. *Darwin: The Life of a Tormented Evolutionist.* Norton, New York.

Dodson, S. 1989. Predator-induced reaction norms. *BioScience* 39: 447–452.

Dohle, W. 1976. Die Bildung und Differenierung des postnauplialen Keimstreifs von Diastylis rathkei (Crustacea, Cumacea). II. Die Differenzierung und Musterbildung des Ektoderms. *Zoomorphologie* 84: 235–277.

Foelix, R. F. 1982. *The Biology of Spiders.* Harvard University Press, Cambridge.

Friedrich, M. and D. Tautz. 1995. Ribosomal DNA phylogeny of the major extant arthropod classes and the evolution of myriapods. *Nature* 376: 165–167.

Garstang, W. 1985. *Larval Forms and Other Zoological Verses.* University of Chicago Press, Chicago.

Gilbert, S. F. 1997. *Developmental Biology,* 5th Ed. Sinauer Associates, Sunderland, MA.

Glenner, H. and J. T. Høeg. 1995. A new motile, multicellular stage involved in host invasion by parasitic barnacles (Rhizocephala). *Nature* 377: 147–150.

Grasshoff, M. 1964. Die Kreuzspinne Araneus pallidus—ihr Netzbau und ihre Paarungsbiologie. *Natur und Museum* 94: 305.

Hand, S. C. and J. F. Carpenter. 1986. pH-induced metabolic transition in *Artemia* embryos mediated by novel hysteretic trehalase. *Science* 232: 1535–1537.

Heymons, R. 1901. Die Entwicklungsgeschichte der *Scolopendra. Zoologica* 13: 1–244.

Høeg, J. T. and J. Lutzen. 1985. *Marine Invertebrates of Scandinavia.* No. 6. Norwegian University Press, Oslo.

Holm, A. 1940. Studien über die Entwicklung und Entwicklungsbiologie der Spinnen. *Zool. Bidrag, Uppsala* 19: 1–214.

Holm, A. 1952. Experimentelle Untersuchungen über die Entwicklung und Entwicklungsphysiologie der Spinnerenembryos. *Zool. Bidrag, Uppsala* 29: 293–424.

Hudinaga, M. 1942. Reproduction, development, and rearing of *Penaeus japonicus* Bate. *Japan J. Zool.* 10: 305–392.

Johannsen, O. A. and F. H. Butt. 1941. *Embryology of Insects and Myriapods.* McGraw-Hill, New York.

Kishinouye, K. 1891. On the development of Araneina. *J. Coll. Sci. Imp. Univ., Tokyo* 4: 55–88.

Kukalová-Peck, J. 1992. The "Uniramia" do not exist: The ground plan of the Pterogota as revealed by Permian Diaphanopterodea from Russia (Insecta: Paleodictyopteroidea). *Can. J. Zool.* 70: 236–255.

Lampert, W., R. Tollrian and H. Stibor. 1994. Chemical induction of defense mechanisms in fresh-water animals. *Naturwissen.* 81: 375–382.

MacBride, E. W. 1914. *Textbook of Embryology.* Macmillan, NY.

Manton, S. M. 1928. On the embryology of a mysid crustacean, *Hemimysis lamornae. Phil. Trans. Roy. Soc. Lond.* B 216: 363–463.

Martinez-Arias, A. and Lawrence, P. A. 1985. Parasegments and compartments in the *Drosophila* embryo. *Nature* 313: 639–642.

Montgomery, T. H.1908. On the maturation, mitosis, and fertilization of the egg of *Theridium. Zool. Jb. Anat.* 25: 237–250.

Panganiban, G., L. Nagy and S. Carroll. 1994. The development and evolution of insect limb types. *Curr. Biol.* 4: 671–675.

Patel, N. H., T. B. Kornberg and C. S. Goodman. 1989. Expression of *engrailed* during segmentation in grasshopper and crayfish. *Development* 107: 201–212.

Paulus, H. F. 1979. Eye structure and the monophyly of arthropoda. In *Arthropod Phylogeny* , A. P. Gupta (ed.). Van Nostrand Reingold, NY; p. 299–383.

Scholz, G. and W. Dohle 1996. Cell lineage and cell fate in crustacean embryos: A comparative approach. *Int. J. Dev. Biol.* 40: 211–220

Sekiguchi, K. 1968. Arachnida. In *Invertebrate Embryology*, M. Kume and K. Dan (eds.). National Library of Medicine, Washington, D.C.

Shiino, S. M. 1968. Crustacea. In *Invertebrate Embryology*, M. Kume and K. Dan (eds.). National Library of Medicine, Washington, D.C.

Telford, M. J. and R. H. Thomas. 1995. Demise of the atelocerata? *Nature* 376: 123–124.

Thomas, J. B., M. J. Bastiani, M. Bate and C. S. Goodman. 1984. From grasshopper to Drosophila: A common plan for neuronal development. *Nature* 310: 203–207.

van der Weele, C.. 1995. *Images of Development: Environmental Causes in Ontogeny.* Elinkwijk, Utrecht.

Vachon, M. 1958. Contribution á l'étude du développement postembryonnaire des araignées. Généralités et nomenclature des stades. Bull. Soc. Zool. France. 82: 337–354

Wedeem, C. J. and D. A. Weisblat. 1991. Segmental expression of an engrailed-class gene during early development and neurogenesis in an annelid. *Development* 113: 805–814.

Whitington, P. M., D. Leech and R. Sandeman. 1993. Evolutionary change in neural development within the arthropods: Axonogenesis in the embryos of two crustaceans. *Development* 118: 449–461.

Whitington, P. M., T. Meier and P. King. 1991. Segmentation, neurogenesis, and formation of early axonal pathways in the centipede *Ethmostigmus rubripes* (Brandt). *Roux's Arch. Dev. Biol.* 199: 349–363.

Winsor, M. P. 1969. Barnacle larvae in the nineteenth century: A case study in taxonomic theory. *J. Hist. Med. Allied Sci.* 24: 294–309.

Woltereck, R. 1909. Weitere experimentelle untersuchungen über das Wesen quantitativer Artunderscheide bei Daphnien. *Versuch. Deutsche. Zool. Ges.* 1909: 110–172.

CHAPTER 14

Arthropods: The Insects

Fritz E. Schwalm

THE MOST SPECTACULAR RADIATION OF SPECIES ON THIS planet has been that of the class Insecta (sometimes referred to as Hexapoda). Nearly one million species of insects have been described, and this number is thought to represent only 2–5% of the actual number of species in this class. Beetles, alone, contain more species (some 65,000) than any other animal phylum with the exception of the molluscs (Brusca and Brusca 1990). However, despite the abundance of species that comprise the class Insecta, we can identify a highly conserved body pattern.

Adult Morphology (the Imago)

Insect segments are combined into three distinct body regions (**tagmata**): the head, thorax and abdomen. Tagmata, especially the head capsule, reveal their segmented origin in the adult to varying degrees. The segmental origin is often more distinct during embryonic or larval stages, especially in the holometabolous orders that undergo complete metamorphosis.

Segments in each tagma show characteristic modifications of appendages and of the derivatives of the nervous system: in the *head* we find one pair of antennae, paired mouthparts represented by clypeolabrum, mandible, maxilla, and labium. Adaptation to any imaginable food source has led to great diversity in the structures of the mouthparts so that they can serve as devices for chewing, sucking, licking, piercing. Paired segmental ganglia in the head region are fused to form a brain. Prominent compound eyes occupy large portions of the head capsule. The *thorax* is highly conserved in the class, with three segments more or less clearly delineated. Paired appendages on each segment can be modified into legs adapted to walking, jumping, swimming, catching of prey. Articulat-ed segments of the legs allow mobility and a high degree of diversity. A pair of wings extends from the dorsal side of the second (**mesothorax**) and the third (**metathorax**). Modifications of the basic wing pattern may be typical for different orders. The *abdomen* is the tagma that most clearly shows external segment borders. Its most prominent taxonomic variations are found in the terminal segments leading to adaptations that serve in depositing eggs (oviposition) and in copulation. Matsuda (1965, 1970, 1976) has treated the variability of tagmata in different orders in great detail.

The chitinous exoskeleton covers the adult and prevents additional growth after the final, adult molt has occurred. Internal organs include a digestive tract into which Malpighian tubules enter at the border of the midgut and the intestine. A dorsal heart and an anterior blood vessel deliver hemolymph into an otherwise open circulatory system. A bilateral tracheal system with branching tracheoles serves as respiratory system. The reproductive system in the female consists of bilateral ovaries with oviducts and accessory glands and with a seminal receptacle connected to a vagina. Male reproductive organs consist of paired testes, vas deferens, and seminal vesicles that are connected to a common ejaculatory duct.

Although the monophyletic origin of Arthropoda is disputed (Manton 1977) as reviewed by Patel (1993) and by the previous chapter, there is sufficient reason to view the class Insecta as a single taxon (see Carroll et al. 1995). A simplified phylogenetic relationship is indicated in Figure 14.1. It is relevant to the discussion of embryogenesis that some of the orders undergo a gradual, **hemimetabolous**, **metamorphosis** from the embryo to the adult stage, while the remaining orders undergo a complete **holometabolous metamorphosis** between larval and adult stages.

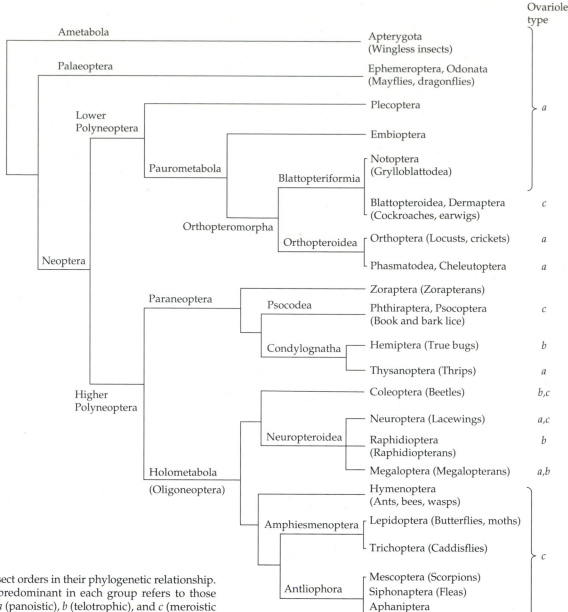

Figure 14.1   Major insect orders in their phylogenetic relationship. The type of ovariole predominant in each group refers to those shown in Figure 14.2: *a* (panoistic), *b* (telotrophic), and *c* (meroistic polytrophic). (After Schwalm 1988.)

Gametogenesis

Paired gonads in males and females reveal an originally segmented arrangement: **ovarioles** in the female, which are covered by a mesodermal peritoneal membrane, are supplied with germ cells from a terminal (proximal) **germarium** (Figure 14.2). An array of ovarioles is found in each side of the abdomen. They are connected to a lateral oviduct, which posteriorly joins a centrally located vagina. Within each ovariole, germ cells are aligned chronologically, with the most mature follicle at the distal end closest to the oviduct. Each gamete is surrounded by mesodermal follicle cells.

Three types of ovarioles can be distinguished in differ-

ent insect orders based on the relationship of the oocyte to other cells in the follicle (Figure 14.2). Oocytes in the **panoistic** ovariole (Figure 14.2*a*) initiate oogenesis as soon as they are surrounded by follicle cells. Gene products to be stored in the oocyte and those required for oocyte metabolism result from activity in the oocyte nucleus, sometimes accompanied by formation of **lampbrush chromosomes**. These chromosomes extend loops of chromatin and are the places of rapid gene transcription. In the **telotrophic** ovariole (Figure 14.2*b*), oogonia divide and the oocyte remains connected to the follicular trophocytes by a **trophic channel**. The oocyte becomes surrounded by follicle cells and moves distally into the ovariole. Stored gene products in the oocyte are produced in the trophocytes as well as in

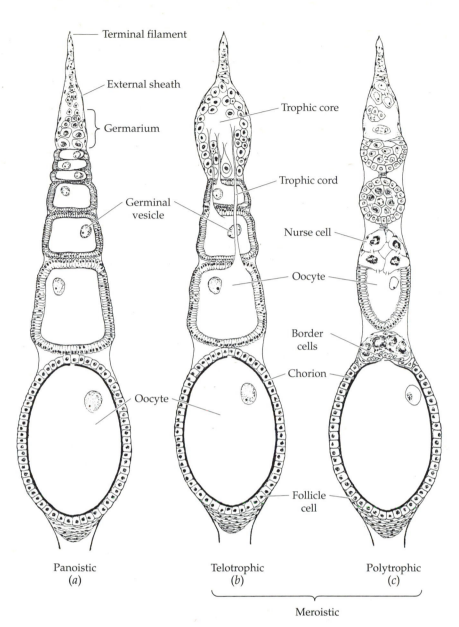

Terminal filament

External sheath

Germarium

Germinal vesicle

Oocyte

Trophic core

Trophic cord

Nurse cell

Oocyte

Border cells

Chorion

Follicle cell

Panoistic
(*a*)

Telotrophic
(*b*)

Polytrophic
(*c*)

Meroistic

Figure 14.2   Ovarioles with different internal organization: (*a*) panoistic, (*b*) telotrophic, (*c*) meroisitic polytrophic.

oocyte nucleus. In the **polytrophic** ovariole (Figure 14.2*c*) trophocytes are enclosed together with the oocyte in the follicle. These trophocytes, also called **nurse cells**, vary in number in different species, grow substantially during the first phases of oocyte maturation, and show a high degree of gene activity. Their products are transferred into the oocyte through channels that have resulted from incomplete cytokinesis between the oocyte and its sibling nurse cells. It appears that trophocyte activity can substitute for gene activity in the oocyte to a large extent, since in some Diptera (flies) the chromatin of the oocyte nucleus can be found to be condensed into a nuclear **endobody** that does not transcribe RNA. Toward the end of oogenesis the majority of the cytoplasm of the nurse cells is transferred into the oocyte, before the eggshell (**chorion**) closes off the anterior end (reviewed in King and Büning 1985). While RNA and proteins in all three types of ovarioles are syn-

thesized in the oocyte and the trophocytes, a major storage protein, **vitellin**, is produced by the fat body and is transported to the egg through the hemolymph. Growing oocytes absorb the protein by pinocytosis, during selective stages of oogenesis (DiMario and Mahowald 1986; Raikhel and Dadhialla 1992).

RNA and protein synthesis in the trophic cells and in the oocyte supply reserve nutrients as well as embryonic determinants. The diversity of these contributions has been analyzed extensively for the formation of the anterior-posterior and dorsal-ventral axes in *Drosophila*. The formation of the anterior-posterior axis is directed by the production of two mRNAs by the nurse cells. The mRNA for bicoid protein is transported through the trophic channels and becomes stored in the portion of the egg adjacent to the nurse cells. The mRNA for the nanos protein becomes localized in that portion of the egg farthest from the nurse cells. When the egg is fertilized, these mRNAs are translated into proteins. Bicoid protein is at one end of the egg, and nanos protein is at the other. When cleavage occurs, those nuclei in the portion of the egg containing bicoid protein are told to make anterior structures (the more bicoid protein in the region, the more anterior the structure), and the cells containing the nanos protein become the posterior cells of the larva (Figure 14.3; St. Johnson and Nüsslein-Volhard 1992; Gilbert 1997).

The formation of the dorsal-ventral axis involves the interaction between the oocyte and the follicle cells. The placement of yolk puts the oocyte nucleus closer to one side of the follicle than the other. The nucleus secretes or localizes compounds that tell the follicle cells close to it to become dorsal follicle cells. These dorsal follicle cells are inhibited from synthesizing particular proteins. However, the ventral follicle cells do secrete these proteins, and these proteins tell the region of the egg nearest to them that it is to be ventral. The most ventral cells are told to become the precursors of the mesoderm, the cells above them are told to become the neuronforming (neurogenic) ectoderm, while the remaining cells will become the precursors of the ectoderm, endoderm, and outer covering (serosa) (reviewed in Bownes 1994; Spradling 1993; Gilbert 1997).

The follicle cells secrete the chorion. The chorion is one of the most important adaptations in the history of life on earth, as it permitted insects to colonize the dry land of the planet. It prevents desiccation, but can allow the passage of gases through small channels called **hydropyles** and **aeropyles**. These are in contact with the **plastron**, a thin layer of air between the ovum and the chorion. But the whole point of having the egg would be abolished if sperm were not allowed access, and sperm cannot penetrate the chorion. Fertilization can only take place where a specific modification of the chorion, the **micropyle**, has been made. These complex pores allow the movement of sperm through the chorion. Micropyles are constructed when follicle cells invade the border between the nurse chamber and the oocyte and become positioned over the anterior center of the oocyte. In lower Diptera, however, the micropyle is assembled anterior to the chamber and becomes secondarily attached to the chorion after all nurse cell contents has been transferred into the oocyte (Sander 1994). Micropyles are found in various positions along the sides of the chorion in other orders, even in the posterior half of the egg. Since the position of the micropyle has been implicated in the pattern formation of the embryo (in stick insects, Sander 1983), the ability of follicle cells to produce micropyles in specific locations may ultimately be the result of a signal from the oocyte, which has directed the follicle cells in the specific location to construct micropyles (St. Johnston and Nüsslein-Volhard 1992). The larva hatches from the chorion by opercula and hatching lines. Other chorion modifications serve as floating devices or to attach eggs to suitable substrates (summarized from Schwalm 1988).

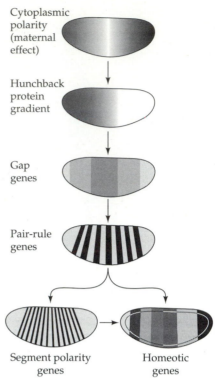

Figure 14.3   Cascade of regulation in the *Drosophila* embryo. (*a*) The fertilized egg initiates two gradients based on the translation of messenger RNAs placed in its cytoplasm by the nurse cells. The bicoid protein diffuses from the anterior end, and the nanos protein diffuses from the posterior end. Bicoid protein stimulates the synthesis of the hunchback protein; nanos inhibits its production. (*b*) This creates a gradient of hunchback protein from anterior to posterior. (*c*) Depending on their relative combinations, nanos, bicoid, and hunchback proteins activate "gap genes." The expression of these genes defines large domains of the embryo. (*d*) The interaction of these genes activate pair-rule genes whose expression domains cover pairs of segments. (*e*) The combination of all these genes then activates both the segment-polarity genes, which define the segmental and parasegmental boundaries, and the homeotic genes that give these segments their fates. (After Gilbert 1997.)

In *Drosophila*, chorion synthesis by the follicle cells is accompanied by amplification of specific genes. In these nuclei, the genes for the chorion proteins are replicated many times more than any other gene. Other taxonomic groups may not show such amplification, as demonstrated for several species of moths (reviewed in Spradling 1993). A detailed description of the shell structures of a wider variety of orders is given in Margaritis (1985).

Eggs of very different cytoplasmic composition are synthesized in the different types of ovarioles and in different taxonomic groups (see Figure 14.1). Panoistic ovarioles produce large eggs, rich in fatty yolk and in most cases requiring periods of 2 weeks to several months for completion of development. Development in these cases is hemimetabolous. Eggs produced in polytrophic ovarioles display a higher proportion of euplasm (nonyolky cytoplasm) and develop into a larval stage (they are mostly found in holometabolous species) in periods that may be as short as one day and rarely exceed a few weeks.

There are numerous modifications of oogenesis throughout the insect class. Some insects undergo **parthenogenesis** ("virgin birth") wherein the species is comprised entirely of females. In the fly *Drosophila mangabeiri*, for eample, the polar body returns and "fertilizes" the egg, both activating it and restoring the diploid chromosome number. In other insects (such as the Orthopteran *Moraba virgo*), the oogonia double their chromosome number prior to meiosis. Meiosis, then, restores the nomal diploid chromosome content. In some aphids, there is no meiosis, and the egg develops from a diploid cell. In these instances, all the offspring are genetically identical to each other and to their mother.

The gonads in males consist of paired testes with varying numbers of follicles in each testis. Spermatogonia in the apical region may form **spermatocysts** after having become surrounded by mesodermal cells. Within a cyst, the spermatogonium may undergo a discrete number of mitotic divisions before all its descendants mature into spermatocytes that subsequently enter into meiosis. For example, in *Drosophila*, four mitotic divisions, followed by meiosis, produce 64 spermatids. These spermatids are initially connected by cytoplasmic canals (a result of incomplete cytokinesis) and are transformed into mature sperm cells in synchrony. The morphogenesis of spermatids into mature sperm affects the nuclear condensation (chromatin becomes hypercondensed), the production of an acrosome at the anterior end of the spermatid, the merging of all mitochondria into a single "**nebenkern**" and the outgrowth of the axoneme from the distal end of the nucleus. The mitochondrial derivatives can become condensed into rodlike structures, which accompany the midsection of the sperm in one or two crystalline structures (Fuller 1993). Transient microtubules emanate from the distal nuclear surface and surround the nebenkern derivatives as well as the nucleus in a **manchette** that purportedly serves to transport cytoplasm and nucleoplasm into the regions of the spermatid that can be sloughed off in the process of size reduction. During these processes, the cyst moves distally toward the vas deferens into which bundles of mature sperm are released. Subsequently, they are stored in (paired) seminal vesicles. One fascinating aspect of insect sperm production concerns the length of Dipteran sperm tails. The sperm of *Drosophila bifurca*, for instance, are 58 mm long. This is 15,000 times the length of the human sperm, and it is 20 times longer than the adult animal that produces it. The reason for manufacturing such enormously elongated sperm is still unknown (Pitnick et al. 1995).

During copulation, large quantities of sperm may be encapsulated into a **spermatophore** for transfer to the female gonopore. Alternatively, sperm may be deposited into a female spermatheca through an extendable penis, equipped with a sperm pump. After sperm has migrated from the spermatophore or the spermatheca to the seminal receptacle of the female it may provide a lifelong supply of gametes for egg fertilization.

Eggs may be deposited individually or in batches, depending on the adaptive strategy of the individual species. Newly hatched larvae depend on immediate food resources and the selection of the appropriate substrate for oviposition is crucial for the survival of many species.

Fertilization

Most insects rely on internal fertilization. As the egg passes through the vagina during oviposition, its micropyle becomes apposed to an opening of the duct that guides sperm cells from the seminal receptacle toward the opening of the micropyle(s). The insemination process is credited with an activating phenomenon that sets developmental processes in motion. These same sequences have been initiated by mechanical stimuli (Went and Krause 1974). Stimuli other than fertilization must account for egg activation in parthenogenetic development. This type of development is common in a variety of species among insects, sometimes being utilized as alternate form of reproduction as seasonal adaptation, or, as in the Hymenoptera (bees, wasps, ants), as mechanism for sex determination. In these cases, the fertilized eggs become females, and the haploid eggs develop into males.

After sperm entry and egg activation, dynamic cytoplasmic dislocations occur in the egg cytoplasm. The oocyte nucleus completes the second meiotic division. The male pronucleus decondenses its chromatin, replacing protamines by histones. Both pronuclei migrate toward an egg region in which **syngamy** (pronuclear fusion) occurs. Pronuclei may undergo an S phase while migrating toward each other, and chromosomes from male and female pronuclei may be separated in groups during the first nuclear division (**gonomery**), as in *Drosophila*. In Coccidae, permanent heterochromatinization of all paternal chromosomes is observed (Brown and Nur 1961). This means that the genes derived from the paternal chromosomes are inactive. Assuming that insemination of the egg occurs at the time of oviposition, the time period between insemination and syngamy varies greatly among different species and ranges from 12 min to 5–6 h among different species (compiled in Schwalm 1988).

Cleavage Pattern and Blastoderm Formation

Since the yolk occupies the central region of the insect cytoplasm, early embryonic development in insects represents **centrolecithal** or **intravitelline** division. Commensurate with the diversity in oogenesis, differences in yolk composition and in the duration of embryonic development are observed. Nuclei divide in rapid cycles while cytokinesis is

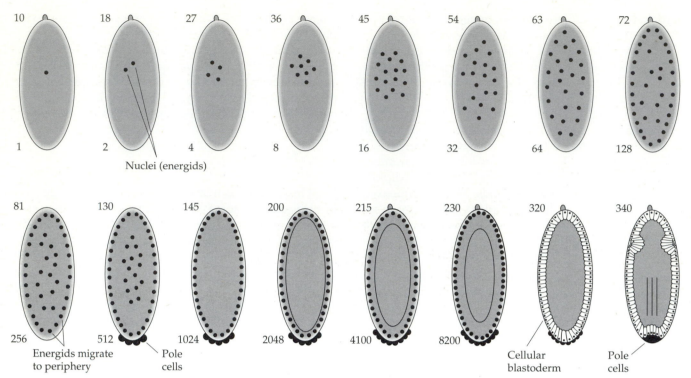

Nuclei (energids)

Energids migrate to periphery

Pole cells

Cellular blastoderm

Pole cells

Figure 14.4 Superficial cleavage in *Drosophila*. The numeral above each embryo corresponds to the number of minutes after oviposition; the number at the bottom represents the number of nuclei present. Pole cells are seen at the 512-nuclei stage even though the cellularization of the blastoderm does not occur for another three hours. The times represent development at room temperature. (After Foe et al. 1993.)

absent (all nuclei share a common reservoir of cytoplasm). During this **syncytial** or **plasmodial phase** of development, nuclear divisions are synchronous and can be as short as five minutes in some Diptera or they may last one or several hours in Coleoptera and Orthoptera. The cells create a **syncytial blastoderm**. The events have been described in greatest detail in *Drosophila* by Foe et al. (1993).

Homologies for eggs from other species with different yolk composition need to be established. During this phase of rapid division (the first 13 cycles in *Drosophila*), nuclei are surrounded by a halo of microtubule-enriched cytoplasm distinct from the yolky components of the egg. This cytoplasm and its respective nucleus are called an **energid**. Energids separate quickly from each other after each nuclear division, and most of them reach the egg surface (Figure 14.4). The rapid karyokinesis is attributed to the precocious replication of the centrosome, its involvement in mitotic spindle assembly, and the assembly of microtubules that transport cytoplasm into the energid and of microtubules that constitute the migration cytaster. The literature on *Drosophila*, reviewed by Foe et al. (1993) indicates that the centrosome originates from the sperm, while Wolf (1980) has observed cytasters in gallmidge (*Wachtliella*) eggs that were unfertilized. Time-lapse cinematography reveals that rhythmic contractions of the entire egg contents occur in synchrony with each nuclear division in eggs that are not heavily encumbered with fatty yolk (Fink and Wieschaus 1991). The length of nuclear division cycles in species for which observations have been reported increases after the 7th to the 13th division cycle when the cellularization of the blastoderm occurs. Nuclei and their accompanying cytoplasm reach the egg surface before the 10th nuclear division. In many species, they form a nucleated rim in the relatively yolk-free **periplasm** of the egg.

Cellularization of the Blastoderm

Nuclei and the associated microtubular arrays that reach the periphery in yolk rich eggs such as from *Drosophila*, cause the oolemma (the cell membrane derived from the oocyte) to bulge outward (termed "budding" by Foe 1993). The buds retract temporarily during each of the 11th, 12th, and 13th nuclear division and reappear over each nucleus during its interphase. Cellularization in *Drosophila* occurs during the 14th interphase. This cell cycle takes a considerably longer time period. Membrane folds originate between the buds that formed after the 13th cycle and extend centripetally into the egg (Figure 14.5). The advancing furrow is broadened and will expand laterally as soon as it reaches the yolky interior egg content. Thus, the membranous furrow will advance toward the center underneath each blastoderm cell, forming the distal cell-membrane and a membrane covering the yolk sac. A narrow canal of cytoplasm between the cell and the yolk sac forms in a yolk stalk. The furrows elongate in close association with microtubules that surround the nucleus in each forming blastoderm cell. Nuclei of the

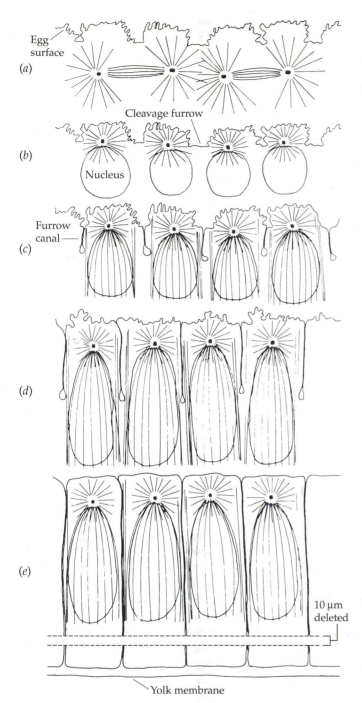

Egg surface

(a)

Cleavage furrow

(b)

Nucleus

Furrow canal

(c)

(d)

(e)

10 μm deleted

Yolk membrane

Figure 14.5 Diagram of nuclear elongation and cellularization in the *Drosophila* blastoderm. (After Fullilove and Jacobson 1971.)

curred and then disperse along the egg surface while they continue to divide. Euplasm is sparse around the nuclei in these eggs and there is not a distinct layer of periplasm underneath the egg surface. The energids under the egg surface will still be unevenly distributed, with higher numbers near the regions of their first arrival, and few, if any, at the distant (pole) regions at the time of cellularization of the blastema.

Germ Band Formation

After the blastoderm has been cellularized, blastomeres are more or less evenly distributed on the egg surface. Subsequent aggregation and heterochronous cell proliferation lead to the formation of the embryonic primordium, called the **germ band** ("keimanlage"). This single layer of columnar cells constitutes the rudiment from which the embryo will be formed. This structure may extend over longer or short distances along the ventral egg surface or may be reduced to a cap of cells covering the posterior pole of the egg. Insect species are often characterized as having one of three types of germ bands: long, intermediate, and short. In species with *short germ bands* (as in many short-horned grasshoppers, Caelifera; Orthoptera) the germ band is a disclike structure that will differentiate into the primordia for the anteriormost segments of the head, while a posterior proliferation zone will gradually supply the cellular substrate for additional segments. Germ bands of *intermediate length* are found in damsel flies (Odonata) and many long-horned Orthoptera (e.g., crickets). These germ bands form from two ventrolateral cell aggregations which cover 30–50% of the egg surface. After the two germinal rudiments have merged ventrally, they provide primordia for head and thorax segments, while additional cell proliferation buds off the more posterior segments (summarized from Schwalm 1988). Species with *long germ bands* utilize a cell layer that extends over 80–100% of the egg surface. The type is best represented by Diptera such as *Drosophila* but is common among Hymenoptera (bees, ants, and wasps) and Lepidoptera (butterfies and moths). Primordia of all segments that will form in the embryo (and adult) are contained in the germ band of these forms. The types of germ band found may vary more in some orders than in others. Coleoptera (beetles) show the greatest diversity within a single order.

Cell Differentiation

Vitellophages

While most of the nuclei migrate to the periphery of the cell, some nuclei remain in the egg interior and become the **yolk nuclei** of the **vitellophages**. Vitellophages extend branched cytoplasmic processes throughout the noncellular yolk, and they break down the yolk so that it can be used for nutrition in the developing embryo (Ikeda et al. 1990). Observations in *Drosophila* indicate that DNA syn-

entire blastoderm expand and elongate during this cell cycle (Fullilove and Jacobson 1971). In yolk-rich large eggs such as those of the cricket *Acheta domesticus* (Orthoptera), energids move further apart than in the smaller dipteran eggs. Large quantities of fatty yolk become interspersed between them, and they may or may not be connected to each other by cytoplasmic bridges while they move through the yolk. They initially arrive at the egg surface closest to the region in which nuclear division first oc-

thesis continues in yolk nuclei in the absence of mitosis. It is expected that these cells become polyploid, each of their nuclei having numerous sets of chromosomes (Foe et al. 1993). Additional yolk nuclei may come from nuclei that are slow in reaching the the egg periphery and from the release of cells from cellular blastoderm into the yolk as secondary vitellophages. Nuclear shape may become highly irregular and yolk nuclei may cluster in groups. In intermediate-length and short germ-band organisms, the cytoplasm immediately underneath the germ band becomes enriched in vitellophages. In embryos of the stick insect (*Carausius morosus*), they form a contiguous sheet of cells, a yolk cell-membrane whose pseudopodial processes extend into the yolk (Fausto et al. 1994).

Germ Plasm

Initial differences in the length of nuclear cycles in different egg regions appear when the most of nuclei have reached the egg surface. In many of the long germ band organisms, cells at the **posterior pole** (so named because the posterior end of the embryo will be located at this pole) will behave differently from the cells of the remaining blastoderm. The **pole plasm** in this region is distinguishable from cytoplasm in other egg regions. It frequently contains unique granular organelles—considered individually as **grana** or as the **oosome** in their entirety—and specific types of messenger RNA that are found only in this region. Under the influence of these components, cells in the posterior egg region segregate rapidly from the syncytial blastoderm while the blasterm nuclei undergo a succession of three or four nuclear division cycles. Instead of reversibly budding out of the egg surface as described earlier for the blastoderm cells, **pole cells** push out of the context of the egg surface and segregate from it during the 10th mitotic cycle in *Drosophila* (Illmensee and Mahowald

1974; Foe et al. 1993). During germ band formation and during organogenesis these pole cells will become integrated into the mesodermal components of the gonads. They constitute the **primary germ cells** and will differentiate into the gametes.

One of the earliest signs of cellular commitment in the blastoderm is the expression of pattern genes. These are the genes (known in *Drosophila* to be regulated by the maternally synthesized determinants mentioned earlier) that determine the fate of the region along the anterior-posterior, dorsal-ventral, and left-right axes. The expression of these genes and the morphogenetic consequences have been studied most intensely in the long germ-band embryo of *Drosophila*, but comparisons to expression in other germ band types have been reported (Patel 1993; Carroll et al. 1995). It appears that many of the genes used to determine the fate of insect segments may be used throughout the insect class. However, the routes to get these genes expressed may differ. Recent evidence from these comparisons suggests that intermediate-length germ band formation is ancestral to the derived forms of the long germ band and the short germ band in more recent taxa (Tautz et al. 1994) and that the the original winged insect had wings on each segment and that wing production was inhibited in most segments as more regulatory genes evolved (Carroll et al. 1995).

Gastrulation, Mesoderm Formation, and Blastokinesis

Long germ-band embryos—exemplified by *Drosophila melanogaster*—will develop from the blastoderm that covers the ventral and lateral surfaces of the egg in a layer of uniformly columnar cells. Cells located along the ventral

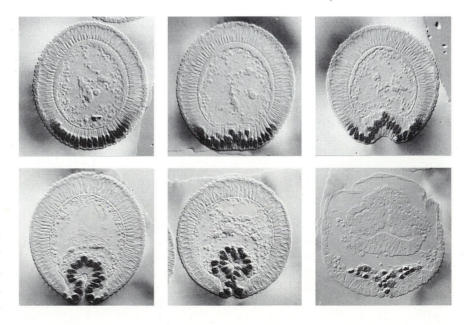

Figure 14.6   Gastrulation in *Drosophila*. The mesodermal cells at the ventral portion of the embryo buckle inward, forming a tube that then generates the mesodermal organs. The nuclei are stained with an antibody to a protein that activates mesodermal-specific genes. (From Leptin 1991.)

(*a*) Germ band of *Drosophila*

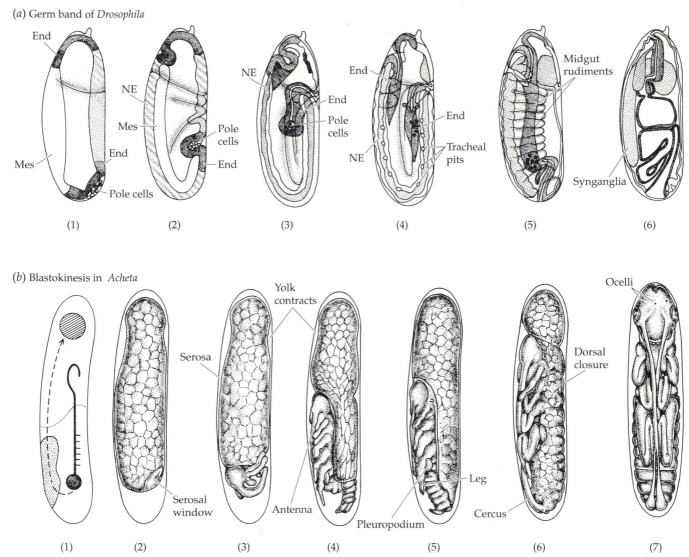

(*b*) Blastokinesis in *Acheta*

Figure 14.7 Movements and major features of embryos during morphogenesis. Lateral view, ventral side faces to the left (except in *b*7). (*a*) Germ band of *Drosophila*. (*a*1–*a*3) Mesoderm (Mes) forms by invagination; neurogenic ectoderm (NE) produces a chain of ventral ganglia. Anterior and posterior midgut rudiments invaginate as endoderm (End); the posterior midgut rudiment extends along the dorsal side and encloses the pole cells into its invagination. (*a*4–*a*5) After reaching its full extension, the germ band retracts. Tracheal pits appear in the posterior segments; midgut rudiments extend from the end regions of the germ band and fuse during retraction. Parasegments (after germ-band extension, *a*4) are replaced by definite segments by the end of retraction (*a*5). (*a*6) Anterior segments are retracted into the thorax segments during head involution. The central nervous system consolidates into synganglia. Dorsal extension of the flanks leads to dorsal closure. (*b*) Blastokinesis in the cricket em-

bryo, *Acheta domesticus*. (*b*1) Idealized tracing of movements of the embryo. The intermediate-length germ band (stippled on the ventral side) becomes immersed in the yolk by moving around the posterior pole, assuming an inverted position (solid outline of the embryo in egg interior). After establishing segmentation, katatrepsis begins (hatched line). The embryo revolves around the posterior end of the yolk and assumes its final position. (*b*2–*b*4) Initially visible at a serosal window, the head emerges first from the yolk and moves anteriorly. While the embryo covers more and more of the yolk sac, tissues differentiate. (*b*5, *b*6) The embryo closes its dorsal flanks. The leading edge contains the cardioblasts. Undulating motion of the edges of the flank create a circulatory movement inside the egg shell. (*b*7) Frontal view of the cricket embryo during completion of organogenesis. (*a* after Leptin 1994; *b* after Schwalm 1988.)

egg surface invaginate rapidly in a distinct morphogenetic movement (Figure 14.6) . The apical surfaces of 4–5 cells along both sides of the ventral midline contract, the cells sink into the yolk, and create a **ventral furrow**. The furrow quickly closes into a mesodermal tube and separates from the cells that remain on the egg surface, most of which are prospective ectoderm.

On the dorsal surface, the anterior end of the tube defines the site of the future **anterior midgut invagination**, the posterior end will form the **proctodeal invagination**. Coinciding with the closure of the mesodermal tube along the ventral midline is a rapid extension of the germ band around the posterior pole of the egg (Figure 14.7). During the extension, the proctodeal invagination moves toward the anterior region, along the dorsal midline of the egg surface. It has engulfed the pole cells and provides the location for the posterior endoderm (prospective midgut) invagination (reviewed in Costa et al. 1993 and Leptin 1994). The pole cells follow this deepening invagination, penetrate its outer layers and eventually become surrounded by mesodermal gonad components.

While the blastoderm forms and gastrulation occurs on the ventral and lateral egg regions, the dorsal surface of the embryo is covered by a serosa. The movement of the germ band during the formation of the basic body pattern is limited (**germ band extension**) but it affects all segments posterior to the head region. This germ band extension moves the region at which the gastrulation fold ends—in a proctodeal pit which surrounds the pole cells at the dorsal side of the posterior egg pole—anteriorly, along the dorsal midline, until it stops approximately 30% of the way from the anterior pole (Figure 14.7a1–3). Yolk movement as well as proliferation of cells within the germ band facilitate the extension. When the extension reaches its full extent, the cells underlying the pole cells proliferate and form the posterior midgut invagination in the region between the anterior head fold and the proctodeum, taking the primary gamete cells into the egg interior.

External delineations become apparent on the embryo surface during the germ band extension. These are the borders of parasegments. These are not the prospective segment borders of the hatching larva, but they are critical in compartmentalizing the embryo during its development (Martinez Arias and Lawrence 1985). The time period for which the germ band remains in the extended position lasts less than 10 percent of the total embryonic development time for *Drosophila* (from 6–8 hours after oviposition). Subsequently, the germ band retracts while the serosal layer covers the middorsal regions of the yolk sac. The posterior end of the embryo will become located at the posterior end of the egg. Differentiation processes will continue after germ band retraction. As the flanks of the embryo extend over the yolk, they will gradually resorb the serosa, which disappears upon dorsal closure when the two flanks merge along the dorsal midline of the egg (Figure 14.7a). In other long germ band embryos, the

invagination of a mesodermal tube can be replaced by the immersion of a median plate into the yolk that results from cell proliferation along a limited band of cells along the median midline of the blastoderm.

Blastokinesis

In intermediate-length and short germ-band embryos, mesoderm forms by cell invagination and delamination of newly divided cells along a **gastral groove** or a **primitive groove** that begins centrally behind the head lobes of these organisms. Since these shorter germ bands represent only anterior rudiments of the prospective embryo, the proliferative zone at their posterior end provides precursors for ectoderm and mesoderm simultaneously.

Germ bands of short and intermediate length follow a pattern of movement in their eggs that is called **blastokinesis** (Figure 14.7b). It consists of an immersion phase (**anatrepsis**) where the embryo extends into the yolk, and an emersion (**katatrepsis**) during which the embryo resumes its position on the egg surface. Long germ bands do not follow the same pattern, but the germ band extension during homologous stages of their development could constitute a remnant of this process.

As the intermediate-length or short germ bands thicken, they produce amniotic folds from the margins of the embryo and from the head lobes. These folds extend medially toward the center of the germ band and result in a cell layer that covers the embryo ventrally. The converging fold will pull the flanking regions of the serosa ventrally, so that the embryo will eventually be covered by two extraembryonic membranes (Figure 14.8). Concurrently with the amnion formation, these embryos extend around the posterior egg pole, directing the embryonic growth zone anteriorly into the egg interior. The head lobe that had been assembled on the ventral surface of the egg moves with its posterior end first, around the posterior pole, and the anterior end will point toward the posterior pole (Figure 14.7b). The embryo is inverted with regard to its position prior to hatching while its basic body pattern is established.

The immersion will be reversed in a movement called katatrepsis, prior to the completion of differentiation. Yolk

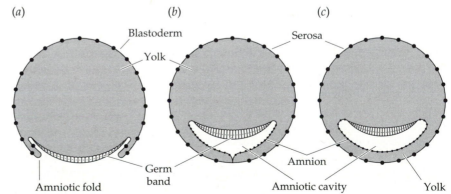

Figure 14.8 Schematic diagram illustrating the development of the amniotic cavity. (*a*) Lateral folds starting at edges of the germ band begin to grow over the germ band. (*b*) The lateral folds meet ventrally beneath the germ band. (*c*) The membranes fuse, causing the separation of the amnion and the serosa, and surrounding the embryo with yolk. (After Chapman 1982.)

contraction in anterior egg regions and changes in cell shape in serosa cells (from squamous to columnar) seem to provide the forces required to let the embryo emerge, head first, from the yolk (Figure 14.7*b*2,3). The serosa breaks at the anterior end of the embryo and pulls rapidly back along the ventral egg surface; the embryo, heading downward toward the posterior egg pole, bends upward and quickly slides over the ventral surface of the yolk sac, covering its sides with the lateral regions of its mesoderm/ectodermal layers (Figure 14.7*b*5,6). During the subsequent stages of differentiation, the flanks of the embryo extend dorsally and eventually complete the dorsal closure, similar to that described for long germ-band embryos.

Organogenesis

Differentiations from the Ectoderm

NEUROGENESIS, CENTRAL NERVOUS SYSTEM. After invagination or delamination of the mesoderm, cell populations can be identified that will differentiate into the ectodermal, mesodermal, and endodermal derivatives. In the long germ-band embryos with distinct gastrulation movement, several rows of cells along the ventral midline have entered into the yolky side of the germ layer. As result of their movement, blastoderm that originally occupied lateral egg regions joins along the ventral midline to form the neurogenic region of the ectoderm. This group of cells is capable of generating neuroblasts, the cells that give rise to neurons, and the pattern of neuroblast positioning is highly conserved among the different orders of insects. Neuroblasts originate from both sides of the ventral midline during the germ band extension of the long germ-band embryo or when the short or intermediate germ band has grown to its full length and is immersed into the yolk.

The number of primary neuroblasts is segment-specific and species-specific. Genes that control which of the neurectoderm cells will differentiate into the neurogenic cells have been identified in *Drosophila* (Doe 1992). The pattern of their distribution in bilaterally symmetric plates is similar to that found in Orthoptera (Thomas et al. 1984). The two lines of cells that meet along the ventral midline as result of the gastrulation movement will give rise to special midline precursor cells that will differentiate into neurons and glial cells. Campos-Ortega (1993) describes the pattern of neuroblast segregation from the neurectoderm in *Drosophila*. Neuroblasts leave the neurectoderm in the direction of the mesoderm at the time when the newly formed mesoderm cells divide mitotically. After an initial growth phase, each neuroblast produces a distinct number of ganglioblasts in teloblastic divisions before it atrophies. Descendants have been traced as "identified neurons" (Bate 1976; Goodman and Doe 1993), indicating that their fate is strictly determined. Neurons forming under the ventrolateral surface of each segment join to

form a ganglion from which axons extend to the adjacent ganglia: two commissures connect the ganglion to its contralateral (opposite side) ganglion in the same segment while connectives reach the ganglia in the anterior and posterior ipsilateral (same side) ganglion. During the advanced stages of embryogenesis and during larval development, ganglia of the ventral nerve cord lose their strictly segmental position. A forward compression of the posterior ganglia into a few synganglia obscure the segmental and bilateral origin of the nerve cord in the later stages.

Neurogenesis in the procephalic lobe of *Drosophila* varies somewhat from the mode described above. Presumptive neuroblasts anterior to the cephalic furrow divide while they are still in the neurectoderm and do not form as distinct an array as they do in the postgnathal (posterior to the jaw) region of the germ band. Unequal (telotrophic) divisions occur in the neuroblasts, resulting in ganglion mother cells without the neuroblast leaving the neurectoderm. The ganglion mother cells can generate a set of neurons that have specific functions. The procephalic region lacks segmental characteristics. The number of neurogenic fields is obscured as result of the absence of segments in most taxa. Generally, three neurogenic regions, neuromeres, can be distinguished (equivalent to the postgnathal ganglia). They are attributed to the intercalary, antennal and the clypeolabral segments. These paired neuromeres merge to form subdivisions of the brain and are in close contact with the optic lobes that form from the anteriormost region, the presumptive acron. Observations on the morphogenesis of the head are further obscured by the fact that the procephalon (forebrain) is expanded to its fullest extent only during the earliest stages of these events. Contraction, dorsal flexure, dorsal closure, and the extreme modifications of the "acephalic" embryo of the Diptera leave few clues about the segmented origin of the organ rudiments during their formation.

SENSORY ORGANS, PERIPHERAL NERVOUS SYSTEM. Neurons from the epidermis extend their axons toward the central nervous system as the buds of antennae, legs and posterior sensory projections (cerci) appear at their respective segments. After the axon has followed the epidermal basal membrane for a short distance it penetrates into the mesoblastic tissues and targets the ganglion (or neuromere) for its input. Secondary neurons use the original pioneer neuron for guidance (Edwards 1977). Such neurons conduct action potentials produced by chemoreceptors or mechanoreceptors of epidermal origin. The sensory organ may be cuticular sensillae, which form in the following sequence. An epidermal cell divides into a sensory cell and a bristle mother cell. The latter divides into a tormogen and a trichogen cell that will produce a bristle. Scale sensillae originate in the same manner (Keil and Steinbrecht 1983). Another type of ectodermal sensory organ is called scolopidium.

These are chordotonal sensory organs that monitor internal body movements, and they are formed from the epidermis without cuticular auxiliary structures. The scolopidium spans a hollow space in the leg, trachea, or antenna; the accompanying sensory cell transmits potentials generated by vibrations to which the scolopidia can respond.

COMPOUND EYE DIFFERENTIATION. The area of the procephalic lobe that will produce the compound eye may form an eye placode in which further differentiation is deferred into the larval stages. (Such a placode forms the eye disc in some holometabolous forms.) The individual units of the compound eye, the ommatidia, differentiate in hemimetabolous insects, such as the stick insect *Carausius*, in the following fashion (as described by Such 1975 and reviewed in Schwalm 1988). A circular array of cells differentiates into a central group of seven or eight photoreceptive retinula cells, surrounded by four crystallogen cells and flanked by two pigment cells. Retinula cells extend axons toward the optic lobes, while their apical surfaces differentiate into light-sensitive rhabdomeres. Each retinula cell is connected to its neighbors by a tight zonula adhaerens junction near its apical surface. As the apical surface is tranformed into a rhabdomere, it becomes involuted toward the center of the ommatidium where the eight rhabdomeres form the rhabdome. Crystalline and pigment cells overgrow and surround the retinula cells and provide the dioptric apparatus. We will see that the differentiation of eye discs in Holometabola follows the same pattern of morphogenesis of ommatidia (described in the paragraph on metamorphosis).

Epidermal Invaginations

The head capsule of embryos of many orders of insects results from secretions of the cuticle by the epidermis of the head segments. During the dorsal closure, regions of the head that had originated anteriorly to the stomodeum on the ventral side of the germ band may turn posteriorly; the portions of the brain that had resulted from neurogenesis in the procephalic lobe come to lie dorsally of the subesophageal ganglia. Invaginations between the segments participate in the construction of a head skeleton and of various glands of the head region. Paired invaginations between the antennal and mandibular segment and between the maxillar and labial segment, respectively, protrude as apodemes into the head interior and fuse internally to form the tentorium, as described for the beetle *Lytta viridana* (Rempel and Church 1971) The location of several apodemes in *Carausius morosus* is shown in Figure 14.9 *a* (Scholl 1969). Shorter apodemes are produced by invaginations from the base of the labrum, the mandible, and the labium. These serve as attachment for head musculature. Ectodermal invaginations from the labium do not produce apodemes but form diverticula, which can differentiate into silk glands or into salivary glands in other forms.

The corpora allata, an important endocrine organ, arises from an infolding of the epidermis between the mandibular and maxillar segments (Cassier 1979). In the dorsal part of the prothorax, papilionidae (swallowtail butterflies) produce an extrudable gland, a bifurcated osmeterium, which originates as an ectodermal invagination. Larvae use this organ as a defense gland when disturbed, using muscle pressure on the hemolymph to cause the pressure to evert it.

HEAD INVOLUTION IN ACEPHALIC LARVAE. A remarkable larval adaptation among the Diptera is the involution of the head region, resulting in an apparently headless "acephalic" larva. The initial pattern of segmentation of the head region is concordant with the phylotypic insect body pattern (Koerpergrundgestalt). Subsequently, the rudiments of the procephalic and the gnathal segments are *retracted* into the thorax region so that the prothorax region becomes the anteriormost segment of the embryo. The retraction is accomplished by several concurrent morphogenetic movements during which the head region becomes compacted and unique larval structures emerge (Jürgens and Hartenstein 1993). During these movements, regions of the ventral segments anterior to the thorax invaginate through the stomodeal opening. The resulting fold is called the atrium, which is delineated dorsally by the pharynx. The opening of the salivary glands, by now already merged into a single duct, opens into the floor of the atrium. Dorsally of the stomodeum, a fold pulls in the structures that had formed from the anterior part of the procephalic lobes, producing a dorsal pouch. The clypeolabrum that had been located at the rostral end of the larva invaginates laterally of the pharynx, thus creating lateral folds that connect the atrium with the dorsal pouch, at least in the anterior parts of the embryo.

During these movements, sensory organs of most of the cephalic segments shift anteriorly so that they surround the opening of the stomodeum. The region of the maxillary segment produces a sclerotized mouth hook and parts of a cephalopharyngeal skeleton. This latter organ increases in complexity during subsequent larval stages. At the end of the morphogenetic movements, only remnants of the retracted head region, especially the sense organs, surround the orifice and together with the mouth hook protrude slightly from the first segment and form a pseudocephalon. At the same time, groups of ectodermal cells that eventually will give rise to the imaginal discs from which the adult head will be constructed are inconspicuous. However, they can be identified by genetic assays that identify active genes within cells (Jürgens and Hartenstein 1993).

(a)

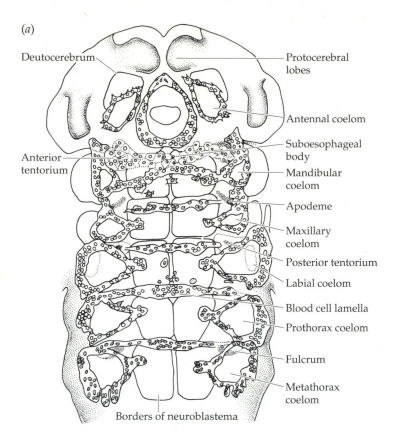

Deutocerebrum

Protocerebral lobes

Antennal coelom

Suboesophageal body

Anterior tentorium

Mandibular coelom

Apodeme

Maxillary coelom

Posterior tentorium

Labial coelom

Blood cell lamella

Prothorax coelom

Fulcrum

Metathorax coelom

Borders of neuroblastema

(b)

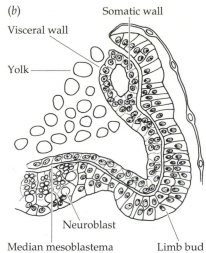

Somatic wall

Visceral wall

Yolk

Neuroblast

Median mesoblastema

Limb bud

Figure 14.9   Somite differentiation into various mesodermal tissues. (a) Head and thorax somites in embryos of *Carausius morosus*. (b) Mesoderm becomes layered into visceral and somatic coelomic walls. (c) Cell groups from the different regions of the somite separate and differentiate into muscle, fat body, and cardioblasts. (a after Scholl 1969; b after Ullmann 1964; c after Ibrahim 1958.)

(c)

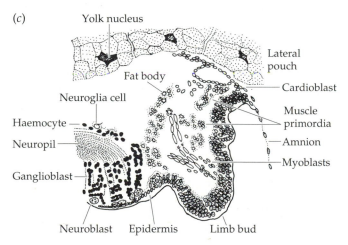

Yolk nucleus

Fat body

Lateral pouch

Cardioblast

Neuroglia cell

Haemocyte

Muscle primordia

Neuropil

Amnion

Myoblasts

Ganglioblast

Neuroblast   Epidermis   Limb bud

TRACHEAL SYSTEMS.   Lateral invaginations in each segment from the second thorax to the eighth abdominal segment show tracheal pits as an indication of tracheal invagination. The pits originate intrasegmentally. A diverticulum extends interiorly, branches when it encounters certain derivatives from the mesoderm, and may extend several branches in different directions, thus clearly establishing metameric tracheal structures. The anteriormost branch from the second thorax primordium will form the ventral head trunk; branches that extend anteriorly and posteriorly and encounter branches from other segments will anastomose with these and eventually estab-

lish a network of tubular tracheae. Other branches extend into the deep tissues in each segment. (Scholl 1969 described this process for the stick insect *Carausius* and Rempel and Church 1972 did so for *Lytta viridana*, a beetle). This pattern is virtually identical to that described for *Drosophila* in which tracheal pits originate at the time of full germ band extension (Figure 14.7a; Leptin 1994). Techniques employing tracheole-specific antibodies allow more detailed analysis than was possible with strictly cytological investigations, and a succession of minute branching patterns has been established for several Diptera.

Despite the fact that initial morphogenesis of the tracheal system is homologous between hemi- and holometabolous orders, major differences are observed, primarily for larval stages in Holometabola. Only a posterior spiracle—from abdominal segment 8 or 9—functions during all larval stages in Diptera. Anterior spiracles function during distinct larval stages and are the only openings in the pupa, while the adult has spiracles in the mesothorax and metathorax as well as in abdominal segments 1–7. Transition from larval to adult stages in holometabolous insects requires major reconstructions of the tracheal system. Transient pupal systems and a very different adult tracheal system rely on the availablility of tracheoblasts in the later stages. The reconstruction is essential for oxygen supply to the adult specifc organs such as the eyes, thoracic flight muscles, ovaries, and to the legs (summarized from Manning and Kraznov 1993).

APPENDAGES. The metameric appendages in embryos with direct development begin to bud out from the ventral surface as a single epidermal layer. The mesodermal tissues extend into these buds as the somites subdivide into their respective derivatives. Antennal buds form early and extend along the ventral side of the embryo. Gnathal appendages develop into highly specialized mouthparts (examples in Schwalm 1988, after Matsuda 1965). Thorax limb buds extend in the posterior direction, developing circular grooves where the articulation will occur upon cuticle secretion (Figure 14.7b). Buds in abdominal segments vary greatly with segment number and among different taxa. The first abdominal homolog to a limb bud becomes a glandular pleuropodium in some orders, purportedly producing hatching enzymes required by the embryo at the end of its development. Several posterior abdominal segments use appendages as copulatory apparatus or for oviposition. The last pair of abdominal appendages forms the cerci, which carry sensillae similar to those found in antennae.

EMBRYONIC CUTICLE FORMATION AND SEGMENTATION. The embryonic epidermis progressively assumes new functions in the organism. As producer of the exoskeleton it is responsible for the external features. Secretion of an embryonic cuticle is the first expression of this function, secretion of denticles in a segment-specific pattern is a second function. The chromosomes of the epidermal cells replicate while neither nuclear nor cell division occur. This combination creates polytene chromosomes, wherein numerous copies of the replicated DNA adhere to each other, forming a thick chromosome. These epidermal cells are retained during subsequent larval instars, and when the cuticle is shed, these cells form it anew. The synthetic cycles result in molting and subsequent synthesis of new cuticles. Another characteristic is that the epidermal cells produce attachment sites for the segmental musculature (Martinez Arias 1993). Precursor cells for adult epidermis are produced at the same time that the epidermis differentiates. In holometabolous insects these cells are interspersed among the larval epidermis (Fristrom and Fristrom 1993).

The most conspicuous epidermal pattern is seen in the establishment of segments that subdivide the embryo into a phylotypic number of metamers. A precise sequence of position-dependent gene expression is required for proper segmentation. This sequence has been elaborated in great detail for *Drosophila* (Pankratz and Jäckle 1993). During germ band extension, mitotic divisions in the differentiating epidermis occur in domains rather than uniformly. The extension results from rearrangements of epidermis cells. Expression of pattern genes in characteristic stripes and the appearance of 14 transient grooves in the ventral and lateral regions of the germ band (Martinez Arias and Lawrence 1985) indicate that cell differentiation occurs in parasegments. Morphological distinction of paraseg-

ments disappears during germ band shortening when definite segments are clearly and permanently established. The establishment of segmental boundaries is accomplished by the interaction of cells at these borders.

Endoderm–Midgut Rudiments

The anterior and posterior midgut primordia originate at the two ends of the ventral furrow in the long germ band embryos. The anterior end of the furrow invaginates or produces a group of cells by telotrophic cleavage wherein one cell remains in place while the other (usually smaller) cell enters the interior. The cells of this group proliferate faster than those of the mesoderm and the ectoderm (Foe et al. 1993). Two groups of cells extend toward the egg interior, one on each side of the yolk sac. The posterior midgut invagination begins as a depression at the posterior end of the ventral furrow into which pole cells (where they exist) are enclosed. As the germ band extends anteriorly, the posterior depression invaginates as a proctodeum. The endodermal cells at the tip of the proctodeum continue to proliferate. Upon germ band retraction they extend anteriorly and encounter the cell strands from the anterior midgut rudiments and fuse with them. Cells spread dorsally and ventrally and eventually enclose the yolk sac. In embryos of insects with germ bands that are short or of intermediate length, the origin of the endoderm follows a similar pattern. Anterior endoderm forms while the posterior segments are being produced, and posterior endoderm can only form after all segments have formed. Rudiments of the endoderm remain at the terminal locations until katatrepsis has occurred, when the bands of endodermal cells begin to proliferate and extend toward the midregion of the embryo, gradually enveloping the yolk in a layer of cells that will differentiate into the midgut.

Stomodeal and proctodeal invaginations may be the place of origin of endoderm formation, or these invaginations make contact with the endodermal cell mass soon after they become established. These invaginations are of ectodermal origin and will differentiate into the foregut and hindgut, respectively (Figure 14.10).

At the interface of hindgut and midgut, stem cells of the malpighian tubules differentiate and proliferate, extending the tubules into the egg interior. The stem cells are considered to be of ectodermal origin, but exceptions have been noted (Nüesch 1988) in Ephemeroptera (*Cloeon dipterum*) where endodermal origin can be demonstrated.

Mesodermal Derivatives

After separation of the mesodermal tube or delamination of mesodermal cells during gastrulation, proliferation increases the cell masses differentially with respect to different regions of the embryo. Mesoderm in the terminal regions (anterior head and the last abdominal segments) may be reduced or absent. Within the other segments, proliferation of cells occurs in the lateral regions, while

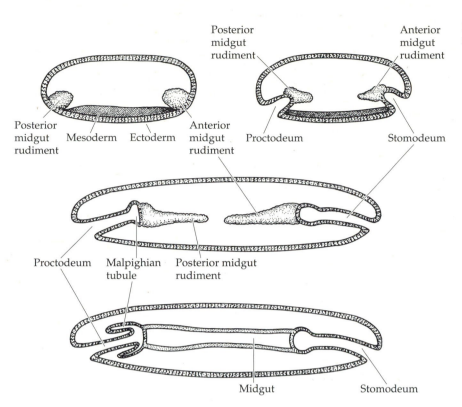

Posterior midgut rudiment

Anterior midgut rudiment

Posterior midgut rudiment   Mesoderm   Ectoderm   Anterior midgut rudiment

Proctodeum

Stomodeum

Proctodeum   Malpighian tubule   Posterior midgut rudiment

Midgut   Stomodeum

Figure 14.10   Diagrammatic representation of midgut development. (After Chapman 1982.)

few cells may reside in the midline region of the embryo. Mesodermal cells may group into distinct lateral somites in each segment, or may form coelomic cavities in each side of a segment, or they may form loosely associated layers of cells that show clumping in a pattern that follows parasegmental arrangements (Bate 1993). Depending on the position in the embryo (segment specificity) and on the position of cell groups within an individual somite, mesodermal regions will differentiate into specific tissues. Figure 14.9 will serve for orientation and allow introduction of some common terminology for a "typical" somite structure. The coelomic sac is delineated by a wall of external somitic mesoderm (lining the ectoderm) and by an internal visceral (or splanchnic) mesoderm adjacent to the yolk sac (Figure 14.9 *b*) or adjacent to the endoderm after this layer has reached the respective segment. Pouches within the coelomic sac may extend medially or laterally and subdivide the coelomic wall into distinct surface areas. Derivatives of these surfaces will then assume specific functions in the developing embryo. Cells that reach the midline of the embryo—either as result of incomplete separation of the coelom during somite formation or as a result of migration into these regions—will constitute the blood-cell lamellae from which haemocytes and plasmatocytes differentiate. Portions of the visceral wall differentiate into midgut muscle. Derivatives from the somatic wall invade the segment-specific appendages to form the muscles (Figure 14.9 *c*). Other portions form longitudinal musculature in the segments or differentiate into fat body cells (reviewed in meticulous detail for *Drosophila* in Bate

1993). Cells from the dorsolateral border of splanchnic and somatic layers differentiate into cardioblasts; these line the flanks of the embryo along the abdominal and thorax segments and produce the blood sinus of the heart and the walls of the cephalic aorta in close contact with the associated segmental muscle fibers.

Gonad Development

Gonads develop from cells of dual origin: the primary germ cells originate independently from mesodermal formations. In many long germ-band embryos, primary germ cells are the first cells to leave the blastoderm syncytium when they form the pole cells. They rarely divide in the early embryonic stages, and remain clustered until they penetrate the posterior midgut rudiment (in *Drosophila*) during germ band retraction. After breaking into two clusters, they become surrounded by splanchnic-wall mesoderm derivatives from the somites in the respective segments and form a primary gonad in each segment. In *Drosophila*, gonad formation seems to be restricted to abdominal segment 5. Detailed analysis has shown that mesoderm from parasegments 10–12 (abdominal segments 5–8) contributes to this organ (Bate 1993; Grieg and Akam 1995). Mesodermal cells will become interspersed between gametes and will produce the follicle cells of the ovarioles and the spermatocysts. They will also anchor the ovarioles (by differentiating into terminal filaments, see Figure 14.2) in the fat body. Ventrally located portions will grow out into strands that differentiate into the oviducts, from each one of the bilateral ovaries, during the larval stages of development (reviewed in Schwalm 1988). The number of gamete cells in embryos that will develop into males and into females does not differ in the first stages of cell segregation but it appears that the male primary gametes in *Drosophila* undergo higher rates of proliferation during the later stages of embryogenesis (Poirié et al. 1995). The segregation of primary germ cells in short and intermediate- length germ bands is less distinct than in the long germ- band embryos. Strips of primary germ cells can be distinguished in segments 2–8 in the abdomen of *Pyrrhocoris* (Hemiptera) and in segments 3–6 in the stick insect *Clitumnus*, where they associate with the splanchnic portion of the mesoderm to form the subdivision of the gonads. The oviduct will extend posteriorly from each portion and its subdivisions will connect to each other before they merge to form a vaginal duct at the posterior end (summarized from Schwalm 1988).

Postembryonic Development and Metamorphosis

Upon completion of the embryonic phase of development, the larva hatches from the egg shell. The escape is facilitated by different mechanisms: hatch lines in the egg shell may break in response to internal pressure, the larva may break the shell with the help of special spines, or the larva may bite its way through the shell. The juvenile will mature into an adult generally by acquiring the typical features of wings, external genitalia, mature gonads, and fully developed eyes. Larval development occurs in defined growth periods, each terminated by ecdysis (the shedding of the exoskeleton). The larva that hatches is called the first instar, and after the first molt comes the second instar, and so forth. After a fixed number of instars (with some plasticity due to environmental needs) there is the final or adult molt.

Two major groups of orders can be distinguished. Those in which the larvae show the characteristic features of the respective order have external wing pads (at least in the later larval instars) and constitute the exopterygota. The endopterygote larvae of the holometabolous orders display unique larval morphologies (details in Sehnal 1985) in adaptation to their habitat , which frequently differs substantially from that of the adult form in the same species. Adult structures emerge rapidly during the pupal stage from imaginal discs and histoblasts. These precursors of adult tissues originate in the embryo and become competent for transformation into the respective structure during larval development.

Origin and Development of Imaginal Discs and Histoblasts

Imaginal discs are flat epithelia that are attached to the larval epidermis by a **peripodial stalk** (Figure 14.11), while histoblasts are cells that occur in groups in distinct regions of the epidermis interspersed among the epidermal cells of the larva. Cells from which imaginal discs or histoblasts will develop become differentially committed during the blastoderm stage of the embryo. Their commitment depends on their anterior-posterior position (which segment they reside in) and their position along the dorsal-ventral axis. After the primordia of each disc have formed, their cells rarely divide during the remainder of embryogenesis, but they proliferate during larval development. In *Drosophila*, there is substantial growth of the discs attributed to cell proliferation in the third (final) larval instar. In the pupal stage, cells in the disc respond to the titer of ecdysteroid hormones in the hemolymph by evaginating through the peripodial stalk (Figure 14.11). In evaginating, the center of the disc moves to the tip of the evaginating structure. This tip will form the distal part of the wing, leg, or antenna. The regions around the center will become the more distal regions and the outer margin of the disc will contribute to the adult epidermis at the base of the respective structure. Histoblasts, which replace the majority of the larval epidermis, are interspersed among the larval epidermis cells (Fristrom and Fristrom 1993). Since the histoblasts remain diploid, they can be distinguished histologically from the polytene larval cells.

CELL NUMBERS AND DISC COMPARMENTALIZATION. The number of cells that form the adult primordia has been assayed by gynandromorph analysis or by mitotic mutations in which X-ray exposure was used to enhance mitotic recombination in *Drosophila* (reviewed by Oberlander 1985). In these analyses, cells of the embryo get marked either by a mutation (whose phenotype is discernible in the adult) or by sex (where there is sexual dimorphism in color, bristle number, or some other marker) at a certain stage of development. The number of cells originally committed to produce a disc range from 2–11 for specific parts of the genital disc and from 10–12 for leg and wing discs. As the number of cells increases during larval development, compartments of the presumptive organ are being established. Techniques similar to those used to determine the number of original cells sequestered in a primordium have shown that clones of cells that descend from a single mutated ancestor cell are limited to specific compartments in the imaginal disc of which they are a part. Clones from cells mutated at an early stage of disc growth contribute to large compartments—anterior, posterior in wing, leg, proboscis in flies; clones that are induced in cells during later stages will be confined to smaller subcompartments within the respective adult organ (reviewed by Oberlander 1985 and Cohen 1993). Immature discs are determined to form their segment- specific adult structure, but they need to reach a certain size to become competent for transformation into the respective structure. This maturation process is largely independent of the position of the disc in the larva and can progress even

(a) Leg disc

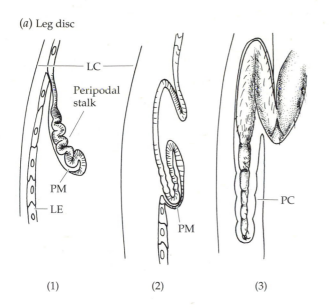

(1)   (2)   (3)

(b) Eye disc

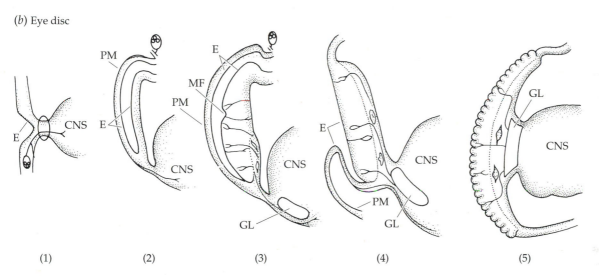

(1)   (2)   (3)   (4)   (5)

Figure 14.11   Imaginal disc evagination in flies. (a) Leg disc. (a1) The leg disc epithelium is attached to the larval epidermis (LE). A peripodial membrane (PM), a squamous epithelium, covers one side of the disc. (a2) The disc evaginates through the peripodial canal while the peripodial membrane contracts and the folds in the disc disappear. (a3) After evagination and differentiation of segments and bristles, a pupal cuticle is secreted by the leg epithelium (PC). In cyclorrhaphous Diptera, this event proceeds inside the last instar larval cuticle (LC), which changes into a puparium. (b) Eye disc. (b1) Embryonic ectoderm (E) folds inward above the cells of the optic stalk which connects the eye to the central nervous system (CNS). (b2) The disc, folded into a saclike structure formed from a single cell layer, is enlarged by cell proliferation during the larval stages. The peripodial membrane (PM) covers the external side of the fold. (b3) The morphogenetic furrow (MF) appears and cells become determined in ommatidial clusters, behind the furrow. (b4) Eversion begins in the pupal stage aided by retraction of the peripodial membrane. The eye surface now becomes exposed to the outside. (b5) As the ommatidia arrange their cells into the typical pattern, photoreceptor axons establish contact with the lamina ganglionaris (GL). (a from Poodry 1992; b from Wolff and Ready 1993.)

in a disc that has been transplanted into an adult abdomen. This observation has facilitated experimental techniques employed to maintain discs in "in vivo" culture for long time periods.

The imaginal discs of legs, antennae, wings, and terminal appendages of the abdomen are derived from individual segments and their transformation into adult structures can be accommodated by the model given for leg metamorphosis (Figure 14.11). Primordia of the adult head are compacted into a group of very closely associated discs. The analysis is complicated by the fact that some of these discs may be of mixed segmental origin, like that established for the eye–antennal disc. Only three pairs of discs can be distinguished clearly, but these discs contain adult primordia from five to six head segments. The tracing of the ontogeny of head structures from these discs is available in a separate treatise for *Drosophila* by Jürgens and Hartenstein (1993). One primordium that has been given much attention in recent studies is the eye disc of

Drosophila, since a wealth of genetic information in this organism can be linked to morphological observations in eye differentiation.

Eye Differentiation in an Imaginal Disc

The eye disc in *Drosophila* originates from 6–23 cells of the dorsolateral ectoderm during the blastoderm stage. This same region had earlier delaminated the cells that form the optic lobe (part of the central nervous system). This ectoderm region folds during head involution (Figure 14.11b). During the first two larval stages, the epithelium increases in size but shows no signs of cell differentiation. Proliferation accelerates during the third larval instar during which a dorsolateral indentation, the morphogenetic furrow (Figure 14.11b3), appears at the apical surface of the disc. In the pupal stage, disc eversion occurs rapidly under retraction of the peripodial membrane. (Compare the leg eversion in Figure 14.11a2 and the eye eversion in Figure 14.11b4). The disc epithelium expands laterally over the optic stalk, while increasing numbers of differentiating cells extend axons toward the lamina ganglionaris, which will eventually become centered beneath the pupal eye (Figure 14.11*b*5). The morphogenetic furrow, mentioned earlier, is the result of the following cellular modifications: Several rows of cells, located parallel to and near the posterior margin of the disc shorten and retract their apical surfaces below the original perimeter of the eye disc (Figure 14.12). The apical surfaces of these cells also become reduced to 1/15 of their original area. After completion of these changes, the cells undergo a cell division. The newly divided cells extrude away from the furrow beyond the posterior margin of the disc. These cells enter the patterning stage of ommatidial differentiation. Thus the **morphogenetic furrow** (Figure 14.11*b*3) separates cells in their proliferative stage from those that are in the patterning stage, within the same disc. Cells anterior to the furrow will undergo the same transition as those that have just completed the division—more cells will be extruded on the posterior side of the furrow. By the time the pupa has reached one-third of its development, all cells in the eye disc have completed this first wave of morphogenetic mitosis that precedes differentiation of ommatidia.

After the cells have been extruded from the furrow, they arrange themselves into cell clusters from which ommatidia will differentiate (Figure 14.12). As additional clusters become interspersed between the initial clusters and the furrow, a gradient of maturity is established, with newly formed clusters leaving the furrow and the most mature clusters becoming located distally from the furrow. By the time when eight to nine new clusters have become interspersed between the furrow and a maturing cluster, some of the cells in the maturing cluster will undergo a second mitotic division, bringing the number of cells needed to construct a complete ommatidium to a full complement. Wolff and Ready (1993) designate the arrays of cells in the immature clusters (before the second mitosis) as "preclusters" and reserve the term ommatidial clusters to the full complement of cells. During the final maturation of an ommatidial cluster, processes similar to those described earlier for the ommatidial differentiation in the stick insect occur. The eight retinula cells in the

inner core are connected to each other by an apical zonula adherens. Retinula cells differentiate into photoreceptors by retracting the apical cell surface at the center of the ommatidium while this surface remains extended at the circumference of the retinula cells: as the apical surfaces differentiate into a rhabdomere, they slip into a common region shared by the eight retinula cells, forming the rhabdome between them. In *Drosophila*, the cell that is to become photoreceptor 8 is the first photoreceptor to mature, and it induces cells on either side of it to become photoreceptors 2 and 5. These interact with the other nearby retinula cells until each of the eight photoreceptors has differentiated and is in its proper orientation with respect to the others. The cone cells (C in Figure 14.12) surround the retinula cells in the original ommatidial cluster and are homologous to the crystallogen cells that Such (1975) described in the eye differentiation of the stick insect (mentioned earlier). These cells move above the retinula cells and form four lenslike structures. Pigment cells surround the aggregate of retinula and cone cells and demarcate adjacent ommatidia from their neighboring receptors. The secretion of a cuticular cornea completes the differentiation of the pupal eye in *Drosophila* (Wolff and Ready 1993).

Summary

I have attempted to account for structural events in insect eggs from oogenesis to the hatching of the adult. In a class as ancient and as diverse as that of the Insecta, comparisons of observations that have been made for different orders will necessarily appear to be forced. For some large taxonomic groups, only a few species have been investigated. Extensive reports that stress detailed differences between groups have been published (Anderson 1972; Haget 1977; Schwalm 1988). This chapter represents a condensed review of significant events reported on various aspects of insect morphogenesis. It has neglected many of the fascinating details of sequential gene expression during oogenesis and embryogenesis of *Drosophila* that are filling volumes of literature. An exemplary collection of reviews of this literature has been edited by Bate

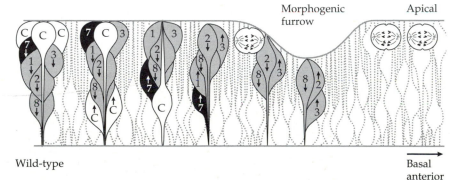

Figure 14.12   Retinal development in *Drosophila*. Longitudinal view of ommatidial assembly in the third instar eye disc of *Drosophila*. The eye disc is a monolayer (except for dividing cells). The sequence of differentiation is shown as the morphogenetic furrow travels towards the anterior of the disc. Dividing cells (at the right) enter the furrow and begin differentiating. Cells R1 and R6, R2 and R5, and R3 and R4 behave as pairs and only one of the pair is represented on this diagram. The cell movements are shown by arrows. Cone cells are represented as C. (After Tomlinson and Ready 1987.)

and Martinez Arias (1993). Reviews by Patel (1993), Beeman et al. (1993), Sander (1994), and Carroll et al. (1995) address comparative aspects of gene expression in the development of different insect orders.

Literature Cited

Anderson, D. T. 1972. The development of hemimetabolous insects. The development of holometabolous insects. In *Developmental Systems: Insects*, vol. 1., S. J. Counce and C. H. Waddington (eds.). Academic Press, London, pp. 95–242.

Bate, C. M. 1976. Embryogenesis of an insect nervous system. I. A map of the thoracic and abdominal neuroblasts in *Locusta migratoria* (Orthoptera, Acrididae). *J. Embryol. Exp. Morph.* 35: 107–123.

Bate, M. 1993. The mesoderm and its derivatives. In *The Development of Drosophila melanogaster*, M. Bate and A. Martinez Arias (eds.). CSH Laboratory Press, New York, pp. 1013–1090.

Beeman, R., S. Brown, J. Stuart and R. Denell. 1993. The evolution of genes regulating development commitments in insects and other animals. In *Evolutionary Conservation Of Developmental Mechanisms*, 50th Symposium SDB, A. C. Spradling (ed.). Wiley-Liss, New York, pp. 71–84.

Bownes, M. 1994. Interactions between germ cells and somatic cells in *Drosophila melanogaster*. *Seminars Dev. Biol.* 5: 31–42.

Brown, S. W. and U. Nur. 1961. Heterochromatic chromosomes in the coccids. *Science* 145:130–136.

Brusca, R. C. and G. J. Brusca. 1990. *Invertebrates*. Sinauer Associates, Sunderland, MA.

Carroll, S. B., S. D. Weatherbee and J. A. Langeland. 1995. Homeotic genes and the regulation of insect wing number. *Nature* 375: 58–61.

Campos-Ortega, J. 1993. Early neurogenesis in *Drosophila melanogaster*. In *The Early Development of Drosophila melanogaster*, M. Bate and A. Martinez Arias (eds.) CSH Laboratory Press, New York, pp. 1091–1131.

Cassier, P. 1979. The corpora allata of insects. *Int. Rev. Cytol.* 57: 1–73.

Chapman, R. F. 1982. *The Insects: Structure and Function*. 3rd Ed., Harvard University Press, Cambridge.

Cohen, S. M. 1993. Imaginal disc development. In *The Development of Drosophila melanogaster*, M. Bate and A. Martinez Arias (eds.). CSH Laboratory Press, New York, pp 747–842.

Costa, M., D. Sweeton and E. Wieschaus. 1993. Gastrulation in *Drosophila*: Cellular mechanisms of morphogenetic movements. In *The Development of Drosophila melanogaster*, M. Bate and A. Martinez Arias (eds.). CSH Laboratory Press, New York, pp. 425–464.

DiMario, P. J. and A. J. Mahowald. 1987. Female sterile (1) yolkless: A recessive female sterile mutation in *Drosophila melanogaster* with depressed numbers of coated pits and coated vesicles within the developing oocytes. *J. Cell Biol.* 105: 199–206.

Doe, C. Q. 1992. The generation of neuronal diversity in the *Drosophilia* embryonic central nervous system. In *Determination of Neuronal Identity*, M. Shankland and E. Macagno (eds.). Academic Press, New York, pp. 149–154.

Edwards, J. S. 1977. Pathfinding by arthropod sensory nerves. In *Identified Neurons and Behavior of Arthropods*, G. Hoyle (ed.). Plenum Press, New York, pp. 483–593.

Fausto, A. M., M. Carcupino, M. Mazzini and F. Giorgi. 1994. An ultrastructural investigation on vitellophage invasion of the yolk mass during and after germ band formation in embryos of the stick insect *Carausius morosus br. Dev. Growth Differ.* 36:197–207.

Fink, R. and E. Wieschaus. 1991. Fruit fly embryogensis. In *A Dozen Eggs. Time Lapse Microscopy of Normal Development*. R. Fink (ed.). Sinauer Associates, Sunderland, MA. A videotape.

Foe, V. E., M. O. Garrett and B. A. Edgar. 1993. Mitosis and morphogenesis in the *Drosophilia* embryo point and coun-

terpoint. In *The Development of Drosophila melanogaster*, Vol. 1. M. Bate and A. Martinez Arias (eds.). CSH Laboratory Press, New York, pp. 149–300.

Fristrom, D. and J. W. Fristrom. 1993. The metamorphic development of the adult epidermis. In *The Development of Drosophila melanogaster*, Vol 2. M. Bate and A. Martinez Arias (eds.). CSH Laboratory Press, New York, pp. 842–898.

Fuller, M. T. 1993. Spermatogenesis. In *The Development of Drosophila melanogaster*. M. Bate and A. Martinez-Arias (ed.). CSH Laboratory Press, New York, pp. 71–147.

Fullilove, S. L. and A. G. Jacobson. 1971. Nuclear elongation and cytokinesis in *Drosophila montana. Dev. Biol.* 26: 560–577.

Gilbert, S. F. 1997. *Developmental Biology*, 5th Ed. Sinauer Associates, Sunderland, MA.

Goodman, C. S. and C. Q. Doe. 1993. Embryonic development of the *Drosophila* nervous system. In *The Development of Drosophila melanogaster*. M. Bate and A. Martinez Arias (eds.). CSH Laboratory Press, New York, pp. 1131–1206.

Greig, S. and M. Akam. 1995. The role of homeotic genes in the specification of the *Drosophila* gonad. *Curr. Biol.* 5: 1057–1062.

Haget, A. 1977. L'embryologie des insectes; In *Grassé, Traité de Zoologie*, Vol. 8, part VB.

Masson, Paris, pp. 1–262; 279–387.

Ibrahim, M. M. 1958. Grundzüge der Organbildung im Embryo von *Tachycines* (Insecta, Saltatoria). *Zool. Jb. Anat. Ont.* 76: 541–594.

Ikeda, M., T. Sasaki and O. Yamashita. 1990. Purification and characterization of protease responsible for vitellin degradation of the silkworm, *Bombyx mori. Insect Biochem.* 20: 725–734.

Illmensee, K. and A. Mahowald. 1974. Transplantation of posterior polar plasm in *Drosophila*. Induction of germ cells at the anterior pole of the egg. *Proc. Nat. Acad. Sci. U.S.A.* 71:1016–1029.

Jürgens, G. and V. Hartenstein. 1993. The terminal regions of the body pattern. In *The Development of Drosophila melanogaster*, M. Bate and A. Martinez Arias (eds.). CSH Laboratory Press, New York, pp. 687–746.

Keil, T. A. and R. A. Steinbrecht. 1984. Mechanosensitive and olfactory sensilla of insects. In *Insect Ultrastructure*, Vol. 2. R. King and H. Akai (eds.). Plenum Press, New York, pp. 477–516.

King, R. and Büning, J. 1985. The origin and functioning of insect oocytes and nurse cells. In *Comprehensive Insect Physiology Biochemistry and Pharmacology*, Vol. 1. G. Kerkut and L. Gilbert (eds.). Pergamon Press, Oxford, pp. 37–82.

Leptin, M. 1991. Mechanics and genetics of cell shape changes during *Drosophila* ventral furrow formation. *In* R. Keller et al. (eds.), *Gastrulation: Movements, Patterns, and Molecules*. Plenum, New York, pp. 199–212.

Leptin, M. 1994. *Drosophila*. In *Embryos* Color Atlas of Development. J. Bard (ed.). Wolfe Mosby Publishing Co., London, pp. 113–134.

Manning, G. and M. Krasnov. 1993. Development of the *Drosophila* tracheal sytem. In *The Development of Drosophila melanogaster*, M. Bate and A. Martinez Arias (eds.). CSH Laboratory Press, New York, pp. 609–685.

Manton, S.M. 1977. *The Arthropoda: Habits, Functional Morphology, and Evolution*. Clarendon Press, Oxford.

Margaritis, L. H. 1985. Structure and physiology of the eggshell. In *Comprehensive Insect Physiology Biochemistry and Pharmacology*, Vol. 1. G. Kerkut and L. Gilbert (eds.). Pergamon Press, Oxford, pp. 153–230.

Martinez Arias, A. 1993. Development and patterning in the larval epidermis of *Drosophila*. In *The Development of Drosophila*, M. Bate and A. Martinez Arias (eds.). CSH Laboratory Press, New York, pp. 517–608.

Martinez Arias, A. and P. A. Lawrence. 1985. Parasegments and compartments in the *Drosophila* embryo. *Nature* 313: 639–642.

Matsuda, R. 1965. *Morphology and Evolution of the Insect Head. Mem. Am. Ent. Inst.*, No. 4. American Entomological Institute, Ann Arbor.

Matsuda, R. 1970. *Morphology and Evolution of the Insect Thorax.* *Mem. Can. Ent. Soc.* No. 76. Entomological Soc. Canada, Ottawa.

Matsuda, R. 1976. *Morphology and Evolution of the Insect Abdomen.* Pergamon Press, Oxford.

Nüesch, H. 1988. Development of the Malpighian tubules in *Cloeon dipterum* (Ephemeroptera) *J. Morphol.* 197: 241–247.

Oberlander, H. 1985. The imaginal discs. In *Comprehensive Insect Physiology, Biochemistry, and Pharmacology*, Vol. 2. G. Kerkut and L. Gilbert (eds.). Pergamon Press, Oxford, pp. 151–200.

Pankratz, M. J. and H. Jäckle. 1993. Blastoderm segmentation. In *The Development of Drosophila melanogaster.* M. Bate and A. Martinez Arias (eds.). CSH Laboratory Press, New York, pp. 467–516.

Patel, N. 1993. Evolution of insect pattern formation: A molecular analysis of short germband segmentation. In *Evolutionary Conservation of Developmental Mechanisms, 50th Symposium in SDB*, A. Spradling, (ed.). Wiley-Liss, New York, pp. 85–110.

Pitnick, S., G. S. Spicer and T. A. Markow. 1995. How long is a giant sperm? *Nature* 375: 109.

Poirié, M., E. Niederer and M. Steinmann-Zwicky. 1995. A sex-specific number of germ cells in embryonic gonads of *Drosophila.* *Development* 121: 1867–1873.

Poodry, C. A. 1992. Morphogenesis of *Drosophila.* In *Morphogenesis*, E. Rossomando, E. and S. Alexander (eds.). Marcel Dekker, Inc. New York, pp. 143–188.

Raikhel, A. and T. Dadhialla. 1992. Accumulation of yolk proteins in insect oocytes. *Annu. Rev. Entomol.* 37: 217–251.

Rempel, J. G. and N. S. Church. 1971. The embryology of *Lytta viridana* Le Conte (Coleoptera: Meloidae). VII. Eighty-eight to 132-h: The appendages, the cephalic apodemes, and head segmentation. *Can. J. Zool.* 49:1571–1581.

Rempel, J. G. and N. S. Church. 1972. The embryology of *Lytta viridana* Le Conte (Coleoptera: Meloidae). VIII. The respiratory system. *Can. J. Zool.* 50: 1547–1554 .

St. Johnston, D. and C. Nüsslein-Volhard. 1992. The origin of pattern and polarity in the *Drosophila* embryo. *Cell* 68: 201–220.

Sander, K. 1983. The evolution of patterning mechanisms: Gleanings from insect embryogenesis and spermatogenesis. In *Development and Evolution*, B. C. Goodwin, N. Holder and C. C. Wylie (eds.). Cambridge University Press, Cambridge, pp. 137–159.

Sander, K. 1994. The evolution of insect patterning mechanisms: A survey. *Development Suppl.* pp. 187–191.

Scholl, G. 1969. Die Embryonalentwicklung des Kopfes und Prothorax von *Carausius morosus* B. (Insecta, Phasmida). *Z. Morph. Ökol. Tiere* 65: 1–142.

Schwalm, F. E. 1988. *Insect Morphogenesis. Monographs in Developmental Biology*, Vol. 20. H.W. Sauer (ed.). Karger, Basel.

Sehnal, F. 1985. Growth and life cycles. In *Comprehensive Insect Physiology, Biochemistry, and Pharmacology*, Vol. 2. G. Kerkut and L. Gilbert (eds.). Pergamon Press, Oxford, pp. 1–86.

Spradling, A.C. 1993. Developmental genetics of oogenesis. In *The Development of Drosophila melanogaster.* M. Bate and A. Martinez Arias (eds.). CSH Laboratory Press, New York, pp. 1–70.

Such, J. 1975. Analyse ultrastructurale de la morphogenèse ommatidienne au cours du développement embryonnaire de l'œil composé chez le phasme *Carausius morosus.* *Br. C.r. hébd. Séanc. Acad. Sci. Paris* 281D: 67–70.

Tautz, D., M. Friedrich and R. Schröder. 1994. Insect embryogenesis—what is ancestral and what is derived. *Development Suppl.* 1994: 193–199.

Thomas, J. B., M. J. Bastiani, M. Bate and C. S. Goodman. 1984. From grasshopper to *Drosophila*—A common plan for neuronal development. *Nature* 310: 203–207.

Tomlinson, A. and D. F. Ready. 1987. Cell fate in the *Drosophila* ommatidium. *Dev. Biol.* 123: 264–275.

Ullmann, S. L. 1964. The origin and structure of the mesoderm and the formation of the coelomic sacs in *Tenebrio molitor* L. (Insecta, Coleoptera). *Phil. Trans. R. Soc. Ser.* B 248: 245–277.

Went, D. F. and G. Krause. 1974. Alternations of egg architecture and egg activation in an endoparasitic hymenopteran as a result of natural or imitated oviposition. *Roux's Arch. Entwmech. Org.* 175: 173–184.

Wolf, R. 1980. Migration and the division of cleavage nuclei in gall midge *Wachtliella persicariae.* II. Origin and ultrastructure of the migration aster. *Roux's Arch. EntwMech. Org.* 188: 65–74.

Wolff, T. and D. F. Ready. 1993. Pattern formation in the *Drosophila* retina. In *The Development of Drosophila melanogaster*, M. Bate and A. Martinez Arias (eds.). CSH Laboratory Press, New York, pp. 1277–1326.

CHAPTER 15

Phoronids, Brachiopods, and Bryozoans, the Lophophorates

Russel L. Zimmer

T HE THREE PHYLA CONSIDERED IN THIS CHAPTER—BRYO-zoa, Brachiopoda, and Phoronida—are recognized as being closely related by most zoologists and are often identified under the collective term lophophorates or tentaculates. A **lophophore** consists of a ring of tentacles that are used in feeding and respiration and are part of a small, specialized, circumoral mesosome, the middle one of three body regions. Adult members of the three phyla share a number of other morphological traits in addition to the lophophore (e.g., three-part body with protosome, mesosome, and metasome each containing a coelom; tentacles surrounding mouth, but not anus; U-shaped gut); collectively, these common features not only unite the three lophophorate phyla, but also suggest an alliance of the lophophorates with the deuterostome phyla. However, the life history patterns of the three phyla are rather divergent and contain a confusing mixture of both protostome and deuterostome traits. To complicate matters further, recent molecular analyses (e.g., Halanych et al. 1995) have suggested that the three lophophorate phyla are most closely linked with the protostome phyla Annelida and Mollusca, that Brachiopoda and Phoronida are more closely related to each other than either is to Bryozoa, and that Brachiopoda may consist of two distantly related groups. Although the Phylum Entoprocta has been allied with Bryozoa, especially by Nielsen (e.g., 1971), and the newly proposed phylum Cycliophora (Funch and Kristensen 1996) is suggested to be related to Ectoprocta and Entoprocta, neither entoprocts or cycliophorans have developmental or adult traits that would ally them with lophophorates.

If one were to devise a hypothetical adult lophophorate (HAL), it would be in the shape of an elongate cylinder with a tuft of tentacles at one end. The animal would be bilateral, triploblastic, coelomate, marine, and sedentary.

Distinguishing characteristics would include a three-part body consisting of a minute preoral **protosome**, a small circumoral **mesosome** bearing the tentacles, and a large postoral **metasome** forming the rest of the body. Each of these units would have paired coeloms, the **proto-**, **meso-**, and **metacoels**. The proto-, meso-, and metasome, each of which is quite different from the others, are not considered segments since they do not represent repeated structures. There are several nonlophophorate phyla that have a similar tri- or oligomerous compartmentalization of the body (that is, with three body regions, preoral, circumoral and postoral). In some of these, trimery is evident in the larva but becomes obscured in the adult (e.g., echinoderms); in others it is retained in both the larva and adult (Hemichordata). These two phyla are usually considered closely related to vertebrates and a logical question for one to ask is, What are the relationships among phoronids, bryozoans, brachiopods, echinoderms, hemichordates, and vertebrates? The developments of echinoderms and hemichordates have a number of parallels with the development of phoronids. On the other hand, the developments of both brachiopods and bryozoans are difficult to correlate with that of any other phylum.

To return to our description of HAL, the mesosome includes a very small portion of the elongate body of the hypothetical ancestor plus the tentacles; the term lophophore is used variably for either the entire mesosome or the tentacles only. Although the tentacles encircle the mouth, they would be in a simple ring only if our hypothetical HAL were a minute individual (as in most bryozoans); in larger forms, the ring of tentacles would be strongly indented on one side so that the tentacles are no longer in a circle but follow the edges of horseshoe (as in brachiopods, phylactolaemate bryozoans, and phoronids). This indentation is called the lophophoral concavity and it is here that the anus and the metanephridia open. (Recall that the mouth and anus are separated by only a very short distance, not at opposite ends of the body!)

To continue our description, the most anterior body region in our HAL, the protosome, is a small, simple, moveable flap that functions much like a trap door over the mouth, hence the name **epistome**. An analogous relationship is that between the epiglottis and glottis of mammals. By far the largest part of the body consists of the postoral metasome or **trunk**. The trunk has a muscular body wall permitting movements and contain the U-shaped gut and the gonads. Our hypothetical form has a closed circulatory system and each of the three body regions has a pair of coeloms. The metacoel is drained by paired metanephridia that serve in both excretion and reproduction.

The nervous system is intraepidermal with a circumesophageal nerve ring from which one or two lateral nerves extend the length of the trunk. That part of the nerve ring between the mouth and anus is a thickened ganglion, the brain of HAL.

You may have noted that the terms anterior, posterior, dorsal, and ventral have been avoided in describing HAL. Suffice it to say here that we usually think of the anus as being at the posterior end and the mouth near the anterior one in metazoans, but in HAL they are both at the lophophoral end separated by only a short distance. That short gap between the mouth and anus represents the dorsal midline, whereas virtually the entire trunk is ventral! This arrangement is so different from what is true in virtually all other bilateral animals that it must have a complicated explanation. How HAL "lost" its anterior-posterior axis and got a very short dorsal midline, but a very long ventral one will become apparent from studying the metamorphosis of phoronids below.

The hypothetical lophophorate just described provides a nearly accurate picture of a phoronid, an adult of which is shown in Figure 15.1; the only significant discrepancy is that none of the coeloms of phoronids is paired embryologically, although the trunk coelom becomes divided into right and left halves following metamorphosis.

Phylum Phoronida

Since both their embryology and adult morphology are considered the most primitive among lophophorates, it is logical for us to study phoronids first. The phylum Phoronida consists of but two genera and only about 15 species. All species are marine and all live in tubes made of their own secretion; a description of a typical adult phoronid was provided above. Despite the small size of the phylum, three patterns of development have been recognized. Two of these involve the production of **actinotroch** larvae (Figure 15.2) that have a long planktotrophic phase; the two variants are distinguished by the facts that, in one, the eggs are about 100 μm in diameter and are brooded to the stage of early larvae whereas, in the other, 60-μm eggs are shed into the seawater where all development takes place. In both patterns, the long-lived actinotroch undergoes a dramatic and rapid metamorphosis (Figure 15.3). Members of the genus *Phoronopsis* have only the nonbrooding pattern, but members of the genus *Phoronis* may have either. The third pattern of development is unique to the phylum's smallest species (*Phoronis ovalis*), but involves the largest eggs (125 μm). These are brooded within the

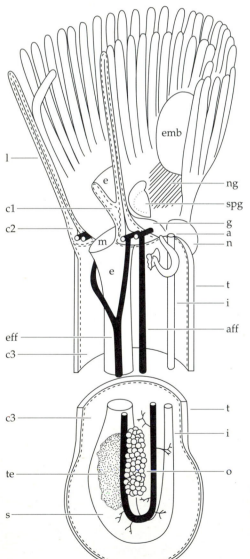

Figure 15.1 Diagrammatic hemisection of an adult phoronid (based on the hermaphroditic species *Phoronis vancouverensis*). The three parts of the body—the epistome (e), lophophore (l), and trunk (t)—contain separate coeloms (c1, c2, and c3, respectively). Because of the length of the trunk, a portion has been deleted. Other abbreviations: a, anus; aff, afferent blood vessel; eff, efferent blood vessel; emb, embryo mass; e, esophagus; g, main ganlion; I, intestine, m, mouth; n, nephridum; ng, nidamental gland; o, ovary; s, stomach; spg, sphermatophoral gland; te, testis. (Redrawn from Zimmer 1991.)

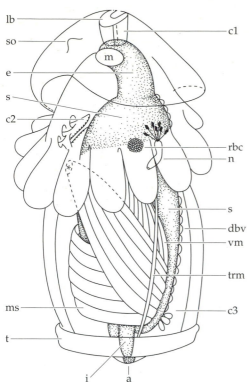

Figure 15.2 Advanced actinotroch larva competent to undergo metamorphosis (based on *Phoronopsis harmeri*). The larva measures about 1 mm. Abbreviations: *, opening to metasomal sac; a, anus; c1, protocoel or preoral hood coelom; c2, mesocoel or collar coelom; c3, metacoel or trunk coelom; dbv, dorsal blood vessel; e, esophagus; lb, larval brain; m, mouth; ms, metasomal sac, n, duct of left protonephridium capped with soleno-cytes; rbc, mass of red blood cells in collar blastocoel; s, stomach; so, sense organ; t, telotroch; trm, trunk retractor muscle on left side. (Redrawn from Zimmer 1991.)

Brief accounts of adult morphology (and development) are available in most recent invertebrate texts; more detailed accounts in English are those of Emig (1982) and Hyman (1959). Silén's 1954 paper is a seminal study on reproductive biology. Recent review articles that cover development and/or specific aspects of reproductive biology include Emig (1977, 1983, 1990) and Zimmer (1987, 1991). Articles that will provide an introduction to the varying interpretations of the phylogenetic position(s) of phoronids and other lophophorates include Zimmer (1973), Siewing (1974), Nielsen (1987), Field et al. (1988), and Halanych et al. (1995).

Gametogenesis

The genus *Phoronis* has both gonochoristic and hermaphroditic members, but *Phoronopsis* has only gonochoristic ones. The gonads of phoronids develop within the metacoel in association with the coelomic lining of blood capillaries attached to the stomach. Nutrients for gametogenesis are accumulated during the nonbreeding season in vasoperitoneal tissues of the same capillaries.

The sperm are highly modified and V-shaped. One limb of the V is the flagellum or tail, whereas the other

parental tube and develop into a short-lived, lecithotrophic, slug-shaped larva. The metamorphosis of this species has not been described, but initially results in a hemispherical mass that remains seemingly unchanged over a considerable period of time.

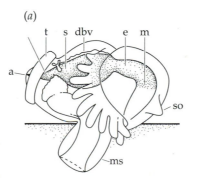

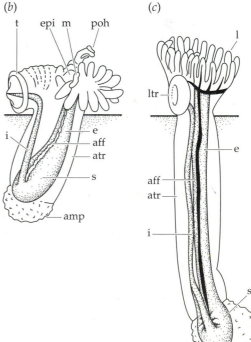

Figure 15.3 Metamorphosis in the Phoronida (based on *Phoronopsis harmeri*). (*a*) Larva with sensory organ (so) protruded and metasomal sac (ms) partly protruded into the sediment (shaded area). Other abbreviations: a, anus; dbv, dorsal blood vessel; e, esophagus; i, larval intestine; m, mouth; s, stomach; t, telotroch. (*b*) Midway (10 minutes) through metamorphosis. The larval trunk is partly collapsed and the adult trunk (atr) is fully everted and contains the now U-shaped gut. The distinctive lower part of the adult trunk is the ampulla (amp). Much of the preoral hood (poh) has degenerated, but one portion will persist as the adult protosome or epistome (epi). The dorsal blood vessel of the larva is now the afferent blood vessel (aff) or "heart-artery" of the adult. (*c*) A 3-day-old juvenile. The collapsed larval trunk (ltr) bears the anus at its center. The larval tentacles, now the adult lophophore (l), are elevated, obscuring the mouth and epistome. Portions of the fully formed closed circulatory system, including the afferent artery (aff), are shown in black. (Redrawn from Zimmer 1991.)

has an elongate nucleus starting at the junction of the two limbs and terminating in a pair of rod-shaped mitochondria. A solid-appearing, membrane-bound rod located at the tip of the V in the nonflagellar limb is a possible acrosome.

Sperm are collected in the metanephridia, which also serve as gonoducts for the female gametes, and are passed as a compact mass through the nephridiopore and into one of a pair of glandular pockets—spermatophoral glands—in the lophophoral concavity. Here the sperm mass is molded into a spermatophore held together by layers of mucus and sometimes provided with a mucoid spiral sail. Spermatophores are released into the ambient seawater and apparently drift in water currents. Probably aided by feeding currents, the spermatophores reach the lophophores of adjacent individuals. In some species, spermatophores are thought to be swallowed and to then enter the visceral coelom through the stomach wall. In *Phoronopsis harmeri*, the spermatophore contacts several tentacles; the mucous envelopes surrounding the sperm mass rupture, and sperm become an ameboid mass. By means which remain unknown, this mass digests a hole through one or more of the tentacles and streams into the mesocoelomic lumen of those tentacles. The still coherent mass of sperm reaches the base of the lophophoral cavity and eventually penetrates the septum separating the mesocoel and metacoel. Once inside the metacoel, the sperm disperse throughout this cavity. Although fertilization is commonly indicated to be external, the eggs are fertilized internally within the metacoel, probably at the time of their release from the ovarian tissue. Spermatophores and spermatophoral glands have been found in all species in which they have been sought, with the exception of *Phoronis ovalis*.

In all phoronids but one, the fertilized eggs are collected in the metanephridia located at the lophophoral end of the animals and, on exit from the nephridiopore, either pass into the surrounding seawater where all development will take place or are brooded. Nonbrooding species produce hundreds to thousands of eggs (average diameter about 60 μm) each day during their reproductive period, which may extend over a month. The length of time from spawning of a zygote to the settlement of the larva is estimated to be about 4 weeks. Except in *Phoronis ovalis* (see following paragraph), the embryos of brooding species are attached to tentacles adjacent to the nephridiopores. These tentacles transitorily develop special nidamental (nest-secreting) glands whose mucous secretions temporarily anchor the young stages to the tentacles. Species that brood their early young in this way produce a few dozen to perhaps a thousand eggs, which average about 100 μm in diameter. The brooding period probably averages about 2 weeks, by which time the zygotes have developed to early actinotrochs capable of capturing microscopic food organisms, including phytoflagellates and diatoms. The larvae probably require an additional 2–4 weeks to complete their development.

A different brooding pattern is found in *Phoronis ovalis*, which releases all of its female gametes at one time through the severed end of the trunk after casting off the lophophoral end! Only about 2 dozen eggs are produced by members of this smallest phoronid species. Interestingly, the eggs are the largest in the phylum (125 μm) and are brooded within the maternal tube. Early stages of development have not been described in detail, but slug-shaped, ciliated larvae eventually crawl from the tube and spend a short time swimming near the bottom. These larvae have only a partial gut and are presumed to be lecithotrophic (nonfeeding). Although so dissimilar externally from an actinotroch, the two larval types have in common three pairs of coelomic chambers and a massive ventral invagination, the metasomal sac (Zimmer 1991).

Cleavage, Blastulation, and Gastrulation

When spawned, the eggs are technically primary oocytes that have already been fertilized and in which the female nucleus has been arrested at metaphase of first meiosis. Soon after spawning, meiosis of the female gamete is completed and early cleavages begin. Cleavage is radial or more properly biradial, although spiral cleavage has been reported by several workers (e.g., Rattenbury 1954). By the 16- or 32-cell stage there is a small blastocoel that becomes proportionately larger with each subsequent cleavage. The ciliated blastula has longer cilia at the former animal pole, and here the larval brain will soon form; gastrulation will occur at the opposite (vegetal) pole.

Gastrulation is by invagination. Initially the archenteron is a shallow cup and the blastopore is circular, or nearly so. The archenteron becomes an L-shaped tube as the lips of the blastopore push toward each other and fuse, starting at the future posterior end of the gastrula. Eventually, the blastopore is reduced to a small anterior remnant; this structure remains open as the larval (and adult) mouth, a clear protostome trait. The developing gut becomes regionated into esophagus, stomach, and blind intestine. The anus forms later and may be a posterior derivative of the blastopore.

The Actinotroch Larva

Several changes soon take place in the epidermis. The larval skin just anterior to the mouth grows ventrally and posteriorly as a two-layered, umbrella-like flap that eventually overhangs the larval mouth. Called the preoral hood, this distinctive feature of the actinotroch represents the protosome of the trimerous larva.

Perhaps the most easily recognized feature of the actinotroch (and the source of its name) is an oblique ring of tentacles. The tentacles originate from a thickened ectodermal ridge that encircles the larva. The first tentacles form on either side of the ventral midline, and additional pairs are added more dorsally until eventually there is a nearly complete ring of tentacles. The tentacle ring marks the posterior boundary of the middle, circumoral collar or mesosome of the actinotroch larva.

The third part of the trimerous larva consists of the rest of the body—that part posterior to the tentacle ring. Properly the metasome, this region is usually identified as the larval trunk. In early actinotrochs, the trunk is little more than a small bump, but it will grow to form the largest part of the fully grown larva (Figure 15.2).

The paired ducts of the larval **protonephridia** originate quite early from a midventral epidermal invagination in the anteriormost portion of the still minute trunk. The **solenocytes**, which may be of mesodermal rather than ectodermal origin, protrude into the blastocoelic space of the collar rather than into the coelomic cavity of the trunk. Some authors identify the possession of protonephridia in actinotroch larvae as a protostome trait since protonephridia occur in a number of protostome larvae and adults, but only in cephalochordates among deuterostomes.

While the preoral hood, tentacle ring, and protonephridial ducts were being elaborated as ectodermal derivatives, mesoderm was forming and differentiating internally. Formation of the mesodermal cells of phoronids involves neither a 4d cell, as in protostomes, nor typical archenteric pouches, as in deuterostomes. The origin of mesoderm is poorly documented, but some mesodermal cells originate from the anterior part of the archenteron as individual ameboid cells; this process is called ingression. The cells ingress either from a complete ring around the future esophageal region near the blastopore or from all but the posterior part of that ring. The initial differentiation of these anterior mesodermal cells into the protocoel has been followed closely only in *Phoronis vancouverensis* (Zimmer 1980). In this species, these cells remain ameboid for a time and migrate to line the blastocoel within the preoral hood. They then unite as an epithelial lining of the preoral hood cavity, forming a closed, fluid-filled vesicle. By definition this is a coelom; as its position is preoral, it corresponds to the **protocoel** of trimerous forms. Although initially extending throughout the preoral hood, the protocoel or **preoral hood coelom** soon becomes reduced to a small vesicle between the larval brain and esophagus in many species and becomes functionally obliterated in others. We will consider the fate of this coelom (and the other two found in actinotrochs) when we describe metamorphosis. Formation of a coelom by ameboid wandering of mesodermal cells may be unique to this phylum.

Although not certain, some of the anterior mesoderm cells probably form muscles and possible blood cells within the blastocoel of the collar and may contribute to the U-shaped coelom formed of that region. This collar coelom could also be called the **mesocoel** or, considering its fate, the **lophophoral coelom**. It forms only very late in larval life and occupies only a small part of the collar cavity, much of which remains as an unlined blastocoel. (However, see Herrmann [1986] for an alternative interpretation of the formation of this and the other two coeloms in *Phoronis muelleri*).

The larval trunk also has a coelom and, again, this has been described in detail only for *Phoronis vancouverensis*. In this species, some 8–10 mesodermal cells form a ventrally interrupted ring around the blind-ending intestine of the incipient actinotroch. The origin of these cells has not been determined, but may involve ingression from the posterior end of the archenteron. Rapid cell division followed by schizocoely establishes a C-shaped balloon curving around the junction of the stomach and intestine. Its two ends abut, forming a two-layered mesentery in the ventral midline of the trunk. Note that, because of its mode of formation, the **trunk coelom** or **metacoel** of the actinotroch has only a midventral mesentery, whereas most bilateral animals have both ventral and dorsal mesenteries since their coeloms originate as paired right and left chambers. In fact, all three coeloms of actinotroch larvae are unpaired, despite the obvious bilateral nature of the larva. Embryologically unpaired coeloms in bilateral animals are found in larvae and adults of some deuterostomes (e.g., the protocoel of enteropneusts), but usually coeloms that appear unpaired (such as that containing our heart and the protocoel of starfish larvae) originate by fusion of right and left members or by loss of one of them.

The development of the heavily ciliated tentacles provides the larva both a means of locomotion and a ciliary mechanism for collecting suspended food items. Actinotrochs feed using a single-band or upstream ciliary system, which is characteristic of deuterostome larvae, rather than a double-band or downstream one employed by protostome larvae. In addition to the well-developed ciliation on their tentacles, all but very early actinotroch larvae are provided with a powerful posterior **telotroch** of fused cilia that propel the larva forward, but play no apparent role in feeding.

About midway through larval life, the skin in the ventral midline just below the tentacle ring thickens and then invaginates into the trunk coelom as an elongate epidermal sac. This **metasomal sac**, a diagnostic feature of the actinotroch, pushes between the right and left leaves of the midventral mesentery of the trunk coelom as it invaginates. Thus the midventral mesentery forms a coelomic covering of the invaginating sac as well as a connection between the sac and the digestive tract. As the larva grows, the metasomal sac increases dramatically in size and its coelomic lining forms an extensive musculature of longitudinal and circular elements. In larvae ready to metamorphose, the heavily muscled sac fills most of the cavity of the trunk not occupied by the gut. The metasomal sac plays a major and surprising role at metamorphosis, as will soon be seen.

In advanced actinotrochs, clumps of hemoglobin-containing blood cells are formed in the collar blastocoel, and incipient blood vessels are formed in the collar and trunk regions. These anticipate the closed circulatory system that will be established during metamorphosis.

Actinotroch larvae also have a complex musculature that differs between species. Of special note is a powerful pair of **trunk retractor muscle**s that run through the trunk coelom (in most, but not all species). The muscles of the actinotroch are all smooth to our knowledge. These muscles play an important role at metamorphosis.

The nervous system of the larva is intraepidermal. The **larval brain** is located at the anteriormost end of the larva in the preoral hood. Near the ganglion is a **larval sense organ**; as this develops only shortly before metamorphosis and is actively protruded during presettlement behavior, it is thought to assist the larva in finding an appropriate place to begin existence as a benthic adult, perhaps functioning as a chemo- or mechanoreceptor. Detailed studies on the larval nervous system include Hay-Schmidt (1989, 1990a, 1990b) and Lacalli (1990).

Metamorphosis

Larvae that are fully differentiated and able to complete the elaborate metamorphosis are said to be competent. The metamorphosis of such larvae can be triggered by a number of materials (Herrmann 1976), including actively growing, gram-positive bacteria and cesium and rubidium ions. The former may help ensure that the larva settles in a biologically rich area; the role of the ions is less clear, but metamorphosis of many marine larvae is enhanced by augmented concentrations of potassium, a related monovalent alkaline metal. In anticipation of metamorphosis, the larvae swim along the surface of the sediment, repeatedly protrude the larval sense organ, and, in many species, strongly arch themselves dorsally.

Metamorphosis of larval phoronids is dramatic and irreversible (Figure 15.3). The transformation from larva to adult begins with eversion of the metasomal sac; in most species this is due to the contraction of the powerful trunk retractor muscles identified earlier. The resulting shortening of the larval trunk apparently increases the pressure within its coelom and forces out the metasomal sac much as the inside-out finger of a rubber glove can be everted. As a consequence, the metasomal sac now exists as an elongate protrusion sticking out at right angles to the shortened larval trunk. The larval digestive tract is pulled into this protrusion, the adult trunk, thanks to the midventral mesentery of the trunk coelom. The larval trunk, no longer containing either the metasomal sac or most of the digestive tract, becomes extremely shortened, placing the anus only a short distance from the mouth.

Shortly after the metasomal sac has everted, the preoral hood turns opaque and starts to degenerate. In addition, parts or all of each larval tentacle may also undergo degenerative changes. As they disintegrate, the cast-off portions of the hood (and tentacles) are swept into the animal's mouth by feeding currents so that the first meal of the incipient adult is its own transitory larval tissues. With collapse of the larval trunk, the (remaining) tentacles become directed anteriorly rather than extending radially or posteriorly as they were in the larva, and the ring of tentacles becomes indented. Thus, the tentacles now have the orientation, position, and horseshoe shape of the adult lophophore and, in fact, are that lophophore. A small part of the preoral hood remains as the primordium of the adult epistome, but the larval brain (and sense organ) have been lost. Whether a portion of the larval protocoel remains as the coelom of the epistome is controversial.

At this point, we should note that the metamorphosing individual already has the shape of an adult with an elongated trunk containing a U-shaped gut and capped with a lophophore. Although it seems illogical, note that the entire adult trunk is a ventral structure. The ventral midline courses from the mouth to the end of the trunk and back to the anus, whereas the dorsal midline of the adult extends only the short distance from the mouth to the anus.

Internally, several important changes have occurred. Within 15–30 minutes, the red blood cells, which had been in one to several compact masses, have broken into individual cells, and these are now circulating through a complex vascular system nearly identical to that of the adult. During metamorphosis, the solenocytes of the protonephridia are lost (and temporarily circulate with the erythrocytes), but the ducts are retained. Eventually the inner end of each duct will connect with a ciliated funnel formed from the lining of the metacoel. In this way the larval protonephridia are transformed into the metanephridia of the adult.

Externally, we can now see that the end of the trunk opposite the lophophore is a thin-walled and somewhat bulbous **ampulla** (this was the innermost end of the metasomal sac). Almost as soon as it is formed, the adult trunk is surrounded by a transparent sheath of secretions from the epidermis just anterior to the ampulla. The sticky secretions accumulate sand grains and other materials and collectively these components harden into a tube within which the adult lives . The ampulla will anchor the adult within this tube and perhaps is used by the early imago to burrow into the sediment.

Experimental Observations, Problems to be Solved, and Conclusions

Phoronids have been the subject of only limited experimental analysis. Zimmer (1973) reported that blastomeres separated from 2- and 4-cell cleavage stages could regulate to form complete (but small) actinotrochs and Herrmann (1976) experimented with metamorphic clues as noted earlier. Recently, Freeman (1991) studied axis formation and regulative capacity in *Phoronis vancouverensis*, provided a fate map for this species, and compared this fate map to those of articulate and inarticulate brachiopods.

A major set of unsolved questions concerning the embryology of Phoronida includes the origin of the mesoderm and its differentiation into coelomic compartments. Even in the best studied species, *Phoronis vancouverensis*, the sources of mesoderm are poorly understood. The available evidence indicates that each of the three coeloms of its larva may form by a different method. As an added complication, patterns of coelom formation may differ among species with different modes of development. Additional areas needing further study include (a) descriptions of early development and metamorphosis in the lecithotrophic larva of *Phoronis ovalis* and (b) determination of the roles of sensory structures and external stimuli in the metamorphosis of actinotroch larvae.

Phoronids (and the other two lophophorate phyla) remain objects of interest because of their controversial relationships. Because phoronids are widely accepted as the least derived of the three phyla covered in this chapter, they are of central importance in understanding the phylogenetic position of lophophorates. Confusion about where to place phoronids exists because they have (a) a number of embryological and adult features that support a deuterostome relationship (cleavage pattern, regulative development, trimery, poor cephalization, an intraepidermal nervous system), (b) at least one embryological characteristic typical of protostomes (the blastopore forms the adult mouth, [although exceptions to this "rule" are known]), and (c) several equivocal features (origin[s] of mesoderm, modes of coelom formation, presence of protonephridia in the larva). Actinotroch larvae share virtually no distinctive features with typical protostome larvae (trochophores) and little other than trimery and mode of feeding with those of deuterostomes. As mentioned in the introduction to the chapter, the available molecular data (most importantly 18S rDNA sequences) suggest that lophophorates have protostomian affinities. Unfortunately, paleontology has not provided us many useful clues as to the affinities of lophophorates to date and may not do so in the future (but see Conway Morris and Peel 1995). Since virtually all of the morphological and embryological criteria used in establishing relationships have been explored for the Phoronida, the resolution of their phylogenetic placement will almost certainly rely on the procurement of additional phylogenies based on "molecular-clock" data for multiple molecules of phylogenetic significance. Should additional molecular phylogenies be congruent with that currently available from 18S rDNA, the importance of embryological and morphological features—especially those which indicate deuterostome affinities—will be questioned. Should the several molecular phylogenies be divergent, it will be important to find how to weight all available characters (embryological, morphological, functional, molecular) in an effort to resolve these phylogenetic incongruities.

Phoronids also provide a classic example of the importance of studying the complete life history of an organism: One is unlikely to ascertain the eventual roles of coeloms, clumps of red blood cells, or the metasomal sac in the larva by studies of that stage alone. On the other hand, one would be most unlikely to unravel how the gut became U-shaped, how the adult trunk formed, or the origin of the adult metanephridia from examination of only postmeta-

morphic individuals. In their ontogeny, phoronids provide an elegant example of how **heterochrony**—changing the times at which structures or features are developed or become functional—can influence life histories. In phoronids, the precocious formation of adult structures during larval life (this form of heterochrony has been termed **adultation**) has permitted the rapid transition from a well-adapted planktonic larva to a well-adapted tubiculous adult. Thus a long period of metamorphosis with ill-adapted or nonfunctional transitional stages has been avoided, enhancing the survival chances of the incipient juvenile. Although quite bizarre, the early stages of the life history of phoronids facilitate an understanding of the structure of the adult. Unfortunately, that is not the case with the other two lophophorate phyla, as we shall soon describe.

Phylum Brachiopoda

The phylum Brachiopoda flourished during the Paleozoic and Mesozoic eras (with about 12,000 fossil species), but is now on the wane with only about 350 living species. The reader is referred to any recent invertebrate text for an introduction to adult morphology of this interesting group. As with phoronids, all members are marine and sedentary, but the body is protected within two valves, one dorsal and one ventral, rather than a tube. The adult is usually anchored to solid substrates or within burrows in softer sediments by an extension of the metasome called the pedicle. The relationships of the brachiopods are highly disputed, and quite divergent classifications have been proposed. There is frequent argument that the two classes of Brachiopods represent separate phyla, a concept supported in part by recent molecular and other data. The classification shown in Table 15.1 is a conservative one that will be followed in this chapter.

Recent review articles on reproductive biology and development include Chuang (1977, 1983a, 1983b, 1990), James et al. (1992), Long and Stricker (1991), and Nielsen (1990b).

TABLE 15.1   A Classification of the Phylum Brachiopoda

Class Articulata (just under 300 living species)

    Order Rhynchonellidae

    Order Terebratulida

    Order Thecideidina

Class Inarticulata (about 45 living species)

    Order Lingulida

    Order Acrotretida

        Suborder Acrotretidina

        Suborder Craniidina

Reproductive Patterns through Gastrulation in Brachiopoda

As reproductive patterns and early development are fairly uniform throughout the phylum, they will be described in a general section. However, separate coverage will be provided for the four distinctive taxon-specific patterns of later development, two of which involve lecithotrophic larvae and two of which involve planktotrophic stages.

Brachiopods are typically gonochoristic, although a few hermaphroditic articulates (*Argyrotheca*, *Platidia*, and *Pumilus*) are known. Most brachiopods freely spawn their gametes, but some articulates are brooders. The majority of these retain their early embryos within the mantle cavity, but *Argyrotheca* broods within the nephridia. In some mantle brooders the embryos are attached to the lophophore, and *Lacazella* has evaginations of the ventral valve within which young are retained.

The gonads are simple in structure, forming from peritoneal cells that line the metacoel. The testes and ovaries of inarticulates except craniids are contained in the spacious main visceral coelom. In contrast, in both craniids and articulates, the perivisceral portion of the metacoel is small, and the gonads are largely found within the tubular extensions of this coelom into the mantle folds. The paired metanephridia normally serve as gonoducts for both sperm and eggs in all species.

The eggs have only a modest range in size, from about 95 μm in *Lingula* to 150 μm in *Terebratalia transversa*. The female gametes typically are shed as primary oocytes either at the germinal vesicle stage or in first prophase of meiosis. They may be enclosed in an envelope of follicle cells that is soon shed.

As correlated with their free-spawning habits and external fertilization (discussed later), brachiopods have sperm of a primitive type. Such sperm typically have a small acrosome, round nucleus, short midpiece with few mitochondria, and a simple flagellum.

Fertilization is external in both free-spawning and mantle-brooding species, but is probably internal in *Argyrotheca*, which is reported to practice nephridial brooding. In some species, fertilization is followed by formation of a fertilization envelope.

Cleavage may be somewhat asynchronous and irregular, but is clearly radial or biradial, holoblastic, and more or less equal. In *Terebratalia transversa* and some others, a sizable portion of the products of third cleavage have the eight cells in a single ring, while the rest have the expected two tiers of four cells each. In either case, however, the 16-

cell stages have two tiers of eight cells each. The 32-cell stage commonly has two tiers, each with 16 cells in a square array. Further cleavages result in a blastula that develops cilia and then undergoes gastrulation by invagination of the vegetal plate. The early cup-shaped archenteron largely obliterates the blastocoel. The blastopore is initially large and circular, but progressively closes, starting from its posterior end (except in *Crania* in which the blastopore is reported to close from the anterior end). In contrast to phoronids, in brachiopods the blastopore eventually closes completely, so that the relationships of the adult mouth and, when present, anus to this structure are very difficult to observe and remain controversial.

The Later Development of Articulate Brachiopods

The life histories of numerous articulates are well documented and comparison indicates that development in the three orders of articulates (Rhynchonellidae, Terebratulida, and Thecideidina) is surprisingly uniform, with only a few minor variations in detail: all articulates studied produce distinctive, three-lobed larvae that are lecithotrophic, short-lived, and undergo a similar metamorphosis. Curiously, these well known larvae do not have a widely accepted descriptive name.

As development through gastrulation has already been described, we resume the story with subsequent stages: In articulate embryos, while the blastopore is closing, or shortly before, the archenteron becomes partitioned into an elongate median cavity (the future gut) and a U-shaped one that partly surrounds the future gut (anteriorly and laterally in most species, but posteriorly and laterally in some). The subdivision of the archenteron is achieved by a U-shaped curtain of cells (Figure 15.4) that extends from the roof (dorsal surface) of the early archenteron ventrally, dividing it into the two components. This curious method by which the archenteron of articulate brachiopods is divided into gut and coelomic parts has recently been reported to occur also in the inarticulate brachiopod *Crania* (Nielsen 1990b; see below). Outside the brachiopods, similar ledges that subdivide

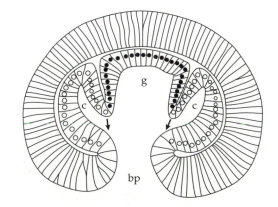

Figure 15.4   Diagrammatic cross section of brachiopod gastrula showing paired lateral ledges that extend in the direction of the arrows and subdivide the archenteron at this level into lateral coelomic compartments (c) and a central gut (g). Cilia of cells are not shown. Nuclei of cells that form the gut and coelomic compartments are shown as filled and open circles, respectively. The drawing is based on *Terebratalia transversa*. Although the ledges dividing the archenteron in this species are a single layer at this stage, they later become double and are shown as such for ease of interpretation. bp, blastopore. (Modified from Long and Stricker 1991.)

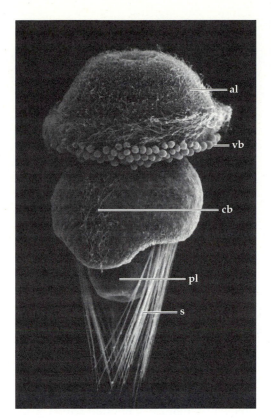

Figure 15.5   SEM of advanced larva of *Terebratalia transversa*. The larval body is about 150 μm in length. Abbreviations: al, apical lobe; cb, midventral band of cilia on the ventral mantle lobe; pl, pedicle lobe; s, larval setae; vb, vesicular bodies of uncertain function. (From Stricker and Reed 1985.)

the larva is uniformly ciliated initially, the cilia on the pedicle and all but a midventral band on the mantle are soon lost (Figure 15.5). The apical lobe retains its overall ciliation, but an apical tuft and a band encircling the posterior margin of the apical lobe become prominent. During the brief larval life, paired lateral and dorsolateral tufts of setae grow from the inner surface of the mantle lobe (Figure 15.5). Although normally directed posteriorly and held against the pedicle lobe, these setae can be splayed in all directions as an apparent defense mechanism if the larva is irritated. Ultrastructural study has shown that these setae are formed exactly as are the setae of polychaetes, each from a single cell with complex apical microvilli (Gustus and Cloney 1972). Although in the same topological position as the mantle setae of adults, these larval setae are transitory structures that will be replaced by adult setae after metamorphosis.

Internally, a somewhat complex musculature develops, but the gut remains a simple sac blind at both ends and no larval excretory organs develop. A simple nervous system including an apical plate forms within the epidermis. **Pigment-cup ocelli** on the apical lobe are reported in one species, and red patches, often referred to as ocelli or eyespots, are found in a corresponding position in other articulate larvae. Although the latter patches seem to involve only epidermal pigmentation, articulate larvae, including some that lack any traces of photoreceptors, are positively phototactic initially, but reverse that response before metamorphosis. In many species, the cells along the posterior margin of the apical lobe each develop a large, vesicle-filled protrusion that remains connected to the cell proper by a narrow stalk. Although their pendant morphology suggests they might be balance organs, these bead-like projections remain of unknown function.

The planktonic life may be only hours long in articulates that brood, but, even in free spawners, is only 4 to 5 days in length. Despite the apparent simplicity of their nervous system, including the lack of obvious sense organs, the larvae are quite selective in choosing the site for metamorphosis.

Metamorphosis is more progressive than dramatic and begins with attachment to a hard substrate by the free end of the pedicle lobe. The mantle lobes, which had surrounded the pedicle, slowly reverse their direction to surround the apical lobe (Figure 15.6). The mantle lobe is now seen to have dorsal and ventral halves; both of these soon begin secretion of the proteinaceous and calcareous layers that form the dorsal and ventral valves of the juvenile. Within a few days, one to three pairs of short **cirri** or **tentacles** develop simultaneously from the apical lobe, which

the archenteron into gut and coeloms are known only in chaetognaths, but coelom formation in these two phyla probably originated independently.

Externally, the incipient larva becomes ciliated and slightly elongated. Two transverse constrictions will partly divide it into three regions called the **apical**, **mantle**, and **pedicle lobes** (anterior to posterior, Figure 15.5). The apical lobe will remain a large, rounded anterior lobe, while the mantle lobe grows posteriorly as a thick fold that largely encloses the small pedicle lobe. Perhaps unexpectedly, these three parts of the larva do not correspond to the proto-, meso-, and metasome of a trimerous organism, but rather become the lophophore, the mantle (which secretes the shells) plus that part of the body that contains the digestive tract and perivisceral coelom or metacoel, and a solid pedicle (or attachment stalk that represents an extension of the trunk), respectively. Further, the U-shaped coelom described above does not become subdivided into the three parts that might be anticipated for a trimerous form. It remains undivided and sends extensions into each of the three larval regions in most species, but becomes divided into two pairs of cavities in some (e.g., *Terebratella*; Percival 1960). The two pairs of coeloms in *Terebratella* are assumed to become the mesocoels and metacoels of the adult; since the epistome or protosome of articulates apparently lacks a coelom (it is filled with connective tissue), the absence of a third coelom may not be surprising.

After hatching, the larva completes its differentiation, but never feeds (development is lecithotrophic). Although

Figure 15.6 Metamorphosis of *Terebratalia transversa*. Diagrammatic sagittal section of recently metamorphosed imago attached to the substratum by its pedicle (pl). The ventral and dorsal mantle lobes have reversed direction and have secreted the pedicle (pv) and brachial (bv) valves, respectively. The interior of the imago contains mesoderm (stippled; coelomic compartments and muscles not shown) and the gut (g). Other abbreviations: al, apical lobe; vb, vesicular bodies.

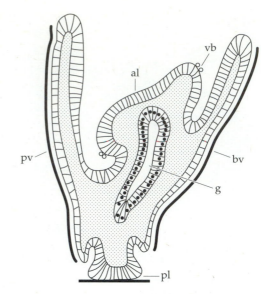

is now contracted against the dorsal valve. An ectodermal invagination or stomodeum forms the mouth and esophagus, permitting juveniles with as few as four cirri to feed. Additional pairs of cirri are sequentially added, allowing the lophophore to take its adult configuration. The differentiation of the gut is not well described, but as adult articulates have a blind-ending gut, an anus never forms. The differentiation of the simple circulatory system, the formation of the one or two pairs of metanephridia, completion of the musculature, addition of the lophophoral skeleton, and the reworking of the coeloms to establish the complex adult system remain largely unstudied. As noted earlier, the site of closure of the blastopore and the relationship of this site to that of the adult mouth remain controversial. A ledge of tissue that overhangs the mouth and corresponds in position and function to the epistome is present in articulates, but the fact that it does not contain a coelomic cavity and has no specific precursor in the larva causes many to doubt its equivalency to the epistome or protosome of trimerous animals.

The Later Development of Lingulids (Class Inarticulata, Order Lingulida)

In contrast to articulate brachiopods that have three-lobed, lecithotrophic larvae and produce shells only after metamorphosis, lingulids have a planktotrophic, shelled larva that develops the pedicle belatedly. The few available studies on development are limited to the genera *Lingula* and *Glottidia*. Following gastrulation and closure of the blastopore, the early embryos of these lingulids become externally regionated into two parts. The more vegetal portion will become the mantle lobe, and this begins growing around the more anterior (animal) portion. The latter is soon recognizable as equivalent to the apical lobe of articulate larvae.

Mesoderm formation and differentiation take place after gastrulation, but are reported to differ between *Lingula* and *Glottidia*. In the former (Yatsu 1902), mesoderm is proliferated from the walls of the archenteron as right and left masses of mesoderm. Ameboid cells from the anterior ends of these masses are said to migrate into the presumptive lophophoral region, but the remaining portions hollow by schizocoely to form a pair of coeloms. In *Glottidia* (Freeman 1995), mesoderm forms as a mass anterior to the blastopore. Although the mode of formation is not reported, a coelom develops within the apical lobe (which will form the lophophore), and mesenchyme cells surround the gut. Unfortunately, further differentiation of the body

cavities is not documented for either species, but eventually there will be paired coeloms within the lophophore, paired coeloms around the viscera with extensions into the mantle lobes, and a single coelom in the pedicle. It is not known whether the cavity of the epistome represents its own coelom or only extensions of the lophophoral coelom. The complex musculature of lingulid adults is developed during larval life.

The early embryo rapidly develops into an incipient larva with two mantle lobes surrounding a body lobe. A stomodeal invagination, reportedly at the site of the closed blastopore, establishes the mouth at one edge of this lobe so that the gut now is complete anteriorly. This opening becomes surrounded by the median tentacle and one pair of cirri, the rudiments of lophophore. (Seemingly, all the extensions of the lophophore could be considered tentacles—as in other lophophorates—or cirri, but in brachiopods the term tentacles is reserved for the median element). The mantle lobe soon produces chitinous dorsal and ventral valves at the posterior end of the embryo. It is at this stage that hatching from the fertilization envelope of the egg occurs in the laboratory, although the youngest stages collected from the plankton have three pairs of cirri. The number of pairs of cirri, the size of the larval valves, and the complexity of the larva increase during the planktotrophic phase (Figures 15.7, 15.8). While the larva is swimming, its cirri radiate outward from the mouth over the edges of the widely gaping valves, much like the spokes of an opened umbrella. The ciliation of the cirri provides locomotion, driving the animal with its mouth forward, as well as a feeding mechanism. The epistome appears at a stage with three pairs of cirri, but the origin of its cavity remains unclear. During further planktonic life, the gut is completed by an anus, rudiments of the nephridia appear, and the addition of concentric bands of juvenile shell material to all but the hinge region converts the embryonic shells to a circular and then to the ellipsoidal shape characteristic of the adult shells. *Lingula* from Japan

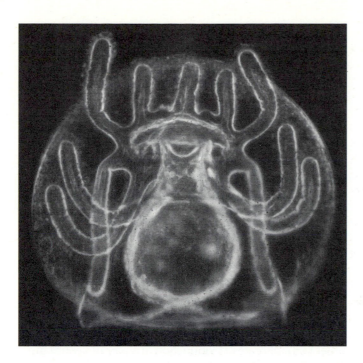

Figure 15.7 Photomicrograph of early *Glottidia albida* larva with still nearly round shells. The five upward-directed structures are the median tentacle and two pairs of cirri. Three additional pairs of cirri are seen on each side of the pear-shaped stomach. The larval mouth is directly below the median tentacle. When the larva is swimming, the tentacle and cirri extend radially outside the larval shells above the mouth.

are estimated to complete embryonic and larval development in 31–32 days (Chuang 1990). Larval life in the lingulid *Glottidia pyramidata* is about 20 days at temperatures of 25–30°C, but the pelagic phase can be markedly prolonged if appropriate substrates are unavailable (Paine 1963). The larvae have a pair of statocysts that presumably are important in orienting the pelagic larvae. Interestingly,

these gravity-sensing organs persist in the benthic and burrow-living adults.

Late in the developmental period, the peduncle of the incipient imago appears and rapidly elongates. The peduncle is technically an outgrowth of the ventral mantle lobe, but, during the pelagic period, the peduncle is coiled between the valves rather than protruding from between them (Figure 15.8).

Competent larvae have almost fully anticipated the imaginal stage with the exception that they are planktonic whereas the juveniles and adults are benthic. Settlement involves little more than protrusion of the pedicle. This muscular structure anchors the juveniles (and adults) in vertical mud burrows within which they are limited to minor vertical shifts up or down. Settlement cues have not been identified, but gregarious settlement must be involved considering that dense aggregations of adults are typical.

It should be obvious that one might question whether *Lingula* and its close relatives have direct or indirect development; that is, should we recognize the free-swimming stage as a larva or as a free-swimming juvenile? Certainly metamorphosis is extremely simple. A few larval tissues or structures (certain muscles and the median tentacle) are lost during larval life or after metamorphosis, indicating that there are at least a few transitory larval structures.

The Later Development of *Discinisca* and *Pelagodiscus* (Class Inarticulata, Order Acrotretida, Suborder Acrotretidina)

As in lingulids, *Discinisca* and *Pelagodiscus* have free-swimming larvae that in many ways are miniature adults. In contrast to lingulids, the fully developed larvae of disciniscids have a maximum of four pairs of cirri in addition to the median tentacle (Figure 15.9). During their ontogeny, these larvae sequentially produce four cohorts of mantle setae, each with a distinctive morphology. It has been argued that the changes in setation should be considered subtle forms of metamorphosis, but it is hard to rationalize that metamorphosis can be a serial process. Discini-

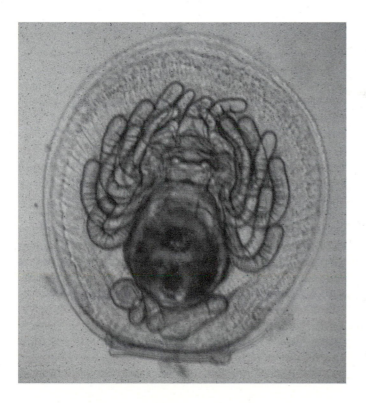

Figure 15.8 Photomicrograph of advanced *Glottidia albida* larva near metamorphosis. Compare with Figure 15.7, noting the increased number of tentacles, the elliptical shape of the shells, and the presence of an elongate pedicle coiled within the shells below the dark stomach. Abundant setae at the edge of each mantle are not readily visible in this micrograph.

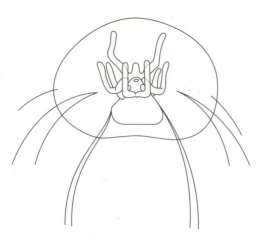

Figure 15.9   Diagram of free-swimming larva of *Discinisca*. Note the similarity with the early *Glottidia* larva (Figure 15.7); however, this larva never has more than four pairs of cirri and is provided with several morphologically distinct setal types. (After Stricker and Reed 1991.)

scid adults attach to hard substrates by a short ventral pedicle that apparently is produced shortly before their settlement.

The Later Development of *Neocrania* (Class Inarticulata, Order Acrotretida, Suborder Craniidinida)

The final known pattern of brachiopod development involves the inarticulate *Neocrania*, which as an adult attaches directly to hard substrata by the ventral valve rather than a pedicle. The early development of this brachiopod has been recently reported by Nielsen (1990b).

All development occurs after free spawning of the 125-μm eggs and results in a larva that is lecithotrophic, in contrast to other inarticulates. In contrast to all other brachiopods, the blastopore of *Neocrania* closes near the posterior rather than the anterior end of the late gastrula. The

larva does not become trilobed as in articulates, but shows a partial division into a rounded anterior head (ciliated and apparently comparable to the apical lobe of articulates) and a dorsoventrally flattened body (Figure 15.10). The absence of a pedicle lobe is understandable, since craniids attach directly the substrate by their ventral valve. Curiously, the larva has four pairs of coelomic sacs, the more posterior three pairs of which are each associated with a pair of dorsolateral setal tufts. Attachment to the substrate at metamorphosis is very close to the posterior end where the blastopore closed, so the adult mouth cannot be related to the blastopore, but the anus might. (Note that the mouth and anus of *Neocrania* are at opposite ends of a more-or-less straight gut, not closely positioned as in other lophophorates.) Contraction of a pair of ventral muscle greatly shortens the ventral side of the larva as it metamorphoses; note that this is opposite to what occurs in phoronids, in which the ventral side becomes greatly elongated at metamorphosis. Shortly, both adult valves will form on the now convex dorsal side of the larva, with the more posterior one (identified as the ventral valve!) anchored directly to the substratum. The fate of the four coeloms is only partially known. The anterior pair is reported to disappear, although it might be expected to form the cavity of the episteome, the second and third pairs are presumed to form the lophophoral and visceral coeloms, respectively, and the fourth is speculated to fuse as the undivided anal chamber that is unique to this group. Presumably, the anterior lobe of the larva becomes the lophophoral region, but its differentiation has not been followed.

Experimental Analysis, Problems to be Solved, and Conclusions

Brachiopod embryology has been subjected to very little experimental analysis. Separation of blastomeres has documented that development is regulative (e.g., Zimmer 1973). Recently, Freeman has analyzed regional specification during embryogenesis in an articulate brachiopod (*Terebratalia transversa*; Freeman 1993) and an inarticulate one (*Glottidia pyramidata*; Freeman 1995) and provided fate maps for these two species.

Although additional comparative studies of the inarticulate groups—including elucidation of the origin of

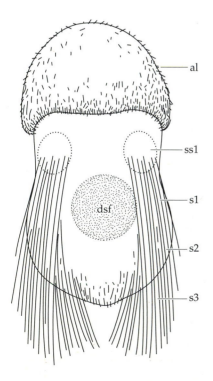

Figure 15.10   Diagram of advanced larva of *Crania anomala* in dorsal view. Abbreviations: al, apical lobe; dsf, dorsal shell field that will secrete the brachial valve after metamorphosis; s1, s2, s3, three setal bundles on right side; ss1, epidermal sac that secreted setal bundle 1. (After Nielsen 1990b.)

the mesoderm and its differentiation—is needed, it is doubtful that this information will clarify our understanding of the relationships of brachiopods with other phyla. However, it should contribute to a resolution of the controversies concerning the possible polyphyletic nature of the group.

Little embryological or morphological evidence exists to relate brachiopods with protostomes, but there are two exceptions. First, the adult mouth may form at the site of closure of the blastopore (a larval mouth is found only in some inarticulates and its formation relative to the blastopore is not documented) and second, the setae and setae-producing cells of larval and adult brachiopods are similar to those of polychaetes even at the ultrastructural level. However, cleavage pattern, regulative potential, and, although not typically enterocoelous, coelom formation all suggest deuterostome affinities. Another deuterostome characteristic, trimery, is evident in all brachiopod adults (although the epistome lacks a coelomic lumen in articulates) and in the larvae of lingulids and disciniscids. In contrast, trimery is falsely represented in the three-part external partitioning in the larvae of articulate brachiopods and is nowhere evident internally in these larvae. The presence of four pairs of coeloms in *Crania* larvae could support trimery if one recognizes that adult craniids have an anal coelomic chamber, which is an apparent subdivision of the metacoel. Unfortunately, the first pair of larval coeloms, which should form the protocoel or the cavity of the epistome in the adult in this scenario, are reported to disappear. However, the formation of neither the epistome (or protosome) nor its cavity (the protocoel) has been elucidated in juvenile craniids; it is conceivable that a reduction in size or transitory collapse of the anteriormost pair of coeloms, as occurs in the protocoel of different phoronid larvae, may have been misinterpreted as loss. Neither the lecithotrophic or planktotrophic larvae of brachiopods can be favorably compared with larvae typical of either protostomes or deuterostomes.

Disciniscids and lingulids are unique among extant brachiopods in having long-swimming larvae that are planktotrophic, traits that we usually consider as primitive ones. However, it must be noted (a) that the larval feeding mechanism is in fact that of the adult not an alternative one as is the case in most taxa and (b) that, because these larvae so closely resemble diminutive adults, adultation, the precocious appearance of adult features in pre-adult stages, must have affected the early stages of both lingulids and disciniscids almost globally. In contrast, articulate brachiopods and craniids have larvae that are lecithotrophic, are only briefly planktonic, and have little if any evidence of adultation other than a crude regionalization into lophophore, mantle, and pedicle precursors. Since lecithotrophic development is usually interpreted as a derived pattern, changes associated with a shift in articulates from planktotrophic development may have masked primitive traits that once existed in the larvae of

ancestral forms. In summary, it is probable that inarticulate larvae as well as articulate ones retain few, if any, original larval features.

Phylum Bryozoa

The phylum Bryozoa (or Ectoprocta) has some 20,000 living and fossil species arranged in three classes with living representatives and several that are extinct. Of the living classes (Gymnolaemata, Phylactolaemata, and Stenolaemata), the gymnolaemates are by far the most successful today in terms of diversity and distribution, as well as abundance. The phylactolaemates are found only in freshwater and not surprisingly have strongly modified reproductive patterns. Curiously, however, in adult morphology, they have closer similarities with phoronids than either of the two other groups that are largely (Gymnolaemata) or entirely (Stenolaemata) marine. One classification of the Bryozoans is shown in Table 15.2.

In contrast to phoronids and brachiopods, bryozoans exist as **colonies** consisting of tens to thousands of individual **zooids** that are interconnected with each other. All the zooids of a colony (with the exception of the sexually produced founding individual or **ancestrula**) originate by asexual budding. Most, if not all, zooids of a colony are capable of feeding themselves and are therefore called **autozooids**. Some zooids, especially of cheilostome gymnolaemates, have become highly specialized for reproductive, defensive, or other functions and are called **heterozooids**. Heterozooids are not capable of feeding and must depend on the colony-wide **funicular system** for their nourishment; although it consists of tube-like structures in only some areas and cordlike strands elsewhere, the funicular system has been compared to a circulatory system. The zooids of both cheilostome gymnolaemates and of stenolaemates have exoskeletons with both calcareous and chitinous components, whereas those of ctenostome gymnolaemates and of phylactolaemates lack the calcified layer. As individual zooids are quite small (about 0.5 mm in length), miniaturization is often listed as a factor to explain the small number of lophophoral tentacles and the absence of nephridia, a typical circulatory system, and, in all but phylactolaemates, the epistome.

TABLE 15.2   *A Classification of the Living Bryozoa*

Class Gymnolaemata (about 3000 extant species)

    Order Ctenostomata

    Order Cheilostomata

Class Stenolaemata

    Order Cyclostomata (about 900 extant species)

Class Phylactolaemata (about 50 extant species)

Information on the relationships and adult anatomy—and to a lesser extent developmental patterns—for this phylum is available in most general invertebrate textbooks and Ryland (1976). More detailed reviews on reproductive biology and development include Hyman (1959), Nielsen (1971, 1990a), Ryland (1976b), Ström (1977), Zimmer and Woollacott (1977a,b), Franzén (1983), and Reed (1987, 1991).

To understand the metamorphosis of most bryozoan larvae, it is important to know two processes that are important in the adult colonies: asexual reproduction and polypide replacement. It will perhaps be easier to first consider polypide replacement: For reasons that should soon be clear, a bryozoan individual or zooid can be considered to consist of two parts, the **cystid** and the **polypide**. In simple terms, the cystid consists of most of the outer body wall and its musculature; technically, it is the nonretractable portions of the metasome, including the epidermis and its underlying mesodermal lining; characteristically, the exoskeleton and main funicular cords are also considered cystid components. The polypide includes the remaining portions, most of which are at least partially moveable, including the tentacle sheath, lophophore, neural ganglion and circumoral connective, gut, and certain muscles (e.g., the retractors). In most bryozoans, there is periodic degeneration and replacement of the polypide, so that the cystid of an individual may house several polypides in sequence. Why replacement of most of the innards of an individual is necessary may be related to the fact that the miniaturized individuals (a) have no excretory system and (b) use the cells lining the stomach for intracellular digestion, resulting in their progressive contamination. Polypide replacement involves degeneration of its multiple components into a compact, globular dark mass, the so-called **brown body**. None of the cells are retained, but presumably there is resorption and reuse of many of the component compounds. With the loss of the polypide, the zooid is reduced to a cystid only. Although consisting only of ectodermal and mesodermal derivatives, this cystid is capable of proliferating a new polypide including a gut. The anlage of the new polypide develops from the body wall at the proximal end of the zooid. Initially only a thickening of the cystids two layers (epidermis and its mesodermal lining), the primordia invaginates to form a simple sac-like structure. The vesicle formed becomes bilobed with an outer portion (nearer the surface) and an inner one. The more superficial part will produce the tentacle sheath, lophophore, mouth and pharynx, while the deeper one differentiates as the remainder of the digestive tract (stomach, caecum, intestine, and anus) (Figure 15.11).

As the polypide matures, its attachment to the surface shifts from near the proximal end of the zooid to near the distal end. New retractor muscles and an operculum must also be produced, but soon the zooid is once again complete and capable of all functions.

Asexual reproduction initially involves the production of a new cystid only, but as should now be obvious, this is capable of forming the polypide needed for a complete zooid. A cystid bud forms as a ballooning out of the body wall, either from the preceding generation of autozooids or from stolon-like individuals (**kenozooids**). As the new cystid nears full size, a perforated partition separates the new cystid from its progenitor, and a polypide bud starts to develop from its frontal surface. The new polypide forms exactly as in polypide replacement. The growth of the asexual bud is achieved thanks to the transfer of nutrients through a system of cords or tubes (the funicular system) that unite the bud and its progenitor.

Important points to note about these related process of asexual budding and polypide replacement are that the unit bryozoan (zooid) can be considered to be composed of a cystid and a polypide, of which the polypide is expendable, since an existing cystid can make a new polypide for itself as well as generating new (daughter) cystids. The ability of epidermal cells (adult skin) to make structures that are considered of both ectodermal and endodermal origins contradicts our usual interpretation that the endoderm of triploblastic animals is an essential germ layer formed only during gastrulation.

Reproductive Biology of Gymnolaemate Bryozoa

Gymnolaemates have varied patterns of sexuality. The large majority of species have colonies in which most of the zooids are reproductive and are hermaphroditic. As one variant, some species have sterile, male, and female zooids, so that, although the colony is hermaphroditic, its individual components are gonochoristic. Finally, there are some species with separate male and female colonies. Commonly, the zooids of hermaphroditic species are protandrous (produce sperm before eggs).

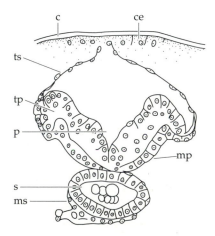

Figure 15.11 Stage in polypeptide differentiation of *Alcyonidium polyoum*. Abbreviations: c, cuticle; ce, cystid epidermis; mp, mesoderm lining pharynx; ms, mesoderm lining stomach; p, pharynx; s, stomach rudiment; tp, tentacle primoridum; ts, tentacle sheath. (After Zschiesch 1909.)

The gametes develop from the peritoneal lining of the metacoel, often in association with the funicular tissue or the body wall or both.

Where known, spermatozoa are released through pores at the tips of some or all tentacles, whereas eggs are released through a small **supraneural pore** between the mouth and anus near the bases of the tentacles. (Bryozoans lack the nephridia that serve as gonoducts in the other lophophorates as well as in many other phyla.) In species with free-spawning of their oocytes (and in a few others), the epidermis of the two tentacles adjacent to the supraneural pore become modified as a vase-shaped chamber through which the eggs must pass after exiting the supraneural pore. This structure, the **intertentacular organ**, is known to serve also as an entry port for spermatozeugmata (see below) in *Membranipora*.

Although found in only a small number of genera, the most primitive developmental pattern within the gymnolaemates (and the phylum) involves the free spawning of small eggs that develop into planktotrophic larvae. Individual zooids of species that reproduce this way can each produce dozens of small eggs daily over many weeks. Their larvae, called **cyphonautes**, are laterally flattened, provided with a pair of triangular valves, and spend perhaps a month or two feeding and growing in the water column. As with other bryozoan larvae, they undergo a dramatic metamorphosis to form a sessile ancestrula that is the progenitor of a colony.

The vast majority of gymnolaemate species brood their young, releasing them as **coronate larvae** that typically lack any trace of a gut and spend but a few hours in the water column before metamorphosing. They are usually identified as being lecithotrophic, but recent studies show they are capable of incorporating at least small quantities of dissolved organics (Jaeckle 1994). Most coronate larvae originate from large, yolk-rich eggs that are retained through embryogenesis in one of several different categories of brood chamber. Most ctenostomes brood their young within the tentacle sheath of the mother, but cheilostomes have a wealth of brooding mechanisms, most of which involve special external chambers. The external brood chambers, which are called **ovicells** or **ooecia**, typically require the participation of the maternal zooid and several additional ones. Except rarely, only a single embryo is brooded at a time; the brooding period is probably two or three weeks in most species. In many species, only a single larva will be produced by an individual, but, in some, two or more larvae are reared in sequence, using the same ovicell. Although the reproductive potential of brooding species might seem limited at first sight, a colony (which could be considered an individual since it arises from a single larva) often consists of thousands of zooids, most of which can reproduce sexually.

There are several other variations in reproductive mode in addition to the two main patterns just described. In a few genera, coronate larvae are produced, but these develop from quite small eggs (around 35 μm); the embryonic stages are brooded in ovicells and there receive extraembryonic nutrients provided by the maternal individual via a placenta-like system derived from her body wall. In all species that have placental nourishment of their embryos, the maternal polypide routinely degenerates during the development of the brooded embryo. (Presumably the polypide provides some of the nutrients needed for growth of the embryo.) However, such degeneration also characterizes species that brood within the tentacle sheath and occasionally occurs in species with other brooding patterns. As a final example of alternative life histories, several genera produce large eggs that are brooded but result in shelled larvae. During their brief planktonic life, these larvae do not feed, but they may have an incomplete gut. Sometimes called **pseudocyphonautes**, these larvae usually share with cyphonautes few features other than the possession of paired valves, but these larvae are also distinctly different from coronate larvae.

Brooded larvae typically are released (or escape) from the brood site shortly after the onset of dawn light (or artificial illumination). The freeing of the larva probably requires participation of both the maternal zooid (e.g., opening of the closure to the brood chamber by muscular action) and the larva (e.g., release of hatching enzymes to weaken the vitelline envelope which surrounded it and active swimming by cilia).

Gametogenesis, Gametes, and Fertilization in Gymnolaemates

The sperm of bryozoans are highly specialized suggesting that fertilization is modified (see below and Franzén 1983). In gymnolaemates, for example, the sperm have a conical head and an elongate midpiece that extends about 50% of the length of the flagellum. The midpiece initially contains two elongate mitochondrial rods, but these become altered beyond recognition during spermiogenesis. Although usually released as individual units, spermatozoa are joined in a bundle as a **spermatozeugma** in *Membranipora*. In this genus, some 32–64 sperm originate by incomplete divisions of a single spermatogonium; as they mature, the sperm become precisely aligned and form a coherent bundle that are freely spawned, but later enter another individual via its intertentacular organ.

Oogenesis results in the simultaneous production of numerous, small, yolk-poor eggs in free-spawning species such as *Membranipora*. In most brooding species, only a single large, yolky egg is ovulated at one time, although several additional generations may mature at intervals of several weeks.

It is commonly reported that fertilization must be external (since there is no obvious route for sperm entry) or involve selfing or both. Silén (1972) provided observations supporting external-fertilization in two species of *Electra* but believed there was cross- rather than self-fertilization.

However, the actual event of sperm entry into a female gamete seems never to have been observed. The identification of sperm nuclei in coelomic eggs of a free-spawning form (*Membranipora*) and in extremely young (previtellogenic) oocytes within the ovary in all brooding species examined suggests internal fertilization is typical (Temkin 1994). In *Membranipora*, fertilization apparently occurs as the individual oocytes (primary oocytes in metaphase of first meiosis) are released from the ovary. In species that brood, fertilization occurs within the ovary, often before vitellogenesis begins. In some species, the last mitotic division of the oogonium is incomplete, resulting in duets of primary oocytes joined by an intracellular bridge. Both cells are in early prophase I of meiosis when one of them is fertilized. The sperm nucleus remains unchanged in the recipient cell, which now undergoes rapid growth. The sister cell remains small and apparently degenerates. Regardless of whether fertilization is early or relatively late, it is internal, but formation of the zygote nucleus does not occur until the oocytes are spawned and have completed their meiotic divisions.

Development of Gymnolaemates with Cyphonautes Larva

Although the early development of cyphonautes larvae is poorly documented in the literature, the more advanced stages are very common in marine plankton and have been well studied (e.g., Stricker et al. 1988a, b). Because the cyphonautes larva is widely considered the most primitive of all bryozoan larvae, we will begin our study of gymnolaemate embryology using it as an example. Later, where appropriate, we will supplement that description with information from coronate development.

Cyphonautes larvae are found in only six genera of gymnolaemates. Its status as the least specialized of all bryozoan larvae is in part supported by the fact that (a) it is the only bryozoan larva that has a functional gut and is planktotrophic, (b) it is found in both cheilostomes and ctenostomes, and (c) the cheilostome genera in which it occurs (e.g., *Membranipora* and *Electra*) are among the most primitive and one of the ctenostome genera (*Alcyonidium*) is relatively primitive. Curiously, however, the two other genera that produce cyphonautes are among the most highly specialized ctenostomes.

Early cleavages have been followed more closely in species with large eggs (that are brooded), but, to the best of our knowledge, the events are similar in forms such as *Membranipora* that freely spawn small eggs. The cleavages are total, synchronous, biradial, and more or less equal. The 16-cell stage has a 4-by-2 array of eight cells in each hemisphere. The fourth division produces two tiers of eight cells each in the animal half and tiers of 12 and four cells in the vegetal one. The fates of cells can be identified for cyphonautes as early as this 32-cell stage. In fact, although cyphonautes and coronate larvae appear to be vastly different, both have the same cell lineages and later have a common suite of structures, most of which represent epidermal derivatives. The eight cells in the most animal tier at the 32-cell stage will form the **apical disc** and **pallial epithelium**, and the second tier of eight cells will form the **corona**. The tier of four cells at the vegetal pole will be internalized at gastrulation, and the remaining tier of 12 cells in the vegetal hemisphere will eventually form the **pyriform complex**, the **larval oral epithelium**, and that part of the epidermis that invaginates as the **internal sac**. The fates of the internal cells are more variable, but all gymnolaemate larvae form larval muscles and undifferentiated primordia of adult structures; only cyphonautes have a functional gut, however. No gymnolaemates have coelomic cavities in the larva, although such cavities are found in the adults and are formed very early in embryogenesis in the other two classes. No adult or larval bryozoans has any type of excretory organ. Table 15.3 documents the fate of different cell groups as general tissues common to all gymnolaemate embryos and identifies the more specific fates of these general tissues in both coronate and cyphonautes larvae. The table identifies the basic fates (column 2) of the four tiers of cells at the 32-cell stage (column 1) as well as the specific fates of the several basic tissues in the larvae of four species. Note that (1) both categories of larvae have the same basic components and (2) cyphonautes larvae are more similar to the coronate larva of *Tanganella* than the coronate larvae of *Bugula* and *Bowerbankia* are to each other. The table also illustrates that the cystid of the ancestrula is formed in three different ways (pallial epithelium plus internal sac, internal sac alone, pallial epithelium alone) and that the formation of the polypide can be quite different in different species, although it is common for the tissue that forms ectodermal and endodermal epithelia to originate from undifferentiated epidermal cells associated with the apical disc.

Gastrulation is quite precocious, occurring at or near the 64-cell stage. In forms with cyphonautes, the cells at the vegetal pole are reported to invaginate, forming an archenteron (see, however, how gastrulation is achieved in coronate larvae below). The blastopore closes at the vegetal pole in forms with coronate larvae, and apparently this closure also occurs in forms with cyphonautes, since their guts are completed by stomodeal (oral) and proctodeal (anal) invaginations. The fate of the internalized cells has not been followed in detail for any species, but these cells undergo rapid proliferation to fill the blastocoel and are thought to represent both endodermal and mesodermal precursors. In cyphonautes, some of the internal cells (presumably endodermal ones) form an ellipsoidal archenteron that will become the stomach, the middle portion of a three-part gut complete with mouth and anus. Unfortunately, the relationship between the blastopore and the larval mouth (or anus) is not clear in any bryozoan; in fact, only cyphonautes and a very few coronate and pseudocyphonautes larvae have mouths. To add further confusion, the mouth and anus of the sexually produced indi-

TABLE 15.3   *Cell Lineages for Cyphonautes and Coronate Larvae*

Origin at 32-cell stage	Basic tissue formed	*Membranipora* (primitive cheilostome with cyphonautes)	*Tanganella* (primitive ctenostome with coronate larva)	*Bugula* (advanced cheilostome with coronate larva)	*Amathia* and *Bowerbankia* (advanced ctenostomes with coronate larva)
1st tier of 8 cells	Apical disc	*Larval brain, epidermal blastema[ep]*	*Larval brain, epidermal blastema[ep]*	*Larval brain; epidermal blastema[ep]*	*Larval brain*
	Pallial epithelium	Extensive, exposed pallial epithelium[ec]; forms paired larval shells	Extensive pallial epithelium[ec], in shallow groove	Small, fully invaginated pallial epithelium	Very extensive, fully invaginated pallial epithelium[ec]
2nd tier of 8 cells	Corona	*Narrow corona, with hundreds of cells in 2 rows*	*Medium-wide corona, with ~40 cells in single row*	*Very wide corona, usually with 32 cells in single row*	*Very wide corona, with ~40 cells in single row*
3rd tier of 12 cells	*Pyriform complex*	*Typical pyriform complex*	*Typical pyriform complex*	*Typical pyriform complex*	*Typical pyriform complex*
	Oral epithelium	*Lining of vestibule, ciliated ridges*	*Ciliated and undifferentiated oral epithelium[ep?,mp?]*	*Ciliated oral epithelium*	Cupiform layer[ep] with external and internal (!) cells; reduced oral ectoderm
	Internal sac	Large, complex internal sac[ec]	Large, complex internal sac[ec]	Very large, complex internal sac[ec]	*Small internal sac with neck region only*
4th tier of 4 cells	*Endodermal cells*	*Functional gut*	*Partial, nonfunctional gut*	*Nutrient-rich cells?*	*Nutrient-rich cells?*
	Mesodermal cells	Mesodermal blastema[mp] associated with apical disc; *complex, striated musculature*; other mesodermal cells[mc?]	Mesodermal blastema[mp] associated with apical disc; *simple, smooth musculature*; other mesodermal cells[mc?]	Mesodermal blastema[mp] associated with apical disc; *simple, smooth musculature*; pigmented cells[mc], other mesodermal cells?	Median cord[mp]; *simple, smooth musculature*; other mesodermal cells[mc?]

Note: Italics identify transitory larval tissues (lost at metamorphosis). Other identification notes follow.

[ep] source of ectodermal and endodermal parts of ancestrular polypide

[mp] source of mesodermal parts of ancestrular polypide

[ec] source of ectodermal part of ancestrular cystid

[mc] source of mesodermal part of ancestrular cystid

? identified fate is not certain

vidual (ancestrular zooid) in all gymnolaemates are formed far from the site of gastrulation.

Other gastrulated cells are presumed to be mesodermal. Some of these form a complex musculature including both striated and smooth elements. In cyphonautes as in coronates, some of the mesodermal cells remain undifferentiated until metamorphosis is initiated and then differentiate as the mesodermal derivatives of the imago or ancestrula. In cyphonautes, one group of such undifferentiated cells underlies the apical disc; this will be the mesodermal blastema of the polypide rudiment. Other mesodermal cells of cyphonautes become lipid-filled and are used as a nutrient store to fuel the complex metamorphosis.

Although initially cylindrical, the growing cyphonautes larva rapidly assumes a laterally flattened, triangu-

lar shape with the apical disc at one apex (Figure 15.12) and with coronal cells ringing the opposing side. The corona surrounds an opening where the entire vegetal hemisphere has been inverted inside the animal hemisphere, forming a cavity called the vestibule. A bilateral pair of shells is secreted by the pallial epidermis that forms the larval surface between the apical disc and corona. The vestibule becomes functionally divided into inhalent and exhalent chambers by a pair of ledges that extend toward each other from the lateral surfaces of the vestibule. Lateral cilia on these ledges draw water into the inhalent chamber and out the exhalent one. Frontolateral cilia capture food particles, perhaps by filtration, and frontal cilia pass these toward the mouth at the apex of the inhalent chamber. The patterning and function of the cilia on the ledges is remarkably similar to that found on adult tentacles (although the frontolaterals may play a more important and different role in larval life).

Figure 15.12 Right lateral view of advanced cyphonautes of *Membranipora membranacea*. The larva is laterally flattened and has right and left shells (not shown) that have the same profile as the larva. The larval measures about 500 mm in height. Abbreviations: a, larval anus; ad, apical disc; adm, main adductor muscle; c, corona; cr, ciliated ridge between inhalent (iv) and exhalent (ev) regions of vestibule; dm, dorsal mucle; e, larval esophagus; i, larval intestine; is, internal sac; m, larval mouths.

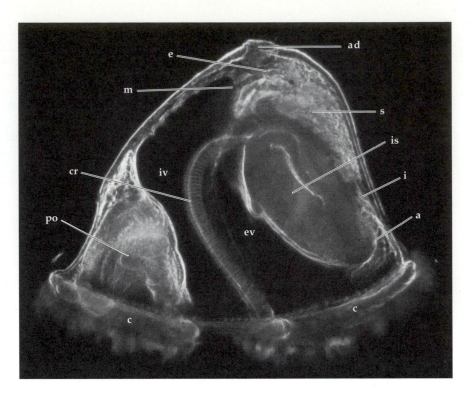

Internally, the gut has been completed, with the anus opening into the exhalent chamber of the vestibule so that feces are carried away. A complex set of striated muscles will develop, but there are no traces of either coelomic cavities or excretory structures.

Two structures characteristic of all gymnolaemate larvae, the pyriform organ and the internal sac, form during larval life of cyphonautes but during embryogenesis in forms with coronate larvae. The first is a glandular-sensory complex just anterior to the entrance to the inhalent chamber of the vestibule. The internal (or adhesive or metasomal) sac forms as a greatly thickened invagination of the vestibular lining (oral epithelium) between the mouth and anus. Both structures play critical roles in metamorphosis, as described below.

In anticipation of metamorphosis, bryozoan larvae seem to crawl along the bottom with the pyriform organ appressed against the substratum. Periodically, the larvae may rotate around a single site and may become temporarily attached by a mucous cord secreted by gland cells of the pyriform organ. The eversion of the metasomal sac signals the onset of permanent attachment. The outermost (neck) region of this sac consists of gland cells that apparently provide a cement for the initial attachment, but then degenerate. The actual anchoring of the juvenile is by the innermost (roof) region of the sac. As the sac spreads against the substratum, the pallial epithelium and the attached larval shells become flattened above it: the positions of the shells of the cyphonautes during larval life and after metamorphosis can be compared with those of the covers of a book when it is closed and opened face down, respectively.

At the boundary between the edges of the everted internal sac and the now flattened pallial epithelium, the corona and the lining of the vestibule (the oral epithelium) have invaginated. The margins of the pallial epithelium and internal sac zip together, forming a continuous flattened vesicle, the epidermis of the cystid. Internally, this epidermis will soon be lined by mesodermal cells, although the origin of these cells is poorly known in cyphonautes (and in most species with coronate larvae). The apical disc soon retracts deep into the interior of the cystid. Note that the apical disc of cyphonautes consists of the larval brain and undifferentiated epidermal blastemal cells (Table 15.4). The sheet of epidermal blastemal cells and a closely fitted layer of mesodermal blastemal cells will form a two-layered cup that is the primordium of the polypide of the future ancestrula. At this stage, which is achieved within a few minutes of the onset of metamorphosis, the individual consists of the body wall or cystid, within which are contained the transitory larval tissues (the larval brain, corona, pyriform organ, larval gut, and vestibular lining) and the still simple rudiment of a polypide (Figure 15.13). Called the **preancestrula,** this stage will change rapidly during the next few days as the cystid takes the characteristic shape and secretes its cuticular (and calcareous) exoskeleton. Internally, the transitory larval tissues undergo degenerative changes and the polypide rudiment grows and differentiates rapidly, forming the lophophore, gut, and other parts of the functional polypide exactly as in polypide replacement and asexual budding. When these changes are completed in one to several days, the lophophore is everted and the new indi-

TABLE 5.4   Comparison of Features of Cyphonautes and Coronate Larvae

Feature	Cyphonautes larvae	Coronate larvae
Shape	Flattened, triangular	Spherical to subspherical
Size of apical disc	Small	Small to large
Pallial epithelium	Forms much of larval surface, secretes paired shells	No shells formed; pallial epithelium invaginated into annular furrow; therefore largely or completely absent from surface proper
Corona	Narrow; hundreds of cells in two rows	Wide to very wide; one row of cells; often only 32 cells.
Oral epithelium	Extensive; invaginated as vestibule with role in feeding	Covers little of surface; not invaginated
Internal sac	Large, complex	Small to very large; simple or complex
Endodermal cells	Form functional gut	Usually no trace of a gut, but remnants may be nutrient-laden cells
Nutrition	Planktotrophic	Lecithotrophic or placental
Musculature	Elaborate, striated	Simple, smooth
Embryonic development	In seawater; 1–2 days from zygote to feeding larva	Brooded; probably about 2 weeks from zygote to (nonfeeding) larva
Length of larval life	4 weeks or so	Few hours

vidual, now called an ancestrula, begins feeding.

Surprisingly, the ancestrula is near the size of zooids in adult colonies and, once functional, never grows further as an individual zooid; however, by forming asexual buds that grow into new zooidal units, it contributes to the establishment and growth of a colony.

Development of Gymnolaemates with Coronate Larvae

A full description of the development and structure of coronate larvae is not necessary since so many of the details are shared with cyphonautes larvae. Therefore, only features that differ significantly between the two forms will be considered below.

Early cleavage is as described for embryos leading to cyphonautes larvae, but the process of gastrulation and the fate of the gastrulated cells are different: In forms with coronate larvae, the four cells adjacent to the vegetal pole divide to produce four internal and four external cells during the sixth cleavage The four internal cells are said to have been gastrulated by primary delamination or more simply to have been internalized by cell division. Interestingly, while still at the 64-cell stage, the four external cells

Figure 15.13 Midsagittal section of just-metamorphosed cyphonautes of *Electra pilosa*. The asterisks (*) identify sites where epidermal tissues from the internal sac (eis) and from pallial tissue (ep) will fuse to complete the cystid epidermis. Other abbreviations: ad, invaginated apical disc with underlying mesodermal cells (polypide rudiment); c, corona; dm, dorsal muscle; e, larval esophagus or pharynx; i, larval intestine; ls, larval shells; po, pyriform organ; s, larval stomach, ve, vestibular epithelium. (After Kupelweiser 1905.)

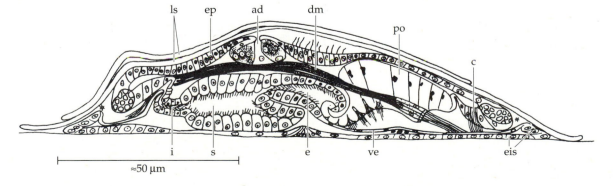

Figure 15.14 Scanning electron micrograph of coronate larva of *Watersipora arcuata* in right lateral view. Abbreviations: ad, apical disc; c, corona; np, neural plate; ois, opening to internal sac; po, pyriform organ. (Micrograph courtesy of Christopher Reed.)

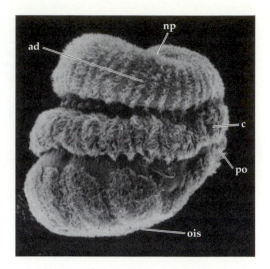

progressively disappear from the surface as they squeeze into the interior of the early embryo. As they become internal, their former site on the surface is overgrown by 8-cells that earlier surrounded them, so that there is never an open blastopore. Thus gastrulation begins and is completed during the 64-cell stage and involves delamination (of the first four cells), ingression (of the other four cells), and epiboly (overgrowth by surrounding cells).

Guts are known in a few coronate larvae, but none function in larval life. Probably all coronate larvae have abundant nutrient-laden cells that may be endodermal remnants. The musculature of coronate larvae is very simple in contrast to that of cyphonautes and consists only of smooth fibers. Undifferentiated cells that will form mesodermally derived structures after metamorphosis and a variety of cell types of uncertain function(s) complete the tissues which completely fill the interior.

Shells, which were present in cyphonautes, are lacking in coronate larvae. The pallial epithelium that secretes the cyphonautes valves is present but is typically invaginated so that, at the larval surface, the corona comes almost into contact with the apical disc. The corona of coronate larvae most commonly consists of but 16 pairs of large cells; these are arranged in a single ring that covers from 25%

(e.g., Watersipora; Figure 15.14) to 90% of the larval surface (e.g., *Bugula* [Figure 15.15] and its relatives and *Bowerbankia* [Figure 15.16] and its relatives).

Since the larvae are usually spherical, it should be apparent that the oral epithelium has not invaginated to form a vestibule as it did in cyphonautes. The metasomal sac invaginates at the site of the vegetal pole, directly opposite the apical disc. In most coronate larvae, the sac is large and complex in cytology as in cyphonautes, similarly functioning in temporary attachment and in production of about half of the cystid epidermis at metamorphosis. The internal sac in *Bugula* and its relatives is especially

Figure 15.15 Diagrammatic sagittal section of larva of *Bugula neretina*. Muscles shown in black; stippled areas represent parenchymal tissue that fills the interior (m). Other abbriviations: c, corona; cc, ciliated cleft; eb, ectodermal blastema of polypide rudiment; igf, inferior glandular field; is, internal sac; mb, mesodermal blastema of polypide rudiment; nn, nerve nodule; np, neural plate; ois, opening to internal sac; pe, pallial epithelium; po, pyriform organ; sgf, superior glandular field; vp, vibratile plume. (After Reed 1991.)

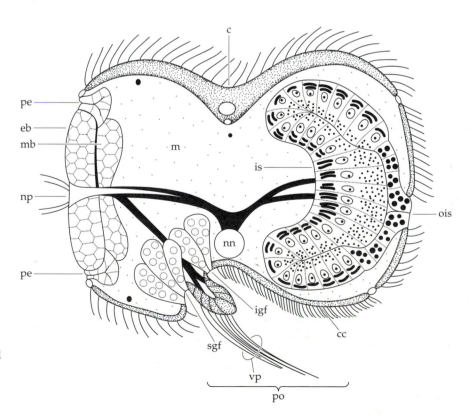

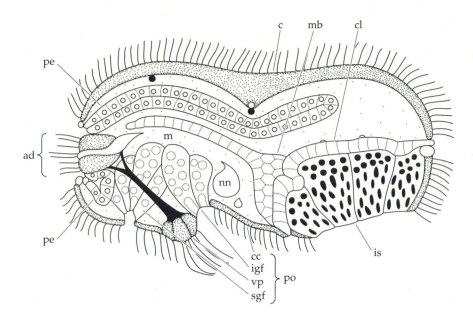

Figure 15.16   Diagrammatic sagittal section of larva of *Bowerbankia gracilis*. Muscles shown in black; stippled areas represent parenchymal tissue that fills the interior (m). Other abbriviations: ad, apical disc with neural plate; c, corona; cc, ciliated cleft; cl, cupiform layer of undifferentiated "ectodermal" cells of polypide rudiment; igf, inferior glandular field; is, internal sac; mb, median band of undifferentiated mesodermal cells of polypide rudiment; nn, nerve nodule; pe, pallial epithelium; po, pyriform organ; sgf, superior glandular field; vp, vibratile plume. (After Reed 1991.)

large, in keeping with the fact it forms *all* of the cystid epidermis (the role of the pallial epithelium becomes very reduced). Interestingly, the sac also serves as the larval part of the placenta-like system in *B. neritina* (and possibly other placental species). In contrast, in a few groups of ctenostome gymnolaemates (e.g., *Bowerbankia* and its relatives), the sac is reduced to the neck region only (which functions only in early attachment before it degenerates) so that none of the cystid epidermis is provided by the internal sac; rather, the pallial epithelium will form all the epidermis of the body wall of the cystid including the permanent attachment disc.

Thus the formation of the cystid epidermis can involve the pallial epithelium and metasomal sac equally or one or the other uniquely. Clearly, joint production of the cystid epidermis is the plesiomorphic (primitive) pattern. The origin of the cystid mesoderm is poorly documented, but it is known to involve pigmented mesodermal cells that were associated with the corona during larval life in *Bugula neritina* and vesiculated cells associated with the pallial epithelium in *Amathia vidovici*.

Curiously, the ectodermal and mesodermal cells that form the polypide rudiment have different origins in different species. The best known examples involve epidermal and mesodermal blastemas associated with the apical discs of coronate larvae of *Bugula* and relatives and of the cyphonautes of *Membranipora* and *Electra*. A quite different origin is reported for *Bowerbankia*; the so-called median band and cupiform layers are believed to form the inner (epidermal) and outer (mesodermal) layers of the polypide rudiment. Unfortunately, the full range of sources for polypide components is as yet unclear.

A majority of coronate larvae are provided with one to many pairs of pigment cup ocelli and other potential sensory structures (chemoreceptors, mechanoreceptors) embedded between adjacent coronal cells. Although cypho-

nautes lack distinctive photic responses, coronate larvae tend to be strongly photopositive and geonegative when first released, but become photonegative or photoneutral and geopositive in a few hours. The initial phase is speculated to enhance distribution during the few hours of larval life, whereas the later, opposite responses are thought to facilitate selection of an appropriate benthic site.

The metamorphosis of coronate larvae is fundamentally identical to that of cyphonautes. A preancestrula is formed in the first few minutes after settlement of the larva and consists of a cystid containing larval transitory tissues and a polypide rudiment. Differentiation of the cystid and polypide lead to the formation of a functional ancestrula that is the progenitor of the colony.

Development of Stenolaemate Bryozoans

Stenolaemate reproduction involves a rare mode of sexual reproduction called **polyembryony**, in which a single zygote produces multiple offspring. Identical twins in humans involve a similar mechanism, but stenolaemates can produce some one hundred to many hundreds of larvae from each zygote. To accommodate such large broods, stenolaemates have unusually large embryo chambers called **gonozooids** produced by the hypertrophy of single female zooids or by the fusion of the chambers of a number of zooids. Since the eggs are quite small, polyembryony also requires an efficient means of providing extraembryonic nutrition to the developing embryos. In contrast to gymnolaemates, only a small percentage of the individuals in a stenolaemate colony reproduce; the fertile individuals are usually gonochoristic, but may be hermaphroditic. The female-acting gonozooids produce a single egg that is fertilized quite early.

Cleavage of the zygote is poorly documented, and typical blastula and gastrula stages do not occur. The early blastomeres commonly separate from each other, but eventually reaggregate to form a hollow embryo that is not a blastula since its walls consist of two cell layers. Presumably, the two layers arose by rearrangement of cells, and this would have to be considered gastrulation. Comparison with other bryozoan developmental patterns suggests that the internal cells are mesodermal and that this stage is a cystid. If these assumptions are true, we should now regard the blastocoel as a coelomic cavity (it would represent the metacoel or visceral coelom, considering subsequent development) and conclude that gastrulation directly results in coelom formation, but not gut formation! This bilayered vesicle is called the **primary embryo** in stenolaemates. The primary embryo increases in size and then develops irregular projections or lobes that pinch off small **secondary embryos**. They are more or less spherical and, again, are hollow and two-layered. The secondary embryos will usually differentiate as larvae, but the fragmentation of the secondary embryos to produce **tertiary embryos** occurs in some species. Differentiation of the secondary (or tertiary) embryos to larvae remains incompletely described.

Before describing larval structure and metamorphosis, we should note that the earlier developmental stages are retained inside the gonozooid. At its inception, the oocyte is associated with a polypide rudiment that soon degenerates. The developing egg is surrounded by primary follicle cells and later stages of embryogenesis will be embedded in a meshwork of cells that presumably are a source of or pathway for nutrients to the embryos. The origins and functions of both the follicular and putative nutrient tissues are obscure, but may form from the membranous sac (a coelomic compartment unique to stenolaemates).

On exposure of fertile colonies to light, swarms of larvae will escape through a specialized opening at the distal end of each gonozooid. The larvae are small (about 100 μm) and spherical (or discus-shaped) with uniformly ciliated surface cells (equivalent to the corona in coronate larvae). In contrast to gymnolaemate larvae, there is no pyriform organ. Although not obvious, there is a pore at both the leading and trailing ends of the larva (Figure 15.17). Both openings are the sites of epidermal invaginations. As one might guess, the posterior opening leads to an internal sac composed of glandular cells. At the other end, a conical sheet of squamous cells leads to a flattened disc of columnar epidermis which is underlain by a layer of cuboidal mesodermal cells. Clearly some epidermal cells of this anterior invagination are pallial epithelium and one would expect there to be an apical disc. No trace of a larval brain is evident, but whether there is an epidermal blastema as seen in many gymnolaemates is controversial (see below). Internally, there is a small population of vacuolated cells (potentially endodermal, but the fate of which is unknown), but no muscles, coeloms, or excretory structures.

After a very brief dispersal period (often only minutes long), metamorphosis is initiated by the eversion of the internal sac; this sac becomes a flattened disc attached to the substratum, anchoring the incipient imago. Simultaneously, those cells invaginated at the opposite pole (all of which have secreted a cuticle at their apical surfaces) evert as a hemispherical dome. The ciliated coronal cells invaginate into the larval interior between this expanding dome and the internal sac. Although widely separated in the larva, the edges of the dome (that corresponds to pallial epithelium of gymnolaemates) and internal sac disc are now adjacent to each other, thanks to internalization of the coronal cells, and they soon fuse with each other, forming the cystid epidermis. The hemispherical individual is called the primary disc in stenolaemates and represents a preancestrula. Internally, the coronal cells degenerate. The origin of the two-layered vesicle that forms inside the preancestrula and will later differentiate as the polypide is controversial. Barrois (1886) deduced from a study of intact imagos that the disc of cuboidal mesodermal cells and the overlying epidermis invaginated to form this structure. However, Nielsen (1970) concluded that the invaginated epidermal cells form only the tentacle sheath and that the cuboidal mesodermal cells proliferate and are the sole source of the two-layered vesicle that is the primordium of the polypide; a similar origin of the polypide was reported by Ostroumoff (1886). Obviously, clarification of this point is important.

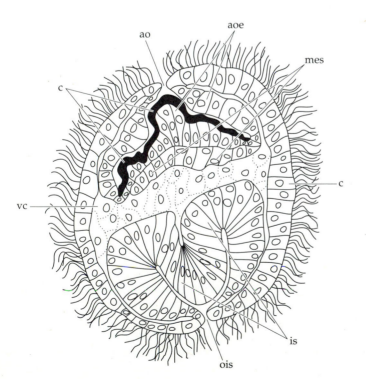

Figure 15.17 Longitudinal section of larva of *Berenicea patina*. Abbreviations: ao, anterior opening; aoe, invaginated aboral epithelium; c, corona; cu, cuticle covering aboral epithelium; is, internal sac; mes, mesodermal cells underlying central part of aboral epithelium; ois, opening to internal sac; vc, vacuolated cells. (After Nielsen 1970.)

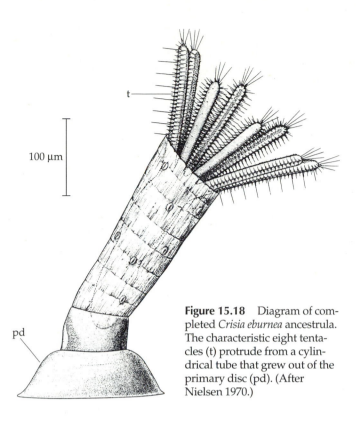

Figure 15.18 Diagram of completed *Crisia eburnea* ancestrula. The characteristic eight tentacles (t) protrude from a cylindrical tube that grew out of the primary disc (pd). (After Nielsen 1970.)

(ovisacs) within which the development stages undergo considerable growth.

As in other bryozoans, the sperm are highly specialized, suggesting the fertilization process is a modified one. Unfortunately, little is known about fertilization, but it is presumed to be internal.

In its maternal functions, an individual will produce but one ovisac containing a single embryo that develops from a small and yolk-poor egg. How the egg is transferred from the ovary into the ovisac remains unclear. Interestingly, up to 50 oocytes surrounded by follicle cells develop within the ovary and continue to grow during early embryogenesis. These oocytes as well as the maternal polypide seem logical sources of extraembryonic nutrients for the growing embryo, but, in at least some species, the degeneration of these structures does not occur until after release of the larva.

Early cleavages are poorly known, but result in an irregular mass of cells that, with further cleavages, hollows as an elongate, single-layered, blastula-like stage. Gastrulation involves the unipolar proliferation of cells into the blastocoel from either the more superficial (distal) end of the blastula or the deeper one. These cells line the blastocoel of the one-layered sac to establish a two-layered one (Figure 15.19). Recall that a similar stage exists in stenolae-

In the meantime, external changes have occurred. The primary disc has developed a calcareous covering, and a short tube now extends from it; the developing polypide will eventually occupy this tube, and the simple lophophore will be protruded from its distal end (Figure 15.18).

It should be obvious that (a) stenolaemate larvae can be interpreted, at least in part, as dramatically simplified versions of gymnolaemate ones and (b) metamorphoses in these two groups have a number of parallels. Considering the mode of polypide formation during the metamorphosis of many cheilostome and during polypide replacement and asexual budding of bryozoans in general, the reported formation of this composite structure from internal (mesodermal?) cells only seems improbable; however, such an origin may be parallel to that seen during the metamorphosis of *Bowerbankia* and its relatives.

Development of Phylactolaemate Bryozoans

Although many mysteries remain about the reproductive stages of stenolaemates, even more surround phylactolaemate development. As with gymnolaemates, most individuals in phylactolaemate colonies are capable of sexual reproduction and are hermaphroditic. In all species, brooding occurs in saclike invaginations of the body wall

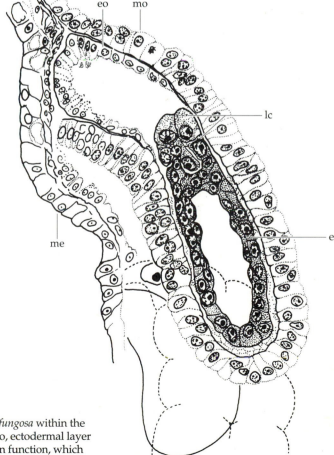

Figure 15.19 Longitudinal section through early embryo of *Plumatella fungosa* within the maternal embryo sac. Abbreviations: e, two-layered embryo (= cystid); eo, ectodermal layer of embryo sac; mo, mesodermal layer of ovisac; lc, large cells of uncertain function, which soon degenerate; me, epidermis of maternal zooid. (After Brien 1953.)

Figure 15.20 Three-dimensional diagram of swimming "larva" of *Plumatella fungosa*, containing two fully differentiated zooids. This stage swims in the direction of the large arrow. Abbreviations: ae, adult or ancestrular epidermis; g, gut of one zooid; l, lophophore of other zooid; le, larval epidermis. (After Brien 1953.)

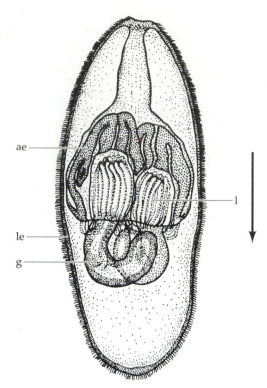

mates (see above) and that this probably represents a cystid containing a coelomic cavity. In a few species, cells presumed to be endodermal are proliferated either internally or *externally* at the distal end. Whether internal or external and whether endodermal or not, these cells apparently make no contribution to the eventual imago.

The bilayered cystid grows, apparently receiving nutrients thanks to a ringlike thickening near the equator of embryo and a corresponding thickening of the adjacent wall of the ovisac that together form a placenta. As it enlarges, the cystid develops a budding zone at one end. Here, depending on the species involved, one to four fully formed polypides will differentiate; in some species, one or more simple polypide buds are also formed. These polypides (and early buds) are formed exactly as those formed during asexual production in adult colonies.

Late in embryogenesis, near the equator of the embryo, a second ringlike thickening of its body wall appears. This thickening develops as a two-layered fold, the mantle, which progressively grows toward the budding end, eventually closing over that end except for a small pore. During its extension, the mantle ruptures the placental connection between the developing individual and its mother. The outer fold of mantle cells and the end of the original cystid that was not overgrown by the mantle develop ciliation and collectively form the surface of the incipient larva.

Eventually, the larva escapes, apparently through rupture of the body wall rather than by escape through the pore leading into the brood cavity. Externally the elongate larva is almost featureless except for the nearly uniform ciliation (Figure 15.20). However, it swims with one end forward; at this forward end the cilia are longer, and both gland and neural cells are found. At the trailing end is the pore that leads inward to the one to four zooids formed at the budding end of the embryo. After a swimming period of a few minutes to a day, the larva settles by its anterior end. The ciliated larval skin progressively contracts to form a compact wad that soon degenerates. During this process, the zooids that had been concealed inside at the trailing end are exposed. A portion of the adult skin will grow toward the substratum, surrounding the shriveled

larval epidermis and providing the definitive attachment of the colony. Within minutes of attachment of the larva, the polypides can be everted, permitting the miniature colony to begin feeding and growth (Figure 15.21).

In summary, phylactolaemate development is significantly different from those of both stenolaemates and gymnolaemates. Although placental relationships exist between embryo(s) and mother in each of the three groups, different components are apparently used in each class. One might propose that the ciliated epidermis of phylacto-

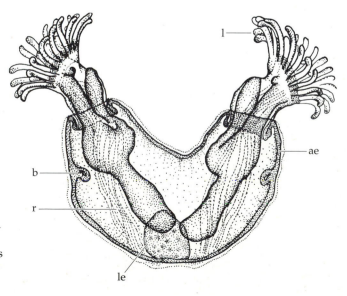

Figure 15.21 Three-dimensional diagram of ancestrula of *Plumatella fungosa*, whose two functional zooids have everted lophophores. Abbreviations: ae, ancestrular epidermis, which now forms body wall of the early colony; b, asexual bud, l, lophophore; le, larval epidermis, which is undergoing degeneration inside ancestrula body cavity or coelom; r, retractor muscles used to retract lophophore. (After Brien 1953.)

laemate larvae represents a corona, that the anterior attachment site shares one function at least with the internal sac, and that the budding zone might be equated with the polypide rudiments found in the other two classes, but it is hard to imagine that cell lineage studies would validate any of these comparisons. It is probably justifiable to call the free-swimming stage a larva since it has a distinctive morphology, is planktonic, and must undergo settlement and a metamorphosis with loss of at least one transitory tissue (the ciliated epidermis) to establish the juvenile colony. An alternative interpretation might be that embryogenesis led to formation of a cystid that was retained within the ovisac, where it developed into a potential colony provided with a secondarily formed, ephemeral locomotory system unrelated to the coronas of other bryozoans. Regardless, it is obvious that extreme adultation has occurred, permitting the almost instantaneous establishment of a miniature functional colony in these freshwater forms. (Although the developmental history of no truly freshwater gymnolaemate species is known, the little information available clearly indicates that the parallels will be with other gymnolaemates not with phylactolaemates.)

Experimental Analysis, Problems to be Solved, and Conclusions

Considering how little work has been completed on the early embryology of bryozoans, it may not be surprising that most studies are observational. The very limited experimental reports include confirmation that cleavage is regulative (such cleavage was documented in a free-spawning gymnolaemate species and clearly occurs normally during polyembryony of stenolaemates), one study in which morphogenetic movements during metamorphosis have been manipulated, and several studies focused on natural and artificial stimuli for metamorphosis.

Two of the most pressing problems in bryozoan life histories are related: one concerns the fate of cells that are internalized during gastrulation and the other has to do with the origin of the inner organs of adults. In other phyla, the gastrulated cells typically differentiate during the developmental stages into readily recognizable mesodermal and endodermal tissues or organs that are retained entirely or in large part through metamorphosis—this is not the case in most bryozoans. Let us first consider the endoderm. Recall (a) that larvae with functional guts are found only in a few genera of gymnolaemates while the others usually have no more than scattered cells of presumed endodermal origin and (b) that in the larvae of the two other bryozoan groups there are *no* clear endodermal cells or tissues. Even when larvae have functional guts, the digestive organs of the juvenile formed at metamorphosis do not originate from the larval gut or from gastrulated cells, but from various parts of the larval epidermis (or possibly from mesodermal cells in stenolaemates). Similarly, the guts of subsequent zooids originate from the body wall epidermis during both asexual bud-

ding in the adult and polypide replacement. Clearly, endoderm is not an "essential" germ layer, as it is often considered. Relative to mesoderm, its formation and differentiation in the larva remain poorly studied. Similarly, the origin(s) of mesodermal structures of the adult (muscles, coeloms, gonads, and funicular tissue) are virtually unknown. During embryogenesis, there is clearly neither a 4d cell as is characteristic of protostomes nor, in any bryozoan, archenteric pouches as in deuterostomes. As noted earlier, with the exception of muscles, gymnolaemate larvae lack coeloms and other differentiated structures of obvious mesodermal origin. The cavity found in the early two-layered cystid-like stages of both stenolaemates and phylactolaemates, if it is truly a mesoderm-lined coelom, arises exceptionally early and perhaps by a method unique in the animal kingdom. Certainly, its mode of formation is unrelated to the schizocoely of protostomes or the archenteric pouching of typical deuterostomes. It seems probable that the cell-lined cavity of phylactolaemate larvae is retained as the visceral coelom of the ancestrular polypides, but in stenolaemate larvae the cavity itself is lost and the fate of the cells that lined it in the early embryo is obscure. Thus, since there is general agreement that larval muscles do not persist through metamorphosis, no mesodermal structures of adult zooids are preformed in the larva, with the probable exception of the metacoel in phylactolaemates. Precursor cells that may be either scattered or organized into compact groupings have been identified (many only tentatively), but their differentiation into muscles, coeloms, funicular tissue, and gonadal tissue of zooids remains virtually unexplored. Since some of the mesodermal primordia are reported to be of epidermal origin, mesoderm in bryozoan may also defy our conventional understanding of the three germ-layer concept.

Additional gaps in our knowledge are abundant. Among gymnolaemates, there are a number of larval types (e.g., pseudocyphonautes) for which detailed studies are needed. Such comparative studies of additional gymnolaemates will contribute significantly to our understanding of the diversity of larval forms and will very probably identify as yet undiscovered primordia of adult structures. However, it is doubtful that the studies will clarify our understanding of the relationships within this class. In all three classes, the biology of fertilization of the oocyte and activation of the zygote await detailed study. Finally, multiple unanswered questions concerning the reproductive biologies and early life histories of stenolaemates and phylactolaemates have been specifically identified earlier in this chapter and will not be repeated here.

Considering those embryological features that are believed to be informative, the only two incontestable ones, radial cleavage and regulative development, are both deuterostome traits. In the vast majority of species, the blastopore closes precociously at the vegetal pole; in the few forms that have a functional gut (or a partial, nonfunctional one) the site of formation of the larval mouth may be near that of blastopore closure, but no direct observations have been made as confirmation. Even when a larval mouth is present, its site and that of the mouth of the ancestrular zooid are unrelated. As mentioned earlier, neither the origin of mesoderm nor the formation of coeloms has clear parallels in either protostomes or deuterostomes. In contrast to both brachiopods and phoronids, bryozoan larvae lack *any* suggestion of trimery and, in parallel with both, are extremely different from either trochophore larvae of protostomes or representative larvae of deuterostomes.

Because of the complexity and heavily degenerative nature of its metamorphosis, the cyphonautes larva, the most primitive of existing bryozoan larvae, appears in fact to be highly specialized rather than being close to an ancestral stage. It is quite possible that the miniaturization of adult individuals, the polypide-cystid duality, and the advancement of regenerative mechanism into the early life history have so distorted developmental stages in this group that embryology provides few reliable clues to the history of the group either with respect to their relationship with other phyla or the mode of development in ancestral forms.

Literature Cited

Afzelius, B. A. and M. Ferraguti. 1978. Fine structure of brachiopod spermatozoa. *J. Ultrasturct. Res.* 63: 308–315.

Barrois, J. 1886. Mémoire sur la métamorphose de quelques Bryozoaires. *Biblitque. Ec. ht. Etud., Paris, Sect. Sci. Nat.,* 32: 1–94.

Bartolomaeus, T. 1989. Ultrastructure and relationship between protonephridia and metanephridia in *Phoronis muelleri* (Phoronida). *Zoomorphology* 109: 113–122.

Brien, P. 1953. Etude sur les Phylactolemates. *Ann. Soc. Roy. Zool. Belg.* 84: 301–440.

Chuang, S. H. 1977. Larval development of *Discinisca* (inarticulate brachiopod). *Amer. Zool.* 17: 39–53.

Chuang, S. H. 1983a. Brachiopoda. In *Reproductive Biology of Invertebrates, Vol. I: Oogenesis, Oviposition, and Oosorption*, K. G. Adiyodi and R.G. Adiyodi (eds.). John Wiley, Chichester, pp. 571–584.

Chuang, S. H. 1983b. Brachiopoda. In *Reproductive Biology of Invertebrates, Vol. II: Spermatogenesis and Sperm Function*, K. G. Adiyodi and R.G. Adiyodi (eds.). John Wiley, Chichester, pp. 514–530.

Chuang, S. H. 1990. Brachiopoda. In *Reproductive Biology of Invertebrates, Vol. IV, Part B: Fertilization, Development and Parental Care*, K.G. Adiyoidi and R.G. Adiyodi, (eds.). John Wiley, Chichester, pp. 211–254.

Conway Morris, S. and J.S. Peel, 1995. Articulated halkieriids from the Lower Cambrian of North Greenland and their role in early protostome evolution. Phil. Trans. R. Soc. Lond. B. 347: 305-358.

Emig, C. C. 1977. The embryology of the Phoronida. *Amer. Zool.* 17: 21–37.

Emig, C. C. 1982. The biology of phoronids. *Adv. Mar. Biol.* 19: 1–90.

Emig, C. C. 1983. Phoronida. In *Reproductive Biology of Invertebrates, Vol. I: Oogenesis, Oviposition, and Oosorbtion*, K. G. Adiyodi and R. G. Adiyodi (eds.). Academic Press, Chichester, pp. 535–542.

Emig, C. C. 1990. Phoronida. In *Reproductive Biology of Invertebrates, Vol. IV, Part B: Fertilization, Development and Parental Care*, K.G. Adiyodi and R.G. Adiyodi (eds.). John Wiley, Chichester, pp. 165–184.

Field, K. G., G. J. Olsen, D. J. Lane, S. J. Giovannoni, M. T. Ghiselin, E. C. Raff, N. R. Pace and R. A. Raff. 1988. Molecular phylogeny of the animal kingdom. *Science* 239: 748–753.

Franzén, Å. 1983. Bryozoa Ectoprocta. In *Reproductive Biology of Invertebrates, Vol. 2: Spermatogenesis and Sperm Function*. K. G. Adiyodi and R.G. Adiyodi (eds.). John Wiley, Chichester, pp. 491–504.

Franzén, Å. and K. Ahlfors. 1980. Ultrastructure of spermatids and spermatozoa in *Phoronis pallida*, Phylum Phoronida. *J. Submicrosc. Cytol.* 12: 585–597.

Franzén, Å. and T. Sensenbaugh. 1983. Fine structure of the apical plate in the larva of the freshwaer bryozoan *Plumatella fungosa* (Pallas) (Bryozoa: Phylactolaemata). *Zoomorphology* 102: 87–98.

Freeman, G. 1991. The bases for and timing of regional specification during larval development in *Phoronis. Dev. Biol.* 147: 157–173.

Freeman, G. 1993. Regional specification during embryogenesis in the articulate brachiopod *Terebratalia. Dev. Biol.* 160: 196–213.

Freeman, G. 1995. Regional specification during embryogenesis in the articulate brachiopod *Glottidia. Dev. Biol.* 172: 15–36.

Funch, P. and R. M. Kristensen. 1996. Cycliophora is a new phylum with affinities to Entoprocta and Ectoprocta. *Nature* (Lond.) 378: 711–714.

Gustus, R. M. and R. A. Cloney. 1972. Ultrastructural similarities between setae of brachiopods and polychaetes. *Acta Zool.* 53: 155–174.

Halanych, K. M., J. D. Bacheller, A. A. Aquinaldo, S. M. Liva, D. M. Hillis and J. A. Lake. 1995. Evidence from 18s ribosomal DNA that the lophophorates are protostome animals. *Science* 267: 1641–1643.

Hay-Schmidt, A. 1987. The ultrastructure of the protonephridia of the actinotroch larva (Phoronida). *Acta Zool.* 68: 35–47.

Hay-Schmidt, A. 1989. The nervous system of the actinotroch larva of *Phoronis muelleri* (Phoronida). *Zoomorphology* 108: 333–351.

Hay-Schmidt, A. 1990a. Distribution of catecholamine-containing, serotonin-like and neuropeptide FMRFamide-like immunoreactive neurons and processes in the nervous system of the actinotroch larva of *Phoronis muelleri* (Phoronida). *Cell Tissue Res.* 259: 105–118.

Hay-Schmidt, A. 1990b. Catecholamine-containing, serotonin-like and FMRFamide-like immunoreactive neurons and processes in the nervous system of the early actinotroch larva of *Phoronis vancouverensis* (Phoronida): Distribution and development. *Can. J. Zool.* 68: 1525–1536.

Herrmann, K. 1976. Untersuchungen über Morphologie, Physiologie und Ökologie der Metamorphose von *Phoronis mülleri* (Phoronida). *Zool. Jahrb. Abt. Anat.* 95: 354–426.

Herrmann, K. 1986. Die Ontogenese von *Phoronis mülleri* (Tentaculata) unter besonderer Berüchsichtigung des Mesodermdifferenzierung und Phylogenese des Coeloms. *Zool. Jahrb. Abt. Anat.* 114: 441–463.

Hyman, L. H. 1959. *The Invertebrates, Vol. VI: Smaller Coelomate Groups: Chaetognatha, Hemichordata, Pogonophora, Phoronida, Ectoprocta, Brachiopoda, Sipunculida: The Coelomate Bilateria.* McGraw-Hill, New York.

Jaeckle, W. B. 1994. Rates of energy consumption and acquisition by lecithotrophic larvae of *Bugula neritina* (Bryozoa, Cheilostomata). *Mar. Biol.* 119: 517–523.

James, M. A., A. D. Ansell, M. J. Collins, G. B. Curry, L. S. Peck and M. C. Rhodes. 1992. Biology of living brachiopods. *Adv. Mar. Biol.* 28:175–387.

Kupelwieser, H. 1905. Untersuchungen über den feineren Bau und die Metamorphose des Cyphonautes. *Zoologica* (Stuttgart) 19: 1–50.

Lacalli, T. C. 1990. Structure and organization of the nervous system in the actinotroch larva of *Phoronis vancouverensis. Philos. Trans. Roy. Soc. London B* 227: 655–685.

Long, J. A. and S. A. Stricker. 1991. Brachiopoda. In *Reproduction of Invertebrates. Vol. VI: Lophophorates and Echinoderms*, A. C. Giese, J.

S. Pearse and V. B. Pearse (eds.). pp. 1–35. Boxwood Press, Pacific Grove, CA.

Mackey, L. Y., B. Winnepenninckx, R. de Wachter, T. Backeljau, P. Emschermann and J. R. Garey. 1996. 18S rRNA suggests that Entoprocta are protostomes, unrelated to Ectoprocta. *J. Mol. Evol.* 42: 552–559.

Nielsen, C. 1970. On metamorphosis and ancestrula formation in cyclostomatous bryzoans. *Ophelia* 7: 217–341.

Nielsen, C. 1971. Entoproct life cycles and the entoproct/ectoproct relationship. *Ophelia* 9: 209–341.

Nielsen, C. 1985. Animal phylogeny in light of the Trochea theory. *Biol. J. Linn. Soc.* 243–299.

Nielsen, C. 1987. Structure and function of metazoan ciliary bands and their phylogenetic significance. *Acta Zool.* 68: 205–262.

Nielsen, C. 1990a. Bryozoa Ectoprocta. In *Reproductive Biology of Invertebrates, Vol. IV, Part B: Fertilization, Development and Parental Care,* K. G. Adiyodi and R. G. Adiyodi (eds.). John Wiley, Chichester, pp. 185–200.

Nielsen, C. 1990b. The development of the brachiopod *Crania (Neocrania) anomala* (O. F. Müller) and its phylogenetic significance. *Acta Zool.* 72: 7–28.

Ostroumoff, A. A. 1886. Contribution à l'étude zoologique et morphologique des bryozoaires du Golfe de Sébastopol. *Arch. Slaves Biol.* 1: 557–569; 2: 184–190; 2: 329–355.

Paine, R. 1963. Ecology of the brachiopod *Glottidia pyramidata. Ecol. Mono.* 33: 187–213.

Percival, E. 1960. A contribution to the life-history of the brachiopod, *Terebratella inconspicua* Sowerby. *Trans. Roy. Soc. New Zealand* 74: 1–23.

Rattenbury, J. C. 1954. The embryology of *Phoronopsis viridis. J. Morphol.* 95: 289–349.

Reed, C. G. 1985. The many motors of morphogenesis: The roles of muscles, cilia, and microfilaments in the metamorphosis of marine bryozoans. In *The Cellular and Molecular Biology of Invertebrate Development, Vol. 15, Belle W. Baruch Series in Marine Biology,* R. H. Sawyer and R. M. Showman (eds.), pp. 197–219. University of South Carolina Press, Columbia.

Reed, C. G. 1987. Bryozoa. In *Reproduction and Development of Marine Invertebrates of the Northern Pacific Coast: Data and Culture Methods for the Study of Eggs, Embryos, and Larvae,* M. F. Strathmann (ed.), pp. 494–510. University of Washington Press, Seattle and London.

Reed, C. G. 1991. Bryozoa. In *Reproduction of Marine Invertebrates, Vol. 6: Lophophorates and Echinoderms,* A. C. Giese and J. S. Pearse (eds.). Boxwood Press, Pacific Grove, CA, pp. 85–245.

Rowell, A. J. 1960. Some early stages in the development of the brachiopod *Crania anomala* (Müller). *Ann. Mag. Nat. Hist.,* Ser. 13, 3: 35–52.

Ryland, J. S. 1976. Behaviour, settlement and metamorphosis of bryozoan larvae: A review. *Thalass. Jugosl.* 10: 239–262.

Siewing, R. 1974. Morphologische Untersuchungen zum Archicoelomatenproblem. 2. Die Körpergliederung bei *Phronis muelleri* de Selys Longchamps. Ontogenese—Larva—Metamorphose—Adultus. *Zool. Jahrb. Abt. Anat.* 93: 275–318.

Silén, L. 1954. Developmental biology of Phoronidea of the Gullmar Fjord area (West coast of Sweden). *Acta Zool.* 35: 215–257.

Silén, L. 1972. Fertilization in the Bryozoa. *Ophelia* 10: 27–34.

Stricker, S. A., C. G. Reed, and R. L. Zimmer. 1988a. The cyphonautes larva of the marine bryozoan *Membranipora membranacea.* I. General morphology, body wall, and gut. *Can. J. Zool.* 66: 368–383.

Stricker, S. A., C. G. Reed and R. L. Zimmer. 1988b. The cyphonautes larva of the marine bryozoan *Membranipora membranacea.* II. Internal sac, musculature, and pyriform organ. *Can. J. Zool.* 66: 384–398.

Ström, R. 1977. Brooding patterns of bryozoans. In *Biology of Bryozoa,* R. M. Woollacott, and R. L. Zimmer (eds.). Academic Press, New York, pp. 23–55.

Temkin, M. H. 1994. Gamete spawning and fertilization in the gymnolaemate bryozoan *Membranipora membranacea. Biol. Bull.* 187: 143–155.

Yatsu, N. 1902. On the development of *Lingula anatina. J. Coll. Sci. Imp. Univ. Tokyo* 17: 1–112.

Zimmer, R. L. 1973. Morphological and developmental affinities of the lophophorates. In *Living and Fossil Bryozoa,* G.P. Larwood (ed.). Academic Press, London, pp. 593–599.

Zimmer, R. L. and R. M. Woollacott, 1977a. Structure and classification of gymnolaemate larvae. In *Biology of Bryozoa,* R. M.Woollacott, and R. L. Zimmer (eds.). Academic Press, New York, pp. 57–89.

Zimmer, R. L. and R. M. Woollacott, 1977b. Metamorphosis, ancestrulae, and coloniality in bryozoan life cycles. In *Biology of Bryozoa,* R. M. Woollacott and R. L. Zimmer (eds.). Academic Press, New York, pp. 91–142.

Zimmer, R. L. 1980. Mesoderm proliferation and formation of the protocoel and metacoel in early embryos of *Phoronis vancouverensis* (Phoronida). *Zool. Jahrb. Abt. Anat.* 103: 219–232.

Zimmer, R. L. 1987. Phoronida. In *Reproduction and Development of Marine Invertebrates of the Northern Pacific Coast,* M. F. Strathmann (ed.). University of Washington Press, Seattle and London, pp. 476–485.

Zimmer, R. L. 1991. Phoronida. In *Reproduction of Marine Invertebrates, Vol. VI: Lophophorates and Echinoderms,* A. C. Giese, J. S. Pearse and V. B. Pearse (eds.). Boxwood Press, Pacific Grove, CA, pp. 1–35.

Zschiesche, A. 1901. Untersuchungen über die Metamorphose von *Alcyonidium mytili. Zool. Jahrb. Abt. Anat.* 28: 1–72.

SECTION VI

Bilateral Animal Phyla: Deuterostome Coelomates

CHAPTER 16

Echinoderms

Gregory A. Wray

ECHINODERMS OCCUPY A UNIQUE PLACE IN EMBRYOLOGICAL studies. Among the very first experimental analyses of development were separations of 2-cell sea urchin blastomeres, carried out independently by Driesch and Fiedler in 1891 (see Driesch 1892). The isolated blastomeres developed into miniature, but otherwise nearly normal, larvae. A century later this result is still amazing, but at the time it created an absolute sensation. Only a few years earlier, Roux had carried out a similar experiment, killing one blastomere of a 2-cell amphibian embryo and observing that the surviving cell developed into a half-larva. Echinoderms had made their dramatic and controversial debut in embryological studies. We now know that Roux's experiment was technically flawed, and that the embryos of many animals (including those of amphibians) possess remarkable regulative abilities, while the embryos of many other animals do not. The discovery of regulation, and cases where regulation did not occur, prompted numerous experimental studies with a wide range of organisms, and development was transformed from a descriptive to an experimental science.

Echinoderm embryos remain an important experimental system in modern developmental biology for several reasons. They are transparent, allowing direct observation of cell divisions and cell movements throughout the embryo and larva. This has been a tremendous advantage for characterizing the events of echinoderm development, as well as for carrying out experiments and monitoring their results in living embryos. The abundant eggs of echinoderms are another important advantage, providing material for biochemical and molecular characterizations, and permitting experiments with very large numbers of synchronously developing embryos. Experimental embryologists also have taken advantage of the ease with which it is possible to physically manipulate echinoderm embryos. This has allowed experiments involving blastomere re-

moval and transplantation, isolation of particular germ layers, and even bisection of entire embryos or larvae. Finally, echinoderms are a morphologically and ecologically diverse group, and they provide important material for analyses of developmental evolution and larval ecology. Despite these advantages, echinoderms have an important limitation as an experimental system: their unsuitability for genetic analyses. Because of these particular strengths and weaknesses, echinoderms have proven most useful for experimental studies of fertilization, cell–cell interactions, and morphogenesis. The strengths of echinoderms as experimental organisms thus complement some major weaknesses of the animals that have proven most useful for genetic analyses: *Drosophila*, *Caenorhabditis*, and the mouse.

Overview of Echinoderm Development

Adult echinoderms are familiar to most people as sea urchins, sea stars, and sea cucumbers (Figure 16.1*b*). These animals are characterized by fivefold **radial symmetry**, a biomineral **endoskeleton**, and the **water vascular system**, a unique organ system that is used for both locomotion and circulation. Echinoderms live exclusively in marine habitats, where they are widely distributed and are often an ecologically significant group. As in many animal phyla, most echinoderm species **free-spawn**, releasing their gametes directly into the seawater, where they are fertilized (Figure 16.2*a*). A series of rapid, synchronous, and stereotypic cell divisions produce a hollow blastula of several hundred cells that swims by means of cilia (Figure 16.2*b–e*). Soon after, the cell movements of gastrulation begin, ultimately establishing an embryo with three distinct layers of cells (Figure 16.2*f–k*). Mesenchyme cells migrate into the

(a)   (b)

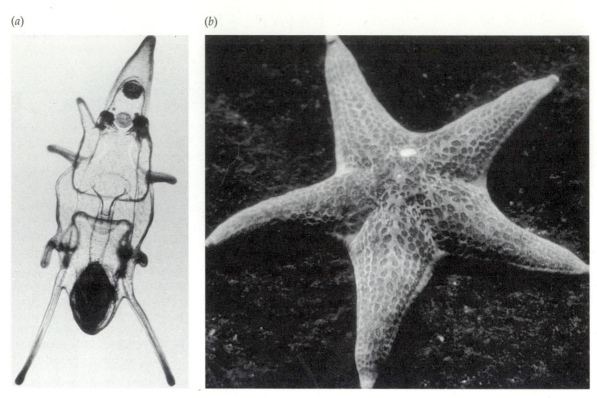

Figure 16.1   Echinoderm larva and adult. In most echinoderms, the embryo develops into a tiny, bilaterally symmetrical larva (*a*) that swims in the plankton, grazing on unicellular algae. (*b*) A drastic metamorphosis later converts the larva into a radially symmetrical adult that lives on the sea floor. The larva and adult shown here are of the starfish *Dermasterias imbricata*.

blastocoel as the archenteron invaginates and elongates. In some species, including most sea urchins, a distinct population of mesenchyme cells ingresses before gastrulation (Figure 16.2*g*). The mesenchyme cells give rise to several differentiated cell types, including myocytes, pigment cells, skeletogenic cells, and "coelomocytes," cells of unknown function. Toward the end of gastrulation, the archenteron buds off paired lateral pouches that become the coeloms and the remainder becomes the gut (Figure 16.2*k*). Eventually, the tip of the archenteron contacts the blastocoelar wall and a perforation appears that becomes the mouth. At this point, the embryo is generally able to feed.

Unlike the embryos of mammals and insects, which develop directly into miniature adults, most echinoderms have an **indirect development**, first producing a tiny larva (Figure 16.1*a*). These larvae swim for several days or weeks in the plankton, grazing on unicellular algae. A loop of ectodermal cells with long cilia is used both for swimming and feeding during this time. Larval development involves an increase in size, elaboration of the ciliated band, and often the growth of "arms." In many species, a larval skeleton supports the arms. Toward the end of larval life, a portion of the coelom and the overlying ectoderm together form a region of rapidly proliferating cells. This cluster of cells, the **imaginal adult rudiment**, will produce almost all of the radially symmetrical adult. A drastic and rapid metamorphosis marks the transition from swimming larva to bottom-dwelling juvenile. The water vascular system is already functional, and the juvenile can crawl about on its locomotory **tube feet**. It is many days, however, before feeding begins; the gut is first completely replumbed, and a new mouth and anus form.

Echinoderm development therefore produces two rather different body organizations in succession: first a bilaterally symmetrical larva and later a radially symmetrical adult. The vast majority of descriptive and experimental studies have examined the first of these processes, and very little is known about metamorphosis or postmetamorphic development. There is also a phylogenetic bias in studies, with nearly all research on echinoderm development coming from a few, relatively closely related species of sea urchins. These biases will be apparent through the remainder of the chapter. Recent reviews of early development of echinoderms include those of Ettensohn and Ingersoll (1992) and Hardin (1994).

Fertilization

Three factors make echinoderms particularly favorable for studying fertilization. First, it is possible to obtain large numbers of gametes with ease, which has permitted extensive biochemical analyses of the molecules involved in fertilization. Second, fertilization occurs outside the body, rather than internally, allowing detailed studies of the

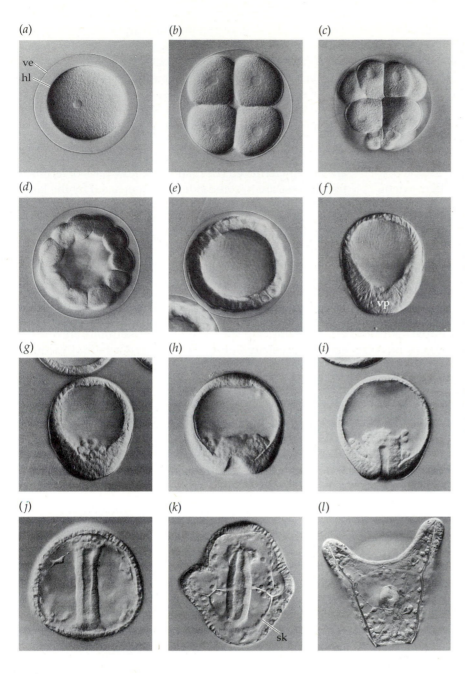

(a) (b) (c) (d) (e) (f) (g) (h) (i) (j) (k) (l)

Figure 16.2 Overview of early development in sea urchins. The course of early development is shown for the sea urchin *Lytechinus variegatus*. (*a*) The fertilized egg is surrounded by a thin vitelline envelope (ve); closely appressed to the cell surface is the hyaline layer (hl). (*b*) The first three cleavage divisions are equal, producing an embryo with eight cells in a cubic arrangement. (*c*) The fourth cleavage is unequal in most sea urchins. From animal pole (top) to vegetal pole, there are eight mesomeres, four macromeres, and four micromeres. (*d*) During cleavage, a blastocoel (bc) develops within the embryo. (*e*) Shortly after the rapid cell divisions of cleavage end, the previously rounded cells form a true epithelium. (*f*) A thickening at the vegetal pole (vp) of the embryo signals the beginning of morphogenesis. (*g*) Primary mesenchyme cells (pmc) ingress before gastrulation in many echinoderms. (*h*) The initial invagination of the archenteron involves a buckling of the thick vegetal plate. (*i*) Archenteron extension begins immediately after the initial invagination. (*j*) Full extension involves a narrowing of the archenteron, driven by cell rearrangements. (*k*) The primary mesenchyme cells have begun to synthesize the larval skeleton (sk). (*l*) In this early larva, the skeleton now extends into the first two arms that will be used for feeding. The ciliated band (cb) is visible as thickened ectoderm stretched between the two arms.

physiological events that attend fertilization. And third, the eggs are relatively transparent, allowing investigators to observe pronuclear migration and spindle assembly with particular clarity. Because of these experimental advantages, more is known about fertilization in echinoderms than perhaps any other group of animals.

Gametes and Fertilization

Echinoderm eggs are large cells, ranging in size from 80 to over 2000 μm in diameter, depending upon the species. Surrounding the egg is a thick, protective **jelly coat**. The yolk protein **vitellogenin** is distributed throughout the cytoplasm in granules, as are numerous small vesicles containing lipid. Together, this yolk and lipid supply the em-

bryo and early larva with energy and precursor molecules prior to feeding (Scott and Lennarz 1989). Also present in the cytoplasm are membrane-bound vesicles called **cortical granules** that release their contents following fertilization. Other vesicles exocytose somewhat later, adding material to the extracellular matrix. Echinoderm eggs also contain unusually high concentrations of materials required for gene expression and cell division: large numbers of ribosomes, prefabricated nuclear pore complexes, tubulin for mitotic spindles, and enzymes and nucleotides required for DNA replication. These stockpiled materials allow unusually rapid cell divisions following fertilization.

Once released into the ocean, echinoderm gametes have a limited lifespan. Sperm swim vigorously for 10–30 minutes, while eggs can be fertilized for perhaps an hour or two. Sperm cells swim using long, whiplike flagella. A sperm cell approaching an egg will burrow into the jelly coat, an action that triggers a dramatic response in the sperm called the **acrosome reaction**. The **acrosomal vesicle**, a large membrane-bound vesicle near the tip of the sperm head, exocytoses, releasing a proteolytic enzyme that digests the jelly coat and allows the sperm to approach the surface of the egg. Meanwhile, actin monomers just under the surface of the acrosomal vesicle polymerize to form a long **acrosomal process** that extends toward the egg surface. As the surfaces of the gametes approach, a specific interaction takes place between the protein **bindin** on the sperm cell (Vacquier and Moy 1977) and a receptor on the egg surface (Foltz et al. 1992). The membranes of the two cells fuse, allowing the sperm pronucleus to begin moving toward the egg pronucleus.

Sperm-egg fusion sets in train a complex series of responses. Among the most important are processes that prevent polyspermy, or fertilization by multiple sperm cells. This condition is almost invariably fatal, as it interferes with normal cell divisions and causes polyploidy. Two complementary processes prevent polyspermy in sea urchins: a rapid, transient depolarization of the egg's plasma membrane and the slower, more permanent production of a physical barrier. The physical barrier is a tough egg shell called the **vitelline envelope**, or fertilization envelope (Figure 16.2*a*). It is produced by the exocytosis of cortical granules, which releases proteins that are rapidly crosslinked by peroxidases. Several other important events are set in train by sperm-egg fusion. Cortical granule exocytosis also releases a sticky protein, **hyalin**, that forms a tough extracellular matrix called the **hyaline layer** (Figure 16.2*a*). The hyaline layer holds the cells of the early embryo together until they develop cell junctions in the blastula. The initial membrane depolarization acts through second messenger systems to activate the egg physiologically: respiration and the rate of translation both increase dramatically. The egg pronucleus completes meiosis (except in sea urchins, where this happens before fertilization), and two pronuclei fuse to form the diploid nucleus of the zygote. The first mitotic division of cleavage begins a few minutes later.

Challenges of External Fertilization

Among animals that live in the ocean, internal fertilization is uncommon. Most marine invertebrates free-spawn, a characteristic that poses several problems that are not experienced by species that copulate. The first problem is simply bringing together the gametes. Given the vastness of the ocean and the tiny size and short life span of a gamete, the chances of bumping into a complementary gamete of the correct species are remote. Echinoderms use several methods to increase these odds. Most species synchronize gametogenesis and confine spawning to a dis-

tinct time of the year. (Somewhat surprisingly, very few echinoderms congregate during spawning, although this is a seemingly simple way to increase the odds of a successful encounter between gametes.) A protein in the jelly coat of sea urchin eggs called **resact** acts as a chemoattractant to sperm (Ward et al. 1985), while an unrelated molecule performs the same function in starfishes (Miller 1985). Direct measures from natural spawnings suggest that anywhere from 1–90% of eggs released by a female are actually fertilized (Levitan 1995). Low rates of fertilization success are to some extent compensated by the large numbers of eggs released by echinoderms, which may be in the hundreds of thousands. Sperm are considerably more numerous, and a single male may release tens or even hundreds of millions.

Another problem that arises with external fertilization is species specificity. A gamete that combines with that of another species is effectively wasted, and free-spawning animals have evolved various ways of preventing this. In echinoderms, the sperm surface-protein bindin confers species specificity during fertilization (Glabe and Vacquier 1977). As part of the biochemical machinery of sperm-egg fusion, bindin is ideally suited to this role. The portion of the bindin molecule that directly binds to the receptor on the egg's surface is highly conserved among different sea urchin species, but adjacent regions evolve very rapidly (Vacquier et al. 1995). Changes in these regions of the bindin protein may be partially responsible for species specificity, in much the same way that small differences in the teeth of key provide specificity for a particular lock. Additional species specificity may be provided by the jelly coat protein resact and by the bindin receptor.

Cleavage, Cell Lineage, and the Blastula

Following fertilization, the zygote begins to divide rapidly. Much of the basic spatial organization of the larva is established during these early cell divisions, through extensive cell interactions and a limited number of molecules stored in specific regions of the egg cytoplasm (Davidson 1989). Both methods of specifying cell fate rely on highly stereotypic cell divisions, which position cells in particular places at particular times.

Cleavage and Early Cell Lineages

Early cell divisions follow a characteristic geometry in echinoderms, although this geometry varies somewhat among species (Figure 16.3). The first two cleavage divisions are longitudinal, intersecting the animal and vegetal poles. These divisions lie at right angles to one another, dividing the embryo into four cells of equal size. The third division is equatorial, lying perpendicular to the animal-vegetal axis and the first two cleavage divisions. Up to this point, cell divisions are equal, and in most groups of

(a)

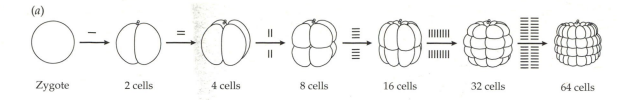

Zygote   2 cells   4 cells   8 cells   16 cells   32 cells   64 cells

(b)

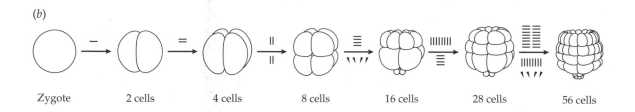

Zygote   2 cells   4 cells   8 cells   16 cells   28 cells   56 cells

(c)

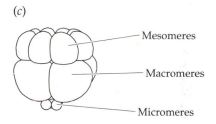

Mesomeres

Macromeres

Micromeres

(d)

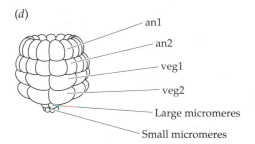

an1

an2

veg1

veg2

Large micromeres

Small micromeres

Figure 16.3 Geometry of cleavage. Within a particular echinoderm species, cleavage geometry is invariant, but patterns differ among species. (a) In most echinoderms, including starfishes and sea cucumbers, cleavage divisions are equal. (b) In most sea urchins, however, unequal fourth and fifth cleavages in the vegetalmost blastomeres (bottom) produce distinct size classes of cells. In all known echinoderms, the first two mitoses are oriented perpendicular to the animal-vegetal axis and the third parallel to it; thereafter cell divisions alternate between perpendicular and parallel to the animal-vegetal axis, with the exception of the unequal cleavages in sea urchins. (The small bars between embryos in this figure indicate the number and orientation of the mitotic axes.) Although cell divisions are initially synchronous, they become asynchronous between tiers following the fourth and fifth cleavages in sea urchins. Because the micromeres do not divide until the macromeres and mesomeres have already divided twice, there are no 32- or 64-cell stages in sea urchins. The unequal cleavages in sea urchin embryos allow one to identify groups of blastomeres with distinct fates. The conventional names of these cells are given for the 16-cell stage (c) and 56-cell stage (d).

echinoderms, they continue to be equal throughout the remainder of cleavage (Figure 16.3a). In most sea urchins, however, the four vegetal cells divide unequally and the animal four cells divide equally at fourth cleavage (Figure 16.3b) to produce a 16-cell embryo with three tiers, each containing blastomeres of a different size (Figure 16.2c). These cells are known, from the animal pole down, as **mesomeres, macromeres**, and **micromeres** (Figure 16.3c). Because they are easily recognized and each has a unique fate in the unperturbed embryo, these cells have been the subject of much experimental inquiry, as described later.

(Note that these cells are not homologous to cells with the same names in other phyla, such as ctenophores and molluscs.) In sea urchins the micromeres divide unequally again during fifth cleavage (Figure 16.3b) to form **large micromeres** and **small micromeres** at the extreme vegetal pole (Figure 16.3d); the mesomeres and macromeres divide equally during the fifth cleavage (Figure 16.3b), and from this point on, all cell divisions are equal.

A variety of methods have been used to study the fates of early blastomeres. Some of the earliest studies were done by Hörstadius (1939), who used the vital dye Nile Blue to mark the surface of individual cells in sea urchin embryos. He found that mesomeres give rise to ectoderm in the larva, that micromeres become skeletogenic cells, and that macromeres produce ectoderm, gut, and a variety of mesenchymal cells. More recent studies have confirmed and considerably extended these observations, using intracellular injections of fluorescent tracer dyes in the sea urchin *Strongylocentrotus purpuratus* (Cameron and Davidson 1991). Sea urchin blastomeres through at least the 32-cell stage have specific, predictable fates (Figure 16.4), a condition common to many phyla called **invariant cell lineage**. Cell lineage has also been examined in the starfish *Astrina pectinifera* (Kominami 1984), but blastomere fates have not been followed beyond the 8-cell stage. Nonetheless, there are a number of similarities between starfish and sea urchin cell lineage. In both cases,

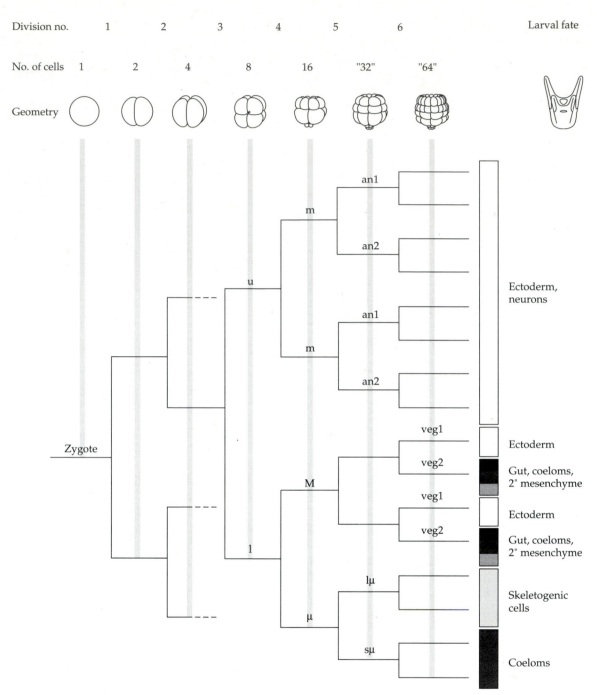

Figure 16.4 Cell lineage of a sea urchin embryo. The embryonic cell lineages of the sea urchin *Strongylocentrotus purpuratus* have been characterized in detail. In this branching diagram, horizontal lines represent cells and vertical lines mitoses. The configuration of the embryo at each round of cell division is indicated above, and linked to the blastomeres present at that time by a vertical gray bar. Beyond the 4-cell stage only one-quarter of the embryo is represented; the cell lineages in the other three quadrants are nearly identical. Note that the horizontal axis is not scaled to represent the relative timing of cell divisions, which become asynchronous after the fourth cleavage. There are no true 32- and 64-cell stages because of this asynchrony, and the diagram indicates the fates of cell lineages after an equivalent number of cell divisions. Cells with distinct names are indicated: u, upper and l, lower blastomeres of the 8-cell embryo; m, mesomere, M, macromere, and μ, micromere of the 16-cell embryo; lμ, large micromere; sμ, small micromere; an1, upper animal tier and an2, lower animal tier of the 28-cell embryo; veg1, upper vegetal tier and veg2, lower vegetal tier of the 56-cell embryo. (See Figure 16.3 for the position of these cells within the embryo.) The fates of these cells in the larva are indicated on the right.

cell lineages are invariant, and the spatial distribution of fates is similar. Cells in the animal half give rise to ectoderm, including ectodermally derived neurons, while cells in the vegetal half produce descendants in all three germ layers. The invariant geometry of cleavage and the predictable relationship between early blastomeres and their differentiated progeny strongly suggest that blastomeres acquire unique and detailed information during early cleavage in echinoderms. This conclusion is supported by analyses of gene expression and by experimental manipulations that are discussed later.

Midblastula Transition

The rapid, invariant, and synchronous divisions of cleavage last from 9 to 11 cell cycles in echinoderms, depending upon the species. An intricate molecular timer involving the protein **cyclin** regulates the pace and synchrony of these divisions (Evans et al. 1983). During cleavage, cell divisions are oriented parallel, rather than perpendicular, to the surface of the embryo, so that all cells remain within a single, curved plane. Blastomeres shift positions somewhat from the idealized arrays shown in Figure 16.3a and b, and close off the "holes" at the animal and vegetal poles. Eventually, a distinct space develops within the embryo, forming a **blastula** (Figure 16.2d). The space inside the blastula is the **blastocoel**, or primary body cavity, of the embryo. The cells that line the blastocoel are initially rounded and held together by the hyaline layer during cleavage. As cleavage ends, several things happen almost simultaneously. The cells become organized into a true epithelium (Figure 16.2e), with permanent cell junctions, a distinct polarity, and complex extracellular matrices on the interior and exterior surface. Each cell forms a single cilium on its outer surface, and these begin to beat. Cell cycles slow down considerably and neither the position nor the timing of cell divisions can be predicted in advance (Matsuda and Sato 1994). In most echinoderms, a proteolytic **hatching enzyme** is released about this time, which ruptures the vitelline envelope, allowing the blastula to hatch.

Early Zygotic Gene Expression

Echinoderms develop through the end of cleavage with little or no transcription of zygotic genes, but they do need to translate maternal mRNAs stored in the egg during oogenesis. This was first demonstrated by raising zygotes exposed to either actinomycin D, an inhibitor of transcription, or to emetine, an inhibitor of translation (Gross and Cousineau 1964). In the presence of actinomycin D, zygotes developed into blastulae, but then arrested, while in the presence of emetine, they did not even complete cleavage. This suggests that the zygotic genome does not play a role in echinoderm development until after most of the important cell lineages have already been established.

The activation of zygotic transcription just after cleavage has been confirmed by analyzing the expression of numerous genes individually (Kingsley et al. 1993). Even the earliest zygotic transcripts accumulate in subsets of cells, rather than throughout the embryo (Coffman and Davidson 1992; Kingsley et al. 1993). In many cases, genes are expressed by the descendants of particular cleavage stage blastomeres (Figure 16.5). For example, the primary mesenchyme cells of sea urchins, which are the exclusive descendants of the large micromeres, uniquely express several genes that are associated with skeletogenesis (Figure 16.5a). Davidson (1989) has proposed that sea urchin

Figure 16.5 Cell lineage-specific gene expression. Many genes in sea urchin embryos are expressed within groups of cells that descend from a particular set of cleavage-stage blastomeres. The expression of such genes is therefore restricted to particular clones of cells within the embryo. (a) *msp130* is expressed only by descendants of the large micromeres, the skeletogenic cells. The protein product is a component of the spicule sheath. (b) *endoV* is expressed only by gut cells, the descendants of some veg2 descendants. Its product is a structural protein. (c) A calmodulin-related gene, *Spec1*, is expressed only by cells of the aboral (dorsal) ectoderm. These cells descend from all the veg1 cells plus a subset of the an1 and an2 cells. Figures 16.3 and 16.4 indicate where and when during cleavage the founder cells for these territories are "born" in the embryo. Panels (a) and (b) are immunolocalizations; panel (c) is an *in situ* hybridization.

(a)

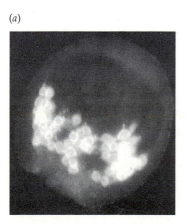

(b)

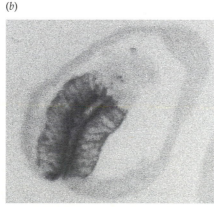

(c)

embryos are composed of five **territories** that are established as clones of cells during cleavage and that express unique sets of genes in later embryos and larvae. These territories are: (1) the **oral ectoderm**, surrounding the mouth; (2) the **aboral ectoderm**, comprising the bulk of ectodermal cells and including the ciliated band that is used by the larva for feeding and locomotion; (3) the **large micromeres**, which ingress as primary mesenchyme and become skeletogenic cells; (4) the **small micromeres**, that form the coelomic pouches; and (5) the **vegetal plate**, a complex mix of cells that will give rise to the endoderm and various secondary mesenchyme cell populations. Although many genes are expressed within one territory, several exceptions are known, involving expression within subregions of a territory, expression in two territories, or both (Kingsley et al. 1993). Even in many of the exceptions, however, gene expression is limited to combinations of particular cell lineages or sublineages of those that make up the five territories.

Differential gene expression is an unambiguous indicator that cells have begun to acquire distinct identities within the embryo. The fact that most differential gene expression in early sea urchin embryos is bounded by large cell clones suggests that spatial information was actually established during cleavage, at the time the founder cells of these clones existed. It further suggests that spatial information is established on a cell-by-cell basis, rather than through the action of global signals or diffusible morphogens.

Positional Information in the Early Embryo

Numerous experiments have taken advantage of the ability to manipulate echinoderm embryos in order to analyze how spatial information is established during development. Indeed no other group of organisms has been as thoroughly analyzed by classical embryological approaches such as egg and embryo bisection and blastomere isolation, transplantation, and recombination (Hörstadius 1973; Davidson 1989). These experiments illustrate the unique power of embryological manipulations for revealing how and when positional information is established within an embryo.

As mentioned earlier, some of the earliest experiments with sea urchin embryos were done by Fielder and Driesch over a century ago. Their blastomere separations were the first of many experiments demonstrating the remarkable ability of echinoderm embryos to compensate for major experimental perturbations. This phenomenon became known as **regulation**. Later experiments have shown that sea urchin embryos bisected during cleavage, or as blastulae, or even larvae, can also regulate (Hörstadius 1973; D. McClay, personal communication). Such extensive regulative capabilities are not incompatible with

the cell lineage and gene expression studies just discussed, which indicate that blastomeres acquire specific spatial information during early cleavage. **Specification**, the process of gaining a distinct identity, is initially labile for most cells in echinoderm embryos. This lability can be demonstrated experimentally, for example by transplanting a cell to a new area within the embryo. Eventually, a cell's specified identity becomes irreversible, a process known as **commitment** (or determination). Commitment is not apparent until the cell undergoes differentiation, and becomes physiologically and morphologically distinct from other cell types.

Specification and Commitment of Axes

Sea urchins provided the first example of a fixed spatial relationship between the position of the polar bodies, cleavage planes, and body axes. In 1901, Boveri marked the transparent jelly coat of sea urchin eggs with ink particles and discovered the **jelly canal**, a narrow passage through the coat. Boveri's careful observations of fertilization and subsequent development demonstrated that the jelly canal marks the future animal pole of the embryo. This simple observation proved that the animal-vegetal (AV) axis is specified maternally, before fertilization. Were this not the case, early cleavage divisions would have a random distribution with regard to the AV axis. In most sea urchins, the AV axis is rather subtle morphologically, but in a few species it is marked by a subequatorial band of pigment granules. In echinoderms other than sea urchins, meiosis is not completed prior to fertilization, and the first polar body frequently remains attached to the unfertilized egg. In these groups, the AV axis is much simpler to spot than in sea urchins, since the polar bodies are extruded at the animal pole.

Experimental manipulations demonstrate that the AV axis of sea urchins is not only specified before fertilization, but also committed. When unfertilized eggs are bisected at the equator, each half can be fertilized and will develop for some time, regardless of whether it contains the oocyte pronucleus or not. Animal fragments develop into hollow, ciliated blastulae, while vegetal fragments become nearly normal larvae, with an apparently full complement of cell types (Hörstadius 1973; Maruyama et al. 1985) (Figure 16.6c). It is likely that a maternal **determinant** (or determinants) localized to the vegetal half of the egg during oogenesis is responsible for both specifying and committing the AV axis, but its molecular identity remains elusive.

The dorsoventral (DV) axis, in contrast, is specified and committed after fertilization. Surface marking with vital dyes and intracellular dye injections have been used to study establishment of the DV axis. In several sea urchins and a starfish, the radial position of the first cleavage plane does not bear a unique relationship to the DV axis of the larva (Kominami 1983; Henry et al. 1992) (see Fig-

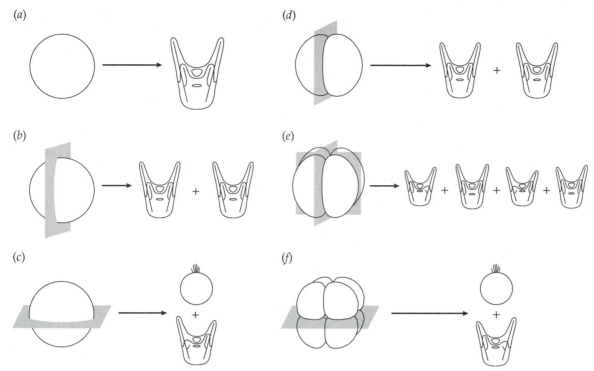

Figure 16.6 Axis specification and commitment. Bisections of unfertilized eggs provide information about the timing of AV and DV axis specification and commitment. Both halves of a bisected egg can be fertilized and will develop to a remarkable degree. If the egg is bisected parallel to the AV axis (*b*), both halves will form nearly normal larvae that are smaller than control larvae (*a*). Since each half-egg regulates to form a new DV axis, this axis is not committed before fertilization. If the egg is bisected perpendicular the AV axis, however, only the vegetal half will develop into a more-or-less normal larva; the animal half does not regulate, but develops into ball of ectodermal cells (*c*). This indicates that the AV axis is both specified and committed before fertilization. Bisections of embryos provide further information about the timing of DV axis commitment. Embryo bisections parallel to the AV axis at the 2- and 4-cell stages result in miniature larvae (*d* and *e*), indicating that the DV axis is still not committed after fertilization. Embryo bisections perpendicular to the AV axis result in unequal development (*f*), indicating a committed AV axis and confirming the result obtained in egg bisections (*c*). Note that the manipulation diagrammed in (*d*) was one of the first experiments in the history of embryology.

ure 16.7), suggesting that the DV axis is specified after fertilization. The first cleavage plane is not randomly oriented, however, but almost invariably occupies one of four positions with respect to the later DV axis: parallel, perpendicular, or oblique (Figure 16.7). This suggests that the DV axis is specified at about the 8-cell stage, when the only cell boundaries that exist occupy these positions. Bisections of unfertilized eggs parallel to the AV axis produce two miniature, but nearly normal larvae (Figure 16.6*b*), in contrast to the equatorial bisections described

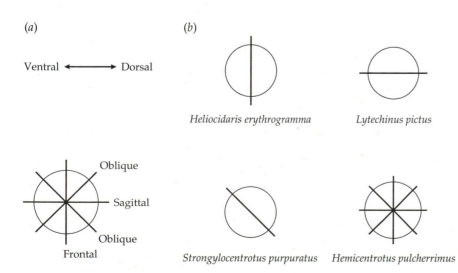

Figure 16.7 Radial position of the first cleavage plane. The first cleavage division in all known echinoderms is oriented perpendicular to the AV axis, but its radial position varies both among and within species. (*a*) Looking down onto the animal pole of an embryo, the first cleavage plane (point of contact between blastomeres) is oriented in one of four positions with respect to the DV axis: parallel, perpendicular, left-oblique, or right-oblique. This orientation has been studied by marking a spot on one blastomere or microinjecting one blastomere just after first cleavage and observing where the mark or labeled cells reside within the larva. (*b*) The position of the first cleavage plane varies among, and often within, species. The line or lines indicate the most common positions of the first cleavage plane in various species.

earlier. This observation indicates that the DV axis is not committed before fertilization, as expected from the later time of specification. The mechanisms of DV axis specification are not known. It is possible that the position of sperm entry establishes the eventual position of either the first cleavage plane or the DV axis in echinoderms, as it does in some other phyla.

Cell Fate Specification and Commitment

Early models of cell fate specification in echinoderm embryos proposed a double gradient of maternal determinants distributed along the AV axis (Child 1916; Hörstadius 1939). More recent interpretations have stressed how little preformed maternal information is present and have emphasized instead the importance of extensive cell interactions during fate specification (Wilt 1987; Davidson 1989). In principle, a cell's fate may be specified in one of two basic ways: through localized maternal determinants or through cell interactions. Because localized maternal determinants do not depend upon other cells in the embryo for their activity, this first method is called **autonomous specification**. In such cases, specification and commitment occur simultaneously. In contrast, **conditional specification** is the result of cell interactions, and the fate of the specified cell is conditional on the presence of the cell that initiates the signal. In these cases, commitment often occurs well after specification, and a cell's fate is subject to respecification until this time. Another feature of conditional specification is that a cell's fate is often refined over time from a general to a specific fate.

The only blastomeres whose fate is specified autonomously in sea urchin embryos are the micromeres. If these cells are removed from 16-cell embryos and cultured in isolation, they will differentiate into skeletogenic cells and synthesize spicules (Okazaki 1975). Since this is their fate in unperturbed embryos, these results indicate that the specification of micromere fate does not depend upon other cells. None of a variety of experimental manipulations has succeeded in altering the fate of micromeres, indicating that this fate is committed as soon as these cells are born. Micromeres thus meet the operational criteria for autonomous specification. In contrast, fates of mesomeres and macromeres are not committed in the early embryo. For example, if the micromeres are removed from a 16-cell embryo, the macromeres will produce differentiated skeletogenic cells (Figure 16.8*b*) (Hörstadius 1939), something they will not do in a unperturbed embryo (Figure 16.4). Similarly, isolated mesomeres, whose progeny would normally differentiate into ectoderm, will produce gut and mesodermal cells as

Figure 16.8 Cell fate specification. A variety of excision and "cut-and-paste" experiments provide information about cell interactions during early sea urchin development. (*a*) During normal development, the mesomeres, macromeres, and micromeres have stereotypic fates in the larva (see Figure 16.5). If the micromeres are excised from a 16-cell embryo, some descendants of the macromeres will adopt their fate to become skeletogenic cells, and will synthesize skeletal rods (*b*). If the micromeres and macromeres are removed, the remaining mesomere "cap" (*c*) will become ectoderm, its normal fate in the embryo. If, however, micromeres are added to a mesomere "cap" (*d*), they will induce some mesomere descendants to produce gut and pigment cells, which normally descend only from macromeres. Finally, an excised pair of mesomeres (*e*), in contrast to an intact "cap" (*c*), will give rise to gut and primary and secondary mesenchyme cells. An example is shown in (*f*) and (*g*), which are two images of the same embryo: (*f*) is a brightfield image, showing a gut and pigmented mesenchyme cells (secondary mesenchyme derivates), while (*g*) shows skeleton glowing under polarized light, indicating the presence of differentiated primary mesenchyme cells. Together, the results of these experiments demonstrate that many cell fates are specified before they are committed in sea urchin embryos, and that cell interactions of both instructive and inhibitory kinds operate during cleavage.

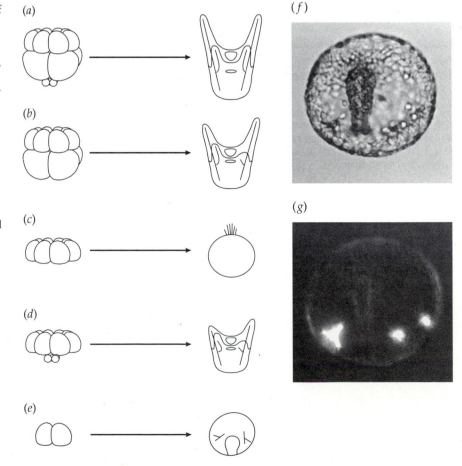

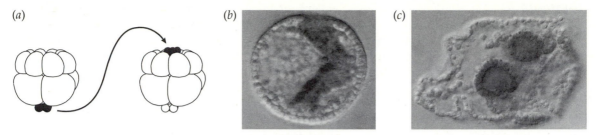

Figure 16.9 Micromere transplantation experiment demonstrates that micromeres have the capacity to induce gut formation in cells that normally form ectoderm. (*a*) Micromeres are labeled and removed from a donor embryo, then placed at the animal pole of a host embryo. (*b*) The resulting chimeric embryo gastrulates from both the animal and vegetal poles. (*c*) Eventually, a double embryo forms, with two digestive systems and two sets of skeletal rods. Because the donor micromeres were labeled prior to transplantation, it is clear that the second gut is formed from the host embryo's mesomeres and not from the transplanted cells. Panels (*b*) and (*c*) are *in situ* hybridizations carried out for a gut-specific marker gene, demonstrating that the induced gut is composed of differentiated cells.

well if isolated at the 16-cell stage (Henry et al. 1989) (Figure 16.8*e*–*g*). Mesomeres and macromeres thus meet the criteria for conditional specification.

The available evidence strongly suggests that the majority of cell fates in sea urchins are specified conditionally. Many of the earliest cell interactions apparently occur between tiers of cells along the AV axis. These interactions were the subject of an extensive series of experiments by Hörstadius (reviewed in 1939 and 1973). Micromeres transplanted to an ectopic region of a host embryo will induce a second AV axis, including a second gut, skeleton, and mouth (Hörstadius 1939; Ransick and Davidson 1993) (Figure 16.9). Similarly, adding micromeres to isolated animal half-embryos will "rescue" their development (Figure 16.8*d*). Instead of producing only ectoderm, the cells of the isolated animal halves differentiate into a variety of cell types, including gut and mesoderm, along an organized AV axis (Hörstadius 1939). These experiments demonstrate that micromeres can organize cell fate specification along the AV axis inductively, and that any other cell type is competent to receive this inductive signal. Further evidence for cell interactions comes from treating embryos with Li$^+$, which interferes with second messenger systems during signal transduction. Embryos treated with Li$^+$ become "**vegetalized**": a greater number of cells adopt vegetal fates such as endoderm, suggesting that normal cell interactions are being disrupted. Some cell interactions along the AV axis are instructive, while others are inhibitory. The operation of inhibitory signals in normal embryos can be seen in two experiments mentioned in the previous paragraph: macromeres will produce some skeletogenic cell progeny following micromere removal (Figure 16.8*b*), and mesomeres can produce gut and skeletogenic cells if isolated

in pairs (Figure 16.8*e*,*f*) but not as an intact group (Figure 16.8*c*). These interactions are summarized in Figure 16.10. To recap, broad specification of cell fates along the AV axis has begun by the 16-cell stage, but any nonmicromere cell remains capable of respecification for some time afterwards.

Much less is known about cell fate specification along the DV axis. In contrast to the AV axis, the DV axis is not autonomously specified. DV axis specification probably occurs at about the 8-cell stage (as discussed earlier). Cell lineage studies suggest that specification of animal tier blastomeres has begun by this time, since these cells have unique ectodermal fates, and the first dorsoventral asymmetries in vegetal cell fates are apparent by the 16-cell stage (Cameron and Davidson 1991). Specification of the DV axis is disrupted if embryos are treated with Ni^{2+}, becoming "**radialized**" with no distinct DV organization of morphology in the larva. As with position along the AV axis, commitment along the DV axis occurs well after specification, since bisected blastulae and gastrulae can regulate to produce small but otherwise nearly normal larvae (D. McClay, personal communication).

Figure 16.10 Summary of cell interactions in sea urchins. A variety of cell interactions occur during early development in sea urchins. The autonomously specified micromeres induce gut cell fates in the macromeres while inhibiting the skeletogenic cell fate. Macromeres in turn inhibit nonectodermal cell fates in the mesomeres, and interactions among the mesomeres reinforce this inhibition.

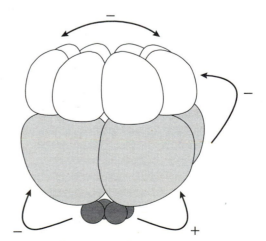

Early Morphogenesis

Once cells have acquired unique identities and begun to express different sets of genes, the stage is set for morphogenesis and differentiation. Morphogenesis begins shortly after cleavage in echinoderms, quickly establishing the three primary germ layers (Figure 16.2*f–j*). The archenteron invaginates from the vegetal pole to become the larval gut, and later the adult gut. Mesodermal cells arise in two ways: as loosely organized **mesenchyme cells** that ingress into the blastocoel from the vegetal plate or invaginating archenteron (Figure 16.2*g, i*), and as **coelomic epithelia** that form as pouches off the sides of the archenteron (Figure 16.2*k*). Morphogenesis has been studied intensively in sea urchins because of their relatively simple anatomy, because it is possible to observe cell movements in living embryos over long periods of time, and because it is possible to experimentally manipulate embryos in various informative ways (Gustafson and Wolpert 1967; Ettensohn and Ingersoll 1992).

Mesenchyme Cell Ingression and Migration

The first overt morphogenetic event in many echinoderms is the ingression of a subset of mesenchyme cells from the vegetal pole region of the late blastula (Figure 16.2*g*). In most sea urchins, these **primary mesenchyme cells** are the descendants of the large micromeres, and become skeletogenic cells. In some brittle stars and sea cucumbers, primary mesenchyme cells are also present. The lineal origin and subsequent fate of these cells have not been examined, however, and it is not clear whether they are homologous to the primary mesenchyme cells of sea urchins. In all echinoderms, mesenchyme cells ingress during archenteron extension (Figure 16.2*i*), and are called **secondary mesenchyme cells**. In sea urchins, these cells are known to form a heterogeneous population, giving rise to brightly colored **pigment cells** that invade the ectoderm, to muscle cells that line the pharynx, to migratory **coelomocytes**, and to other, less well characterized cell populations.

Mesenchyme cell ingression has been studied in detail in sea urchins using a variety of approaches. During ingression, primary mesenchyme cells become round, retract their cilia, and migrate into the blastocoel. Using an assay that measures the strength of cell binding, Fink and McClay (1985) demonstrated that presumptive primary mesenchyme cells loose their affinity for the apical lamina and for their epithelial neighbors, and gain an affinity for the basal lamina lining the blastocoel. These changes in adhesion presumably allow the cells to ingress. Once inside the blastocoel, primary mesenchyme cells migrate seemingly at random for a brief period, then congregate in the vegetal half of the embryo in a ring with two ventrolateral clusters (Figure 16.2*k*), where they later synthesize skeletal rods (Figure 16.2*l*). If primary mesenchyme cells

are experimentally transferred to the animal pole region of a host embryo, they join the host's cells in forming the ring and clusters. If pigment cells, representing another population of mesenchyme, are transplanted, they do not migrate to the ring and cluster (Ettensohn and McClay 1986).These results, and much earlier experiments by Driesch (1896), suggest that primary mesenchyme cells do not follow a hard-wired path during migration, but instead depend upon directional cues in their environment.

This is confirmed by Ni^{2+} treatments that radialize and Li^+ treatments that vegetalize embryos. Primary mesenchyme cells do not migrate to their normal positions in such embryos. Rather, they form up to eight clusters without any dorsoventral organization in radialized embryos, and the ring forms closer to the equator in vegetalized embryos. These altered patterns of migration suggest that cues external to the primary mesenchyme cells guide their migration, and that these cues are directly tied to the AV and DV axes. These cues are already present before the primary mesenchyme cells ingress, since cells transferred to early blastulae will migrate towards the vegetal pole (Ettensohn and McClay 1986). Time-lapse analyses suggest that migrating cells follow global rather than local directional cues (Malinda and Ettensohn 1994). The molecular nature of these cues is not well understood, but they are likely to reside within the basal lamina lining the blastocoel. At least some molecules in the basal lamina are distributed asymmetrically with respect to the AV axis (Solursh and Katow 1982; Ingersoll and Ettensohn 1994), but the distribution of no known molecule corresponds exactly to the ring-and-cluster formed by the primary mesenchyme cells.

Gastrulation

The ectoderm surrounding the vegetal pole thickens to form the **vegetal plate** just before gastrulation begins (Figure 16.2*f*). Archenteron formation in sea urchins begins with a buckling, or **invagination**, of the vegetal plate (Figure 16.2*h*). This seemingly simple process is poorly understood. If blastulae are bisected at or below the equator, vegetal plate buckling occurs on schedule in the isolated vegetal hemispheres (Moore and Burt 1939; Ettensohm 1984), indicating that neither global ectodermal forces nor changes in blastocoelar pressure are required. An intriguing possibility is that the extracellular matrix secreted by cells at the vegetal pole bends as it expands, much as a bimetallic strip does when heated (Lane et al. 1993). Accompanying the buckling of the vegetal plate is a limited amount of **involution**, or flow of cells over the lip of the blastopore (Burke et al. 1991). Involution does not continue past this initial phase of gastrulation, however, and during most of its elongation the archenteron has a fixed volume (Hardin 1989). The product of this initial invagination and involution is a short, wide, and thick-walled archenteron.

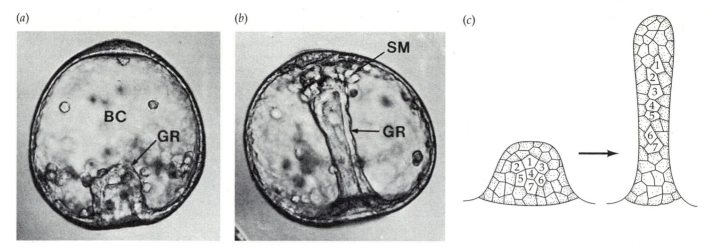

(a)　　　　　　　　　　　*(b)*　　　　　　　　　　　*(c)*

Figure 16.11　Cell rearrangements during gastrulation. During gastrulation, the archenteron is transformed from a low, wide dome (*a*) into a long, narrow tube (*b*). (*c*) The number of cells that make up the circumference of the archenteron drops, and the number of cells that span its length increases. Given that cells do not appreciably elongate during gastrulation, the only way this can happen is for cells to undergo rearragements with their neighbors. The micrographs in (*a*) and (*b*) are of the sea urchin *Lytechinus variegatus*. BC, blastocoel; GR, archenteron; SM, secondary mesenchyme cells.

The next phase of gastrulation (Figure 16.2*i, j*), wherein the short, wide archenteron becomes long and narrow, is more thoroughly understood. Several lines of evidence indicate that **cell rearrangements** are a crucial part of this process. The diameter of the archenteron decreases as it elongates (see Figure 16.11*a,b*), and this is accompanied by a decrease in the number of cells composing the circumference of the archenteron (Ettensohn 1985). The only way this change can occur is by rearrangements of cells (see Figure 16.11*c*). These rearrangements have been observed directly with time-lapse microscopy (Hardin 1989). Changes in cell shape also occur during this time and probably contribute to archenteron extension. As the archenteron is elongating, secondary mesenchyme cells ingress from its tip (Figure 16.11*b*). These cells send out long processes called **filopodia** that contact the inner wall of the blastocoel, suggesting a possible role for filopodial traction in archenteron extension. During the first two-thirds of gastrulation, these cells apparently play no significant role, since extension to about two-thirds the normal length occurs in **exogastrulae** (experimentally produced or rare naturally abnormal embryos where the archenteron pokes out, rather than into the blastocoel) and in embryos where the filopodia are ablated with a laser (Hardin 1988). During the last third of gastrulation, however, laser ablation of filopodia halts archenteron extension, suggesting that filopodial traction plays a role during this final phase (Hardin 1988).

Formation of the Mouth and Coeloms

As the tip of the archenteron approaches the animal pole of the blastocoel, it begins to bend ventrally, towards the prospective oral region. Careful observations and experiments demonstrate that the secondary mesenchyme cells guide this movement. The filopodia of secondary mesenchyme cells are highly dynamic, but spend more time in contact with basal lamina near the animal pole than other regions, and eventually form stable contacts on the inner surface of the prospective oral ectoderm (Hardin and McClay 1990). If the animal pole of an early gastrula is pushed down within premature reach of the filopodia, they form stable contacts earlier than in control embryos; conversely, squeezing the embryo to move the animal pole farther away results in a prolonged period of filopodial activity (Hardin and McClay 1990). As the archenteron approaches the prospective oral region, the ectoderm invaginates to meet it (Gustafson and Wolpert 1963). This invagination is not the result of an inductive interaction triggered by the archenteron, since it will occur on schedule in exogastrulae, where the tip of the archenteron is very far away. The oral and archenteron epithelia make contact, and a hole appears, which is the larval mouth. The **blastopore**, or original site of invagination, becomes the larval anus, as in most deuterostome phyla. Just before the archenteron makes contact with the prospective oral field, another important morphogenetic movement, coelom formation, takes place (Figure 16.2*k*).

In most echinoderms, the coeloms form as pouches from the tip, and in some species, the sides of the archenteron. The pouches pinch off, forming sacs, and these eventually subdivide to form three pairs, as in many other deuterostome groups. The coeloms primarily contribute to adult structures, so their subsequent morphogenesis and fates will be discussed in connection with metamorphosis. The cells of the archenteron remaining after coelom formation become the larval gut. Soon after gastrulation is complete, the gut differentiates morphologically and biochemically into a muscular pharynx, a large digestive stomach, and a narrow intestine. The embryo is now properly called a larva, and it is ready to begin eating.

Larvae and Metamorphosis

Echinoderm larvae do not resemble the adults to which they ultimately give rise (Figure 16.1), and have distinct names reflecting this difference: **pluteus** in sea urchins and brittle stars, **bipinnaria** in sea stars, and **auricularia** in sea cucumbers. Although it does not appear so to the casual eye, much about the morphology of these tiny larvae (Figure 16.12 and front cover) helps them obtain food, and they have been described as "feeding machines" (Hart and Strathmann 1995). The **ciliated band**, a loop of cells with exceptionally long cilia, traces a convoluted path along the body edges and arms surrounding the mouth. In the pluteus larvae of sea urchins and brittle stars, an elaborate endoskeleton composed of long rods supports the arms. Microscopic water currents set up by coordinated beating within the ciliated band move water past the oral region. When a food item, usually a unicellular alga, bumps into a cilium, the individual cilium immediately reverses its beat and flicks the meal toward the mouth, where shorter cilia sweep it in. A simple nervous system (Bisgrove and Burke 1987) coordinates these activities. Larvae drift in the plankton, relentlessly grazing on algae and growing. Initially, energy derived from feeding goes into expanding the size of the larval body, which in turn facilitates feeding (Hart and Strathmann 1995). Later, energy is diverted into the adult rudiment, a group of cells that will eventually form the adult body through a complex metamorphosis.

Skeletogenesis

The skeleton of both larval and adult echinoderms is synthesized internally by mesenchyme cells, making it a true endoskeleton. In sea urchins, skeletogenic cells are the exclusive descendants of the large micromeres, but in the other echinoderm groups, micromeres do not form during cleavage and the lineal origin of skeletogenic cells is not known. The skeletogenic cells of sea urchins begin to

form syncytia as soon as they congregate in the ring and ventrolateral clusters during gastrulation. The first skeletal material is deposited as triradiate spicules within membrane-bound vesicles by cells lying in the ventrolateral clusters (Figure 16.2k). The skeleton itself is a complex biomineral matrix. The organic component includes several proteins, of which five relatively small and highly acidic proteins make up the bulk of material (Benson et al. 1986). The inorganic component of the skeleton is the mineral **calcite**, composed primarily of $CaCO_3$ with small amounts of $MgCO_3$. It is deposited along a single crystal axis (Okazaki and Inoué 1976), making the skeleton glow under polarizing light (Figure 16.8g).

The larval skeleton of sea urchins is intricate. A variety of experiments have been used to ask whether the skeletogenic primary mesenchyme cells are preprogrammed or rely on external information for synthesizing this complex structure. The skeletons of miniature larvae resulting from blastomere isolations at the 2- or 4-cell stage are scaled to the smaller size of the resulting larvae (Figure 16.6d,e) (Driesch 1892; Hörstadius 1973). Similarly, if primary mesenchyme cell number is altered, either by removing some cells or by adding cells from a sibling, the size of the skeleton is unaffected (Ettensohn and Malinda 1993). These results demonstrate that information external to the primary mesenchyme cells is used for scaling the skeleton. The situation is more complex for the source of information used to shape the skeleton. Primary mesenchyme cells cultured *in vitro* can synthesize skeletal spicules, but these do not resemble ones produced in intact embryos (Okazaki 1975), suggesting that external cues are required to synthesize the correct shape. Another experiment, however, suggests that some spatial patterning information is intrinsic to the primary mes-

Figure 16.12 Brittle star larva. This pluteus illustrates the basic anatomy of an echinoderm larva. The mouth (m) and anus (a) mark the two ends of the simple digestive system. The ciliated band (cb), a sinuous loop of ectodermal cells with long cilia, runs along the body edges and the arms. The beating of these cilia produces tiny currents that are used for feeding and swimming. A skeleton (s) is present in the larvae of sea urchins and brittle stars. The skeleton supports the arms, improving feeding efficiency. The larvae of starfishes have arms as well (Figure 16.1a), but they are not supported by a skeleton. New material is added to the skeleton at the distal tips of the rods (*).

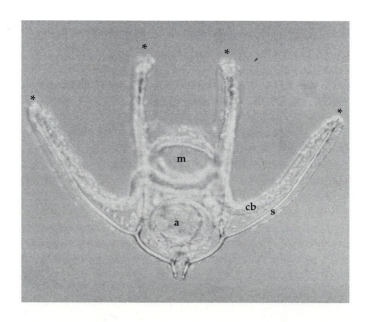

(a)

(b)

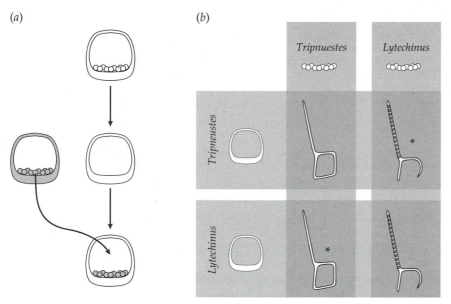

Tripnuestes *Lytechinus*

Tripneustes

Lytechinus

Figure 16.13 Primary mesenchyme cells can be transplanted between embryos of species that have different skeletal morphology. (*a*) In one interesting experiment, the primary mesenchyme cells are removed from a host embryo, and replaced with the labeled primary mesenchyme cells of a donor embryo. (*b*) These hybrid embryos develop into larvae with skeletons characteristic of the donor rather than the host species; the hybrids are indicated by asterisks. This experiment indicates that some aspects of skeletal morphology are "hard-wired" into the primary mesenchyme cells and are not dependent on external cues. It further suggests that the general cues used to guide the migration of primary mesenchyme cells are conserved between species.

enchyme cells. These cells can be transplanted between embryos of species with different larval skeletal morphology. When the host embryo's primary mesenchyme cells are removed and replaced with those of another species (Figure 16.13*a*), the transplanted cells migrate to the correct locations and synthesize a skeleton that appears normal. The skeleton produced by the transplanted cells, however, is characteristic of the donor, and not the host, species (Figure 16.13*b*) (Armstrong and McClay 1994). This result suggests that primary mesenchyme cells use general cues for migration that are conserved among species, but that information for details of skeletal structure is autonomous, rather than based on external information.

Primary mesenchyme cells continue to add skeletal material through most of larval life. New matrix is added at the tips of the skeletal rods (Figure 16.12), where clusters of primary mesenchyme cells reside just under the ectoderm. If this arm tip ectoderm is ablated on one side of the larva, the primary mesenchyme cells stop depositing matrix in that arm, but not in the unoperated control arm of the same embryo. This observation suggests that the arm-tip ectoderm provides a local signal that is required for continued skeletal growth (Ettensohn and Malinda 1993).

The rate of skeleton elongation is not limited by the number of primary mesenchyme cells, since it is unaffected by the addition of two or even four times the normal number of primary mesenchyme cells (Ettensohn and Malinda 1993). Ultimately, the rate at which larvae feed determines the rate of skeletal, and indeed overall, growth. Larvae raised on algal diets of various concentrations show a tight correlation between food intake and body size, including skeleton size (Strathmann et al. 1992) (Figure 16.14). In ad-

Figure 16.14 Effect of diet on larval growth and development. The amount of food available to a larva determines not only how quickly it grows, but also how soon it begins to build an adult rudiment relative to arm growth. Larvae in a low-food regime are smaller than those on normal or high-food regimes at any given time. More surprisingly, formation of the adult rudiment begins after the arms are significantly longer in reduced-food larvae than normal or enhanced-food larvae. Thus low-food larvae on day 19 had longer arms and no rudiment compared to high-food larvae on day 9, which had shorter arms and large rudiments. These are larvae of the sea urchin *Paracentrotus lividus*.

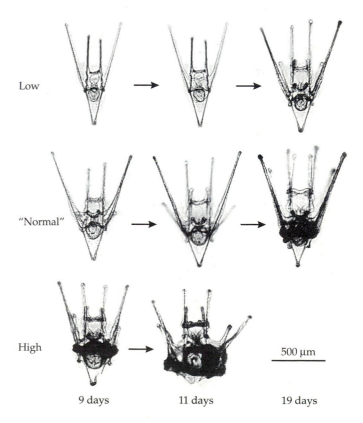

Low

"Normal"

High

500 µm

9 days   11 days   19 days

dition to setting growth rates, food levels also influence the shape of the skeleton: well-fed larvae produce arms sooner than partially starved siblings, but arm growth ceases at a shorter overall length (Figure 16.14). This plasticity in larval morphogenesis is probably adaptive: the slower growing, but ultimately longer, arms of hungry larvae allow them to feed more efficiently, while well-fed larvae do not need to divert energy towards making a more efficient feeding apparatus and instead begin to prepare for metamorphosis (Strathmann et al. 1992).

The Imaginal Adult Rudiment

Late in larval life, a fascinating sequence of events establishes a tiny **imaginal rudiment**, usually on the left side of the larva, that will eventually become the adult animal

(Bury 1895; Burke 1989). The rudiment is derived in large part from the coeloms, whose origin was discussed earlier. The coelomic pouches divide (or, in some species, independently bud off the archenteron) to form three pairs of sacs lying along the gut: anterior **axocoels**, intermediate **hydrocoels**, and posterior **somatocoels** (these correspond to the protocoels, mesocoels, and metacoels of other deuterostome phyla) (Figure 16.15a–c). The morphology and fate of the coelomic sacs vary among echinoderms. In general, the left hydrocoel, and often one or both axocoels, make up the bulk of the water vascular system of the adult; the somatocoels generally become the mesenteries and coeloms that line the body cavity of the adult; and the right hydrocoel and the right axocoel often degenerate. The peculizar asymmetries in coelomic fates are directly

(a)

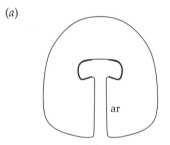

(d)

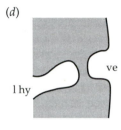

(b)

(e)

Figure 16.15 Coelom and rudiment formation. The coeloms are intimately involved in the development of the imaginal adult rudiment. (a) They form as pouches on the sides of the archenteron (ar) toward the end of gastrulation. (b) These pouches eventually pinch off to form right and left coelomic sacs (rc and lc). (c) In the feeding larva, the coelomic sacs on the right and left sides divide to form three separate regions: axocoel (ax), hydrocoel (hy), and somatocoel (so). Note that this is an idealized configuration; the actual location and relative sizes of the coelomic sacs vary considerably among echinoderm species. (d) As the larva feeds, the left hydrocoel (l hy) enlarges through cell proliferation and interacts with the overlying ectoderm, which thickens and, in many species, invaginates to form a vestibule (ve). (e) Active cell division and morphogenetic movements in the left hydrocoel soon establish the beginnings of the water vascular system (wvs). The first five tube feet (tf) form from branches of the water vascular system and the overlying ectoderm. Skeletogenic cells move into the area and begin to synthesize skeletal plates and spines (sk). The adult mouth will eventually form in the center of the five radially organized tube feet.

(c)

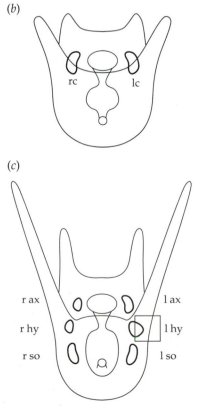

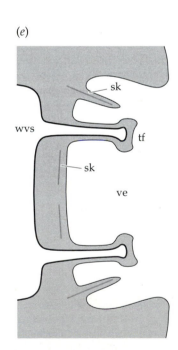

connected to the development of the imaginal rudiment on the left side of the larva. The left hydrocoel expands through cell proliferation and the ectoderm to the left of the mouth invaginates to meet it (Figures 16.15d, 16.16b). The hydrocoel continues to grow, forming five lobes that are the first morphological manifestation of the adult fivefold symmetry. The overlying ectoderm also proliferates and begins to change shape, eventually surrounding the hydrocoel lobes to form the first **tube feet**, which are locomotory organs attached to the water vascular system (Figure 16.15e). In many echinoderms, the developing rudiment forms within a depression in the ectoderm called the **vestibule.** The coeloms continue to expand during late larval life, forming the principle conduits of the water vascular system. Meanwhile, skeletogenic cells migrate into the developing rudiment and begin to synthesize the first skeletal plates of the adult, also organized in fivefold symmetry (Gordon 1929). Through all of this growth and activity in the rudiment, the larva continues to swim and feed (Figure 16.16c).

Surprisingly little experimental research has analyzed rudiment formation. An important early event appears to be an inductive interaction between the left hydrocoel and the overlying ectoderm. If the left hydrocoel is ablated by an ultraviolet microbeam, the ectoderm does not undergo the cell proliferation and morphogenetic movements characteristic of rudiment assembly (Czihak 1971). Not surprisingly, the amount of food available to the larva influences the time at which the rudiment is formed, as well as

how quickly it grows (Strathmann et al. 1992). The patterning mechanisms that establish fivefold symmetry within the rudiment have not been investigated at all, despite their obvious developmental and evolutionary significance (Raff and Kaufman 1983). Based on the sequence of morphological events, it seems likely that radial symmetry is first established in the hydrocoel during the early phase of cell proliferation, and that this spatial organization is propagated through subsequent inductive interactions to the ectoderm and skeletogenic cells. Rapid progress during the past few years towards understanding the molecular basis of pattern formation in *Drosophila* and other genetically manipulable organisms may eventually provide tools for understanding how radial symmetry is established in echinoderms.

Metamorphosis and Beyond

During **overt metamorphosis**, a bilaterally symmetrical echinoderm larva converts within a few minutes to a radially symmetrical juvenile that looks completely different. In perhaps no other phylum is this process so dramatic. Indeed, when echinoderm larvae were first discovered they were thought to be microscopic organisms that had never been encountered before, rather than the larvae of well-known creatures. Several larvae were even described as new species and assigned formal Latin names, and it was not until metamorphosis was directly observed that a connection to the much larger and morphologically distinct adults was realized. A rapid and dramatic overt metamorphosis is possible in echinoderms because extensive preparations are made within the imaginal rudiment during late larval life. Overt metamorphosis essentially consists of an eversion of the rudiment, after which the individual is called a **juvenile** rather than a larva. The future oral surface of the adult lies at the center of the everting rudiment, and in most echinoderm groups, comes to lie

Figure 16.16 Rudiment in sea urchin larvae. These photographs illustrate the formation of the imaginal adult rudiment in the sea urchin *Strongylocentrotus droebachiensis*. (a) The left hydrocoel (l hy) has begun to grow in this eight-armed larva, and a tiny vestibular (ve) invagination of ectoderm is visible just to the right. (b) A close-up of the left hydrocoel and vestibule, which are now in contact. (c) In this late larva, the rudiment (ru) has grown considerably, and is apparent as the large dark mass.

(a)

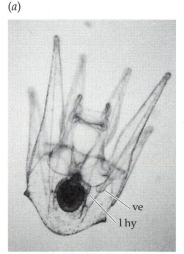

(b)

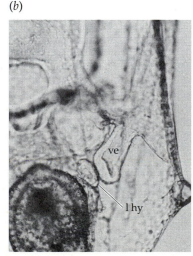

(c)

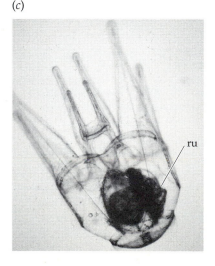

facing down. The ectoderm surrounding the mouth expands to produce much of the adult ectoderm, while the bulk of the larval ectoderm is histolyzed (Cameron and Hinegardner 1978). In most echinoderms, the larval mouth and anus are lost during metamorphosis, and the gut retracts and undergoes proliferation before forming a new mouth and anus. Replumbing the digestive system may take days or weeks, and recently metamorphosed juveniles must rely on larval nutrient reserves for nourishment during this time.

The adult skeleton of echinoderms is composed of the same biomineral matrix as the larval skeleton, but is considerably more complex. The first few elements of the adult skeleton are initiated within the rudiment during late larval life, but the overall configuration of plates and spines takes shape during the weeks or months following metamorphosis (Gordon 1929). Throughout postmetamorphic life, new skeletal plates are added at five **growth zones**, while existing plates behind the growth zones are enlarged and resculpted (Mooi et al. 1994). Recently metamorphosed echinoderms grow considerably before they reach a size where reproduction begins, a process that takes from one to several years.

Overt metamorphosis is not only a dramatic morphological transformation, it is also a critical ecological shift from the plankton to the sea floor called **settlement**. The tube feet, which are already functional before overt metamorphosis, contact the substratum, and the juvenile literally walks away. Larvae that are competent to undergo metamorphosis will generally not do so unless presented with a suitable substrate, such as sediment or rocks collected from a site where adults of the same species live. This observation suggests that the timing of metamorphosis is not hard-wired, and larvae rely on environmental cues to determine when and where to settle. Aqueous extracts of sediment inhabited by adult sand dollars will cause larvae of the same species to metamorphose within minutes (Burke 1984), while larvae of other species respond to algae-encrusted rocks. These observations suggest that larvae respond to diffusible cues in choosing a settlement site.

Evolution of Echinoderm Development

Developmental processes, particularly those of embryogenesis, are usually described as if they were uniform throughout large groups of organisms. Careful comparisons between species, however, make it clear that substantial variation in developmental processes can exist within a phylum, sometimes even between very closely related species. A few cases have already been mentioned in this chapter. For example, a three-tiered 16-cell embryo is limited to the sea urchins (Figure 16.3), suggesting that the central role the micromeres play in cell fate specification in sea urchins is not present in other echinoderm

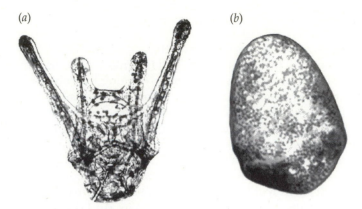

Figure 16.17 Feeding and nonfeeding larvae. The feeding (a) and nonfeeding (b) larvae of sea urchins differ in general appearance. These are the larvae of two closely related species, *Heliocidaris tuberculata* and *H. erythrogramma*. The nonfeeding larva lacks a mouth and anus, and the larval skeleton is reduced to the point that there are no arms.

groups. Many other developmental processes differ among species, including the radial position of the first cleavage plane (Figure 16.7), the timing of mesenchyme cell ingression, spatial patterns of skeletogenesis, cell movements during gastrulation, methods of coelom formation, and morphogenesis of the adult rudiment. Some of these differences may be required to construct the rather different adult morphologies of sea stars, sea urchins, and so forth.

The most dramatic evolutionary differences in developmental process in echinoderms have nothing to do with adult morphology, however, but are instead associated with a modified larval lifestyle. On several separate occasions, echinoderm larvae have lost the ability to feed, and instead develop from large, nutrient-rich eggs. The larvae of these species are highly modified morphologically (Figure 16.17). The development of one such species, the sea urchin *Heliocidaris erythrogramma*, has been studied in detail. This species differs in numerous and rather substantive ways from the "typical" development of sea urchins presented throughout most of this chapter (Wray and Raff 1991). Among other modifications, the DV axis is committed by the 2-cell stage, there are no unequal cell divisions during cleavage, cell fates are specified at different times (Figure 16.18), gastrulation involves extensive involution, no larval mouth forms, and the timing of the expression of several early zygotic genes is altered. Less is known about the many other echinoderms with nonfeeding larvae. However, the available data suggests that their development is also highly modified, often in strikingly parallel ways that suggest specific adaptations to the loss of larval feeding (Wray and Bely 1994). Such cases remind us that model organisms, for all their practical advantages, can mislead us into thinking that we understand *the* way a group of organisms develops.

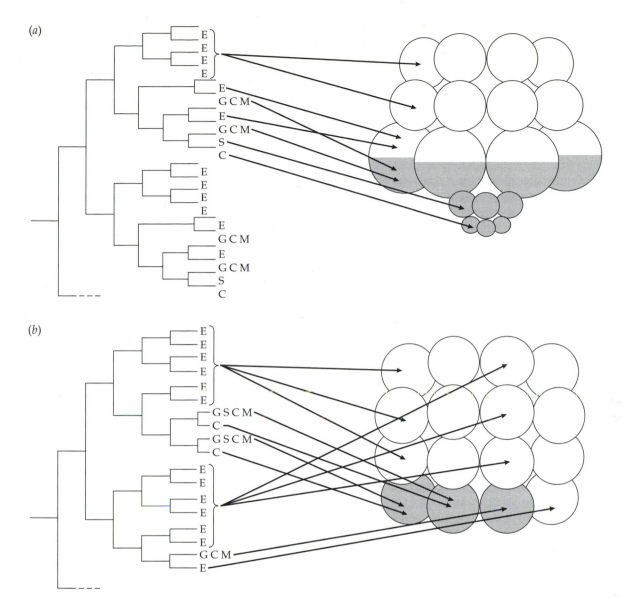

Figure 16.18 Evolutionary changes in sea urchin cell lineage. The characteristic geometry of early cleavage found in most sea urchins is altered in most species with nonfeeding larvae, which have no unequal cleavages and therefore superficially resemble starfish embryos. This figure compares the embryonic cell lineages (left) and fate maps (right) of (*a*) *Strongylocentrotus purpuratus*, a species with feeding larvae, and (*b*) *Heliocidaris erythrogramma*, a species with nonfeeding larvae. Several alterations in cell lineage are evident, with many cell lineages being specified at different times and in a different order (compare cell lineage diagrams) or in different places (compare fate maps). Two quadrants of each embryo are shown for comparison. In *S. purpuratus*, each quadrant has a nearly identical early cell lineage (for clarity, only one quadrant is mapped onto the fate map). In *H. erythrogramma*, there are radial differences in cell lineages between quadrants. Abbreviations: E, ectoderm; G, gut; C, coelom; M, muscle, pigment cells, and coelomocytes; S, skeletogenic cells.

Literature Cited

Armstrong, N. and D. R. McClay. 1994. Skeletal pattern is specified autonomously by the primary mesenchyme cells in sea urchin embryos. *Dev. Biol.* 162: 329–338.

Benson, S. C., N. Crise Benson and F. Wilt. 1986. The organic matrix of the skeletal spicule of sea urchin embryos. *J. Cell Biol.* 102: 1878–1886.

Bisgrove, B. W. and R. D. Burke. 1987. Development of the nervous system of the pluteus larva of *Strongylocentrotus droebachiensis*. *Cell Tissue Res.* 248: 335–343.

Boveri, T. 1901. Über die Polaritat von Ovocyte, Ei und Larve des *Strongylocentrotus lividus. Zool. Jb. (Abt. Anat.)* 14: 630–653.

Burke, R. D. 1984. Pheromonal control of metamorphosis in the Pacific sand dollar, *Dendraster excentricus. Science* 225: 442–443.

Burke, R. D. 1989. Echinoderm metamorphosis: comparative aspects of the change in form. *Echinoderm Studies* 3: 81–108.

Burke, R. D., R. L. Meyers, T. L. Sexton and C. Jackson. 1991. Cell movements during the initial phase of gastrulation in the sea urchin embryo. *Dev. Biol.* 146: 542–557.

Bury, H. 1895. The metamorphosis of echinoderms. *Quart. J. Microsc. Sci.* 29: 409–136.

Cameron, R. A. and E. H. Davidson. 1991. Cell type specification during sea urchin development. *Trends Genet.* 7: 212–218.

Cameron, R. A. and R. T. Hinegardner. 1978. Early events in sea urchin metamorphosis: Description and analysis. *J. Morphol.* 157: 21–32.

Child, C. M. 1916. Experimental control and modification of larval development in the sea urchin in relation to axial gradients. *J. Morphol.* 28: 65–133.

Coffman, J. A. and E. H. Davidson. 1992. Expression of spatially regulated genes in the sea urchin embryo. *Curr. Opin. Genet. Dev.* 2: 260–268.

Czihak, G. 1971. Echinoids. In *Experimental Embryology of Marine Invertebrates*, G. Reverberi (ed.). North Holland, Amsterdam, pp. 383–506.

Davidson, E. H. 1989. Lineage-specific gene expression and the regulative capacities of the sea urchin embryo: A proposed mechanism. *Development* 105: 421–445.

Davidson, E. H. 1990. How embryos work: A comparative view of diverse modes of cell fate specification. *Development* 108: 365–389.

Driesch, H. 1892. The potency of the first two cleavage cells in echinoderm development: Experimental production of double and partial formations. Reprinted in *Foundations of Experimental Embryology*, B. H. Willier and J. M. Oppenheimer (eds.). Hafner, New York.

Driesch, H. 1896. Die taktische Reizbarkeit von Mesenchymzellen von *Echinus tuberculatus. Roux Arch. Entwick.* 3: 362–380.

Ettensohn, C. A. 1984. Primary invagination of the vegetal plate during sea urchin gastrulation. *Amer. Zool.* 24: 571–588.

Ettensohn, C. A. 1985. Gastrulation in the sea urchin embryo is accompanied by the rearrangement of invaginating epithelial cells. *Dev. Biol.* 112: 383–390.

Ettensohn, C. A. and E. P. Ingersoll. 1992. Morphogenesis of the sea urchin embryo. In *Morphogenesis: An Analysis of the Development of Biological Form*, E. F. Rossomondo and S. Alexander (eds.). Marcel Dekker, New York, pp.189–262.

Ettensohn, C. A. and K. M. Malinda. 1993. Size regulation and morphogenesis: A cellular analysis of skeletogenesis in the sea urchin embryo. *Development* 119: 115–167.

Ettensohn, C. A. and D. R. McClay. 1986. The regulation of primary mesenchyme cell migration in the sea urchin embryo: Transplantations of cells and latex beads. *Dev. Biol.* 117: 380–291.

Evans, T., E. T. Rosenthal, J. Youngbloom, D. Distel and T. Hunt. 1983. Cyclin: A protein specified by maternal mRNA in sea urchin eggs that is destroyed at each cleavage divisions. *Cell* 33: 389–396.

Fink, R. D. and D. R. McClay. 1985. Three cell recognition changes accompany the ingression of sea urchin primary mesenchyme cells. *Dev. Biol.* 107: 66–74.

Foltz, K. R., J. S. Partin and W. J. Lennarz. 1992. Sea urchin egg receptor for sperm: Sequence similiarity of binding domain and hsp70. *Science* 259: 1421–1425.

Glabe, C. G. and V. D. Vacquier. 1977. Species-specific agglutination of eggs by bindin isolated from sea urchin sperm. *Nature* 267: 836–838.

Gordon, I. 1929. Skeletal development in *Arbacia, Echinarachnius,* and *Leptasterias. Philos. Trans. Roy. Soc. London* 217: 289–334.

Gross, P. R., and G. H. Cousineau. 1964. Macromolecular synthesis and the influence of actinomycin D on early development. *Exp. Cell. Res.* 33: 368–395.

Gustafson, T. and L. Wolpert. 1963. Studies on the cellular basis of morphogenesis in the sea urchin embryo: Formation of the coelom, the mouth, and the primary pore-canal. *Exp. Cell Res.* 29: 561–582.

Gustafson, T. and L. Wolpert. 1967. Cellular movement and contact in sea urchin morphogenesis. *Biol. Rev.* 42: 442–498.

Hardin, J. D. 1988. The role of secondary mesenchyme cells during sea urchin gastrulation studied by laser ablation. *Development* 103: 317–324.

Hardin, J. D. 1989. Local shifts in position and polarized motility drive cell rearrangement during sea urchin gastrulation. *Dev. Biol.* 136: 430–445.

Hardin, J. 1994. The sea urchin. In *Embryos: Color Atlas of Development,* J. Bard (ed.). Mosby-Year Book Europe, London, pp. 37–54.

Hardin, J. D. and D. R. McClay. 1990. Target recognition during gastrulation by the archenteron during sea urchin gastrulation. *Dev. Biol.* 142: 86–102.

Hart, M. W. and R. R. Strathmann. 1995. Mechanisms and rates of suspension feeding. In *Larval Ecology of Marine Invetebrates*, L. R. McEdward (ed.). CRC Press, Boca Raton, pp. 193–222.

Henry, J. J., S. Amemiya, G. A. Wray and R. A. Raff. 1989. Early inductive interactions are involved in restricting cell fates of mesomeres in sea urchin embryos. *Dev. Biol.* 136: 140–153.

Henry, J. J., K. M. Klueg and R. A. Raff. 1992. Evolutionary dissociation between cleavage, cell lineage, and embryonic axes in sea urchin embryos. *Development* 114: 931–938.

Hörstadius, S. 1939. The mechanics of sea urchin development, studied by operative methods. *Biol. Rev.* 14: 132–179.

Hörstadius, S. 1973. *Experimental Embryology of Echinoderms.* Clarendon, Oxford University Press, London.

Ingersoll, E. P. and C. A. Ettensohn. 1994. An N-linked carbohydrate-containing extracellular matrix determinant plays a key role in sea urchin gastrulation. *Dev. Biol.* 163: 351–366.

Kingsley, P. D., L. M. Angerer and R. C. Angerer. 1993. Major temporal and spatial patterns of gene expression during differentiation of the sea urchin embryo. *Dev. Biol.* 155: 216–234.

Kominami, T. 1983. Establishment of the embryonic axes in larvae of the starfish, *Asterina pectinifera. J. Embryol. Exp. Morphol.* 75: 87–100.

Kominami, T. 1984. Allocation of mesendodermal cells during early embryogenesis in the starfish, *Asterina pectinifera. J. Embryol. Exp. Morphol.* 84: 177–190.

Lane, M. C., M. A. R. Koehl, F. Wilt and R. Keller. 1993. A role for regulated secretion of apical extracellular matrix during epithelial invagination in the sea urchin. *Dev. Biol.* 117: 1049–1060.

Levitan, D. R. 1995. The ecology of fertilization in free-spawning invertebrates. In *Ecology of Marine Invertebrate Larvae*, L. R. McEdward (ed.). CRC Press, Boca Raton, pp. 123–156.

Malinda, K. M. and C. A. Ettensohn. 1994. Primary mesenchyme cell migration in the sea urchin embryo: Distribution of directional cues. *Dev. Biol.* 164: 562–578.

Maruyama, Y. K., Y. Nakaseko and S. Yagi. 1985. Localization of cytoplasmic determinants responsible for primary mesenchyme formation and gastrulation in the unfertilized eggs of the sea urchin *Hemicentrotus pulcherrimus. J. Exp. Zool.* 236: 155–163.

Masuda, M. and H. Sato. 1984. Asynchronization of cell division is concurrently related with ciliogenesis in sea urchin blastulae. *Dev. Growth Differ.* 26: 281–294.

Miller, R. L. 1985. Sperm chemo-orientation in the metazoa. In *Biology of Fertilization*, Vol. 2, C. B. Metz and A. Monroy (eds.). Academic Press, New York, pp. 275–337.

Mooi, R., B. David and D. Marchand. 1994. Echinoderm skeletal homologies: Classical morphology meets modern phylogenetics. In *Echinoderms Through Time*, B. David, A. Guille, J.-P. Féral and M. Roux (eds.). Balkema, Amsterdam, pp. 87–95.

Moore, A. R. and A. S. Burt. 1939. On the locus and nature of the forces causing gastrulation in the embryos of *Dendraster excentricus. J. Exp. Zool.* 82: 159–171.

Okazaki, K. 1975. Spicule formation by isolated micromeres of the sea urchin embryo. *Amer. Zool.* 15: 567–581.

Okazaki, K. and S. Inoue. 1976. Crystal property of the larval sea urchin spicule. *Dev. Growth Differ.* 18: 413–434.

Raff, R. A. and T. C. Kaufman. 1983. *Embryos, Genes, and Evolution.* Macmillan, New York.

Ransick, A. and E. H. Davidson. 1993. A complete second gut induced by transplanted micromeres in the sea urchin embryo. *Science* 259: 1134–1138.

Scott, L. B. and W. J. Lennarz. 1989. Structure of a major yolk glycoprotein and its processing pathway by limited proteolysis are conserved in echinoids. *Dev. Biol.* 132: 91–102.

Solursh, M. and H. Katow. 1982. Initial characterization of sulfated macromolecules in the blastocoels of mesenchyme blastulae of *Strongylocentrotus purpuratus* and *Lytechinus pictus. Dev. Biol.* 94: 326–336.

Strathmann, R. R., L. Fenaux and M. F. Strathmann. 1992. Heterochronic developmental plasticity in larval sea urchins and its implications for evolution of nonfeeding larvae. *Evolution* 46: 972–986.

Vacquier, V. D. and G. W. Moy. 1977. Isolation of bindin: the protein responsible for adhesion of sperm to sea urchin eggs. *Proc. Natl. Acad. Sci. USA* 74: 2456–2460.

Vacquier, V. D., W. J. Swanson, and M. E. Hellberg. 1995. What have we learned about sea urchin sperm bindin? *Dev. Growth Differ.* 37: 1–10.

Ward, G. E., C. J. Brokaw, D. L. Garbers and V. D. Vacquier. 1985. Chemotaxis of *Arbacia punctulata* spermatozoa to resact, a peptide from the jelly layer. *J. Cell Biol.* 101: 2324–2329.

Wilt, F. H. 1987. Determination and morphogenesis in the sea urchin embryo. *Development* 100: 559–575.

Wray, G. A. and A. E. Bely. 1994. The evolution of echinoderm development is driven by several distinct factors. *Development* 1994 (Supplement): 97–106.

Wray, G. A. and R. A. Raff. 1991. The evolution of developmental strategy in marine invertebrates. *Trends Ecol. Evol.* 6: 45–50

CHAPTER 17

Tunicates

William R. Jeffery and Billie J. Swalla

T UNICATES ARE MARINE ANIMALS COMMONLY KNOWN AS "sea squirts." These animals have been popular subjects in developmental biology for more than a century. The French biologist L. Chabry (1887; see Fischer 1992) conducted the first experiment in embryology on a tunicate. Chabry destroyed one blastomere of a 2-cell embryo, and the remaining blastomere developed into an incomplete larva. This pioneering experiment demonstrated that tunicate embryos cannot compensate for missing parts and launched a new field: experimental embryology.

Tunicates are soft-bodied animals encased in a gelatinous or leathery tunic (Figure 17.1). These animals were known to Aristotle in ancient Greece, who called them the "Thalia," a name still in use for a class of tunicates. When naturalists began to categorize animals, the tunicates were considered to be molluscs because of their soft bodies and filter-feeding life styles. However, in 1866, the Russian biologist A. Kowalevsky showed that tunicates have a tailed larva with a dorsal nervous system and notochord—the unmistakable features of a chordate (see Figure 17.2; also see Figure 17.10). In most tunicates, these chordate features disappear during metamorphosis, and the adult bears no resemblance to other chordates. Kowalevsky's important discovery showed that the tunicates are members of the phylum Chordata and sparked interest in their development.

This chapter begins with a discussion of the general aspects of tunicate biology, including their morphology, life cycles, and phylogeny. The attributes of the ascidians, one of the tunicate classes, as an experimental system in developmental biology are also described. We then discuss gametogenesis, fertilization, embryogenesis, metamorphosis and adult development, and conclude with a discussion of the evolution of ascidian development.

What Is a Tunicate?

The hallmarks of a tunicate are (1) a tunic, (2) a perforated pharynx or branchial sac, and (3) a notochord. These chordate features are usually present during a particular portion of the life cycle. For example, in some tunicates the notochord is present only in the larva and the branchial sac is present only in the adult. The tunicates are classified as the subphylum Tunicata (or Urochordata) in the phylum Chordata. This subphylum consists of three classes: (1) the Ascidiacea, or ascidians, (2) the Larvacea, or larvaceans, and (3) the Thaliacea, or thaliaceans (from Aristotle's original name for all tunicates).

Ascidians are sessile marine animals attached to rocks, shells, docks, and other firm objects at the seashore, or submersed in mud or sand flats on the ocean floor. The morphology of an adult ascidian is shown in Figure 17.1a. Like all other tunicates, ascidians are surrounded by a tunic, which is composed of a cellulose-like compound, tunicine, secreted by the epidermis. The body wall or mantle is located immediately beneath the tunic. The mantle consists of the epidermis and a layer of muscle. The feeding apparatus begins with the oral siphon or mouth, and ends with the anal siphon or anus. Food particles are filtered from the oral siphon into the branchial sac, where they are trapped in a coat of mucus and moved into the digestive tract. Food is digested in the stomach and intestine, and excrement is discarded through the anal siphon. There is no coelom and the circulatory system is open except for a small tubular heart, which pumps blood cells in alternating directions. The nervous system is relatively simple, consisting of a ganglion located between the siphons, which radiates fine peripheral nerves to the siphon and body wall muscle. Most ascidians are hermaphrodites, with gonads contain-

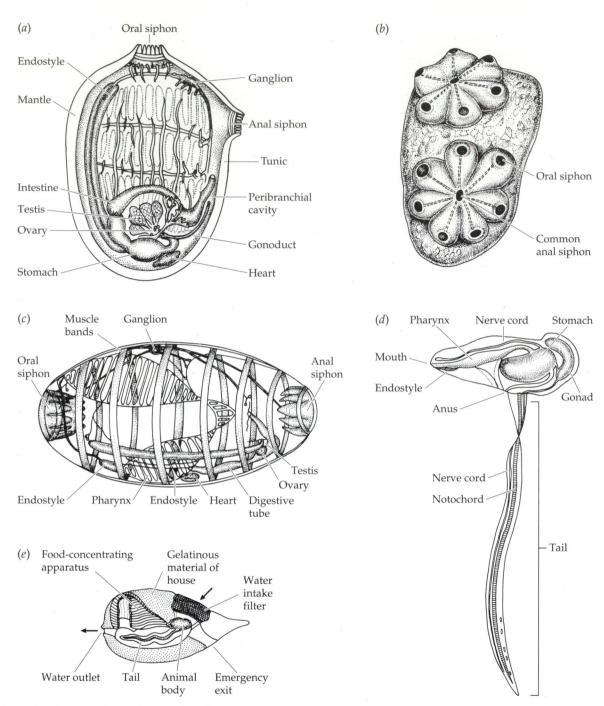

Figure 17.1 Tunicates. (*a*) The internal organization of a solitary ascidian (*Molgula*). (*b*) The external organization of a compound ascidian (*Botryllus*). (*c*) The internal organization of a thaliacean (*Doliolum*). (*d*) The internal organization of a larvacean (*Oikopleura*). (*e*) A larvacean within its tunic, or house. (*c* after Berrill 1950; *d* after Holland et al. 1994.)

ing distinct regions devoted either to sperm or egg production. Solitary ascidians live as single individuals, reproducing exclusively by sexual reproduction, whereas compound ascidians live in colonies (Figure 17.1*b*), reproducing by both sexual and asexual reproduction. Each individual, or **zooid**, in a colony has its own mouth but

shares an anus with several neighboring individuals.

The thaliaceans and larvaceans are pelagic tunicates. The thaliaceans consist of three related groups: the salps, doliolids, and pyrosomes. The morphology of an adult doliolid is shown in Figure 17.1*c*. The oral and anal siphons are present at opposite ends of the body. The mantle is ringed by contractile muscle bands, which produce water currents used in propulsion, feeding, and respiration. The structure of an adult larvacean is shown in Figure 17.1*d*. Larvaceans are unusual in that a tail is present in the adult. The tail does not serve a locomotory function, as in other tunicates, but instead is used to propel food particles through the branchial sac. The larvacean tunic is a highly specialized structure encasing the body in a gelatinous "house" (Figure 17.1e), which filters and concentrates plankton (Fenaux 1986).

Tunicate Life Cycles

Most tunicates exhibit a life cycle with larval and adult stages. This is called indirect development. The life cycle of a solitary ascidian is shown in Figure 17.2. The adult produces both sperm and eggs. In some species, eggs can be fertilized with sperm of the same individual, but other species are self-sterile: only gametes from different individuals can undergo fertilization. After fertilization, the zygote cleaves rapidly, and the embryo develops into a tadpole-shaped larva. The ascidian tadpole larva does not feed and is adapted for dispersal. Depending on the species, ascidian tadpoles swim for only a few hours or for several days before settling and beginning metamorphosis. During metamorphosis, the larva is completely remodeled. The tail is retracted into the head, tail tissues are destroyed, and new adult tissues and organs are formed from undifferentiated cells in the head. The juvenile feeds, grows, and eventually develops gonads. In compound ascidians, the life cycle also includes asexual reproduction by budding.

In some ascidian and thaliacean species, the larval phase of the life cycle is skipped. This is called direct development. In the larvaceans, the adult phase of the life

cycle contains chordate features that are present only in the larval phase of other tunicates. It has been postulated that larvaceans evolved from an ascidian or thaliacean ancestor by neoteny (Garstang 1928), a process in which juvenile features are retained in the adult.

Tunicate Phylogeny

The relationship between the tunicates and other chordates has fascinated biologists since the time of Kowalevsky. The other chordate subphyla are the Cephalochordata (*Amphioxus*) and the Vertebrata. The Tunicata and Cephalochordata are known as protochordates because of their simple body plans relative to those of the Vertebrata. Which of the protochordates is most closely related to the vertebrates? Great strides have been made in reconstructing the evolutionary history of organisms using DNA sequences. A chordate evolutionary tree based on DNA sequences is shown in Figure 17.3. According to this tree, the tunicates diverged from the chordate line before the appearance the cephalochordates and vertebrates (Wada and Satoh 1994; Turbeville et al. 1994). Therefore, the cephalochordates appear to be most closely related to the vertebrates, but the tunicates are the next closest vertebrate relative.

The evolutionary relationship among the tunicates can also be inferred from DNA sequences. According to the evolutionary tree (Figure 17.3), the common ancestor of the tunicates diverged first to form the larvaceans and later to form the ascidians and thaliaceans (Wada and

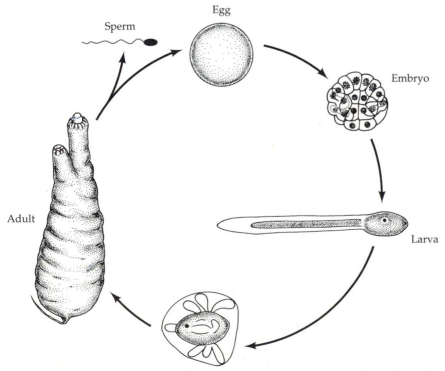

Figure 17.2 The life cycle of a solitary ascidian.

Figure 17.3 A chordate evolutionary tree inferred from small ribosomal DNA sequences.

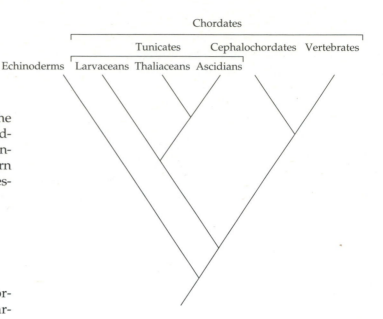

Satoh 1994). This evolutionary tree does not support the idea that the larvaceans evolved by neoteny from an ascidian or thaliacean ancestor (Garstang 1928). Instead, the ancestral tunicate may have been organized like the modern larvacean or ascidian tadpole larva, implying that the sessile life style appeared later during tunicate evolution.

Ascidians: An Experimental System in Developmental Biology

The class of tunicates that has been studied most thoroughly is the ascidians. In contrast, thaliacean and larvacean development is poorly understood. The major obstacle in studies of thaliacean and larvacean development is that these animals inhabit the open ocean and are difficult to collect and transport to laboratories. The ascidians are easy to collect because they are sessile and usually live near the sea shore. Therefore, in this chapter we emphasize ascidian development.

There are many reasons why ascidians are popular as research animals in developmental biology. First, ascidian embryos and larvae have a small number of cells. At the beginning of gastrulation, ascidian embryos contain only about 110 cells, whereas amphibian gastrulae contain about 10,000 cells. The tadpole larva consists of a few thousand cells and only six different tissues. The low number of cells allows each blastomere to be accounted for during early development. Second, ascidians develop rapidly. In many species, swimming larvae hatch only 12 to 18 hours after fertilization. Rapid development is an advantage for experimental analysis. Third, some ascidians have eggs with colored cytoplasmic regions. During cleavage these regions are distributed to specific blastomeres. This feature allows cell fates to be followed by routine microscopy. Fourth, as simple chordates, ascidians may provide basic information about the development of more complex chordates, such as the vertebrates. Fifth, ascidians have small genomes that facilitate cloning genes involved in developmental processes. Sixth, some closely related ascidian species have evolved different modes of development. This attribute permits these species to be used to study the evolution of development. Other attributes of ascidians will become evident as we describe their development.

There are also some limitations in using ascidians as an experimental system. One limitation is that most ascidians have restricted breeding seasons, and living embryos can be obtained only at certain times of the year. Another limitation is that genetic analysis is usually not available, as it is in *Drosophila, Caenorhabditis elegans*, and the mouse. The genetic approach has been developed in some of the compound ascidians (Rinkevich and Weissman 1987).

There are 13 families of ascidians. Various species in these families are used for research in developmental biology. These species and the principal features that make them attractive research subjects are listed in Table 17.1.

Gametogenesis

The first step in development is the production of the gametes. Since most ascidians are hermaphrodites, their gonads have specialized regions devoted to the production of sperm and eggs. The gonads contain germ cells, which are the precursors of the gametes, and somatic cells, which sup-

TABLE 17.1 *Ascidians Used in Developmental Biology*

Genus/Species	Family	Major attribute
Ciona sp.	Cionidae	Rapid development; long breeding season
Phallusia mammallata	Ascidiidae	Transparent embryos
Ascidia sp.	Ascidiidae	Transparent embryos
Styela sp.	Styelidae	Colored cytoplasms
Botryllus sp.	Botryllidae	Colony formation; asexual reproduction
Boltenia villosa	Pyuridae	Colored cytoplasms
Halocynthia roretzi	Pyuridae	Large embryos
Molgula citrina, M. oculata, M. occulta	Molgulidae	Evolutionary studies

port gamete development. The origin of the germ cells is poorly understood. In *Ciona*, the germ cells appear to be derived from wandering blood cells (Sugino et al. 1987). These cells originate from mesenchyme cells during adult development. After entering the gonads, the germ cells give rise to both sperm and eggs (Okada and Yamamoto 1991).

Sperm

Ascidian sperm are adapted for motility, the delivery of a haploid genome to the egg, and a short life span. The sperm of most animals contain three parts: (1) the head, (2) the midpiece, and (3) the tail. Ascidian sperm are unique in containing a head and tail but no midpiece (Figure 17.4). The sperm head contains the nucleus, an acrosome, centrosomal material, and a single large mitochondrion, which is displaced to one side of the nucleus (Cloney and Abbott 1980; Fukumoto 1990). As described below, the mitochondrion is discarded during fertilization and does not enter the egg. The sperm tail contains an axoneme, exhibiting the typical 9+2 pattern of microtubules (Woollacott 1977). The sperm swims by the coordinated beating of the tail. During sperm movement, the microtubules slide over each other and the axoneme bends. These processes are mediated by the hydrolysis of ATP, which is catalyzed by the microtubule-associated ATPase dynein.

Sperm development occurs in the testes, a specialized region of the gonad (or **ovotestis**; Figure 17.1*a*). The testes is composed of bundles of seminiferous tubules containing sperm precursor cells at all stages of development. The developing sperm are ordered in sequence from the outer to the inner region of the seminiferous tubule according to their stage in maturation. The wall of the seminiferous tubule contains the germ cells, which produce spermatogonia by mitosis. During spermatogenesis, diploid spermatogonia are converted into haploid spermatids by the meiotic reduction divisions. After completing meiosis, the

spermatids differentiate into sperm. During spermiogenesis, the spermatid nucleus and cytoplasm are reorganized (Woollacott 1977). Some of the structural and biochemical changes that occur during ascidian spermiogenesis are indicated below.

A Golgi complex is present in the anterior region of the spermatid (Kubo et al. 1978), which probably produces the acrosome. The single sperm mitochondrion is formed by the fusion of numerous individual mitochondria in the spermatid cytoplasm. As the volume of the cytoplasm is reduced, the nucleus flattens and elongates (Franzen 1976). The change in nuclear shape is mediated by bundles of microtubules, which are located in the cytoplasm adjacent to the nuclear envelope. Within the nucleus, the dispersed chromatin becomes condensed and compacted into lamellae. The compaction process is controlled by a switch in chromosomal proteins from histones to protamines (Chiva et al. 1992). The **protamines** are small basic nuclear proteins that crosslink DNA into tight packages. The mature sperm are stored in the lumen of the seminiferous tubules and pass out of the testes through the sperm duct during spawning.

Eggs

The egg is a specialized cell that contributes half of the zygotic genome and all of the cytoplasm to the embryo. Although egg size varies in different ascidian species, most ascidian eggs are small, ranging from about 100 to 200 μm in diameter. The eggs are produced in a region of the gonad called the ovary (Figure 17.1*a*).

Ascidian eggs contain an internal region, or endoplasm, which is packed with yolk granules. The endoplasm is surrounded by a thin cortical shell enriched in mitochondria, pigment granules, endoplasmic reticulum, and ribosomes. These organelles are tethered in the cortex by a meshwork of cytoskeletal filaments, which is linked to the plasma membrane (Figure 17.5). The complex of cortical organelles and cytoskeletal elements defines a unique cytoplasmic region known as the **myoplasm**. Eggs are surrounded by a **vitelline envelope**, which is retained throughout embryogenesis. The vitelline envelope consists of: (1) the follicle cells, (2) the chorion, and (3) the test cells (Figure 17.6; also see Figure 17.9*a*). The follicle cells are present on the outside of the vitelline envelope, the acellular **chorion** is present between the **follicle cells** and the **test cells**, and the test cells, which often adhere to the surface of the egg, are located inside the chorion. The space between the chorion and the egg surface, known as the **perivitelline space**, also contains an extracellular matrix. The origin of the follicle and test cells is controversial. Some investigators think that they are part of the germ line (sister cells of the oocyte), while others believe that they are somatic cells that have migrated into the ovary. The follicle cells secrete the chorion, which consists of several layers of glycoprotein filaments. In some species, the follicle cells also produce chemotactic substances that at-

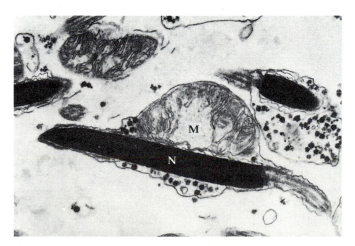

Figure 17.4 An electron micrograph showing the head of an ascidian sperm. N, nucleus; M, mitochondrion. (From Fukomoto 1990.)

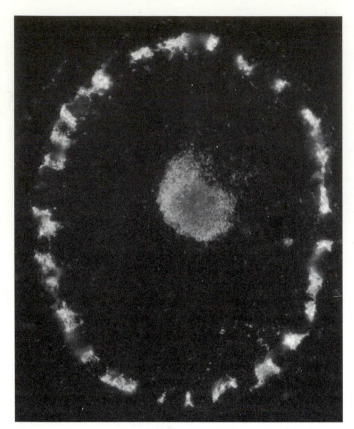

Figure 17.5 An ascidian oocyte. The germinal vesicle (bright area in center) and myoplasm (discontinuous bright region at edge) are stained with an antibody that recogizes the cytoskeletal protein p58.

tract the sperm to the egg (Miller 1982) or produce large gas-filled vacuoles that cause eggs to float in sea water (Lambert and Lambert 1978). Other functions of the accessory cells will be described in the sections on fertilization and embryogenesis.

Germ cells within the ovary produce oogonia by mitotic divisions. The oogonia undergo oogenesis. Oogenesis is divided into three stages based on the timing of yolk deposition: (1) previtellogenesis, (2) vitellogenesis, and (3) postvitellogenesis. Oocytes in different stages of oogenesis are present in the ovary. During previtellogenesis, the oocyte grows and increases its cytoplasmic volume. Although many cytoplasmic components are elaborated at this stage, the most prominent is a mass of lipid pigment granules and mitochondria located on the vegetal side of the **germinal vesicle**, or oocyte nucleus. This region is thought to be the precursor of the myoplasm (Hsu 1963). During vitellogenesis, yolk granules are laid down in the cytoplasm, causing a large increase in oocyte volume. The yolk granules are produced by intracellular processes, but there is also an external contribution, as indicated by the fusion of pinocytotic vesicles with the growing yolk granules (Kessel 1966). During vitellogenesis, or yolk deposition, the mass containing mitochondria and pigment granules is translocated from its original location beneath the germinal vesicle to the cortex. During postvitellogenesis, the oocyte becomes full size and ready to initiate maturation. The postvitellogenic oocyte is organized with the germinal vesicle, in the central region, positioned toward the animal pole. The yolky endoplasm surrounds the germinal vesicle, accounting for most of the cytoplasmic mass. The myoplasm is located in the cortex.

During maturation, the postvitellogenic oocyte undergoes the first meiotic division, producing a secondary oocyte and a polar body. The secondary oocyte initiates the second meiotic division, but it arrests at metaphase II. Thus, the unfertilized egg is actually a secondary oocyte arrested at metaphase of the second meiotic division. Unfertilized eggs pass out the ovary via the oviduct. In some species, a large number of eggs are stored in the oviduct during the breeding season.

Breeding Seasons and Spawning

Ascidians spawn sperm and eggs every day during the breeding season. Breeding seasons differ among ascidian species. Some species have breeding seasons restricted to the summer or winter months, whereas other species breed throughout the year. The gametes are released into the sea water through the anal siphon. Spawning is triggered by changes in the photoperiod. In some ascidian species, a short pulse of light following an extended dark period triggers spawning (Whittingham 1967). These species spawn in the morning. In other species, spawning is initiated by a longer light period following darkness (West and Lambert 1976) . These species spawn at dusk. The natural spawning conditions can be mimicked in the laboratory by adjusting the light-dark cycle. In this manner, sperm and eggs can be obtained for experimental purposes at any time during the breeding season (West and Lambert 1976).

Fertilization

Fertilization is divided into the following successive events: (1) sperm attraction to the egg, (2) sperm activation, (3) sperm binding and penetration of the chorion, and (4) sperm-egg fusion (Figure 17.7). Ascidian eggs have mechanisms to prevent the entry of multiple spermatozoa, a potentially lethal condition known as **polyspermy**. The events of fertilization and the mechanisms that prevent polyspermy and self-fertilization are described in this section.

The vitelline envelope plays a major role in fertilization. As mentioned earlier, the follicle cells secrete materials that attract sperm to the egg (Miller 1982). The molecules responsible for sperm chemotaxis are unknown. The follicle cells also function in sperm activation. That follicle cells are required for fertilization can be demonstrated by inseminating eggs with and without follicle cells (Fuke 1983). The results show that eggs with follicle cells can be

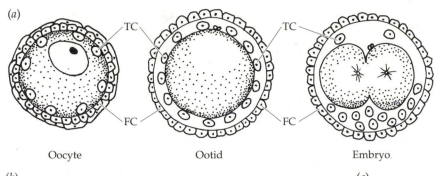

Oocyte   Ootid   Embryo

Figure 17.6 The structure of the vitelline envelope. *(a) Styela* oocytes, ootids, and 2-cell embryos showing the follicle cells (FC) and test cells (TC). The chorion is the thin layer between the follicle cells and test cells. *(b)* Scanning electron micrograph of a *Styela* egg showing the layer of follicle cells. *(c)* Scanning electron micrograph of a *Styela* egg with part of the follicle cell layer removed, showing the underlying chorion.

(b)

(c)

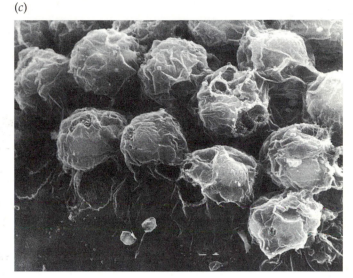

fertilized, but those lacking follicle cells cannot be fertilized. However, *Phallusia* eggs lacking the vitelline envelope can be fertilized if sperm are first exposed to other eggs with follicle cells (Speksnijder et al. 1989). Sperm activation may be achieved by interaction of the sperm with the follicle cells or with a substance they secrete into seawater.

The sperm must bind and penetrate the vitelline envelope to reach the egg surface. Sperm binding occurs at clefts between the bases of the follicle cells (Figure 17.6), where parts of the underlying chorion are exposed to the environment (Bates 1980). The general mechanism of sperm binding is the formation of a macromolecular complex between a sperm surface enzyme and a sugar residue in the chorion. The sugar-binding sites are probably parts of larger glycoprotein molecules. In *Ciona*, sperm bind to L-fucose residues in the chorion (Rosati and DeSantis 1978). The role of L-fucose has been demonstrated by experiments in which the ability of eggs to bind sperm was tested in the presence of different sugars. The rationale for these experiments is that the addition of sugars that are the same or similar in structure to the natural sperm bind-

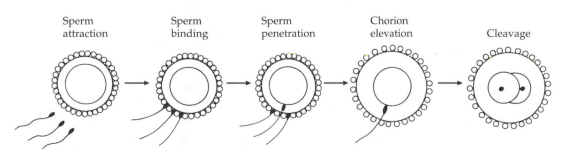

Sperm attraction   Sperm binding   Sperm penetration   Chorion elevation   Cleavage

Figure 17.7 A summary of the steps of ascidian fertilization. (After Satoh 1994.)

ing sites should prevent fertilization by interacting with sperm before they reach the chorion. In contrast, sugars that are unrelated to the natural binding sites should have no effect on fertilization. The results show that fertilization is prevented by L-fucose, but not other sugars such as D-fucose, mannose, *N*-acetylglucosamine, or *N*-acetylgalactosamine. The binding of sperm to L-fucose residues is mediated by the formation of a complex with α-L-fucosidase, which is present on the sperm surface (Hoshi et al. 1985). In *Phallusia* and *Ascidia*, the sperm binding site is *N*-acetylglucosamine. Accordingly, fertilization in these species is inhibited by treatment of eggs with *N*-acetylglucosamine, but not unrelated sugars (Honegger 1986; Lambert 1986). In these species, the *N*-acetylglucosamine groups are complexed with the sperm surface enzyme *N*-acetylglucosamine (Godknecht and Honegger 1991).

Many sperm are released in close proximity to the eggs during spawning. These eggs seldom contain more than a single sperm nucleus, suggesting that there are effective means to prevent polyspermy. In *Ascidia*, *N*-acetylglycosaminidase located on the egg surface is released into seawater as a response to the first sperm that binds to the chorion (Lambert 1989). This enzyme destroys the unoccupied *N*-acetylglucosamine binding sites in the chorion, prohibiting the binding of additional sperm. An additional block to polyspermy may be mediated by electrical depolarization of the egg plasma membrane (Goudeau et al. 1994). The relationship between the two blocks to polyspermy remains to be investigated.

After the sperm binds to the vitelline envelope, it must penetrate the chorion to reach the egg surface. Sperm penetration is mediated by proteolytic enzymes which digest a hole in the chorion. The role of proteases in sperm penetration was first suggested by an experiment in which treatment with the protease inhibitors leupeptin and chymostatin prevented fertilization (Hoshi et al. 1981). Subsequently, specific proteases were isolated from sperm (Sawada et al. 1982; 1983; Pinto et al. 1990). As mentioned previously, the head of ascidian sperm contains a small **acrosome** (Fukumoto 1990). During the acrosome reaction, the outer membrane of the acrosome fuses with the sperm plasma membrane, and the acrosome contents are exposed to the environment. Some of the acrosomal materials, such as the enzymes involved in sperm binding, may remain associated with the tip of the sperm, while others, such as the proteases involved in sperm penetration, are released into the environment. As the sperm penetrates the chorion, the mitochondrion slides posteriorly out of the head and along the axoneme to the tip of the tail, where it is discarded (Figure 17.8). The sliding mitochondrion may propel the sperm through the chorion (Lambert and Koch 1988). Ascidians are one of the few animals in which sperm mitochondria do not enter the egg. Thus, the mitochondria of the fertilized ascidian egg are exclusively of maternal origin.

Numerous hermaphrodites have evolved specific

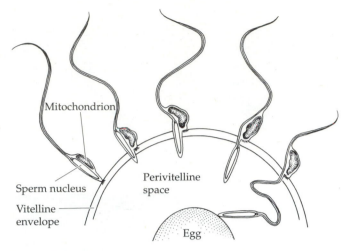

Figure 17.8 Movement of the mitochondrion during sperm penetration of the chorion. (After Lambert and Koch 1988.)

mechanisms to prevent self-fertilization. Self-fertilization is undesirable because two copies of potentially deleterious alleles could be transmitted to the offspring. The ascidian *Ciona* exhibits **self-sterility**. In a classic experiment, T. H. Morgan (1939) removed the vitelline envelope of *Ciona* eggs and showed that the naked eggs are capable of fertilization by sperm of the same animal. These results suggested that self-fertilization is controlled by the vitelline envelope. However, the results could not distinguish whether the follicle cells or the chorion was the agent that prevented self-fertilization. This experiment was repeated with eggs lacking the entire vitelline envelope or only the follicle cells (Rosati and DeSantis 1978). The results showed that eggs lacking the vitelline envelope, but not those lacking only the follicle cells, were capable of self-fertilization. The chorion appears to be responsible for self-sterility. Self-sterility is established late during oogenesis, accompanied by the appearance of a thin layer of dense material on the outer surface of the chorion (Pinto et al. 1995). Unfortunately, little is known about the molecules that are responsible for preventing self-fertilization. Some ascidians, including viviparous solitary and colonial species, are self-fertile. The manner in which these species evade the deleterious effects of homozygosity is currently being investigated.

After penetrating the chorion, sperm enter the perivitelline space and eventually bind and fuse with the egg. The events of sperm-egg interaction are complicated by the vitelline envelope. As in the case of sea urchins, however, ascidian sperm protrude an **apical process** that binds to the egg surface. The apical process probably forms as the final event in the **acrosome reaction** (Fukumoto 1990). During sperm-egg fusion the centrosomal material enters the egg cytoplasm along with the nucleus. The centrosome later generates the sperm aster (see below) and is required for cleavage. Eggs artificially activated by calcium ionophore undergo some of the activation events, but do not form a sperm aster or cleave (Jeffery 1982). There is no elevation of a fertilization membrane, such as in fertilized sea urchin eggs. However, in some species the perivitelline space expands or the chorion is hardened after fertilization. These events may also be related to the prevention of polyspermy.

Embryogenesis

Ascidian embryogenesis is divided into four phases: (1) egg maturation and ooplasmic segregation, (2) cleavage, (3) gastrulation and neurulation, and (4) morphogenesis and cell differentiation. The four phases of embryogenesis are shown in Figure 17.9.

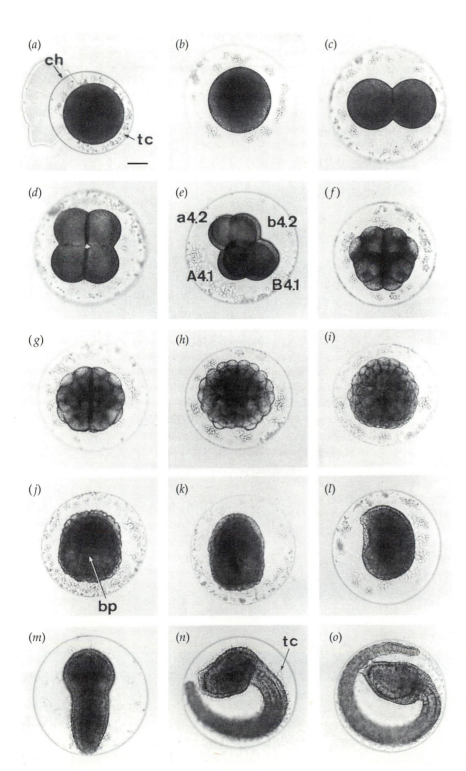

Figure 17.9 Embryonic development in the ascidian *Halocynthia roretzi*. (*a*–*b*) Maturation and ooplasmic segregation. (*c*–*i*) Cleavage showing 2-, 4-, 8-, 16-, 32-, 64-, and 110-cell embryos, respectively. (*j*–*k*) Gastrulation and (*l*) neurulation. (*m*–*o*) Morphogenesis and cell differentiation. (m) Early, (n) mid-, and (*o*) late tailbud stage; ch, chorion; tc, test cells. Scale bar: 100 μm. (From Satoh et al. 1990.)

As a prelude to discussing embryogenesis, we will first describe the tadpole larva (Figure 17.10). The tadpole is divided into a head and a tail. The larva contains six different types of tissue: epidermis, nervous system, notochord, muscle, mesenchyme, and endoderm (Katz 1983). At the anterior end of the larva, there is an adhesive organ, which is used for attachment during metamorphosis. The larval head contains the anterior portion of the nervous system, known as the brain. The brain contains two pigmented sensory organs, the otolith and ocellus. The otolith consists of a single melanocyte suspended in a vesicle, which is surrounded by sensory neurons. The **otolith** may function in orientation of the swimming larva. Although variable in different species, the ocellus usually consists of a single cell containing small melanin granules and a lenslike structure. The **ocellus** is a photoreceptor. *Molgula* tadpoles have an otolith but no ocellus. At its posterior end, the brain tapers into the spinal cord, which runs through the dorsal side of the tail. The larval head also contains the rudiments of adult organs and tissues, such as the branchial sac, gut, heart, and body-wall muscle. These organs and tissues are formed from endoderm and mesenchyme cells, which remain undifferentiated during embryogenesis. The larval tail contains the notochord running through its center, which is flanked by bands of striated muscle cells. A strand of endoderm runs through the ventral portion of the tail. The entire larva is covered by the epidermis. The number of cells that comprise each of the larval tissues is shown in Table 17.2. The tadpole consists of about 2500 cells. We will now describe the development of the tadpole larva.

Maturation and Ooplasmic Segregation

The fusion of the sperm and egg results in the activation of development. Four critical processes occur during egg activation: (1) the completion of meiosis, (2) the development of the sperm pronucleus, (3) ooplasmic segregation, and 4) syngamy, or pronuclear fusion. These processes are required for subsequent cleavage and the initiation of development.

Although the sperm is a haploid cell, the egg must still complete the second meiotic division to become haploid. During the second maturation division, the haploid female pronucleus is formed, and another polar body appears at the animal pole. Meanwhile, the sperm nucleus is undergoing changes leading to the formation of the male pronucleus in the vegetal hemisphere. The sperm can enter anywhere on the egg surface and is swept into the vegetal hemisphere during **ooplasmic segregation** (Bates and Jeffery 1988; Speksnijder et al. 1989). Ooplasmic segregation, a dramatic shift in egg cytoplasmic regions, is initiated by sperm entry. The events of ooplasmic segregation are spectacular in *Styela* and *Boltenia* eggs, which have colored myoplasms (Conklin 1905; Jeffery 1982). The segregation of the myoplasm is shown in Figure 17.11a.

There are two phases of ooplasmic segregation (Sardet et al. 1989; Jeffery and Swalla 1990). During the first phase,

TABLE 17.2 *Approximate Cell Numbers in Tissues of Ascidian Larvae*

Tissue	Cell number
Endoderm	500
Epidermis	800
Nervous system	300
Notochord	40
Muscle	38
Mesenchyme	900

Source: From Monroy 1979.

(a)

(b)

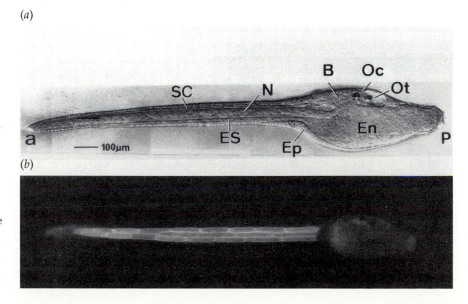

Figure 17.10 The ascidian tadpole larva. (*a*) The larva of *Halocynthia roretzi*. (*b*) A tadpole stained with a flourescent label highlighting the tail-muscle cells. Ep, epidermis; P, adhesive papillae; B, brain; Ot, otolith; Oc, ocellus; En, endoderm; N, notochord; SC, spinal cord; ES, endodermal strand. Scale bar: 100 μm. (From Satoh et al. 1990.)

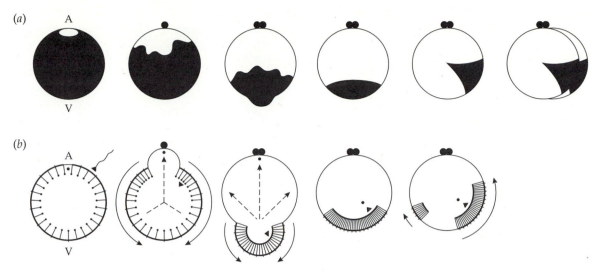

Figure 17.11 Ooplasmic segregation. (*a*) The myoplasm (dark area) is translocated first to the vegetal pole and then into a crescent at the future posterior pole of the embryo. (*b*) The myoplasmic cytoskeleton mimics segregation of the myoplasm. A small amount of myoplasm is also translocated into the anterior pole region. A, animal pole; V, vegetal pole. (From Jeffery and Swalla 1990.)

the myoplasm is translocated into the vegetal hemisphere, forming a cap of colored cytoplasm near the vegetal pole (Figure 17.11*a*). Although not shown in Figure 17.11*a*, the endoplasm and ectoplasm (the cytoplasmic regions responsible for forming the endoderm and ectoderm, respectively) are also rearranged during ooplasmic segregation. The endoplasm is translocated into the animal hemisphere and the ectoplasm into the vegetal hemisphere, where it is layered above the cap of myoplasm. The myoplasm is a cytoskeletal domain consisting of a superficial layer of actin filaments and a deeper cytoskeletal matrix (Jeffery and Meier 1983). Mitochondria, pigment granules, and other organelles are associated with the inner layer of the cytoskeleton. The myoplasmic cytoskeleton and its movements during ooplasmic segregation are shown in Figure 17.11*b*. The first phase of ooplasmic segregation is mediated by the contraction of actin filaments in the myoplasm. The role of actin filaments in ooplasmic segregation was demonstrated by an experiment in which cytochalasin B, an inhibitor of actin polymerization, was shown to prevent the translocation of the myoplasm to the vegetal pole region (Sawada and Osanai 1981). The myoplasm is intimately associated with the egg plasma membrane. Thus, during the first phase of ooplasmic segregation, membrane proteins and microfilaments cosegregate into the vegetal hemisphere (Figure 17.12*a*, *b*). In *Phallusia* eggs, the focal point of ooplasmic segregation, which is usually (but not always) localized near the vegetal pole (Roegiers et al. 1995), is defined by a contraction pole consisting of a tuft of microvilli (Figure 17.12*c*). The aggregation of these microvilli during ooplasmic segregation may be mediated by the underlying cytoskeleton.

During the second phase of ooplasmic segregation, the myoplasm shifts away from the vegetal pole region, forming a crescent on one side of the egg (Figure 17.11*a*). In some species, the pigment granules associated with the myoplasm form the famous yellow crescent, first described in *Styela* eggs by E. G. Conklin (1905). A smaller part of the myoplasm also shifts in the opposite direction (Sardet et al. 1989). In the remainder of this chapter, however, we will refer to the yellow crescent cytoplasm as the myoplasm. The ectoplasm is also shifted during the second phase of ooplasmic segregation, first to the equatorial region near the yellow crescent, and then into the animal hemisphere, where it displaces part of the endoplasm into the vegetal hemisphere. The sperm aster, a radial system of microtubules, forms in the vegetal hemisphere during the second phase of ooplasmic segregation. The second phase is sensitive to colchicine, an inhibitor of microtubule polymerization, but not to cytochalasin, indicating that it is controlled by microtubules (Sawada and Schatten 1989). The sperm aster is probably responsible for localizing the yellow crescent during the second phase of ooplasmic segregation. As ooplasmic segregation is being completed, the female and male pronuclei fuse to form the diploid zygotic nucleus.

An important consequence of ooplasmic segregation is the appearance of the embryonic axes. The unfertilized egg shows **radial symmetry**: any plane through the animal-vegetal axis separates it into two equal parts. After the completion of ooplasmic segregation, however, the fertilized egg shows **bilateral symmetry**: only one plane (the plane of the future left-right axis of the embryo) can bisect it into two equal parts. The three axes of the embryo are: (1) the dorsoventral axis, (2) the anteroposterior axis, and (3) the right-left axis (Figure 17.13). The dorsoventral axis

Figure 17.12   Accumulation of plasma membrane proteins (*a*), microfilaments (*b*), and microvilli (*c*) at the contraction pole during the first phase of ooplasmic segregation in *Phallusia* eggs. (*a*) Staining with a fluorescent dye that labels macromolecules on the cell surface. (*b*) Microfilament staining with fluorescent phalloidin. (*c*) Scanning electron micrograph of the tuft of microvilli (arrowheads). c, contraction pole or focal point for ooplasmic segregation; a, animal pole; F, fertilized egg; NF, unfertilized egg. (From Roegiers et al. 1995.)

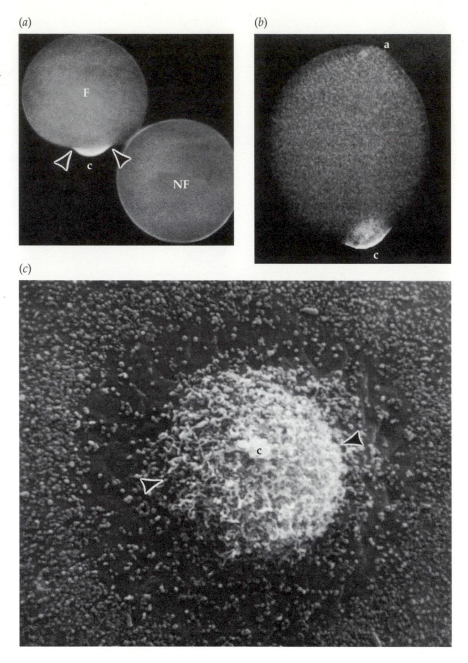

approximates the animal-vegetal axis, and the vegetal pole region becomes the dorsal side of the embryo. The antero-posterior axis lies perpendicular to the dorsoventral axis, and the yellow crescent region becomes the posterior pole of the embryo.

Cleavage

After maturation, ooplasmic segregation, and pronuclear fusion are completed, a series of rapid cleavages is initiated, which divide the zygote into a multicellular embryo. The first cleavage occurs about an hour after fertilization in most ascidian species. As in other animals, ascidian cleavage is characterized by short, synchronous cell cycles. The cleavage pattern is bilateral.

The pattern of the cleavage up to the 64-cell stage is shown in Figures 17.14 and 17.15. The nomenclature for ascidian blastomeres was introduced by E. G. Conklin (1905). Two letters are used in Conklin's system: A and B (Figure 17.15). Lowercase letters (a and b) are used for blastomeres in the animal hemisphere, and uppercase letters (A and B) for blastomeres in the vegetal hemisphere. The letters are followed by numbers: the first digit indicates the cell generation, increasing by one after each division, and the second digit (after a decimal point) indicates the number of cells in each quadrant of the embryo. Cells on the right and left halves of the bilaterally symmetric embryo are designated by the same letters, but cells on the right are underlined. For example, the A4.1 blastomere is

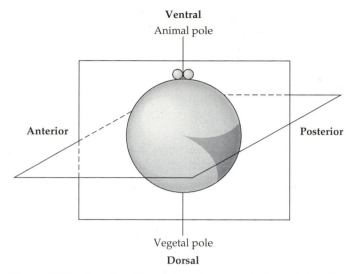

Figure 17.13 Polarity of the uncleaved zygote after completion of the second phase of ooplasmic segregation. The dorsoventral and anteroposterior axes are indicated. The yellow crescent is shaded.

located in the anterior quadrant of the vegetal hemisphere on the right side of the 8-cell embryo, and is a member of the fourth generation of blastomeres (Figure 17.15c). Likewise, the b6.8 blastomere is located in the posterior quadrant of the animal hemisphere on the left side of the 32-cell embryo, and is a member of the sixth generation of blastomeres (Figure 17.15f). Using this nomenclature, we will now describe ascidian cleavage.

The uncleaved zygote is called the A1 cell. The first cleavage plane bisects the A1 cell through the animal-vegetal axis and yellow crescent, establishing the right and left sides of the embryo. The first two cells are the AB2

Figure 17.14 Scanning electron micrographs of cleaving *Halocynthia roretzi* embryos. Upper row: (a) 1-cell zygote; (b) 2-cell embryo; (c) 4-cell embryo; (d) 8-cell embryo. Center row: 16-cell embryos viewed from the animal (e) and vegetal (f) poles; 32-cell embryos viewed from the animal (g) and vegetal (h) poles. Lower row: 64-cell embryos viewed from the animal (i) and vegetal (j) poles; 110-cell embryos viewed from the animal (k) and vegetal (l) poles. pb, polar body; tc, test cell. (From Satoh 1979.)

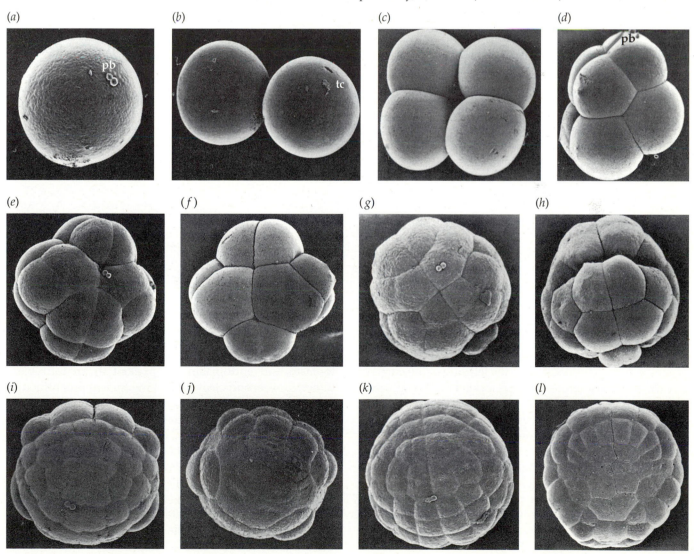

Figure 17.15 Cleavage of the ascidian embryo showing blastomere nomenclature. (*a–c*) 2-cell through 8-cell embryos. (*d–e*) 16-cell embryos viewed from the animal (*d*) and vegetal (*e*) poles. (*f–g*) 32-cell embryos viewed from the animal (*f*) and vegetal (*g*) poles. (*h–i*) 64-cell embryos viewed from the animal (*h*) and vegetal (*i*) poles. (From Venuti and Jeffery 1989.)

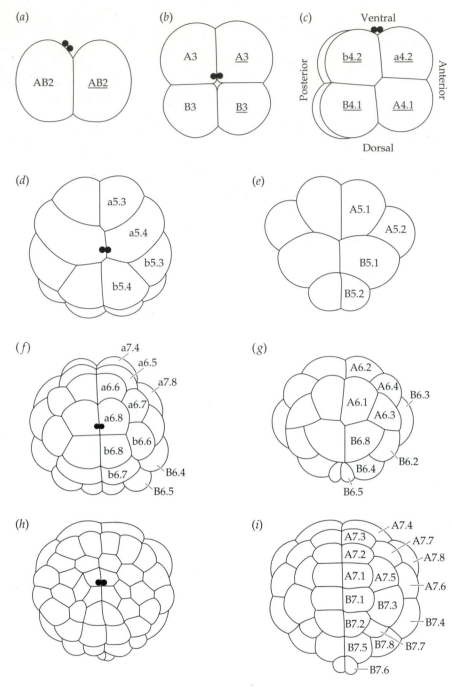

blastomere, which is located on the left side of the embryo, and the AB2 blastomere, which is located on the right side of the embryo (Figure 17.15*a*). The ectoplasm, endoplasm, and myoplasm are divided equally between the AB2 and AB2 blastomeres. The second cleavage plane also cuts through the animal-vegetal axis, but it is perpendicular to the plane of the first cleavage, resulting in four approximately equal-sized blastomeres. The A3 blastomere generates the anterior-right quadrant, the A3 blastomere the anterior-left quadrant, the B3 blastomere the posterior-right quadrant, and the B3 blastomere the posterior-left quadrant of the embryo (Figure 17.15*b*). Although the ec-

toplasm and endoplasm are present in all four blastomeres, at second cleavage the myoplasm is confined to the posterior (B3 and B3) blastomeres. The third cleavage plane cuts through the equator of the embryo, forming four animal and four vegetal blastomeres. The slanting of these cleavage planes results in a trapezoid-shaped 8-cell embryo, the four animal cells lying slightly anterior to the four vegetal cells (Figures 17.9*e*, 17.14*d*, and 17.15*c*). The animal hemisphere contains the a4.2, a4.2, b4.2, and b4.2 blastomeres, and the vegetal hemisphere contains the A4.1, A4.1, B4.1, and B4.1 blastomeres. At third cleavage, the myoplasm enters the B4.1 and B4.1 cells, most of the endoplasm enters the four vegetal cells, and most of the ectoplasm enters the four animal cells. Beginning at fourth cleavage, the cleavage planes are oriented differently in the animal and vegetal hemispheres, and after fifth cleavage, cell division synchrony begins to fade. The late cleavages are not described here, but can be inferred from Figures 17.14 and 17.15. By the 64-cell stage, the cytoplasmic regions of the egg have been distributed to different blastomeres. The ectoplasm is located in the animal-hemisphere blastomeres, the endoplasm in the vegetal-hemisphere blastomeres, and the myoplasm is located in the posterior vegetal-hemisphere blastomeres.

Cleavage continues through the 110-cell stage, when gastrulation begins. As result of the early cleavages, the fertilized egg is converted into a blastula. The blastomeres are packed closely together within the blastula, and there is no blastocoel. The interval between the first and eighth cleavages is about 3–4 hours.

Fate Maps and Cell Lineages

Determination is the process in which the fate of an embryonic cell or region is decided, but is yet to be established by differentiation. After the early cleavages are completed, the three germ layers of the embryo (ectoderm, mesoderm, and endoderm) are formed, and cell fates begin to be determined.

An understanding of regional and cell fates is an absolute requirement for studying the mechanism of embryogenesis. A **fate map** is a description of the developmental history of embryonic regions or cells and a prediction of the tissues they will form in the larva. The ascidian fate map is accurately known, due to the small number of embryonic cells. Fate maps are constructed by marking a portion of the egg or early embryo and determining the location of the marker at a later stage in development. The ascidian fate map has been constructed with several different markers. E. G. Conklin (1905) made a crude fate map of the *Steyla* embryo using the colored cytoplasmic regions as markers. Other investigators have made fate maps by staining with vital dyes or by placing small chalk particles on the egg surface (Ortolani 1955; Zalokar and Sardet 1984; Bates and Jeffery 1987). Fate maps have also been constructed by following cleavages with the scanning electron microscope (Nicol and Meinertzhagen 1988). The most detailed fate map has been made by microinjecting horseradish peroxidase, an enzyme used as an intracellular marker, into specific blastomeres (Nishida and Satoh 1983, 1985; Nishida 1987).

Fate maps of the *Halocynthia* egg and 8-cell embryo are shown in Figure 17.16*a*. The ectoderm is located in the animal hemisphere, the mesoderm is located in the upper portion of the vegetal hemisphere, and the endoderm is located in the lower portion of the vegetal hemisphere (Figure 17.16*a*). The epidermis and nervous system are derived from the ectoderm. The notochord, muscle, and mesenchyme cells are derived from the mesoderm. The

endoderm initiates gastrulation at the end of the cleavage phase but otherwise remains undifferentiated during embryogenesis. After metamorphosis, the endoderm will differentiate into the adult tissues and organs.

The fate map of a single blastomere is called a **cell lineage**. The cell lineage of the *Halocynthia* embryo is shown in Figure 17.16*b*. The cell lineage shows the number of cell divisions of each blastomere and the location of its progeny in the embryo. It can be concluded from the cell lineage that most larval tissues originate from more than one quadrant of the ascidian embryo. This principle is illustrated by considering the origin of the tail muscle cells (Figure 17.17). *Halocynthia* tadpoles contain 42 tail muscle cells, 21 cells on each side of the tail. On the left side of the tail, 14 anterior cells are derived from the B4.1 blastomere (posterior-vegetal quadrant), 2 posterior cells are derived from the A4.1 blastomere (anterior-vegetal quadrant), and 5 cells at the tip of the tail are derived from the b4.2 blastomere (posterior-animal quadrant) of the 8-cell embryo. The muscle cells derived from the B4.1 blastomeres are called primary muscle cells, whereas the muscle cells derived from the A4.1 and b4.2 blastomeres are called secondary muscle cells. Likewise, the notochord, endoderm, and mesenchyme cells originate from cell lineages in different quadrants (Figure 17.16*b*). Later in this chapter, we will discuss the mechanisms of tissue determination.

Gastrulation and Neurulation

During the cleavage period, the blastomeres do not change their position in the embryo. Beginning about 4–5 hours after fertilization, however, a coordinated series of cell movements begins in the three germ layers, which convert the blastula into a multilayered embryo. These movements are called gastrulation and neurulation. Gastrulation involves three processes: (1) invagination, (2) involution, and (3) epiboly (Figure 17.18). The endoderm undergoes invagination, the mesoderm undergoes involution, and the ectoderm undergoes epiboly.

The first step in gastrulation is **invagination**, the internal movement of a sheet of cells (the endoderm) based on changes in their shape. The ascidian blastula is slightly flattened in the vegetal hemisphere (Figure 17.18*a*). At the beginning of gastrulation, the vegetal endoderm cells contract at their basal margins and expand at their apical margins, changing from a columnar to a wedge shape. The most extreme changes occur in the four largest endoderm cells (B7.1, B7.1, A7.1, and A7.1), which appear to lead the process of invagination. Once inside the embryo, the invaginated endoderm cells continue to migrate anteriorly beneath the ectoderm. After invagination is complete, the blastopore is bordered by mesoderm cells on all sides (Figure 17.18*b*). The fate map shows that the presumptive notochord cells are located at the anterior lip of the blastopore, the presumptive mesenchyme cells at the lateral lips of the blastopore, and the presumptive muscle cells at the posterior lip of the blastopore.

Figure 17.16 Ascidian fate maps and cell lineage. (*a*) Fate maps of the 1-cell zygote (left, after completion of ooplasmic segregation) and 8-cell embryo (right). (*b*) The cell lineage of the *Halocynthia roretzi* embryo. Only the left side of the bilaterally symmetric embryo is indicated. The cleavage stage and time elapsed since fertilization are indicated at the top. (From Satoh 1994.)

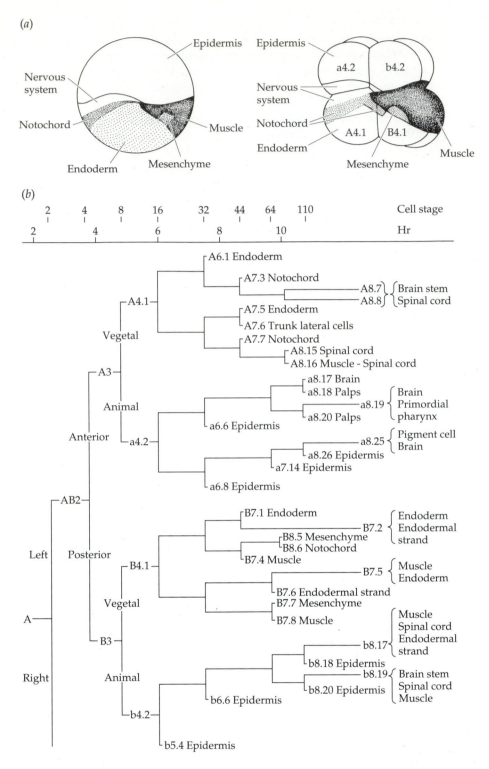

The second step in gastrulation is **involution**, the process in which a sheet of migratory cells (the mesoderm) moves over a sheet of stationary cells (the ectoderm). The mesoderm cells involute over the lips of the blastopore in an anterior to posterior sequence: the presumptive notochord cells disappear into the blastopore first, followed by the presumptive mesenchyme cells, and then by the presumptive muscle cells. After the completion of involution, the mesoderm is entirely internalized. As the endoderm and mesoderm cells enter the blastopore, **epiboly**, the third step in gastrulation, begins in the ectoderm. During epiboly, the columnar ectoderm cells flatten and spread into the vegetal hemisphere. Similar to involution, epiboly is initiated in the anterior region of the

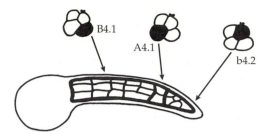

Figure 17.17 The three sources of tail muscle cells. (From Venuti and Jeffery 1989.)

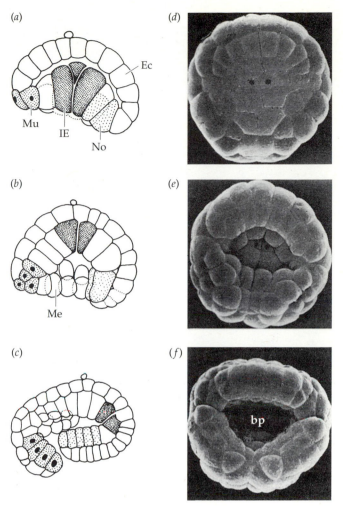

Figure 17.18 Gastrulation. (*a–c*) Diagrams of cross sections. The posterior poles are on the left. (*b*) A mid-gastrula showing involution of the mesoderm. (*d–f*) Scaninng electron micrographs of embryos viewed from the vegetal pole. Black dots indicate the position of two invaginating endoderm cells. Ec, ectoderm; Mu, presumptive muscle cells; IE, invaginating endoderm; No, presumptive notochord; Me, presumptive mesenchyme cells; bp, blastopore. (*a–c* from Jeffery 1992; *d–f* from Satoh 1978.)

gastrula and proceeds posteriorly, as indicated by the keyhole shape of the blastopore during the late gastrula stage (Figure 17.18*f*). Eventually, the entire surface of the embryo is covered by ectoderm. At the end of gastrulation, the embryo begins to elongate along its anteroposterior axis. The elongation process involves the extension of the neural plate (several rows of columnar ectoderm cells on the dorsal side of the embryo), the anterior migration of the endoderm cells, and the posterior migration of the mesoderm cells.

Neurulation is the process in which the neural tube is formed on the dorsal side of the embryo (Figure 17.19). The neural plate folds inward along the anteroposterior axis, and the neural folds eventually meet to form the neural tube. The folding of the neural plate occurs by a process that resembles invagination. The neural-plate cells contract at their basal margins and expand at their apical margins, and thus sink into the interior of the embryo (Satoh 1978). The closure of the neural tube occurs in a posterior to anterior wave (Figure 17.19). The **neuropore**, an orifice leading into the neural tube, is visible at the end of neurulation. The anterior part of the neural tube will expand and become the brain, and the posterior part will become the spinal cord.

Morphogenesis and Cell Differentiation

The next step in embryogenesis is the formation of the tadpole larva. Larval development involves an integrated set of morphogenetic movements followed by cell differentiation. The neural tube elongates and differentiates into the brain and spinal cord, the otolith and ocellus differentiate in the brain, and the epidermis forms the adhesive organs and secretes the inner layer of the larval tunic. In contrast, the outer portion of the tunic, which forms the tail fins in the posterior region of the larva, is secreted by the test cells. If the test cells are removed, the embryo develops into a normal larva, but it lacks tail fins (Cloney and Cavey 1982). The most prominent event during larval development is tail formation.

Tail formation involves morphogenetic movements and subsequent differentiation of the presumptive notochord, muscle, spinal cord, and tail epidermis cells. The force that drives tail formation is the elongation of the notochord

(Miyamoto and Crowther 1985). The presumptive notochord cells are located in the anterior half of the gastrula. During the neurula and subsequent tailbud stages, these cells migrate posteriorly and interdigitate to form the notochord, a single row of 40 cells. Cell lineage analysis has shown that notochord cells interdigitate randomly (Nishida and Satoh 1985): notochord cells derived from the right and left sides of the embryo appear in an unordered sequence along the notochord (Figure 17.20*a*). After the completion of interdigitation, the notochord cells begin to differentiate (Cloney 1990). Notochord differentiation involves the secretion of an extracellular matrix, which is initially packed into spaces between the cells, but later becomes continuous through the center of the notochord (Figure 17.20*b*). The extracellular matrix is responsible for

(a)

(b)

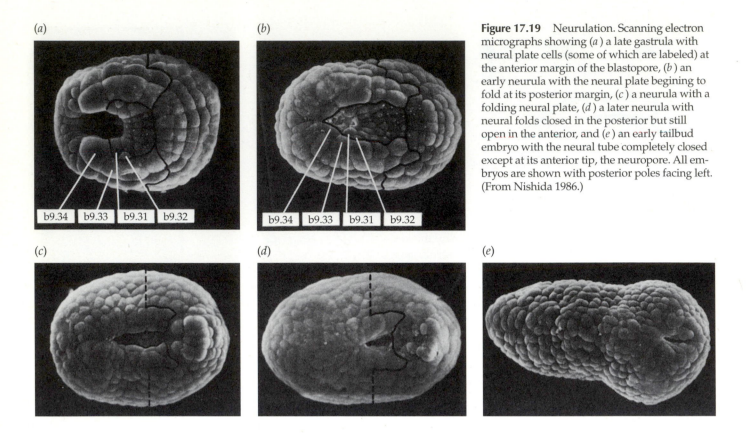

| b9.34 | b9.33 | b9.31 | b9.32 |

| b9.34 | b9.33 | b9.31 | b9.32 |

(c)

(d)

(e)

Figure 17.19 Neurulation. Scanning electron micrographs showing (a) a late gastrula with neural plate cells (some of which are labeled) at the anterior margin of the blastopore, (b) an early neurula with the neural plate begining to fold at its posterior margin, (c) a neurula with a folding neural plate, (d) a later neurula with neural folds closed in the posterior but still open in the anterior, and (e) an early tailbud embryo with the neural tube completely closed except at its anterior tip, the neuropore. All embryos are shown with posterior poles facing left. (From Nishida 1986.)

the flexibility of the notochord during larval swimming.

As the notochord is forming, presumptive muscle cells differentiate into tail muscle. Muscle cell differentiation involves the synthesis of actin, myosin, and other specific proteins that comprise the myofibrils. Ascidian muscle cells resemble those of vertebrates in being striated and containing a specific muscle actin protein (Kovilur et al. 1994). However, in contrast to vertebrate skeletal muscle cells, which fuse into myotubes, ascidian tail muscle cells are unicellular. Tail muscle differentiation is completed just before hatching, when the tail begins to contract and twitch in the chorion.

Figure 17.20 Morphogenesis and differentiation of the notochord. (a) Cell interdigitation during notochord development. Presumptive notochord cells originating from one side of the plane of bilateral symmetry are indicated by shading. (b) Differentiation of the notochord. The areas of secreted extracellular matrix are indicated by dark shading. Development of the notochord progresses from the top to the bottom of the figure. (a after Jeffery 1992; b after Cloney 1990.)

(a)

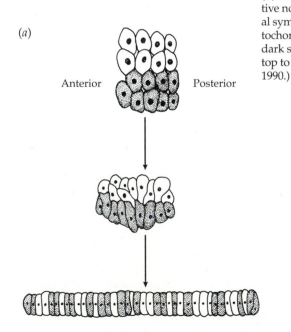

Anterior   Posterior

(b)

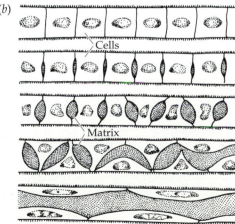

Cells

Matrix

Control of Embryogenesis

Having described embryogenesis, we will now describe how the egg is converted into the larva. As in other animals, ascidian embryogenesis is a progressive process in which successively smaller and smaller embryonic regions are determined before the final steps of cell differentiation. The experiments of L. Chabry showed that determination is initiated at the time of first cleavage. When one blastomere of a 2-cell embryo was destroyed, the remaining blastomere continued to cleave as if it were still half of the embryo, forming a half-larva (Chabry 1887). The same experiment done at the 4-cell stage also produced a incomplete larva. The results of these experiments were different from those obtained in similar experiments carried out with sea urchin and other "regulative" embryos, in which miniature whole larvae were produced from isolated blastomeres of the same 2- and 4-cell embryos. Thus, ascidian embryos were regarded as the prototype of "mosaic" embryos, which cannot compensate for lost parts. Unfortunately, "mosaic" development is sometimes misinterpreted to mean that all the instructions for embryogenesis exist in the egg. We now understand that some larval tissues are determined *intrinsically* by factors present in the egg cytoplasm, but other tissues are determined *extrinsically* by cell interactions during embryogenesis. As examples of the control of embryogenesis, we will describe axis and tail muscle cell determination, which occur by intrinsic mechanisms, and notochord determination, which occurs by an extrinsic mechanism.

Axis Determination

As described earlier, the axes of the ascidian embryo are apparent in the uncleaved zygote after the second phase ooplasmic segregation. By first cleavage, the three axis are irreversibly established. Thus, the initial aspects of axis determination must be controlled by an intrinsic mechanism. Factors present in the egg cytoplasm that act to determine embryonic regions or cell fates by an intrinsic mechanism are called cytoplasmic determinants.

The first axis to appear in the zygote is the dorsoventral axis, which is defined by the cap of myoplasm in the vegetal hemisphere. Fate-mapping experiments have shown that the vegetal pole region of the *Styela* zygote is the site of invagination during gastrulation and the future dorsal pole of the embryo (Bates and Jeffery 1987). To determine whether a cytoplasmic region is responsible for determining the site of gastrulation, small parts of the vegetal pole cytoplasm were removed by microsurgery between the first and second phase of ooplasmic segregation (Bates and Jeffery 1987). The results showed that zygotes lacking the vegetal pole, but not other regions of the egg, did not gastrulate or establish a dorsoventral axis (Figure 17.21). Despite lacking a dorsoventral axis, these embryos produced muscle cells, suggesting that the anteroposterior axis was specified normally. When the same experiment was done

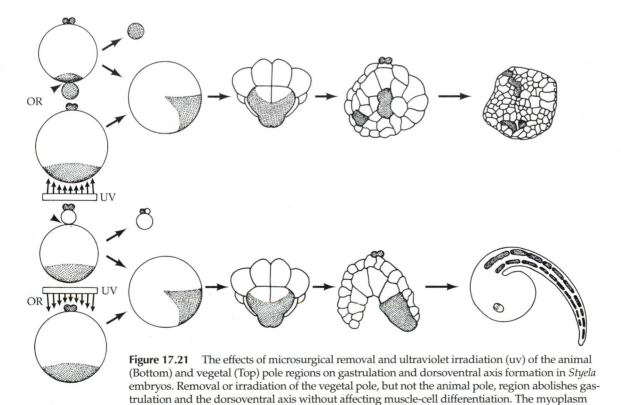

Figure 17.21   The effects of microsurgical removal and ultraviolet irradiation (uv) of the animal (Bottom) and vegetal (Top) pole regions on gastrulation and dorsoventral axis formation in *Styela* embryos. Removal or irradiation of the vegetal pole, but not the animal pole, region abolishes gastrulation and the dorsoventral axis without affecting muscle-cell differentiation. The myoplasm and muscle cells are shaded. (From Jeffery and Swalla 1990.)

(a)   (b)

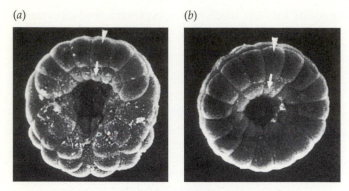

Figure 17.22 Scanning electron micrographs of gastrulae produced from normal zygotes (left) and zygotes lacking part of the posterior cytoplasm (right). The zygotes lacking posterior cytoplasm are radialized, and cells surrounding the blastopore exhibit the size and shape of cells normally found only at the anterior lip of the blastopore. The embryos are viewed from the vegetal pole. (From Nishida 1994.)

with unfertilized eggs, and the eggs were subsequently fertilized, there was no effect on gastrulation or the dorsoventral axis.

In a subsequent series of experiments, Nishida (1996) confirmed these results in *Halocynthia*. In addition, he showed that transplantation of the vegetal pole region to the animal pole of another egg induced an ectopic site of gastrulation at the animal pole. The results suggest that cytoplasmic determinants placed near the vegetal pole during the first phase of ooplasmic segregation specify gastrulation and the dorsoventral axis. It was also shown that irradiation of the vegetal pole of *Styela* eggs with ultraviolet light at wavelengths that preferentially inactivate nucleic acids produces the same effects on gastrulation and dorsoventral-axis determination as removal of the vegetal pole region (Figure 17.21; Jeffery 1990a). One of the targets of ultraviolet light is an mRNA encoding a 30-kilodalton protein (Jeffery 1990b). The translation of this protein during embryogenesis may initiate a developmental pathway that determines the site of gastrulation and the dorsoventral axis.

The second axis to appear in the zygote is the anteroposterior axis. The anteroposterior axis is defined by the myoplasmic yellow crescent, which is located at the future posterior pole after the completion of the second phase of ooplasmic segregation. A series of microsurgical experiments with *Halocynthia* eggs have shown that cytoplasmic determinants localized in the myoplasmic crescent determine the anteroposterior axis (Nishida 1994). When part of the myoplasm was removed after completion of the second phase of ooplasmic segregation, embryos continued to cleave and were able to gastrulate, but formed radially symmetric gastrulae and larvae without an anteroposterior axis. The posterior blastomeres cleaved identically to the anterior blastomeres, and at the gastrula stage, presumptive notochord cells, usually found only at the ante-

rior lip, ringed the blastopore (Figure 17.22). Presumptive muscle cells did not appear in the radialized embryos.

Other experiments showed how the anteroposterior determinants may function (Nishida 1994). In these experiments, the posterior cytoplasmic region was removed and transplanted (1) to the anterior region of a zygote lacking posterior cytoplasm or (2) to the anterior region of a normal zygote. The first experiment resulted in the formation of an anteroposterior axis and muscle cells, but their orientations were reversed relative to the original axis: the anterior region became posterior, and vice versa. This result shows that anteroposterior determinants specify the posterior region and muscle cells. The second experiment, which produced embryos with posterior cytoplasm at both poles, enhanced the development of posterior features at the expense of anterior features. The results suggest that anteroposterior determinants localized in the myoplasmic crescent determine (1) the posterior pole, (2) the posterior cleavage pattern, and (3) the muscle cell precursors. In contrast, the absence of anteroposterior determinants at the opposite pole of the embryo determines the anterior pole and subsequent anterior cleavage pattern, as a default pathway.

The dorsoventral and anteroposterior axes are specified by the activation of cytoplasmic determinants during ooplasmic segregation (Figure 17.23a). After the first phase of ooplasmic segregation, dorsoventral determinants are activated in the cap of myoplasm located near the vegetal pole. These determinants define an **organization center** that specifies the site of gastrulation and the future dorsal pole of the embryo. During the second phase of ooplasmic segregation, the anteroposterior determinants are activated in the myoplasmic crescent. These determinants define a second organization center, which specifies the posterior region of the embryo and initiates muscle cell determination. The two organization centers appear to function independently.

Muscle Determination

Muscle cell determination is also mediated by cytoplasmic determinants. Several different experiments have been done indicating that muscle cells are determined by an intrinsic mechanism (Figure 17.24). These experiments were carried out using the following markers for differentiating muscle cells: (1) acetylcholinesterase (a muscle-specific enzyme), (2) muscle actin mRNA, and (3) myosin heavy-chain protein.

The cleavage-arrest method was used to assess muscle cell determination in ascidian embryos (Whittaker 1973). In a cleavage arrest experiment, embryos are treated with cytochalasin, which inhibits cytokinesis, at specific cleavage stages. The arrested embryos are "frozen" at the cleavage stage in which they were initially treated with the drug, nuclear division continues, and muscle cell markers are expressed as they would be in normal embryos. In embryos arrested at the 4-, 8-, 16-, 32-, and 64-cell stages,

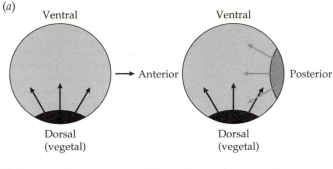

(b)

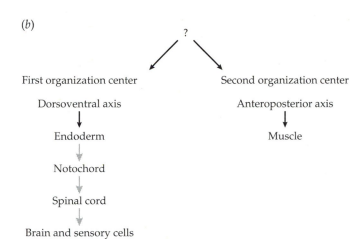

Figure 17.23 The two-organization-center model for ascidian development. (*a*) The two organization centers containing cytoplasmic determinants that specify the dorsoventral and anteroposterior axes. (*b*) The determinant activities and inductive cascades responsible for development of the tadpole larva. Determinant activites are shown by dark lines with arrows, and inductive activities by shaded lines with arrows. For simplification, some tissues have been omitted. The question mark indicates the initial determinant in the sequence.

acetylcholinesterase appears in only 2, 2, 4, 6, or 8 posterior cells, respectively (Whittaker 1973), as predicted by the distribution of myoplasm and the primary muscle cell lineage (Figure 17.16*b*). These results suggest that acetylcholinesterase development is specified by cytoplasmic determinants that are present in the primary muscle lineage.

The existence of muscle determinants in *Styela* embryos was subsequently investigated by the following experiment (Whittaker 1980). As described earlier, the third cleavage normally occurs in the equatorial plane, distributing the myoplasm to B4.1 and B4.1 blastomeres. These two blastomeres are the only cells of the 8-cell embryo that contain the yellow crescent and produce acetylcholinesterase in cleavage-arrested embryos (Whittaker 1973). When *Styela* embryos were compressed under glass cover slips, the third cleavage plane was horizontal, distributing the yellow crescent to four rather than two blastomeres of the 8-cell embryo (Figure 17.24*a*). When the compressed embryos were cleavage-arrested, up to four blastomeres

produced acetylcholinesterase, suggesting that the muscle cell lineage was increased. The results indicate that muscle cells are determined by the cytoplasm they obtain during cleavage.

If muscle cells are specified by intrinsic factors, determination should be cell autonomous, that is, muscle cell markers should be expressed in blastomeres that develop in isolation from other embryonic cells. The following experiment was carried out to investigate muscle cell autonomy (Jeffery 1993). At the 2-, 4-, and 8-cell stage, *Styela* embryos were dissociated into individual cells, and when the dissociated cells divided, the daughter cells were also dissociated, and so forth, resulting in a culture of single cells (Figure 17.24*b*). When the control embryos developed muscle cells, the cell cultures were assayed for muscle cell determination using muscle actin mRNA as a marker. The results showed that single cells derived from embryos dissociated as early as the 2-cell stage produced muscle actin mRNA, and were already determined to be muscle cells. Thus, muscle determination is cell autonomous and does not require the presence of other embryonic cells.

The final experiment demonstrating intrinsic muscle cell determination is shown in Figure 17.24*c*. In this experiment, small anucleate fragments containing the myoplasm were removed from *Halocynthia* eggs or uncleaved zygotes and transplanted to a blastomere in the 8-cell embryo (Nishida 1992a). The myoplasm-containing fragments were fused to the a4.2 or a4.2 cells, which are members of the epidermal cell lineage. When myoplasm was introduced into one of these cells, it attained the ability to produce two muscle cell markers: acetylcholinesterase and myosin heavy-chain. The results demonstrate that myoplasm can cause a nonmuscle blastomere to develop muscle cell features. Therefore, muscle cell development is controlled by cytoplasmic determinants localized in the myoplasm.

In this section, we have described primary muscle cell determination. Ascidian embryos also have secondary muscle cells. Although experimental analysis of secondary muscle cell determination is still at an early stage, cell isolation experiments suggest that extrinsic mechanisms are involved (Nishida 1992b). Thus, the primary and secondary muscle cells are determined by different mechanisms.

Notochord Determination

The notochord consists of a single row of 40 cells in the center of the larval tail (Figure 17.25 top), which are derived from the A4.1/A4.1 and B4.1/B4.1 quadrants of the embryo. Extrinsic determination of the notochord was demonstrated by two kinds of experiments (Nakatani and Nishida 1994). In the first set of experiments, the presumptive notochord cells were isolated from embryos and cultured. Later in development, the isolates were assayed for notochord differentiation using Not-1, a notochord specific antibody (Figure 17.25, top). The presumptive no-

Figure 17.24 Three experiments showing that muscle cells are determined by intrinsic factors. (*a*) Compression of *Styela* embryos between cover slips at third cleavage distributes the yellow crescent to four rather than two blastomeres of an 8-cell embryo and expands the capacity to produce acetylcholinesterase from two to four cells. (*b*) Dissociation of blastomeres from early cleaving *Styela* embryos and continued dissociation of the subsequent division products shows autonomous expression of muscle actin mRNA in muscle lineage cells. (*c*) Epidermal lineage cells fused with anuclear portions of 1-cell zygotes containing the myoplasm obtain the capacity to produce acetylcholinesterase and the heavy chain of myosin.

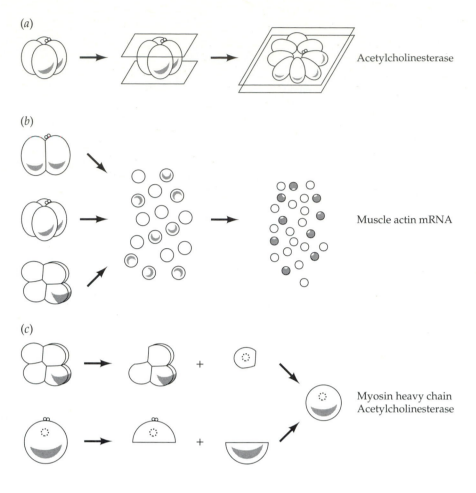

tochord cells do not differentiate into notochord under these conditions, showing that interactions with other cells are necessary to promote notochord differentiation. Thus, notochord determination is not cell autonomous. In the second set of experiments, single presumptive notochord blastomeres were co-isolated from 32-cell embryos in combination with adjacent endoderm, ectoderm, or other presumptive notochord blastomeres and cultured. The partial embryos containing a single presumptive notochord cell and ectoderm cells did not develop into differentiated notochord cells (Figure 17.25*e–f, k–l*). Thus, the presence of ectoderm cells is insufficient to induce notochord differentiation. In contrast, partial embryos containing single presumptive notochord cells co-isolated with endoderm cells were capable of differentiating into notochord cells (Figure 17.25*a–b, g–h*). This result suggests that notochord determination is controlled by an inductive signal produced by the endoderm cells. Notochord determination is complicated, however, by the following additional result. When partial embryos containing multiple presumptive notochord cells but no endoderm cells were cultured, they also produced differentiated notochord cells (Figure 17.25 *c–d, i–j*). Thus, other presumptive notochord cells, as well as endoderm cells, can induce notochord differentiation.

The following sequence of events may occur during notochord determination. First, presumptive notochord cells are induced by a signal emanating from the adjacent endoderm cells. Second, the initial inductive signal is enhanced by a redundant signal from the surrounding notochord cells. Recent experiments show that human basic fibroblast growth factor is capable of inducing notochord precursor cells to differentiate into notochord. Thus, a related growth factor may operate to induce the notochord in ascidian embryos (Nakatani et al. 1996).

Determinants, Inductive Cascades, and Regulatory Genes

The induction of the notochord is probably the first event in a cascade of cell–cell interactions that generate cell fates along the dorsoventral axis (Jeffery 1994). The sequence of events in the induction cascade is shown in Figure 17.23*b*. According to this model, dorsoventral specification involves the following steps: (1) dorsoventral determinants specify the endoderm, (2) the endoderm induces the notochord, (3) the notochord induces the spinal cord, and (4) the spinal cord induces the brain and sensory organs. Some of these inductive events have been demonstrated experimentally (e.g., endoderm specification by determinants: Nishida 1993; spinal cord induction of neural tis-

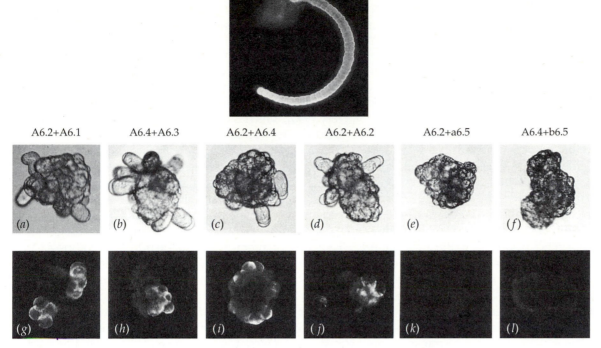

| A6.2+A6.1 | A6.4+A6.3 | A6.2+A6.4 | A6.2+A6.2 | A6.2+a6.5 | A6.4+b6.5 |

Figure 17.25 Induction of the notochord during *Halocynthia* embryogenesis. (Top) A late-tailbud embryo showing the notochord specifically stained by the Not-1 antibody. (Bottom) Partial embryos obtained in coisolation experiments. Light (*a–f*) and immunofluorescence (*g–l*) micrographs of the same partial embryos. Above the micrographs, the notochord blastomere is indicated to the left of the + sign and the coisolated blastomere of other lineage on the right of the + sign. When A6.2 or A6.4 (presumptive notochord) blastomeres are coisolated with A6.1 or A6.3 (presumptive endoderm) blastomeres or A6.2 or A6.4 blastomeres (presumptive notochord), the Not-1 antigen is expressed (*g–j*). When A6.2 (presumptive notochord) blastomeres are co-isolated with a6.5 (presumptive brain) and b6.5 (presumptive epidermis) blastomeres, the Not-1 antigen is not expressed (*k,l*). (From Nakatani and Nishida 1994.)

sues: Nishida 1991), while others are currently speculative (notochord induction of spinal cord). The dorsoventral system appears to be regulated independently of the anteroposterior system, which is also involved in muscle cell determination. Inactivation or removal of dorsoventral determinants does not affect muscle development.

Drosophila embryogenesis is controlled by specific regulatory genes. The proteins encoded by these genes function as transcription factors that turn on or off other regulatory genes. In this way, the activation a small number of genes controls the expression of a larger number of regulatory genes in different parts of the embryo. It is a remarkable that some of the genes that regulate fly development are conserved and have similar functions in vertebrates (Scott 1994). Although analysis of developmental regulatory genes is at an early stage in ascidians, a few of the genes that may function in ascidian development have now been identified.

Gastrulation and neurulation are critical morphogenetic events because they bring together cells from distant parts of the embryo for inductive signaling. In vertebrates,

the *Msx* genes are expressed at sites of morphogenetic movements and cell interactions. The *Msx* proteins are transcription factors containing a specific DNA-binding region called the **homeodomain**. An *Msx* gene has been identified in the ascidian *Molgula* (Ma et al. 1996). The ascidian *Msx* gene is expressed in the involuting mesoderm cells during gastrulation and later in the neural plate and neural folds during neurulation. The homeobox gene *Hrlim* is expressed in endoderm cells of *Halocynthia* embryos beginning at the 32-cell stage, and expression ceases just before these cells invaginate during gastrulation (Wada et al. 1995). The homeobox gene *orthodenticle* is also expressed in the invaginating endoderm, as well as in the involuting mesoderm, during *Halocynthia* gastrulation (Wada et al. 1996). Finally, *Manx*, another *Molgula* gene, is expressed in the involuting mesoderm, the neural plate, and neural folds during gastrulation, and also in the dorsal ectoderm during epiboly (Swalla et al. 1993). The *Manx* protein contains a zinc finger and other potential DNA-binding domains, and may be a transcription factor. The *Msx*, *Hrlim*, *orthodenticle*, and *Manx* genes may control cell movements during gastrulation and neurulation. As described later, *Manx* also plays a critical role in the evolution of ascidian development.

The *T* or *Brachyury* gene encodes a transcription factor that is expressed during early development of the notochord and other mesoderm derivatives in vertebrates. Mutations in the *T* gene result in mice lacking tails. A *T* gene has recently been identified in *Halocynthia* (Yasuo and Satoh 1994). The ascidian *T* gene is expressed in the presumptive notochord cells beginning at the 64-cell stage (Figure 17.26), after the notochord is induced (Nakatani and Nishida 1994). Interestingly, the ascidian *T* gene is expressed exclusively in the notochord. Thus, expression of *T* in other mesodermal derivatives in vertebrate embryos may be related to their more elaborate body plan.

Muscle differentiation in vertebrates and other animals is mediated by several genes belonging to the *MyoD* family. The *MyoD* family genes contain a helix-loop-helix DNA-binding domain and function as transcription factors. Loss of function of various *MyoD* family genes results in defective muscle cell differentiation. A *MyoD*-related gene called *AMD1* has been identified in *Halocynthia* (Araki et al. 1994). The *AMD1* gene may be transcriptionally activated either directly or indirectly by the muscle determinant.

Thus far, we have considered regulatory genes that are expressed during embryogenesis. In *Drosophila*, maternal genes encode mRNAs that are localized in different regions of the egg. These localized mRNAs encode transcription factors that regulate the expression of downstream genes in the zygote. In ascidians, we would expect the RNA products of maternal regulatory genes to accumulate in different cytoplasmic regions of the egg, such as the myoplasm. Recently, an RNA called *YC* (for yellow crescent) has been identified in *Styela* eggs (Swalla and Jeffery 1995), which is localized in the myoplasm and distributed to the muscle cells during cleavage (Figure 17.27). The association of *YC* RNA with the myoplasm is mediated by binding to the underlying cytoskeletal domain. The *YC* RNA may play a role in the early aspects of muscle cell determination. Interestingly, *YC* RNA does not encode a protein. Instead, it may function as an RNA molecule. The *YC* RNA has recently been shown to contain a region that is capable of interacting with the mRNA encoding a DNA replication factor (Swalla and Jeffery 1996a). Thus, *YC* RNA is a candidate for regulating the posterior cleavage pattern in the ascidian embryo.

An mRNA called *posterior end mark* is localized in the myoplasm of *Halocynthia* eggs (Yoshida et al. 1996). The *posterior end mark* RNA may encode a protein involved in determining anteroposterior polarity in the embryo. At present, no localized mRNAs or other factors are known to be involved in the muscle-determining function of the myoplasm.

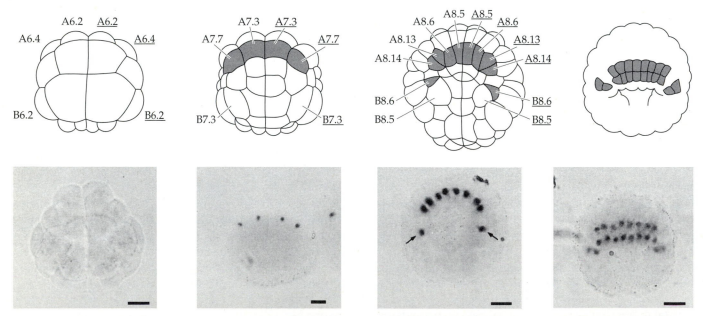

Figure 17.26 Expression of the *T* gene in the presumptive notochord cells of *Halocynthia* embryos. (Upper row) The notochord lineage in 32-cell, 64-cell, early-gastrula, and late-gastrula stage embryos. (Lower row) Accumulation of *T* mRNA in the presumptive notochord cells at the same stages of development as in the top row. The *T* mRNA was detected by whole mount in situ hybridization. (From Yasuo and Satoh 1994.)

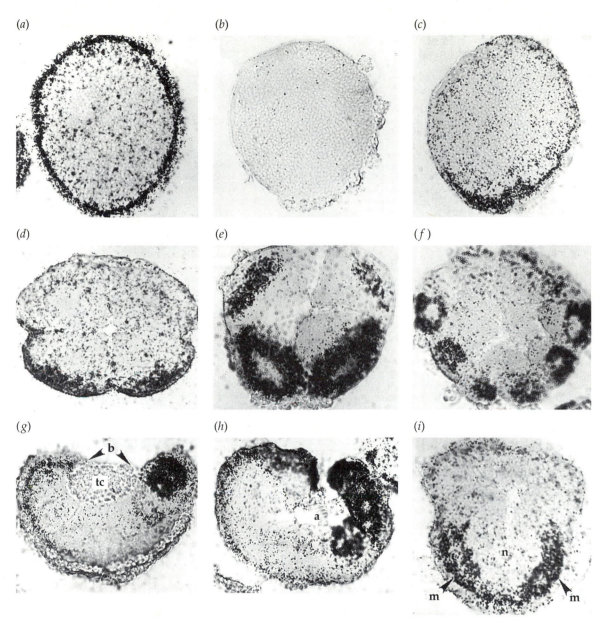

Figure 17.27 Sections of *Styela* eggs and embryos showing localization of *YC* RNA in the myoplasm and presumptive muscle cells. The *YC* RNA is represented by the black dots in the in situ hybridizations. (*a*) An unfertilized egg with *YC* RNA in the cortical myoplasm. (*b*) An unfertilized egg that served as a control and is lacking in black dots. (*c*) A zygote after completion of the first phase of ooplasmic segregation with *YC* RNA localized in the myoplasmic cap. (*d*) A 4-cell embryo with *YC* RNA localized in the myoplasm of the B3 and <u>B3</u> cells. (*e*) An 8-cell embryo with *YC* RNA localized in B4.1 cells (bottom) and to some extent in A4.1 cells (top). (*f*) A 32-cell embryo with *YC* RNA localized in six presumptive muscle cells (bottom). (*g*–*h*) Early (*g*) and late (*h*) gastrulae with *YC* RNA localized in presumptive muscle cells at the posterior lip of the blastopore (b). (*i*) An early tailbud stage embryo with *YC* RNA localized in two rows of differentiating muscle cells. tc, test cells; a, archenteron; m, muscle cells ; n, notochord. (From Swalla and Jeffery 1995.)

Metamorphosis

After embryogenesis is complete, the tadpole larva hatches from the chorion. Hatching usually takes place about 12 to 36 hours after fertilization and is mediated by prote-olytic enzymes secreted by the larva. Depending on the species, the tadpoles swim for a few minutes or several days before settling and beginning metamorphosis (Berrill 1935). The ascidian larva is a nonfeeding larva that is important for dispersal of the sessile adult. As you will see in the last section, the ascidian tadpole larva can be evolutionarily reduced or eliminated in the ascidian life cycle because, as long as the embryo undergoes metamorphosis into an adult, it will be able to reproduce. Metamorphosis is a complex process in which the larval tissues are destroyed or remodeled and replaced by adult tissues and organs (Cloney 1978, 1982). Most importantly, the adult develops gonads for sexual reproduction of the next generation. First, we describe the process of metamorphosis and then discuss what is known about how this critical phase of the life cycle is controlled.

Morphological Events during Metamorphosis

Metamorphosis from a tadpole larva to a feeding adult usually takes at least 5 days. In this time, the individual undergoes dramatic morphological changes as tissue rearrangement occurs. This process can be divided into four separate steps (Figure 17.28). The first event that takes place is larval attachment to a suitable substrate. Next, the tail is resorbed by contraction of the tail epidermis. Then the larva rotates 90 degrees and begins tissue remodeling. After several days of development, the newly formed siphons open and the juvenile begins feeding. It will take several months of growth before the ovotestes mature and the adult becomes capable of reproduction.

Metamorphosis begins with larval attachment to a substrate by **adhesive papillae** at the anterior end of the tadpole larva (Figure 17.28a; Cloney 1978). The tadpole larva is the dispersal phase of the ascidian life cycle (see Figure 17.2), because the adult never moves after metamorphosis. Ascidian larvae do not feed while they swim in the ocean; they rely on energy stores from the egg for tail muscle contraction. If the head and tail of the ascidian tadpole larva are severed, the head can metamorphose into a complete adult, while the tail continues to swim for several hours (B. J. Swalla, unpublished data). These experiments suggest that the entire presumptive adult tissues are contained in the head. The length of time spent swimming varies depending on the species (Berrill 1935), and whether some of the adult structures have already formed in the head, a process called **adultation** (see next section; Figure 17.31). Some larvae may move far from the adults that spawned the sperm and eggs before finding a suitable substrate to which to attach, while others may settle near the parents.

After attachment at the anterior of the head, the tail is retracted and the entire body is rotated and remodeled (Figure 17.28). The tail undergoes programmed cell death (**apoptosis**) and the cells of the tail are absorbed and probably used as an energy source for metamorphosis (Figure 17.28b, c). The tail is contracted by a variety of mechanisms, depending on the species (Cloney 1978), but usually either the tail epidermis or the central notochord becomes contractile to facilitate tail retraction and resorption. These properties were discovered by excising the tail from the swimming larva and showing that the epidermis or notochord contracted in isolation from the rest of the larva. The 90 degree rotation of the presumptive siphons

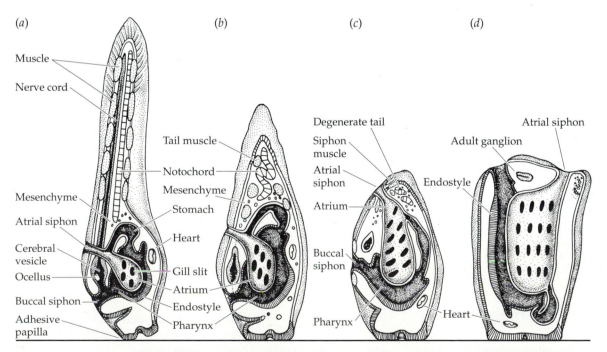

Figure 17.28   Metamorphosis. (a) The larva attaches to a suitable substrate at the anterior end. (b) The tail is retracted and resorbed. (c) The larval organs rotate 90 degrees and tissue remodeling continues. (d) After several days of development, the functional siphons open and the juvenile begins feeding.

and other organs ensures that the siphons will be pointing away from the substratum (Figure 17.28). Little is known about the cues or mechanism(s) underlying this process (Cloney 1978).

After attachment and tail resorption, the metamorphosing individual forms long epidermal structures called **ampullae** (Figure 17.29). These are transient structures, continually being formed and retracted during the process of tissue remodelling. Recent studies suggest that the ampullae express a homeobox gene called *distalless* during outgrowth. Use of an antibody to distalless protein showed that the gene is expressed during outgrowth of appendages and related structures in many protostome and deuterostome phyla (Panganiban et al. 1997).

The ampullae have a cavity through which blood cells can be seen to move, and time-lapse studies show that there are rhythmic contractions that occur along a single ampullae (DeSanto and Dudley 1969; Torrence and Cloney 1981). These structures appear to help anchor the individual to the substrate, but are also believed to be important for respiration and blood flow until the heart is developed (Torrence and Cloney 1981) and also for immune response (Bates 1991). It has been shown that individuals that settle near the ampullae of another individual may die or detach, suggesting that blood cells carried by the ampullae may have a negative immune response with other settling individuals.

Little is known about the factors that trigger metamorphosis of the ascidian larva, although it has been extensively investigated. Larvae tend to settle near adults or juveniles (recently metamorphosed adults) of the same species. In the laboratory a number of diverse substances trigger the attachment of larvae to substrates, including dyes, metals, and oxidizing reagents (Lynch 1961). It has also been shown that larvae show specific preferences for substrates in controlled experiments. However, it is not clear whether these factors are important in the ocean, the larval natural environment.

Adult Development

It has been suggested that the first vertebrate evolved from an ancestor similar to an ascidian, but the adult stage got shorter and shorter until the larva contained both larval and adult tissues. Because of this theory, people have looked for organs in the ascidian adults to see if there are similar organs to those found in the vertebrates. After discussing a few of the organs found in an ascidian and what their functions appear to be, we will look at colonial ascidians and discuss some of the unique features that are found in this ascidian lifestyle.

Solitary Ascidians

Figure 17.1a shows a cross section through an adult ascidian with the main organs labeled. The adult is firmly attached to the substrate by the strong tunic secreted during metamorphosis, which is characteristic of the Tunicata. The continual flow of water through the two siphons brings plankton and oxygen into the adult. The oral siphon opens into a chamber surrounded by the branchial basket called the peribranchial cavity. Water is filtered through the branchial basket and oxygen is absorbed, while carbon dioxide is expelled into water flowing out the anal siphon.

The ascidian heart is a two-chambered heart that periodically reverses blood flow. The muscle composing the heart is similar to heart muscle found in vertebrates, and its function is nearly identical. The blood of ascidians has a similar function to that of vertebrates, but there are a

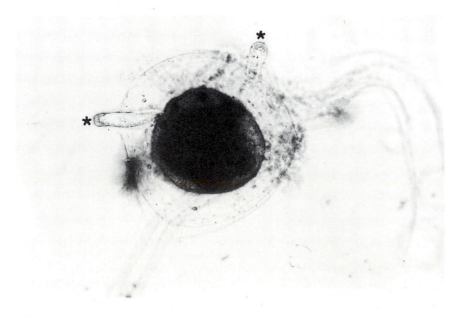

Figure 17.29 Photograph of *Molgula citrina* individual during mid-metamorphosis, two days after larval attachment. The tail has been retracted (tunic remnant to the right), and three ampullae are seen protruding from the test. The two ampullae in focus are marked with an asterisk; the third extends to the bottom of the photo, out of the plane of focus.

large number of different blood cells. The blood and muscle cells come from the mesenchyme cells located at the base of the larval head and they begin to wander freely once metamorphosis begins (Figure 17.28).

The nervous system is relatively simple and consists of a central cylindrical ganglion that regulates contraction of the siphons. From each end arise a variable number of nerves. Anteriorly, the lateral nerves supply the buccal siphon; posteriorly the nerves enervate the anal siphon, the branchial sac, and the visceral organs.

Colonial Ascidians

Colonial ascidians live in a colony made of multiple zooids that share a vascular system (Figure 17.30). Colonial ascidians may reproduce asexually, by budding, or sexually by producing gametes (Satoh 1994). The larva attaches and develops into an **oozoid** (zooid from an egg). This oozoid can bud asexually, but will never produce gametes. In contrast, once the oozoid produces a **blastozooid** (zooid from a bud), the blastozooid can now reproduce both sexually and asexually. There are two functional types of budding in colonial ascidians: propagative and survival buds. **Propagative budding** allows for colony growth, while **survival budding** allows the adult to survive adverse environmental conditions (Nakauchi 1982). The process of budding requires that certain cells of the oozoid are multipotential, or able to give rise to more than one cell type. Each bud shows bilateral asymmetry; usually the heart is on the right side of the organism and the digestive tract is on the left side (Figure 17.30). However, if the asymmetry is experimentally reversed, the pattern is transmitted to the future asexual zooids (Sabbadin et al. 1975). Recently, it was shown that retinoic acid can cause duplication of bud axes (Hara et al. 1992). This is remarkable because retinoic acid has been shown to cause duplication of the body axes in vertebrates as well (Tabin 1991). The mechanism by which this duplication occurs is not

yet known, but it is intriguing that the asexual zooid is patterned by a mechanism similar to vertebrates.

Evolution of Ascidian Development

Now that our description of larval and adult development is complete, we will discuss the ways in which ascidian development has changed during evolution. Here, we describe three evolutionary changes in the conventional mode of development: (1) adultation, (2) caudalization, and (3) anural development (Figure 17.31; also see Figure 17.32). Each of these changes has resulted in dramatic alterations in the morphology of the tadpole larva.

Adultation: Superimposing the Larva and Adult

Adultation is a mode of development in which adult tissues and organs differentiate precociously in the head of the tadpole (Figure 17.31c–e). The evolutionary advantage of adultation is that it reduces the time between metamorphosis and feeding. Adultation can be minimal (Figure 17.31c), with the early appearance of one or both siphons, a partial digestive tract, a few gill slits in the branchial sac, and a rudimentary heart, or it may be more extreme (Figure 17.31d, e). In the extreme cases, a miniature juvenile with a digestive tract, a branchial sac with numerous gill slits, a beating heart, and an asexual bud develops within the larval head. Ascidians with adultation resemble larvacean tunicates in that they have adult organs and a tail;

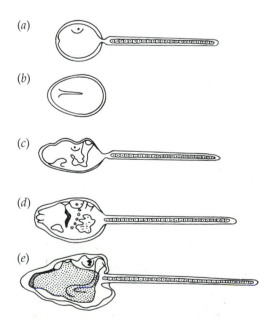

Figure 17.31 Adultation and anural development. (*a*) A conventional tadpole larva with an undifferentiated head, containing an otolith and a tail. (*b*) Anural development, in which the otolith and larval tail are lost. (*c*–*e*) Increasingly extreme cases of adultation, in which adult tissues and organs develop precociously in the head. (From Jeffery and Swalla 1992.)

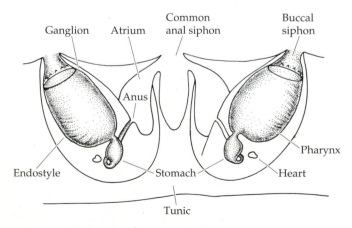

Figure 17.30 Representation of two individuals in a *Botryllus* colony, showing the asymmetry of the individuals.

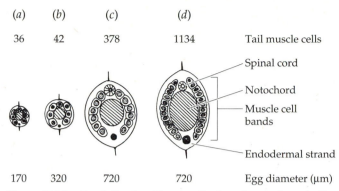

(a) (b) (c) (d)

36   42   378   1134   Tail muscle cells

Spinal cord

Notochord

Muscle cell bands

Endodermal strand

170   320   720   720   Egg diameter (μm)

Figure 17.32 Caudalization. Top row displays the number of tail-muscle cells; middle row shows cross sections through the tail; bottom row gives egg diameters in μm. (*a*) *Ciona intestinalis*; (*b*) *Halocynthia roretzi*; (*c*) *Stolonica socialis*; (*d*) *Ecteinascidia turbinata*. (From Jeffery and Swalla 1992.)

however, they lack mature sexual organs and lose their tails during metamorphosis.

Adultation is an example of **heterochrony**, an evolutionary change in developmental timing. Heterochrony is a common way in which the morphology of a species can change during evolution. The mechanism of heterochrony is related to changes in the timing of gene expression. The ascidian genome contains several different muscle actin genes, some of which are expressed during embryonic development and others in the adult (Beach and Jeffery 1992; Kusakabe et al. 1992). A muscle actin gene that is normally expressed in the adult muscle cells of *Styela*, which has a conventional tadpole larva, is expressed in the mesenchyme cells during larval development of *Molgula citrina*, a species with adultation (Swalla et al. 1994). The expression period of the muscle actin gene is shifted to the larval stage to promote rapid muscle differentiation during adult development. Regulatory genes that function upstream of structural genes such as the muscle actin genes may be responsible for shifting the timing of development. Earlier we described the ascidian *Msx* gene, which encodes a homeodomain transcription factor. The *Msx* gene is known to be involved in heart development. The heart is one of the adult organs that begins to differentiate precociously during adultation. Thus, the *Msx* gene is expressed in the heart primordium during the larval stage of *Molgula citrina*, a species with adultation (Ma et al. 1995). Further studies on ascidian adultation will contribute to our general understanding of heterochrony.

Caudalization: Remodeling the Tail

Caudalization is a mode of development in which additional muscle cells are formed in the tadpole tail without changing the cell number in other tissues. The advantage of caudalization is that it produces a more robust tadpole that is capable of enhanced swimming and dispersal. Caudalization usually occurs in the same species with adultation, and the robust tail may have evolved to propel the

enlarged larval head. Similar to adultation, caudalization can be minimal or extreme (Figure 17.32). In cases of minimal adultation, the three rows of muscle cells flanking the notochord are increased in length by adding a few muscle cells to the tip of the tail. For example, *Halocynthia* larvae have 42 tail muscle cells (Nishida and Satoh 1983), rather than the conventional 36–38 muscle cells present in other ascidian species (Figure 17.32*b*). The additional muscle cells are derived from the b4.1/<u>b4.1</u> quadrant, suggesting that *Halocynthia* produces more muscle cells because of enhanced induction of secondary muscle cells. In cases of extreme caudalization, however, both the length and the width of the muscle cell rows increase in the larval tail (Figure 17.32*c*–*d*). For example, *Stolonica* tadpoles have 378 muscle cells and *Ecteinascidia* tadpoles have 1134 muscle cells, although there are only 40 notochord cells in the tail of each species. In species with extreme caudalization, the additional muscle cells are produced by extra cell divisions before terminal differentiation in the muscle cell lineage (Berrill 1935). Whether the number of cell divisions is increased in the primary muscle lineage, the secondary muscle lineage, or both the primary and secondary lineages is unknown.

The extent of caudalization is correlated with egg size: the larger the egg, the greater the number of tail muscle cells. This relationship to egg size also holds for adultation (Berrill 1935). Thus, adultation and caudalization may be controlled by the extent of the maternal contribution to embryogenesis. Ascidians with caudalization provide a model system for investigating the evolution of cell cycle control during development.

Anural Development: Loss of the Tadpole Larva

In **anural development**, the larval phase of the life cycle is lost or extensively modified (see Figure 17.31b). Some species with anural development have retained a tailless larva, whereas others have completely lost the larva, resulting in direct development. Anural species live in spatially restricted environments, such as sand and mud flats on the ocean floor, or shorelines with strong currents or wave action. The loss of the swimming larva is an advantage in these habitats because dispersal to undesirable locations would be prevented. Most species with anural development are members of the family Molgulidae. The molgulid evolutionary tree (Hadfield et al. 1995) shows that anural species were derived from ancestors with tadpole larvae and that anural development evolved at least four separate times (Figure 17.33).

The mechanism of anural development has been studied in two closely related species: *Molgula oculata*, which has a tadpole larva, and *Molgula occulta*, which has a tailless, or anural larva (Figure 17.34a). Because of their similar scientific names, we refer to these animals as the tailed (or urodele) species (*M. oculata*) and the tailless (or anural) species (*M. occulta*). The tailless species does not differentiate a brain sensory organ, a notochord, or muscle cells

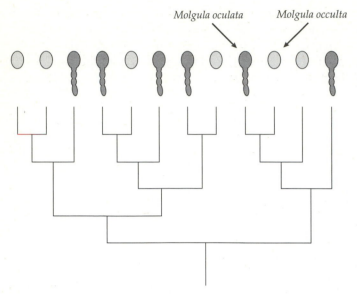

Figure 17.33 Evolutionary tree of anural development in *Molgula* species. Tailed (dark) larvae are species with urodele development. Tailless (shaded) larvae are species with anural development. (Data from Hadfield et al. 1995.)

(Jeffery and Swalla 1991). The tail does not form because the presumptive notochord and muscle cells do not undergo morphogenetic movements or differentiation. The egg size and cleavage patterns are identical in the tailed and tailless species, suggesting that anural development is controlled by mechanisms that are different from those that regulate adultation and caudalization.

The close relationship between the tailed and tailless species has allowed interspecific hybridization experiments to be carried out (Swalla and Jeffery 1990). Interspecific hybridization experiments can determine whether anural development is controlled by maternal processes, zygotic processes, or both maternal and zygotic processes. The two *Molgula* species can be crossed in both directions: eggs of the tailed species can be fertilized with sperm of the tailless species, and vice versa. When eggs of the tailed species were fertilized with sperm of the tailless species, the hybrid embryos developed into tadpole larvae that were morphologically identical to those of the tailed species (Figure 17.34b). This result shows that the genome of the tailless species does not produce factors that inhibit the development of tadpole larvae. When eggs of the tailless species were fertilized with sperm of the tailed species, however, the hybrid embryos developed into larvae with a short tail, a notochord, and a brain sensory organ (Figure 17.34b). Thus, the tadpole features that were lost during evolution were recovered by introducing the genome of the tailed species into the eggs of the tailless species. The restoration of the tadpole larval features suggests that anural development evolved by loss of function mutations in zygotic genes. However, the tail muscle cells were not rescued in these hybrids, suggesting that mater-

nal changes, which the genome of the tailed species cannot compensate for, are also responsible for the evolution of anural development. Therefore, both maternal and zygotic changes contribute to the evolution of anural development. Below we describe some of these maternal and zygotic changes.

The most significant maternal change is in the composition of the myoplasm (Swalla et al. 1991). Urodele ascidians have a 58-kilodalton cytoskeletal protein (**p58**) in their myoplasm (see Figure 17.5). Although p58 is present in previtellogenic oocytes of the tailless species, it fails to be localized in the myoplasm, and is degraded later in oogenesis. Thus, mature eggs of the tailless species lack, or contain reduced quantities, of p58. When p58 was investigated in several species that evolved anural development independently (see Figure 17.33), the protein was also missing in their eggs. It is possible that p58 is required for localization of muscle determinants in the myoplasm.

A critical zygotic change in the tailless species is the lack of notochord differentiation. As described earlier, the

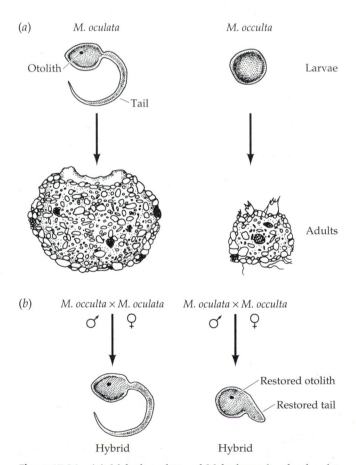

Figure 17.34 (*a*) *Molgula oculata* and *Molgula occulta*, closely related species with different modes of development. Larvae and adults of both species are shown; *M. oculata* (left) and *M. occulta* (right). (*b*) Interspecific hybridization of *M. oculata* and *M. occulta*. (Left) *M. oculata* eggs fertilized with *M. occulta* sperm yield hybrid tadpole larvae. (Right) *M. occulta* eggs fertilized with *M. oculata* sperm yield hybrid larvae with a restored otolith and tail. (After Jeffery 1994.)

presumptive notochord cells are induced by endoderm during the late cleavage stages, initiating an inductive cascade resulting in differentiation of the brain and its sensory organs. Therefore, the lack of tail and otolith formation in the tailless species may be caused by defective notochord induction. While altered notochord induction remains to be proved experimentally, one of the genes involved in the loss of the tail and otolith has been identified. Earlier we discussed the *Manx* gene, which may encode a zinc finger transcription factor. The *Manx* gene is expressed in the mesoderm and ectoderm cells during gastrulation and neurulation, but is also active in the elongating notochord, the migrating muscle cells, and the tail epidermis during later development (Swalla et al. 1993). Three lines of evidence suggest that inactivation of the *Manx* gene is responsible for anural development. First, although *Manx* is strongly expressed in embryos of the tailed species, its expression is reduced in embryos of the tailless species. Second, *Manx* expression is enhanced in hybrids with restored tadpole larval features. Third, when *Manx* gene expression is specifically inhibited in these hybrids, they are unable to restore the tadpole larval features (Swalla and Jeffery 1996b). Therefore, the *Manx* gene may regulate the development of the tadpole larva.

Summary

The tunicates are a group of sessile invertebrates with definitive chordate features, including a dorsal nervous system, a notochord, and gill slits. The life cycle of most tunicates includes larval and adult phases. The ascidians, one of the three classes of tunicates, are used as a model system in developmental biology. Among the attributes of ascidians are rapid development, small embryonic and larval cell numbers, and eggs and embryos with colored cytoplasmic regions. The tadpole larva has only six different types of tissues. Ascidians are hermaphrodites with gonads containing both sperm and eggs. After the completion of spermiogenesis and oogenesis, the mature gametes are shed into seawater. Fertilization is a complex process because of the vitelline envelope surrounding the mature egg. After fertilization, the zygote completes maturation and undergoes ooplasmic segregation: a dramatic series of cytoplasmic movements that determine the embryonic axes. The cleavages are rapid and bilateral. Gastrulation occurs by invagination, involution, and epiboly, and is followed by neurulation. After neurulation the embryo elongates and undergoes morphogenetic movements and cell differentiation leading to the formation of the tadpole larva. Ascidian embryogenesis is controlled by intrinsic and extrinsic mechanisms. The embryonic axis and muscle cells are specified by cytoplasmic determinants, which act intrinsically, whereas the notochord is determined by inductive cell interactions, which act extrinsically. Regulatory genes are beginning to be identified that may control embryonic development. After the larva is formed, it hatches, disperses by swimming, then settles and undergoes metamorphosis. Metamorphosis involves the destruction of most of the larval tissues and their replacement with adult tissues and organs, which develop from undifferentiated cells in the larval head. In colonial ascidians, new adults are also formed by asexual budding. Different modes of development have evolved in ascidians, including heterochronic shifting of adult tissues and organs into the larval phase of the life cycle (adultation), remodeling of the larval tail (caudalization), and loss of the tadpole larva (anural development). As the most primitive chordates, tunicates offer a chance to learn how developmental processes evolved in more complex chordates, such as the vertebrates.

Literature Cited

Araki, I., H. Saiga, K. W. Makabe and N. Satoh. 1994. Expression of *AMD1*, a gene for a MyoD1-related factor in the ascidian *Halocynthia roretzi*. *Roux's Arch. Dev. Biol.* 203: 320–327.

Bates, W. R. 1980. A scanning electron microscope study of the structure and fertilization of an ascidian egg, *Ciona intestinalis*. *Can. J. Zool.* 58: 1942–1946.

Bates, W. R. 1991. Ampullar morphogenesis in anural and urodele molgulid ascidians. *Dev. Growth Differ.* 33: 401–411.

Bates, W. R. and W. R. Jeffery. 1987. Localization of axial determinants in the vegetal pole region of ascidian eggs. *Dev. Biol.* 124: 65–76.

Bates, W. R. and W. R. Jeffery. 1988. Polarization of ooplasmic segregation and dorso-ventral axis determination in ascidian embryos. *Dev. Biol.* 130: 98–107.

Beach, R. L. and W. R. Jeffery. 1992. Multiple actin genes encoding the same alpha-muscle isoform are expressed during ascidian development. *Dev. Biol.* 151: 55–66.

Berrill, N. J. 1935. Studies in tunicate development. Part III. Differential retardation and acceleration. *Phil. Trans. Roy. Soc. London.* B 255: 266–326.

Berrill, N. J. 1950. *The Tunicata with an Account of the British Species.* Ray Society, London.

Chabry, L. 1887. Contribution á l'embryologie normale et térotologique des Ascidies simples. *J. Anat. Physiol.* 23: 167–319.

Chiva, M., F. LaFargue, E. Rosenberg and H. E. Kasinsky. 1992. Protamines, not histones, are the predominant basic proteins in sperm nuclei of solitary tunicates. *J. Exp. Zool.* 263: 338–349.

Cloney, R. A. 1978. Ascidian metamorphosis: Review and analysis. In *Settlement and metamorphosis of marine invertebrate larvae*, F. S. Chia and M. E. Rice (eds.). Elsevier, New York, pp. 255–282.

Cloney, R. A. 1982. Ascidian larvae and the events of metamorphosis. *Am. Zool.* 22: 817–826.

Cloney, R. A. 1990. Urochordata-Ascidiacea. In *Reproductive Biology of Invertebrates*, K. G. Adiyodi and R. G. Adiyodi (eds.). Oxford and IBH, New Dehli, pp. 361–451.

Cloney, R. A. and L. C. Abbott. 1980. The spermatozoa of ascidians: Acrosome and nuclear envelope. *Cell Tissue Res.* 206: 261–270.

Cloney, R. A. and J. M. Cavey. 1982. Ascidian larval tunic: Extracellular structures influence morphogenesis. *Cell Tissue Res.* 222: 547–562.

Conklin, E. G. 1905. The organization and cell lineage of the ascidian egg. *J. Acad. Nat. Sci. Philadelphia* 13: 1–119.

DeSanto, R. S. and P. L. Dudley. 1969. Ultramicroscopic filaments in the ascidian *Botryllus schlosseri* (Pallas) and their possible role in ampullar contractions. *J. Ultrastruct. Res.* 28: 259–274

Fenaux, R. 1986. The house of *Oikopleura dioca* (Tunicata, Appendicularia): Structure and functions. *Zoomorphol.* 106: 224–231.

Fischer, J. L. 1992. The embryological oeuvre of Laurent Chabry. *Roux's Arch. Dev. Biol.* 201: 125–127.

Franzen, A. 1976. The fine structure of spermatid differentiation in a tunicate, *Corella parallelogramma* (Muller). *Zoon* 4: 115–120.

Fuke, T. M. 1983. Self and non-self recognition between gametes of the ascidian, *Halocynthia roretzi*. *Roux's Arch. Dev. Biol.* 192: 347–353.

Fukumoto, M. 1990. Morphological aspects of ascidian fertilization. *Zool. Sci.* 7: 989–998.

Garstang, W. 1928. The morphology of the Tunicata, and its bearing on the phylogeny of the chordata. *Quart. J. Micr. Sci.* 72: 51–187.

Godknecht, A. and T. G. Honneger. 1991. Isolation, characterization, and localization of a sperm-bound *N*-acetylglucosamidiase that is indispensible for fertilization in the ascidian *Phallusia mammillata*. *Dev. Biol.* 143: 398–407.

Goudeau, H., Y. Dpresle, L. Rosa and M. Goudeau. 1994. Evidence by a voltage clamp study of an electrically mediated block to polyspermy in the egg of the ascidian *Phallusia mammallita*. *Dev. Biol.* 166: 489–501.

Hadfield, K. A., B. J. Swalla and W. R. Jeffery. 1995. Multiple origins of anural development in ascidians inferred from rDNA sequences. *J. Mol. Evol.* 40: 413–427.

Hara, K., S. Fujiwara and K. Kawamura. 1992. Retinoic acid can induce a secondary axis in developing buds of a colonial ascidian, *Polyandrocarpa misakiensis*. *Dev. Growth Differ.* 34: 437–445.

Holland, P. W. H., J. Garcia-Fernandez, N. A. Williams and A. Sidow. 1994. Gene duplications and the origins of vertebrate development. *Development Suppl.*1994:124–133.

Honneger, T. G. 1986. Fertilization in ascidians: Studies on the egg envelope, sperm and gamete interactions in *Phallusia mammillata*. *Exp. Cell Res.* 138: 446–451.

Hoshi, M., T. Numakunai and H. Sawada. 1981. Evidence for participation of sperm proteases in fertilization of the solitary ascidian, *Halocynthia roretzi*: Effects of protease inhibitors. *Dev. Biol.* 86: 117–121.

Hoshi, M., R. DeSantis, M. Pinto, F. Cotelli and F. Rosati. 1985. Sperm glycosidases as mediators of sperm-egg binding in the ascidian. *Zool. Sci.* 2: 65–69.

Hsu, W. S. 1963. The nuclear envelope in developing oocytes of the tunicate *Boltenia villosa*. *Z. Zellforsch.* 58: 660–678.

Jeffery, W. R. 1982. Calcium ionophore polarizes ooplasmic segregation in ascidian eggs. *Science* 216: 545–547.

Jeffery, W. R. 1984. Pattern formation by ooplasmic segregation in ascidian eggs. *Biol. Bull.* 166: 277–298.

Jeffery, W. R. 1990a. Ultraviolet irradiation during ooplasmic segregation prevents gastrulation, sensory cell induction, and axis formation in the ascidian embryo. *Dev. Biol.* 140: 388–400.

Jeffery, W. R. 1990b. An ultraviolet sensitive maternal mRNA encoding a cytoskeletal protein may be involved in axis formation in the ascidian embryo. *Dev. Biol.* 141: 141–148.

Jeffery, W. R. 1992. A gastrulation center in the ascidian egg. *Development Suppl.* 1992: 53–63.

Jeffery, W. R. 1993. Role of cell interactions in ascidian muscle and pigment cell specification. *Roux's Arch. Dev. Biol..* 202: 103–111.

Jeffery, W. R. 1994. A model for ascidian development and developmental modifications during evolution. *J. Mar. Biol. Assoc. U.K.* 74: 35–48.

Jeffery, W. R. and S. Meier. 1983. A yellow crescent cytoskeletal domain in ascidian eggs and its role in early development. *Dev. Biol.* 96: 125–143.

Jeffery, W. R. and B. J. Swalla. 1990. The myoplasm of ascidian eggs: A localized cytokeletal domain with multiple roles in embryonic development. *Seminars Dev. Biol.* 1: 253–261.

Jeffery, W. R. and B. J. Swalla. 1991. An evolutionary change in the muscle lineage of an anural ascidian embryo is restored by interspecific hybridization with a urodele ascidian. *Dev. Biol.* 145: 328–337.

Jeffery, W. R. and B. J. Swalla. 1992. Evolution of alternate modes of development in ascidians. *BioEssays* 14: 219–226.

Katz, M. C. 1983. The comparative anatomy of the tunicate tadpole *Ciona intestinalis*. *Biol. Bull.* 164: 1–27.

Kessel, R. G. 1966. Electron microscope studies on the origin and maturation of yolk in oocytes of the tunicate *Ciona intestinalis*. *Z. Zellforsch.* 71: 525–544.

Kovilur, S., R. L. Beach, J. W. Jacobson, W. R. Jeffery and C. R. Tomlinson. 1994. Evolution of the chordate muscle actin gene. *J. Mol. Evol.* 36: 361–368.

Kowalevsky, A. O. 1866. Entwickungsgeschichte der einfachen Ascidien. *Mém. Acad. Sci. St. Petersb.* 10: 1–19.

Kubo, M., M. Ishikawa and T. Numakunai. 1978. Differentiation of apical structures during spermiogenesis and fine strucutre of the spermatozoon in the ascidian *Halocynthia roretzi*. *Acta Embryol. Exp.* 1978: 283–295.

Kusakabe, T., K. W. Makabe and N. Satoh. 1992. Tunicate muscle actin genes: structure and organization as a gene cluster. *J. Mol. Biol.* 227: 955–960.

Lambert, C. C. 1986. Fertilization induced modification of chorion *N*-acetylglucosamine groups in the ascidian chorion, decreasing sperm binding and polyspermy. *Dev. Biol.* 116: 168–173.

Lambert, C. C. 1989. Ascidian eggs release glycosidase activity which aids in the block against polyspermy. *Development* 105: 415–420.

Lambert, C. C. and R. A. Koch. 1988. Sperm binding and penetration during ascidian fertilization. *Dev. Growth Differ.* 30: 325–336.

Lambert, C. C. and G. Lambert. 1978. Tunicate eggs utilize ammonium ions for floating. *Science* 200: 64–65.

Lynch, W. F. 1961. Extrinsic factors influencing metamorphosis in bryozoan and ascidian larvae. *Am. Zool.* 1: 59–66.

Ma, L., B. J. Swalla, J. Zhou, J. Chen, J. R. Bell, S. L. Dobias, R. Maxson and W. R. Jeffery. 1996. Expression of an *Msx* homeobox gene in ascidians: Insights into the archtypal chordate expression pattern. *Dev. Dynam.* 205: 308–318.

Miller, R. L. 1982. Sperm chemotaxis in ascidians. *Am. Zool.* 22: 827–840.

Miyamoto, D. M. and R. J. Crowther. 1985. Formation of the notochord in living ascidian embryos. *J. Embryol. Exp. Morphol.* 86: 1–17.

Monroy, A. 1979. Introductory remarks on the segregation of cell lines in the embryo. In *Cell Lineage, Stem Cells and Cell Determination*, N. Le Douarin (ed.), Elsevier, Amsterdam, pp. 3–13.

Morgan, T. H. 1939. The genetic and physiological problems of self-sterility in *Ciona*. III. Induced self-fertilization. *J. Exp. Zool.* 80: 19–54.

Nakatani, Y. and H. Nishida. 1994. Induction of notochord during ascidian embryogenesis. *Dev. Biol.* 166: 289–299.

Nakatani, Y., H. Yasuo, N. Satoh and H. Nishida. 1996. Basic fibroblast growth factor induces notochord formation and the expression of *As-T*, a *Brachyury* homolog, during ascidian embryogenesis. *Development* 122: 2023–2031.

Nakauchi, M. 1982. Asexual development of ascidians: Its biological significance, diversity, and morphogenesis. *Am. Zool.* 22: 753–63.

Nicol, D. R. and I. A. Meinertzhagen. 1988. Development of the central nervous system of the larva of the ascidian *Ciona intestinalis* L. I. The early lineages of the neural plate. *Dev. Biol.* 130: 721–736.

Nishida, H. 1986. Cell division patterns during gastrulation of the ascidian *Halocynthia roretzi*. *Dev. Growth Differ.* 28: 191–201.

Nishida, H. 1987. Cell lineage analysis in ascidian embryos by intracellular injection of a tracer enzyme. III. Up to the tissue restricted stage. *Dev. Biol.* 121: 526–541.

Nishida, H. 1991. Induction of brain and sensory pigment cells in the ascidan embryo analyzed by experiments with isolated blastomeres. *Development* 112: 389–395.

Nishida, H. 1992a. Developmental potential for tissue differentiation by fully dissociated cells of the ascidian embryo. *Roux's Arch. Dev. Biol.* 201: 81–87.

Nishida, H. 1992b. Regionality of egg cytoplasm that promotes muscle differentiation in embryo of the ascidian *Halocynthia roretzi*. *Development* 116: 521–529.

Nishida, H. 1993. Localized regions of egg cytoplasm that promote expression of endoderm-specific alkaline phosphatase in embryos of the ascidian *Halocynthia roretzi*. *Development* 118: 1–7.

Nishida, H. 1994. Localization of determinants for formation of the anterior-posterior axis in eggs of the ascidian *Halocynthia roretzi*. *Development* 120: 3093–3104.

Nishida, H. 1996. Vegetal egg cytoplasm promotes gastrulation and is responsible for specification of vegetal blastomeres in embryos of the ascidian *Halocynthia roretzi*. *Development* 122: 1271–1279.

Nishida, H. and N. Satoh. 1983. Cell lineage analysis in ascidian embryos by intracellular injection of a tracer enzyme. I. Up to the eight-cell stage. *Dev. Biol.*. 99: 382–394.

Nishida, H. and N. Satoh. 1985. Cell lineage analysis in ascidian embryos by intracellular injection of a tracer enzyme. II. The 16- and 32-cell stages. *Dev. Biol.* 101: 440–454.

Okada, T. and Y. Yamamoto. 1991. Immunoelectron microscopic study of the germ cells in an ascidian, *Ciona savignyi*. *Zool. Sci.* 8: 1067.

Ortolani. G. 1955. The presumptive territory of the mesoderm in the ascidian germ. *Experientia* 11: 445–446.

Panganiban, G., S.M. Irvine, C. Lowe, H. Roehl, L. S. Corley, B. Sherbon, J. Grenier, J. F. Fallon, J. Kimble, M. Walker, G. Wray, B. J. Swalla, M. Q. Martindale and S. B. Carroll. 1997. The origin and evolution of animal appendages. *Proc. Natl. Acad. Sci. U.S.A.* In Press.

Pinto, M. R., M. Hoshi, R. Marino, A. Moroso and R. DeSantis. 1990. Chymotrypsin-like enzymes are involved in sperm penetration through the vitelline coat of *Ciona intestinalis* egg. *Mol. Reprod. Dev.* 26: 319–323.

Pinto, M. R., R. De Santis, R. Marino and N. Usui. 1995. Specific induction of self-discrimination by follicle cells in *Ciona intestinalis* oocytes. *Dev. Growth Differ.* 37: 287–291.

Rinkevich, B. and I. L. Weissman. 1987. A long-term study on fused subcolonies in the ascidian *Botryllus schlosseri*: The resorption phenomenon (Protochordata: Tunicata). *J. Zool. (Lond.)* 213: 717–733.

Rosati, F. and R. DeSantis. 1978. Studies on fertilization in the ascidians. I. Self-sterility and specific recognition between gametes of *Ciona intestinalis*. *Exp. Cell Res.* 112: 111–119.

Roegiers, F., A. McDougall and C. Sardet. 1995. The sperm entry point defines the orientation of the calcium-induced contraction wave that directs the first phase of cytoplasmic reorganization in the ascidian egg. *Development* 121: 3457–3466.

Sabbadin, A., G. Zaniolo and F. Majone. 1975. Determination of polarity and bilateral asymmetry in palleal and vascular buds of the ascidian *Botryllus schlosseri*. *Dev. Biol.* 46: 79–87.

Sardet, C., J. E. Speksnijder, S. Inoué and F. Jaffe. 1989. Fertilization and ooplasmic movements in the ascidian egg. *Development* 105: 237–249.

Satoh, N. 1978. Cellular morphology and architecture during early morphogenesis of the ascidian egg: An SEM study. *Biol. Bull.* 155: 608–614.

Satoh, N. 1979. Visualization with scanning electron microscopy of cleavage pattern of the ascidian eggs. *Bull. Mar. Biol. Stat. Asamushi, Tohoku Univ.* 16: 169–178.

Satoh, N. 1994. *Developmental Biology of Ascidians*. Cambridge University Press, Cambridge, UK.

Satoh, N., T. Deno, H. Nishida, T. Nishikata and K. W. Makabe. 1990. Cellular and molecular mechanisms of muscle cell differentiation in ascidian embryos. *Int. Rev. Cytol.* 122: 221–258.

Sawada, T. and K. Osanai. 1981. The cortical contraction related to the ooplasmic segregation in *Ciona intestinalis* eggs. *Roux's Arch. Dev. Biol.* 193: 127–132.

Sawada, T. and G. Schatten. 1989. Effects of cytoskeletal inhibitors on ooplasmic segregation and microtubule organization during fertilization and early development in the ascidian *Molgula occidentalis*. *Dev. Biol.* 79: 181–198.

Sawada, H., H. Yokosawa, M. Hoshi and S. Ishii. 1982. Evidence for acrosin-like enzyme in sperm extract and its involvement in fertilization of the ascidian, *Halocynthia roretzi*. *Gamete Res.* 5: 291–301.

Sawada, H., H. Yokosawa, M. Hoshi and S. Ishii. 1983. Ascidian sperm chymotrypsin-like protease: Participation in fertilization. *Experientia* 39: 377–278.

Scott, M. P. 1994. Intimations of a creature. *Cell* 79: 1121–1124.

Speksnijder, J. E., L. F. Jaffe and C. Sardet. 1989. Polarity of sperm entry in the ascidian egg. *Dev. Biol.* 133: 354–378.

Sugino, Y. M., A. Tominaga and Y. Takashima. 1987. Differentiation of the accessory cells and structural regionalization of the oocytes in the ascidian *Ciona savignyi*. *J. Exp. Zool.* 242: 205–214.

Swalla, B. J. and W. R. Jeffery. 1990. Interspecific hybridization between an anural and urodele ascidian: Differential expression of urodele features suggests multiple mechanisms control anural development. *Dev. Biol.* 142: 319–334.

Swalla, B. J. and W. R. Jeffery. 1995. A maternal RNA localized in the yellow crescent is segregated to the larval muscle cells during ascidian development. *Dev. Biol.* 170: 353–364.

Swalla, B. J. and W. R. Jeffery. 1996a. PCNA mRNA has a 3' UTR antisense to yellow crescent RNA and is localized in ascidian eggs and embryos. *Dev. Biol.* 178: 23–34.

Swalla, B. J. and W. R. Jeffery. 1996b. Requirement of the *Manx* gene for restoration of ancestral chordate features in a tailless ascidian larva. *Science* 274: 1205–1208.

Swalla, B. J., M. R. Badgett and W. R. Jeffery. 1991. Identification of a cytoskeletal protein localized in the myoplasm of ascidian eggs: Localization is modified during anural development. *Development* 111: 425–436.

Swalla, B. J., K. W. Makabe, N. Satoh and W. R. Jeffery. 1993. Novel genes expressed differentially in ascidians with alternate modes of development. *Development* 119: 307–318.

Swalla, B. J., M. E. White, J. Zhou and W. R. Jeffery. 1994. Heterochronic expression of an adult muscle actin gene during ascidian development. *Dev. Genet.* 15: 51–63.

Tabin, C. J. 1991. Retinoids, homeoboxes, and growth factors: toward molecular models for limb development. *Cell* 66: 199–217.

Torrence, S. A. and R. A. Cloney. 1981. Rhythmic contractions of the ampullar epidermis during metamorphosis of the ascidian *Molgula occidentalis*. *Cell Tissue Res.* 216: 293–312.

Turbeville, J. M., J. R. Schultz and R. A. Raff. 1994. Deuterostome phylogeny and the sister group of the chordates: Evidence from molecules and morphology. *Mol. Biol. Evol.* 11: 648–655.

Venuti, J. M. and W. R. Jeffery. 1989. Cell lineage and determination of cell fate in ascidian embryos. *Int. J. Dev. Biol.* 33: 197–212.

Wada, H. and N. Satoh. 1994. Details of the evolutionary history from invertebrates to vertebrates, as deduced from the sequences of 18S rDNA. *Proc. Nat. Acad. Sci. U.S.A.* 91: 1801–1804.

Wada, S., Y. Katsuyama, S. Yasugi and H. Saiga. 1995. Spatially and temporally regulated expression of the LIM class homeobox gene *Hrlim* suggest multiple distinct functions in development of the ascidian *Halocynthia roretzi*. *Mech. Dev.* 51: 115–126.

Wada, S., Y. Katsuyama, Y. Sato, C. Itoh and H. Saiga. 1996. *Hroth*, an *orthodenticle*-related homeobox gene of the ascidian *Halocynthia roretzi*: Its expression and putative roles in the axis formation during embryogenesis. *Mech. Dev.* 60: 59–71.

West, A. B. and C. C. Lambert. 1976. Control of spawning in the tunicate *Styela plicata* by variations in a natural light regime. *J. Exp. Zool.* 195: 263–270.

Whittaker, J. R. 1973. Segregation during ascidian embryogenesis of egg cytoplasmic information for tissue-specific enzyme development. *Proc. Nat. Acad. Sci. U.S.A.* 70: 2096–2100.

Whittaker, J. R. 1980. Acetylcholinesterase development in extra cells caused by changing the distribution of myoplasm in ascidan embryos. *J. Embryol. Exp. Morphol.* 55: 343–354.

Whittingham, D. G. 1967. Light induction of shedding of gametes in *Ciona intestinalis* and *Molgula manhattensis*. *Biol. Bull.* 132: 292–298.

Woollacott, R. M. 1977. Spermatozoa of *Ciona intestinalis* and analysis of ascidian fertilization. *J. Morphol.* 152: 77–88.

Yasuo, H. and N. Satoh. 1994. An ascidian homologue of the mouse *Brachyury* (*T*) gene is expressed exclusively in notochord cells at the fate-restricted stage. *Dev. Growth Differ.* 36: 9–18.

Yoshida, S., Y. Marukawa and N. Satoh. 1996. *posterior end mark*, a novel maternal gene encoding a factor localized in the ascidian embryo. *Development* 122: 2005–2012.

Zalokar, M. and C. Sardet. 1984. Tracing of cell lineage in embryonic development of *Phallusia mammillata* (Ascidia) by vital staining of mitochondria. *Dev. Biol.* 102: 195–204.

CHAPTER 18

Cephalochordates, the Lancelets

J. R. Whittaker

L ANCELETS, SO NAMED BECAUSE THEIR SHAPE RESEMBLES the old style of surgeon's lance, are a small and homogeneous group of seemingly primitive and exclusively marine invertebrate chordates that comprise the subphylum **Cephalochordata** (Acrania). They are named formally from two morphological features: a notochord that runs forward above the sensory vesicle of the nervous system to the very anterior of the animal (*cephalo* = head; *chorda* = notochord) and the fact that they have virtually no delineated cranial region (*a* = without; *crania*). The adults are of a laterally flattened cylindrical shape, not usually more than 6 cm in length and 0.5–1 cm in diameter. They are fishlike in their general appearance and even to the taste (Light 1923). There are metameric muscle structures along the body length that enable them to swim and burrow. Dorsal, ventral, and caudal fins assist in the swimming. They burrow upright in sand with a protruding rostral region and feed by straining phytoplankton through a complex, ciliated, pharyngeal filtering apparatus.

These lancelets, also called amphioxus and amphioxids, are of enduring zoological fame because their chordate body structures seem intermediate in complexity between those found in the swimming tadpole larvae of tunicates (Urochordata) and those in modern day jawless (agnathan) vertebrates, the hagfishes and lampreys. Lampreys, in fact, retain a burrowing **ammocete** larva, which is similar in many aspects of its morphology and feeding mechanism to the lancelet. The lancelets thus appear to some observers as a modern invertebrate survivor of a successful and surprisingly stable ancestral stage in the evolution of vertebrates, remaining essentially unchanged "living fossils" of an ancient prevertebrate form.

Amphioxus, a formerly used genus name referring correctly to the animal being pointed at both ends (*amphis* = both; *oxys* = sharp), now remains only a common name for the group. In accordance with taxonomic rules, the *Amphioxus* name was displaced by an earlier-used genus name, *Branchiostoma*. Gabriele Costa in 1834 was the first to recognize the approximate phylogenetic position of amphioxus as a fishlike creature allied to the hagfishes and lampreys. Costa, unfortunately, misinterpreted the circlet of tentacle-like processes surrounding the mouth (stoma) as respiratory filaments (branchiae) and in so designating the genus as *Branchiostoma* has left us with a permanent misnomer. Later investigators, including the renowned Johannes Müller in 1841/1844, confirmed the chordate nature of amphioxus but recognized that it differed from fish to an even greater extent than fish themselves differ from amphibians. The first person to describe amphioxus and who gave it the name *Limax lanceolatus*, Peter Simon Pallas in 1778, mistook it for a molluscan slug.

For over a hundred years, amphioxus has been a familiar animal in both the classroom and research laboratory for studies of comparative anatomy and comparative embryology with respect to vertebrate ancestry (Willey 1894). Although amphioxus is known to be abundant in only a few accessible locations, and has an inconveniently short breeding season, commercial collectors and biological supply houses can still readily provide preserved specimens and mounted embryonic stages on demand, and the more determined research embryologist can travel to the collecting sites during the breeding season. In the last few decades amphioxus has remained of continued intellectual interest because its small size and epithelial nature lend themselves well to electron microscopic studies of its adult and larval morphology. Recent molecular genetic studies have focused a renewed attention on the animal and reopened many of the older and still fascinating questions about its development.

Chordate Features

Examined in lateral aspect in a purely linear fashion, one is struck forcibly by the general similarity of the amphioxus body plan to that of proper lower vertebrates. It fits closely to what might be construed as an archetypic body plan of an early vertebrate ancestor (Figure 18.1). An obvious notochord runs dorsally along the full length of the animal, extending anteriorly beyond the end of the nerve cord. This notochord is surrounded by a thick and collagenous extracellular fibrous matrix, the notochordal sheath (Eakin and Westfall 1962a). Immediately above the notochord there is a dorsal tubular nervous system, with only a slightly enlarged cerebral vesicle at its most anterior end (Figure 18.2). Beneath the notochord is an endodermally derived digestive tract connected anteriorly to a pharyngeal area with gill slits and ciliated gill bars; posteriorly, the digestive tract exits at an anal opening, beyond which occurs a short postanal tail. Lateral contractile muscle blocks, 50 to 75, occur along each side of the body, and are arranged segmentally as **myotomes** or **myomeres** (Figures 18.1 and 18.3), but actually not in corresponding register on the two sides. The monographic classics by Willey (1894), Franz (1927), and Drach (1948) are still the best compendia of adult morphology and natural history. Young (1981), Jefferies (1986), and Ruppert and Barnes (1994) offer excellent summary accounts of amphioxus biology.

Amphioxus suspends its pharynx and reproductive organs in an atrium that runs ventrally from the mouth region along two-thirds the length of the animal, with a distal atriopore opening. The atrium forms during development by the fusion of two lateral folds. After entering the mouth, water passes into the atrium through the ciliated and mucous-invested gill slits of the pharynx, where small food particles are removed, and exits through the atriopore. This flow of water is apparently sufficient for gas exchange. When gonads (Figure 18.3) beneath the coelomic membrane lining the atrial wall mature, the gametes burst through the wall and also exit the atriopore.

The circulatory system is highly developed and has a general pattern resemblance to the large blood vessels one finds in fishes. It has been described in some detail after intravascular injection with Indian ink. There is no heart in amphioxids, propulsive contractility being achieved by three of the major ventral vessels being lined with myogenically active myoepithelial cells. A colorless blood does contain proteins but no circulatory cells, hemoglobin, or other respiratory protein. Unlined tissue channels make up a closed circulatory circuit along with the smaller vessels.

One well-marked feature of amphioxus is a midgut diverticulum, also called a **hepatic caecum**, which extends forward into the atrium along the right side of the pharynx (Figure 18.3). This structure is a digestive gland that secretes enzymes into the intestine and is also the site of both intracellular digestion and nutrient storage. It is not a liver in a formal functional sense but is undoubtedly a primitive homologue of the vertebrate liver.

The pharynx has a ventrally located ciliated groove or **endostyle** that produces mucous in conjunction with the filter feeding mechanism. A similarly organized endostyle occurs in both the urochordate ascidian larva and the ammocoete larva of lamprey. In each case, iodine accumulat-

Figure 18.1 Morphology of the adult *Branchiostoma*. (*a*) Side view. (*b*) Internal view with body wall removed.

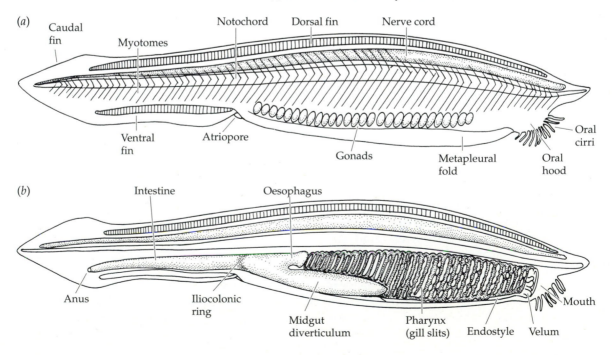

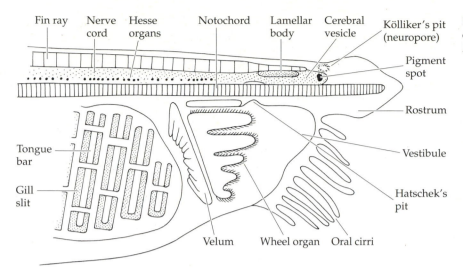

Figure 18.2 Specialized structures in the cephalic and branchial regions of the adult *Branchiostoma*.

ed by cells of the endostyle is incorporated into the thyroxine substances tri- and tetraiodothyronine. During metamorphosis of the ammocoete larva, parts of the endostyle become integrated into the developing thyroid gland—one of the more convincing examples of the probable origin and subsequent homology of a complex vertebrate organ from a simple epithelial structure. Similarly,

Hatschek's pit, a mucus-secreting epithelial structure in the oral vestibular roof of amphioxus (Figure 18.2), has long been suspected on morphological grounds of being a homologue of the vertebrate adenohypophysis. Recent results with immunohistological reagents indicate that the structure indeed secretes pituitary-like hormones (Nozaki and Gorbman 1992).

The marked segmentation-metamerism of the muscle blocks to form somites (myometamery) is a very vertebrate-like character, as is also the lined coelomic cavities. Similarly the outgrowth of nerves is segmental in arrangement, and gonads occur segmentally in rows on both sides of the atrium in *Branchiostoma* species, but only on the right side of *Asymmetron* (*Epigonichthys*) species. The gill slits follow quite a different form of segmentation, branchiomery, involving paired excretory organs and the dorsal aortae of the circulatory system.

Cephalochordates, however, lack certain other features of close vertebrate affinity and could not be classified as vertebrates. A cephalochordate has no cartilaginous or bony endoskeleton, no indication of a three-part brain structure with a cranial housing and vertebral column through which the nerve cord passes, and in fact no significant development of a separate head, except for a certain anteriorization (the rostrum) of structures associated with feeding (Figure 18.2). Amphioxus also lacks separate eyes, nasal openings, ears or jaws, nor does it show any traces or indications in its development of once having had such structures. The relative "headlessness" and "brainlessness" of amphioxus have also contributed to recurring dis-

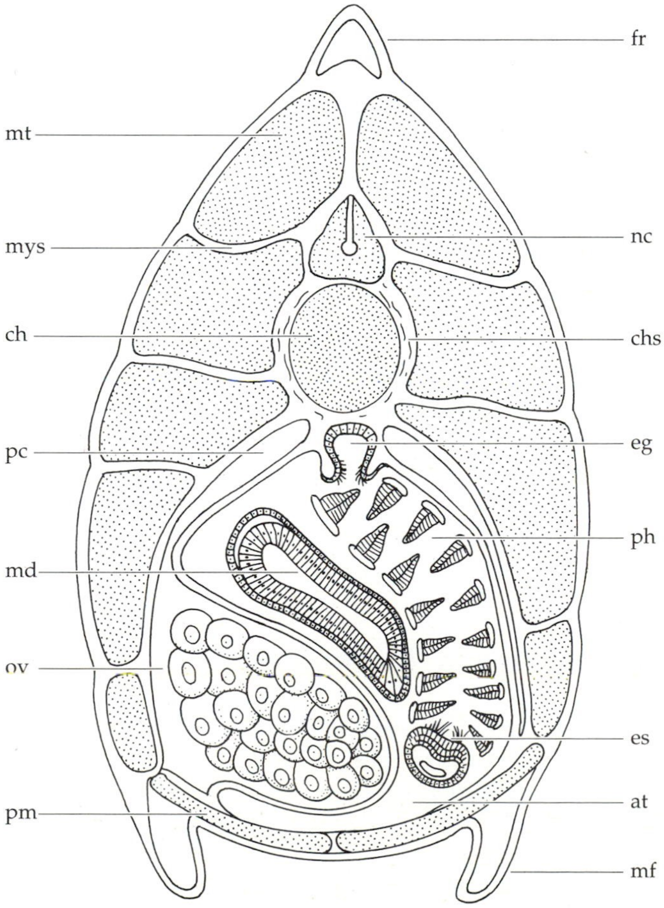

Figure 18.3 Cross section of the branchial region in an adult *Branchiostoma*. Abbreviations: at, atrium; ch, notochord; chs, notochordal sheath; eg, epipharyngeal groove; es, endostyle; fr, fin ray; md, midgut diverticulum; mf, metapleural fold; mt, myotome; mys, myoseptum; nc, nerve cord; ov, ovary; pc, periviscer al coelom; ph, pharynx; pm, pterygial muscle. (After Drach 1948.)

cussions about the vertebrate head problem. Amphioxids also appear to lack a neural crest, that *sine qua non* of vertebrate evolution, a middorsal neural tube tissue region of migratory cells, the elaborations of which contribute to many new structures in vertebrates, especially in the head region (Gans and Northcutt 1983).

Cephalochordate Peculiarities

Amphioxus has a number of anatomic oddities that have contributed to almost a century of argument about the animal's presumed primitiveness. Certainly, their distinct chordate features noted above, along with their advanced state of muscle metamerism, qualify them for consideration as vertebrate relatives. Do they remain, however, in a primordial ancestral condition similar to that from which the first fishlike vertebrates evolved or are they a very highly specialized and divergent animal that bears no such close ancestral relationship? There are four such conditions or structures that most authorities readily agree have contributed a share to these uncertainties: (1) notochord structure and its forward extension, (2) brain and spinal nerve organization, (3) body asymmetries of larva and adult, and (4) the nature of the excretory organs. All of these encompass interesting developmental problems, which as yet have hardly been addressed by researchers.

The structure of the amphioxus notochord is like that of no other chordate. It consists of a longitudinal series of disc-shaped cells or plates, each having unusually thick myosin filaments running transversely from left to right and from which paramyosin (tropomyosin A) can be extracted (Flood, Guthrie, and Banks 1969). These filaments prove to be contractile, and the notochord is a "muscular organ" and hydrostatic skeleton that contributes as an antagonist to the lateral muscles when amphioxus swims or burrows (Guthrie and Banks 1970). This extreme specialization of the amphioxus notochord detracts from the sometime argument that amphioxus is a degenerate rather than specialized form. The forward extension of the notochord beyond the sensory vesicle, which will be discussed later, is perhaps closely related to its hydrostatic skeletal functions.

Numerous attempts to homologize the rostral or cerebral vesicle part of the spinal cord with the craniate brain have been provocative but inconclusive (Lacalli et al. 1994). In addition to its rudimentary brain, amphioxus has spinal nerves that seems primitive in comparison to lower vertebrates. There is no clear indication that any of the rostral spinal nerves are cranial nerves in the later vertebrate sense (Fritzsch and Northcutt 1993). Spinal ganglia are also absent, structures which in vertebrates arise from neural crest tissues. Dorsal spinal roots are similar to the mixed sensory and motor dorsal roots of many cranial nerves in lampreys, but all spinal nerves in amphioxus lack a separate ventral motor root, which is a constant feature of all spinal motor roots in lampreys and other vertebrates. Processes originating from myotomal muscles form the equivalent of a "ventral root," which makes synaptic contact with somatic motoneurons in the spinal cord (Flood 1966). Similarly, there is a peculiar "medial root" (horn) formed by processes of the contractile notochord cells (Flood 1970) that makes cholinergeric synapses at the spinal cord with motoneurons of unknown origin. Consensus tends to favor the judgment of a primitive nervous system from which later vertebrate neural structures could be readily derived.

Perhaps the most unusual and puzzling feature of amphioxus is its right-left asymmetry, which arises in very extreme form at the beginning of larval development and persists in minor part even in characters of the adult. Oddly, this most extraordinary feature of the animal is only rarely discussed or even mentioned in textbooks and other summary literature. The adult amphioxus is quite noticeably asymmetric in the oblique positioning of its muscle segments, nerves, and gonads. The right and left sets of each are a half-segment out of register with its opposite counterpart. Although the atriopore forms medioventrally, the adult anus is displaced slightly to the left of the ventral midline and develops originally on the right of the larval midline. Opinion is about equally divided on whether this asymmetry is archaic or highly specialized. This question is relevant to cephalochordate ancestry and will be addressed further in describing larval development.

The excretory organs of amphioxus consist of about 90 pairs of **nephridia** located dorsally in the pharyngeal region and associated with the two dorsal aortae. For over 70 years they were believed to be essentially marine invertebrate in character. For osmoregulatory reasons, amphioxus was dismissed as too specialized a marine organism to be a likely invader of fresh water, where the first vertebrates were thought to have arisen. Meanwhile, opinion has shifted back to a marine origin of vertebrates, and electron microscopy reveals a more nephron-like nature to the amphioxus organs. More specifically a "glomerular" vascular filtration surface on extensions of the dorsal aortae is overlain with a kind of podocyte (modified coelomic epithelial cell) similar to those lining the renal capsule of vertebrates. The amphioxus organs are now regarded as primitive potential precursors to vertebrate nephrons (Ruppert 1994).

Organism and Techniques

At recent count there are about 25 demonstrably distinct worldwide amphioxus species of two genera, *Branchiostoma* and *Epigonichthys* (= *Asymmetron*). The familiar *Branchiostoma lanceolatum* (Pallas) is found in the temperate regions of Europe; a *B. belcheri* Gray occurs in the temperate regions of China and Japan. Most of the other known species are distributed tropically and subtropically, but all seem restricted to relatively shallow water (0.25–14 m) and a distinct kind of substratum consisting of coarse and current-swept sand. Except for a major study by Webb and Hill (1958) on an West African species, *B. nigeriense* Webb, the ecology of amphioxus has been little examined. Information about larval and adult behavioral ecology is largely anecdotal, although several studies of larval feeding (Bone 1958; Webb 1969) and larval hovering (Stokes and Holland 1995a) are noteworthy. In spite of their relatively few species and their narrow habitat specialization, the amphioxus species can become immensely abundant and prolific in certain locations. For example, a population of *B. senegalense* Webb in a region along the Spanish Sahara coast (at Cape Blanc) produces an estimated total of 1.1 million tons of planktonic larvae each year (Flood et al. 1976).

Two investigators have examined the age structure of amphioxus populations. The *B. belcheri* of the Xiamen region of China have a life span of aobut 3 years (Chin 1941). Off the southern coast of France, *B. lanceolatum* lives to about 4 years of age; those at Helgoland, in the very much colder North Sea, grow at only about half the annual growth rate of the Mediterranean population, but survivors live up to 6 years (Courtney 1975).

In North America an exceptionally large and easily accessible population of *B. floridae* Hubbs, formerly called *B. caribaeum*, occurs in Old Tampa Bay, Florida where they can be collected from a 1-meter depth by means of shovel and sieve (Nelson 1969). Their spawning season is from early August to early September. *B. lanceolatum*, studied in the first part of the century from the Bay of Naples and the Straits of Messina, is no longer found in abundance at those locations. Nonetheless, these spring-spawning animals can still be obtained in useful numbers at various locations in the vicinity of marine laboratories along the European coastline and in the Mediterranean basin. Large populations of *B. belcheri* var. *tsingtauense* occur along the China coastline near Xiamen and Quingdao and are harvested commercially in the Xiamen (Amoy) region (Light 1923; Chin 1941). They spawn from mid-June to mid-July and are in easy reach of a research facility at Quingdao, where they have been much studied and even cultured through complete life cycles (Wu et al. 1994).

There are two absolute requirements for serious modern ontogenetic studies of any organism, and not until quite recently have these been met by amphioxus: (1) reliable and reproducible artificial fertilizations and (2) precise embryonic and larval staging series at given regulated temperatures. The first has been met by the spawning methods described by Holland and Holland (1993) for *B. floridae*, and good staging series are also now available (Holland and Holland 1993; Stokes and Holland 1995b). Table 18.1 and Figure 18.7 illustrate times and stages for *B. floridae*. Other necessary techniques for handling, culturing, and rearing embryos are included in these papers and that of Wu et al. (1994).

TABLE 18.1 *Development schedule (at 25°C) for Branchiostoma floridae*

Time	Stage
0 h	Insemination
55 min	2-cell
75 min	4-cell
95 min	8-cell
3.5 h	Blastula
5.5 h	Gastrula
10.0 h	Neurula hatches
38.0 h	Larval mouth opens
18 d	Metamorphosis begins
23 d	Metamorphosis complete

Source: From Holland and Holland 1993.

Embryonic and Larval Development

Understanding of the mode of development in amphioxus began with discoveries by Alexander Kowalevsky in 1867 and 1877 that the development of *B. lanceolatum*, particularly in embryonic stages and with respect to formation of its nervous system, was remarkably similar to that of lower vertebrates. These discoveries were at first challenged but soon widely accepted as evidence of vertebrate affinity. Later investigators, particularly Hatschek (1893), Cerfontaine (1906), and Conklin (1932) examined the embryology of this species with ever more precision; their descriptions (and figures) have formed the canonical textbook accounts of amphioxus development. Only quite recently has this collective sketch been improved upon by studies on *B. belcheri* and *B. floridae* with the transmission and scanning electron microscopes (Hirakow and Kajita 1990, 1991, 1994; Stokes and Holland 1995b). There have been no major surprises. The accounts of normal embryogenesis given below are based on these recent publications.

Sexual Dimorphism

The sexes of amphioxus are readily separable as they begin to develop their rows of creamy-white testes or lemon-yellow ovaries; otherwise, no simple external morphological traits are helpful. At the time of spawning, individual females produce thousands of eggs. While it seems probable that an "X and Y" sex determination mechanism is functioning, investigations of *B. floridae* and *B. lanceolatus* have failed to identify obvious X and Y chromosomes; the diploid chromosome number for each species is 38 autosomes (see Colombera 1982). Some indication of a biological tentativeness may be indicated by how frequently hermaphrodite individuals have been reported, accounts summarized only in part by Wickstead

(1975). Hermaphroditism occurs as a normal state in most tunicate species; it is common as an occasional condition in hagfish and lampreys.

Fertilization

A large literature is devoted to matters of sexual reproduction, gonadal development, gametogenesis, and fertilization of the amphioxus egg, as reviewed extensively by Wickstead (1975). Much of it is in need of reinvestigation by modern technology with present-day questions in mind. Recent publications by Nicholas and Linda Holland based on observations with transmission electron microscopy are the most reliable sources of information on the topic of gametogenesis, and readers are referred to those works (Holland and Holland 1989a, 1991); the subject will not be reviewed here. Highlights of cytological events before, during, and immediately after fertilization, as outlined below for *B. floridae,* are described by Holland and Holland (1989b, 1992).

The spawned, mature amphioxus egg rests in a condition of second meiotic metaphase, with the first polar body attached to the animal pole surface of the egg. After sperm penetration, no marked ooplasmic segregation takes place. Sheets of dense granules and associated endoplasmic reticulum do, however, aggregate along with numerous mitochondria into whorls and accumulate in a yolk-free zone near the vegetal pole. These whorls make up the vegetal pole cytoplasm, which possibly becomes segregated into the germs cells during subsequent cleavages.

In the 45 seconds (at 25°C) after sperm penetration, a cortical granule exocytosis occurs around the whole egg surface to create (in cooperation with a coarsely granular vitelline layer) a **fertilization envelope**, which gradually forms a large perivitelline space between egg and envelope and lifts away the first polar body. After 2 minutes, second meiotic telophase begins. By 16 minutes, male and female pronuclear fusion occurs to form the $2n$ zygote nucleus and the second polar body has been extruded from the egg. The second polar body replaces the first at the egg surface and marks approximately the animal pole of the animal-vegetal embryo axis. Ascidians, with which they are most frequently compared, have significantly different egg membrane structures, no fertilization envelope, no vegetal pole cytoplasm, and very pronounced ooplasmic segrega-

tion of ultrastructurally different zygotic regions of cytoplasm. Cephalochordates do not in these respects resemble lower vertebrates any more closely than they do ascidians.

Cleavage Stages

The first cleavage furrow in amphioxus passes meridionally through the animal-vegetal axis denominated approximately by the position of the attached second polar body, to produce two equal-sized blastomeres. The second cleavage plane is also meridional, but at right angles to the first, resulting (for *B. belcheri*) in four adjacent cells of equal volume (Hirakow and Kajita 1990). *B. lanceolatum* may differ, since both Cerfontaine (1906) and Conklin (1932), but not Hatschek (1893), indicate that one pair is just barely larger and hence second cleavage is slightly asymmetric. Yet all four agree that third cleavage, which is equatorial and separates a top and bottom quartet of cells, results in an animal quartet slightly but noticeably smaller than the bottom four (Figure 18.4*a*). This difference seems to be more pronounced in *B. lanceolatum.*

Cleavages are roughly synchronous in time through seventh cleavage (up to the 128-cell stage), at which stage it becomes difficult to observe and keep track of all the cells. The cell size difference noted at the 8-cell stage (Figure 18.4*a*) is equally pronounced at the 16-cell stage, when the fourth cleavage planes have become meridional again to produce top and bottom octets of cells. Even after fifth cleavage, which is meridional and now creates four discernibly stacked octets of cells at the 32-cell stage (Figure 18.4*b*), one can note size differences between animal and vegetal ends of the embryo, and these persist through blastula and gastrula formation. The four octets of cells at the 32-cell stage are labeled an_1, an_2, veg_1, and veg_2 in Figure 18.4*b*; cells from these distinctive layers can be isolated surgically with accuracy of provenance.

At the 64-cell stage, individual blastomeres remain spherical and loosely connected, but already there is a sig-

Figure 18.4 Early embryonic stages in *Branchiostoma.* (*a*) 8-cell stage, (*b*) 32-cell stage. an_1, animal pole outer octet of blastomeres; an_2, animal pole inner octet of blastomeres; veg_1, vegetal pole inner octet of blastomeres; veg_2, vegetal pole outer octet of blastomeres. (*c*) Blastula (broken open) of *B. belcheri.* (*d*) Mid- to late-gastrula (optical midsection) of *B. lanceolatum.* (*a–c* after Hirakow and Kajita 1990; *d* after Conklin 1932.)

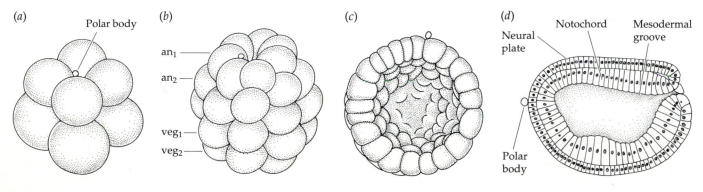

(*a*) Polar body   (*b*) an_1 an_2 veg_1 veg_2 (*c*)   (*d*) Neural plate   Notochord   Mesodermal groove   Polar body

nificant blastocoel cavity. Later, at the 128-cell (seventh cleavage) and "256-cell" (eighth cleavage) stages when the cells are more adherent, the blastocoel is pronounced. The eighth cleavage marks the beginning of the blastula stage (Figure 18.4c); the blastocoel wall is now thinner, and the outer and inner surfaces of the wall cells are distinctive in appearance.

In its early embryology up to the blastula stage, amphioxus development is generic deuterostomal: cleavage is arguably radial, and the embryology is at least superficially indistinguishable from an echinoderm's. Gastrulation and neurulation first begin to demonstrate their chordate affinities.

Gastrula and Neurula

At the beginning of gastrulation there is a flattening of the vegetal zone cells of the blastula into an endodermal plate. A deep sinking or ingression of this flattened plate soon occurs, which slowly eliminates the blastocoelic space and causes the embryo to become cap-shaped with a deepened archenteron. Initially, a large blastopore is formed with thickened lips, the dorsal lip made up of presumptive notochord cells and the lateral lips of mesodermal precursor cells. Progressive invagination of these dorsal and lateral lip cells forms an internal notochordal plate with mesodermal grooves of each side of it as the roof of the archenteron. By the late gastrula stage the embryo has become ovoid and slightly flattened, with a small posteriodorsal blastopore (Figure 18.4d). The flattened dorsal side of the embryo is the neural plate, which becomes a broad shallow groove.

By about the time of late gastrulation, cells of the ectoderm and neuroectoderm have each developed a single cilium. These cilia propel the embryo within its fertilization envelope in an anticlockwise rotation, as seen from behind (Hatschek 1893). After hatching, which occurs during the neurula stage, the embryos and larvae swim by ciliary movement in the water column. Larvae retain these cilia until metamorphosis and employ them in hovering and feeding (Stokes and Holland 1995b). At metamorphosis most of the surface cilia are lost. Different commentators have remarked that the presence of cilia on the surface of amphioxus larvae attests to their affinities with ancient ciliated deuterostome larvae.

Neurulation begins with enclosure of the neural plate. The ectodermal layer at the edges of the neural plate dissociates from the plate edges and from the deeper tissue layer below it, and these ectodermal edges both move transversely toward the dorsal midline and fuse, creating a smooth ectodermal surface layer and in the process distorting the neural plate below into a more pronounced V shape. The V ridges of the sunken neural plate gradually coalesce from the rear forwards, creating a closed neural tube that opens anteriorly at the neuropore and posteriorly by the neurenteric canal connecting (temporarily) with the gut cavity (Figure 18.5a).

During the time of neural plate enclosure and neural tube formation, the notochord and mesoderm are devel-

oping from the adjacent **chordamesodermal plate** that constitutes the roof of the archenteron just below. Their formation is not complete until a late-neurula–early-larva stage at about the time of hatching (Figure 18.5b). The notochord rudiment forms by a pouching upwards of midline cells along of the chordamesodermal plate (Figures 18.6a,b). Edges of this plate pouch upwards dorsolaterally to create the rudiments of the first few of the dorsolateral series muscle blocks. This **enterocoelic** or pouchlike formation of the myocoel in the anterior few somites (Figure 18.6a) is an embryologic curiosity of the amphioxids; coelomic cavities in the posterior somites arise by a splitting process called **schizocoely**, a more deuterostomal and vertebrate characteristic. The more ventral parts of the archenteron lining are the endodermal tube rudiment, and they fuse along the midline to form the pharyngeal tube and gut continuum.

One further oddity of somite formation in amphioxus is the fate of the anterior parts of the archenteron, which are regarded by some as forming "premandibular" somites that are the actual first pair. As the notochord begins to grow anteriorly (Figures 18.5b,c), two right and left pockets, the gut diverticulae, form. The left diverticulum forms the preoral pit (see below); the right one contributes the anterior coelom of the larva and later rostral coelom of the adult.

A comparison of amphioxus gastrular and neurular development with that of the ascidian larva reveals remarkable similarities between the two taxa. Given the probable necessary adjustments for the smaller cell numbers involved in ascidian development at these stages (gastrulation begins just after the 64-cell stage), ascidian gastrulation and neurulation (Swalla 1993) do have a reasonably close similarity to what is observed in amphioxus. There is also little question about the homology to obviously comparable processes in amphibian development, but the correspondence to ascidians seems closer.

Organization of the Neurula

At the end of neurulation and the beginning of larval formation the animal has become a near-perfect representative of the common, generalized, primitive embryonic body form of all vertebrate embryos, which according to Nelsen (1953) has: (1) an elongated structure, cylindrical in shape, and is somewhat compressed laterally; (2) four basic organ-forming epithelial tubes, epidermal, neural, endodermal, and mesodermal, oriented around a primitive axis, the notochord; that is, a triploblastic "tube within tube" structure augmented by a neural tube and notochord; (3) distinct body regions of (a) "head," (b) pharyngeal area, (c) trunk, and (d) tail.

Amphioxus is also a predominantly epithelial animal. A particular example of this characteristic is the structure of the adult skin, which remains a mucous-secreting single-layered epidermis (Olsson 1961). Any study of amphioxus morphogenesis beyond gastrulation is an extended exercise in tracing the development of its organs and

(a)

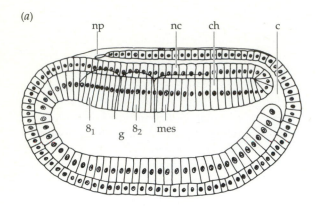

Figure 18.5   Later embryonic and larval stages of *Branchiostoma lanceolatum*: (*a*) Neurula stage (2 somite pairs). (*b*) Newly hatched embryo (9 somite pairs). (*c*) One-day larva (15 somite pairs; 1 gill slit). (*d*) Two-day larva (20–22 somite pairs; 2 gill slits). (*e*) Five-day larva (35 somite pairs; 3 gill slits). Abbreviations: ap, anterior process of first somite; ba, branchial anlage; c, neurenteric canal; cg, club-shaped gland; ch, notochord (or notochordal territory); cv, cerebral vesicle; es, endostyle; gs1–gs4, numbered gill slits; hc, cells of Hesse organ; i, intestine; ld, left anterior gut diverticulum (Hatschek's pit); m, mouth; mc, myocoel; mes, unsegmented mesoderm; nc, nerve cord; np, neuropore (Kölliker's pit); p, pigmented spot; S1–S2, numbered somites. (*a,b* based on Kellicott's [1913] modification of figures from Hatschek 1893; *c,d* after Conklin 1932; *e* after Lankester and Willey 1890.)

(b)

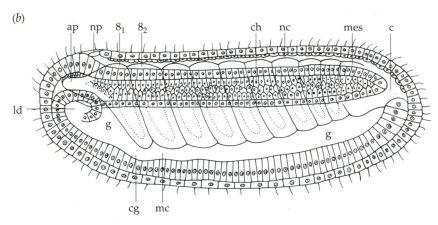

(c)

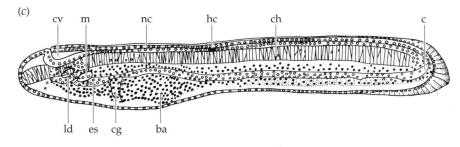

(d)

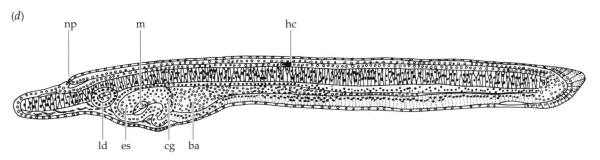

(e)

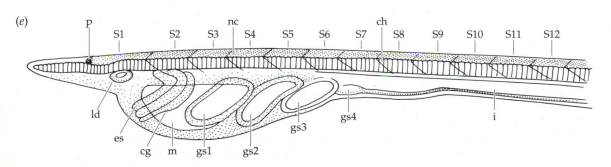

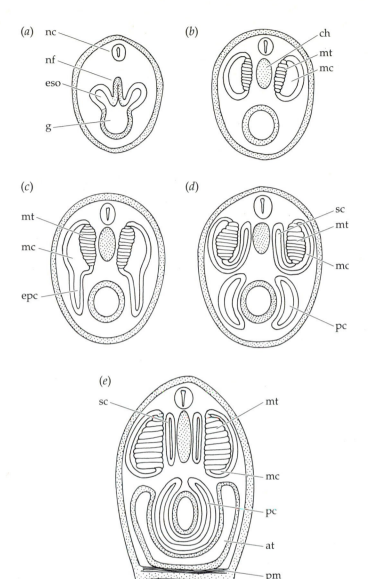

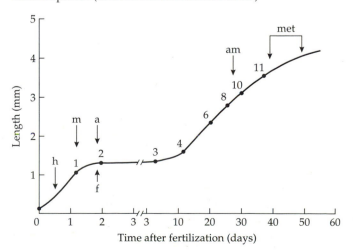

Figure 18.6 Major muscle blocks and coeloms of amphioxus. Diagrammatic cross sections through the postbranchial intestinal region at various developmental stages. (*a*) Neurula stage corresponding to Figure 18.5*a*. (*b*) Hatched embryo corresponding to Figure 18.5*b*. (*c*) One-day larva corresponding to Figure 18.5*c*. (*d*) Two-day larva corresponding to Figure 18.5*d*. (*e*) Postmetamorphic larva (6 weeks) corresponding to Figure 18.5*d*. Abbreviations: at, atrium; ch, notochord; epc, evaginating periviseral coelom; eso, evaginating somite; g, gut cavity; mc, myocoel; mt, myotome; nc, nerve cord; nf, notochordal fold; pc, periviseral coelom; pm, pterygial muscle; sc, schlerocoel. (After Holland et al. 1995.)

The left gut diverticulum (Figures 18.5*c*,*d*) begins to produce the preoral pit which opens to the left exterior (Figures 18.8*a*,*b*), and after metamorphosis this structure contributes cells to Hatschek's pit and the **buccal wheel organ** (Figure 18.2); its strictly larval functions remain unknown. An endodermal thickening in the left wall of the archenteron heralds the position of mouth formation; similarly, a thickening in the right wall gives rise to the club-shaped gland that develops on the right side of the anterior pharynx but whose duct opens to the left. The structure of this gland is now well known (Olsson 1983), but its larval function remains a mystery; it disappears at metamorphosis. The mouth opening, which quickly enlarges (Figure 18.5*d*,*e*) and remains substantial during larval stages, eventually migrates during metamorphosis to the velar region (Figure 18.2) as the adult mouth.

Because the anlagen of the mouth and club-shaped gland are bilaterally situated and each has initially the essential distinctive properties of gill-pouches, they have been regarded by some as derivatives of a primordial "first" pair of gill slits (van Wijhe 1913). Hence, the larval

Figure 18.7 Growth and development curve for *Branchiostoma floridae* from hatching to metamorphosis at 22.5°C. The numbered points along the curve indicate the advent of new gill slits. Other events shown by abbreviations: h, time of hatching; m, mouth forms; a, anus forms; f, time of first feeding; am, anus migrates; met, metamorphosis. (From Stokes and Holland 1995b.)

structures from simple epithelial beginnings that barely move beyond that state. For this reason alone amphioxus has much to recommend it as a very early "vertebrate" embryo prototype. Yet amphioxus and its embryos do differ profoundly from vertebrates: they lack the massive mesenchymal tissues of typical vertebrates, and their larva has no detectable mesenchymal cells.

Larva and Metamorphosis

The three larvae illustrated in Figures 18.5*c*–*e* show stages of rapid change in their tissue and organ structures (Figure 18.7; Table 18.1). The larva depicted by Figure 18.5*c* already has approximately 15 somites. At about this general time there is a shifting of symmetry as the intersomitic boundaries shift a half-segment between right and left sides. This asymmetry persists into the adult. At the anterior of the larva a number of even more profound changes are taking place.

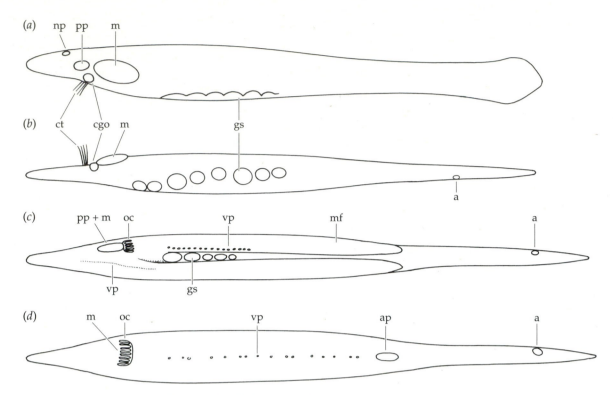

Figure 18.8 Major structural changes in *Branchiostoma floridae* at the time of metamorphosis. Late premetamorphic larvae in left-lateral (*a*) and ventral view (*b*). Mid-metamorphic stage (*c*) and early juvenile stage (*d*), both shown in ventral view. Abbreviations: a, anus; ap, anterior process of first somite; atp, atriopore; cgo, club-shaped gland opening; ct, ciliary tuft; m, mouth; mf, metapleural fold; np, neuropore; oc, oral cirri; pp, preoral pit; vp, ventral pit cells. (After Stokes and Holland 1995b.)

and subsequently adult mouth is actually an enlarged gill slit. One further reason for believing so is the presence of Hatschek's nephron in the dorsal branchial vicinity of the mouth. Other nephrons eventually occur in pairs located between lateral pairs of ultimately differentiated gill slits. What might have happened to the original primordial mouth, however, is not yet obvious from the results of any embryological study. In vertebrates the middle ear originates from elements of the first gill-pouch. Thus, amphioxus, lacking auditory organs, and having mislaid its mouth, might be said to eat with its left ear.

What is nominally the first morphological gill slit in amphioxus develops from a branchial anlage of endoderm situated below the third somite. It develops simultaneously with the mouth, although feeding does not begin until the second gill slit and anus open about 18 h later (Figure 18.7). Gill-slit formations are yet another of the bizarre larval asymmetries. The first gill slit belongs to the left side but forms midventrally and migrates to the right side where it fuses with ectoderm and breaks through the body wall; others form similarly, but only the foremost several migrate toward the right, whereas the others remain midventrally located. The anus initially develops at the right of the midline, perhaps reflecting the axial distor-

tion created by the positioning of the left gill slits.

In B. lanceolotum, about 16 "left" larval gill slits develop, but at the time of metmorphosis (days 40–50; Figure 18.7), eight of these abort and the remaining eight actually migrate to the left side. Meanwhile, during metamorphosis eight new gill slits of the proper right side appear there (Willey 1894; van Wijhe 1913). During metamorphosis the gill slits become enclosed into a peribranchial cavity by the midventral fusion of two metapleural folds, which form a posterior atripore for the evacuation of water. While the "left" gill slits are migrating to the left, where they apparently belong, the anus also migrates from the right to the left side of the midline. The changes appear to be quite similar in B. floridae (Stokes and Holland 1995a). After metamorphosis, branchial metameres of the juveniles become secondarily divided and gill slits multiply with age until there are 180 or more pairs. The juveniles adopt an adult life style of feeding and behavior; with increasing size their bilateral rows of coelomically derived gonads develop and mature.

Some amphioxus larvae fail to undergo metamorphosis and seem to reside indefinitely in the plankton; they acquire greater size, develop more numerous gill slits and myotomes, the metapleural folds remain unfused, and they are sometimes found with gonadal rudiments containing spermatocytes and oocytes (Wickstead 1964, 1975). These were initially classified as pelagic species of the genus *Amphioxides*. "Amphioxides" are no longer regarded as distinct from other species, but are now thought to arise by paedogenesis—that is, sexual maturity occurs in what is clearly a larval organism.

The Significance of the Asymmetries

Long ago, Willey (1894) favored the somewhat neo-Lamarckian viewpoint that the developmental mechanics accompanying the forward extension of the notochord into a rostral region dislocated the primitive symmetry of the larva, the notochordal extension being adaptive and advantageous, whereas the accompanying asymmetries were neither. Two later explanations which might be designated the historical relics theory and the special adaptations theory have been treated more seriously.

Medawar (1951) has been one of the more recent proponents of the amphioxus asymmetries being historical relics of axial disturbances arising from neotenization of a tunicate-like larva in the ancestral origins of the amphioxus animal. According to this view, the adult amphioxus is like a larval tunicate, and he sees the asymmetries, especially the pharyngeal ones, as a residual part of the axial torsion that ascidian larvae undergo during metamorphosis. Bone (1958), however, favors an adaptationist explanation that the asymmetries have no phylogenetic significance and are optimizations related solely to larval feeding. There is little question that the larvae are well adapted to their present feeding strategy (Stokes and Holland, 1995a), but one might equally suspect that such finely tuned optimizations of feeding strategy could as readily have improved upon a necessity imposed originally by the early organization of the neotenate.

Medawar's idea has the charm of being a deducible consequence of the perhaps most widely accepted theory of cephalochordate origins, namely, neotenization of a tunicate-like tadpole larva, permitting it to retain the tail structures while developing adult feeding and reproductive features. Berrill (1987) points out another salient but little known characteristic of ascidian postmetamorphic change. In the most primitive families of solitary ascidians there is also a secondary set of "metamorphic" changes during which a primary three-siphoned oozooid slowly reorganizes its branchial region to become a secondary two-siphoned ascidiozooid. Oddly, amphioxus also has two equivalent sets of "metamorphic" change, the first occurring during original pharyngeal development after neurulation, and a second set of adjustments to symmetry during its own metamorphosis.

Primitive "Eye" Structures

Amphioxids lack paired bilateral eyes in their cephaplic region, yet ammocoete larvae and adult cyclostomes have primitive paired cephalic eyes typical of the more advanced vertebrates. How might such eyes have originated? Several presumably photoreceptive structures do occur in amphioxus (Lacalli et al. 1994). These include the frontal eye or "eye spot" at the anterior end of the slightly enlarged cerebral vesicle, as well as two other structures positioned in the dorsal wall of the cerebral vesicle: a pineal-like mid-dorsal lamellar body (Figure 18.2), and Joseph cells. The frontal eye and lamellate body have ciliary photoreceptors; the Joseph cells appear to be microvillar. Additionally, a series of **cup organs of Hesse** (Figure 18.2) are distributed lateroventrally within the neural tube (Hesse 1898). The Hesse organ is two-celled: a spherical photoreceptor cell (also microvillar) capped with a cup-shaped black melanocyte (Eakin and Westfall 1962b). Lancelets respond kinetically to light, actively swimming away from it (Parker 1908).

On the basis of tissue arrangements and overall structure, as revealed by three-dimensional reconstructions, Lacalli (1996) deduced that the frontal eye represents an unpaired homolog of the two lateral eyes in vertebrates. The frontal eye in amphioxus specimens is sometimes divided in two (Ayers 1890). Recently developmental expression of *AmphiOtx*, the amphioxus member of the *Otx* homeobox class of regulatory genes, was observed in the anterior mesendoderm and overlying neuroectoderm of both vertebrates and cephalochordates, also lending support to the suggestion that the frontal and vertebrate paired eyes are homologs (Williams and Holland 1996).

The diversity of amphioxus photosensory structures might prove valuable for investigating the evolution of eyes and for addressing the now-confusing concept of eye homology. That amphioxus should have four separate photosensory structures seems to affirm the likelihood of frequent recurrent independent eye evolution, as predicted by the calculations of Nilsson and Pelger (1994). One wonders if each of the amphioxus organs will prove to employ the same *Pax-6* regulatory gene expression found in the anatomically different eye structures of very distantly related phyla (Tomarev et al. 1997). If all animal photoreceptors are homologous as cells, then one might expect the various amphioxus structures to share a common conserved *Pax-6*-dependent mechanism. There is no present basis for regarding the four separate amphioxus bodies as homologous organs.

The Missing Neural Crest

The lateroventrally distributed cup organs of Hesse are located in the neural tube next to the neural cavity in linear clusters that are a half-segment out of register on the two sides, similar to the asymmetry of the myotomes. Melanocytes of the Hesse organ can be seen quite clearly in living or fixed specimens.

The first of these melanocytes develops in the 1-day larva (Figure 18.5c) at about the level of the fifth somite. Very gradually others develop all along the tube and forward to almost the level of the cerebral vesicle (Figure

18.2), until there are about 1500. These melanocytes originate within the neural tube, but their initial location and manner of development remain unknown. In vertebrates, melanocytes are unquestionably a neural crest derivative originating in a middorsal part of the neural tube; they are arguably its first such derivative. Results of experiments with different types of neural crest cells in cell and tissue culture indicate that melanocytogenesis is a common default condition in the differentiation of neural crest cells, as one might predict if melanocytes were the original neural crest primordium (Weston 1991). One need not be much of a contrarian to suspect that the neural crest literature is quite mistaken about the lack of a primordium or trace of neural crest antiquity in amphioxids. Hesse organ melanocyte differentiation merits a modern careful investigation with this discordant thought in mind.

Further Analysis of Histodifferentiations

As yet there are relatively few ultrastructural studies of tissue differentiations in amphioxus development but many notable examples of the analysis of adult and larval structures. Ultrastructural analysis of embryos, of which Hirakow and Kajita (1994) provide an good example, must be complemented with studies of molecular expressions to be fully revealing. An illustration is offered as follows: In the early larval stage corresponding to Figure 18.5c, myofibrils are found in the muscle cells of the myotome and paramyosin is seen in the notochordal cells (Hirakow and Kajita 1994). At neurula stages having 4–5 somites and corresponding to Figures 18.5a,b, messenger RNA (mRNA) for an alkali myosin light-chain gene (MLC-alk) can already be seen by in situ hybridization techniques to be localized in the myotomes of the forming somites, and it continues thereafter to be expressed strongly in all of the developing muscle cells of the embryo (L. Z. Holland et al. 1995). In larval stages some cells of the notochord (in cross-section) can be seen reacting faintly with this probe; it is not a strong indicator of the contractile elements in notochord but more or less correlates with the advent of muscular twitching in the larva.

The murine Brachyury (T) gene, an apparent regulatory gene active in the mouse notochord during embryogenesis, is also functional in mesoderm formation. It appears to be involved in controlling the differentiation and fate of notochord cells. Brachyury homologues of this gene (AmBra-1 and AmBra-2) isolated from B. floridae are also expressed actively during notochord (and mesoderm) formation in their embryos as early as the late gastrula and early neurula stages (P W. H. Holland et al. 1995).

Other interesting vertebrate regulatory genes (transcription factors) have also been isolated and their expressions studied in amphioxus, notably the axial-regulating Hox genes, which affect the nervous system as well as other structures (see final section). Although there has been much excellent investigation of the morphology of

the nervous system in amphioxus (see Lacalli et al. 1994 and Fritzsch and Northcutt 1993), one hopes that the isolation of additional genes concerned with neuronal differentiation will help unravel the functions of the amphioxus nervous system. AmphiPax-1 is a regulatory gene in amphioxus that is expressed in pharyngeal endoderm during the formation of the mouth and first two gill slits and may play a role in pharyngeal patterning (N. D. Holland et al. 1995). In vertebrates it is also expressed in endoderm of pharyngeal pouches.

Experimental Embryology

Attempts to analyze the nature of the developmental processes in amphioxus by cell and tissue manipulations, or what has been called classical experimental embryology, are remarkably few in the history of amphioxus study. This lack results in large part from the purely practical difficulties of short breeding seasons and limited animal accessibility. For successful experimental embryology, in addition to good staging series at given regulated temperatures and reproducible artificial fertilizations, one requires a relatively antiseptic technique for culturing demembranated partial embryos and blastomere recombinations. Bacterial and protozoan infestations of cultures, their activities and toxins, will cause general abnormalities in the already stressed embryos. In avoiding further unnecessary stress to embryos, one must also employ conditions of culture that are optimal rather than marginal for temperature, salinity, pH, and oxygen tension. Few of the above conditions were met in the earliest reported experiments; this failure to meet optimal conditions may explain the inconsistency of results.

Cell Lineage

Among the central questions about egg and embryo organization that experimental embryology addresses is the fate of early and later cells and when (and how) their fates become fixed. Many invertebrate embryos, including that of amphioxus, have a determinate cleavage pattern, whereby under the normal undisturbed circumstances of embryonic development individual cells in the patterns can be traced to a distinct position and tissue in the resulting embryo. The fixed relationship of cells in these patterns of cleavage is called a cell lineage.

If one marks regions of ascidian and amphioxus eggs and embryos with adherent particles in the case of ascidian or by vital dye staining in the case of amphioxus (Tung et al. 1962a), cell lineage maps can be constructed that indicate a striking similarity between the two (Figure 18.9). This similarity undoubtedly confirms the relatively close, but much argued, phylogenetic relationship between them. The distinct cytoplasmic regions of the ascidian egg and embryo identified by these marking methods can also be seen clearly by simple light microscopic techniques.

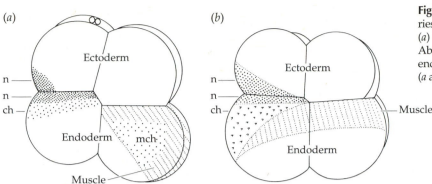

Figure 18.9 Comparison of the cell lineage territories at the 8-cell stage in the ascidian *Ciona intestinalis* (a) and the amphioxus *Branchiostoma belcheri* (b). Abbreviations: ch, notochordal territory; mch, mesenchyme tissue territory; n, neural tissue territory. (*a* after Ortolani 1954; *b* after Tung et al. 1962b.)

Conklin (1933) was quite convinced that regionally distinct cytoplasmic differences corresponding to those in ascidians occurred also in *B. lanceolatum*, but modern investigators using electron microscopy have not confirmed this for either *B. belcheri* (Hirakow and Kajita 1990) or *B. floridae* (Holland and Holland 1992). Given the considerable flexibility in development of amphioxus blastomeres to be discussed below, it seems inherently unlikely that amphioxids have the same physically distinct regions as those in ascidians.

Considering the importance of comparative embryology to questions of phylogenetic relationship, amphioxus cell lineages should really be carefully reinvestigated using the more precise modern methods applied in recent years to ascidian embryos. The diameter of the living *B. floridae* egg is about 140 μm; other species are similar, judging from measurements reported only for formalin-fixed eggs. This size is sufficiently large to accommodate the same blastomere microinjection techniques for introducing horseradish peroxidase or fluorescent dyes coupled to microspheres into the cells.

Blastomere Isolations and Cell Fate

In the ascidian embryo, to which amphioxus embryos are most frequently compared, a century of investigation has led to the general conclusion that fates of their cells are determined by two broad mechanisms: autonomous specification and conditional specification (Davidson 1990). **Autonomous specification** is created by the presence and localization of specific maternal cytoplasmic factors in the egg that get segregated during subsequent cleavages into specific regions and tissues of the embryo. These cytoplasmic "determinants," which directly or indirectly may be of gene regulatory character, then presumably influence the differentiation fates of the cells in which they come to reside. **Conditional specification** is the alteration or adjustment of cell fates resulting from contact interactions between cells of different lineages. In the ascidian, the balance between these two mechanisms is strongly but not exclusively on the side of autonomous specification. But so much so, that blastomeres separated at the 2-cell stage each give rise to essentially lateral half-larvae, and cleav-

age cells or cell combinations isolated from later stages produce only partial embryos, with expressions of differentiation related to normal positional fates of the cells in the whole embryo (as shown by cell lineage maps).

Contrary to what was observed with ascidians, Wilson (1892, 1893), Morgan (1896), and Conklin (1933) all found that isolated 2-cell-stage blastomeres from *B. lanceolatum* adopted normal cleavage and gastrular patterns and could give rise to smaller but essentially normal and complete larvae. Similarly, lateral 2/4 half-embryos from the 4-cell stage sometimes produced half-sized complete larvae. Wilson and Morgan differed from Conklin in their results with single blastomere isolates from the 4-cell stage. Usually these produced only incomplete embryos, but both Wilson and Morgan found cases of more or less complete larvae that convinced them that development was "regulative," or as we would now say, conditionally specified, up to the 4-cell stage.

Many years later, these questions were reinvestigated using *B. belcheri* by Professor and Mrs. T. C. Tung and their associate, Dr. S. C. Wu at Quingdao. They confirmed Wilson's conclusion that essentially complete small larvae could be obtained frequently from one-quarter isolates (Tung et al. 1958). Aseptic techniques of culture, with antibiotics included, were used in their laboratory. Isolates of animal and vegetal quartets at the 8-cell stage (Figure 18.4a) and of the various octets of cells taken from the 32-cell stage (Figure 18.4b) developed only partial expressions now related to their normal fates as indicated by their cell lineages; these studies would seem to indicate that at stages beyond the 4-cell an autonomous specification was becoming largely responsible for cell fate (Tung et al. 1958, 1960a). Perhaps each cell no longer received a full complement of the necessary cytoplasmic factors.

This conclusion is confounded by the results with recombinations of the octet layers taken from the 32-cell stage (Tung et al. 1960a). All four possible combinations that included one animal (an_1 or an_2) and one vegetal (veg_1 or veg_2) octet produced a typical larva with normal structures. Similarly, normal larvae resulted when one of the octets was eliminated, leaving the three others. One might only conclude from this that cell specification in amphioxus has become extremely conditional at later stages. The final truth undoubtedly, is exactly the same as for ascidian: both modes of specification are operative, except in amphioxus conditional specification becomes very dominant and overriding for a period of time.

Regional differences in cell specification are illustrated by the findings of two other experiments. Cell lineage marking indicates that the anterior pair of animal quartet cells in an undisturbed 8-cell stage (Figure 15.9b) normally contribute a part of the larval neural tissue (Tung et al. 1960a). When the animal quartet is rotated front-to-back (180°), presenting different cell lineages to ultimately receive a neural inductive stimulus, normal larvae still result (Tung et al. 1960b). In a different kind of experiment, using cells from 32-cell stages (Figure 18.4b), when one or two cells from the an_1 ectodermal layer were substituted for one or two cells of the veg_2 endodermal layer, the an_1 cells were usually converted into endodermal tissues in the subsequent larva (conditional specification). Conversely, veg_2 cells substituted into an_1 locations generally retained their endodermal fate (autonomous specification).

These results, like others noted above, deserve to be reinvestigated and extended. Given the exceptional conditional specification displayed in embryonic cleavage stages, one naturally wonders about the regenerative potential of later embryonic stages, larvae, and adults. What few experiments have been done on regeneration are inconclusive. From the summaries given by Wickstead (1975), infections and poor culture conditions seem to have been a major impediment to obtaining results.

Neural Induction

The embryos of both ascidians and vertebrates show a common general mechanism in forming the neural tube, as discussed in a previous section. There is also a common form of neural induction, whereby the presumptive ectoderm is conditionally specified for its final fate by contact interaction with cells in the roof of the respective archenterons. In ascidians the transformation can be induced experimentally with either notochordal (chordal) or endodermal tissues, which are the tissues in contact during normal induction. The interaction in amphioxus appears to come only from presumptive notochordal tissues in the archenteron roof. The dorsal lip of the young gastrula will induce a secondary neural system when it is transplanted into another young gastrula. Similarly, notochord, but neither mesodermal nor endodermal cells taken from an already invaginated plate, will induce secondary neural structures when transplanted into a young gastrula (Tung et al. 1962b).

The tissue causing neural induction in amphibian vertebrates is chordamesodermal tissue from the dorsal lip of the blastopore and the archenteron roof. This tissue itself requires interaction with endoderm before it becomes an inducer, although endodermal tissues themselves are not neural-inductive.

So Singular a Little Fish

In the 155 years since affinity of living amphioxus to the primitive jawless fishes was first recognized independently by John Goodsir, Heinrich Rathke, and Johannes Müller, the principal recurring question about the animal is whether its apparently primitive nature and primordial features are historic relics of some ancient prevertebrate form or are secondary specializations acquired over time. Some of the more unique and distinguishing features of amphioxus lend themselves to this speculation, particularly the absence of any proper head, the cerebral vesicle being an almost negligible anterior swelling of the nerve cord ("brainless"), and the notochord extending past the cerebral vesicle to the anteriormost tip of the body. We have already discussed their odd excretory structures and larval asymmetry in other sections.

In the face of countless millennia of natural selection, does any living animal form exhibit a mere survival of a corresponding degree of simplicity in their remote ancestors? Some authorities regard this as highly improbable. In 1875 Dohrn, in a view endorsed subsequently by many others, speculated that amphioxus simply originated by degeneration from a fishlike vertebrate, which in some ways it resembles. Yet the surprising fact is that a few modern survivors, like amphioxus, may indeed be primitive relics after all.

Fossils unearthed in recent decades are suggestive of the primordial nature of amphioxus. *Pikaea gracilens*, found in the Burgess Shale deposits of the Middle Cambrian (520 million years ago), is an amphioxus-like primitive chordate. It has a gross morphology resembling that of modern *Branchiostoma* in having a pointed fore-end, no obviously differentiated "head" region, a dorsal longitudinal notochord-like structure, and open V-shaped muscle bands (Briggs et al. 1994). A detailed description of *Pikaea* and its structures is expected to resolve many questions about cephalochordate-vertebrate affinities. This work has been in preparation for well over a decade by Prof. Simon Conway-Morris and must now be one of the most breathlessly awaited events in the history of paleontology!

Animals with an amphioxus-like body plan have seemingly been present since the great Lower to Middle Cambrian (510–525 Mya) evolutionary explosion (Briggs et al. 1994; Shu et al. 1996). No vertebrate-like fossils occur until traces of an ostracoderm (*Anatolepis*) and the conodont animal occur in the Upper Cambrian and Lower Ordovician strata. Much later, in the Early Permian period (286 Mya), a seemingly "amphioxus" fossil specimen (*Palaeobranchiostoma*) occurred, with its notochord projecting to the anterior end (Oelofsen and Loock, 1981). Results of actual experiments by Briggs and Kear (1994) to evaluate decay and soft-tissue preservation using modern *Branchiostoma* specimens suggest that fossil imprints interpreted as notochord are probably such, although some other interpretations of structure remain equivocal.

Evidence of another sort concerning *Hox* genes, as reviewed by Holland and Garcia-Fernàndez (1996), more directly refutes Dohrn's suggestion of a degenerative origin for amphioxus. *Hox* genes are a related cluster of homeobox-containing regulatory genes known in insects and vertebrates to act in controlling the specification of segment identity and linear organization of body plan. In *Branchiostoma* they occur in a single archetypal gene cluster, but duplications or multiple *Hox* gene clusters (up to four) are present in the genomes of all vertebrates. Whole duplicate sets of these genes seem unlikely to have become lost, especially when they have been preserved so faithfully in diverse vertebrates. This argument is strengthened by the finding that another quite unrelated gene, the alkali myosin light-chain gene involved in muscle differentiation (L. Z. Holland et al. 1995), occurs as only a single gene in amphioxus, but as multiple isoforms in vertebrates.

The rostral extension of the notochord in modern amphioxids, as well as its contractile nature, correlate with the rapid burrowing behavior of the adult in coarse sand. The notochord also extends to the most caudal end of the animal but not usually beyond the limit of the myotomes. Amphioxids seem to burrow with equal facility from either end of the body. Natural selection obviously does influence the degree of this extension since even among the few living amphioxids, two unusual forms occur. In one of these, *Branchiostoma (Dolichorhyncus) indicus* (Willey 1901), the notochord is prolonged anteriorly into a severely elongated cephalic fin. In another, *Epigonichthys lucayanum* (Andrews 1893), the notochord and caudal fin extend far behind the posterior limit of the myotomes. It is, therefore, difficult to know whether the usual rostral extension is primordial based on a very conserved life history or is a recent specialization.

Willey (1894) has claimed that rostral notochord development is histologically and temporally distinguishable from that of the remaining notochord, which indicates that it might be a later evolutionary innovation. Modern technology offers a different suggestion. In amphioxus, *Brachyury* is a regulatory gene associated with the differentiation of notochord cells. Its mRNA transcripts are distributed throughout the length of the early notochord with no gross qualitative or temporal differences of expression between rostral and other notochord (P. W. H. Holland et al. 1995). One might have expected obvious differences if the rostral extension were indeed a recent innovation.

The results of similar experiments with actual tissue expressions of certain anterior-regulating *Hox* genes in amphioxus, as discussed by Holland and Holland (1996), have indicated that the primitive cerebral vesicle of amphioxus may be equivalent to the hindbrain part of the vertebrate brain. These are both examples of how molecular probes prepared by the techniques of recombinant DNA technology can now be applied embryologically to investigating evolutionary questions about amphioxus.

Finally, the phylogenetic relationship of *Branchiostoma* to members of the Craniata (= vertebrates plus hagfishes and lampreys) has been explored by nucleotide sequence analysis of 18S rRNA genes (Turbeville et al. 1994; Wada and Satoh 1994). Evolutionary distances calculated on the basis of molecular site comparisons confirm the deduced kinship supported by anatomical embryological details. Unfortunately, a sister group relationship only indicates with varying degrees of probability that the two share a common ancestry. Close affinity does not necessarily establish a cephalochordate as a direct craniate ancestor.

Many scientists continue the hundred-year fascination with amphioxus as a primitive chordate from which we may yet learn much about evolutionary processes that shaped our own origin as vertebrates. Nonetheless, there still remains strong resistance to the idea that vertebrates are derived from cephalochordates. Yet its value is such that one textbook (Neal and Rand 1936) has paraphrased what Voltaire said about God, remarking that if amphioxus had not been discovered it would have to have been invented.

Acknowledgments

This contribution is dedicated to the memory of my friend Dr. Shan-chin Wu (1921–1988). Preparation of this chapter was supported by a grant from NSERC Canada.

Literature Cited

Andrews, E. A. 1893. An undescribed acraniate: *Asymmetron lucayanum. Johns Hopkins Univ. Stud. Biol. Lab.* 5: 213–247.

Ayers, H. 1890. Concerning vertebrate cephalogenesis. *J. Morphol.* 4: 221–245.

Berrill, N. J. 1987. Early chordate evolution. Part 2. Amphioxus and ascidians. To settle or not to settle. *Int. J. Invert. Reprod. Dev.* 11: 15–28.

Bone, Q. 1958a. Observations upon the living larva of amphioxus. *Pubbl. Staz. Zool. Napoli* 30: 458–471.

Bone, Q. 1958b. The asymmetry of the larval amphioxus. *Proc. Zool. Soc. London* 130: 289–293.

Briggs, E. G., D. H. Erwin and F. J. Collier. 1994. *The Fossils of the Burgess Shale.* The Smithsonian Institution Press, Washington, D.C.

Briggs, D. E. G. and A. J. Kear. 1994. Decay of *Branchiostoma*: Implications for soft-tissue preservation in conodonts and other primitive chordates. *Lethaia* 26: 275–287.

Cerfontaine, P. 1906. Recherches sur le développement de l'*Amphioxus. Arch. Biol.* (Liège) 22: 229–418.

Chin, T. G. 1941. Studies on the biology of Amoy amphioxus *Branchiostoma belcheri* Gray. *Philippine J. Sci.* 75: 369–424.

Colombera, D. 1982. New developments in vertebrate cytotaxonomy 6. Cytotaxonomy and evolution of lower chordates. *Genetica* 58: 97–102.

Conklin, E. G. 1932. The embryology of amphioxus. *J. Morphol.* 54: 69–151.

Conklin, E. G. 1933. The development of isolated and partially separated blastomeres of amphioxus. *J. Morphol.* 64: 303–375.

Courtney, W. A. M. 1975. The temperature relationships and age-structure of North Sea and Mediterranean populations of *Branchiostoma lanceolatum. Symp. Zool. Soc. Lond.* 36: 213–223.

Davidson, E. H. 1990. How embryos work: A comparative view of diverse models of cell fate specification. *Development* 108: 365–389.

Drach, P. 1948. Embranchement des céphalocordés. In *Traité de Zoologie. Anatomie, Systématique, Biologie. Vol. 11 (Échinodermes, Stomocordés, Procordés)*, P.-P. Grassé (ed.). Masson et Cie, Paris, pp. 931–1037.

Eakin, R. M. and J. A. Westfall. 1962a. Fine structure of the notochord of amphioxus. *J. Cell Biol.* 12: 646–651.

Eakin, R. M. and J. A. Westfall. 1962b. Fine structure of photoreceptors in amphioxus. *J. Ultrastruct. Res.* 6: 531–539.

Flood, P. R. 1966. A peculiar mode of muscle innervation in amphioxus. *J. Comp. Neurol.* 126: 181–218.

Flood, P. R. 1970. The connection between spinal cord and notochord in amphioxus (*Branchiostoma lanceolatum*). *Z. Zellforsch.* 103: 115–128.

Flood, P. R., D. M. Guthrie and J. R. Banks. 1969. Paramyosin muscle in the notochord of amphioxus. *Nature* 222: 87–88.

Flood, P. R., J. G. Braun and A. R. de Leon. 1976. On the annual production of amphioxus larvae (*Branchiostoma senegalense* Webb) off Cap Blanc, Northwest Africa. *Sarsia* 61: 63–70.

Franz, V. 1927. Morphologie der Akranier. *Ergebnisse der Anatomie und Entwicklungsgeschichte* 27: 464–692.

Fritzsch, B. and R. G. Northcutt. 1993. Cranial and spinal nerve organization in amphioxus and lampreys: Evidence for an ancestral craniate pattern. *Acta Anat.* 148: 96–109.

Gans, C. and R. Northcutt. 1983. Neural crest and the origin of vertebrates: A new head. *Science* 220: 268–274.

Guthrie, D. M. and J. R. Banks. 1970. Observations on the function and physiological properties of a fast paramyosin muscle—the notochord of amphioxus (*Branchiostoma lanceolatum*). *J. Exp. Biol.* 52: 125–138.

Hatschek, B. 1893. *The Amphioxus and its Development*, James Tuckey (transl. and ed.). Swan Sonnenschein & Co., London.

Hesse, R. 1898. Untersuchungen über die Organe der Lichtemfindung bei neideren Thieren. IV. Die Sehorgane des Amphioxus. *Z. wiss. Zool.* 63: 456–464.

Hirakow, R. and N. Kajita. 1990. An electron microscopic study of the development of amphioxus, *Branchiostoma belcheri tsingtauense*: Cleavage. *J. Morphol.* 203: 331–344.

Hirakow, R. and N. Kajita. 1991. Electron microscopic study of the development of amphioxus, *Branchiostoma belcheri tsingtauense*: The gastrula. *J. Morphol.* 207: 37–52.

Hirakow, R. and N. Kajita. 1994. Electron microscopic study of the development of amphioxus, *Branchiostoma belcheri tsingtauense*: the neurula and larva. *Acta Anat. Nippon* [Kaibogaku Zasshi] 69: 1–13.

Holland, L. Z. and N. D. Holland. 1992. Early development of the lancelet (=Amphioxus) *Branchiostoma floridae* from sperm entry through pronuclear fusion: Presence of vegetal pole plasm and lack of conspicuous ooplasmic segregation. *Biol. Bull.* 182: 77–96.

Holland, L. Z. and N. D. Holland. 1996. Expression of *AmphiHox-1* and *AmphiPax-1* in amphioxus embryos treated with retinoic acid: Insights into evolution and patterning of the chordate nerve cord and pharynx. *Development* 122: 1829–1838.

Holland, L. Z., D. A. Pace, M. L. Blink, M. Kene and N. D. Holland. 1995. Sequence and expression of amphioxus alkali myosin light chain (*AmphiMLC-alk*) throughout development: Implications for vertebrate myogenesis. *Dev. Biol.* 171: 665–676.

Holland, N. D. and L. Z. Holland. 1989a. The fine structure of the testis of a lancelet (= Amphioxus), *Branchiostoma floridae* (Phylum Chordata: Subphylum Cephalochordata = Acrania). *Acta Zool.* (Stockholm) 70: 211–219.

Holland, N. D. and L. Z. Holland. 1989b. Fine structural study of the cortical reaction and formation of the egg coats in a lancelet (= Amphioxus), *Branchiostoma floridae* (Phylum Chordata: Subphylum Cephalochordata = Acrania). *Biol. Bull.* 176: 111–122.

Holland, N. D. and L. Z. Holland. 1991. The fine structure of the growth stage oocytes of a lancelet (= Amphioxus), *Branchiostoma lanceolatum. Int. J. Invert. Reprod. Dev.* 19: 107–122.

Holland, N. D. and L. Z. Holland. 1993. Embryos and larvae of invertebrate deuterostomes. In *Essential Developmental Biology. A Practical Approach*, C. D. Stern and P. W. H. Holland (eds.). IRL Press, Oxford, pp. 21–32.

Holland, N. D., L. Z. Holland and Z. Kozmik. 1995. An amphioxus *Pax* gene, *AmphiPax-1*, expressed in embryonic endoderm, but not in mesoderm: Implications for the evolution of class I paired box genes. *Mol. Mar. Biol. Biotechnol.* 4: 206–214.

Holland, P. W. H. and J. Garcia-Fernández. 1996. *Hox* genes and chordate evolution. *Dev. Biol.* 173: 382–396.

Holland, P. W. H., B. Koschorz, L. Z. Holland and B. G. Herrmann. 1995. Conservation of *Brachyury* (*T*) genes in amphioxus and vertebrates: Developmental and evolutionary implications. *Development* 121: 4283–4291.

Jefferies, R. P. S. 1986. *The Ancestry of the Vertebrates*. British Museum, London.

Kellicot, W. E. 1913. *Outlines of Chordate Development*. Henry Holt and Company, New York.

Lacalli, T. C. 1996. Frontal eye circuitry, rostral sensory pathways, and brain organization in amphioxus larvae: Evidence from 3D reconstructions. *Phil. Trans. R. Soc. Lond. B.* 351: 243–263.

Lacalli, T. C., N. D. Holland and J. E. West. 1994. Landmarks in the anterior central nervous system of amphioxus larvae. *Phil. Trans. R. Soc. London* B 344: 165–185.

Lankester, E. R. and A. Willey. 1890. Development of the atrial chamber of amphioxus. *Quart. J. Microsc. Sci.* 31: 445–466.

Light, S. F. 1923. Amphioxus fisheries near the University of Amoy, China. *Science* 58: 57–60.

Medawar, P. B. 1951. Asymmetry of larval amphioxus. *Nature* 167: 852–853.

Morgan, T. H. 1896. The number of cells in larvae from isolated blastomeres of amphioxus. *Wilhelm Roux Arch. EntwMech. Org.* 3: 269–294.

Neal, H. V. and H. W. Rand. 1936. *Comparative Anatomy*. P. Blakiston's Son & Co., Philadelphia.

Nelsen, O. E. l953. *Comparative Embryology of the Vertebrates*. McGraw-Hill, New York.

Nelson, G. E. 1968. Amphioxus in old Tampa Bay, Florida. *Quart. J. Florida Acad. Sci.* 31: 93–100.

Nilsson, D.-E. and S. Pelger. 1994. A pessimistic estimate of the time required for an eye to evolve. *Proc. R. Soc. Lond. B.* 256: 53–58.

Nozaki, M. and A. Gorbman. 1992. The question of functional homology of Hatschek's pit of amphioxus (*Branchiostoma belcheri*) and the vertebrate adenohypophysis. *Zool. Sci.* 9: 387–395.

Oelofsen, B. W. and J. C. Loock. 1981. A fossil cephalochordate from the Early Permian Whitehall formation in South Africa. *S. Afr. J. Sci.* 77: 178–180.

Olsson, R. 1961. The skin of amphioxus. *Z. Zellforsch.* 54: 90–104.

Olsson, R. 1983. Club-shaped gland and endostyle in larval *Branchiostoma lanceolatum* (Cephalochordata). *Zoomorphol.* 103: 1–13.

Ortolani, G. 1954. Risultati definitivi sulla distribuzione dei territori presuntivi degli organi del germe die Ascidie allo stadio VIII, determinati con le marche al carbone. *Pubbl. Staz. Zool. Napoli* 25: 161–187.

Parker, G. H. 1908. The sensory reactions of amphioxus. *Proc. Amer. Acad. Arts Sci.* 43: 415–455.

Ruppert, E. E. 1994. Evolutionary origin of the vertebrate nephron. *Amer. Zool.* 34: 542–553.

Ruppert, E. E. and R. D. Barnes. 1994. *Invertebrate Zoology*. 6th ed. Saunders College Publishing, Fort Worth, TX.

Shu, D.-G., S. Conway Morris and Z.-L. Zhang. 1996. A *Pikaia*-like chordate from the Lower Cambrian of China. *Nature* 384: 157–158.

Stokes, M. D. and N. D. Holland. 1995a. Ciliary hovering in larval lancelets (= Amphioxus). *Biol. Bull.* 188: 231–233.

Stokes, M. D. and N. D. Holland. 1995b. Embryos and larvae of a lancelet, *Branchiostoma floridae*, from hatching to metamorphosis: Growth in the laboratory and external morphology. *Acta Zool.* (Stockholm) 76: 105–120.

Swalla, B. J. 1993. Mechanisms of gastrulation and tail formation in ascidians. *Microsc. Res. Tech.* 26: 274–284.

Tomarev, S. I., P. Callaerts, L. Kos, R. Zinovieva, G. Halder, W. Gehring and J. Piatigorsky. 1997. Squid *Pax-6* and eye development. *Proc. Natl. Acad. Sci. U.S.A.* 94: 2421–2426.

Tung, T. C., S. C. Wu and Y. Y. F. Tung. 1958. The development of isolated blastomeres of amphioxus. *Scientia Sinica* 7: 1280–1320.

Tung, T. C., S. C. Wu and Y. Y. F. Tung. 1960a. The developmental potencies of the blastomere layers in Amphioxus egg at the 32-cell stage. *Scientia Sinica* 9: 119–141.

Tung, T. C., S. C. Wu and Y. Y. F. Tung. 1960b. Rotation of the animal blastomere in amphioxus egg at the 8-cell stage. *Science Record* 4: 389–394.

Tung, T. C., S. C. Wu and Y. Y. F. Tung. 1962a. The presumptive areas of the egg of amphioxus. *Scientia Sinica* 11: 639–644.

Tung, T. C., S. C. Wu and Y. Y. F. Tung. 1962b. Experimental studies on the neural induction in amphioxus. *Scientia Sinica* 11: 805–820.

Tung, T. C., S. C. Wu and Y. Y. F. Tung. 1965. Differentiation of the prospective ectodermal and entodermal cells after transplantation to new surroundings in amphioxus. *Scientia Sinica* 14: 1785–1794.

Turbeville, J. M., J. R. Schulz and R. A. Raff. 1994. Deuterostome phylogeny and the sister group of the chordates: Evidence from molecules and morphology. *Mol. Biol. Evol.* 11: 648–655.

Wada, H. and N. Satoh. 1994. Details of the evolutionary history from invertebrates to vertebrates, as deduced from the sequences of 18S rDNA. *Proc. Natl. Acad. Sci. U.S.A.* 91: 1801–1804.

Webb, J. E. 1969. On the feeding behavior of the larva of *Branchiostoma lanceolatum*. *Mar. Biol.* 3: 58–72.

Webb, J. E. and M. B. Hill. 1958. The ecology of Lagos Lagoon. IV. On the reactions of *Branchiostoma nigeriense* to its environment. *Phil. Trans. Roy. Soc. London B* 241: 355–391.

Weston, J. A. 1991. Sequential segregation and fate of developmentally restricted intermediate cell populations in the neural crest lineage. *Curr. Topics Dev. Biol.* 25: 133–153.

Wickstead, J. H. 1964. On the status of the "amphioxides" larva. *J. Linn. Soc. (Zool.)* 45: 201–207.

Wickstead, J. H. 1975. Chordata: Acrania (Cephalochordata). In *Reproduction of Marine Invertebrates*, Vol. 2. A. C. Giese and J. S. Pearse (eds.). Academic Press, New York, pp. 283–319.

Wijhe, J. W. van. 1913. On the metamorphosis of *Amphioxus lanceolatus*. *Proc. Sect. Sci. Kon. Ned. Akad. Wetensch.* 16: 574–583.

Williams, N. A. and P. W. H. Holland. 1996. Old head on young shoulders. *Nature* 383: 490.

Wilson, E. B. 1892. On multiple and partial development in amphioxus. *Anat. Anz.* 7: 732–740.

Wilson, E. B. 1893. Amphioxus and the mosaic theory of development. *J. Morphol.* 8: 579–638.

Willey, A. 1894. *Amphioxus and the Ancestry of the Vertebrates.* Macmillan, New York.

Willey, A. 1901. *Dolichorhyncus indicus*, n. g., n. sp. A new acraniate. *Quart. J. Micr. Sci.* 2 ser. 44: 269–271.

Wu, X. H., S. C. Zhang, Y. Y. Wang, B. L. Zhang, Y. M. Qu and X. J. Jiang. 1994. Laboratory observation on spawning, fecundity and larval development of amphioxus (*Branchiostoma belcheri tsingtauense*). Chinese J. Ocean. Limn. 12: 289–294.

Young, J. Z. 1981. *The Life of Vertebrates*, 3rd ed. Clarendon Press, Oxford.

CHAPTER 19

Fishes

James A. Langeland and Charles B. Kimmel

THE TERM *FISH* REFERS TO THREE EXISTING VERTEBRATE classes: primitive jawless fish (Agnatha), the cartilaginous fish (Chondrichthyes), and the bony fish (Osteichthyes) (Figure 19.1). These classes comprise over 20,000 described species of fish, more than all other vertebrates combined.

Despite their broad radiation, there are several basic embryological characteristics that unite the fish classes as a group, as well as with all other vertebrates. The fundamental similarity of fish and other vertebrate embryos has been appreciated since the early nineteenth century when von Baer noted that all vertebrate embryos pass through a common developmental stage, with the distinguishing characteristics of different vertebrate classes becoming apparent only at later stages (Figure 19.2). This common, or phylotypic stage has since been termed the pharyngula (Ballard 1981) due to the prominent series of arches surrounding the pharynx. In addition to these pharyngeal arches, the embryos of fish and other vertebrates also possess a distinct brain and sensory organs, a segmental series of somites extending along much of the length of the body, as well as the general chordate characters of a notochord and a dorsal nerve tube (Figure 19.3). Given these shared characteristics and their basal position in the vertebrate lineage, fish are particularly well suited as model systems for the study of many aspects of vertebrate embryogenesis. Additionally, most fish embryos develop externally. This allows for direct examinations of living embryos, revealing cellular aspects of development at a level of detail that is not possible in many higher vertebrates.

This chapter will focus on the development of teleost fish, the relatively advanced ray-finned bony fish in which most of the embryological investigations in fish have been undertaken. Our primary reference species is the zebrafish [*Danio (Brachydanio) rerio*], with occasional use of other model teleosts. It is important to bear in mind that although all fish exhibit basic morphological similarities, only a handful of species have been investigated in detail. Exceptional modes of development are bound to occur in some particular group, making generalized descriptions of any developmental process difficult.

Overview of Fish Embryogenesis

Figure 19.4 provides an overview of the first day of zebrafish embryogenesis. This early developmental profile is broadly representative of other teleosts and can be subdivided into discrete stages. Following fertilization, the one celled zygote (Figure 19.4a) enters a period of rapid, synchronous divisions. The early cleavages (Figures 19.4b,c) result in stereotypical arrays of blastomeres that are perched on top of a nondividing region containing the yolk. Divisions continue in a less orderly manner to produce the blastula, a ball of largely undifferentiated cells (Figure 19.4d). Following the midblastula transition, cells of the blastoderm become motile and spread thinly over the yolk cell during the process of epiboly (Figures 19.4e,f). During the gastrula period, the major cell rearrangements of involution and convergent extension occur to establish distinct germ layers and the primary embryonic axis (Figures 19.4g,h). The segmentation period is defined by the sequential appearance of bilaterally paired somites along the embryonic axis (Figures 19.4i,j,k,l). During the segmentation period there is also a variety of morphogenetic processes that give rise to the central nervous system, the neural crest, and the major sensory organs, thereby firmly establishing the body plan of the pharyngula (Figure 19.4l). Thus, in the short

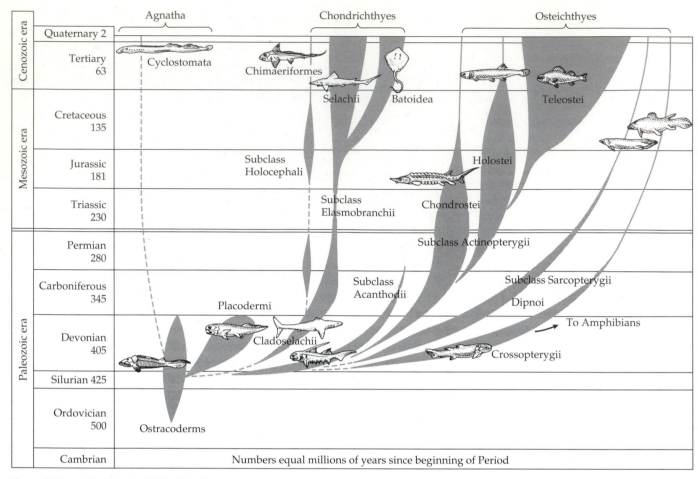

Figure 19.1   A cladogram of fish. This diagram illustrates the evolutionary relationships of the major groups of fish and their relative abundance in the fossil record. Note that teleosts are a subset of the bony fish (Osteichthyes). They arise relatively late and are the largest group of modern fish. (After Villee et al. 1978.)

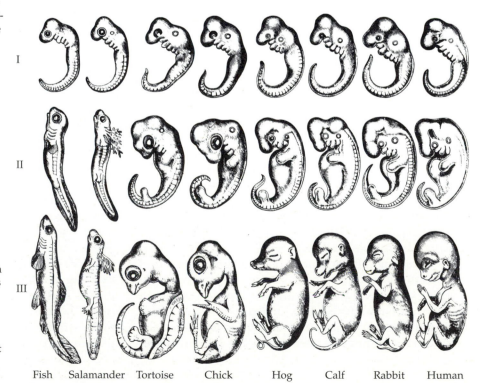

Figure 19.2   A conserved developmental stage for all vertebrates. As described by von Baer, at an early stage all vertebrate embryos are very similar and exhibit the general features of the vertebrate subphylum (I). This has since been termed the phylotypic stage. Differences between the various vertebrate groups become progressively more apparent at later developmental stages (II and III). (From Romanes 1901.)

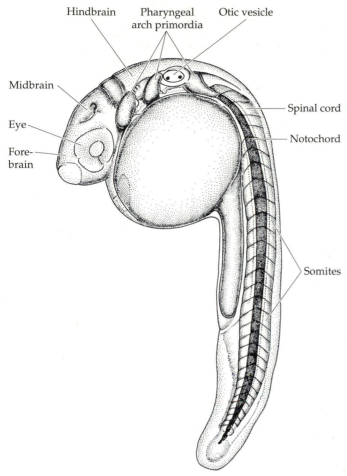

Figure 19.3 The early zebrafish pharyngula. This drawing depicts an approximately 24-hour-old zebrafish and illustrates the prominent features of the vertebrate phylotypic stage. The embryo is supported by the notochord, which extends most of the length of the animal. In the trunk and tail, the notochord is surrounded by the segmented series of somites that will give rise to trunk muscle and vertebrae. The brain is prominent and well sculpted and sensory organs such as the eyes, ears (otic vesicles), and nose (olfactory epithelium) are present. The primordia of the pharyngeal segments, or arches, for which the developmental stage is named, are just beginning to form.

span of 24 hours, the zebrafish zygote is transformed into a clearly recognizable vertebrate possessing all of the characters of the phylotypic stage. What makes this orderly progression all the more striking in teleosts is the optical clarity of the embryos, which allows the embryogenesis of internal structures to be observed in real time with nothing more than a simple dissecting microscope.

Producing the Zygote

Reproductive Strategies

Perhaps nowhere is the diversity of fish more evident than in their reproductive strategies. While the vast majority of teleosts are oviparous, with eggs fertilized externally after deposition by females, both ovoviviparity (internally hatch-

ing eggs) and full viviparity (true live birth) occur in some species (Bond 1979). Eggs may be produced seasonally or year round, and may be either free floating (pelagic) or sedimenting (demersal). The degree of parental care ranges from elaborate to nonexistent, and clutch size from tens to tens of thousands of eggs. In general among oviparous teleosts, freshwater species tend to produce fewer, larger eggs than their marine counterparts. For practical reasons, oviparous freshwater species, with demersal eggs, such as the zebrafish, rainbow trout (*Salmo giardneri*), killifish (*Fundulus heteroclitus*), rosy barb (*Barbus conchonius*), and medaka (*Oryzias latipes*) have been the most accessible experimentally and therefore the most well-characterized developmentally.

Gametogenesis and Fertilization

Modes of oogenesis and spermatogenesis are highly dependent upon general reproductive strategies. In species capable of several annual broods such as zebrafish, repeated waves of oogenesis take place. Oocytes in all stages of growth can be observed in the ovaries, although each progressive batch of eggs develops synchronously (Hisaoka and Firlit 1962). The mature, unfertilized zebrafish egg is approximately 1 millimeter in diameter, although many teleosts (e.g., *Salmo*) have eggs up to several millimeters in diameter. The egg proper is telolecithal, the plasma membrane surrounding a cytoplasm rich with membrane-bound yolk vesicles. An outer or vitelline membrane is the site of osmoregulation for the egg, and the entire structure is supported by the surrounding chorion.

Spermatogenesis in zebrafish males also proceeds in waves. The development of spermatogenic clones proceeds within specialized cysts within the testis and consists of 5 or 6 synchronous spermatogonial generations before final differentiation into spermatids. After being released into the sperm ducts, the mature spermatozoa mix with fluid secretions to produce the milt that is released at spawning (Ewing 1972).

Oviparous eggs are fertilized after deposition. Typically the male sheds mature sperm over the eggs immediately after they are laid, and the spermatozoa swim only a short distance to the eggs. Sperm gain access to the egg surface by passing through a specialized opening above the animal pole of the chorion known as the micropyle (Hart and Donovan 1983); following fertilization, the micropyle becomes plugged, perhaps as an adaptation to prevent polyspermy. Membrane fusion between the fertilizing sperm and the egg occurs on a distinct tuft of microvilli on the egg plasma membrane (Wolenski and Hart 1987).

Making the Blastula

Early Cleavage

Very shortly after fertilization, waves of contractile forces cause the homogeneous mixture of yolk granules and cy-

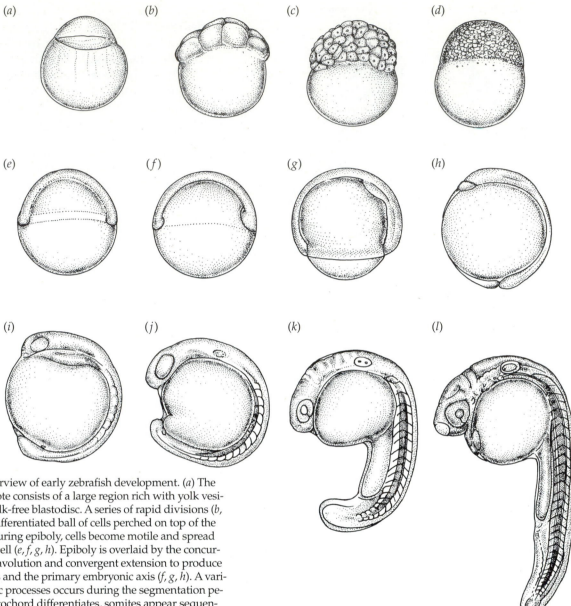

(a) (b) (c) (d)

(e) (f) (g) (h)

(i) (j) (k) (l)

Figure 19.4 An overview of early zebrafish development. (a) The newly fertilized zygote consists of a large region rich with yolk vesicles and a smaller yolk-free blastodisc. A series of rapid divisions (b, c, d) produce an undifferentiated ball of cells perched on top of the nondividing yolk. During epiboly, cells become motile and spread thinly over the yolk cell (e, f, g, h). Epiboly is overlaid by the concurrent movements of involution and convergent extension to produce different germ layers and the primary embryonic axis (f, g, h). A variety of morphogenetic processes occurs during the segmentation period (i, j, k, l). The notochord differentiates, somites appear sequentially along the axis, the tail extends dramatically, and the central nervous system and sensory organs become prominent. By roughly 24 hours, the zebrafish embryo possesses all the characters of the vertebrate phylotypic stage. (Adapted from Kimmel et al. 1995.)

toplasm to separate (Figure 19.5). The nonyolky cytoplasm is squeezed out of the vegetal parts of the egg and comes to occupy a small yolk-free zone called the blastodisc (Lewis and Roosen-Runge 1942, 1943). The presence of the large yolk mass restricts the mitotic apparatus and cleavage furrows to the blastodisc and thus has a pronounced effect on the style of early cleavage. The blastodisc divides and gives rise to the embryo proper, while the remainder of the zygote does not divide and becomes the yolk sac. This style of cleavage is termed discoidal. Furthermore, the early division furrows do not undercut the blastodisc and the early blastomeres remain in cytoplasmic continuity with the yolk and each other. Zebrafish

cleavage is thus also meroblastic or partial, as opposed to holoblastic.

The meroblastic nature of early zebrafish cleavage can be visualized by injecting dye molecules into early blastomeres. Relatively small molecules such as Rhodamine-Dextran (MW 17 KDa) pass freely from one blastomere to another as well as to and from the yolk sac (Kimmel and Law 1985). Early blastomeres are known to be coupled via gap junctions (Dasgupta and Singh 1982), but this dye is too large to pass them and must therefore travel through direct cytoplasmic bridges. These cytoplasmic bridges are maintained for several cleavage cycles (see below). There is a size limit to these links however, as larger dyes (2000KDa Texas Red-Dextran; Strehlow and Gilbert 1993) cannot pass between even the earliest blastomeres or into the yolk sac, and therefore remain confined to single blas-

(a)   (b)

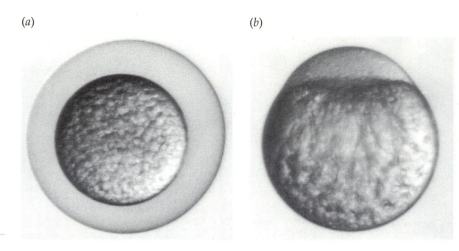

Figure 19.5   Segregation of the blastodisc. (*a*) The newly fertilized zygote (shown within the chorion) is seemingly homogeneous. (*b*) Within minutes of fertilization, waves of contractile forces cause yolk-free cytoplasm to segregate to the animal pole and form the blastodisc (chorion removed and shown at higher magnification). (From Kimmel et al. 1995.)

tomeres and their progeny. The use of these various dyes for tracing cell lineage and constructing embryonic fate maps is discussed in a later section.

The cleavage stages in the zebrafish are marked by a rapid increase in cell number with no significant increase in the size of the organism (Hisaoka and Battle 1958; Kimmel et al. 1995). The first six divisions in the zebrafish are synchronous, take place at regular 15-minute intervals, and result in fairly stereotyped arrays of blastomeres (Figure 19.6). The first cleavage furrow is vertical and divides the blastodisc into two equal blastomeres (Figure 19.6*b*). The second cleavage occurs at a right angle to the first to produce a 2×2 array of blastomeres (Figure 19.6*c*). The third cleavage consists of two furrows parallel to the first, producing a 2×4 array of cells when viewed from the animal pole (Figure 19.6*d*). The fourth cleavage also consists of two furrows, parallel to the second furrow. This produces a 4×4 array of cells (Figure 19.6*e*). The four interior blastomeres of the array are now completely undercut by cleavage furrows, and even 17 KD dyes injected into these blastomeres will not pass to other cells. In contrast, the 12 outer cells of this array retain their cytoplasmic links to the yolk and are termed marginal blastomeres. In all subsequent divisions, resulting marginal cells will retain such cytoplasmic links, while nonmarginal cells will not. The fifth cleavage consists of four furrows, again parallel to the first. The blastomeres lie in a single plane that is partially curved around the yolk, and frequently, but not exclusively, they form a regular 4×8 array (Figure 19.6*f*).

The sixth cleavage is the first that is horizontal rather than vertical. The furrow passes through the entire blastoderm producing two tiers of cells. Note that just as the fourth cleavage segregated marginal from nonmarginal blastomeres, this cleavage results in blastomeres that are now completely covered by other blastomeres. Blastomeres on the upper or superficial layer together with the marginal blastomeres of the lower layer are all exposed and collectively cover or envelope the nonmarginal cells of the lower layer. These two populations of cells are termed the **enveloping layer** (**EVL**) and the deep cells. Initially,

cells of the EVL divide to produce two types of daughter cells, both deep cells and more EVL cells. Eventually, however, EVL cell divisions occur only in the plane of the EVL, and thus daughters will contribute only to the EVL and not the deep-cell population.

Beyond the sixth cleavage, it becomes progressively more difficult to discern any stereotypical array of blastomeres. Cleavage planes are no longer regularly patterned and the cell cycles, rather than being absolutely synchronous, become metasynchronous, with waves of mitosis passing through the blastoderm. The steady increase in cell number combined with the tight association of marginal cells with the yolk sac causes the blastoderm to become a roundish ball perched on top of the yolk sac (Figures 19.7*a* and 7*b*). This constitutes the early blastula. Note that as cell number increases, the deep cells come to greatly outnumber those of the EVL. Additionally, the marginal blastomeres also begin to lose their lower borders and become more obviously joined with the yolk sac. Here they form a distinctive region of the yolk sac, the **yolk syncytial layer** (**YSL**), a structure unique to teleosts. Importantly, once a cell becomes part of the YSL, none of its progeny is known to contribute to the overlying blastoderm, or later to the embryo proper.

All teleost blastulae look superficially the same, with the blastoderm consisting of a few thousand cells, covered by the EVL, and nested on top of the YSL (Figure 19.7*e*). However, the eggs of various species vary dramatically in their sizes, reflecting the amount of yolk present. In general, species with large eggs, such as *Salmo* and *Fundulus* generally develop more slowly than those with relatively small yolks, such as zebrafish.

The Midblastula Transition

As in many other metazoans, the zebrafish blastula undergoes a series of cellular changes that collectively are termed the **midblastula transition** (**MBT**). In general, the MBT is characterized by an increase in cell cycle time, a loss of cell cycle synchrony, the initiation of zygotic transcription, and the beginning of cell motility. By the tenth

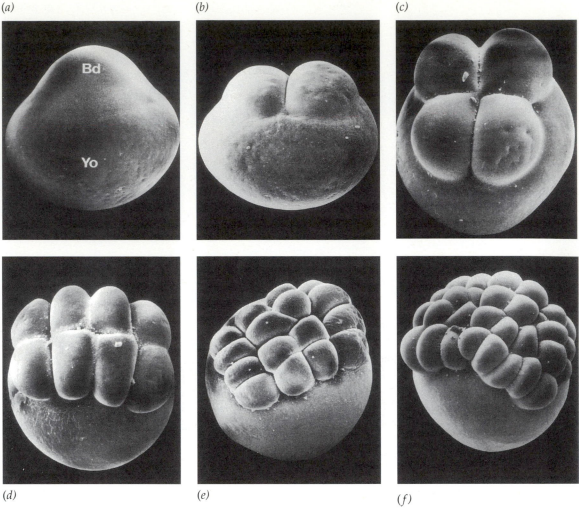

(a)   *(b)*   *(c)*

(d)   *(e)*   *(f)*

Figure 19.6   Stereotyped early-cleavage pattern in the zebrafish. Scanning electron micrographs of early-cleavage stage zebrafish embryos. (*a*) The segregated blastodisc (Bd) is perched on top of the yolk region (Yo). (*b*) The blastodisc divides to form two blastomeres, while the yolk region does not divide. (*c*) The second-division furrow occurs at a right angle to the first, to produce four blastomeres. (*d*) The third cleavage consists of two furrows, parallel to the first, to produce eight blastomeres. (*e*) By the fourth cleavage, a 4×4 array of 16 blastomeres is formed, which consists of two furrows parallel to the second furrow. This is the first cleavage that results in distinct populations of marginal and interior blastomeres. (*f*) The fifth cleavage consists of four furrows parallel to the first furrow, to generate 32 blastomeres in a 4×8 array. Note that each of the first five cleavages are vertical, and the resulting blastomeres all lie in a single plane, partially curved around the yolk region. Each blastomere is also in cytoplasmic contact with the yolk cell. (From Beams and Kessel 1976; courtesy of R. G. Kessel.)

cell cycle in zebrafish (around 1000 cells), cell cycles begin to lose synchrony and are measurably longer than previous cycles. This is the beginning of the MBT in zebrafish. Although the MBT normally begins during the tenth cycle, it is does not appear to be initiated by a strict cell cycle clock. For example, haploid fish embryos commence the MBT one cycle late, while tetraploid embryos begin one cycle early, relative to the normal, diploid state (Kane and Kimmel 1993), just as previously known for frogs.

The transition therefore seems to be a function of nucleo-cytoplasmic ratio. As blastomeres become progressively smaller, the amount of nuclear material relative to cytoplasm increases and when some critical threshold is reached, the MBT is triggered.

The Beginning of Cell Motility: Epiboly

In addition to altered cell cycles, the MBT is characterized by the acquisition of cell motility. In teleost embryos, the first major cell movement to occur is epiboly, which results in the rearrangement of the blastoderm relative to the yolk. At the onset of epiboly, the blastoderm rests on top of the yolk and consists of large number of deep cells overlaid by the EVL, and ringed by the YSL (Figures 19.7*b,e*). In the initial phase of epiboly, the blastoderm flattens onto the yolk cell, generating a fairly smooth sphere. The blastoderm continues to thin and spread over the yolk cell in a process that has been likened to pulling a knitted ski cap over one's head (Figures 19.7*c, d, f*). As the blastoderm flattens around the yolk, inner blastoderm cells intercalate outward among those of more superficial layers (Warga and Kimmel 1990; Helde et al. 1994). This is termed radial intercalation.

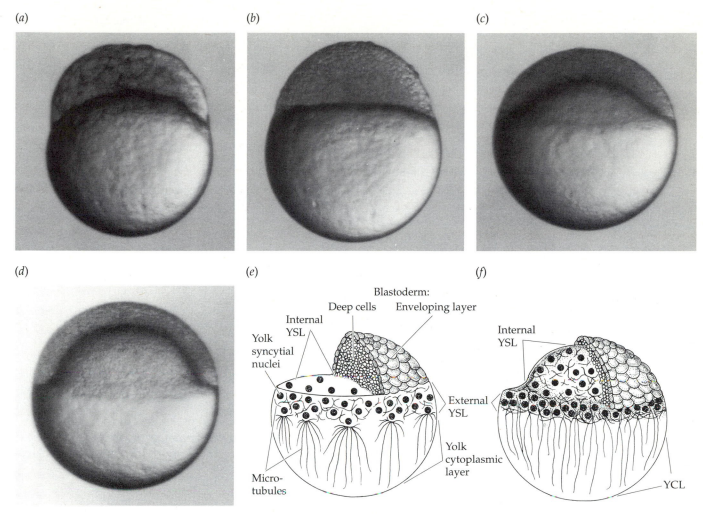

(a) (b) (c)

(d) (e) (f)

Figure 19.7 The zebrafish blastula and its rearrangement during early epiboly. (*a*) 256-cell stage. (*b*) Late blastula, just prior to the onset of epiboly. Continued cell divisions have produced a dense ball of cells still perched on top of the yolk cell. (*c*) 30% epiboly. The yolk cell has domed up toward the animal pole and the blastoderm now covers 30% of the yolk cell. (*d*) 50% epiboly. The blastoderm continues to advance and now covers half of the yolk cell. (*e* and *f*) Schematic illustrations of cellular organization in the late blastula (*e*) and at 30% epiboly (*f*). Prior to the onset of epiboly (*e*), deep cells of the blastoderm are surrounded by the enveloping layer (EVL). The animal surface of the yolk cell is flat and contains nuclei of the yolk syncytial layer (YSL). Microtubule networks extend through the external YSL and also extend vegetally through the yolk cell. At 30% epiboly (*f*), doming of the yolk cell and contraction of the external YSL cause deep cells of the blastoderm to repack by radial intercalation and spread thinly over the yolk. (*a, b, c, d* from Kimmel et al. 1995; *e* and *f* after Solnica-Krezel and Driever 1994.)

The process of epiboly in teleosts is dramatic and fascinating to observe. Moreover, it represents one of the simplest cases of morphogenesis in any vertebrate system, providing unique opportunities to examine the cellular interactions that underlie morphogenesis. Studies in *Fundulus* (Trinkaus 1984, 1992) established that a motive force for epiboly appears to be an autonomously expanding yolk syncytial layer. The enveloping layer is tightly attached to the YSL and is pulled along with it. The deep cells of the blastoderm then move to fill the space between the YSL and EVL during their epibolic expansion. In a striking demonstration of this process, if the attachments between the YSL and the EVL are broken, the blastoderm springs back on top of the yolk, while the YSL continues to expand around the yolk.

More recently in zebrafish, the role of the cytoskeleton in the cell movements of epiboly has been examined (Strahle and Jesuthasan 1993; Solnica-Krezel and Driever 1994). Labeling experiments with antitubulin antibodies reveal two distinct arrays of microtubules in the zebrafish blastula. One array forms a dense network through the YSL, while the other extends from the YSL toward the vegetal pole of the yolk (Figure 19.7*e*). As epiboly proceeds, the microtubule network of the YSL expands, while the array oriented along the animal-vegetal axis shortens (Figure 19.7*f*). Furthermore, treatment of embryos with either UV light (known to damage microtubules), or with chemical agents that cause the depolymerization of microtubules results in the inhibition of epiboly. It appears likely then, that epiboly is dependent upon some type of microtubule-mediated motor activities, perhaps similar to those involved in the motion of the spindle apparatus during mitosis.

The cell movements of epiboly are continuous and eventually result in an expansion of the blastoderm completely around the yolk cell. As epiboly progresses, a concurrent set of cell movements, those of gastrulation, begin to produce distinctive germ layers and define the primary embryonic axes.

Gastrulation

Commitment and the Fate Map

The movements of gastrulation are complex and mark the beginning of the wholesale transformation of the undifferentiated blastula into a highly patterned embryo. Importantly, it is during gastrulation that cells first appear *specified* and then *committed* to a particular fate. The term specification is used to describe a cell's initial embarkment on a developmental pathway leading from an undifferentiated state to a particular developmental fate (see Davidson 1990). Commitment, on the other hand, describes the actual restriction of cell potential such that a cell will autonomously give rise to one particular type and not another (see Ho and Kimmel 1993). Specification and perhaps commitment can be somewhat plastic, however, and may change if conditions in the embryo are perturbed. As revealed by comparing findings from cell lineage analysis and cell transplantation experiments in zebrafish, the process of specification and commitment may occur in discrete stages.

Given the combination of external development and optical clarity, the use of vital dyes to trace cell fate in teleost embryos has proven extremely powerful. Marking single cells at nearly any embryonic stage by microinjection with fluorescent dyes is quite easy to do with fish embryos. If a single dividing cell is so injected, its progeny can be identified for many cell cycles, thus generating a clone of labeled cells. By injecting progenitor cells at various stages of embryogenesis and examining the tissues that its clonal progeny come to occupy, fundamental observations have been made about how the early cells contribute to the body plan (Kimmel and Warga 1987; see Figure 19.8).

First, although the early cleavages produce highly stereotyped arrays of cells, no aspect of the final body plan appears to be specified by these arrays. For example, one might presume that the planes of a 2×4 or 4×8 array of early blastomeres correlate with one of the major body axes such as dorsal-ventral, anterior-posterior, or right-left, as in fact was claimed by one group of workers (Strehlow and Gilbert 1993, Strehlow et al. 1994). However, by labeling any of the blastomeres during early cleavage and observing the clonal progeny of the injected cell after the body plan is established, it is clear that neither the major body axes, nor specific tissue types, correlate with the position of any early blastomere (Figure 19.8a). In particular, the relationship between any of the early cleavage planes and the later mid-sagittal plane of the developing embryo is random (Helde et al. 1994), or very nearly so (Abdelilah et al. 1994).

Indeed, if later blastomeres are labeled, even as late as the midblastula transition, the clones scatter extensively such that one finds descendants of a single blastula cell among different tissues, different organs, and even different embryonic germ layers (Figure 19.8b). Given this cell scattering and lack of clonal restriction, deep cells of the blastula must be largely unspecified as to their eventual fate. Very active scattering of cells actually begins during the late blastula. As the blastoderm thins during epiboly, cells intercalate radially and repack in a manner that appears to preclude any stereotyped relationship between a cell's position in the blastula and the position of its descendants in the gastrula. The cells will then differentiate according to their positions in the gastrula. Careful examinations of cell scattering during epiboly reveal that cells in the central blastoderm mix extensively, while cells near the margin mix very little (Helde et al. 1994). This may be important, because marginal cells form the fish organizer (see below) and may not tolerate the widespread mixing that occurs with the more central cells.

An extreme example of such a lack of positional specification in the blastula occurs in the development of certain African and South American annual fishes. Annual fish are species that maintain permanent populations in bodies of water that are present only seasonally. The eggs of such species can undergo prolonged and reversible developmental arrest to await favorable conditions. Associated with such an early arrest of development, the blastoderm dissociates into a disorderly array of single cells. Development later resumes, with apparently random reaggregation of blastomeres (Carter and Wourms 1991).

Radial intercalations in late blastula are confined to the deep cells and do not affect the enveloping layer (EVL). EVL cells thus remain morphogenetically segregated, within the outer epithelial monolayer of blastoderm cells. This different behavior of the EVL and deep cells during epiboly has distinct consequences in terms of early clonal restriction in cell fate. Unlike the deep cells, when a single EVL cell is marked with dye in the late blastula, it will generate a clone that is entirely confined to a single tissue. The tissue arising from the EVL is the periderm, the outer covering of the embryo.

Fate restriction of clones of labeled deep cells begins a few cell cycles later than the EVL restriction to the periderm. Just as gastrulation commences, when nearly all of the deep cells are in cycle 15, the progeny of single labeled cells are often restricted to single tissues, particularly those contributing to the head or body trunk (Figure 19.8c). Such restricted behavior provides evidence that specification of cell fate is underway in the early gastrula. Additionally, because clonal progeny from single deep cells are not widely scattered among different tissues, it is now possible, for the first time, to construct of a detailed fate map of the embryo—in other words, a map of how the position of a cell in the early gastrula correlates with the tissue it will eventually generate (Kimmel et al. 1990). By understanding this

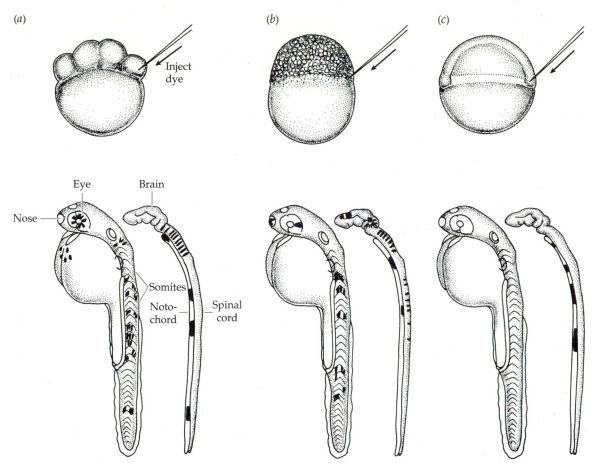

Figure 19.8 Tissue-restricted clones in zebrafish are first seen in the early gastrula. Schematic diagrams illustrating the results of cell-marking experiments at various stages of zebrafish development. Individual cells are injected with a lineage tracing dye at either early-cleavage (*a*), late-blastula (*b*), or early-gastrula stage (*c*). The animals are allowed to develop and are then examined to determine which tissues the labeled progeny of the injected cell occupy. Embryos are shown in two cutaway views, with the central nervous system and notochord on the right, and overlying, more lateral tissues on the left. Labeled tissues are indicated by black patches. Cells labeled at any time during cleavage stages (*a*) through late blastula (*b*) give rise to progeny that are scattered among several different tissues. In addition to the lack of tissue restriction, neither anterior-posterior, dorsal-ventral, nor right-left polarity correlates with any clones labeled at cleavage stages. However, fate restriction of labeled clones is evident in the early gastrula (*c*). Here, the progeny of single labeled cells are restricted to single tissues (example shown is the notochord.) This observation indicates that specification of cell fate is underway and allows for the construction of a detailed fate map of the gastrula. (Adapted from Kimmel and Warga 1987.)

early fate map, it will be easier to follow and interpret the cellular reorganizations that constitute gastrulation.

Figure 19.9 shows a lateral view of the zebrafish gastrula fate map and the tissue type that is predicted for various positions. The map has distinct and reproducible domains along both the animal-vegetal axis, and around the circumference of the gastrula. A dorsal view would reveal that the map is bilaterally symmetric.

Fates are distributed along the animal-vegetal axis of the early gastrula according to the germ layer the descendent cells will occupy after gastrulation. Thus, cells along the extreme margin (at the most vegetal position before any involution occurs) will give rise to classical endodermal fates such as the epithelial linings of the visceral organs such as the pharynx, liver, and intestines. A ring of cells just towards the animal pole above the margin contain mesodermal precursors, including tissues such as notochord, both head and trunk muscle, kidney, heart, and blood. Cells located above this ring, in an "animal cap" located towards the animal pole and including the animal pole will produce ectodermal derivatives such as epidermis, neural crest, sensory organs, brain, and spinal cord. Recently, detailed fate maps have been described for the zebrafish neural plate (Woo and Fraser 1995; Papan and Campos-Ortega 1994) as well as the embryonic shield (Shih and Fraser 1995; Melby et al. 1996).

The zebrafish fate map is topologically similar, in broad outline at least, to those of the other vertebrates classes, and protochordates. Despite the fact that amphibian and ascidian blastomeres can be fate mapped much earlier than their fish counterparts, and that these animals display very different early cleavage patterns, the fate maps for these

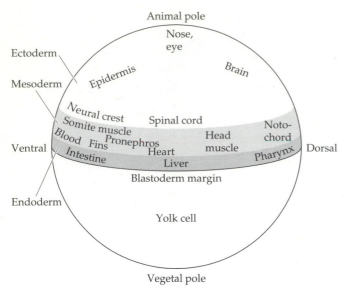

Figure 19.9 Fate map of the zebrafish early gastrula. Lateral view showing the correlation of cell position with the tissue its labeled progeny will generate. Not all organ rudiments are shown, for the sake of clarity. Ectodermal fates (including the central nervous system, epidermis, and sensory organs) map nearest the animal pole. Mesodermal fates (such as notochord and head and trunk muscle) map to a broad ring near the margin. Endodermal fates (such as pharynx, liver, and intestine) map at the margin. Note that all known embryonic fates derive from deep cells of the blastoderm. Enveloping layer (EVL) cells contribute exclusively to the periderm (not shown) and no fates derive from the yolk cell. (From Kimmel et al. 1995.)

embryos are broadly similar, and many topological homologies can be discerned. For example, if one conceptually opens up a frog blastula at the vegetal pole and stretches this opening into a marginal ring, the resulting fate map looks rather like that of the zebrafish at 50% epiboly.

There is presently some controversy about how similar the zebrafish fate map is to those for other fishes, notably other teleosts with much larger eggs such as the trout and suckers (Ballard 1973). Fate maps for these species were made long ago, and techniques available then, such as deposition of fine carbon particles among blastoderm cells, were less refined, and perhaps less reliable than the newer intracellular dye-injection methods. In any case, fate maps with different topologies than the zebrafish one were obtained. Thus, a fate map for the late blastula of the salmon shows the presumptive germ layers already present in a layered arrangement of deep cells: endoderm the deepest, ectoderm on the outside, and mesoderm largely in between. This is quite different than in the animal-to-vegetal arrangement described above for zebrafish. At present we do not know whether the newer methods of cell lineage tracing would confirm this very curious arrangement of the germ-layer precursors. If so, this would indicate a very distinctive kind of cellular morphogenesis during gastrulation for these large yolky eggs (Ballard 1982).

Even though cell fate is already specified in the early gastrula and labeled progenitor cells will develop tissue-

restricted clones, it does not necessarily follow that cells are committed at this stage. Indeed, the results of single-cell transplantation experiments from one blastoderm location to another indicate that commitment occurs later than clonal restriction. When transplants are made at early gastrula stages, the relocated cells develop entirely in accordance with their new surroundings. This shows clearly that they have not yet been formally committed to their normal fates (Ho and Kimmel 1993). If however, the donor cells come from a late gastrula stage, a different result is observed. Such cells migrate away from their new "foreign" neighborhood and take up a blastoderm position seemingly more in accordance with that from which they were obtained. Here they differentiate appropriately for their environment. These findings suggest that, although specified, early gastrula cells remain pluripotent and that commitment becomes firmly established only after the midgastrula period of development.

The Mechanics of Early Gastrulation

Gastrulation begins in a formal sense with the establishment of the germ ring. Typically, this occurs at some point after epiboly begins to stretch the blastoderm around the yolk cell. In zebrafish, the germ ring is established when roughly 50% of the yolk cell is covered (50% epiboly). Gastrulation begins much earlier in the process of epiboly in species with larger eggs (such as *Salmo* and *Fundulus*; Trinkaus 1984). The germ ring is a distinctive thickening that circumscribes the entire margin and consists of two cell layers (Figure 19.10). The more superficial layer is termed the epiblast, and the inner layer is termed the hypoblast. Interestingly, teleosts appear to employ a variety of mechanisms to generate these cell layers. In the zebrafish (Warga and Kimmel 1990; Schmitz and Campos-Ortega 1994), and the Rosy Barb (*Barbus conchonius*; Wood and Timmermans 1988), the formation of the hypoblast has been described to occur by involution. During involution (as shown in Figure 19.10), deep cells of the blastoderm move vegetally toward the margin, then tuck under the blastoderm to occupy a position below their starting point, and proceed to move back toward the animal pole. This movement is continuous such that cells physically pass around the margin before entering the hypoblast. In contrast, in *Fundulus*, the hypoblast appears to be established by ingression (Trinkaus 1996). Rather than passing around the margin, ingressing cells simply delaminate from their original superficial position near the margin, to a deeper position. Although mechanistically distinct, involution and ingression are similar in that the hypoblast is generated by the inward movement of cells that originally were more superficial. However, in trout, it has been reported that neither involution nor ingression occurs (Ballard 1966); rather, the hypoblast may be formed by an outward movement of the most central deep cells. The significance of these differing mechanisms is unclear. It remains to be seen which is more common in teleosts and

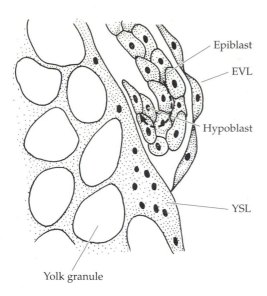

Epiblast

EVL

Hypoblast

YSL

Yolk granule

Figure 19.10 Formation of the germ ring in zebrafish. Schematic diagram of a cross section near the margin of the zebrafish gastrula just after formation of the germ ring. Cells of the blastoderm involute near the margin to create a deeper cell layer. This new cell layer is termed the hypoblast, and the layer superficial to it is termed the epiblast. Once formed, the hypoblast moves away from the margin toward the animal pole. Both germ layers remain covered by the EVL. Note that in some species of teleosts, ingression, and not involution may give rise to the germ ring. EVL, enveloping layer; YSL, yolk syncytial layer.

whether some species employ more than one of them during the initial stages of gastrulation.

Convergence and Extension: Making the Body Axis

Involution begins on the dorsal side of zebrafish embryos (Schmitz and Campos-Ortega 1994), but occurs all around the margin such that the germ ring is initially fairly uniform. However, the movements of involution (or ingression) are quickly overlaid by movement of deep cells of both the hypoblast and epiblast to one side of the embryo. This movement is termed convergence and results in the formation of a localized thickening of the germ ring, the embryonic shield. The appearance of the embryonic shield is important in that it unambiguously identifies the dorsal side of the embryo. Indeed, if the shield is removed and ectopically grafted onto a host embryo, a secondary axis is induced (Oppenhiemer 1936; Ho 1992; Shih and Fraser 1996). The shield is therefore formally homologous to the dorsal blastopore lip of amphibia and is considered the **organizer** of the fish embryo.

The mechanisms of convergence have been best described in *Fundulus* (Trinkaus and Erickson 1983; Trinkaus et al. 1992). In contrast to the seemingly passive movement of EVL cells during epiboly (where the YSL pulls them along), during convergence, clusters of deep cells appear to actively migrate dorsally by extending cytoplasmic processes (Figure 19.11). Although apparently not playing a direct role in cell movement during convergence, the YSL has been implicated in determining the orientation of the dorso-ventral axis, and thus the location of the embry-

onic shield. For example, when the blastoderms of rainbow trout (*Salmo giardneri*) gastrulae were removed from the yolk cell and replaced by very young blastoderms, the new blastoderms adopted the dorso-ventral axis of their host (Long 1983).

As more cells converge upon the embryonic shield, mediolateral intercalations cause the shield to narrow and elongate toward the animal pole. This process is termed extension and morphologically defines anterior-posterior axis of the embryo. Although difficult to conceptualize, the cell movements of epiboly, involution, convergence, and extension all occur simultaneously (Figure 19.12). Thus, as the blastoderm is expanding around the yolk, it is also involuting to form the germ layers. Both the epiblast and the hypoblast converge to the dorsal side (embryonic shield) as well as extend along the anterior-posterior axis.

Convergence and extension can be fairly accurately visualized by examining gene expression patterns at successive stages of gastrulation. For example, transcripts of the gene *no tail* (*ntl*), the homolog of the mouse *Brachyury* or *T* gene, while initially detected uniformly throughout the

Figure 19.11 Deep cells of *Fundulus* during convergence. A scanning electron micrograph of a landscape of deep cells clinging to the internal yolk syncytial layer from which the blastoderm has been removed. The area is just beneath the embryonic shield. Note the cytoplasmic processes (filopodia) extended by the cells during their migration. Scale bar = 100 μm. (From Trinkaus and Erickson, 1983.)

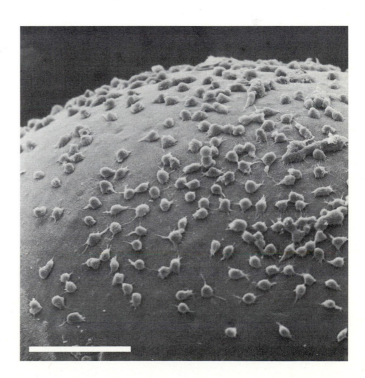

germ ring, accumulate in the embryonic shield hypoblast (Figure 19.12*b*). At progressively later stages (Figure 19.12*c*), *ntl* expression becomes narrow and elongated along the dorsal midline of the hypoblast (Shulte-Merker et al. 1992). This expression represents the primordium of the notochord and is termed the chordamesoderm. Similarly, the gene *snail1* (*sna1*) is also detected uniformly throughout the germ ring. However, during convergence and extension, expression is absent from the central portion of the shield, but persists in two parallel domains adjacent to the chordamesoderm (Figures 19.12*d,e*; Thisse et al. 1993). These rows of cells are termed adaxial cells and represent a subset of the paraxial mesoderm. Still later, the somites, also derivatives of the paraxial mesoderm, express *sna1* (not shown). It appears then, that the cell movements of gastrulation rearrange the early gastrula fate map (refer to Figure 19.9) in an orderly manner. In the early gastrula, cells of the presumptive notochord lie in a dorsal patch and are flanked on either side by presumptive head and trunk muscle. This relationship is maintained as presumptive notochord cells extend along the axis and as presumptive head and trunk muscle converge and extend along either side.

It is important to note the limitations of examining gene expression to address cell movements and lineage. Gene expression can be dynamic, with no assurance that expression will persist in a given cell or its progeny. The definitive means to address cell fate during gastrulation is to construct fate maps at successive stages. In a recent study of epiblast fates during gastrulation (Woo and Fraser 1995), cells fated to become the various brain regions were shown to be initially spread broadly across the epiblast. During convergence, these cell fates accumulate along the dorsal midline and eventually coalesce into the neural keel.

THE TAIL BUD. When epiboly is complete, the yolk cell is completely covered by blastoderm, and the combined movements of involution and convergent extension have established the rudiments of the head and trunk. The tail bud initially forms just after epiboly is complete, and as embryogenesis proceeds, the tail bud extends away from the yolk cell to produce the embryonic tail (see Figure 19.14*a*; Kanki and Ho 1997). Morphogenesis of the tail bud therefore results in the continued lengthening of the anterior posterior axis of the embryo. In a sense, the formation and growth of the tail bud represent a continuation of the basic convergence and extension movements of gastrulation after they have largely ceased in the head and trunk.

Elaborating the Body Plan

Soon after gastrulation establishes the germ primordia and the embryonic axes, a variety of concurrent morphogenetic processes operate to elaborate the distinct features of the embryonic body design. One of the most evident of

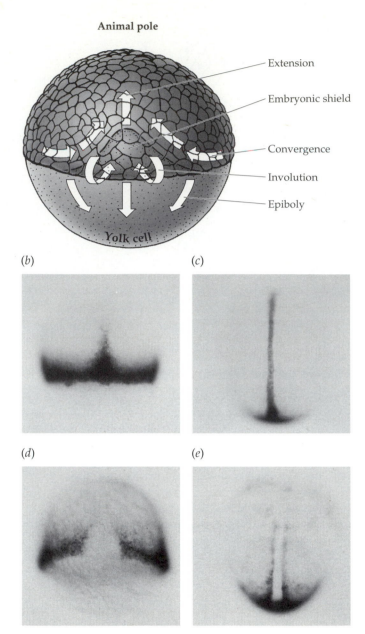

Figure 19.12 Convergence and extension in the zebrafish gastrula. (*a*) Diagram illustrating the many concurrent cell movements occurring during teleost gastrulation, dorsal view. As epiboly spreads the blastoderm over the yolk, involution (or ingression) generates the hypoblast. Convergence brings cells of both the hypoblast and epiblast toward the dorsal side to form the embryonic shield. Within the shield, cells intercalate and extend toward the animal pole to produce the primary embryonic axis. (*b, c, d, e*). Expression patterns of the *no tail* (*ntl, b, c*) and *snail1* (*sna1, d, e*) genes during convergent extension in zebrafish, dorsal views. Both transcripts are initially detected throughout the germ ring. During convergence, *ntl* transcripts accumulate in the midline, while *sna1* transcripts disappear from the central part of the shield. At later stages, *ntl*-expressing cells extend along the midline toward the animal pole to define the axial mesoderm (*c*), while *sna1*-expressing cells form two parallel rows flanking the axial mesoderm. Note that in *c* and *e*, the margin is at the vegetal pole and epiboly is complete. (*a* from Gilbert 1997: *b, c* courtesy of W. Talbot; *d, e* courtesy of C. Thisse, B. Thisse, and J. Postlethwait.)

these is the appearance of somites, the primordia of the segmented body wall muscle and this period is referred to as the segmentation period (Figure 19.13).

Major morphogenetic events that occur during the segmentation period in teleosts include differentiation of the notochord, the formation and differentiation of the brain and sensory organs, and the specification and migration of neural crest. In general, there is an anterior-posterior wave to morphogenesis during this period such that it is often possible to view a wide range of morphogenetic events at one time point in a single embryo. Thus, when head structures are at a relatively advanced stage of morphogenesis, the trunk is just beginning to undergo these processes, and tail structures may be completely undifferentiated.

The Meaning of Vertebrate Segmentation

Segments are present in many animal phyla. The evolutionary significance of segments during vertebrate em-

bryogenesis is hotly debated, but there can be little doubt that certain features of the vertebrate body pattern are arranged and develop segmentally. These include not only the somites, but also the hindbrain (which during the segmentation period becomes partitioned into about seven units called rhombomeres) and the pharyngeal arches. These arches, present peripherally to the hindbrain, will form the jaws and support the iterated series of gills. Each of these structures will be considered in more detail later. Fish contribute an important understanding to the general issue of vertebrate segmentation and development of segmental body plans, primarily because of the relatively simple organization of the fish's early embryonic central nervous system.

As defined more than 100 years ago by Bateson (1894), segments need not be overt, morphologically distinctive packages of body parts (such as in annelid worms), but rather segmentation should be viewed as a repetition of

(a)

(d)

(b)

(c)

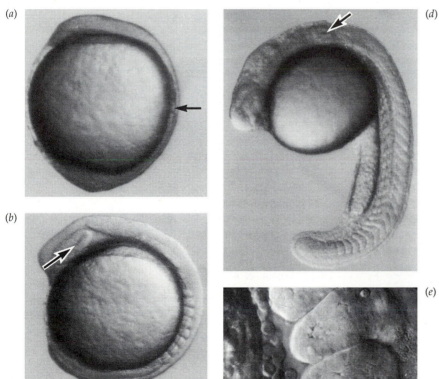

(e)

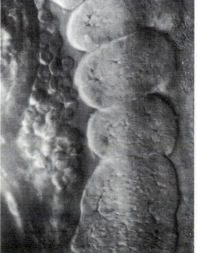

Figure 19.13 Somite formation defines the segmentation period. (a) A zebrafish embryo at the 2-somite stage. The second somite is entirely pinched off from the segmental plate (arrow indicates its posterior boundary). (b) 8-somite stage. Note that the optic primordium is now present (arrow). (c) 15-somite stage. The tail bud is now evident (arrow) and will continue to elongate and add somite pairs. (d) 25-somite stage. The tail is now quite extended and is partially flanked along its ventral side by an extension of the yolk. The arrow indicates the otic vesicle. (e) Somite morphogenesis is shown in this lateral view of the last three somites of a 19-somite-stage embryo. Somite 20 (bottom) is only just beginning to pinch off while somite 17 (top) is taking on a chevron shape. (After Kimmel et al. 1995.)

pattern elements along the major body axis. By this definition, more subtle features of the zebrafish embryonic hindbrain, as well as the spinal cord, have elaborate segmental patterning. As we shall see, the rhombomeres each develop families of individually recognizable neurons, some of which are identical from segment to segment and others of which vary systematically. Moreover, the ventral spinal cord, which does not make any overt swellings, develops repeated sets of primary motoneurons deep to each overlying somite. Each primary motoneuron has a characteristic cell body position within the set, and each will innervate a characteristic subset of muscle cells. Hence, segmentation in the spinal cord very intimately corresponds with that in the muscular periphery. The same may be true in the head, where cranial nerves emerging from the rhombomeres stereotypically innervate pharyngeal muscles in particular pharyngeal segments. Considered in this light, the developmental mechanisms that underlie segmentation in patterning the vertebrate axis become of central interest.

The Notochord

During the segmentation period, the notochord differentiates from the tightly packed mesenchyme of the chordamesoderm (Figures 19.14a,b). In the initial stages of differentiation, the notochord primordium resembles a fairly neat stack of pennies as the central cells begin to vacuolate and swell (Figure 19.14b). However, this appearance is only transient as cells continue to enlarge to produce a stiff rod of large, vacuolated cells surrounded by an epithelial sheath (Figure 19.14c). Importantly, the notochord never shows any hint of segmentation.

Figure 19.14   Morphogenesis of the notochord during tail development. Lateral views of 21-somite stage (a) , 24-hour (b) and 42-hour (c) zebrafish embryos. (a) The primordium of the notochord differentiates from the tightly packed mesenchyme of the tail bud (arrow). (b) The notochord primordium (n) has a transient appearance resembling a stack of pennies. Note also that the tail bud is much smaller and is surrounded by the median fin fold. (c) Notochord cells vacuolate in an anterior to posterior gradient. The tail bud has disappeared entirely and rays of actinotrichia fan out into the fin fold.

Somitogenesis

Beginning shortly after epiboly is complete, bilateral pairs of somites arise sequentially in the segmental plate, the paraxial mesoderm of the trunk and tail. The question of whether paraxial mesoderm in the head is segmented (before stages when neural crest migrates into the region) in fish or any vertebrate is controversial. It has been known for some time (e.g., de Beer 1922) that in sharks, a set of distinctive vesicles form in the head paraxial mesoderm; these have been proposed to represent early segmentation of the head mesoderm. No such distinctive vesicles are present in teleosts (Horder et al. 1993). However, in one teleost species, somitomeres, or transient swellings in the head mesoderm, were described using scanning electron microscopy (SEM; Martindale et al. 1987). Head somitomeres have not been identified by any technique other than SEM in any vertebrate, and there is currently much debate about their significance.

In contrast, somites along the rest of the body are clearly important for the vertebrate embryonic design. Somites are initially distinguished by the appearance of furrows in the segmental plate, with successive furrows demarcating the posterior limits of the preceding somite (Figure 19.13). An epithelial layer develops around each somite, and the interior is an undifferentiated mesenchyme. The total number of somites formed in the zebrafish is between 30 and 34, but this number may vary among different teleosts.

There are two known derivatives of the zebrafish somite, the myotome and the sclerotome. Although it has not been formally demonstrated as in tetrapods, it is likely that the somites in teleosts also will give rise to a dermatome. Most of the somite mesenchyme will differentiate into myotome and give rise to body-muscle segments. The paired somites lie to either side of the notochord, and the cells of the myotome positioned immediately adjacent to the notochord (the so-called adaxial cells) express several distinctive gene markers (such as *sna1*, see Figure 19.12; Thisse et al. 1993). After somite formation, these cells migrate radially away from the notochord and differentiate into a distinctive superficial layer of muscle cells (Figure 19.15b; Devoto et al. 1996). This superficial layer develops as the fish's slow muscle, while the deeper

(a)

(b)

(c)

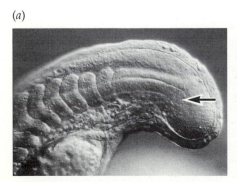
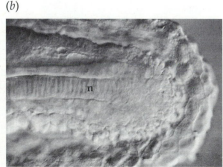

(a)

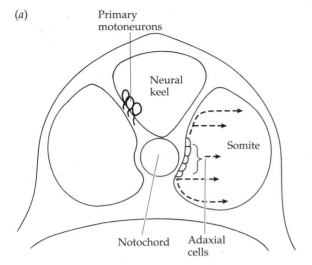

(b)

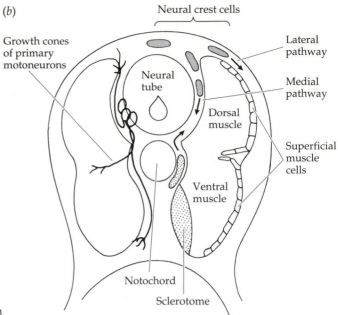

Figure 19.15   Morphogenetic events occur following somite formation. (*a*) Schematic transverse sections of the zebrafish trunk
shortly after somite formation, and (*b*) at the same axial level,
roughly six hours after somite formation. Features are illustrated in
either the right or left half of each section, but are present bilaterally.
Three (and sometimes four) primary motoneurons develop adjacent
to each somite and are uniquely identifiable by their position in the
neural keel (*a*). Growth cones extend from each primary motoneuron and follow stereotyped pathways to innervate either dorsal or
ventral body muscle (*b*). Adaxial cells are a distinct population
along the medial edge of each somite, adjacent to the notochord (*a*).
These cells migrate radially through the somite and differentiate
into a superficial layer of muscle cells (*b*). While most of the somite
becomes myotome and differentiates as body muscle, a ventral medial portion becomes sclerotome and migrates dorsally to form the
rudiment of the vertebral cartilages (*b*). Neural crest cells segregate
from the dorsal edge of the neural keel and migrate along specific
pathways either lateral or medial to the somitic muscle (*b*). Note
that many of the cells movements and axonal projections occur
along similar pathways. (Adapted from Westerfield and Eisen 1988,
Devoto et al. 1996, and Morin-Kensicki and Eisen 1997.)

cells form fast muscle. The myotomes become separated
from one another by a layer of connective tissue termed
the transverse myoseptum and eventually take on a characteristic chevron shape, with the V pointing anteriorly.
Cells of the sclerotome are located ventromedially (Figure
19.15*b*) and will eventually migrate dorsally over the notochord and give rise to vertebral cartilage. The actual formation of this cartilage occurs rather late in development,
well into the larval stages (Morin-Kensicki 1994).

Neurulation

Neurulation refers to the set of morphological processes
that establish the precursor of the brain and spinal cord. In
the earliest stages of neurulation in zebrafish, the presumptive neurectoderm in the gastrula epiblast converges to
form the neural plate (Figure 19.16*a*). By the end of gastrulation, the neural plate can be distinguished from the surrounding ectoderm as a visible thickening, the result of cells
adopting a columnar rather than cuboidal shape. During

the early segmentation period, the neural plate begins to
condense and infold to form the neural keel (Figure 19.16*b*).
The keel then rounds up into a solid cylindrical neural rod
(Figure 19.16*c*). Cell labeling experiments (Papan and Campos-Ortega 1994) show a strong correlation between the
mediolateral position of cells in the neural plate and the
ventrodorsal location of their progeny in the rod. Cells
along the lateral edge of the neural plate move medially toward the midline and come to occupy a dorsal position in
the neural rod, while cells in the center of the neural plate
are enveloped and take up a ventral position.

These morphogenetic movements rearrange the neural-plate cells in a manner that is broadly similar to that described for neurulation in other vertebrates. However, in
the "primary" neurulation in tetrapods, the infolding
edges of the neural plate directly form a hollow nerve
tube, rather than a solid rod. In contrast, the neural rod
formed in teleosts such as the zebrafish and trout (Ballard
1973) is initially solid, and transforms into a tube by a later
process of cavitation (Figure 19.16*d*). Such "secondary"
neurulation also occurs in tetrapods, but is limited to the
caudal spinal cord.

Morphogenesis of the Central Nervous System

Extensive morphogenesis of the central nervous system
continues during the segmentation period (Figure 19.17).
As the neural rod is formed, it is nearly uniform along its
length, with only a slight enlargement demarcating the
brain rudiment from that of the spinal cord (Figure
19.17*a*). However, shortly after it is formed, and even before the neurocoele appears, the neural rod becomes progressively subdivided by constrictions that appear along
its length (Kimmel 1993). In the zebrafish at the 18-somite

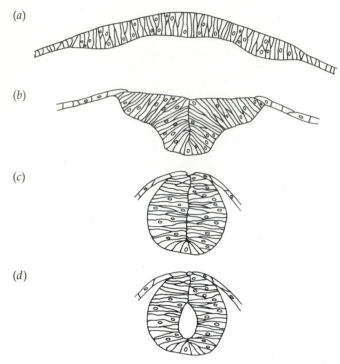

Figure 19.16 Early morphogenesis of the teleost neural primordium, shown in transverse sections. (*a*) The neural plate consists of a distinct population of columnar cells. (*b*) Infolding of the neural plate at the midline forms the neural keel. (*c*) The keel continues to infold and rounds up into the cylindrical neural rod. (*d*) Finally, the neural tube is formed by cavitation. (After Papan and Campos-Ortega 1994.)

stage (18 hours), these constrictions produce 10 distinct swellings termed neuromeres (Figure 19.17*b*). The most rostral three are larger and represent the primordia of the forebrain (telencephalon and diencephalon) and the midbrain (mesencephalon). The more caudal neuromeres subdivide the hindbrain (rhombencephalon) and are termed rhombomeres (Hanneman et al. 1988). By the beginning of the pharyngula period at 24 hours, brain morphogenesis is advanced (Figure 19.17*c*).

Concurrent with this overt morphogenesis of the CNS, the first or primary neurons differentiate within the neural tube. Fish are particularly useful organisms in which to examine this process as it is possible to uniquely identify several neurons in living embryos on the basis of both position of the cell body and the axonal trajectories. Due to the less complicated structure of the spinal cord, the patterns of primary neuron formation are easiest to document there. If we consider one myotome to define a segmental unit of the spinal cord it surrounds, then each left or right spinal cord hemisegment produces three (and sometimes four) unique primary motoneurons (Eisen et al. 1990). The axons of each segmental set of neurons on each side of the spinal cord will share a spinal ventral root that leads to the somite-derived muscle. Each axon then follows stereotyped pathways to innervate either dorsal or ventral body wall muscle (Figure 19.15; Westerfield and Eisen 1988).

The patterns of neurogenesis are more complex in the brain (Hanneman et al. 1988; Chitnis and Kuwada 1990; Ross et al. 1992). However, as in the spinal cord, several neurons in the zebrafish hindbrain have been individually identified (e.g., Metcalfe et al. 1986). There is a clear segmental arrangement of these neurons, and certain serial homologs such as reticulospinal neurons also have segment-specific morphological and presumably functional identities (Figure 19.18). As morphogenesis proceeds in the brain, rhombomeres eventually lose their overt distinction, but can still be recognized by the characteristic patterns of neurons they contain. Less is known of neurogenesis in the forebrain, and the question of whether the forebrain is segmented (as is true of the hindbrain and spinal cord) remains unclear (Ross et al. 1992; Kimmel 1993).

As the developing neuromeres become distinct, the optic primordium develops as a lateral outgrowth of the walls of the diencephalon (Schmitt and Dowling 1994).

Figure 19.17 The brain rudiment is partitioned into neuromeres. Diagrams of zebrafish brains at 12 hours (*a*), 18 hours (*b*), and 24 hours (*c*). A. Shortly after the neural anlage becomes evident, there are no morphological subdivisions. (*b*) At this stage the brain is subdivided into about ten neuromeres, representing the primordia of the telencephalon (T), diencephalon (D), midbrain (M) and seven hindbrain segments or rhombomeres (r1–r7). (*c*) Brain morphogenesis is advanced. The epiphysis (E) and cerebellum (C) are distinctive. In the ventral midline, the floor plate (FP) extends from the rostral midbrain through the hindbrain and spinal cord. (From Kimmel 1993; reprinted with permission from Annual Reviews, Inc.)

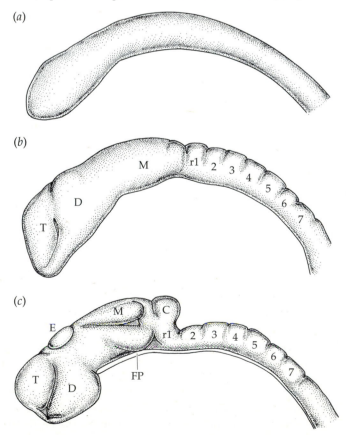

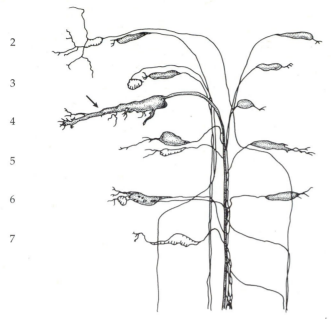

Figure 19.18   Segmental arrays of reticulospinal neurons in the zebrafish hindbrain. Diagram of identified neurons in the larval zebrafish hindbrain. Dorsal view with hindbrain segments numbered. Each cell is shown on only one side, but is bilaterally paired. There is a clear segmentally repeated pattern to these neurons, yet each neuron also has segment-specific features that include cell size, shape, and dendrite branching patterns. For example, the large cell in the fourth rhombomere (arrow) is the famous Mauthner cell. (After Metcalfe et al. 1986.)

This primordium will develop into the optic cup, which consists of an inner cell layer of neural retina, and an outer layer of pigmented retina (Schmitt and Dowling 1994). The lens of the eye develops in conjunction with the optic cup, but does not form from the brain proper. Rather, it is derived from a region of specialized ectoderm termed the lens placode. In addition to the eye, each of the major sensory organs (nose, ear, lateral line), descriptions of which follow later in this chapter, develops in large part from ectodermal placodes.

Neural Crest and Pharyngeal Arches

Of all the major vertebrate embryonic tissues, the neural crest is perhaps the most fascinating. From their initial specification along the dorsolateral edge of the neural keel, neural crest cells migrate ventrally along specific pathways and differentiate into a variety of sensory, pigment, and connective tissue cell types. Neural crest is thought to be unique to vertebrates and the acquisition of neural crest is considered to be of fundamental importance in the evolution of vertebrates from their invertebrate ancestors (Gans and Northcutt 1983).

The behavior of neural crest cells in teleosts appears broadly similar to that of both birds and mammals (see Eisen and Weston 1993). However, since there is no infolding of the neural plate to form a tube, there is no true "crest" in fish as in higher vertebrates. Fish neural crest

cells are derived from the same general precursors as the neural tube, the neurectoderm. Shortly after the neural keel is formed, neural crest cells segregate from the epithelium along the dorsolateral edge of the neural keel. They then take on a mesenchymal appearance and occupy a space between the neural keel and the epidermal epithelium. Fish neural crest cells tend to be comparatively larger and fewer in number than other vertebrates.

In zebrafish embryos, neural crest cells are first recognizable as a distinct population in the head region at around the 6-somite stage (Schilling and Kimmel 1994). Neural crest segregation then continues in a rostrocaudal sequence down the body axis (Raible et al. 1992). Whereas the segregation and dispersal of neural crest cells can be observed in living embryos, the gradient of these processes along the embryo is most evident with scanning electron microscopy (Figure 19.19; Raible et al. 1992). In this figure, the epithelium has been removed to expose the neural crest, the neural tube, and somites. At the caudal end (developmentally youngest) neural crest cells are not yet segregated (Figure 19.19*a*). At more rostral positions, neural crest has clearly segregated (Figure 19.19*b*), and still more rostrally (Figure 19.19*c*), the neural crest has begun its ventral migration. Although neural crest forms along the length of the neural keel, there are clear distinctions between trunk and head (or cranial) crest, both in terms of migration behavior and cellular fate. Bearing in mind that cranial neural crest segregates and migrates prior to that in the trunk, we will first consider trunk neural crest, as it presents a simpler case.

Trunk Neural Crest

At a given position in the trunk, neural crest segregation lags behind the formation of somites. Thus, for example, neural crest cells are not distinctly segregated from the neural keel adjacent to somite 8 until somite 12 is formed. Trunk neural crest cells can be observed across the dorsal and lateral surfaces of the neural keel and are apparently not restricted segmentally (Raible et al. 1992). Shortly after becoming a distinct population of cells, trunk neural crest cells begin to migrate ventrally along one of two distinct pathways (Figure 19.15*b*). The first cells to migrate descend between the neural tube and the somite, along the medial pathway. Later, other cells migrate between the somite and the overlying ectoderm, along the lateral pathway.

After migration, trunk neural crest cells differentiate into a variety of cell types, including neurons of sensory and sympathetic ganglia, Schwann cells, and pigment cells (Raible and Eisen 1994). By labeling premigratory neural crest cells and following their progeny, it has been demonstrated that cells that follow the medial pathway produce all of these cell types, while cells that follow the lateral pathway produce only pigment cells. Furthermore, when labeled prior to migration, most cells produce only one type of progeny, indicating that neural crest cells are

(a) *(b)* *(c)*

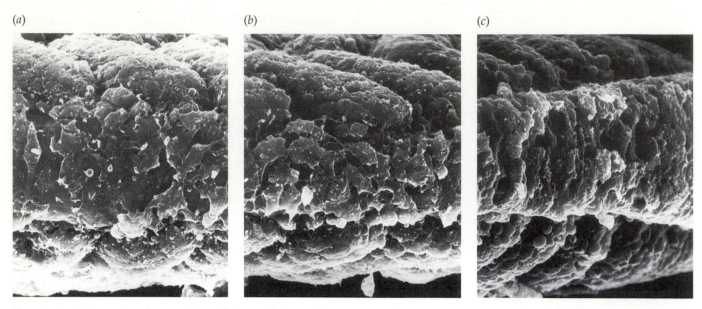

Figure 19.19 A gradient of segregation and dispersal of neural crest cells in the zebrafish trunk. Scanning electron micrographs showing dorsal views of the same 20-somite-stage embryo at different axial levels. The skin is removed to reveal the somites, neural tube, and neural crest. (*a*) In rostral areas of the trunk (somite pairs 4 and 5), neural crest cells have already segregated and are beginning to penetrate the medial pathway between the neural tube and the somites. (b) Around somites 10 and 11, neural crest cells have segregated and overlie the neural rod. They have not begun to migrate. (*c*) Caudal of somite 17, neural crest cells are just beginning to segregate from the neuroepithelium. (From Raible et al. 1992.)

already specified as to their eventual fate even before they begin migrating (Raible and Eisen 1994).

Cranial Neural Crest

Unlike in the trunk, newly segregated neural crest in the head of zebrafish embryos does not cover the dorsal midline and remains on the lateral edges of the neural keel (Schilling and Kimmel 1994). This premigratory mass probably also includes the rudiments of sensory placodes (Landacre 1910). Initially, there is no obvious segmental organization to cranial neural crest. However, as they migrate ventrally, cranial neural crest cells group together, beginning just behind the eye, and migrate adjacent to specific hindbrain segments (rhombomeres; Figure 19.20). The first group forms adjacent to rhombomeres 1 and 2, the second adjacent to rhombomere 4, and the third adjacent to rhombomere 6. Although difficult to detect in the living embryo, these early migrating pathways can be visualized using gene-expression markers. Assuming that all crest cells express such markers, there is a clear segmented pattern of the cranial neural crest.

The distinct segments of migrating cranial neural crest enter and form an integral part of the pharyngeal arches. Pharyngeal arches, of which there are seven in teleosts, are a defining characteristic of vertebrate embryos and are

comprised of tissue from all three primary germ layers. In addition to the ectodermally derived neural crest, both head paraxial mesoderm, as well as some of the endodermal epithelium, contribute to the pharyngeal arches. The neural crest component of the arches differentiates to form pigment cells, sensory elements of the peripheral nervous system, as well as cartilage that will support the gills and jaws. Importantly, agnathan fish such as lamprey do not develop jaws (or paired appendages). Nonetheless, cranial neural crest does contribute to several tissues, including the cartilaginous gill supports that derive from the pharyngeal arches (Newth 1956; Langille and Hall 1986). By 24 hours, the zebrafish pharyngeal arch primordia are

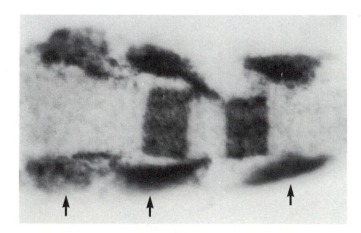

Figure 19.20 Cranial neural crest migrates in segmental pathways. Dorsal view of the hindbrain region of an 18-somite stage zebrafish embryo showing the expression patterns of the *krox20* gene (in rhombomeres r3 and r5) and the *dlx2* gene (arrows). The *dlx2* expression reveals a segmental pattern of cranial neural-crest migration and prefigures the formation of the pharyngeal arches. (Photo courtesy of C. Moens.)

clearly established and extend ventrally below the hindbrain. The definitive morphogenesis and cellular differentiation of arch derivatives, however, do not occur until significantly later (see the later discussion).

As in the trunk, cell labeling experiments indicate that there is extensive patterning information in premigratory cranial neural crest cells (Schilling and Kimmel 1994). When individual neural crest cells are labeled before migration, their progeny generally remain confined to a single arch primordium and form only one cell type. In neither the trunk nor the head are the mechanisms behind the differentiation of neural crest cells understood.

Sensory Placodes

Sensory placodes are a series of specialized regions of the head ectoderm that differentiate into components of all of the major sensory organs of the fish. We have already encountered sensory placodes in describing the development of the fish eye lens. The olfactory (nose) placodes are located at the anterior borders of the eyes and are first recognizable as pits of cells containing beating cilia. The placode gives rise to chemosensory cells, as well as to neurons that send axons into the telencephalon to establish the rudimentary olfactory apparatus (Hansen and Zeiske 1993). Small trigeminal placodes that will form sensory ganglia arise just behind the eyes, and the otic placodes form beside the fifth rhombomere and represent the primordia of the inner ear. The otic placode matures into the otic vesicle, which further develops into the semicircular canals, otolith organs, and receptors for both hearing and balance. A sensory system unique to fish, the lateral-line system, also develops from placodes. The lateral-line primordia arise from lateral placodes, and after forming a ganglion, migrate the length of the embryo, leaving a line of neuromasts in their wakes (Figure 19.21; Metcalfe 1985). This series of mechanosensory cells enables the fish to detect movement in the water, which is thought to aid in escape and the detection of prey, as well as in the social interactions of schooling.

Given that they arise from the ectoderm, produce a variety of neuronal and nonneuronal cell types, and in some cases migrate along specific pathways, sensory placodes share many characteristics with neural crest. Additionally, since they arise just lateral to the edge of the neural keel, it is perhaps useful to consider sensory placodes as the outer edge of a neuroectodermal continuum consisting of a neural keel, neural crest, and sensory placodes.

From Pharyngula to Fish

The segmentation period of embryogenesis ends, at about one day after fertilization in zebrafish, with the establishment of the full complement of somites. At this stage, the embryo is supported along its axis by a differ-

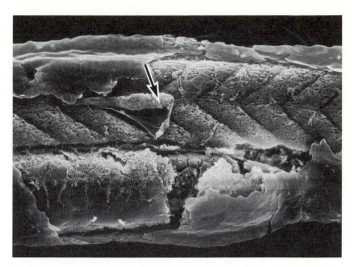

Figure 19.21 Migration of the lateral-line primordium. Scanning electron micrograph of a zebrafish embryo with its skin removed. From its placodal origins near in the head ectoderm, the lateral-line primordium (arrow) migrates over the apex of the underlying chevron-shaped myotomes. (Photo courtesy of W. Metcalfe.)

entiated notochord. Brain morphogenesis is advanced, with the hindbrain fully segmented into rhombomeres. The primordia of the pharyngeal arches are just recognizable, partitioned by a series of grooves in the pharyngeal wall. In the trunk, somite-derived segmental muscle blocks, or myotomes, are innervated by the axons of primary motoneurons, and spontaneous activity in these neurons mediates vigorous muscular contractions. Primary sensory neurons in both the body trunk and head have grown axons to the skin and have connected centrally in the brain and spinal cord. Primary interneurons have made long central axonal pathways, some of these cells projecting between the sensory and motoneurons so as to form sensory motor circuits. Indeed, shortly after the segmentation period is over, a light touch to the embryo will elicit the first reflexive contractile responses.

The embryo at this stage can now properly be termed a pharyngula. Although the pharyngula consists of several well-differentiated body structures that make it clearly recognizable as a vertebrate, significant morphogenesis must still occur to transform the pharyngula into a freely swimming and feeding larva. The number of body structures elaborated during the pharyngula period are numerous; we will only describe the morphogenesis of a few critical structures required for swimming, feeding, and respiring.

Fins

Fins are perhaps the most conspicuous feature of the body of an adult teleost. There are two kinds of fins, the unpaired median fins (including dorsal fins, the caudal or tail fin, and the anal fin), and the bilaterally paired pectoral and pelvic fins. The paired fins are homologous to the paired limbs of tetrapods, and hence their development is of considerable evolutionarily interest (see below). However, since features of the teleost paired fins are quite specialized, their use as a model for the evolutionary precursor of a tetrapod limb is limited to more proximal structures (Coates 1994).

The median and paired fins have different origins. The unpaired fins all derive from a single embryonic median fin fold that develops as a continuous structure along the dorsal, caudal, and ventral midline surface of the early pharyngula (see Figure 19.14c). In contrast, the paired fins each derive from individual fin buds that develop at precise locations along the body flank (Figure 19.22). The pectoral fins begin to develop in the pharyngula, and they function as paddles by the time of hatching. In zebrafish the pelvic fins develop in the larva only some weeks later (Sordino et al. 1995), and hence have not been so thoroughly studied.

In spite of the difference in the form of their rudiments, there are great similarities in how the median and paired fins develop. The earliest morphological sign, in both cases, is the appearance of a thickened ridge of ectodermal cells. The ridge runs along the midline in the case of the median fins and across the apex of a developing mound of mesenchyme in the case of the pectoral fin bud. The latter is the equivalent of the apical ectodermal ridge (AER) of the tetrapod limb bud, which is known to provide signals required for limb outgrowth to the underlying bud mesenchyme. The fish and tetrapod apical ridges express some of the same patterning genes, and hence they may play similar functional roles in fin development. The gene-expression patterns are largely shared as well between the paired and median fin rudiments (Akimenko et al. 1994, 1995).

Fins generally have two types of skeletal elements. The basal or proximal bones are endoskeletal and form by replacing cartilage models that form early in fin morphogenesis. This is also the case for tetrapod limbs. The caudal

fin is supported by specialized bones deriving from tail vertebrae, and the paired fins are supported by girdle elements that originate, along with the fin muscle, in the fin bud mesenchyme. In zebrafish the endoskeletal cartilages of the pectoral fin girdle arise near the end of the pharyngula period and are among the very first cartilages to develop anywhere in the body. The more distal skeletal elements of fins are exoskeletal and form from the fin fold, rather than from the fin bud.

The formation of the fin fold from the fin bud is illustrated in Figure 19.23. Initially, the fold is entirely ectodermal (i.e., the basal surfaces of the folded epithelium meet one another in the manner illustrated), but later mesenchyme invades between the epithelial layers. The ridge-to-fold transition occurs in the development of both the paired and median fins of teleosts and other fish, but there is no hint of such a fold in a tetrapod limb bud. This is a fundamental difference between fins and limbs and makes sense, because the fold develops the paddle, unique to fins. Long collagenous fibers, or actinotrichia, condense from the fin-fold mesenchyme and fan out into the growing fins. These embryonic structural supports are eventually replaced in the adult by bone to produce lepidotrichia, the segmented bony-fin rays.

Importantly, there is never a cartilage model of the fin exoskeleton; the lepidotrichia develop directly from the fin blade's dermal mesenchyme. This is also how the dermal bones of the skull develop in all vertebrates, but no bones seem to develop this way in the tetrapod trunk or tail. Moreover, although cephalic neural crest forms the dermal bones of the skull, mesoderm is thought to give rise to all other bone in the trunk and tail in all vertebrates. Recently, cell labeling experiments suggest that the mesenchyme that produces the caudal fin exoskeleton in zebrafish is derived from neural crest (Smith et al. 1994). If the finding is correct, it would upset the long-held dogma that only cephalic and not trunk or tail neural crest has skeletogenic potential.

Fin structure distinguishes the ray-finned bony fishes, including teleosts such as zebrafish, from their relatives, the lobe-fin, or fleshy-fin fishes. In ray-finned fish, most of the fin develops as the exoskeleton-containing paddle. In contrast, in lobe-finned fish (such as the lung fish and coelacanth), most of the fin develops as the endoskeletal and muscular part of the appendage that drives the pad-

Figure 19.22 Morphogenesis of the pectoral fin bud. Anterior is to the left in all views; arrows indicate the boundary between the inner mesenchyme and the epithelium of the bud. After the initial formation of the fin bud (*a*), the apical ectodermal ridge becomes prominent along the distal edge of the fin (*b*). In *c* the bud begins to curve and taper into the fin fold. Note that an endoskeletal cartilage core has begun to differentiate within the bud mesenchyme (arrow heads). (From Kimmel et al. 1995.)

(*a*)

(*b*)

(*c*)

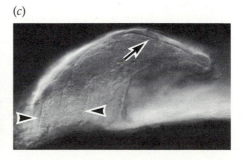

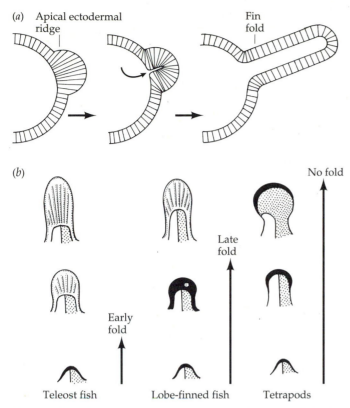

Figure 19.23 The fin fold and the evolution of the vertebrate limb. (*a*) Diagram of the ridge-to-fold transition. The fin bud initially forms with an apical ectodermal ridge, typical of all vertebrate limb buds. In fins, a notch develops in the ridge. It extends into a complete apical ectodermal fold, from which the fin exoskeleton will condense. (*b*) Hypothetical explanation of the developmental differences between teleosts, lobe-finned fish, and tetrapods. In teleost fish, the transition from apical ectodermal ridge to fold takes place early, with a resultant small endoskeleton and extended exoskeleton. In lobe-finned fish (such as the coelacanth), a late ridge-to-fold transition results in more endoskeleton and relatively less exoskeleton. In tetrapods, no such transition occurs, allowing extensive proliferation of the endoskeleton producing mesenchyme. (Adapted from Thorogood 1991, and Sordino et al. 1995.)

dle. Thorogood (1991) has proposed an elegant hypothesis that explains this difference, involving the timing of the ridge-to-fold transition during fin development. The transition would be relatively early in ray fins, and late in lobe fins (Figure 19.23*b*). The ridge is known to be required (at least in tetrapods) to maintain outgrowth of the bud, from which the endoskeleton arises. Hence an early fold transition would promote more exoskeleton, but would do so at the expense of endoskeleton because the bud outgrowth promoting signals from the ridge would be lost. A delay in transition, to provide for lobe-fin development might have been an important prerequisite for the evolution of tetrapods, which need a relatively strong endoskeleton to support body weight on land. Tetrapod limbs have no exoskeleton at all and appear to have lost the ridge to fold transition entirely.

Although the fin fold is unique to fish, there are basic and early similarities between the fish fin bud and the limb bud of tetrapods. Just as mentioned for the AER, very similar gene expression patterns are found in the deep mesenchyme of the fish fin and tetrapod limb buds. One example is *sonic hedgehog*, a secreted protein that is expressed in the zone of polarizing activity in both fin and limb bud mesenchyme. Another intriguing set of genes are those of the *HoxD* cluster, which appear particularly relevant to the evolutionary derivation of tetrapod limbs from fish fins (Sordino et al. 1995). *HoxD* transcripts are present in a proximodistal sequence, but along the posterior side of the limb bud in fish, as well as in birds and mammals. In zebrafish, *HoxD* gene expression persists only in the posterior mesenchyme as the fin bud grows out. However, in the mouse, *HoxD* gene expression relocates from the posterior part to the distal end of the bud, as if territory that was initially posterior now becomes distal. This finding complements morphological evidence that the tetrapod digits derive from the posterior bud.

Heart Development

The pharyngula period of fish development includes the differentiation of the heart tube into distinct chambers. Cell labeling experiments in zebrafish demonstrate that cardiac precursors are first specified in the gastrula (Stanier et al. 1993; Lee et al. 1994). As in other vertebrates, these precursors migrate towards the embryonic axis after involuting, and they coalesce into two tubular primordia, which eventually fuse to form the single heart tube. In zebrafish, the tube is formed and is beating by 24 hours. It lies between the yolk sac and the embryo proper; the portion of the tube that will form the atrium sits anterior to the presumptive ventricle (Figure 19.24*a*; Stanier and Fishman 1992). This curious arrangement is unique to fish and is the opposite of the initial orientation of the heart tube of both the chick and the mouse. However, as described first for the shad (Senior 1909), the polarity of the teleost heart tube switches during development. At the atrial end, the heart tube remains attached to the yolk, while at the ventricular end, it is attached to the embryo near its head. As the embryo grows during the pharyngula period, the yolk becomes depleted and the head rotates dorsally around the yolk. These movements pull the tube such that at later stages, venous return is to the posterior and arterial outflow is to the anterior of the heart tube (Figure 19.24*b*). Thus, the topology of the fish heart tube during later embryogenesis is the same as other vertebrates, and it becomes evident that the early reverse orientation is due to physical distortion caused by the presence of the yolk. Roughly halfway through the pharyngula period (36 hours in zebrafish), there are clear indentations that mark the four chamber boundaries, the sinus venosus, the atrium, the ventricle, and the bulbus arteriosus.

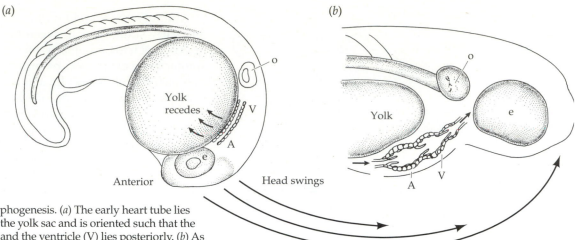

Figure 19.24   Heart morphogenesis. (*a*) The early heart tube lies between the embryo and the yolk sac and is oriented such that the atrium (A) lies anteriorly and the ventricle (V) lies posteriorly. (*b*) As the heart chambers differentiate, the polarity of the heart tube changes. The yolk recedes, pulling the atrium posteriorly, and the head swings dorsally, pulling the ventricle anteriorly. The result is that blood flows through the heart in a normal posterior-to-anterior direction. (After Stanier and Fishman 1992.)

Jaws and Gills

As previously mentioned, the pharyngeal arch primordia are the precursors of both jaws and gills in gnathostome fish. The fact that jaws derive from the same series of embryonic precursors as gills contributes strongly to the theory that jaws evolved from the anterior gill supports of primitive, jawless fish.

Morphogenesis of the pharyngeal arches is extensive. It involves mesenchyme derived from both the mesoderm and the neural crest, with cells from these two sources contributing to different components of the final product. The neural crest gives rise to the series of pharyngeal cartilages and probably to their associated connective tissues. Head mesoderm gives rise to the blood vessels that course through each arch (the aortic arches) and to a complex set of jaw and gill muscles (Schilling and Kimmel 1994). These muscles are innervated by the cranial motoneurons residing in the hindbrain rhombomeres. The system is functional by the time of hatching, such that the young larva is able to capture food and to pump water past its gills. Sensory innervation of the mouth, pharynx, and the operculum (the bony cover for the gills) also comes from cranial nerves. The sensory neurons, derived from either cranial neural crest or from placodes beside the brain, are housed in a series of ganglia associated with the arches themselves.

The two anterior arches, the mandibular and hyoid, form the jaw skeleton and support for the mouth. Two cartilages, a ventral and a dorsal element, form in each of the first two arches during embryonic development (Figure 19.25). The ventral cartilages are thick cylinders curving beneath the pharynx towards the midline. The first one (the mandibular or Meckel's cartilage) forms the lower jaw and the second one (ceratohyal) supports the gular (mouth) cavity. The corresponding but more flattened dorsal elements form the upper jaw (the quadrate cartilage from the first arch) and its support (the hyosymplectic) from the second (Figure 19.25). The jaw cartilages grow rapidly during the last hours before hatching. This

Figure 19.25   Jaw and gill cartilages are serial homologs. Drawing depicts the shape and position of the cartilage derivatives of the pharyngeal arches in a larval zebrafish. The first pharyngeal arch gives rise to a dorsal (quadrate) and ventral (Meckel's) cartilage that form the rudiment of the upper and lower jaw, respectively. The second pharyngeal arch gives rise to the hyosymplectic and ceratohyal cartilages that form essential jaw supports. The third through seventh pharyngeal arches produce ceratobranchial cartilages, all but the last of which will support gills. The eye and otic vesicle (ov) are depicted for perspective. (Original drawing courtesy of G. Kruse.)

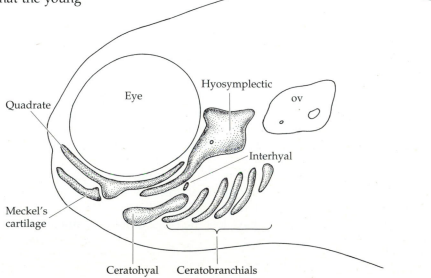

growth serves to reposition the mouth, bringing it ahead of the eyes, and at the same time causing the mouth to gape open widely. The aortic arches within the mandibular and hyoid pharyngeal arches deliver blood to head (including, in particular, the brain) and to the developing operculum. Three of the fish jaw cartilages, Meckel's, the quadrate, and the hyosymplectic, have a particularly interesting evolutionary history in tetrapods. The homologous embryonic elements in mammals develop not as jaws, but rather as the bony ossicles of middle ear, respectively, the malleus, incus, and stapes.

The next four arches will bear the gills. A single cartilage bar (a ceratobranchial) develops initially in each one (Figure 19.25), and by the time of hatching, the first gill filaments appear. Endothelial cells of the aortic arches in these segments develop an elaborately branching network of vessels that supply the gills. The last arch (the seventh) in teleosts does not develop gills and includes no aortic arch. Rather the cartilage in this arch supports pharyngeal teeth opposing the palate and used for food grinding. Most fishes develop teeth on their jaws (for food capturing) and also teeth associated with a number of pharyngeal bones that eventually replace the embryonic cartilages. However, this

is not the case for the zebrafish, as the teeth in the last arch are the only ones it will ever have. Where the cells that form the teeth originate is unknown.

Zebrafish Developmental Mutants

Our knowledge of embryology proceeds along three fronts. One is the careful description of normal developmental processes. A second involves the manipulation of cells via transplantations or ablations to determine the roles individual cells have during development. And a third is the examination of embryos that are defective in one or more genes required for normal development. In addition to being a very useful system for descriptive and manipulative embryology, the zebrafish has recently emerged as a powerful genetic system for understanding development. Hundreds of zebrafish mutants have recently been described, which affect the morphogenesis of discrete structures of the embryonic body plan (Haffter et al. 1996; Driever et al. 1996; reviewed in Eiger 1996).

One example of a zebrafish developmental mutant is shown in Figure 19.26. Embryos lacking *no tail (ntl)* gene function do not have tails (Figure 19.26*a*) and conspicuously lack a notochord (compare Figures 19.26*b* and *c*).

(*a*)

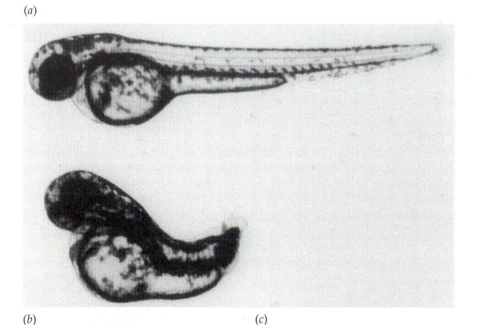

(*b*)

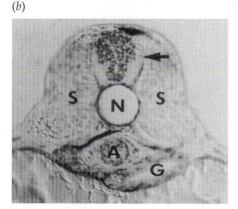

(*c*)

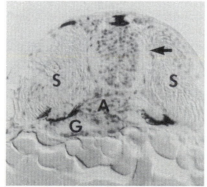

Figure 19.26 An example of a zebrafish developmental mutant. (*a*) A *no tail* (*ntl*) mutant embryo and its wild-type sibling at two days of development. *ntl* mutants are conspicuously lacking extended tails. (*b, c*) Transverse sections through the trunk of a 24-hour wild-type (*b*) and *ntl* mutant embryo (*c*). The notochord (N) is completely absent from *ntl* mutant embryos, while the spinal cord (arrow) and somites (S) are present. A, dorsal aorta; G, gut. (From Halpern et al. 1993.)

This gene is known to be the homolog of the *Brachyury* of *T* gene mouse (Shulte-Merker et al. 1992), which produces the same mutant phenotype as in the fish. In addition to indicating that the *ntl* gene product is required for notochord formation, this mutation also allows one to determine what effect notochord removal has on the embryo. For example, *ntl* mutant embryos have abnormal myotomes and lack muscle pioneers. Transplantation of wild-type notochord precursor cells into a *ntl* mutant host rescues the myotomal phenotype. These findings provide evidence that cells of the notochord provide signals required to properly pattern the somites.

The number of zebrafish developmental mutants is large and growing. Our understanding of embryology will soon benefit from descriptions of mutations affecting discrete processes such as epiboly, gastrulation, somitogenesis, brain development, craniofacial development, and many others. By comparing the development of these embryos with normal embryos, it should one day be possible to understand how individual genes function at the cellular level to transform a single celled zygote into a complex, free-swimming fish.

Literature Cited

Abdelilah, S., L. Solnica-Krezel, D. Y. R. Stanier and W. Driever. 1994. Implications for dorsoventral axis determination from the zebrafish mutation janus. *Nature* 370: 468–471.

Akimenko, M.-A., M. Ekker, J. Wegner, W. Lin and M. Westerfield. 1994. Combinatorial expression of three zebrafish genes related to *distal-less*. Part of a homeobox gene code for the head. *J. Neurosci.* 14: 3475–3486.

Akimenko, M.-A., S. L. Johnson, M. Westerfield and M. Ekker. 1995. Differential induction of four msx homeobox genes during fin development and regeneration in zebrafish. *Development.* 121: 347–357.

Ballard, W. W. 1966. Origin of the hypoblast in *Salmo* II. Outward movement of the deep central cells. *J. Exp. Zool.* 161: 211–220.

Ballard, W., W. 1973. A new fate map for *Salmo giardneri. J. Exp. Zool.* 184: 49–74.

Ballard, W. W. 1981. Morphogenetic movements and fate maps of vertebrates. *Amer. Zool.* 21: 391–399.

Bateson, W. 1894. *Materials for the Study of Variation.* Macmillan and Co., London.

Beams, H. W., and Kessel, R. G. 1976. Cytokinesis: A comparative study of cytoplasmic divisions in animal cells. *Amer. Sci.* 64 (3): 279–290.

Bond, C. E. 1979. *The Biology of Fishes.* W. B. Sanders, Philadelphia.

Carter, C. A. and J. P. Wourms, J. P. 1991. Cell behavior during early development in the South American annual fish of the genus *Cynolebias. Journ. of Morph.* 210: 247–266.

Chitnis, A. B. and J. Y. Kuwada. 1990. Axonogenesis in the brain of zebrafish embryos. *J. Neurosci.* 10: 1892–1905.

Coates, M. I. 1994. The origin of vertebrate limbs. *Development (Suppl.)* 169–180.

Dasgupta, J. D. and U. N. Singh. 1982. Spatio-temporal distribution of gap junctions in zebrafish embryos. *Roux's Arch. Dev. Biol.* 191: 378–380.

Davidson, E. H. 1990. How embryos work: A comparative view of diverse modes of cell fate specification. *Development* 108: 365–389.

Eisen, J. S. and J. A. Weston. 1993. Development of the neural crest in the zebrafish. *Dev. Biol.* 159: 50–59.

Eisen, J. S., S. H. Pike and B. Romancier. 1990. An identified motoneuron with variable fates in embryonic zebrafish. *J. Neurosci.* 10(1): 34–43.

Ewing, H. 1972. Spermatogenesis in the zebrafish *Brachydanio rerio. Anat. Rec.* 172: 308.

Gans, C. and R. G. Northcutt. 1983 Neural crest and the evolution of vertebrates: A new head. *Science* 220: 268–274.

Gilbert, S. F. 1997. *Developmental Biology*, 5th Ed. Sinauer Associates, Sunderland, MA.

Halpern, M. E., R. K. Ho, C. Walker and C. B. Kimmel. 1993. Induction of muscle pioneers and floor plate is distinguished by the zebrafish *no tail* mutation. *Cell* 75: 99–111.

Hanneman, E., B. Trevarrow, W. K. Metcalfe and C. B. Kimmel. 1988. Segmental development of the spinal cord and hindbrain of the zebrafish embryo. *Development.* 103: 49–58.

Hansen, A. and E. Zeiske. 1993. Development of the olfactory organ in the zebrafish, *Brachydanio rerio. J. Comp. Neurol.* 333(2): 289–300.

Hart, N. H. and M. Donovan. 1983. Fine structure of the chorion and site of sperm entry in the egg of *Brachydanio. J. Exp. Zool.* 227: 277–296.

Helde, K. A., E. T. Wilson, C. J. Cretekos and D. J. Grunwald. 1994. Contribution of early cells to the fate map of the zebrafish gastrula. *Science* 265: 517–520.

Hisaoka, K. K. and H. I. Battle. 1958. The normal developmental stages of the zebrafish *Brachydanio rerio. J. Morph.* 102: 311–328.

Hisaoka, K. K. and C. F. Firlit. 1962. Ovarian cycle and egg period in the zebrafish *Brachydanio rerio. Copeia.* 4: 788–792.

Ho, R. 1992. Axis formation in the embryo of the zebrafish *Brachydanio rerio. Sem. Dev. Biol.* 3: 53–64.

Ho, R. K. and C. B. Kimmel. 1993. Commitment of cell fate in the early zebrafish embryo. *Science* 261: 109–111.

Horder, T. J., R. Presley and J. Slipka. 1993. The segmental bauplan of the rostral zone of the head in vertebrates. *Func. Dev. Morphol.* 3: 79–89.

Kane, D. A. and C. B. Kimmel. 1993. The zebrafish midblastula transition. *Development* 119: 447–456.

Kimmel, C. B. 1993. Patterning the brain of the zebrafish embryo. *Annu. Rev. Neurosci.* 16: 707–732.

Kimmel, C. B. and R. D. Law. 1985 Cell lineage of zebrafish blastomeres. I. Cleavage pattern and cytoplasmic bridges between cells. *Dev. Biol.* 108: 94–101.

Kimmel, C. B. and R. M. Warga. 1987. Indeterminate cell lineage of the zebrafish embryo. *Dev. Biol.* 124: 269–280.

Kimmel, C. B., R. M. Warga and T. F. Schilling. 1990. Origin and organization of the zebrafish fate map. *Development* 108: 581–594.

Kimmel, C. B., W. W. Ballard, S. R. Kimmel, B. Ullmann and T. F. Schilling. 1995. Stages of embryonic development of the zebrafish. *Dev. Dyn.* 203: 253–310.

Landacre, F. L. 1910. The origin of the cranial ganglia in *Ameiurus. J. Comp. Neurol.* 20: 309–411.

Langille, R. M. and B. K. Hall. 1986. Evidence of cranial neural crest contribution to the skeleton of the sea lamprey, *Petromyzon marinus. Prog. Clin. Biol. Res.* 217: 263–266.

Lee, R. K. K., D. Y. R. Stanier, B. Weinstein and M. C. Fishman. 1994. Cardiovascular development in the zebrafish. II. Endocardial progenitors are sequestered within the heart field. *Development* 120: 3361–3366.

Lewis, W. H. and E. C. Roosen-Runge. 1942. The formation of the blastodisc in the egg of the zebrafish *Brachydanio rerio. Anat. Rec.* 84: 13–14.

Lewis, W. H., and E. C. Roosen-Runge. 1943. The formation of the blastodisc in the egg of the zebrafish *Brachydanio rerio. Anat. Rec.* 85: 38.

Long, W. L. 1983. The role of the yolk syncytial layer in determination of the plane of bilateral symmetry in rainbow trout (*Salmo gairdneri* Richardson). *J. Exp. Zool.* 228: 91–97.

Martindale, M. Q., S. Meier and A. G. Jacobson. 1987. Mesodermal metamerism in the teleost, *Oryzias latipes* (the medaka). *J. Morph.* 193: 241–252.

Metcalfe, W. K. 1985. Sensory neuronal growth cones comigrate with posterior lateral line primordial cells in zebrafish. *J. Comp. Neurol.* 238: 218–224.

Metcalfe, W. K., B. Mendelson, B. and C. B. Kimmel. 1986. Segmental homologies among reticulospinal neurons in the hindbrain of the zebrafish larva. *J. Comp. Neurol.* 251: 147–159.

Morin-Kensicki, E. M. 1994. Metameric relationship of the somitic sclerotome to elements of the peripheral nervous system and the vertebral column during zebrafish development. Ph. D. Thesis. University of Oregon.

Newth, D. R. 1956. On the neural crest of the lamprey embryo. *J. Emb. Exp. Morph.* 4: 358–375.

Oppenheimer, J. 1936. Transplantation experiments on developing teleosts (*Fundulus* and *Perca*). *J. Exp. Zool.* 72: 409–437.

Papan, C. and J. Campos-Ortega. 1994. On the formation of the neural keel and neural tube in the zebrafish. *Roux's Arch. Dev. Biol.* 203: 178–186.

Raible, D. W. and J. S. Eisen. 1994. Restriction of neural crest cell fate in the trunk of the embryonic zebrafish. *Development* 120: 495–503.

Raible, D. W., A. Wood, W. Hodson, P. Henion, J. A. Weston and J. S. Eisen. 1992. Segregation and early dispersal of neural crest cells in the embryonic zebrafish. *Dev. Dyn.* 195: 29–42.

Romanes, G. J. 1901. *Darwin and After Darwin.* Open Court Publishing, London.

Ross, C. S., T. Parrett and S. S. Easter, Jr. 1992. Axonogenesis and morphogenesis in the embryonic zebrafish brain. *J. Neurosci.* 12: 467–482.

Schilling, T. F. and C. B. Kimmel. 1994. Segment and cell type lineage restrictions during pharyngeal arch development in the zebrafish embryo. *Development* 120: 483–494.

Schmitt, E. A. and J. E. Dowling, J. E. 1994. Early eye morphogenesis in the zebrafish, *Brachydanio rerio. J. Comp. Neurol.* 344: 532–542.

Schmitz, B. and J. Campos-Ortega. 1994. Dorsoventral polarity of the zebrafish embryo is distinguishable prior to the onset of gastrulation. *Roux's Arch. Dev. Biol.* 203: 374–380.

Senior, H. D. 1909. The development of the heart in shad. *Amer. J. Anat.* 9: 212–276

Shih, J. and S. E. Fraser. 1995. Distribution of tissue progenitors within the shield region of the zebrafish gastrula. *Development* 121: 2755–2765.

Shulte-Merker, S., R. K. Ho, B. G. Herrmann and C. Nusslein-Volhard. 1992. The protein product of the zebrafish homologue of the mouse T gene is expressed in nuclei of the germ ring and notochord of the early embryo. *Development* 116: 1021–1032.

Smith, M., A. Hickman, D. Amanze, A. Lumsden and P. Thorogood. 1994. Trunk neural crest origin of caudal fin mesenchyme in the zebrafish *Brachydanio rerio. Proc. R. Soc. Lond.* 256: 137–145.

Solnica-Krezel, L. and W. Driever. 1994. Microtubule arrays of the zebrafish yolk cell: organization and function during epiboly. *Development* 120: 2443–2455.

Sordino, P., F. van der Hoeven and D. Duboule. 1995. *Hox* gene expression in teleost fins and the origin of vertebrate digits. *Nature* 375: 678–681.

Stanier, D. Y. R. and M. C. Fishman. 1992. Patterning the zebrafish heart tube: Acquisition of anteroposterior polarity. *Dev. Biol.* 153: 91–101.

Stanier, D. Y. R., R. K. Lee, and M. C. Fishman. 1993. Cardiovascular development in the zebrafish. I. Myocardial fate map and heart tube formation. *Development.* 119: 31–40.

Strahle, U. and S. Jesuthasan. 1993. Ultraviolet irradiation impairs epiboly in zebrafish embryos: evidence for a microtubule dependent mechanism of epiboly. *Development* 119: 909–919.

Strehlow, D. and W. Gilbert. 1993. A fate map for the first cleavages of the zebrafish. *Nature* 361: 451–453.

Strehlow, D., G. Heinrich and W. Gilbert. 1994. The fates of the blastomeres of the 16-cell zebrafish embryo. *Development* 120: 1791–1798.

Thisse, C., B. Thisse, T. F. Schilling and J. H. Postlethwait. 1993. Structure of the zebrafish *snail 1* gene and its expression in wild-type, *spadetail,* and *no tail* mutant embryos. *Development* 119: 1203–1215.

Thorogood, P. 1991. The development of the teleost fin and implications for our understanding of tetrapod limb evolution. In *Developmental Patterning of the Vertebrate Limb,* J. R. Hincliffe (ed.). Plenum, New York.

Trinkaus, J. P. 1984. Mechanism of *Fundulus* epiboly: A current view. *Amer. Zool.* 24: 673–688.

Trinkaus, J. P. 1992. The midblastula transition, the YSL transition, and the onset of gastrulation in *Fundulus. Development (Suppl.)* 75–80.

Trinkaus, J. P. 1996. Ingression during early gastrulation of *Fundulus. Dev. Biol.* 177: 356–370.

Trinkaus, J. P. and C. A. Erickson. 1983. Protrusive activity, mode and rate of locomotion, and pattern of adhesion of *Fundulus* deep cells during gastrulation. *J. Exp. Zool.* 228: 41–70.

Trinkaus, J. P., M. Trinkaus and R. D. Fink. 1992. On the convergent cell movements of gastrulation in *Fundulus. J. Exp. Zool.* 261: 40–61.

Warga, R. M. and C. B. Kimmel. 1990. Cell movements during epiboly and gastrulation in zebrafish. *Development* 108: 569–580.

Westerfield, M. and J. S. Eisen. 1988. Neuromuscular specificity: pathfinding by identified motor growth cones in a vertebrate embryo. *Trends Neurosci.* 11: 18–22.

Wolenski, J. S. and N. H. Hart. 1987. Scanning Electron Microscope studies of sperm incorporation into the zebrafish (*Brachydanio*) egg. *J. Exp. Zool.* 243: 259–273.

Woo, K. and S. E. Fraser. 1995. Order and coherence in the fate map of the zebrafish nervous system. *Development* 121: 2595–2609.

Wood, A. and L. P. M. Timmermans. 1988. Teleost epiboly: A reassessment of deep cell movement in the germ ring. *Development* 102: 575–585.

CHAPTER 20

Amphibians

Richard Elinson

MANY FUNDAMENTAL CONCEPTS OF VERTEBRATE DEVELopment, such as embryonic induction, axis determination, and morphogenetic mechanisms, are derived from amphibian embryos. The reason for the prominence of the amphibian embryo is that in many ways, it is designed to accommodate the experimental embryologist. The eggs are very large, they develop in pond water, and they divide completely as the embryo forms. Their size makes transplantations, microinjections, and dissections easy, compared to small mammalian embryos. Manipulations of amphibian embryos are unhampered by the large, uncleaved yolk masses or oil droplets found in birds and teleost fish. Culturing of embryos and tissues does not require specialized environments, such as those provided by the extraembryonic membranes in amniotes or the uterus in mammals. It is simple to obtain hundreds of synchronized amphibian embryos; because of the large size of the embryos, isolation of RNAs and other molecules is straightforward.

There are admittedly some characteristics that amphibians lack. Standard genetic analysis is not practical, as the fastest generation times are more than a year. Also, the egg is opaque, so internal changes cannot be observed in intact, living embryos. Amphibian embryologists dream of finding a frog with a three-month generation time and transparent eggs, but in the meantime, they continue to exploit the obvious advantages.

Amphibians are a diverse group of animals and are classified into three orders (Table 20.1). The order Anura consists of the tailless amphibians such as frogs and toads. The South African clawed frog *Xenopus laevis* is the current workhorse of developmental biologists, while in the past, various species of *Rana* and *Bufo* were used. The order Caudata, also known as the urodeles, consists of the tailed newts and salamanders, such as the axolotl *Ambystoma*

mexicanum, the French favorite *Pleurodeles waltl*, and the Japanese *Cynops pyrrhogaster*. The order Gymnophiona are the apodans or caecilians. These legless, wormlike creatures are usually ignored by developmental biologists (Brauer 1897; Wake 1977; Delsol et al. 1981), so there is too little information on them to discuss here. Our generalized view of amphibian development is derived mainly from the above species, with a heavy current emphasis on *X. laevis*. This emphasis creates bias, as *X. laevis* develops very rapidly and shows some developmental differences when comparative data is available.

The question as to whether the three orders of amphibians evolved from a single type of fish that came onto land (monophyly) or whether they evolved from different groups of fishes (polyphyly) has been a favorite topic of discussion over the years. Significant differences in the embryology of anurans and urodeles have provided fuel for the debate, but they have not provided the resolution (Hanken 1986). Compared to fish, amphibian embryos are most similar to embryos of sturgeons (Ballard and Ginsberg 1980; Bolker 1993) and lungfish (Kemp 1981) and quite dissimilar from teleosts (Collazo et al. 1994). Eggs of amphibians, sturgeons, and lungfish all divide completely to produce a blastula, and during gastrulation their embryos show similar patterns of movements to yield the larval body. In this chapter, a generalized view of the amphibian embryo will be presented, but as many variations as possible from this view will be pointed out.

Oocyte Development

It is logical to start a description of development with the egg because the properties of the egg foreshadow the em-

TABLE 20.1 Classification of Amphibians[a]

Order Anura (frogs and toads)

    Suborder Archaeobatrachia

        Superfamily Discoglossoidea
(*Ascaphus, Discoglossus*)

        Superfamily Pipoidea (*Xenopus*)

        Superfamily Pelobatoidea (*Scaphiopus*)

    Suborder Neobatrachia

        Superfamily Bufonoidea (*Rheobatrachus,
Ceratophrys, Lepidobatrachus, Bufo,
Eleutherodactylus, Nectophrynoides,
Rhinoderma*)

        Superfamily Microhyloidea (*Flectonotus,
Gastrotheca*)

        Superfamily Ranoidea (*Rana*)

Order Caudata (urodeles; newts and salamanders)

    Suborder Cryptobranchoidea (*Hynobius,
Cryptobranchus*)

    Suborder Sirenoidea

    Suborder Salamandroidea (*Cynops, Pleurodeles,
Taricha, Triturus, Notophthalmus, Ambystoma*)

Order Gymnophiona (caecilians)

Source: Duellman and Trueb 1986.

[a]All genera mentioned in the text as well as a few others commonly used by developmental biologists are listed.

bryo. Although the sperm is useful genetically, a complete embryo can develop from an egg in the absence of sperm. The volume of the 1.3-mm egg of *X. laevis* is equivalent to five million normal-sized cells, so there is more than enough material in the egg to construct a sensing, responding, swimming embryo, with a complete set of vertebrate organs. All the amphibian egg needs is an appropriate activation stimulus, usually but not necessarily provided by the sperm, to produce the embryo.

The amphibian egg is a highly polarized cell, with an animal and a vegetal half (Figure 20.1). The animal half contains the nucleus and in most species is recognized by darker pigmentation. The vegetal half has fewer pigment granules and is filled with large yolk platelets. In development, most of the organs, including the nervous system, come from the animal half, giving the impression that the animal half is the active half. In fact, development is driven by information, mostly in the form of stored mRNAs, in the vegetal half. An understanding of amphibian development requires an understanding of oogenesis.

Oogenesis

Cells that form sperm or eggs are germ cells, and the mitotically active germ cells in the ovary are called oogonia. Oogonia become oocytes upon entering meiosis. They arrest in diplotene of prophase I and enter into a growth and differentiation phase. The chromosomes become highly extended into lampbrush chromosomes (Figure 20.2), and they are very active in RNA synthesis. In many species, genes for ribosomal RNA are selectively replicated, a

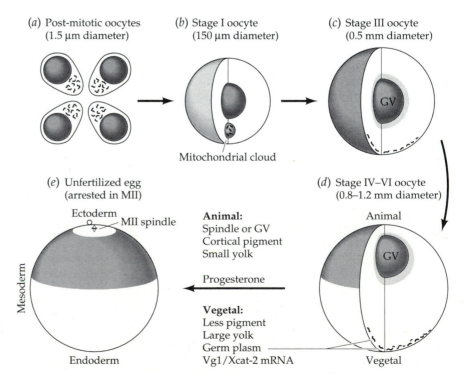

Figure 20.1 Oogenesis in *Xenopus laevis*. Formation of the amphibian oocyte involves growth and polarization. Growth is due mainly to uptake of yolk, while polarization creates an animal-vegetal axis. The animal half is marked by pigment granules and contains a huge nucleus, called the germinal vesicle (GV). The vegetal half has large yolk platelets with localized RNAs and germ plasm at the vegetal cortex. MII, metaphase II. (After Gard 1995.)

(a) Post-mitotic oocytes (1.5 μm diameter)

(b) Stage I oocyte (150 μm diameter)

Mitochondrial cloud

(c) Stage III oocyte (0.5 mm diameter)

GV

(d) Stage IV–VI oocyte (0.8–1.2 mm diameter)

Animal

GV

Vegetal

Animal:
Spindle or GV
Cortical pigment
Small yolk

Progesterone

Vegetal:
Less pigment
Large yolk
Germ plasm
Vg1/Xcat-2 mRNA

(e) Unfertilized egg (arrested in MII)

Ectoderm

MII spindle

Mesoderm

Endoderm

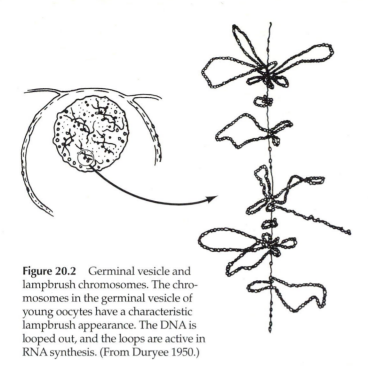

Figure 20.2 Germinal vesicle and lampbrush chromosomes. The chromosomes in the germinal vesicle of young oocytes have a characteristic lampbrush appearance. The DNA is looped out, and the loops are active in RNA synthesis. (From Duryee 1950.)

process known as **gene amplification**, and appear as small nuclear bodies known as **nucleoli** (Gall 1969). This amplification of ribosomal genes allows synthesis of sufficient ribosomes for protein synthesis to supply the growing oocyte and to carry the embryo for the first few days of development. The lampbrush chromosomes and nucleoli are contained in a large nucleus, called the germinal vesicle. As the oocyte starts to grow, mitochondria accumulate on one side of the germinal vesicle, forming the **Balbiani body** or **mitochondrial cloud** (Figure 20.1). In *X. laevis*, the mitochondrial cloud breaks up into smaller clumps and migrates from the germinal vesicle to one surface of the growing oocyte (Heasman et al. 1984). The migration occurs when the oocyte is 400 μm in diameter, and this is the first morphological indication of the future vegetal half of the oocyte. The mitochondria in the cloud are associated with a fibrillar granular material, which can be followed into early development. In the fertilized egg and early embryo of anurans, mitochondria-rich islands of this material are called **germ plasm**, as they end up in a few cells that will become the germ cells of the embryo. These cells in the embryo are called primordial germ cells, and they are the precursors to the sperm or eggs that the animal will eventually produce. The migration of the mitochondrial cloud is obviously an important event because it establishes the polarity of the oocyte and localizes germ plasm to one region. It is not known what initiates the migration or what controls its direction.

The oocyte grows mainly by uptake of yolk, a process called **vitellogenesis**. Yolk precursors, composed of phospholipoproteins, are made in the liver, and are picked up from the circulating blood by the oocytes. The oocyte captures yolk, by pinching off little pieces of membrane, a

mechanism known as pinocytosis. Yolk accumulates as crystalline yolk platelets, which grow larger in the vegetal half than in the animal half. The oldest platelets lie deep within the vegetal mass, while the newest ones are all around the periphery (Figure 20.3) (Danilchik and Gerhart 1987). Shortly after active yolk uptake begins, pigment granules are synthesized within the oocyte and accumulate near the surface of the animal half.

An important molecular accumulation occurs, early in vitellogenesis. While most mRNAs in the oocyte are distributed throughout the cytoplasm, a few are specifically localized to the vegetal cortex, the outer rim of cytoplasm of the vegetal half (Figure 20.1). In effect, the vegetal cortex becomes a source of developmental information for the embryo. These RNAs include Vg1, Xwnt-11, and Xcat-2 (Melton 1987; Ku and Melton 1993; Forristall et al. 1995) and are mentioned here because they are frequently referred to in the analysis of *X. laevis* development. Some of these RNAs are important for the germ plasm, described above, while others may control formation of the embryo's dorsal axis.

The product of oogenesis is a huge, polarized cell, the oocyte. The size of the oocyte depends on the species. Most laboratory species have oocytes which are 1.3–2 mm in diameter. *Nectophrynoides occidentalis* has one of the smallest oocytes at 0.6 mm. This frog is viviparous; the embryos develop in the ovisac of the mother, and they may receive some nutrition from her (Xavier 1977). At the other extreme, some species of marsupial frogs have oocytes that approach 1 cm in diameter (del Pino 1989)! The volume of one of these huge eggs is more than 4000 times greater than that of *N. occidentalis*. The final size of the oocyte seems to be controlled by the surrounding follicle cells, as *X. laevis* oocytes without follicle cells grow larger in culture (Wallace et al. 1981).

Figure 20.3 Oocyte growth. Yolk phospholipoproteins are taken up by the oocyte from the blood and used to make yolk platelets. The oldest and largest platelets, which come from yolk taken up during stages III and IVe, are found deep in the vegetal half, while younger, smaller platelets, formed at stage V, are found around the periphery. (After Danilchik and Gerhart 1987.)

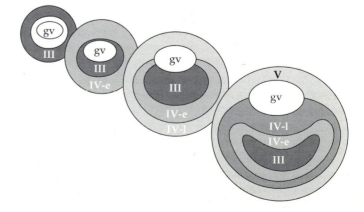

Oocyte Maturation

Full-grown oocytes in the ovary are still diploid and must wait for a hormonal cue from the pituitary gland to resume meiosis. Luteinizing hormone stimulates the follicle cells surrounding the oocyte to produce progesterone, which in turn causes oocyte maturation. Luteinizing hormone also causes ovulation, the release of the oocyte from the ovary.

With oocyte maturation, meiosis resumes. The germinal vesicle breaks down, and a small first meiotic metaphase spindle forms at the animal pole. The oocyte divides, producing a tiny first polar body, and proceeds to metaphase II, where it arrests again (Figure 20.1). Besides this meiotic maturation, the oocyte undergoes cytoplasmic maturation, which includes in *X. laevis* loss of many membranous organelles and changes in the cytoskeleton (Imoh et al. 1983; Klymkowsky and Karnovsky 1994; Gard 1995). Some of the RNAs localized to the vegetal cortex are released and diffuse into the vegetal half, while others remain in small islands, like germ plasm (Forristall et al. 1995).

Oocyte maturation takes about six hours in *X. laevis*, but more typical maturation times are one to two days. With maturation, the oocyte acquires the ability to respond to a sperm or another stimulus and to initiate embryonic development. The matured oocyte is usually called an egg, although technically, it has not completed meiosis and is a secondary oocyte.

Variations

While the goal of meiosis is to reduce the diploid number of chromosomes to the haploid number, some amphibians have bizarre routes to this end. *Ascaphus truei* oocytes start with eight germinal vesicles as a result of three nuclear divisions without cell division in the oogonia (Macgregor and Kezer 1970). Different species of marsupial frogs have 4–3000 germinal vesicles, with *Flectonotus pygmaeus* being the extreme case. These oocytes arise by fusion of many small oocytes into one (del Pino 1989). By the time oocytes of these species are full grown, there is only one germinal

vesicle left, but how this one was selected is unknown.

The European edible frog, *Rana esculenta*, also handles its nucleus in an unusual way, in a mode of reproduction called **hybridogenesis**. *R. esculenta* is actually a hybrid between two species of frogs, *R. lessonae* and *R. ridibunda*. Prior to meiosis, the haploid set of chromosomes from *R. lessonae* is selectively eliminated (Tunner and Heppich-Tunner 1991). The *R. ridibunda* haploid set is duplicated and proceeds through the usual meiotic reduction, so that the hybrid female produces genetically *R. ridibunda* eggs. She then mates with a *R. lessonae* male, thus recreating the *ridibunda–lessonae* hybrid state in the zygote.

Fertilization and Gray Crescent Formation

The general patterns of fertilization are distinctly different in anurans and urodeles. Anuran eggs are inseminated after being laid, while urodele eggs are inseminated in the female's cloaca. Sperm enter the animal half only of the anuran egg, but enter both animal and vegetal halves of the urodele egg. Finally, only one sperm enters the anuran egg (monospermy), while several sperm enter the urodele egg (physiological polyspermy). Because of these differences, fertilization in the two orders will be considered separately.

Anuran Fertilization

The anuran egg is surrounded by jelly coats, secreted around the egg as it passes down the oviduct (Figure 20.4). These coats protect the egg but also provide factors needed by the sperm for fertilization. As the eggs are laid in the water, sperm are deposited over them. The sperm swim through the jelly and digest a hole in the vitelline envelope to reach the surface of the egg. They fuse with the plasma membrane, but only in the animal half. As in most animals, sperm entry activates the egg.

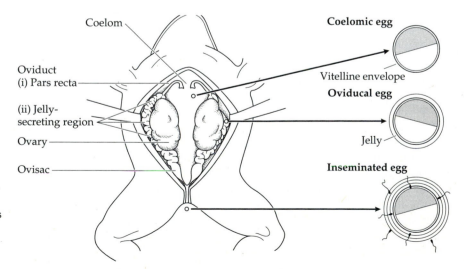

Figure 20.4 Frog reproductive system. Oocytes develop in the ovary. With hormonal stimulation, they are released into the body cavity (coelom) and enter the oviducts, where they acquire jelly coats required for fertilization. Eggs accumulate in the ovisac and are inseminated after laying. Size of sperm is greatly exaggerated. (After Elinson 1986.)

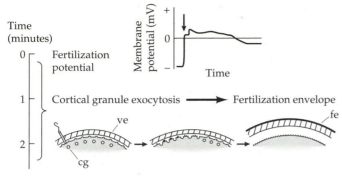

Figure 20.5 Anuran blocks to polyspermy. Within milliseconds of sperm entry of an anuran egg, the fertilization potential serves as a fast block to polyspermy. The fertilization envelope (fe), formed due to cortical granule (cg) exocytosis, provides a slower, more permanent block to polyspermy. (After Elinson 1986.)

The first event of activation is a depolarization of the egg membrane (Figure 20.5). This change in membrane potential serves as a **fast block to polyspermy**, preventing entry of additional sperm (Cross and Elinson 1980). In addition to this change in membrane potential, a second **slow block to polyspermy** is initiated as calcium ions are released in a wave around the egg, starting from the point of

sperm entry (Kubota et al. 1987). The Ca^{2+} causes cortical contraction and exocytosis of the cortical granules. Materials from the cortical granules interact with the vitelline envelope and the inner jelly to produce the **fertilization envelope** (Figure 20.5), and the fertilization envelope physically prevents sperm from reaching the egg (Schmell et al. 1983; Hedrick and Nishihara 1991). Sperm are unable to digest a hole in this modified envelope. Some cortical granule materials are trapped between the egg surface and the fertilization envelope, causing water to be driven osmotically into the space beneath the envelope. As a result, the fertilization envelope lifts off the egg surface, and the fertilized egg is free to rotate with respect to gravity. The dense vegetal half rotates downwards, so that all of the fertilized eggs end up with their animal poles uppermost. This rotation of activation, which usually occurs within a half hour of sperm entry, is the clearest sign of fertilization.

The migration of the sperm nucleus from the cortex into the cytoplasm is due to the **sperm aster**, formed by microtubules growing from a small organelle called the **centrosome** (Manes and Barbieri 1977; Stewart-Savage and Grey 1982). The sperm nucleus stays in the center of the aster, as astral growth pushes the sperm nucleus radially from the surface. The path of the sperm nucleus is marked by pigment granules, and the initial radial movement is called the **penetration path** (Figure 20.6). The sperm nucleus then turns towards a point about one-third

Fkigure 20.6 Pronuclear migration. The male and female pronucleus meet in the center of the sperm aster, and the gray crescent will tend to form on the side opposite to the sperm entry point (SEP). (After Hausen and Riebesell 1991.)

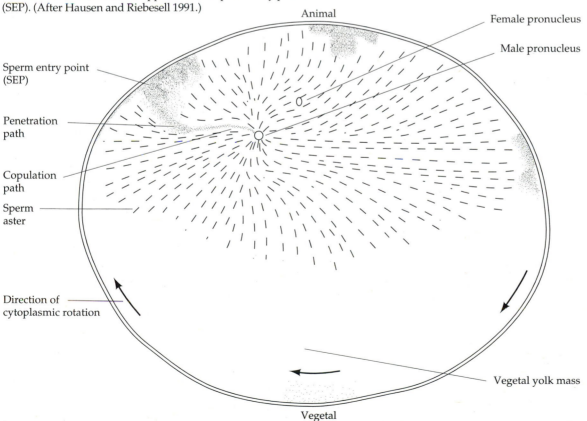

of the egg diameter beneath the animal pole, and this movement is known as the **copulation path**. The change in direction is probably due to astral pushing against the vegetal yolk mass, as well as against the animal cortex. The female nucleus, arrested at metaphase II, resumes meiosis due the Ca^{2+} inactivation of **cytostatic factor** (CSF) (Masui 1992). The second polar body is released, and the female nucleus migrates from the surface into the cytoplasm. Guided by the sperm aster, the female nucleus encounters the sperm nucleus in the vastness of the cytoplasm.

Urodele Fertilization

Urodeles have an elaborate courtship dance. A male attaches a spermatheca, consisting of a clump of jelly with a tuft of sperm on top, to the substrate. He leads the female over the spermatheca, and she picks it up with her cloaca, where the eggs are inseminated. The eggs are then laid in the water.

Sperm swim through the jelly and enter the egg in both the animal and vegetal halves. There is no membrane depolarization, no wave of Ca^{2+}, no cortical granules, and no fertilization envelope (Iwao 1985; Elinson 1986; Grandin and Charbonneau 1992). Several sperm enter since there is no external block to polyspermy, as generated in the anuran egg. Sperm entry points are marked by accumulations of pigment granules and small pits.

The entering sperm all produce asters and start migration, but one, called the **principal sperm nucleus**, reaches the female nucleus in advance of the others. The accessory sperm nuclei are pushed peripherally and vegetally by the dominant aster of the principal sperm nucleus, and the principal sperm nucleus joins the female one to produce the zygote nucleus. The centrosome, derived from the principal sperm nucleus, replicates, and the daughter centrosomes form the poles of the metaphase spindle. The spindle segregates the zygotic mitotic chromosomes as usual and stimulates the cleavage furrow of the first division. Meanwhile the accessory nuclei fail to duplicate their centrosomes and produce a small aster, called a monaster, instead of a spindle with an aster at each pole. The accessory nuclei become condensed and disappear as the egg divides. This unusual elimination of the accessory sperm nuclei is thought to be due to localization of cell cycle factors, near the zygote nucleus (Iwao et al. 1993).

Gray Crescent Formation

About an hour or two after fertilization in anurans, the egg cortex rotates 30 degrees relative to the inner cytoplasm (Figure 20.7). The cortical rotation, distinct from the rotation of activation involving the entire egg, specifies dorsoventral polarity of the embryo (Gerhart et al. 1989). The site of second polar body formation at the animal pole moves towards the **sperm entry point** (SEP), while the SEP moves vegetally. Vegetal cortex moves animally on the side of the egg *opposite* the SEP, and this cortical shift

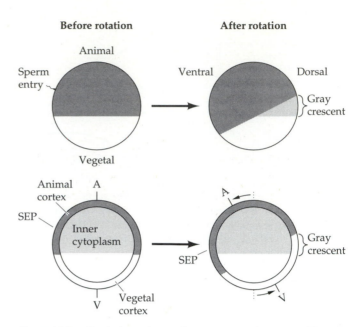

Figure 20.7 Cortical rotation and gray crescent formation. The egg cortex, exaggerated in this illustration, rotates by about 30 degrees relative to the cytoplasm. The cortical rotation forms gray crescent and specifies dorsoventral polarity. SEP, sperm entry point. (After Elinson and Rowning 1988.)

brings about the formation of the gray crescent. The gray crescent is due to the overlying of pigmented animal cytoplasm by nonpigmented vegetal cortex, and the gray crescent marks the future dorsal side of the embryo.

Although a cortical rotation occurs in *X. laevis*, a gray crescent is not visible, as both the animal and vegetal cytoplasm have little pigment. Concomitant with cortical rotation in *X. laevis*, the cytoplasm undergoes extensive rearrangements. The expanding sperm aster pushes a yolk-poor area to one side and a tongue of equatorial cytoplasm swirls towards the animal pole along the dorsal side (Figure 20.8). These movements may be due to relative cytoplasmic viscosities, as they can be mimicked in simple physical models (Danilchik and Denegre 1991).

In anurans, the first cleavage division occurs 1.5–3.5 hours after fertilization at 20°C, depending on the species. The first furrow usually passes through the animal pole and the SEP and bisects the gray crescent (Figure 20.9), since the sperm aster is normally the major influence on the direction of the cortical rotation. Urodele eggs also form a gray crescent, but it is not known whether the aster of the principal sperm nucleus provides the directional cue for cortical rotation. The first cell cycle in urodele eggs characteristically is much longer than that in anuran eggs, typically taking 6–10 hours.

Variations

It is generally true that anuran fertilization occurs externally and is monospermic, while urodele fertilization is internal and polyspermic. There are, however, exceptions. Some

Fertilized

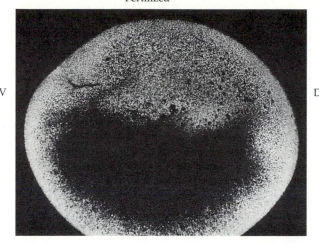

V   D

First cleavage

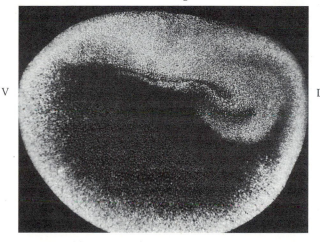

V   D

Fourth cleavage

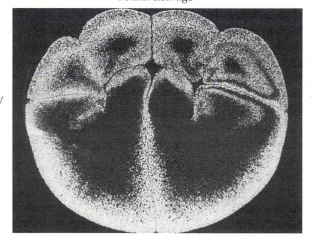

V   D

Figure 20.8 Cytoplasmic rearrangements. The cytoplasm rearranges extensively between fertilization and first cleavage as a result of cortical rotation. There are obvious differences between the dorsal (D) and ventral (V) sides at first cleavage. With each cleavage furrow, cortical cytoplasm is drawn inwards along the furrow. (From Danilchik and Denegre 1991.)

which functions as a fast block to polyspermy (Iwao 1989). Polyspermic fertilization has been reported in the anurans *Discoglossus pictus* and *E. coqui* (Talevi 1989; Elinson 1987a).

A more unusual variant of fertilization occurs among mole salamanders (*Ambystoma* spp.). Some species are triploids and only females are found. The females produce triploid eggs, which are fertilized by normal sperm of neighboring diploid species. The sperm stimulate development and provide the centrosomes for mitosis, but their nuclei all degenerate, and they do not contribute genetically to the resulting triploid female offspring. Use of sperm only to initiate development is called **gynogenesis**, a term indicating that all genes are from the mother. To confuse matters further, a sperm nucleus may survive if the mating pond is warm, and a hybrid salamander with the tetraploid number of chromosomes results (Bogart et al. 1989). The ability to develop without a sperm nucleus illustrates the egg's primary importance in producing the embryo.

Cleavage and Blastula Formation

In the idealized cleavage pattern, the first three furrows are perpendicular to each other (Figure 20.10). Both the first and second cleavage furrow start at the animal pole and proceed vegetally, while the third cleavage is latitudi-

Figure 20.9 Gray crescent. The first cleavage furrow tends to bisect the gray crescent in *Rana pipiens* eggs.

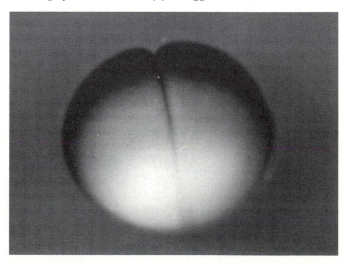

primitive urodeles have external fertilization, while the anurans *Ascaphus truei* and *Eleutherodactylus coqui* have internal fertilization (Duellman and Trueb 1986; Townsend et al. 1981). Similarly, the urodele *Hynobius nebulosus* is monospermic and has a change in fertilization potential,

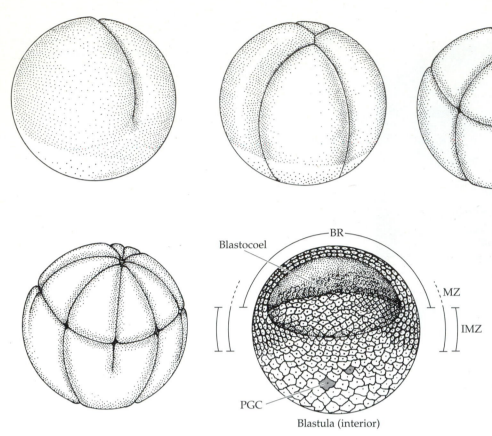

Blastocoel

BR

MZ

IMZ

PGC

Blastula (interior)

Figure 20.10   Cleavage of eggs of most amphibians is regular and synchronous, yielding the blastula. The blastocoel roof (BR) covers the blastocoel and overlaps with the marginal zone (MZ), which gives rise to much of the mesoderm. The involuting marginal zone (IMZ) is that part of the embryo that will involute during gastrulation. The lower boundary of the IMZ is the site of blastoporal lip formation, and the upper boundary closes down to become the posterior point of the embryo at the end of gastrulation. Primordial germ cells (PGC) in anurans are found among the vegetal yolky cells. (After Huettner 1949.)

nal, i.e., parallel to the equator. The third cleavage is closer to the animal pole than the vegetal pole, so the four animal blastomeres at the 8-cell stage are smaller than the four vegetal blastomeres. The animal-vegetal fourth cleavage (Figure 20.10) and the latitudinal fifth cleavage produce a 32-cell embryo with four tiers of eight cells each. Deviations from this idealized pattern are common, and they have no developmental consequences.

As cleavage proceeds, the original surface of the egg remains on the surface, while new membrane makes up most of the internal cleavage furrows (Byers and Armstrong 1986). Besides this internalization of membrane, egg cortical cytoplasm is drawn inwards along the cleavage furrows, so that most blastomeres contain a mixture of egg cortical and internal cytoplasm (Figure 20.8) (Danilchik and Denegre 1991). Cleavage obviously causes considerable rearrangement of surface and cytoplasm, leading to new associations of major egg components. As early as first cleavage, a gap forms between the new furrow membranes in the animal half. The gap expands into a cavity, known as the blastocoel (Figure 20.10), as cleavage proceeds (Kalt 1971). Blastocoel fluid has a salt concentration more than 10 times higher than pond water (Gillespie 1983) and maintains a distinct environment for the internal cells.

Cleavage divisions are synchronous and rapid, each taking half an hour in *Xenopus*, about an hour in *Rana*, and around 1.5 to several hours in various urodeles. Syn-

chrony is lost at the 12th division, the important time known as the **midblastula transition** (MBT) (Signoret and Lefresne 1971; Newport and Kirschner 1982). Prior to MBT, cells go from DNA synthesis (S) to mitosis (M) without intervening gaps (G1, G2), normally found in interphase. With MBT, the G1 period becomes significant, embryonic transcription begins, and development starts to pass from maternal to zygotic control.

The blastula, which results from cleavage, has a characteristic arrangement of cells surrounding the blastocoel (Figure 20.10). The blastocoel roof is several cells thick. The outer cells retain the original egg surface and form an epithelial layer, while the inner cells form looser layers. Pieces of blastocoel roof, symmetric around the animal pole, are frequently dissected for experiments in *X. laevis*. These pieces are called **animal caps**, and they are defined by their size and embryonic stage of origin. When animal caps are placed experimentally in direct contact with vegetal cells, the animal cells form mesoderm such as blood and muscle, instead of ectoderm, such as skin and nerves (Nieuwkoop 1969; Dale and Slack 1987). This signaling from vegetal cells telling animal ones to form mesoderm is called **mesoderm induction** and illustrates another role of the blastocoel. The blastocoel separates animal cells from vegetal ones, except in the marginal zone, so signaling to produce mesoderm is restricted to that zone of contact.

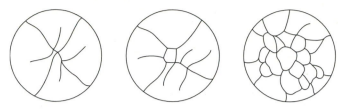

Figure 20.11 Irregular cleavage. Large eggs, such as those of *Gastrotheca riobambae*, often show irregular cleavage, in contrast to that seen in most other amphibians. (From del Pino and Loor-Vela 1990.)

The vegetal half consists of larger, yolky cells, and interspersed among them in anurans are a few dozen primordial germ cells (Figure 20.10). The primordial germ cells are the precursors to the germ cells, and they contain germ plasm, which was originally located at the vegetal surface. The ingression of cortical cytoplasm along cleavage furrows probably accounts for the internalization of the germ plasm and the embedding of primordial germ cells deep within the mass of yolky cells (Savage and Danilchik 1993).

Besides the blastocoel roof and the vegetal half, an area near the equator is defined as the **marginal zone**. The vegetal limit of the marginal zone corresponds approximately with the gray crescent/vegetal boundary (Figure 20.10). During gastrulation, **bottle cells** form near the boundary, marking the vegetal limit of involution as discussed in the next section. The animal limit of the marginal zone is defined by fate and cell behaviors rather than by morphology and extends into the blastocoel roof (Keller et al. 1985; Shi et al. 1987; Johnson et al. 1992; Keller and Jansa 1992). A better-defined region is the **involuting marginal zone** (IMZ), marked by the animal and vegetal limits of involution during gastrulation. The IMZ contains the prospective mesoderm, and movements of these cells drive gastrulation.

Variations

Cleavage in commonly studied amphibians looks similar, even to the extent that MBT was originally described in the 2-mm egg of the urodele *Abystoma mexicanum* (Signoret and Lefresne 1971) and subsequently investigated in the 1.3-mm egg of the anuran *Xenopus laevis* (Newport and Kirschner 1982). In larger eggs (greater than 3 mm), the cleavage furrows are less symmetrical, producing an irregular pattern of cells near the animal pole (Figure 20.11) (Smith 1912; Gitlin 1944; del Pino and Loor-Vela 1990). The furrows progress slowly through the vegetal

half, eventually cellularizing that region. Eggs as large as 7 mm have complete (holoblastic) cleavage (Elinson 1987b), and this is an invariant feature of amphibian development. The large eggs (3 mm) of *Gastrotheca riobambae* have slow, asynchronous cleavage. Nucleoli are visible as early as the 8-cell stage, indicating that these embryos do not have an MBT (del Pino and Loor-Vela 1990).

Gastrulation

Fate Maps

Gastrulation is a complicated rearrangement of regions of the blastula, and it generates the basic arrangement of tissues in the embryo (Figure 20.12). The complexities of gastrulation make it useful to construct fate maps in order to see how the different regions of the blastula contribute to the body plan. Two types of fate maps have emerged. The classical fate map of Vogt (1929), based on several species

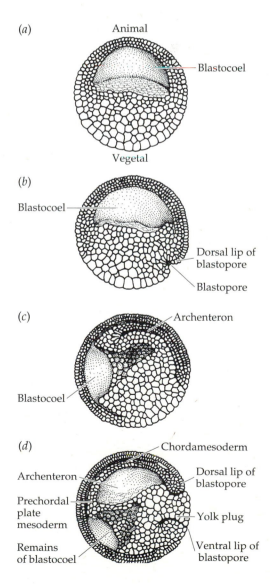

Figure 20.12 Gastrulation. Gastrulation is a complex movement of cells, setting up the tissue patterns of the embryo. Movements start on the dorsal side, seen from the surface as the dorsal lip of the blastopore. A new internal cavity, the archenteron or primitive gut, forms, while the old cavity, the blastocoel is squeezed out. (After Balinsky 1981.)

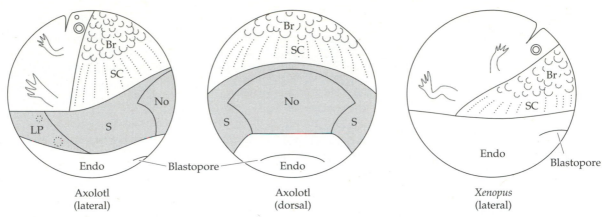

Axolotl
(lateral)

Axolotl
(dorsal)

Xenopus
(lateral)

of urodeles and anurans has mesoderm on the surface, while the *X. laevis* fate map lacks surface mesoderm (Figure 20.13) (Keller 1975). *X. laevis* appears distinctive in this regard, as the surface provides varying mesodermal contributions in all other anurans and urodeles which have been examined (Vogt 1929; Pasteels 1942; Smith and Malacinski 1983; Brun and Garson 1984; Lundmark 1986; Delarue et al. 1992; Purcell and Keller 1993).

Gastrulation is easier to follow in *X. laevis* than in other amphibians because of this surface peculiarity, and the surface and internal movements can be considered separately. The *X. laevis* blastula surface, fated to be ectoderm and endoderm, remains as a surface through gastrulation (Figure 20.14). The endoderm surface is folded inside to line an inner cavity, the archenteron, while the ectoderm surface expands over the exterior to replace the departing endoderm surface. Later, the ectoderm forms epidermis and neurectoderm, and the neurectoderm surface lines another internal cavity, the **neurocoel** of the neural tube.

The internal prospective mesoderm is a ring of tissue in the marginal zone. The ring turns itself inside-out and

Figure 20.13   Fate maps. The animal surface of the *Ambystoma mexicanum* (axolotl) embryo forms the epidermis, covering the whole body, as well as the brain (Br) and spinal cord (SC). Mesoderm (shaded) forms the notochord (No), somites (S), and lateral plate (LP). The mesodermal component of the limbs forms from lateral plate mesoderm (dotted circles). Endoderm is found above and below the blastopore. Endoderm above the blastopore forms the archenteric roof, while that below, forms the archenteric floor. In contrast to most other amphibians, the surface of the *Xenopus laevis* gastrula is fated to form ectoderm (epidermis and nervous system) and endoderm, but not mesoderm. *X. laevis* mesoderm forms from internal cells only. (After Pasteels 1942; Keller 1975.)

Figure 20.14   *Xenopus laevis* gastrulation pattern. (Top) Surface cells remain on the surface through gastrulation. Prospective endoderm (dotted) moves inside to line an internal cavity, the archenteron. Prospective ectoderm (heavy line) stretches to cover the embryo as the endoderm goes inside. (Bottom) The mesoderm (stippled) is an internal ring, which turns inside out and stretches along the dorsal side. Anterior tissues, such as heart, start on the vegetal edge of the mesodermal ring. Bl, blastocoel; Br, brain; SC, spinal cord; No, notochord.

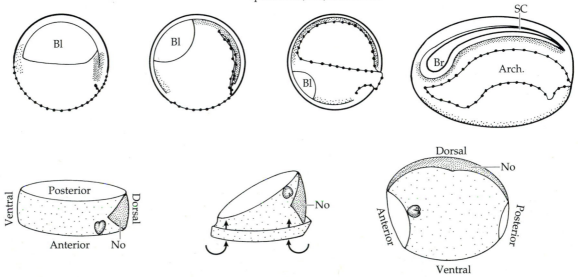

elongates along its dorsal side (Figure 20.14). The elongated side forms the notochord and somitic mesoderm, which make up the mesoderm of the dorsal axis. Combining the surface movements with those of the internal mesodermal ring make it relatively easy to visualize gastrulation. The resulting embryo has an outer ectoderm, a middle notochord and mesoderm, and an inner endoderm surrounding the archenteron.

This view of *X. laevis* gastrulation highlights the difficulty in following gastrulation in other amphibians. Since mesoderm is on the surface originally, it will form the archenteric roof in place of endoderm if no other migrations occur. Consequently, the mesoderm must leave the surface layer during gastrulation, adding to the complexity of movements. In the urodele *A. mexicanum*, prospective somitic cells leave the surface by ingression, which involves a change in cell shape. Their apices contract, and they elongate into bottle cells (Lundmark 1986). The prospective somite cells pass through the blastopore, where they migrate along the blastocoel roof between the ectoderm and endoderm. The prospective notochord cells pass inside at the blastopore, but unlike the somitic cells, they stay on the surface as part of the archenteric roof. Their apical surfaces then contract as the notochord rod forms, and endoderm on either side of the forming notochord joins together to create a continuous sheet of endoderm (Brun and Garson 1984). In the frog *Ceratophrys ornata*, surface cells of both prospective notochord and somite ingress from the archenteric roof to form a middle layer of mesoderm (Purcell and Keller 1993).

Gastrulation Movements

The overall pattern of gastrulation (Figure 20.12) is due to local changes in cell behavior. The first external sign of gastrulation is the formation of the dorsal lip of the blastopore. A group of cells, lying near the boundary of the gray crescent and the vegetal half, constrict their apical surface and elongate into a bottle shape. The formation of these bottle cells causes a pigmented surface depression, which becomes the dorsal lip of the blastopore (Figure 20.15). As gastrulation proceeds, bottle cells continue to form around the embryo's surface to produce lateral lips and finally the ventral lip of the blastopore. Bottle cells were thought to drive gastrulation, as transplantation of bottle cells into a layer of endoderm causes an invagination (Holtfreter 1944). Gastrulation, however, can proceed reasonably nor-

mally in *X. laevis* following removal of the bottle cells, demonstrating that they are not the main engine of gastrulation (Keller 1981). Formation of the archenteron can be followed with the bottle cells, as they are located at the boundary between the roof and the floor of the archenteron (Figure 20.12).

Led by the bottle cells, the prospective endodermal surface leaves the surface at the blastopore lip. With the movement of the endoderm inside, the ectoderm expands in a process known as epiboly. Epiboly involves two changes as the ectoderm becomes thinner during its expansion. Outer ectodermal cells change their shape, while cells of several inner ectodermal layers interdigitate to produce a single inner layer (Figure 20.15) (Keller 1980). Inside the gastrula, cells of the marginal zone migrate up along the blastocoel roof (Figures 20.12, 20.14). As with invagination at the blastopore lip, this migration begins earlier and is much stronger on the dorsal side than elsewhere, but eventually migration occurs all the way around. Cells in the marginal zone initially move vegetally and then turn animally up the roof. This turning is

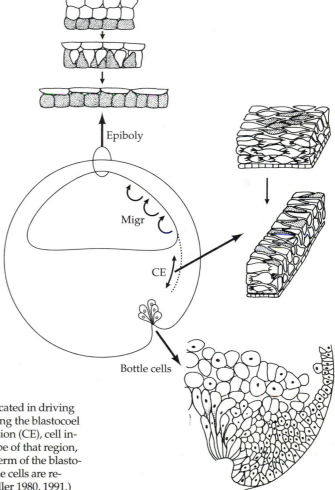

Figure 20.15 Gastrulation mechanisms. Several forces have been implicated in driving gastrulation. With directed mesodermal migration (Migr), migration along the blastocoel roof pulls tissues towards the animal pole. With convergence and extension (CE), cell interdigitation in the dorsal marginal zone narrows and elongates the shape of that region, driving movements in other parts of the embryo. Stretching of the ectoderm of the blastocoel roof is known as epiboly and results from cell rearrangements. Bottle cells are responsible for invagination at the blastopore lip. (After Balinsky 1981; Keller 1980, 1991.)

called involution and corresponds to the inside-out turning of the mesodermal ring (Figure 20.14).

Two forces drive gastrulation: directed cell migration and convergence and extension (Figure 20.15). With the former, cells migrate towards the animal pole due to interactions with extracellular matrix (Johnson et al. 1992). This migration pulls the bottle cells and archenteron towards the future anterior end of the embryo. With convergence and extension, cells of the marginal zone repack along the dorsal midline (Figure 20.15). Interdigitation of these cells causes elongation of the dorsal midline, in a direction perpendicular to movement of individual cells. The extension of the dorsal midline drives the archenteron forward towards the future anterior end of the embryo and drives the blastoporal lip backwards over the yolk mass (Keller and Winklbauer 1992).

The relative importance of directed cell migration and of convergence and extension in the formation of the dorsal axis appears to vary among species. Disruption of interactions between migrating cells and the extracellular matrix in the blastocoel halts gastrulation in *Pleurodeles waltl*, *Ambystoma mexicanum*, and *Rana pipiens* (Johnson et al., 1992, 1993), suggesting that directed cell migration drives gastrulation in these species. In *X. laevis*, however, much of gastrulation, including archenteron formation and blastopore closure, can occur in the absence of a blastocoel roof, indicating that mesodermal migration is not

required for these events (Keller and Jansa 1992).

The end of gastrulation is indicated by movement of all of the endodermal surface to the interior and blastopore closure. In many species, blastopore closure occurs before any sign of neural development is apparent, while in others, the neural plate forms before closure (Ishikawa 1908; Smith 1912). As mentioned above, blastopore closure can occur without a blastocoel roof in *X. laevis*, showing that neither ectodermal epiboly nor mesodermal migration is required. One possible mechanism is that continued convergence and extension pulls the lips over the yolk, forming the yolk plug (Keller and Jansa 1992). There are likely other mechanisms, however. Embryos that lack any dorsal axial development and that probably also lack convergent extension activities nonetheless close the blastopore (Malacinski et al. 1977).

Variations

As already discussed, a number of differences distinguish gastrulation in different species. Among these are the contributions of surface cells to mesoderm, the associations between migrating marginal zone cells and the blastocoel roof, and the timing of blastopore closure. Epibolic thinning of the blastocoel roof is so extreme in embryos from large eggs that a conspicuous transparent window into the blastocoel is created (Figure 20.16). Nonetheless, most gastrulae examined look similar in terms of blastopore formation and closure, with the exception of *Gastrotheca riobambae* (Figure 20.16) (del Pino and Elinson 1983; Elinson and del Pino 1985). Its gastrula differs by having an inconspicuous blastopore lip displaced towards the vegetal pole. Gas-

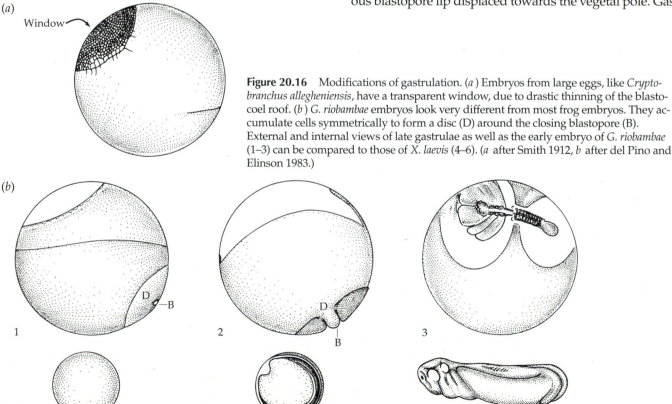

(a)

Window

(b)

1   2   B   3

4   B   5   6

Figure 20.16   Modifications of gastrulation. (*a*) Embryos from large eggs, like *Cryptobranchus alleghaniensis*, have a transparent window, due to drastic thinning of the blastocoel roof. (*b*) *G. riobambae* embryos look very different from most frog embryos. They accumulate cells symmetrically to form a disc (D) around the closing blastopore (B). External and internal views of late gastrulae as well as the early embryo of *G. riobambae* (1–3) can be compared to those of *X. laevis* (4–6). (*a* after Smith 1912, *b* after del Pino and Elinson 1983.)

trulation produces a distinct disc of small cells, overlying a small central archenteron, which later expands anteriorly. Closure of the blastopore in *G. riobambae* is uncoupled from anterior extension of axial tissue, suggesting that convergence and extension plays a minor role in closure.

Spemann's Organizer and Neural Development

Spemann's Organizer

The amphibian embryo develops by successive cell interactions known as inductions. The cortical rotation in the first cell cycle forms the gray crescent and specifies the dorsal axis of the embryo (Gerhart et al. 1989). At the 32-cell stage in *X. laevis*, eight cells along the dorsal side (two cells of each tier) carry dorsal information, with the highest activity in the third tier (Figure 20.17) (Kageura 1990). The dorsal cells of the third and fourth tier and their descendants have been named the **Nieuwkoop center** (Gerhart et al. 1989). The Nieuwkoop Center induces more animal cells to form dorsal mesoderm, while forming endoderm itself. The signals involved in this mesoderm induction include growth factors such as activin and fibroblast growth factor, the latter in conjunction with members of the *wnt* signaling pathway (Kessler and Melton 1994; Elinson and Holowacz 1995).

These early inductions during the blastula stage lead to the formation of Spemann's organizer in the gray crescent region (Figure 20.17) (Spemann and Mangold 1924). Spemann's organizer is centered at the dorsal lip of the blastopore and has powerful inductive properties. Its cells undergo the most extreme convergence and extension movements during gastrulation, and they later form dorsal axial structures such as notochord, somitic mesoderm, and pharyngeal endoderm. Spemann's organizer can induce mesoderm to form more dorsal mesoderm, such as somites, and it induces ectoderm to form neurectoderm of the central nervous system. This latter activity is called **neural induction**.

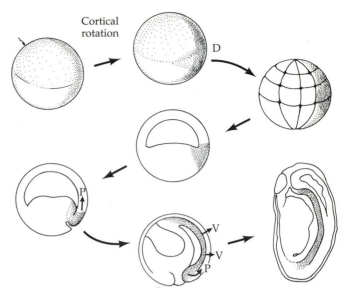

Figure 20.17 Spemann's organizer. The cortical rotation forms the gray crescent, specifying the dorsal (D) side. Dorsal information is spread along one side of the cleaving embryo and becomes concentrated in a region of the blastula (stippled) to form Spemann's organizer. Spemann's organizer signals in a planar (P) and in a vertical (V) direction during gastrulation to induce the ectoderm to form the central nervous system. (From Elinson and Holowacz 1995.)

Until recently, the boundaries of Spemann's organizer were recognized experimentally as an area with induction activity. Inducing cells do not differ morphologically from noninducing neighboring cells. A number of genes, however, are expressed specifically in the dorsal lip in precisely the area defined experimentally as Spemann's organizer (De Robertis 1995). Detection of RNAs from these genes by in situ hybridization effectively "paints" Spemann's organizer (Figure 20.18).

There are two signaling paths from Spemann's organizer involved in neural induction, and these are called vertical and planar (Figure 20.17). In **vertical signaling**, the involuted mesoderm or the archenteric roof induces the overlying ectoderm to form neurectoderm. This induction occurs as

(a)

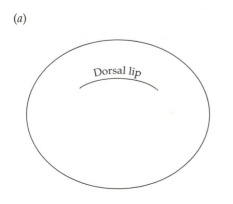

Dorsal lip

(b)

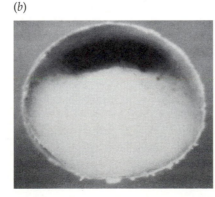

Figure 20.18 Gene expression in Spemann's organizer. In this dorsal view of a *X. laevis* gastrula, *Xnot* RNA is detected in the region defined embryologically as Spemann's organizer. (Photograph courtesy of Linda Gont.)

the mesoderm migrates anteriorly during gastrulation. In **planar signaling**, signals from the dorsal lip pass through the ectoderm towards the animal pole. The degree of anterior neural specification increases with increasing distance from the dorsal lip. Both of these paths exist in experimental situations, although experiments showing planar signaling from Spemann's organizer have only been done in *X. laevis* (Doniach et al. 1992; Doniach 1993; Ruiz i Altaba 1993).

Neurulation

As a result of neural induction, the ectoderm overlying the dorsal axial mesoderm forms the neural plate (Figure 20.19). Ectodermal cells become columnar, and further cell shape changes, involving constrictions, produce neural folds at the boundary between the epidermis and neurectoderm (Figures 20.19 and 20.20). Neural folds are prominent in urodeles, but very weak and difficult to judge in *X. laevis*. The columnarization and constrictions of the neural cells are accompanied by changes in the cells's cytoskeleton (Burnside 1973). Microtubules run the length of the cells, and their polymerization may drive or stabilize the columnar shape. Microfilaments form bands around the apices of the cells, and they may cause constriction of those ends. Disruptions of the cytoskeleton with drugs can alter cell shape and progress of neurulation (Karfunkel 1974; Brun and Garson 1983).

The neural plate takes on a broad keyhole shape, wide at the head end where the forebrain will form and narrow posteriorly where the spinal cord forms (Figure 20.21). Cell elongation and apical constriction is more active in the head end of the plate (Gordon and Jacobson 1978). In addition to changes of cell shape, the keyhole morphology depends on extreme narrowing and elongation of the posterior neural plate, as best illustrated by comparison of neural fate maps at the beginning of gastrulation and at the neural plate stage (Figure 20.21) (Suzuki and Harada 1988; Eagleson and Harris 1990; Keller et al. 1992). The elongation is most active in the **notoplate**, which is the midline of the neural plate immediately over the forming notochord (Gordon 1985), and several mechanisms have been proposed.

The notochord and neural plate adhere to each other, suggesting that elongation of the notochord could cause corresponding stretching of the neural plate. Elongation in culture of the neural plate isolated from *A. mexicanum* required an associated notochord (Jacobson and Gordon 1976; Gordon 1985). In contrast, posterior neurectoderm from *X. laevis* elongated under different culturing conditions without an underlying notochord. This autonomous

Figure 20.19   Neurulation. As a result of induction, the neurectoderm forms a neural plate, bounded by neural folds. The folds arch up and fuse along the embryo's dorsal midline to create the neural tube. (After Balinsky 1981.)

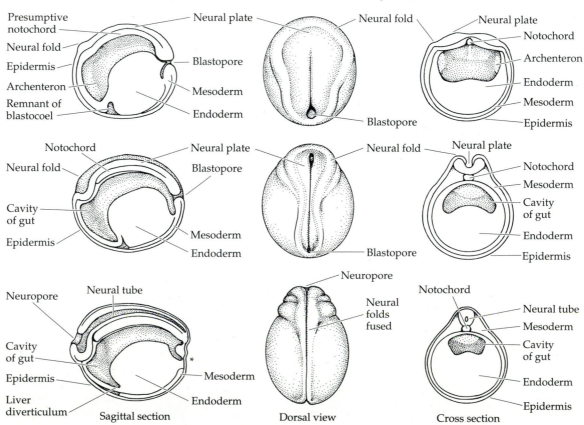

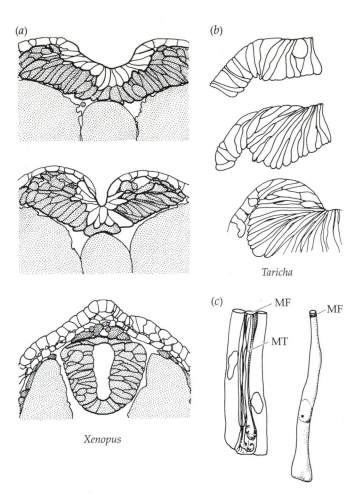

Taricha

Xenopus

Figure 20.20 Cell shape changes in neurulation. (*a*) The neural plate of the anuran *X. laevis* has an outer epithelial layer and an inner sensorial layer (shaded), both of which contribute to the neural tube. (*b*) In contrast, the neural plate of the urodele *Taricha torosa* is one-cell-layer thick. During neurulation, the neurectodermal cells elongate and undergo contrictions at one end, associated with changes in their cytoskeleton. (*c*) Microtubules (MT) align with the long axis of the cells, while microfilaments (MF) are found at constricted ends. (After Schroeder 1970; Jacobson et al. 1986; Burnside 1973.)

ectoderm elongation occurs by convergence and extension, similar to that found in dorsal mesoderm during gastrulation (Keller et al. 1992). These differences between *A. mexicanum* and *X. laevis* could be due to the different experimental conditions, the two species themselves, or an anuran–urodele difference. *X. laevis* appears to make more use of convergence and extension in gastrulation, and its rapid development compared to *A. mexicanum* (15 hours versus 2.5 days for the neural plate stage) may require

more reliance on certain mechanisms. There may however be anuran–urodele differences. Although only a limited number of species have been examined, urodele neurectoderm is one-cell-layer thick (Jacobson et al. 1986; Imoh 1988), while anuran neurectoderm consists of an outer epithelial layer and an inner sensorial layer (Knouff 1935; Schroeder 1970). This difference in cell composition of the neurectoderm (Figure 20.20) could easily affect morphogenesis of the neural plate.

The neural plate forms into the neural tube by rolling up of the neural folds and fusing at the dorsal midline (Figure 20.22). As with formation and elongation of the neural plate, many theories have been proposed for neural tube formation (Gordon 1985; Jacobsen et al. 1986; Schoenwolf and Smith 1990; Clausi and Brodland 1993). Most of the theories include a continuation of processes, such as cell shape change and tissue elongation, that produced the neural plate. A major unsolved problem is whether rolling up of the neural folds is due to forces generated within the neural plate or whether forces from neighboring epidermis or underlying mesoderm are involved (Jacobson and Gordon 1976; Schoenwolf and Smith 1990; Clausi and Brodland 1993).

Besides the brain and spinal cord, another important population of cells results from neural induction, and these are known as neural crest cells. Neural crest cells appear at the boundary of the neural plate and the epidermis (Figure 20.22), and migrate down away from the neural folds to populate different parts of the body. As discussed

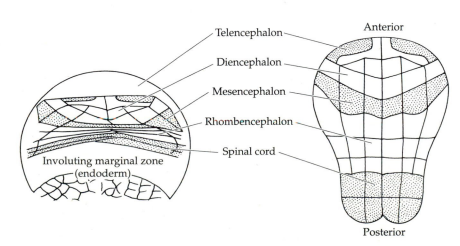

Figure 20.21 Neural plate formation. Prospective neurectoderm at the beginning of gastrulation (left) is rearranged to a typical keyhole shape by the middle of neurulation (right). (After Keller et al. 1992.)

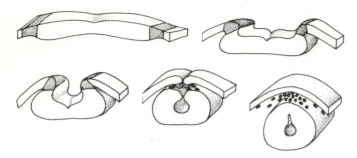

Figure 20.22   Neural crest. Neural crest forms from cells at the juncture between the neurectoderm and the epidermis. (After Balinsky 1981.)

in the next section, neural crest cells play a major role in formation of the head and they form many different cell types, including pigment cells and sensory neurons.

Head Development

As a result of gastrulation and neurulation, all of the tissues are arranged in the order and pattern of the animal's prospective organs. The head of the embryo has very different developmental patterns and tissue arrangements from the trunk and tail, so it will be considered separately. The head contains the brain, the sense organs, and the jaws; the origin of all of these parts depends on neural induction. The brain forms from the broad anterior neural plate; the sense organs originate as thickenings adjacent to the neural plate known as **ectodermal placodes**, and most of the jaw and face cartilage is derived from cranial neural crest.

With neural fold closure, the anterior end undergoes extensive morphogenesis to produce the face. This modeling can be seen in *X. laevis* by following the formation of two glands, derived from the outer epithelial layer of ectoderm (Figure 20.23) (Drysdale and Elinson 1991). The **hatching gland**, used for hatching from the jelly capsule, comes from the anterior neural folds, while the **cement gland**, used for attaching the hatched embryo to plants

and other surfaces, develops just anterior to the transverse (rostral) neural fold. The glands end up encircling the developing nose and mouth, while the eyes lie just outside the hatching gland.

Brain

As in other vertebrates, the forming brain has various constrictions and bulges, allowing several regions to be identified (Figure 20.24). The most anterior region is the forebrain, or **prosencephalon**, which is further divided into the **telencephalon** and **diencephalon**. These are followed by the midbrain, or **mesencephalon**, and the hindbrain, or **rhombencephalon**. The telencephalon expands bilaterally to produce the left and right cerebral hemispheres and olfactory lobes (Figure 20.24). Amphibian cerebral hemispheres remain small, unlike in mammals where they make up most of the brain. The diencephalon produces the pineal gland, posterior pituitary gland, hypothalamus, and optic vesicles. The pineal forms from the roof (dorsal surface) of the diencephalon, and an extension to the epidermis becomes the light-sensitive pineal end organ (Figure 20.24). This so-called third eye can be seen as a spot between a frog's eyes.

The pituitary gland consists of the anterior pituitary (adenohypophysis), derived from mouth ectoderm, and the posterior pituitary (neurohypophysis), derived from the floor of the diencephalon. The pituitary gland is under control of the hypothalamus, which develops adjacent to it from the diencephalon. The types of hormone secreted and the connections to the hypothalamus differ between the anterior and posterior pituitary, reflecting their different embryonic origin. The anterior pituitary secretes luteinizing hormone, thyrotrophin, and prolactin, among others, in response to hypothalamic peptides delivered via blood vessels. The posterior pituitary secretes vasotocin and oxytocin, delivered to it via hypothalamic neurons, which produce the hormones.

Optic vesicles grow out from the diencephalon and contact the ectoderm, where a lens vesicle appears (Figure 20.25). The optic vesicle becomes cup-shaped and develops into the pigmented and neural retina of the eye, while

Figure 20.23   Face morphogenesis. The formation of the *X. laevis* face from the neural plate stage (left) to a swimming embryo (right) can be followed by determining the origin of different surface cell types. The cells used are hatching gland (HG), cement gland (CG), and zones with ciliated (CZ) and nonciliated (NZ) cells. The NZ contains the two nose primordia (N) and a depression, the stomodeum (S), which will form the mouth. The eyes (E) lie outside of the NZ. (From Drysdale and Elinson 1991.)

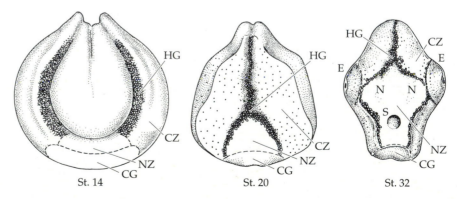

St. 14   St. 20   St. 32

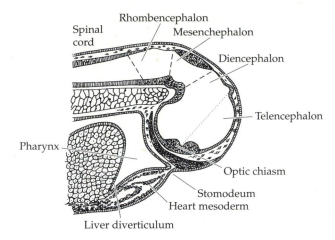

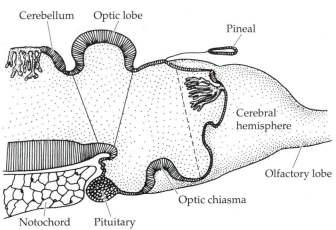

Figure 20.24 Brain regions. (Top) The embryo's brain is initially divided into telencephalon, diencephalon, mesencephalon, and rhombencephalon. (Bottom) The tadpole's brain has small cerebral hemispheres with a large olfactory and optic lobe. (After Rugh 1951.)

the lens vesicle detaches from the ectoderm to form the lens. The coordinate development of the lens and retina is due to induction of head ectoderm by the optic cup. Head ectoderm in turn is sensitized to respond in this way by

prior inductions (Saha et al. 1989); this sensitivity illustrates the concept of **competence**. Competence refers to the ability of a tissue to respond to an induction. The blastocoel roof, discussed earlier, is competent to respond to mesoderm inducing signals, and gastrula ectoderm is competent to respond to neural inducing signals. Both an inducer and a competent responding tissue are required for induction to occur.

With continued development of the eye, the neural retina sends axons back to the brain via the optic stalk. The optic nerves cross at the optic chiasm on the ventral surface of the brain, just anterior to the pituitary. They enter the mesencephalon, where they connect to the optic tectum and generate a precise retinotectal map. When an optic nerve regenerates in amphibians, nerves from each point on the retina reconnect with their corresponding points in the tectum. The optic tectum or optic lobes are large, prominent bulges of the frog brain.

The final part of the brain, the rhombencephalon, connects to the spinal cord and forms the cerebellum and the medulla oblongata. In the past, the anatomical regions of the brain were recognized by morphology. Recently, it has been found that genes are expressed at particular boundaries in the developing brain, and detection of these gene expressions constitutes a molecular anatomy (Figure 20.26).

Placodes

Besides the lens, other ectodermal thickenings, known as placodes, are derived from tissue at the periphery of the neural plate (Figure 20.27) (Knouff 1935; Hausen and Riebesell 1991). The olfactory (nasal) placode gives rise to the sensory olfactory epithelium. The olfactory nerve grows from the epithelium back to the telencephalon, connecting the nose to the brain. The auditory (otic) placode (Figure 20.27) gives rise to the labyrinth of the inner ear and connects to the rhombencephalon via the auditory nerve. An evagination from the otic vesicle produces the endolymphatic duct and sac, which stores calcium. Calcium is required for bone formation at metamorphosis, and the endolymphatic sacs are particularly prominent in em-

Figure 20.25 Eye development. The optic cup grows out from the brain and induces the formation of a lens from the ectoderm. The optic cup becomes the pigmented and neural retina. (From Rugh 1951.)

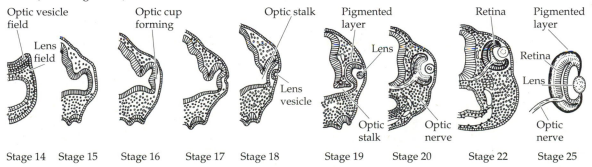

Stage 14   Stage 15   Stage 16   Stage 17   Stage 18   Stage 19   Stage 20   Stage 22   Stage 25

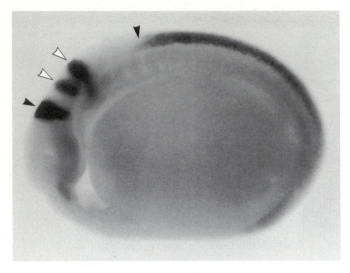

Figure 20.26 Molecular anatomy of the developing brain. Localized gene expression can be used to identify regions of the developing *X. laevis* brain. *En-2* is expressed at the midbrain-hindbrain boundary (left solid arrow), *Krox-20* in two regions of the hindbrain (open arrows), and *XlHbox 6* throughout the spinal cord (starting at right solid arrow). (Photograph courtesy of Tabitha Doniach.)

bryos of *Eleutherodactylus coqui*, a frog that develops without a tadpole (Townsend and Stewart 1985).

Finally, amphibians being aquatic, possess a lateral line system for detecting water currents. Lateral line placodes, arising near the ear, migrate in prescribed paths along the head and body (Figure 20.28). During migration, they deposit cells, which differentiate as mechanoreceptive **neuromasts**. As with the nose and ear, the lateral line placodes

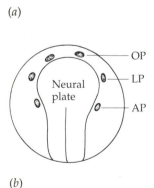

(a)

OP
LP
AP

Neural plate

Figure 20.27 Placodes. (*a*) Placodes, which will form various sense organs, appear at the periphery of the neural plate. AP, auditory placode; LP, lens placode; OP, olfactory placode. (*b*) The auditory placode starts as an ectodermal thickening, which invaginates to form the otic vesicle. The otic vesicle becomes the inner ear. (*a* after Hausen and Riebesell 1991; *b* after Rugh 1951.)

(b)

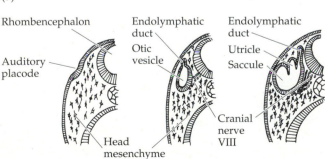

Rhombencephalon

Auditory placode

Endolymphatic duct

Otic vesicle

Endolymphatic duct

Utricle

Saccule

Cranial nerve VIII

Head mesenchyme

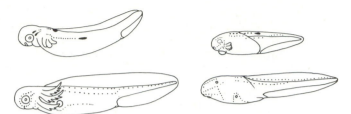

Figure 20.28 Lateral line. Lateral line primordia, derived from placodes, migrate in prescribed pathways and deposit neuromasts, organs for sensing water currents. (Left) *Ambystoma mexicanum* showing early migration and final positions of neuromasts. (Right) *Rana pipiens* showing similar stages. (From Smith et al. 1988.)

produce sensory receptors that connect to the brain (Lannoo and Smith 1989).

Cranial Neural Crest

Much of the cartilage of the head, and hence most of the head structure, is derived from cranial neural crest (Figure 20.29) (Stone 1926, 1929). Three large streams of neural crest cells begin migrating ventrally under the ectoderm around the time of neural fold closure (Figure 20.30) (Stone 1929; Jacobson and Meier 1984; Sadaghiani and Thiebaud 1987). Neural crest arises from the inner sensorial ectoderm in *X. laevis* (Schroeder 1970; Sadaghiani and Thiebaud 1987), but must ingress from the single ectodermal layer in urodeles. It should be emphasized that neural crest migration is a massive movement of tissue. The streams cause bulges under the ectoderm, and these bulges are easily seen on living embryos.

The **mandibular crest**, from the mesencephalic region, forms the most anterior stream (Figure 20.30). It flows around the eye towards the front of the face, where it surrounds the mouth. The mandibular crest forms part of the anterior part of the neurocranium and jaw cartilages, such as the supra- and infrarostrals, the quadrate, and Meckel's cartilage (Sadaghiani and Thiebaud 1987; Seufert and Hall 1990). The flow of **hyoid crest** starts anterior to the ear (Figure 20.30) and forms the ceratohyal cartilage, which underlies the floor of the mouth. The **branchial crest** begins posterior to the ear and forms the cartilage of the branchial arches. There are four branchial arches, with clefts between them allowing water flow from the pharynx to the outside. External gills develop from the branchial arches. In anuran tadpoles but not in urodele larvae, the external gills are covered over by the operculum and are replaced by internal gills, also from the branchial arches.

The extensive contributions of the cranial neural crest to cartilage (Figure 20.29), which is normally derived from mesoderm, support the fascinating hypothesis that the vertebrate head was an evolutionary addition to the body plan of a protochordate (Gans and Northcutt 1983). The cranial neural crest represents the critical tissue that al-

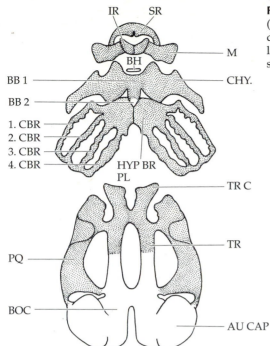

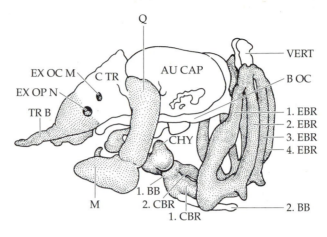

Figure 20.29 Head cartilage derived from cranial crest. Ventral (top left) and dorsal (bottom left) views of the skull of a *Rana palustris* tadpole. All stippled cartilages are derived from neural crest, including the suprarostral (SR) and infrarostral (IR) cartilages of the jaw. A lateral view (right) of the skull of an *Ambystoma punctatum* larva shows a similar large contribution (stippled) of cranial crest. (After Stone 1926, 1929.)

lowed the addition to occur, as it provided a novel source of cartilage. If this conjecture is so, the anuran tadpole may represent a continuation of head evolution, since the tadpole has two additional jaw cartilages, the suprarostral and infrarostral (Figure 20.29). These cartilages are the most anterior ones, and they support the tadpole's jaw. Neither urodeles nor any other vertebrate has these cartilages, and they are lost when the tadpole metamorphoses into a frog.

Other Parts

The head also contains noncrest mesoderm, mostly derived from the prechordal plate. The cells of the prechordal plate migrate along the blastocoel roof during gastrulation (Figure 20.12). They lead the prospective notochord and undergo spreading rather than convergence and extension. The anterior end of the notochord underlies the hindbrain, so the prechordal plate underlies most of the brain. This location suggests that prechordal plate could be important in induction of the forebrain and midbrain. The main skull element from noncrest head mesoderm is the posterior portion of the neurocranium (Figure 20.29).

Finally, the head has a mouth. An initial invagination, the stomodeum, marks the site where the ectoderm and

endoderm fuse to form the mouth (Figure 20.23). This region is notable for the lack of intervening mesoderm, which allows contact between the ectoderm and the endoderm (Figure 20.24). Teeth in urodeles are derived from cranial neural crest as in other vertebrates, but this is not the case in anuran tadpoles. The latter possess keratinous beaks and teeth, rather than true teeth. Presumably, neural crest-derived tooth progenitors are held in reserve until metamorphosis, when the frog's teeth appear.

There are several unusual variations in mouth formation. Tadpoles of *X. laevis* are filter feeders and lack the keratinous beaks and teeth that are used by most tadpoles to scrape plant surfaces. Tadpoles of *Lepidobatrachus laevis* are carnivorous and have a huge mouth without a beak (Ruibal and Thomas 1988). They have a short gut, unlike the long coiled gut required for herbivory in most tadpoles. Finally, there are variations in mouth structure within populations of tadpoles or larvae, depending on what they eat. Some *Scaphiopus multiplicatus* (spadefoot toad) tadpoles have a flattened head and reduced gut, and are carnivorous. They convert from the typical herbivorous form, when shrimp are available to eat (Pfennig 1992). Similarly, some *A. tigrinum* (tiger salamander) larvae have wide mouths and elongated teeth. They are cannibalistic and will eat their more normal relatives along with other pray (Lannoo and Bachmann 1984).

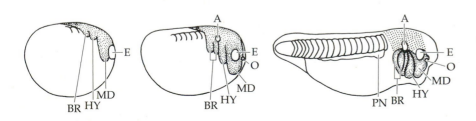

Figure 20.30 Cranial neural crest migration. Cranial crest migrates as three massive streams: mandibular (MD), hyoid (HY), and branchial (BR). E, eye; O, olfactory placode; A, auditory placode; PN, pronephros. (After Stone 1929.)

Trunk and Tail Development

Embryos of all vertebrates pass through a stage in which they look similar to each other (Duboule 1994). Ballard (1981) called this the **pharyngula stage**, and the tissues of the trunk are arranged in a characteristic vertebrate pattern (Figure 20.31). The ectoderm has produced epidermis, neural tube, and trunk neural crest. The mesodermal ring from dorsal to ventral consists of notochord, somites, intermediate mesoderm, and lateral plate mesoderm. The endoderm surrounds the archenteron. The dorsal axial structures, namely the spinal cord, the notochord, and the somites, are basically similar from one end of the trunk to the other. Anteroposterior differences develop in the remaining tissues to give rise to such organs and structures as heart, gonads, kidneys, limbs, lungs, and liver. Organ development has been investigated intensively in previous decades (Huettner 1949; Rugh 1951; Nelsen 1953; Balinsky 1981), but less has been discovered recently. The recent success at molecular analysis of development promises to re-open questions of organogenesis in the future.

Dorsal Axial Structures

The trunk neural tube forms spinal cord and trunk neural crest. The latter migrates from the neural tube to produce pigment cells, sensory neurons, dorsal fin, and other cell types. Unlike cranial crest, trunk crest does not form cartilage. The notochord consists of vacuolated cells with a surrounding sheath. The notochord serves as a flexible rod, supporting the embryo's body. Paraxial mesoderm, which lies on either side of the neural tube and notochord, separates into blocks, known as somites, which repeat

down the axis (Figure 20.31). Each somite consists of sclerotome, dermatome, and myotome. The dermatome gives rise to back dermis underlying the epidermis; the myotome forms back striated muscles, and the sclerotome differentiates into cartilage. The cartilage is the precursor to the vertebral column, which will replace the notochord in supporting the body and protecting the spinal cord.

While the structure of the somite is similar in various amphibians, many developmental paths can lead to that structure (Hanken 1986; Radice et al. 1989; Gatherer and del Pino 1992). In several urodeles, somitic cells surround a cavity in a rosette arrangement. Cells subsequently elongate and fuse to form multinucleated myofibers that run parallel to the notochord. In *X. laevis*, cells originally perpendicular to the notochord rotate to a parallel position, remaining mononucleated throughout. Other anurans exhibit neither rosette formation nor rotation. The significance of this variation in developmental paths is unknown.

Heart and Circulatory System

The heart is of course considered a mesodermal organ of the trunk, but its unusual developmental history makes it distinct from the rest of the trunk mesoderm. Prospective heart tissue maps to the leading edge of the mesodermal ring during gastrulation, lying on either side of the pre-

Figure 20.31 Pharyngula stage and the vertebrate basic body plan. Following neurulation and the beginning of organ formation, the frog embryo is organized similarly to most other vertebrates. A cross section (right) illustrates the basic vertebrate body plan, with the neural tube flanked by somites and underlain by the notochord. A coelom will develop between the somatic and splanchnic layers of lateral plate mesoderm surrounding the endoderm. (After Huettner 1949; Balinsky 1981.)

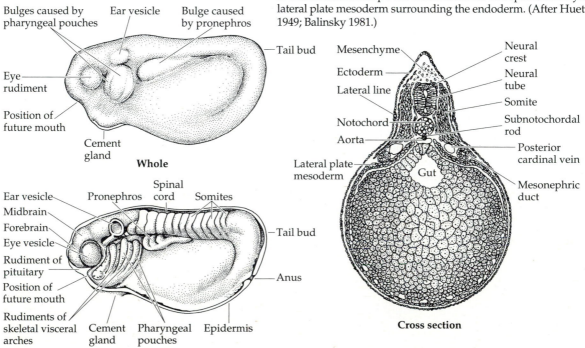

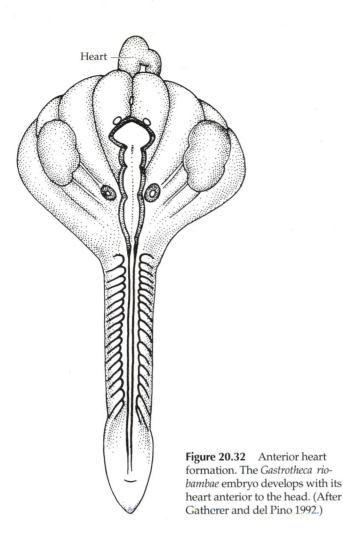

Figure 20.32 Anterior heart formation. The *Gastrotheca riobambae* embryo develops with its heart anterior to the head. (After Gatherer and del Pino 1992.)

chordal plate mesoderm (Figure 20.14). This mapping makes the heart one of the most anterior tissues embryologically, even though it ends up in the trunk rather than in the head. Two observations reinforce this view. First, in *G. riobambae* whose embryo is flattened on a large yolk mass, the heart develops anterior to the head and is secondarily folded into the trunk (Figure 20.32) (del Pino and Escobar 1981). Second, when embryos are treated to enhance anterior development at the expense of posterior development, a central beating heart is present along with extra eyes and other head structures (Kao and Elinson 1988).

As in other vertebrates, there are two lateral regions of prospective heart tissue, which fuse at the ventral midline to create the heart primordium (Figure 20.33). The heart is initially two tubes, an inner endocardium and an outer myocardium. The tubular heart loops into an S-shape, creating a left–right asymmetry that leads to the formation of the ventricle on the left side. This asymmetry depends on an interaction occurring as the heart primordia migrate along the blastocoel roof (Yost 1992).

Other parts of the circulatory system also originate in the mesoderm. The first embryonic blood cells appear in **blood islands**, which form on the surface of the belly in the most ventral mesoderm (Figure 20.33). Blood vessels first form from condensations of cells known as **angioblasts**, which appear in the mesenchyme. Mesenchyme is a loose collection of cells, as distinguished from an epithelium, in which cells form layers. In addition to the heart, amphibians have other beating structures, known as lymph hearts (Duellman and Trueb 1986; Drysdale et al. 1994). These pump lymph but little is known about their development.

Figure 20.33 Heart formation. Prospective heart, on both sides of the embryo (*a*, *b*), migrates to the ventral midline, where the heart forms (*c*, *d*, *e*). Blood first arises in ventral mesoderm (*f*). (After Copenhaver 1955; Huettner 1949.)

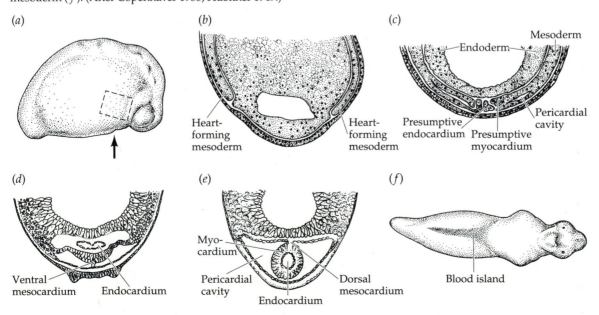

Figure 20.34 Excretory system and gonads. (*a*) The pronephros develops from intermediate mesoderm at the anterior end. The pronephric duct, later taken over by the mesonephros, enters the cloaca, as does the Mullerian duct. (*b*) Primordial germ cells migrate from the yolky endoderm along the dorsal mesentery. They populate the genital ridges, thus forming the gonads. (After Balinsky 1981; Rugh, 1951.)

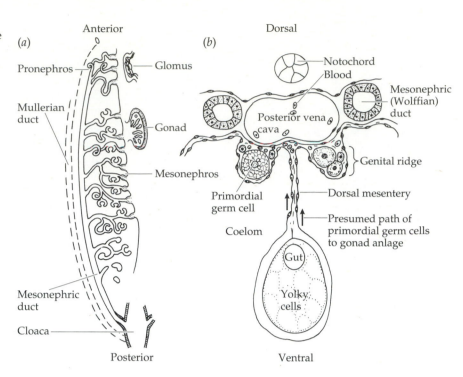

Excretory and Reproductive Systems

The excretory and reproductive systems develop primarily from intermediate mesoderm. The initial kidney is called the **pronephros**, and it functions during larval or tadpole life. Pronephric tubules for collecting wastes form in the anterior intermediate mesoderm, at the level of the forelimb (Figure 20.34). The tubules enter a collecting duct, known as the pronephric duct. The pronephric ducts extend by migrating posteriorly until they join the cloaca, the enlarged ending of the hindgut (Poole and Steinberg 1982) Excretory and digestive wastes, as well as sperm and eggs pass through the cloaca on their way out. A second set of collecting tubules, the **mesonephros**, develop posterior to the pronephros and also empty into the pronephric duct. With metamorphosis, the pronephros degenerates, and the remaining duct becomes the mesonephric duct.

Lateral plate mesoderm splits into an inner **splanchnic layer** and an outer **somatic layer**, and the body cavity or **coelom** develops between the two layers. The somatic layer gives rise to ventral striated muscle, and the splanchnic layer produces mesentery and smooth muscle. Gonads originate as thickenings, known as **genital ridges**, at the dorsal corners of the coelom (Figure 20.34). The genital ridges constitute the somatic (body) part of the gonad, while the germ line, which will form the eggs or sperm, is derived from primordial germ cells.

In anurans, the primordial germ cells develop from cells that incorporate germ plasm, described earlier. They migrate from the yolky endoderm along the dorsal mesentery to populate the genital ridges (Figure 20.34). The origin of primordial germ cells in urodeles contrasts with that in anurans. In urodeles, germ plasm is not seen in the egg, and the germ cells appear to arise from lateral

mesoderm by induction (Nieuwkoop and Sutasurya 1979; Michael 1984; Maufroid and Capuron 1985). Primordial germ cells are the most important cells in the embryo as far as the species is concerned, since they are required for reproduction and passage of genes to the next generation. The dramatic difference in development of primordial germ cells in anurans and urodeles has been a compelling argument for a diphyletic evolution of these two orders from different fish.

In addition to the gonads, there are two pairs of ducts that form regardless of sex. The mesonephric (Wolffian) ducts carry sperm in males as well as urine. The Mullerian ducts arise parallel to the mesonephric ducts in the intermediate mesoderm. They become the oviducts in females for transport of eggs to the cloaca, and they persist in males of some species (Pace 1974).

Limbs

Limbs arise at four sites from the somatic layer of lateral plate mesoderm (Figure 20.35). The forelimbs form posterior to gills and ventral to pronephros, while hindlimbs arise near the anus. The first morphological sign of limbs is an epithelial to mesenchymal transition of somatic mesoderm (Figure 20.35). Some cells lose their connections to each other and become loosely packed. This mass of mesenchyme causes a bulge on the surface called the limb bud. The limb bud grows out, forming first a paddle and then digits.

Limb development in amniotes (birds, reptiles, mammals) depends on the formation of a specialized ectodermal structure at the tip of the limb bud, called the **apical ectodermal ridge** (AER). Reciprocal interactions between

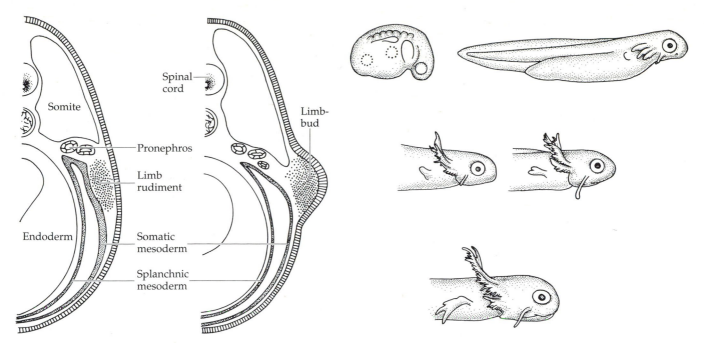

Figure 20.35   Limb development. (Left) Mesenchymal cells released from somatic mesoderm form the limb bud. (Right) Four areas, two on each side (dotted circles), are the presumptive limbs. Limb buds form, grow out, and form digits, as illustrated in this urodele embryo. Development of the balancer at the corner of the mouth and the gills is also clear. (After Balinsky 1981; Nicholas 1955.)

the AER and the limb mesoderm are required for outgrowth and development of the limb. The question of the existence of an AER in amphibians is not completely resolved. Morphologically, an AER has been found in some anurans but not in urodeles (Hanken 1986). Development of limbs depends on an AER in *X. laevis* (Tschumi 1957) but not in the urodele *Pleurodeles waltl* (Lauthier 1985). It

will probably take examination of AER-specific gene expressions to resolve whether urodele limb buds differ from those in anurans and amniotes.

Another obvious difference between anurans and urodeles is the timing of limb bud formation and development. Urodele limb buds are obvious in the embryo (Figure 20.35). The legs grow and differentiate and are functional during larval life. Anuran limb buds usually cannot be seen until the tadpole has begun feeding. Hindlimbs grow slowly and are not functional in the tadpole, while forelimbs are hidden by the opercular fold.

Endoderm

Endoderm surrounds the archenteron and will form the gut and a variety of organs (Figure 20.36). The endoderm

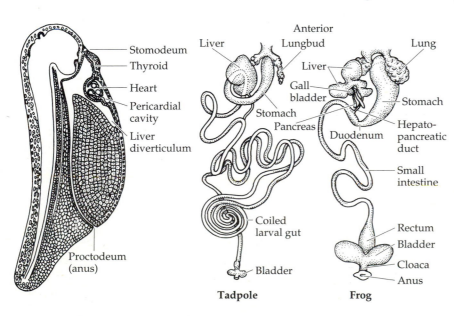

Tadpole   **Frog**

Figure 20.36   Endoderm development. The archenteron becomes a tube with outpocketings, such as the liver diverticulum, and forms the digestive tract and endodermal organs. The long, coiled gut of the herbivorous tadpole is greatly shortened at metamorphosis to the gut of the carnivorous frog. (After Rugh 1951.)

Figure 20.37   Tail development. The *X. laevis* tail forms as a posterior extension of the dorsal lip of the closed blastopore. (After Gont et al. 1993.)

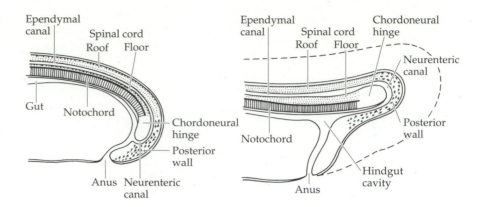

fuses with the ectoderm in the head to create the mouth, and either the blastopore or a secondary opening, created by another endoderm-ectoderm fusion, produces the anus. A large evagination from the floor of the foregut is known as the **liver diverticulum**, and the liver and pancreas will develop from it. Other evaginations develop into lungs and gall bladder.

Tail

The tail consists primarily of spinal cord, notochord, and somites. It forms by an extension posteriorly of these tissues that lie dorsal to the anus (Figure 20.37) (Gont et al. 1993; Tucker and Slack 1995). The last point of involution at the blastopore is marked by the juxtaposition of neural tube and notochord, and this so-called **chordaneural hinge** moves posteriorly with tail elongation. At the start of tail formation, the initial extension is called the tailbud, suggesting that the tissue is similar to a limb bud with a mesenchymal mass. That does seem to be the situation in chicks and mammals. Those amniotes undergo secondary neurulation, where formation of tail neural tube occurs by a hollowing out of a neural rod rather than rolling up of a neural plate (Schoenwolf and Smith 1990; Griffith et al.

1992). In amphibians however, the neural tube and notochord of the tail form the same as in the trunk.

Life Histories

Amphibians have complex life histories, forming first a larva or a tadpole that later metamorphoses into an adult (Figure 20.38) (Duellman and Trueb 1986). Urodele larvae look similar to adults, so their metamorphosis is not as radical as that of the anuran tadpole to a frog. The embryo of anurans has to produce both tadpole structures as well as precursors for adult structures.

Urodele larvae have legs and are carnivorous, and they share these characters with adults. Larva are aquatic, so they have external gills and tail fins not present in adults. Metamorphosis involves changes in the skin, loss of gills and tail fin, and some jaw remodeling. The adult attains sexually maturity, but under some circumstances larvae can also become sexually mature. This is called **neoteny**,

Figure 20.38   Life histories. Anurans have a radical body change as the tadpole metamorphoses into a frog. The transition from a urodele larva to an adult involves less change.

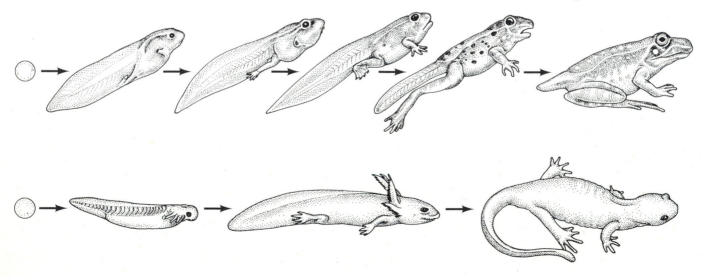

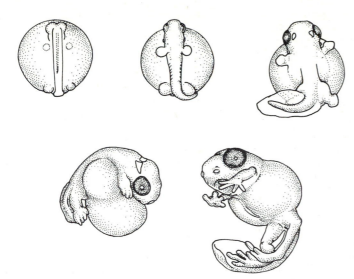

Figure 20.39 Direct development. *Eleutherodactylus coqui* develops into a frog directly from the egg, without a tadpole stage. (After Townsend and Stewart 1985.)

and the best-known case is the axolotl (*Ambystoma mexicanum*), originally found in Mexican lakes but now more common in laboratories. Given thyroid hormone, *A. mexicanum* will metamorphose into an adult form. Different populations and species of Ambystoma show varying tendencies to neoteny, and those tendencies are affected by the environment (Sprules 1974).

Anuran tadpoles look very different from frogs and from any other vertebrate. Their bulbous body accommodates a long coiled gut, needed for digesting vegetable material. Most have keratinous teeth and beaks for scraping plants, as well as extra cartilages to support these jaws, as already mentioned. Internal gills and developing forelimbs are covered over by the operculum, while the hind legs remain small and tucked behind the body at the base of the tail. No tadpole is reproductively mature. With metamorphosis, the jaw is restructured, as is the gut; the tail is histolyzed; and the legs grow. These changes are all triggered by thyroid hormone (Tata 1993; Atkinson 1994).

The tadpole can be viewed as an evolutionary insertion in the frog life history. If we start with a urodele pattern, the larva and the adult would each undergo structural changes. Those in the larva would be towards herbivory, involving changes in jaw and gut, and suppression of limb development. It would not matter what changes are made to tadpole structures as long as tadpole-specific structures degenerate in response to thyroid hormone and adult structures do not depend on those of the preexisting tadpole.

While there is no fossil record of the evolution of the tadpole, the existence of direct development argues that adult structures are independent of tadpole-specific ones. In at least a dozen lineages, tadpoles have been eliminat-

ed, and the embryo develops directly to an adult morphology (Figure 20.39). The direct developer *E. coqui* lacks such tadpole structures as beak, keratinous teeth, lateral line organs, cement and hatching gland, and supra- and infrarostral cartilages (Elinson 1990). Limb buds appear shortly after neurulation, and legs develop continously through embryonic life.

The development on land of *E. coqui* is contrary to the name "amphibian," which suggests that part of the animal's life should be in water. In fact, many amphibians have devised ways of keeping their embryos out of the water (Duellman and Trueb 1986; Duellman 1992). Marsupial frogs keep their embryos in a pouch on their back, while larvae develop in the oviducts of the viviparous frog *Nectophrynoides occidentalis*. More bizarre brooding sites are the vocal sacs of male *Rhinodema darwinii* and the stomach of the Australian *Rheobatrachus silus*. These unusual patterns of reproduction are probably associated with changes in development, but this problem is unexplored.

Conclusion

We know more about development of amphibians than of almost any other animal. This intense scrutiny has suggested that there are important differences between the development of anurans and urodeles. They appear to differ in mechanisms of fertilization, gastrulation, neurulation, primordial germ cell and limb formation, and metamorphosis. Only a few species in each order have been carefully investigated, however, so the extent of variability of developmental mechanisms within each order is not known. As molecular advances are made, comparative analysis will become easier, and a more complete picture of amphibian development will emerge.

Literature Cited

Atkinson, B. G. 1994. Metamorphosis: Model systems for studying gene expression in postembryonic development. *Dev. Gen.* 15: 313–319.

Balinsky, B. I. 1981. *An Introduction to Embryology*, 5th ed. Saunders College Publishing, Philadelphia, PA.

Ballard, W. W. 1981. Morphogenetic movements and fate maps of vertebrates. *Amer. Zool.* 21: 391–399.

Ballard, W. W. and A. S. Ginsberg. 1980. Morphogenetic movements in acipenserid embryos. *J. Exp. Zool.* 213: 69–103.

Bogart, J. P., R. P. Elinson and L. E. Licht. 1989. Temperature and sperm incorporation in polyploid salamanders. *Science* 246: 1032–1034.

Bolker, J. A. 1993. Gastrulation and mesoderm formation in the white sturgeon, *Acipenser transmontanus. J. Exp. Zool.* 266: 116–131.

Brauer, A. 1897. I. Beiträge zur Kenntniss der Entwicklungsgeschichte und der Anatomie der Gymnophionen. *Zool. Jahrb.* 10: 389–472.

Brun, R. B. and J. A. Garson. 1983. Neurulation in the Mexican salamander (*Ambystoma mexicanum*): A drug study and cell shape analysis of the epidermis and the neural plate. *J. Embryol. Exp. Morphol.* 74: 275–295.

Brun, R. B. and J. A. Garson. 1984. Notochord formation in the Mexican salamander (*Ambystoma mexicanum*) is different from notochord formation in *Xenopus laevis*. *J. Exp. Zool.* 229: 235–240.

Burnside, B. 1973. Microtubules and microfilaments in amphibian neurulation. *Amer. Zool.* 13: 989–1006.

Byers, T. J. and P. B. Armstrong. 1986. Membrane protein redistribution during *Xenopus* first cleavage. *J. Cell Biol.* 102: 2176–2184.

Clausi, D. A. and G. W. Brodland. 1993. Mechanical evaluation of theories of neurulation using computer simulations. *Development* 118: 1013–1023.

Collazo, A., J. A. Bolker and R. Keller. 1994. A phylogenetic perspective on teleost gastrulation. *Amer. Nat.* 144: 133–152.

Copenhaver, W. M. 1955. Heart, blood vessels, blood, and endodermal derivatives. In *Analysis of Development*, B. H. Willier, P. A. Weiss and V. Hamburger (eds.). W. B. Saunders Co., Philadelphia, pp. 440–461.

Cross, N. L. and R. P. Elinson. 1980. A fast block to polyspermy in frogs mediated by changes in the membrane potential. *Dev. Biol.* 75: 187–198.

Dale, L. and J. M. W. Slack. 1987. Regional specification within the mesoderm of early embryos of *Xenopus laevis*. *Development* 100: 279–295.

Danilchik, M. and J. C. Gerhart. 1987. Differentiation of the animal-vegetal axis in *Xenopus laevis* oocytes. I. Polarized intracellular translocation of platelets establishes the yolk gradient. *Dev. Biol.* 122: 101–112.

Danilchik, M. V. and J. M. Denegre. 1991. Deep cytoplasmic rearrangements during early development in *Xenopus laevis*. *Development* 111: 845–856.

Delarue, M., S. Sanchez, K. E. Johnson, T. Darribere and J.-C. Boucaut. 1992. A fate map of superficial and deep circumblastoporal cells in the early gastrula of *Pleurodeles waltl*. *Development* 114: 135–146.

del Pino, E. M. 1989. Modifications of oogenesis and development in marsupial frogs. *Development* 107: 169–187.

del Pino, E. M. and R. P. Elinson. 1983. A novel development pattern for frogs: Gastrulation produces an embryonic disk. *Nature* 306: 589–591.

del Pino, E. M. and B. Escobar. 1981. Embryonic stages of *Gastrotheca riobambae* (Fowler) during maternal incubation and comparison of development with that of other egg-brooding hylid frogs. *J. Morphol.* 167: 277–296.

del Pino, E. M. and S. Loor-Vela. 1990. The pattern of early cleavage of the marsupial frog *Gastrotheca riobambae*. *Development* 110: 781–789.

Delsol, M., J. Flatin, J.-M. Exbrayat and J. Bons. 1981. Développement de *Typhlonectes compressicaudus*, amphibien apode vivipare. Hypothèses sur sa nutrition embryonnaire et larvaire par un ectotrophoblaste. *C. R. Acad. Sci. Paris* 293: 281–285.

De Robertis, E. M. 1995. Dismantling the organizer. *Nature* 374: 407–408.

Doniach, T. 1993. Planar and vertical induction of anteroposterior pattern during the development of the amphibian central nervous system. *J. Neurobiol.* 24: 1256–1275.

Doniach, T., C. R. Phillips and J. C. Gerhart. 1992. Planar induction of anteroposterior pattern in the developing central nervous system of *Xenopus laevis*. *Science* 257: 542–545.

Drysdale, T. A. and R. P. Elinson. 1991. Development of the *Xenopus laevis* hatching gland and its relationship to surface ectoderm patterning. *Development* 111: 469–478.

Drysdale, T. A., K. F. Tonissen, K. D. Patterson, M. J. Crawford and P. A. Krieg. 1994. *Xenopus* cardiac troponin I is an embryonic heart-specific marker: Expression during abnormal heart morphogenesis. *Dev. Biol.* 165: 432–441.

Duboule, D. 1994. Temporal colinearity and the phylotypic progression: A basis for the stability of a vertebrate Bauplan and the evolution of morphogenesis through heterochrony. *Development* 1994 Supplement: 135–142.

Duellman, W. E. 1992. Reproductive strategies of frogs. *Sci. Amer.* 267: 80–87.

Duellman, W. E. and L. Trueb. 1986. *Biology of Amphibians*. McGraw-Hill Book Co., New York.

Duryee, W. R. 1950. Chromosomal physiology in relation to molecular structure. *Ann. N. Y. Acad. Sci.* 50:920–953.

Eagleson, G. W. and W. A. Harris. 1990. Mapping of the presumptive brain regions in the neural plate of *Xenopus laevis*. *J. Neurobiol.* 21: 427–440.

Elinson, R. P. 1986. Fertilization in amphibians: The ancestry of the block to polyspermy. *Int. Rev. Cytol.* 101: 59–100.

Elinson, R. P. 1987a. Fertilization and aqueous development of the Puerto Rican terrestrial-breeding frog, *Eleutherodactylus coqui*. *J. Morphol.* 193: 217–224.

Elinson, R. P. 1987b. Change in developmental patterns: embryos of amphibians with large eggs. In *Development as an Evolutionary Process*, R. A. Raff and E. C. Raff (eds.). Alan R. Liss, New York, pp. 1–21.

Elinson, R. P. 1990. Direct development in frogs: Wiping the recapitulationist slate clean. *Seminars Dev. Biol.* 1: 263–270.

Elinson, R. P. and E. M. del Pino. 1985. Cleavage and gastrulation in the egg-brooding, marsupial frog, *Gastrotheca riobambae*. *J. Embryol. Exp. Morphol.* 90: 223–232.

Elinson, R. P. and T. Holowacz. 1995. Specifying the dorsoanterior axis in frogs: 70 years since Spemann and Mangold. *Curr. Top. Dev. Biol.* 30: 253–285.

Elinson, R. P. and B. Rowning. 1988. A transient array of parallel microtubules in frog eggs: Potential tracks for a cytoplasmic rotation that specifies the dorso-ventral axis. *Dev. Biol.* 128: 185–197.

Forristall, C., M. Pondel, L. Chen and M. L. King. 1995. Patterns of localization and cytoskeletal association of two vegetally localized RNAs, Vg1 and Xcat-2. *Development* 121: 201–208.

Gall, J. 1969. The genes for ribosomal RNA during oogenesis. *Genetics* Suppl. 61: 121–132.

Gans, C. and R. G. Northcutt. 1983. Neural crest and the origin of vertebrates: A new head. *Science* 220: 268–274.

Gard, D. L. 1995. Axis formation during amphibian oogenesis: Reevaluating the role of the cytoskeleton. *Curr. Top. Dev. Biol.* 30: 215–252.

Gatherer, D. and E. M. del Pino. 1992. Somitogenesis in the marsupial frog *Gastrotheca riobambae*. *Int. J. Dev. Biol.* 36: 283–291.

Gerhart, J. C., M. Danilchik, T. Doniach, S. Roberts, B. Rowning and R. Stewart. 1989. Cortical rotation of the *Xenopus* egg: Consequences for the anteroposterior pattern of embryonic dorsal development. *Development* 107 Suppl. 37–51.

Gillespie, J. I. 1983. The distribution of small ions during the early development of *Xenopus laevis* and *Ambystoma mexicanum* embryos. *J. Physiol.* 344: 359–377.

Gitlin, D. 1944. The development of *Eleutherodactylus portoricensis*. *Copeia* 1944: 91–98.

Gont, L. K., H. Steinbesser, B. Blumberg and E. M. De Robertis. 1993. Tail formation as a continuation of gastrulation: The multiple cell populations of the *Xenopus* tailbud derive from the late blastopore lip. *Development* 119: 991–1004.

Gordon, R. 1985. A review of the theories of vertebrate neurulation and their relationship to the mechanics of neural tube birth defects. *J. Embryol. Exp. Morphol.* 89 Suppl. 229–255.

Gordon, R. and A. G. Jacobson. 1978. The shaping of tissues in embryos. *Sci. Amer.* 238: 106–113.

Grandin, N. and M. Charbonneau. 1992. Intracellular free Ca^{2+} changes during polyspermy in amphibian eggs. *Development* 114: 617–624.

Griffith, C. M., M. J. Wiley and E. Sanders. 1992. The vertebrate tail bud: Three germ layers from one tissue. *Anat. Embryol.* 185: 101-113.

Hanken, J. 1986. Developmental evidence for amphibian origins. In *Evolutionary Biology*, M. K. Hecht, B. Wallace and G. T. Prance (eds.). Plenum, New York, 20: 389–417.

Hausen, P. and M. Riebesell. 1991. *The Early Development of Xenopus laevis.* Springer-Verlag, Berlin.

Heasman, J., J. Quarmby and C. C. Wylie. 1984. The mitochondrial cloud of *Xenopus* oocytes: The source of germinal granule material. *Dev. Biol.* 105: 458–469.

Hedrick, J. L. and T. Nishihara. 1991. Structure and function of the extracellular matrix of anuran eggs. *J. Elect. Microsc. Tech.* 17: 319–335.

Holtfreter, J. 1944. A study of the mechanics of gastrulation, Part II. *J. Exp. Zool.* 95: 171–212.

Huettner, A. F. 1949. *Comparative Embryology of the Vertebrates.* Macmillan, New York.

Imoh, H. 1988. Formation of germ layers and roles of the dorsal lip of the blastopore in normally developing embryos of the newt *Cynops pyrrhogaster. J. Exp. Zool.* 246: 258–270.

Imoh, H., M. Okamoto and G. Eguchi. 1983. Accumulation of annulate lamellae in the subcortical layer during progesterone-induced oocyte maturation in *Xenopus laevis. Dev. Growth Differ.* 25: 1–10.

Ishikawa, C. 1908. Ueber den Riesensalamander Japans. *Mitteilungen der Deutschen Gesellschaft fur Natur- und Volkerkunde Ostasiens Tokyo* 11: 259–280.

Iwao, Y. 1985. The membrane potential changes of amphibian eggs during species- and cross-fertilization. *Dev. Biol.* 111: 26-34.

Iwao, Y. 1989. An electrically mediated block to polyspermy in the primitive urodele *Hynobius nebulosus* and phylogenetic comparison with other amphibians. *Dev. Biol.* 134: 438–445.

Iwao, Y., N. Sakamoto, K. Takahara, M. Yamashita and Y. Nagahama. 1993. The egg nucleus regulates the behavior of sperm nuclei as well as cycling of MPF in physiologically polyspermic newt eggs. *Dev. Biol.* 160: 15–27.

Jacobson, A. G. and R. Gordon. 1976. Changes in shape of the developing nervous system analyzed experimentally, mathematically and by computer simulation. *J. Exp. Zool.* 197: 191–246.

Jacobson, A. G. and S. Meier. 1984. Morphogenesis of the head of a newt: Mesodermal segments, neuromeres, and distribution of neural crest. *Dev. Biol.* 106: 181–193.

Jacobson, A. G., G. F. Oster, G. M. Odell and L. Y. Cheng. 1986. Neurulation and the cortical tractor model for epithelial folding. *J. Embryol. Exp. Morphol.* 96: 19–49.

Johnson, K. E., J.-C. Boucaut and D. W. DeSimone. 1992. Role of the extracellular matrix in amphibian gastrulation. *Curr. Top. Dev. Biol.* 27: 91–127.

Johnson, K. E., T. Darribere and J.-C. Boucaut. 1993. Mesodermal cell adhesion to fibronectin-rich fibrillar extracellular matrix is required for normal *Rana pipiens* gastrulation. *J. Exp. Zool.* 265: 40–53.

Kageura, H. 1990. Spatial distribution of the capacity to initiate a secondary embryo in the 32-cell embryo of *Xenopus laevis. Dev. Biol.* 142: 432–438.

Kalt, M. R. 1971. The relationship between cleavage and blastocoel formation in *Xenopus laevis.* I. Light microscopic observations. *J. Embryol. Exp. Morphol.* 26: 37–49.

Kao, K. R. and R. P. Elinson. 1988. The entire mesodermal mantle behaves as Spemann's organizer in dorsoanterior enhanced *Xenopus laevis* embryos. *Dev. Biol.* 127: 64–77.

Karfunkel, P. 1974. The mechanics of neural tube formation. *Int. Rev. Cytol.* 38: 245–271.

Keller, R. E. 1975. Vital dye mapping of the gastrula and neurula of *Xenopus laevis* I. Prospective areas and morphogenetic movements in the superficial layer. *Dev. Biol.* 42: 222–241.

Keller, R. 1980. The cellular basis of epiboly: an SEM study of deep-cell rearrangement during gastrulation in *Xenopus laevis. J. Embryol. Exp. Morphol.* 60: 201–234.

Keller, R. 1981. An experimental analysis of the role of bottle cells and the deep marginal zone in gastrulation of *Xenopus laevis. J. Exp. Zool.* 216: 81–101.

Keller, R. 1991. Early embryonic development of *Xenopus laevis.* In *Methods in Cell Biology*, B. K. Kay and H. B. Peng (eds.). Academic Press, San Diego, 36: 62–113.

Keller, R. and S. Jansa. 1992. *Xenopus* gastrulation without a blastocoel roof. *Dev. Dynam.* 195: 162–176.

Keller, R. and R. Winklbauer. 1992. Cellular basis of amphibian gastrulation. *Curr. Top. Dev. Biol.* 27: 39–89.

Keller, R. E., M. Danilchik, R. Gimlich and J. Shih. 1985. The function of convergent extension during gastrulation of *Xenopus laevis. J. Embryol. Exp. Morphol.* 89 Suppl: 185–209.

Keller, R., J. Shih and A. Sater. 1992. The cellular basis of the convergence and extension of the *Xenopus* neural plate. *Dev. Dynam.* 193: 199–217.

Kemp, A. 1981. Rearing of embryos and larvae of the Australian lungfish, *Neoceratodus forsteri*, under laboratory conditions. *Copeia* 1981: 776–784.

Kessler, D. S. and D. A. Melton. 1994. Vertebrate embryonic induction: Mesodermal and neural patterning. *Science* 266: 596–604.

Klymkowsky, M. W. and A. Karnovsky. 1994. Morphogenesis and the cytoskeleton: Studies of the amphibian embryo. *Dev. Biol.* 165: 372–384.

Knouff, R. A. 1935. The developmental pattern of ectodermal placodes in *Rana pipiens. J. Comp. Neurol.* 62: 17–71.

Ku, M. and D. A. Melton. 1993. *Xwnt-11*: A maternally expressed *Xenopus* wnt gene. *Development* 119: 1161–1173.

Kubota, H. Y., Y. Yoshimoto, M. Yoneda and Y. Hiramoto. 1987. Free calcium wave upon activation in *Xenopus* eggs. *Dev. Biol.* 119: 129–136.

Lannoo, M. J. and M. D. Bachmann. 1984. Aspects of cannabalistic morphs in a population of *Ambystoma t. tigrinum* larvae. *Amer. Midl. Nat.* 112: 103–110.

Lannoo, M. J. and S. C. Smith. 1989. The lateral line system. In *Developmental Biology of the Axolotl*, J. B. Armstrong and G. M. Malacinski (eds.). Oxford University Press, New York, pp. 176–186.

Lauthier, M. 1985. Morphogenetic role of epidermal and mesodermal components of the fore- and hindlimb buds of the newt *Pleurodeles waltlii* Michah. (Urodela, Amphibia). *Arch. Biol.* 96: 23–43.

Lundmark, C. 1986. Role of bilateral zones of ingressing superficial cells during gastrulation of *Ambystoma mexicanum. J. Embryol. Exp. Morphol.* 97: 47–62.

Macgregor, H. C. and J. Kezer. 1970. Gene amplification in oocytes with 8 germinal vesicles from the tailed frog *Ascaphus truei* Stejneger. *Chromosoma* 29: 189–206.

Malacinski, G. M., A. J. Brothers and H.-M. Chung. 1977. Destruction of the neural induction system of the amphibian egg with ultraviolet irradiation. *Dev. Biol.* 56: 24–39.

Manes, M. E. and F. D. Barbieri. 1977. On the possibility of sperm aster involvement in dorso-ventral polarization and pronuclear migration in the amphibian egg. *J. Embryol. Exp. Morphol.* 40: 187–197.

Masui, Y. 1992. Towards understanding the control of the division cycle in animal cells. *Biochem. Cell Biol.* 70: 920–945.

Maufroid, J.-P. and A. P. Capuron. 1985. A demonstration of cellular interactions during the formation of mesoderm and primordial germ cells in *Pleurodeles waltlii. Differentiation* 29: 20–24.

Melton, D. A. 1987. Translocation of a localized maternal mRNA to the vegetal pole of *Xenopus* oocytes. *Nature* 328: 80–82.

Michael, P. 1984. Are the primordial germ cells (PGCs) in urodela formed by the inductive action of the vegetative yolk mass? *Dev. Biol.* 103: 109–116.

Nelsen, O. E. 1953. *Comparative Embryology of the Vertebrates*. Blakiston Co., New York.

Newport, J. and M. Kirschner. 1982. A major developmental transition in early *Xenopus* embryos. I. Characterization and timing of cellular changes at the midblastula transition. *Cell* 30: 675–686.

Nicholas, J. S. 1955. Limb and Girdle. In *Analysis of Development*, B. H. Willier, P. A. Weiss and V. Hamburger (eds.). W. B. Saunders Co., Philadelphia, pp. 429–439.

Nieuwkoop, P. D. 1969. The formation of mesoderm in urodelean amphibians. I. Induction by the endoderm. *Roux Arch. Entwick. Org.* 162: 341–373.

Nieuwkoop, P. D. and L. A. Sutasurya. 1979. *Primordial Germ Cells in the Chordates*. Cambridge University Press, Cambridge.

Pace, A. E. 1974. Systematic and biological studies of the leopard frogs (*Rana pipiens* complex) of the United States. *Misc. Pub. Mus. Zool. Univ. Mich.* 148: 1–140.

Pasteels, J. 1942. New observations concerning the maps of presumptive areas of the young amphibian gastrula (*Ambystoma* and *Discoglossus*). *J. Exp. Zool.* 89: 255–281.

Pfennig, D. W. 1992. Polyphenism in spadefoot toad tadpoles as a locally adjusted evolutionarily stable strategy. *Evolution* 46: 1408–1420.

Poole, T. J. and M. S. Steinberg. 1982. Evidence for the guidance of pronephric duct migration by a craniocaudally traveling adhesion gradient. *Dev. Biol.* 92: 144–158.

Purcell, S. M. and R. Keller. 1993. A different type of mesoderm morphogenesis in *Ceratophrys ornata*. *Development* 117: 307–317.

Radice, G. P., A. W. Neff, Y. H. Shim, J.-J. Brustis and G. M. Malacinski. 1989. Developmental histories in amphibian myogenesis. *Int. J. Dev. Biol.* 33: 325–343.

Rugh, R. 1951. *The Frog, Its Reproduction and Development*. McGraw Hill Book Co., New York.

Ruibal, R. and E. Thomas. 1988. The obligate carnivorous larvae of the frog, *Lepidobatrachus laevis* (Leptodactylidae). *Copeia* 1988: 591–604.

Ruiz i Altaba, A. 1993. Induction and axial patterning of the neural plate: Planar and vertical signals. *J. Neurobiol.* 24: 1276–1304.

Sadaghiani, B. and C. H. Thiebaud. 1987. Neural crest development in the *Xenopus laevis* embryo, studied by interspecific transplantation and scanning electron microscopy. *Dev. Biol.* 124: 91–110.

Saha, M. S., C. L. Spann and R. M. Grainger. 1989. Embryonic lens induction: More than meets the optic vesicle. *Cell Differ. Dev.* 28: 153–172.

Savage, R. M. and M. V. Danilchik. 1993. Dynamics of germ plasm localization and its inhibition by ultraviolet irradiation in early cleavage *Xenopus* embryos. *Dev. Biol.* 157: 371–382.

Schmell, E. D., B. J. Gulyas and J. L. Hedrick. 1983. Egg surface changes during fertilization and the molecular mechanism of the block to polyspermy. In *Mechanism and Control of Animal Fertilization*, J. F. Hartmann (ed.). Academic Press, New York, pp. 365–413.

Schoenwolf, G. C. and J. L. Smith. 1990. Mechanisms of neurulation: Traditional viewpoint and recent advances. *Development* 109: 243–270.

Schroeder, T. E. 1970. Neurulation in *Xenopus laevis*. An analysis and model based upon light and electron microscopy. *J. Embryol. Exp. Morphol.* 23: 427–462.

Seufert, D. W. and B. K. Hall. 1990. Tissue interactions involving cranial neural crest in cartilage formation in *Xenopus laevis* (Daudin). *Cell Differ. Dev.* 32: 153–166.

Shi, D.-L., M. Delarue, T. Darribère, J.-F. Riou and J.-C. Boucaut. 1987. Experimental analysis of the extension of the dorsal marginal zone in *Pleurodeles waltl* gastrulae. *Development* 100: 147–161.

Signoret, J. and J. Lefresne. 1971. Contribution a l'étude de la segmentation de l'ouef d'axolotl. I. Definition de la transition blastuléene. *Ann. Embryol. Morphogen.* 4: 113–123.

Smith, B. G. 1912. The embryology of *Cryptobranchus alleghaniensus*, including comparisons with some other vertebrates. II. General embryonic and larval development, with special reference to external features. *J. Morphol.* 23: 455–580.

Smith, J. C. and G. M. Malacinski. 1983. The origin of the mesoderm in anuran, *Xenopus laevis*, and a urodele, *Ambystoma mexicanum*. *Dev. Biol.* 98: 250–254.

Smith, S. C., M. J. Lannoo and J. B. Armstrong. 1988. Lateral-line neuromast development in *Ambystoma mexicanum* and a comparison with *Rana pipiens*. *J. Morphol.* 198: 367–379.

Spemann, H. and H. Mangold. 1924. Über Induktion von Embryonalanlagen durch Implantation artfremder Organisatoren. *Roux Arch. Entwick. Org.* 100: 599–638.

Sprules, G. W. 1974. Environmental factors and the incidence of neoteny in *Ambystoma gracile* (Baird) (Amphibia: Caudata). *Can. J. Zool.* 52: 1545–1552.

Stewart-Savage, J. and R. D. Grey. 1982. The temporal and spatial relationships between cortical contraction, sperm trail formation and pronuclear migration in fertilized *Xenopus* egg. *Roux's Arch. Dev. Biol.* 191: 241–245.

Stone, L. S. 1926. Further experiments on the extirpation and transplantation of mesectoderm in *Amblystoma punctatum*. *J. Exp. Zool.* 44: 95–131.

Stone, L. S. 1929. Experiments showing the role of migrating neural crest (mesectoderm) in the formation of head skeleton and loose connective tissue in *Rana palustris*. *Roux Arch. Entwick. Org.* 118: 40–77.

Suzuki, A. S. and K. Harada. 1988. Prospective neural areas and their morphogenetic movements during neural plate formation in the *Xenopus* embryo. II. Disposition of transplanted ectoderm pieces of *X. borealis* animal cap in prospective neural areas of albino *X. laevis* gastrulae. *Dev. Growth Differ.* 30: 391–400.

Talevi, R. 1989. Polyspermic eggs in the anuran *Discoglossus pictus* develop normally. *Development* 105: 343–349.

Tata, J. R. 1993. Gene expression during metamorphosis: An ideal model for post-embryonic development. *BioEssays* 15: 239–248.

Townsend, D. S. and M. M. Stewart. 1985. Direct development in *Eleutherodactylus coqui* (Anura: Leptodactylidae): A staging table. *Copeia* 1985: 423–436.

Townsend, D. S., M. M. Stewart, F. H. Pough and P. F. Brussard. 1981. Internal fertilization in an oviparous frog. *Science* 212: 469–471.

Tschumi, P. A. 1957. The growth of the hindlimb bud of *Xenopus laevis* and its dependence upon the epidermis. *J. Anat.* 91: 149–173.

Tucker, A. S. and J. M. W. Slack. 1995. The *Xenopus laevis* tail-forming region. *Development* 121: 249–262.

Tunner, H. G. and S. Heppich-Tunner. 1991. Genome exclusion and two strategies of chromosome duplication in oogenesis of a hybrid frog. *Naturwissenschaften* 78: 32–34.

Vogt, W. 1929. Gestaltungsanalyse am Amphibienkeim mit örtlicher Vitalfärbung. II Teil. Gastrulation und Mesodermbildung bei Urodelen und Anuren. *Roux Arch. Entwick. Org.* 120: 384–706.

Wake, M. H. 1977. The reproductive biology of caecilians: An evolutionary perspective. In *The Reproductive Biology of Amphibians*, D. H. Taylor and S. I. Gutman (eds.). Plenum, New York, pp. 73–101.

Wallace, R. A., Z. Misulovin and L. D. Etkin. 1981. Full-grown oocytes from *Xenopus laevis* resume growth when placed in culture. *Proc. Natl. Acad. Sci. USA* 78: 3078–3082.

Xavier, F. 1977. An exceptional reproductive strategy in Anura: *Nectophrynoides occidentalis* Angel (Bufonidae), an example of adaptation to terrestrial life by viviparity. In *Major Patterns in Vertebrate Evolution*, M. K. Hecht, P. C. Goody and B. M. Hecht (eds.). Plenum, New York, pp. 545–552.

Yost, H. J. 1992. Regulation of vertebrate left-right asymmetries by extracellular matrix. *Nature* 357: 158–161.

CHAPTER 21

Reptiles and Birds

Gary C. Schoenwolf

THIS CHAPTER DISCUSSES THE DEVELOPMENT OF REPTILES and birds. Because comparatively little is understood about reptilian development, my approach will be to focus on avian development and to point out information of special interest relating to reptilian development. Also, what is known about reptilian development suggests that reptilian and avian development occur similarly. Thus the reader can assume, unless stated otherwise, that the description given of avian development is generally applicable to reptilian development.

I will emphasize early development, especially that occurring prior to laying of the avian egg (that is, during "in utero" development) and during the egg's first three days of incubation after laying. A series of whole mounts of the chick blastoderm from the stage the egg is laid through three days of incubation is shown in Figure 21.1. Excellent classical references discuss avian development and can be consulted for additional information (Hamilton 1952; Romanoff 1960; Patten 1971; see also Schoenwolf 1995, 1997 for detailed descriptions of avian embryonic morphology).

This chapter begins with a brief discussion of reproductive strategies utilized by reptiles and birds. Then I will discuss the six so-called phases of development. The first phase is gametogenesis. In males, gametogenesis is called more specifically spermatogenesis; it results in the generation of the sperm. In females, gametogenesis is called more specifically oogenesis; it results in the generation of the ovum. Embryonic development per se begins with the second phase: fertilization. During fertilization the ovum and sperm are united, establishing a new individual with its own unique set of genes. Fertilization activates further development of this new organism, and it quickly results in a series of rapid mitotic divisions during the third phase of development: cleavage. The embryo becomes multicel-

lular as a result of cleavage, and the planes of the future body axes become established. The cells formed during cleavage are called blastomeres. Blastomeres are reshuffled and brought into new combinations during the fourth phase of development: gastrulation. Gastrulation establishes the three primary germ layers, the ectoderm, the mesoderm, and the endoderm. These layers undergo morphogenesis, or form-shaping movements, during the fifth phase of development: formation of the primary organ rudiments. The primary organ rudiments consist of the neural tube; mesodermal subdivisions including the heart, notochord, and somites; and the primitive gut, which forms in three distinct subdivisions: the foregut, midgut, and hindgut. Also during the stages of formation of the primary organ rudiments, the tube-within-a-tube body plan of vertebrate embryos is established, and the body proper of the embryo progressively separates from its extraembryonic (supporting) membranes. Finally, primary organ rudiments undergo further development and give rise to organ systems containing specialized tissues. This sixth phase of development, which occupies the vast majority of the developmental period, is called organogenesis. It consists of two processes: growth and differentiation. Here I will give only a brief overview of organogenesis, owing to space constraints. I will conclude this chapter with a discussion of hatching, a summary, and a bibliography.

To aid researchers in describing embryonic development precisely, stage series have been established for organisms commonly studied experimentally. For avian embryos (particularly chick and quail embryos), two stage series have been most useful. Stages of development prior to and immediately after laying of the egg are defined by the Eyal-Giladi and Kochav (1976) stage series (stages designated as

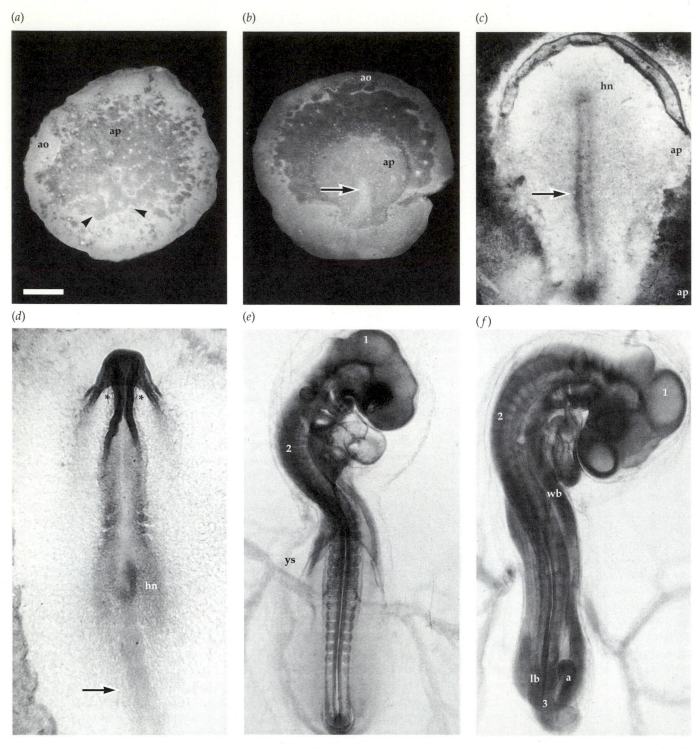

Figure 21.1 Whole mounts of the chick blastoderm from the stage the egg is laid through three days of incubation. (*a*) Blastoderm at Eyal-Giladi and Kochav (1976) stage XI (about 2 hours of incubation). (*b–f*) Blastoderms and embryos; stages per Hamburger and Hamilton (1951). (*b*) Blastoderm stage 3 (about 12 hours of incubation). (*c*) Blastoderm stage 4 (about 18 hours of incubation). (*d*) Embryo stage 8 (about 24 hours of incubation). (*e*) Embryo stage 15 (about 48 hours of incubation). (*f*) Embryo stage 17 (about 72 hours of incubation). Abbreviations: a, allantois; ao, area opaca; ap, area pellucida; hn, Hensen's node; lb, leg bud; wb, wing bud; ys, yolk sac and vitelline blood vessels; arrows, primitive streak; arrowheads, forming hypoblast; asterisks, level of fusion of neural folds; 1, level of cephalic flexure; 2, level of cervical flexure; 3, level of tail flexure. (*a, b, e, f* from Schoenwolf 1995; *c* from Schoenwolf 1997; *d* from Schoenwolf 1991.) Bar = 350 μm.

stages I–XIV). Stages encompassing the entire 21 days of incubation are defined by the Hamburger and Hamilton (1951) stage series (stages designated as stages 1–46). For reptilian embryos, and in particular snapping turtle embryos, the stage series of Yntema (1968) is most useful (stages designated as stages 0–26). Yntema's stage series encompasses the entire 20 weeks of development (at 20°C).

Reptilian embryos and especially avian embryos have been used for experimental embryology to analyze mechanisms underlying development. Techniques of experimental embryology for snapping turtles and chicks have been published (Hamburger 1960; Yntema 1964; Schoenwolf 1995; Darnell and Schoenwolf 1996). The main advantages of using reptiles and birds (especially chicks) for these studies are that eggs can be obtained cheaply and in large numbers; unlike mammals, embryos develop outside of the uterus, providing good access to the embryo throughout the entire period of development; and both reptilian and avian embryos tolerate well experimental manipulation, including microsurgery. Finally, reptiles have the additional advantage that they develop relatively slowly, allowing complex developmental sequences to be dissected out and identified.

Reproductive Strategies

Both reptiles and birds have developed a mode of reproduction that is compatible with life on land. Reptiles and birds lay their eggs on land, unlike most amphibians, which deposit their eggs in a aqueous environment, and most mammals, whose eggs are retained within their mother so that embryos develop within her body. Eggs that are laid on land are called **cleidoic** (Latin: boxlike), that is, they comprise a closed system which is self-sufficient for development (with the exception of gas exchange). Reptiles are cold-blooded organisms, whereas birds are warm blooded. After laying their eggs, reptiles go on their merry way, without further concern for parental care. Thus the rate of development of reptilian eggs varies with the temperature of their external environment (20–30°C is compatible with normal development). Environmental temperature can vary markedly over the light-dark cycle of each day but is mitigated in part by a reproductive behavior of the female who buries her eggs in sand, grass, or suitable material at the time of laying. (The variation in temperature of the external environment during incubation is used by the embryo in an interesting and important way; see the later discussion on packaging the egg and oviposition.) By contrast, birds build nests for their eggs and then sit on their nests to incubate their eggs, thereby maintaining a relatively constant temperature (about 38°C) that is generally higher than that of the external environment. This higher temperature is necessary for further development; little or no development occurs at "room" temperature.

Gametogenesis

Gametogenesis, the formation of the gametes, is commonly thought to begin in adult organisms after the onset of puberty. In reality, gametogenesis is initiated much earlier when the future "father" or "mother" is still an embryo his/herself. During embryonic development of the future parents, specialized cells—called **primordial germ cells**—migrate toward the developing gonads and invade them. In birds, the functioning gonads consist of the paired testes in the male and the single ovary in the female (two ovaries develop in females, but in most birds, including the domestic hen, the right ovary remains rudimentary). After the primordial germ cells invade the gonads they undergo rapid mitotic divisions, increasing in number. Within the gonads these rapidly dividing cells are called **spermatogonia** in the male and **oogonia** in the female. Before hatching of the male embryo, spermatogonia slow down their mitotic divisions considerably. Before hatching of the female embryo, all oogonia completely cease mitosis and become surrounded by a single layer of cells, called **follicle cells**, derived from the ovary. Concomitantly, oogonia initiate meiosis, a process during which the diploid number of chromosomes, the number characteristic of the somatic cells that compose all tissues of the body, becomes reduced to the haploid number, the number characteristic of gametes (i.e., sperm and ova). Each oogonium after the onset of meiosis is called a **primary oocyte**, and each primary oocyte and its surrounding follicles cells is called a **primary follicle**. At this point in time, gametogenesis in both sexes is placed on hold until the onset of puberty, after which gametogenesis resumes with a passion.

Spermatogenesis

Gametogenesis restarts in both males and females at puberty. Throughout the entire life of males, some spermatogonia will continue to divide mitotically, producing additional spermatogonia. After the onset of puberty, some spermatogonia will cease their mitotic divisions and initiate meiosis, after which they are called **primary spermatocytes**. Each primary spermatocyte completes the first meiotic division and forms two **secondary spermatocytes**. Each of these in turn undergoes the second meiotic division and forms two spermatids. Each of the four spermatids generated from each spermatogonium through the process of meiosis contains the haploid number of chromosomes. Within the testes and the gonoducts, the passageways that allow the transport of the sperm to the exterior during copulation, the immature spermatids undergo **spermiogenesis**, a maturation process, after which they are called mature sperm.

Spermatogenesis cannot occur normally at the elevated body temperatures of warm-blooded birds and mammals. Hence both of these classes of vertebrates have devised mechanisms for cooling their testes below normal body temperature. In most mammals this is accomplished by the scrotum, an external sac that houses the testes and contains a special system of veins (the pampiniform plexus) that keeps the testes at the appropriate temperature for spermatogenesis (1–8°F cooler than the temperature of the body cavity). The dartos muscle can raise the scrotum closer to the body (as when the scrotum is in cold water) or lower it (as in fevers). In birds, a scrotum is not present. Instead, the testes are contained within the body proper but are cooled by neighboring air sacs.

Oogenesis

The process of gametogenesis occurs somewhat differently in females than in males. As stated above, by the time of hatching all oogonia have ceased mitosis, have initiated meiosis, and have become surrounded by follicle cells; thus, the ovary contains only **primary oocytes**, not oogonia, and no further increase in the number of potential ova-generating cells can occur.

The reproductive system consists of two main components: an ovary and an oviduct. The ovary is the source of ova, which are generated during oogenesis and are shed from the ovary during the process of ovulation. The oviduct is like a factory assembly line, designed for the packaging of the egg. We will return to the role of the oviduct in packaging later; here I will focus on production of ova during oogenesis.

Recall that oogenesis begins in the female embryo during its own development prior to hatching. In the left ovary of the embryonic hen, cells called oogonia undergo rapid mitotic division. These cells, as well as the comparable cells in the male (called spermatogonia; see above), arise even earlier in development during stages of gastrulation from other cells called primordial germ cells; we will return to these cells in the section on organogenesis. Through mitotic divisions, the number of oogonia increases prior to hatching of the hen to hundreds or thousands. Because these divisions are mitotic rather than meiotic, the diploid number of chromosomes is maintained. Shortly before hatching of the embryonic hen, however, mitotic divisions cease and each oogonium enlarges and initiates the first meiotic division. These cells are now called primary oocytes; additionally, each becomes surrounded by a single layer of **follicle cells**, and each primary oocyte and surrounding cluster of follicle cells is now called a **primary follicle**. Thus at the time the embryonic hen hatches from the egg, her left ovary contains numerous primary follicles, each of which contains a cell that has initiated meiosis.

Primary follicles remain dormant until the growing chick becomes an adult, enters puberty, and becomes a mature hen. At this time, a hormone is produced by the anterior pituitary gland and is secreted into the blood stream. This hormone, called follicle stimulating hormone, causes yolk to be synthesized by the hen's liver. From there it is transported through the blood stream to the primary folli-

cles, where it accumulates and results in their rapid growth. As each primary follicle grows, its cytoplasm and nucleus become localized to one pole of the egg, called the animal pole (the opposite yolky pole is called the vegetal pole). At the animal pole, the sequestered nucleus and cytoplasm form a disclike structure called the **blastodisc**. The blastodisc is the "living" portion of the egg.

Ovulation of the egg (actually, ovulation of the secondary oocyte; see below) from the ovary occurs in response to the secretion of another hormone from the anterior pituitary gland, called luteinizing hormone. A surge of luteinizing hormone is secreted into the blood stream daily in mature hens during their reproductive cycle. (In commercial egg-producing hens, this cycle essentially occurs daily throughout the productive two or three years of the hen's life; she takes relatively few days off during the year from her arduous task of being an "egg machine.") Each day this surge causes one of the largest of the growing follicles to complete its first meiotic division and to form two cells: a **secondary oocyte** and a nonfunctional, much smaller cell, the **first polar body**. Also, the surge of luteinizing hormone causes this same secondary oocyte (and its encasing follicle cells) to be ovulated from the ovary, after which it quickly enters the ostium (opening) of the oviduct.

The oviduct in birds is considered to have five subdivisions (Figure 21.2). Beginning at the side closest to the ovary, these five subdivisions are as follows: the **infundibulum**, **magnum**, **isthmus**, **shell gland**, and **vagina**. After entering the ostium of the oviduct, each secondary oocyte resides for a short period of time in the infundibulum. Further development of the oocyte requires fertilization, and is discussed in the next section. In the absence of fertilization, the egg is ultimately passed through the oviduct and laid. If cracked open and examined, such eggs will be seen to consist of a yolk, surrounded by egg white (albumen), and containing at one pole (the animal pole) a small white dot (about 3–4 mm in diameter), the blastodisc. The unfertilized egg thus contains an uncleaved blastodisc and remains at the secondary oocyte stage.

Fertilization

Fertilization of the ovulated egg occurs within the infundibulum of the oviduct before most of the coverings of the egg have been added (see next section). In birds, the hen and rooster copulate and sperm are deposited by the rooster into the cloaca of the hen. The process of copulation in birds is referred to as the cloacal kiss, because during copulation the cloaca of the male and female are brought into apposition. For the next several days following copulation, a small packet of sperm is released each day, concomitant with ovulation, from a specialized region of the cloaca. Sperm migrate up the various subdivisions of the oviduct to the infundibulum where they en-

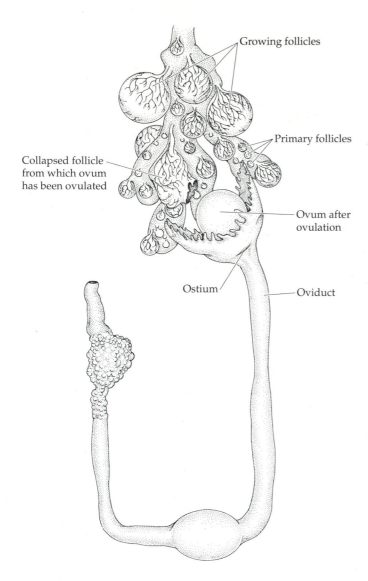

Growing follicles

Primary follicles

Collapsed follicle from which ovum has been ovulated

Ovum after ovulation

Ostium

Oviduct

Figure 21.2 Drawing of the functional reproductive system of the mature hen. The ovary is the lobulated structure at the top of the drawing; the subdivisions of the oviduct are illustrated below it. The magnum is greatly distended by the passage of an ovum. (After Schoenwolf 1995.)

now called the fertilized egg or **zygote**. The few sperm that entered the egg's cytoplasm with the fertilizing sperm but did not participate further in the development of the egg, form accessory nuclei, which can be seen for the first few cleavages. Their ultimate fate is unknown, but they are believed to degenerate and contribute nothing to the developing individual or its membranes.

Packaging the Egg and Oviposition

I think if required on pain of death to name instantly the most perfect thing in the universe, I should risk my fate on a bird's egg. (T. W. Higginson 1863; cited by Romanoff and Romanoff 1949.)

After fertilization occurs the vitelline membrane is secreted (actually, only the outer vitelline membrane is added; an inner vitelline membrane is added in the primary follicle prior to ovulation). The egg remains within the infundibulum for about 15 minutes. Within the magnum of the oviduct, the egg white or albumen is added. This requires about three hours. Within the isthmus the shell membranes (inner and outer) are added. This requires about one hour. Within the shell gland the shell is added. In birds, the shell is calcified and, consequently, hard; in many reptiles, the shell is leathery and far more malleable (in some reptiles the egg is calcified as in birds). The formation of the shell requires the greatest amount of time "on the assembly line," that is, about 20 hours. After formation of the shell is completed, the egg is passed to the vagina, where it is expelled through the cloaca to the outside; that is, the egg is laid or oviposition occurs. Because of the times just listed, the hen can make only one egg per day. One primary follicle grows to the appropriate size each day, one secondary oocyte is generated and ovulated each day, and the egg is packaged in about 24 hours and 15 minutes (i.e., 15 minutes in the infundibulum plus three hours in the magnum plus one hour in the isthmus plus 20 hours in the shell gland). The actual time needed to produce an egg exceeds 24 hours, so each subsequent egg produced by a particular hen will be laid progressively later during the day until afternoon is reached. After several days of laying, she takes a break and skips one ovulation (and, consequently, one day later skips laying one egg). She then starts laying eggs again, one each day, beginning in the morning, thereby repeating the cycle. Each group of eggs that is laid without a break is called a clutch.

For additional details see Bellairs (1964).

counter the freshly ovulated egg. At the time of fertilization, several sperm surround and a few penetrate the plasmalemma overlying the blastodisc. The vitelline membranes provide the first block to polyspermy, that is, the fertilization of the egg by more than one sperm. During fertilization, sperm must pass along tortuous routes through pores in the vitelline membranes to reach the egg's plasmalemma. Most fail in this endeavor. Of the few that enter the egg's cytoplasm only one enlarges as the **male pronucleus**. Fertilization stimulates the completion of meiosis and makes the egg refractory to further penetration by additional sperm, thereby providing the second block to polyspermy. The secondary oocyte at fertilization undergoes the second meiotic division and forms two cells: the **ovum** (containing the haploid number of chromosomes) and the nonfunctional, much smaller **second polar body**. The nucleus of the ovum enlarges into a structure called the **female pronucleus**. Fertilization is completed when the male and female pronuclei fuse and the diploid number of chromosomes is restored in what is

Cleavage and Axis Determination

Cleavage begins while the egg is still within the oviduct of the hen (either the isthmus or shell gland), about five hours after fertilization (Figures 21.3 and 21.4). The mode of cleavage of reptilian and avian eggs is referred to as **partial** (or **meroblastic**) **cleavage** because only a portion of the egg cleaves, and also as **discoidal**, because cleavage is restricted to the disc of cytoplasm at the animal pole of the egg. Both reptilian and avian eggs are highly polarized and are referred to as **telolecithal eggs**, based on the extremely asymmetrical localization of the cytoplasm in relation to the yolk. During cleavage, the initial cleavage furrows only partially separate the blastodisc (which is now called a blastoderm with formation of multiple cells or blastomeres) into adjacent blastomeres; that is, cytoplasmic continuity is retained between adjacent blastomeres (and between some blastomeres and the yolk mass) for a period of time. The first cleavage furrow partially separates the blastodisc into two blastomeres. The second cleavage furrow forms about 15 minutes after the first, is perpendicular to the first, and establishes four partially separated blastomeres. The third cleavage occurs about 1 hour after the second. Two third-cleavage furrows form during the third cleavage. These furrows form parallel to the first and partially separate the blastoderm into eight blastomeres. The fourth and subsequent cleavage furrows begin to completely separate **central blastomeres** from peripheral **marginal blastomeres**. In addition, some of these furrows form horizontally within the blastoderm rather than vertically; these furrows progressively separate the central blastomeres from the underlying **yolk** and establish a fluid-filled cavity called the **subgerminal cavity**. At the time of laying, the avian blastoderm consists of approximately 60,000 cells (i.e., about 14 divisions of the zygote occur prior to laying). Moreover, at this time two distinct zones can be seen within the blastoderm: the central and clearer **area pellucida**, and the peripheral and less translucent **area opaca**. The area pellucida is about 6 cell-layers thick; the area opaca progressively tapers to a single layer as it spreads centrifugally over the yolk.

The **body axes** of the future embryo first become visible during cleavage stages. Three axes define the coordinates of the embryonic body: the **rostrocaudal (head-tail** or **anterior-posterior) axis**, the **dorsoventral axis**, and the **right-left axis**. Shortly before the egg is laid, a layer of large, yolky cells begins to form beneath the area pellucida. This layer is called the **hypoblast**. It serves to define the three future body axes. First, the hypoblast begins its development at what will be the future caudal end of the area pel-

Figure 21.3   Dorsal views of the chick blastoderm during cleavage. Numbers indicate cleavage furrows. (After Patten 1971.)

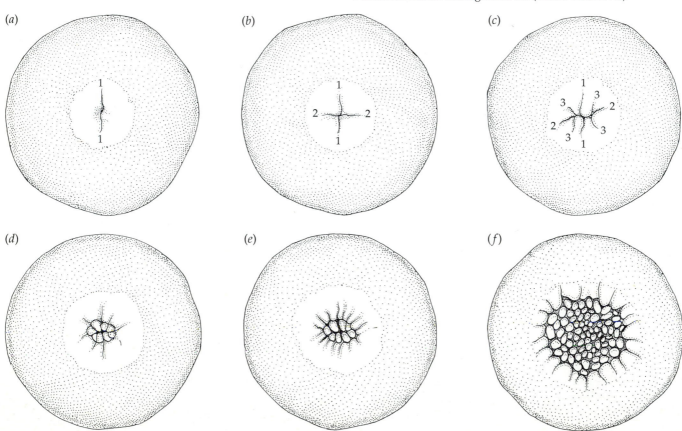

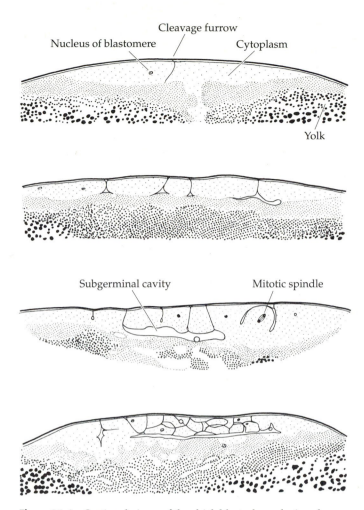

Figure 21.4 Sectional views of the chick blastoderm during cleavage.

Gastrulation

Gastrulation consists of a complex series of morphogenetic (i.e., form-generating) movements in which the three **primary germ layers** are established and during which cells are brought together in new combinations allowing cell–cell inductive interactions to occur (Figures 21.5–21.7).

Recall that immediately preceding gastrulation, the blastoderm is bilaminar, consisting of a dorsal **epiblast** and a ventral **hypoblast**. With the onset of gastrulation, rapid movements begin within the epiblast. These movements are whirl-like in nature and have been referred to as Polonaise movements based on the swirling motion of women's dresses during Polish folk dancing. When viewed from the dorsal surface of the blastoderm, cells swirl counterclockwise on the left side of the blastoderm and clockwise on the right side of the blastoderm as they move toward the caudal midline where they pile up. In birds as cells accumulate, they form a wedge of cells, whose apex points rostrally and whose base lies caudally. This wedgelike structure is the initial **primitive streak**; it elongates and becomes cordlike over the next few hours of development. In reptiles, development occurs somewhat differently: cells pile up at the caudal, median area pellucida but they do not extend rostrally as a primitive streak. Instead, they form a **blastopore**. Subsequently, cells turn inward, or involute, over the lip of the blastopore. Thus gastrulation in reptiles is reminiscent of gastrulation in both birds (in which the blastoderm is flat) and amphibians (in which a blastopore forms). Moreover, a tunnel-like opening in the vicinity of the blastopore extends through the entire thickness of the blastoderm. This opening, called the chordamesodermal canal, eventually connects the cavity of the caudal neural tube to the cavity of the yolk sac. Hence at a later stage it is renamed as the neurenteric canal. A similar neurenteric canal has been described in gastrulating/neurulating human embryos.

The primitive streak of birds and mammals is the site at which cells leave the epiblast and migrate into the interior of the blastoderm during formation of the three **primary germ layers**, the **ectoderm**, the **mesoderm**, and the **endoderm**. Movement of cells through the streak is termed ingression. The first cells to ingress through the primitive streak form endoderm and extraembryonic mesoderm. The prospective endodermal cells within the epiblast move toward the primitive streak—in particular through its rostral half—ingress through it, and enter the hypoblast layer (Figure 21.5). As they enter the hypoblast, they displace the hypoblast cells rostrally toward the interface between the area pellucida and area opaca at the extreme rostral end of the blastoderm. The hypoblast remains in this area and eventually makes two contributions: it forms part of the extraembryonic mesoderm contributing to two extraembryonic membranes and it gives rise to primordial germ cells. Meanwhile, prospective extraembryonic mesodermal cells within the epiblast also

lucida, thereby defining the rostrocaudal axis of the future embryo. Second, the hypoblast delaminates from a more superficial position subjacent to the vitelline membrane to a deeper position adjacent to the subgerminal cavity overlying the yolk, thereby defining the dorsoventral axis of the future embryo. The hypoblast marks the ventral extent of the blastoderm and the cells lying above it (i.e., more superficially), which collectively are now called the **epiblast**, mark the dorsal extent of the blastoderm. This initial dorsoventral polarity of the blastoderm will be modified during gastrulation (see next section) and body folds. Finally, formation of the hypoblast establishes the right-left axis of the future embryo by defining median (midline) and lateral (off the midline) positions. The hypoblast initially forms at the future caudal end of the area pellucida, spanning the midline. Thus median and lateral coordinates can be estimated and, consequently, the future right-left axis of the embryo can be identified.

For additional details see Kochav and Eyal-Giladi (1971); Khaner et al. (1985); Khaner and Eyal-Giladi (1986, 1989); Eyal-Giladi and Khaner (1989); Khaner (1993).

Figure 21.5 Prospective fate maps of the chick area pellucida showing the formation of the ingressed endoderm and the displacement of the hypoblast. The epiblast is shown on the left side of each diagram, but not on the right side. Arrows indicate the direction of displacement of the hypoblast in the initial primitive streak stage, and directions of migration of the ingressed endodermal cells in the intermediate primitive streak stage and definitive primitive streak stage. At all stages illustrated, prospective mesoderm is contained within the epiblast. By the intermediate primitive streak stage, mesoderm is beginning to migrate into the interior of the blastoderm, between the overlying epiblast and underlying hypoblast/endoderm. For the sake of clarity in illustrating the formation of the endoderm, prospective mesoderm and ingressed mesoderm have been omitted from this figure. The numbers used here correspond to those used in Figures 21.6 and 21.7. 1, prospective endoderm; 2, endoderm; 3, hypoblast; 4, primitive streak; 5, germ cell crescent; 6, notochord. (After Schoenwolf 1995.)

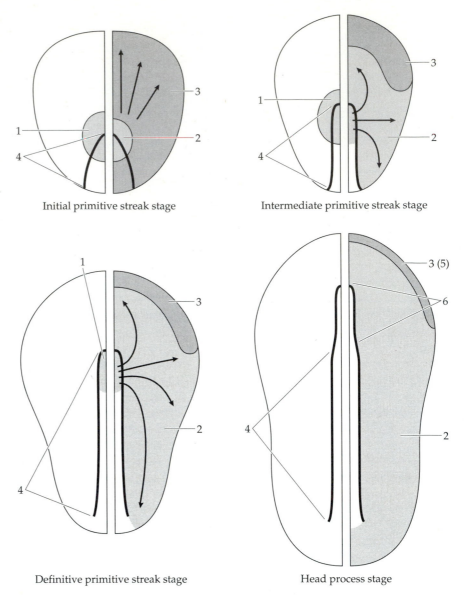

Initial primitive streak stage   Intermediate primitive streak stage

Definitive primitive streak stage   Head process stage

Figure 21.6 Prospective fate map of the chick epiblast. The approximate site of formation of the primitive streak is indicated by dashed lines. Prospective intermediate mesoderm arises from the interface between prospective somitic mesoderm and prospective lateral plate mesoderm, but it is too small and ill-defined to illustrate; it ingresses in conjunction with these two mesodermal subdivisions. The numbers correspond to those used in Figures 21.5 and 21.7. 1, prospective endoderm, 7, prospective extraembryonic mesoderm; 8, prospective heart mesoderm; 9, prospective head mesenchyme, prospective notochord; 10, prospective somitic mesoderm; 11, prospective lateral plate mesoderm; 12, prospective neural plate; 13, prospective skin ectoderm. (After Schoenwolf 1995.)

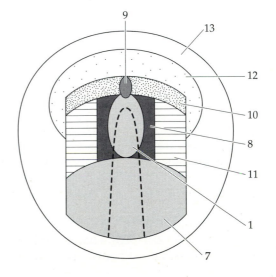

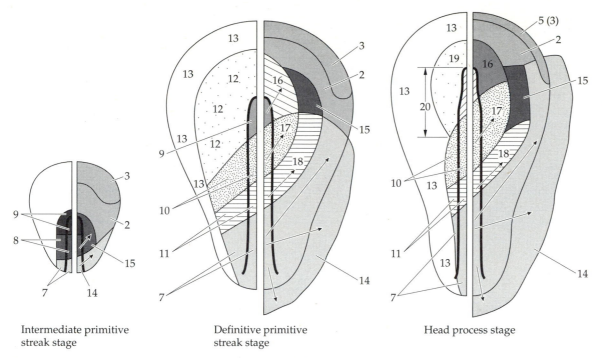

Intermediate primitive
streak stage

Definitive primitive
streak stage

Head process stage

Figure 21.7 Prospective fate maps of the chick area pellucida showing the formation of
the subdivisions of the ingressed mesoderm. The epiblast is shown on the left side of
each diagram, but not on the right side. Arrows indicate the directions of migration of the
ingressed mesodermal cells. The boundaries of the primitive streak and notochord are indicated by a heavy black line. The numbers correspond to those used in Figures 21.5 and
21.6. 2, endoderm; 3, hypoblast; 5, germ cell crescent; 7, prospective extraembryonic
mesoderm; 8, prospective heart mesoderm; 9, prospective head mesenchyme, prospective
notochord; 10, prospective somitic mesoderm; 11, prospective lateral plate mesoderm; 12,
prospective neural plate; 13, prospective skin ectoderm; 14, extraembryonic mesoderm;
15, heart mesoderm; 16, head mesenchyme; 17, somitic mesoderm; 18, lateral plate mesoderm; 19, prosencephalon level of neural plate; 20, mesencephalon, rhombencephalon,
and spinal cord levels of neural plate. (After Schoenwolf 1995.)

move toward the primitive streak—in particular toward
its caudal half—ingress through it, and enter the space between the epiblast and forming endoderm (Figure 21.7).
The extraembryonic mesoderm contributes to all four extraembryonic membranes. With formation of the extraembryonic mesoderm the formerly bilaminar blastoderm
now becomes trilaminar.

The primitive streak continues to elongate during subsequent development until it extends about two-thirds to
three-quarters of the rostrocaudal extent of the area pellucida. Prospective mesodermal cells within the epiblast
continue to move toward the primitive streak, ingress
through it, and enter the space between the epiblast and
the forming endoderm. Different groups of prospective
mesodermal cells originate from different spatially localized regions of the epiblast. This fact was determined by
fate mapping studies in which different regions of the epiblast were labeled precisely with dyes and followed over
time to see what they became. Such studies allowed researchers to construct **prospective fate maps**, drawings
that show the locations of various prospective groups of

cells within the epiblast prior to and during ingression
(Figures 21.5–7). Such maps have revealed in great detail
the origins and movements of prospective ectodermal,
mesodermal, and endodermal cells during gastrulation of
avian embryos. In particular they show that prospective
heart mesoderm ingresses prior to other subdivisions of
prospective *embryonic* mesoderm, and that prospective
heart mesoderm ingresses through the middle third or so
of the primitive streak. At slightly later stages, prospective
head mesenchyme, prospective notochord, prospective
somite, and prospective lateral plate mesoderm undergo
ingression, with prospective head mesenchyme and
prospective notochord ingressing through the rostral end
of the streak, and prospective somites and prospective lateral plate mesoderm ingressing through the middle of the
primitive streak. (Prospective extraembryonic mesoderm
continues to ingress through the caudal portion of the
primitive streak; additionally, prospective intermediate
mesoderm, also known as prospective nephrotome, ingresses in concert with prospective somite and prospective lateral plate mesoderm, but its surface distribution is

very small and its exact origins remain unclear; consequently, prospective intermediate mesoderm has not been indicated on the maps shown in Figures 21.5–7).

As ingression occurs, the primitive streak begins to shorten or regress. During regression, a specialized portion of the primitive streak—**Hensen's node**, the rostral end of the streak—moves caudally leaving the notochord behind in its wake. Hensen's node plays an important role in inducing the **neural plate**; we will return to Hensen's node in the next section. Over time, progressively less of the area pellucida is involved in formation of the germ layers and progressively more becomes involved in morphogenesis to from the **primary organ rudiments**. At the completion of gastrulation, all prospective endodermal and mesodermal cells have ingressed and the primitive streak has largely disappeared. (Its remnants form the **tail bud**, which contributes cells to the neural tube and somites of the tail in a later developmental period called **secondary body development**). The epiblast now consists of only ectodermal cells. Similar events occur during gastrulation in reptilian embryos except that cells grow rostrally from the blastopore to form the embryo; that is, regression does not occur. Thus by the end of gastrulation, three distinct and separate germ layers have formed; during the next phase of development—formation of the primary organ rudiments—each of these layers is sculpted into separate organ rudiments.

For additional details see Spratt (1942, 1952); Rosenquist (1966); Nicolet (1971); Vakaet (1984); Stern and Canning (1988); Stern (1990); Schoenwolf et al. (1992); Garcia-Martinez and Schoenwolf (1993); Garcia-Martinez et al. (1993).

Formation of the Primary Organ Rudiments

Figure 21.8 shows selected transverse (cross) sections through a chick embryo after one day of incubation at 38°C. It illustrates a stage in formation of the primary organ rudiments; thus, it can be examined to see early events in formation of the neural tube, heart, gut, somites, and body folds.

Neurulation

The process of neurulation establishes the **neural tube**, one of the three major defining structures characteristics of chordates (that is, the presence of a dorsal, hollow nerve cord; the other two characteristics are the presence of gill slits and a notochord). The neural tube is the rudiment of the entire adult central nervous system (see later overview of organogenesis). In addition, cells emigrating from the roof of the neural tube, called **neural crest cells**, make major contributions to the peripheral nervous system (see the organogenesis discussion). The neural tube forms from ectoderm, which becomes induced by

Hensen's node to form a thickened layer called the neural plate. Hensen's node, which contains principally prospective notochordal cells, is considered to be the **organizer** of the avian and mammalian embryo. (In the latter class of vertebrates Hensen's node is generally called merely the node, rather than Hensen's node; Professor Hensen described this structure in the chick.) If transplanted to an ectopic site, Hensen's node can organize a complete additional embryo.

It is believed that during normal development Hensen's node, or its derivative cell populations, secrete growth factors that neuralize the ectoderm, that is, that alter the fate of the ectoderm from skin to nerve cells. Thus there are cell–cell inductive interactions that occur that involve the secretion of growth factors by the inducer cells and the presence of appropriate receptors on the responding cells. Induction of the neural plate is followed by three major morphogenetic events: shaping of the neural plate, bending of the neural plate, and closure of the neural groove. During the first of these events, the overall shape of the neural plate is altered. Prior to shaping, the neural plate consists of a single layer of cuboidal cells; the rostrocaudal extent of the neural plate is relatively short and its width (i.e., mediolateral extent) is relatively broad. During shaping, the cells of the neural plate become taller, increasing its thickness; the width of the plate narrows; and the length of the plate increases. Narrowing and lengthening of the plate result chiefly from two cell behaviors: oriented mitosis and the active rearrangement of cells. These two behaviors act in concert and generate a movement typical of morphogenesis called **convergent-extension**. Convergent-extension is best understood with an analogy. Imagine you are waiting in line for a rock concert. The line is flanked on its two sides by barricades. On the average, six people span the width of the line. Suppose that the flanking barricades are now moved closer so that the width of the line narrows to three people. What will happen to the length of the line? It will double (as its width is halved). This coordinated narrowing and lengthening is convergent-extension.

As shaping of the neural plate is underway, bending of the neural plate is initiated and formation of the **neural groove** ensues. During bending, the lateral edges of the neural plate and the adjacent medial edges of the skin ectoderm are thrown into folds called the **neural folds**. Each of the paired neural folds moves dorsally or elevates and then moves medially or converges to meet its partner in the dorsal midline. What causes neural folds to form, and what causes the neural plate to bend and to form a neural groove? Neural folds form as a result of a localized interaction occurring at the interface between their two components: lateral neural plate and medial skin ectoderm. This interaction is autonomous to the interface and can occur when the interface is cultured in isolation from surrounding tissues or when a new interface is created experimentally. However, bending of the neural plate is a more global

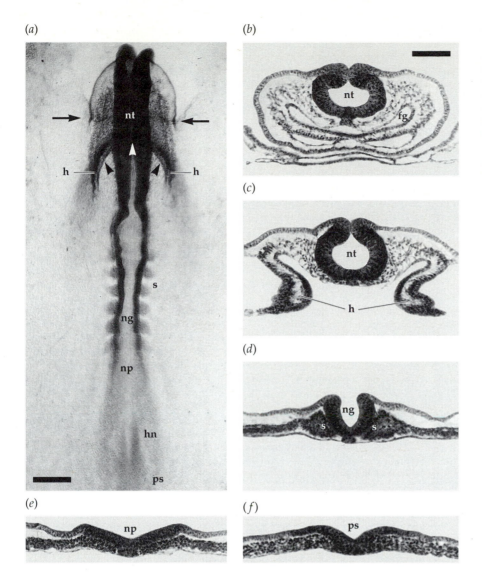

(a) *(b)* *(c)* *(d)* *(e)* *(f)*

Figure 21.8 Whole mount (*a*) and transverse sections (*b–f*) of the chick embryo at Hamburger and Hamilton (1951) stage 9 (about 30 hours of incubation). Transverse sections are lettered in rostral-to-caudal order. fg, foregut; h, heart rudiments; hn, Hensen's node; ng, neural groove; np, neural plate; nt, neural tube; ps, primitive streak; s, somites; arrows, head fold of the body; arrowheads, opening of the foregut. Bars = 150 μm (*a*), 75 μm (*b–f*). (*a* from Schoenwolf 1991; *b–f* from Schoenwolf 1997.)

phenomenon. Cell behaviors within both the neural plate and skin ectoderm—such as mitosis, rearrangement, and change in shape—generate forces required for the uplifting of the neural folds and the formation of the neural groove.

As a result of elevation of the neural folds and their convergence toward the dorsal midline, the paired neural folds are brought into apposition. Subsequently, fusion of the neural folds closes the neural groove and establishes a **neural tube**, a structure now covered dorsally by **skin ectoderm**. During subsequent development, ectodermal cells generated from the roof of the neural tube migrate laterally as paired streams of cells called the **neural crest**. As state above, the neural tube will form the entire central nervous system during subsequent development, whereas the neural crest will form a major component of the peripheral nervous system. (In addition, neural crest cells in the head contribute to the bones of the skull and the associated visceral skeleton within the branchial arches—the

rudiments of the craniofacial region; how individual neural crest cells become committed to one of several possible fates remains a major question in the field of developmental biology.) Failure of the neural groove to close results in an opening of the neural tube, specifically, an opening within a portion of the brain or spinal cord onto the dorsal side of the embryo; that is, the neural tube lumen remains open to the outside, and the nervous system does not become covered with skin. Such defects, called neural tube defects, are generally incompatible with life (unless there is surgical intervention to repair the defect partially, as is done in the case of newborn humans).

For additional details see Weston (1963); Karfunkel (1974); Noden (1975); Bancroft and Bellairs (1976); Anderson and Meier (1981); LeDouarin (1982); Tosney (1982); Bronner-Fraser (1986); Hall and Horstadius (1988); Schoenwolf and Smith (1990); Schoenwolf (1991, 1994); Jacobson (1994).

Body Folding and Early Development of the Primitive Heart Tube and Gut

The **bilaminar blastoderm** (i.e., the blastoderm consisting of epiblast and hypoblast) and **trilaminar blastoderm** (i.e., the blastoderm consisting of ectoderm, mesoderm, and endoderm) are flat, pizza-like structures, which are essentially two-dimensional and which float upon a massive "sea" of yolk. During body folding, regions of the area pellucida begin to fold and to separate off from other areas of the area pellucida that will give rise to extraembryonic membranes. Body folding thus serves to give the embryo a three-dimensional shape—characteristically, **a-tube-within-a-tube body plan**—and to separate the embryonic body from its extraembryonic (supporting) membranes. In addition, body folding is directly involved in formation of the primitive heart tube and gut.

Three types of body folds form during development. The first body fold, called the **head fold of the body**, forms just rostral to the developing brain; initially, it is shaped like a crescent, opening caudally. From its region of origin it expands caudally, undercutting the developing head and forming two spaces lined with epithelium: an ectodermally lined space, the **subcephalic pocket**, and an endodermally lined space, the **foregut**. The second body fold is a paired fold called the **lateral body folds**. The lateral body fold forms in the **somatopleure** on each side of the head and each fold is continuous medially with one side of the crescent-shaped head fold of the body.

Recall that the ingressed mesoderm becomes subdivided into several components. The **notochord** lies in the midline. On each side of the midline, the mesoderm becomes subdivided into the following areas listed in medial-to-lateral order: **somitic mesoderm**, **intermediate mesoderm**, **lateral plate mesoderm**, and **extraembryonic mesoderm**. The notochord acts as a primitive skeletal rod, providing support for the elongating embryo. Ultimately, it contributes to the intervertebral discs of the adult vertebral column. The somitic mesoderm becomes segmented into the somites, each of which becomes further subdivided. The intermediate mesoderm contributes to the urogenital system; namely, the kidneys and their ducts, and the gonads and their ducts. The lateral plate mesoderm and the extraembryonic mesoderm split in a plane perpendicular to their dorsoventral axis into a dorsal and a ventral layer. The upper (dorsomost) layer is called somatic mesoderm and the lower (ventromost) layer is called splanchnic mesoderm. Somatic mesoderm lies just ventral to the ectoderm and splanchnic mesoderm lies just dorsal to the endoderm. Collectively, the ectoderm and adjacent somatic mesoderm are called **somatopleure** and the endoderm and adjacent splanchnic mesoderm are called **splanchnopleure**. The somatopleure forms the body wall, including its paired outgrowths, the limb buds. The splanchnopleure forms the heart and gut walls.

Each lateral body fold consists of two components: so-matopleure and splanchnopleure. At the level of the forming heart, the splanchnic mesodermal component of each lateral body fold buds off cells that coalescence to form a tube, called the **endocardial tube**; after formation of this tube, the remaining splanchnic mesodermal layer at the level of the forming heart is called the **myocardium**. The splanchnopleuric component on each side fuses together with its partner in the ventral midline with expansion of the lateral body folds. As a result of this fusion, the endoderm pinches off as the **foregut**, whose walls consist of an inner layer of endoderm and an outer covering of splanchnic mesoderm, and a primitive, midline **heart tube** is established by fusion of the paired endocardial tubes. Formation of the heart tube requires additional description. After fusion of the endoderm to form the foregut (and at more caudal levels to form the midgut), the two endocardial tubes are free to approach one another and to fuse in the midline, forming a single tube: specifically, the **endocardium** of the primitive heart tube. Thereafter, the adjacent myocardial portion of the splanchnic mesoderm on each side is wrapped around the endocardium to form three structures: **a dorsal mesocardium**, a mesentery connecting the primitive heart tube to the ventral gut wall; the **myocardium**, a layer of the heart that eventually forms cardiac muscle; and a second mesentery, the **ventral mesocardium**, connecting the heart ventrally to the ventral body wall. The dorsal and ventral mesocardia are transient structures that are partially eliminated during looping of the heart.

The final body fold is called the **tail fold of the body**. It forms just caudal to the developing **tail bud** and extends rostrally. As it does so it establishes two spaces: an ectodermally lined space, called the **subcaudal pocket**, and an endodermally lined space, called the **hindgut**. Eventually, all three types of body folds join together around a stalk that connects the midgut to the yolk sac (see next section). This connection is called the **yolk sac stalk**; through it, the embryo remains attached to its **extraembryonic membranes** until the time of hatching.

As a result of body folding, the so-called tube-within-a-tube vertebrate body plan is established. What this means is that in transverse section the embryo is seen to consist of two major tubes: an outer tube, composed of skin ectoderm, the future skin of the embryo; and an inner tube, the gut, formed from endoderm. Vertebrate embryos also have an additional inner tube, the neural tube, which is formed from the ectodermal neural plate during neurulation (see above).

For additional details see Miller (1982).

Extraembryonic Membranes

Four extraembryonic membranes form during avian development: the **amnion, chorion, allantois**, and **yolk sac** (Figure 21.9). Two of these membranes, the chorion and allantois, fuse together during late embryogenesis and form a new structure called the **chorioallantoic membrane** (see below).

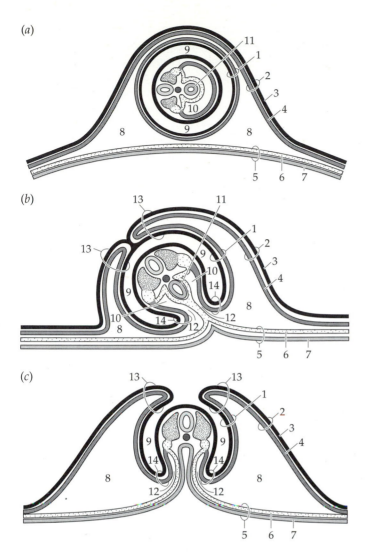

(a)

(b)

(c)

Figure 21.9 Schematic drawings of transverse sections of two-day chick embryos showing the formation of the amnion, chorion, and yolk sac. (a) is more rostral than is (b), which in turn is more rostral than (c). Torsion of the embryo occurs as the lateral amniotic folds elevate and fuse. 1, amnion; 2, chorion; 3, ectoderm; 4, somatic mesoderm; 5, yolk sac; 6, splanchnic mesoderm; 7, endoderm; 8, extraembryonic coelom; 9, amniotic cavity; 10, intraembryonic coelom; 11, gut endoderm; 12, continuity between intraembryonic and extraembryonic coeloms; 13, lateral amniotic fold; 14, lateral body fold. (After Schoenwolf 1995.)

embryo and are called the lateral amniotic folds. These four folds ultimately fuse together to enclose the embryo in an amnion, consisting of ectoderm nearest the embryo and an investing layer of somatic mesoderm; and an outer covering of chorion, consisting of somatic mesoderm nearest the amnion and an investing layer of ectoderm. The space between the amnion and the embryo is called the amniotic cavity; it is directly continuous with the subcephalic (beneath the head) and subcaudal (beneath the tail) pockets, and it is filled with amniotic fluid. The space between the amnion and chorion is called the extraembryonic coelom; it will later be obliterated by expansion of allantois and formation of the chorioallantoic membrane (see below). Smooth muscle fibers develop within the somatic mesodermal layer of the amnion. Contraction of these muscles gently rocks the embryo within the amniotic cavity, preventing the formation of adhesions between the embryo and its amnion.

The yolk sac forms from the endoderm, which was originally located adjacent to the yolk, and from an investing layer of splanchnic mesoderm. During development, the yolk sac expands over the surface of the yolk and ultimately completely encloses it and the surrounding albumen. Just prior to hatching, the yolk sac is withdrawn into the gut (midgut) of the embryo, and its enclosed yolk provides a food source to the hatchling for the next couple of days.

The allantois develops as an outgrowth of the hindgut. Recall that owing to formation of the tail fold of the body, the hindgut consists of an endodermal pocket with an investing layer of splanchnic mesoderm. The allantois is essentially nonfunctional until it fuses with the chorion to form the chorioallantoic membrane.

The chorioallantoic membrane is formed by fusion of two extraembryonic membranes during the latter half of incubation: the chorion and the allantois. The chorioallantoic membrane becomes highly vascular, with three major functions. First, it provides a large reservoir for the storage of nitrogenous wastes. Second, it transports calcium from the shell back to the embryo where it is utilized in bone formation. Third, it functions in respiration; namely, the transport of CO_2 from the embryo through pores in the shell to the outside environment, and the transport of O_2 in the opposite direction.

For additional details see Miller et al. (1994).

All four of the extraembryonic membranes develop from mesoderm and from one other germ layer. The mesoderm is derived from the prospective extraembryonic mesoderm. Recall that during subdivision of the mesoderm following its ingression, the lateral plate mesoderm and the extraembryonic mesoderm split into an upper layer called the somatic mesoderm and a lower layer called the splanchnic mesoderm. Somatic mesoderm lies just ventral to the ectoderm and splanchnic mesoderm lies just dorsal to the endoderm. Collectively, the ectoderm and adjacent somatic mesoderm are called somatopleure and the endoderm and adjacent splanchnic mesoderm are called splanchnopleure.

The amnion and chorion develop from somatopleure, whereas the allantois and yolk sac develop from splanchnopleure. The amnion and chorion develop together as four folds of somatopleure. One fold arches over the head of the embryo and is called the head fold of the amnion. A second fold arches over the tail of the embryo and is called the tail fold of the amnion. The third and fourth folds arch bilaterally over the lateral sides of the

Somitogenesis

One of the most fascinating processes to watch in time-lapse films of living embryos is the formation of the **somites**. Somites are serially iterated, paired blocks of mesoderm that flank the neural tube. During development each of them subdivides into three components: **dermatome**, **myotome**, and **sclerotome**. These give rise, respectively, in later development to the skeletal elements of the vertebral column, the associated skeletal muscle, and the overlying dermis of the skin.

Somites form from the somitic mesoderm at the average rate of one pair about every 90 minutes. Segmentation of the somites occurs uniformly, yielding serially repeated, paired units, and a characteristic number of somites forms in each species at each level (i.e., cervical, thoracic, lumbar, sacral, and coccygeal) of the future vertebral column. The mechanisms underlying segmentation of somites are unclear and many models have been proposed. Recently, it has been discovered that certain classes of genes, called patterning genes, are expressed within the somitic mesoderm in a spatially and temporally localized pattern during somitogenesis. Such genes are homologues of similar genes expressed in the embryo of the fly *Drosophila* (e.g., the homeobox-containing genes). These genes are known to play an important role in segmentation of the fly embryo. Most patterning genes encode DNA-binding proteins, proteins that bind to DNA and likely regulate the transcription of cascades and/or families of other genes; therefore, these proteins are putative transcription factors. An example of genes likely involved in segmentation of the somites are the *Pax* family of genes, genes that encode proteins containing two DNA-binding sites: a homeodomain and a paired-box domain.

After segmentation of the somitic mesoderm has occurred, individual somites become subdivided as stated above into three components called the dermatome, myotome, and sclerotome. Formation of these subdivisions requires cell–cell interactions with adjacent tissues including the neural tube and notochord. Also unique patterns of gene expression occur in the three components of the somites as they are forming.

For additional details on mesodermal subdivisions and somitogenesis see Jurand (1962); Langman and Nelson (1968); Lipton and Jacobson (1974); Bancroft and Bellairs (1976); Packard and Jacobson (1976, 1979); Meier (1979, 1980, 1981); Chernoff (1985); Sausedo and Schoenwolf (1993).

Brief Overview of Organogenesis

Nervous System

As stated above, the nervous system is derived in large part from two ectodermal rudiments. The **neural tube** forms the entire adult **central nervous system**, and the **neural crest** forms much of the adult **peripheral nervous system**. Additional contributions of ectoderm to the peripheral nervous system come from thickenings associated with the branchial arches in the developing craniofacial region. These thickenings, or epibranchial placodes, give rise to cells that migrate away from the thickenings and join other migratory cells derived from the neural crest.

The central nervous system, derived from the neural tube, consists of the **brain** and **spinal cord**. The future brain level of the early neural tube becomes subdivided as development proceeds, first into three primary vesicles, called the **prosencephalon**, the **mesencephalon**, and the **rhombencephalon**, and then into five secondary vesicles derived in the following manner. The prosencephalon subdivides into the **telencephalon** and **diencephalon**. The former forms the adult **cerebrum** and its **cerebral hemispheres**, the **olfactory lobes**, and the **corpora striata**, and the latter forms the adult **thalamus**, **epithalamus**, **hypothalamus**, **pineal gland**, and **posterior pituitary gland**. The mesencephalon does not subdivide into additional vesicles. It forms the adult **superior** and **inferior colliculi** (i.e., the **corpora quadrigemina**) and **the cerebral peduncles**. The rhombencephalon subdivides into the **metencephalon** and **myelencephalon**. The former forms the **cerebellum** and **pons**, and the latter forms the **medulla**.

Within each of the five secondary vesicles of the brain and within the spinal cord, further subdivision occurs. The presence of these subdivisions, called **neuromeres** (also called **prosomeres** in the forebrain and **rhombomeres** in the hindbrain), indicate that the early neural tube is patterned in a segmental fashion. Such patterning is believed to be imposed upon the spinal cord by the pattern of the segmented somites, and subsequently on the neural crest as they leave the spinal cord and aggregate into **spinal ganglia** (see below). However, in the rhombencephalon this segmental pattern is believed to originate in the neural tube with the expression of certain **homeobox-containing genes**, called *Hox* genes.

As just described, the primitive neural tube initially is essentially a straight tube consisting of the primary (and at later stages, secondary) brain vesicles and the spinal cord. These vesicles (and subsequently the neuromeres) are arrayed like beads on a string. The adult configuration of the various regions of the central nervous system, which have specific spatial relationships to one another, is achieved through a process called torsion and through the formation of four flexures. During torsion, the embryonic axis, including the neural tube, twists—beginning rostrally and continuing caudally—such that the embryo now lies on the yolk sac on its left side rather than on its ventral side. During flexion, the neural tube bends back upon itself at four different levels giving rise to four flexures called the cephalic (or cranial) flexure, the cervical flexure, pontine flexure, and tail flexure. As a result of flexion, the central nervous system becomes folded in a manner reminiscent of the closing of an accordion. Thus considerable growth of the central nervous system can occur during development despite the fact that elongation of the neural tube is restricted by its encasement in the developing skull and vertebral **column**.

The peripheral nervous system is derived from the neural crest and epibranchial placodes. It subdivides into two main components: the **somatic nervous system**, consisting of nerves (and associated ganglia) innervating skin (i.e., sensory innervation) and mainly voluntary muscles (i.e., motor innervation), and the **autonomic nervous system**, consisting of nerves (and associated ganglia) innervating mainly involuntary muscles and organs (e.g., the heart and viscera). The somatic nervous system contains nerves and ganglia associated into two groups: the **cranial nerves** and **ganglia**, associated with the brain, and the **spinal nerves** and **ganglia**, associated with the spinal cord. The autonomic nervous system consists of two divisions typically having opposite functions: the **sympathetic division** and the **parasympathetic division** (e.g., one division slows heart rate, the other accelerates it).

For additional details see Kallen (1953); Martin and Langman (1965); Watterson (1965); Langman et al. (1966); Gallera (1971); Revel and Brown (1976); Goodrum and Jacobson (1981); Schoenwolf and Desmond (1984a, b); Couly and LeDouarin (1985, 1987); Hatta et al. (1987); Balaban et al. (1988); Gardner et al. (1988); Lumsden and Keynes (1989); Keynes and Lumsden (1990); Flynn et al. (1991); Glover (1991); Guthrie and Lumsden (1991); Guthrie et al. (1991); Lim et al. (1991); Sechrist and Bronner-Fraser (1991); Bally-Cuif et al. (1992); Alvarado-Mallart (1993); Goulding et al. (1993); Manner et al. (1993, 1995); Pikalow et al. (1994); Millet and Alvarado-Mallart (1995).

Sensory Organs: Eyes and Ears

Sensory organs such as the eyes and ears develop in close association with the neural tube. The eyes begin their development as outgrowths of the prosencephalon level of the neural tube (future diencephalon level). These paired outgrowths are called the **optic vesicles**. Each optic vesicle subsequently folds upon itself producing a double-layered cup, called the **optic cup**. The optic cup will form the retina of the eye.

Recall that the neural tube is covered by skin ectoderm. During development of the eyes, the optic vesicles interact with the overlying skin ectoderm. This interaction results in the formation of the **lens** of the eye, which subsequently becomes nestled within the optic cup, and the **cornea** of the eye.

The **optic nerves** form from the original connections of the optic vesicles to the prosencephalon. These nerves carry impulses from the retina back to the various regions of the brain where vision is interpreted.

The ears start their development alongside the myelencephalon. Skin ectoderm on each side adjacent to the myelencephalon thickens owing to cell–cell interactions between it and the myelencephalon and adjacent head mesenchymal cells. This thickening on each side is called the **otic placode**. During subsequent development, each otic placode folds upon itself to form an **otic vesicle**, which pinches off from the adjacent skin ectoderm. The otic vesicle forms the inner ear, including its auditory and vestibular components. The remainder of the ear (outer and middle ears), develop from components of the branchial arches.

For additional details see Meyer and O'Rahilly (1959); O'Rahilly and Meyer (1959); Hunt (1961); Coulombre (1964); Bancroft and Bellairs (1977); Meier (1978a, b); Hilfer and Yang (1980); Hilfer et al. (1981, 1989); Piatigorsky (1981); Van Rybroek and Olsen (1981); Bard and Ross (1982a, b); Brady and Hilfer (1982); Yang and Hilfer (1982); Hilfer (1983); Garcia-Porrero et al. (1984); Van De Kamp and Hilfer (1985); Bard and Bansal (1987); Calvente et al. (1988); Sinning and Olsen (1988); Alvarez and Navascues (1990); Hilfer and Randolph (1993).

Cardiovascular System

The **cardiovascular system** comprises the **heart** and **embryonic** and **extraembryonic blood vessels** and **lymphatics**. The blood vessels are comprised of **arteries**, **arterioles**, **capillaries**, **venules**, and **veins**. Embryonic blood vessels develop within the body proper of the embryo; extraembryonic blood vessels develop within two of the extraembryonic membranes: the yolk sac and allantois.

Recall that the primitive heart consists of a straight tube with an inner layer of endocardium and an outer layer of myocardium. The heart is suspended within the **body cavity** or **coelom** (actually, within the future **pericardial cavity**) by the dorsal and ventral mesocardia (the coelom forms as a space between the somatic and splanchnic mesodermal components of the lateral plate mesoderm). The initial space between the endocardium and myocardium becomes filled with a gelatinous **extracellular matrix** called **cardiac jelly**. Later an additional layer forms in the heart. This layer, the **epicardium** or **visceral pericardium**, is derived from the **dorsal mesocardium**.

The straight heart tube consists of three subdivisions. Beginning rostrally, is the **bulbus cordis**. The bulbus cordis is continuous caudally with the primitive **ventricle**. The ventricle is continuous caudally with the **sinoatrial region**. Embryonic and extraembryonic **blood vessels** attach to the arterial (bulbus cordis) and venous (sinoatrial) ends of the heart. With the onset of beating of the heart and circulation of the blood, blood flows into the sinoatrial region from the yolk sac, through the paired right and left **vitelline veins**, and from the body of the embryo, through the paired right and left **common cardinal veins**. The vitelline veins branch extensively within the yolk sac and function in transporting yolk to the growing embryo. Similarly, each common cardinal vein splits into a **cranial cardinal vein (precardinal vein)**, which drains the head and rostral trunk, and the **caudal cardinal vein (postcardinal vein)**, which drains the caudal trunk and tail. Blood flows out of the heart through the paired right and left **ventral aortae**, which lie ventral to the foregut. Each ventral aorta connects to a **first aortic arch**, which extends from ventral-to-dorsal along the lateral side of the foregut, and in turn each first aortic arch connects to a **dorsal aorta**, which lies dorsal to the foregut. Each dorsal aorta extends rostrally to supply the head as an **internal carotid artery**, and caudally to supply the trunk and tail. Later a series of **aortic arches** will develop, one pair within each pair of branchial arches. The aortic arches undergo complex transformations to give rise to major arteries of the thorax.

In the adult bird, the heart is a four-chambered structure. To acquire this configuration two processes must occur: looping of the straight heart tube, and partitioning of the looped heart tube. During looping, the heart first bends to the right and then the caudal end of the heart moves rostrally and dorsally. This places the bulbus cordis rostral and ventral; the primitive ventricle, caudal; and the sinoatrial region, which has by now subdivided into the common **atrium** and **sinus venosus**, rostral and dorsal. During partitioning, partitions form within the primitive ventricle to subdivide it into **right** and **left ventricles** and in the common atrium to subdivide it into **right** and **left atria**.

For additional details see Rychter (1962); DeHaan (1965, 1968); Stalsberg and DeHaan (1969); Hay and Low (1972); Manasek et al (1972); Arguello et al. (1975); Hendrix and Morse (1977); Ho and Shimada (1978); Markwald et al. (1978, 1979); Morse and Hendrix (1980); Hirakow and Hiruma (1981); Kirby et al. (1983); Hay et al. (1984); Pardanaud et al. (1987, 1989); Coffin and Poole (1988); Icardo (1988); Kirby (1988); Poole and Coffin (1988, 1989, 1991); Flamme (1989); Hiruma and Hirakow (1989); Easton et al. (1992).

Digestive and Respiratory System

Recall that the gut develops in three distinct units: the foregut, midgut, and hindgut. The **foregut** gives rise to several components of the adult digestive (gastrointestinal) system as well as to the respiratory system. The latter develops as a ventral outgrowth that forms the **larynx**, **trachea**, and **lungs**. Components of the digestive system derived from the foregut include the **pharynx**, **esophagus**, **stomach**, and part of the **duodenum** (the first region of the **small intestine**). The **liver** and **pancreas** develop as outgrowths of the foregut.

The **midgut** gives rise to the remainder of the **small intestine** (i.e., part of the **duodenum**, the **jejunum**, and the **ileum**) as well as to part of the **large intestine** (i.e., the **ascending colon** and part of the **transverse colon**).

The **hindgut** gives rise to the remainder of the **large intestine** (i.e., part of the **transverse colon**, the **descending colon**, **rectum**, and part of the **anus**). Both the **anus** and **mouth** develop from a cooperation between ectodermal depressions (**proctodeum** and **stomodeum**, respectively) and endoderm of the gut (hindgut and foregut, respectively).

For additional details see Locy and Larsell (1916a, b); Bellairs (1953a, b, 1955, 1957); Kingsbury et al. (1956);

Stalsberg and DeHaan (1968); Tsai and Overton (1976); Lim and Low (1977); Nogawa (1981); Nogawa and Mizuno (1981); Hilfer and Brown (1984); Overton and Meyer (1984); Abbott et al. (1991); Yokouchi et al. (1995).

Limbs

Recall that the limbs are paired structures derived from somatopleure, and hence consist of both skin ectoderm and somatic mesoderm. Each limb begins its development as a **limb bud**. Somatic mesoderm forms the loosely packed cellular core of the limb bud and is capped by skin ectoderm. At the apex of each limb bud, the skin ectoderm is thickened as the **apical ectodermal ridge**. Experiments have shown that interactions between the limb bud somatic mesoderm and the apical ectodermal ridge are necessary for normal limb outgrowth. Thus reciprocal interactions occur.

First, the somatic mesoderm induces formation of the apical ectodermal ridge. Then, the ridge acts back on the somatic mesoderm and causes the subridge mesoderm (the so-called **progress zone**) to remain proliferative, thereby resulting in limb outgrowth.

Each limb has a distinctive morphology and a polarized axis that is readily identifiable. For example, the upper and lower limbs of each organism are morphologically (and often functionally) characteristically different from one another, and the right and left members of each pair of limbs have mirror-image orientations. How these morphological differences and polarized axes are established is still not fully understood and the complexity that is known is daunting. Nevertheless, it is clear that a specialized area of the somatic mesoderm of each limb bud, called the **zone of polarizing activity** or **ZPA**, plays an important role. The ZPA is located near the caudal border of each limb bud. If this region is transplanted more rostrally, the experimental limb bud now forms a duplicated limb in which the twins have mirror-image symmetry. Thus the ZPA is a signaling region involved in establishing the limb axes. Several molecules seem to be involved in the signaling process including retinoic acid, growth factors and their receptors, and transcription factors encoded by homeobox-containing genes.

For additional details see Fallon and Caplan (1982); Kelley et al. (1982); Fallon et al. (1992a,b).

Urogenital System

With discussion of the development of the urogenital system we come full circle in our story of embryonic development. Recall that the kidneys, gonads, and their ducts develop from the intermediate mesoderm. The kidneys pass through three phases in their development. The first phase involves formation of the **pronephros** (i.e., the **pronephric kidney**). The pronephros on each side consists of two components: a **pronephric duct** and multiple **pronephric cords**. In birds, the pronephros is a primitive, nonfunctional kidney. The second phase involves formation of the **mesonephros** (i.e., the **mesonephric kidney**). The mesonephros consists of three components: a **mesonephric duct** (the pronephric duct is renamed as it elongates and becomes more robust); **mesonephric cords**, each of which subsequently hollows out and becomes a **mesonephric tubule** (the pronephric cords, formed earlier, degenerate); and **glomeruli**—tufts of blood vessels that become nestled within mesonephric tubules. The mesonephros functions during much of embryonic development. Nitrogenous wastes within the blood pass through the walls of the glomeruli and into the lumina of the mesonephric tubules. From here, they pass to the mesonephric ducts and eventually to the cloaca (and chorioallantoic membrane). The third phase involves formation of the **metanephros** (i.e., **metanephric kidney**). The metanephros forms from two rudiments (plus glomeruli): the **nephrogenic cord** (derived from caudal intermediate mesoderm) and the **ureteric bud**, paired outgrowths of the caudal levels of the mesonephric ducts. Reciprocal inductive interactions occur between these two rudiments during their development. The nephrogenic cord forms the **secretory tubules** of the metanephros, whereas each ureteric bud forms a **ureter**, **renal pelvis**, **minor** and **major calyces**, and **collecting tubules** of the metanephros. The metanephros functions during late incubation and throughout posthatching life. The **urinary bladder** forms from the cloaca, near its connection to the allantois.

Development of the reproductive system involves the formation of common rudiments in both sexes; the stage at which these common rudiments are present is called the indifferent stage. At the indifferent stage, the embryos of both sexes contain paired **mesonephric ducts**, paired **paramesonephric ducts** (also called **Müllerian ducts**; ducts that develop from intermediate mesoderm and run a course that is parallel to that of the mesonephric ducts), and paired **genital ridges** (thickenings of the intermediate mesoderm and its overlying covering of coelomic epithelium).

From the indifferent stage, the embryo progressively develops its characteristic male or female phenotype depending on the genotype of the embryo and on the type of hormones circulating within its blood. In birds, in contrast to mammals, the male is the homogametic sex (it is ZZ; that is, it has two large Z sex chromosomes), whereas the female is the heterogametic sex (it is Zw; that is, it has one large Z sex chromosome and one small w sex chromosome). Under the influence of circulating hormones (either steroids or proteins; these hormones remain unknown) the gonads begin to differentiate from the genital

ridges according to the genotype of the embryo. In males, the early testes secrete a substance (anti-Müllerian hormone) that causes the paramesonephric ducts to regress, as well as testosterone, which acts on several rudiments to promote the development of male secondary sex characteristics (e.g., the development of the comb and wattle). Each of the two persisting mesonephric ducts in the male forms the **epididymis**, **ductus deferens**, and **ejaculatory duct**; the **seminal vesicle** forms as outgrowths from each mesonephric duct. In females, the early ovary (recall that only the left one forms) secretes **estrogen**, which also acts on several rudiments to promote the development of female secondary sex characteristics (e.g., in female duck embryos, estrogen is responsible for differentiation of the female syrinx and genital tubercle, two early somatic sex characteristics). Only the left paramesonephric duct persists in female avian embryos. It forms the **oviduct**.

The testes in the male house the developing sperm-generating cells undergoing spermatogenesis, and in the female, the left ovary houses the developing ova-generating cells undergoing oogenesis. Recall that these gamete-generating cells are derived from the primordial germ cells. Primordial germ cells arise from the epiblast at the time the egg is laid, and as described above, with formation of the hypoblast these cells become incorporated into the hypoblast. Finally during gastrulation stages, primordial germ cells become localized to the germ cell crescent with formation of the endoderm and the displacement of the hypoblast; ultimately, the hypoblast and its contained primordial germ cells form a localized region of the yolk sac endoderm.

After the onset of circulation, primordial germ cells enter the extraembryonic (vitelline) veins of the yolk sac where they circulate to the trunk of the embryo near the forming gonads (Figure 21.10). They leave the circulation

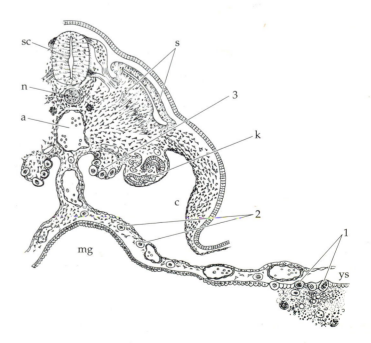

Figure 21.10 Drawing of a transverse section through the trunk of a midincubation chick embryo, illustrating the sequence of migration (1–3) of primordial germ cells from the endoderm of the yolk sac, through the gut wall, and to the developing gonads. Abbreviations: a, aorta; c, coelom; k, kidney; mg, midgut; n, notochord; s, somite; sc, spinal cord; ys, yolk sac (After Patten 1971.)

passing through blood vessel walls, and enter the developing gonads. Experiments suggest that primordial germ cells are drawn to the developing gonads through chemoattraction. Once they enter the gonads, the primordial germ cells are renamed as oogonia in the female and spermatogonia in the male, and oogenesis and spermatogenesis, respectively, ensue.

For additional details see Gallien (1967); Kannankeril and Domm (1968); Clawson and Domm (1969); Swartz and Domm (1972); Poole and Steinberg (1984); Narbaitz and Kapal (1986); Rodemer-Lenz (1989); Maraud et al. (1990); Stoll et al. (1990); Elbrecht and Smith (1992); Jarzem and Meier (1987); Urven et al. (1988, 1989); Zaccanti et al. (1990); Ukeshima and Fujimoto (1991); Ukeshima et al. (1991); Weniger (1991a, b); Weniger et al. (1991); Ukeshima (1992, 1994); Kuwana (1993); Yoshinaga et al. (1993); Austin (1995); Chang et al. (1995).

In several species of turtles, sex determination is dependent upon the temperature of incubation of the egg during a critical time period. For example, in some species incubation at 20°C or at 30°C during stages 14–16 of Yntema (1968) results in the development of only female embryos in all eggs, and incubation at intermediate temperatures results in principally male embryos (in other species of turtles, females develop only at high temperatures and males at low temperatures). The reason for this temperature effect on sex determination is not clear but it has been suggested that at extreme temperatures, testosterone is inactivated and, therefore, cannot act on the hormonally-dependent, indifferent urogenital rudiments, which require testosterone to differentiate according to the male phenotype (the default pathway seems to be female). At intermediate temperatures, testosterone acts on the indifferent rudiments and directs their development toward the male phenotype. Regardless of the exact mechanism, this phenomenon of temperature-dependent sex determination is a remarkable example of adaptation to environmental changes in a cold-blood organism whose eggs develop in absence of maternal care. Why such an adaptation would be evolutionarily advantageous also remains unclear.

For additional details see Yntema (1979, 1981).

Hatching

Unlike mammals, in which birth of the new individual is controlled in large part by the mother and contractions of her uterus, hatching of reptilian and avian embryos can only occur through the concerted efforts of the embryo involved. Either the embryo has the strength and wherewithal to free itself from the confines of the egg, or its development was for naught because it will soon die.

Birds develop a unique behavior that allows them to escape from the egg. During late incubation, they wiggle their way into a position called the hatching position. If the embryo fails to acquire this position, or even if it exhibits minor variation in position, its likelihood of hatching is greatly reduced. When in the hatching position, the beak—with its calcified specialization at its tip, called the egg tooth—is pointed toward the blunt end of the egg. Moreover, the head is tucked beneath the right wing, and the neck is rotated so that the beak is pointed upwards beyond the back of the embryo. Similarly, the legs are flexed at the major joints; during extension of the legs, the feet are pushed firmly against the shell, particularly its pointed end. Muscular extension of the neck, largely through the action of an enlarged dorsal neck muscle, called the hatching muscle, slams the egg tooth against the shell near the blunt end of the egg.

At about 19 or 20 days of incubation, the embryo pips the shell; that is, the egg tooth pierces the shell membranes and the shell at the blunt end of the egg. This allows the embryo to begin respiring through its lungs rather than through the chorioallantoic membrane. The yolk sac begins to withdraw into the midgut approximately concomitant with pipping. The embryo begins hatching about one day later. During this process, the egg tooth acts like a can opener to cut along the circumference of the egg. Beginning at the point of pipping, the embryo begins to rotate in the egg along the egg's long axis, thrusting the egg tooth against the shell repeatedly and pushing the two ends of the shell apart. About one day later, if all is successful, an exhausted but undaunted chick emerges.

For additional details see Pohlman (1919); Oppenheim (1966, 1972a, b); Hamburger and Oppenheim (1967); Brooks and Garrett (1970).

Summary

To me, two of the most remarkable aspects of development are that it usually occurs stereotypically from embryo to embryo and it occurs with rapidity. Here I have only scratched the surface in describing what is known about the development of reptiles and birds. A thorough account of all that is known would far exceed the space allotted by this single volume. Moreover, I would wager that if all that is known about the development of a single organism—for example, about the chick embryo—were to be written in one place, it would probably take the average reader considerably longer to read, to digest, and to understand than it would for the embryo to pass from the zygote to the hatchling (in the case of the chick, this takes only 21 days). Moreover, despite the many advances in developmental biology over the last few decades, much more still remains to be understood. In this chapter, I have tried to give the reader a taste of what is known about the key developmental events of reptiles and birds. I wish to emphasize that it is my hope that this introductory chapter will serve to stimulate the reader to delve deeper into the mystery and wonder of the development of these unique organisms.

Acknowledgments

I wish to thank Dr. Diana K. Darnell for assistance with preparation of Figures 21.1 and 21.8, and Dr. David Packard for directing me to the appropriate literature on temperature effects on sex determination during reptilian development. The original work in the author's laboratory from which some results have been described herein was supported in part by Grant No. NS 18112 from the National Institutes of Health and through Cooperative Agreement HD 28845, as part of a Multicenter Agreement for Studying Neural Tube Defects in Mutant Mice funded by the National Institutes of Health and Human Development, NIH.

Literature Cited

Abbott, L. A., S. M. Lester and C. A. Erickson. 1991. Changes in mesenchymal cell-shape, matrix collagen and tenascin accompany bud formation in the early chick lung. *Anat. Embryol.* 183: 299–311.

Alvarado-Mallart, R.-M. 1993. Fate and potentialities of the avian mesencephalic/metencephalic neuroepithelium. *J. Neurobiol.* 24: 1341–1355.

Alvarez, I. S. and J. Navascues. 1990. Shaping, invagination and closure of the chick embryo otic vesicle: Scanning electron microscopic and quantitative study. *Anat. Rec.* 228: 315–326.

Anderson, C. B. and S. Meier. 1981. The influence of the metameric pattern in the mesoderm on migration of cranial neural crest cells in the chick embryo. *Dev. Biol.* 85: 385–402.

Arguello, C., M. V. De La Cruz and C. S. Gomez. 1975. Experimental study of the formation of the heart tube in the chick embryo. *J. Embryol. Exp. Morphol.* 33: 1–11.

Austin, H. B. 1995. DiI analysis of cell migration during Müllerian duct regression. *Dev. Biol.* 169: 29–36.

Balaban, E., M.-A. Teillet and N. LeDouarin. 1988. Application of the quail-chick chimera system to the study of brain development and behavior. *Science* 241: 1339–1342.

Bally-Cuif, L., R.-M. Alvarado-Mallart, D. K. Darnell and M. Wassef. 1992. Relationship between *Wnt-1* and *En-2* expression domains during early development of normal and ectopic met-mesencephalon. *Development* 115: 999–1009.

Bancroft, M. and R. Bellairs. 1976a. The neural crest cells of the trunk region of the chick embryo studied by SEM and TEM. *Zoon* 4: 73–85.

Bancroft, M. and R. Bellairs. 1976b. The development of the notochord in the chick embryo, studied by scanning and transmission electron microscopy. *J. Embryol. Exp. Morphol.* 35: 383–401.

Bancroft, M. and R. Bellairs. 1977. Placodes of the chick embryo studied by SEM. *Anat. Embryol.* 151: 97–108.

Bard, J. B. L. and M. K. Bansal. 1987. The morphogenesis of the chick primary corneal stroma. I. New observations on collagen organization in vivo help explain stromal deposition and growth. *Development* 100: 135–145.

Bard, J. B. L. and Λ. S. A. Ross. 1982a. The morphogenesis of the ciliary body of the avian eye. I. Lateral cell detachment facilitates epithelial folding. *Dev. Biol.* 92: 73–86.

Bard, J. B. L. and A. S. A. Ross. 1982b. The morphogenesis of the ciliary body of the avian eye. II. Differential enlargement causes an epithelium to form radial folds. *Dev. Biol.* 92: 87–96.

Bellairs, R. 1953a. Studies on the development of the foregut in the chick blastoderm. 1. The presumptive foregut area. *J. Embryol. Exp. Morphol.* 1: 115–124.

Bellairs, R. 1953b. Studies on the development of the foregut in the chick blastoderm. 2. The morphogenetic movements. *J. Embryol. Exp. Morphol.* 1: 369–385.

Bellairs, R. 1955. Studies on the development of the foregut in the chick. III. The role of mitosis. *J. Embryol. Exp. Morphol.* 3: 242–250.

Bellairs, R. 1957. Studies on the development of the foregut in the chick embryo. IV. Mesodermal induction and mitosis. *J. Embryol. Exp. Morphol.* 5: 340–350.

Bellairs, R. 1964. Biological aspects of the yolk of the hen's egg. *Adv. Morphogen.* 4: 217–272.

Brady, R. C. and S. R. Hilfer. 1982. Optic cup formation: A calcium-regulated process. *Proc. Nat. Acad. Sci. USA* 79: 5587–5591.

Bronner-Fraser, M. E. 1986. Analysis of the early stages of trunk neural crest migration in avian embryos using monoclonal antibody HNK-1. *Dev. Biol.* 115: 44–55.

Brooks, W. S. and S. E. Garrett. 1970. The mechanism of pipping in birds. *Auk* 87: 458–466.

Calvente, R., R. Carmona, F. Abadia-Molina and F. Abadia-Fenoll. 1988. Stereological study on the mode of optic cup expansion and the accumulation of mitoses in the early stages of chick embryo development. *Anat. Rec.* 222: 401–407.

Chang, I.-K., A. Tajima, T. Chikamune and T. Ohno. 1995. Proliferation of chick primordial germ cells cultured on stroma cells from the germinal ridge. *Cell Biol. Int.* 19: 143–149.

Chernoff, E. A. G. 1985. Cell movement and contraction in somite development. *Scann. Electr. Micros.* 257–267.

Clawson, R. C. and L. V. Domm. 1969. Origin and early migration of primordial germ cells in the chick embryo: A study of the stages definitive primitive streak through eight somites. *Amer. J. Anat.* 125: 87–112.

Coffin, J. D. and T. J. Poole. 1988. Embryonic vascular development: Immunohistochemical identification of the origin and subsequent morphogenesis of the major vessel primordia in quail embryos. *Development* 102: 735–748.

Coulombre, A. J. 1964. Problems in corneal morphogenesis. In *Advances in Morphogenesis*, M. Abercrombie and J. Brachet (eds.). Academic Press, New York.

Couly, G. F. and N. M. LeDouarin. 1985. Mapping of the early neural primordium in quail-chick chimeras. I. Developmental relationships between placodes, facial ectoderm, and prosencephalon. *Dev. Biol.* 110: 422–439.

Couly, G. F. and N. M. LeDouarin. 1987. Mapping of the early neural primordium in quail-chick chimeras. II. The prosencephalic neural plate and neural folds: Implications for the genesis of cephalic human congenital abnormalities. *Dev. Biol.* 120: 198–214.

Darnell, D. K. and G. C. Schoenwolf. 1996. Modern techniques for cell labeling in avian and murine embryos. In: *Molecular and Cellular Methods in Developmental Toxicology*, G. P. Daston (ed.). CRC Press, Boca Raton.

DeHaan, R. L. 1965. Morphogenesis of the vertebrate heart. In *Organogenesis*, R. L. DeHaan and H. Ursprung (eds.). Holt, Rinehart and Winston, New York.

DeHaan, R. L. 1968. Emergence of form and function in the embryonic heart. *Dev. Biol.* Suppl. 2: 208–250.

Easton, H., M. Veini and R. Bellairs. 1992. Cardiac looping in the chick embryo: the role of the posterior precardiac mesoderm. *Anat. Embryol.* 185: 249–258.

Elbrecht, A. and R. G. Smith. 1992. Aromatase enzyme activity and sex determination in chickens. *Science* 255: 467–470.

Eyal-Giladi, H. and O. Khaner. 1989. The chick's marginal zone and primitive streak formation. II. Quantification of the marginal zone's potencies—temporal and spatial aspects. *Dev. Biol.* 134: 215–221.

Eyal-Giladi, H. and S. Kochav. 1976. From cleavage to primitive streak formation: A complementary normal table and a new look at the first stages of the development of the chick. I. General morphology. *Dev. Biol.* 49: 321–337.

Fallon, J. F. and A. I. Caplan. 1982. *Limb Development and Regeneration. Part A: Proceedings of the Third International Conference on Limb Morphogenesis and Regeneration*. Alan R. Liss, New York.

Fallon, J. F., P. F. Goetinck, R. O. Kelley and D. L. Stocum. 1992a. *Limb Development and Regeneration. Part A: Proceedings of the Fourth International Conference on Limb Development and Regeneration*. Wiley-Liss, New York.

Fallon, J. F., P. F. Goetinck, R. O. Kelley and D. L. Stocum. 1992b. *Limb Development and Regeneration. Part B: Proceedings of the Fourth International Conference on Limb Development and Regeneration*. Wiley-Liss, New York.

Flamme, I. 1989. Is extraembryonic angiogenesis in the chick embryo controlled by the endoderm? *Anat. Embryol.* 180: 259–272.

Flynn, M. E., A. S. Pikalow, R. S. Kimmelman and R. L. Searls. 1991. The mechanism of cervical flexure formation in the chick. *Anat. Embryol.* 184: 411–420.

Gallera, J. 1971. Primary induction in birds. *Adv. Morphogen.* 9: 149–180.

Gallien, L. 1967. Developments in sexual organogenesis. In *Advances in Morphogenesis*, M. Abercrombie and J. Brachet (eds.). Academic Press, New York.

Garcia-Martinez, V. and G. C. Schoenwolf. 1993. Primitive-streak origin of the cardiovascular system in avian embryos. *Dev. Biol.* 159: 706–719.

Garcia-Martinez, V., I. S. Alvarez and G. C. Schoenwolf. 1993. Locations of the ectodermal and non-ectodermal subdivisions of the epiblast at stages 3 and 4 of avian gastrulation and neurulation. *J. Exp. Zool* 267: 431–446.

Garcia-Porrero, J. A., E. Colvee and J. L. Ojeda. 1984. The mechanisms of cell death and phagocytosis in the early chick lens morphogenesis: A scanning electron microscopy and cytochemical approach. *Anat. Rec.* 208: 123–136.

Gardner, C. A., D. K. Darnell, S. J. Poole, C. P. Ordahl and K. F. Barald. 1988. Expression of an *engrailed*-like gene during development of the early embryonic chick nervous system. *J. Neurosci. Res.* 21: 421–437.

Glover, J. C. 1991. Inductive events in the neural tube. *TINS* 14: 424–427.

Goodrum, G. R. and A. G. Jacobson. 1981. Cephalic flexure formation in the chick embryo. *J. Exp. Zool.* 216: 399–408.

Goulding, M. D., A. Lumsden and P. Gruss. 1993. Signals from the notochord and floor plate regulate the region-specific expression of two Pax genes in the developing spinal cord. *Development* 117: 1001–1016.

Guthrie, S. and A. Lumsden. 1991. Formation and regeneration of rhombomere boundaries in the developing chick hindbrain. *Development* 112: 221–229.

Guthrie, S., M. Butcher and A. Lumsden. 1991. Patterns of cell division and interkinetic nuclear migration in the chick embryo hindbrain. *J. Neurobiol.* 22: 742–754.

Hall, B. K. and S. Horstadius. 1988. *The Neural Crest*. Oxford University Press, Oxford.

Hamburger, V. 1960. *A Manual of Experimental Embryology*, Rev. Ed. University of Chicago Press, Chicago.

Hamburger, V. and H. L. Hamilton. 1951. A series of normal stages in the development of the chick embryo. *J. Morphol.* 88: 49–92.

Hamburger, V. and R. W. Oppenheim. 1967. Prehatching motility and hatching behavior in the chick. *J. Exp. Zool.* 166: 171–204.

Hamilton, H. L. 1952. *Lillie's Development of the Chick*, 3rd Ed. Holt, Rinehart and Winston, New York.

Hatta, K., S. Takagi, H. Fujisawa and M. Takeichi. 1987. Spatial and temporal expression pattern of N-cadherin cell adhesion molecules correlated with morphogenetic processes of chicken embryos. *Dev. Biol.* 120: 215–227.

Hay, D. A. and F. N. Low. 1972. The fusion of dorsal and ventral endocardial cushions in the embryonic chick heart: A study in fine structure. *Amer. J. Anat.* 133: 1–24.

Hay, D. H., R. R. Markwald and T. P. Fitzharris. 1984. Selected views of early heart development by scanning electron microscopy. *Scann. Electr. Microsc.* 1984/IV: 1983–1993.

Hendrix, M. J. C. and D. E. Morse. 1977. Atrial septation. I. Scanning electron microscopy in the chick. *Dev. Biol.* 57: 345–363.

Hilfer, S. R. 1983. Development of the eye of the chick embryo. *Scann. Electr. Microsc.* 1983/III: 1353–1369.

Hilfer, S. R. and J. W. Brown. 1984. The development of pharyngeal endocrine organs in mouse and chick embryos. *Scann. Electr. Microsc.* 1984/IV: 2009–2022.

Hilfer, S. R. and G. J. Randolph. 1993. Immunolocalization of basal lamina components during development of chick otic and optic primordia. *Anat. Rec.* 235: 443–452.

Hilfer, S. R. and J.-J. W. Yang. 1980. Accumulation of CPC-precipitable material at apical cell surfaces during formation of the optic cup. *Anat. Rec.* 197: 423–433.

Hilfer, S. R., R. C. Brady and J.-J. W. Yang. 1981. Intracellular and extracellular changes during early ocular development in the chick embryo. In *Ocular Size and Shape Regulation During Development*, S. R. Hilfer and J. B. Sheffield (eds.). Springer-Verlag, New York.

Hilfer, S. R., R. A. Esteves and J. F. Sanzo. 1989. Invagination of the otic placode: Normal development and experimental manipulation. *J. Exp. Zool.* 251: 253–264.

Hirakow, R. and T. Hiruma. 1981. Scanning electron microscopic study on the development of primitive blood vessels in chick embryos at the early somite-stage. *Anat. Embryol.* 163: 299–306.

Hiruma, T. and R. Hirakow. 1989. Epicardial formation in embryonic chick heart: Computer-aided reconstruction, scanning, and transmission electron microscopic studies. *Amer. J. Anat.* 184: 129–138.

Ho, E. and Y. Shimada. 1978. Formation of the epicardium studied with the scanning electron microscope. *Dev. Biol.* 66: 579–585.

Hunt, H. H. 1961. A study of the fine structure of the optic vesicle and lens placode of the chick embryo during induction. *Dev. Biol.* 3: 175–209.

Icardo, J. M. 1988. Cardiac morphogenesis and development. *Experientia* 44: 909–1032.

Jacobson, A. G. 1994. Normal neurulation in amphibians. In *Neural Tube Defects*, Wiley, Chichester, pp. 6–24.

Jarzem, J. and S. P. Meier. 1987. A scanning electron microscope survey of the origin of the primordial pronephric duct cells in the avian embryo. *Anat. Rec.* 218: 175–181.

Jurand, A. 1962. The development of the notochord in chick embryos. *J. Embryol. Exp. Morphol.* 10: 602–621.

Kallen, B. 1953. On the significance of the neuromeres and similar structures in vertebrate embryos. *J. Embryol. Exp. Morphol.* 1: 387–392.

Kannankeril, J. V. and L. V. Domm. 1968. Development of the gonads in the female Japanese quail. *Amer. J. Anat.* 123: 131–146.

Karfunkel, P. 1974. The mechanisms of neural tube formation. *Intl. Rev. Cytol.* 38: 245–271.

Kelley, R. O., P. F. Goetinck and J. A. MacCabe. 1982. *Limb Development and Regeneration. Part B: Proceedings of the Third International Conference on Limb Morphogenesis and Regeneration*. Alan R. Liss, New York.

Keynes, R. and A. Lumsden. 1990. Segmentation and the origin of regional diversity in the vertebrate central nervous system. *Neuron* 2: 1–9.

Khaner, O. 1993. Axis determination in the avian embryo. *Curr. Topics Dev. Biol.* 28: 155–180.

Khaner, O. and H. Eyal-Giladi. 1986. The embryo-forming potential of the posterior marginal zone in stages X through XII of the chick. *Dev. Biol.* 115: 275–281.

Khaner, O. and H. Eyal-Giladi. 1989. The chick's marginal zone and primitive streak formation. I. Coordinative effect of induction and inhibition. *Dev. Biol.* 134: 206–214.

Khaner, O., E. Mitrani and H. Eyal-Giladi. 1985. Developmental potencies of the area opaca and the marginal zone areas of the early chick blastoderms. *J. Embryol. Exp. Morphol.* 89: 235–241.

Kingsbury, J. W., M. Alexanderson and E. S. Kornstein. 1956. The development of the liver in the chick. *Anat. Rec.* 124: 165–187.

Kirby, M. L. 1988. Role of extracardiac factors in heart development. *Experentia* 44: 945–951.

Kirby, M. L., T. F. Gale and D. E. Stewart. 1983. Neural crest cells contribute to normal aorticopulmonary septation. *Science* 220: 1059–1061.

Kochav, S. and H. Eyal-Giladi. 1971. Bilateral symmetry in chick embryo determination by gravity. *Science* 171: 1027–1029.

Kuwana, T. 1993. Migration of avian primordial germ cells toward the gonadal anlage. *Dev. Growth Diff.* 35: 237–243.

Langman, J. and G. R. Nelson. 1968. A radioautographic study of the development of the somite in the chick embryo. *J. Embryol. Exp. Morphol.* 19: 217–226.

Langman, J., R. L. Guerrant and B. G. Freeman. 1966. Behavior of neuro-epithelial cells during closure of the neural tube. *J. Comp. Neurol.* 127: 399–412.

LeDouarin, N. M. 1982. *The Neural Crest.* Cambridge University Press, London.

Lim, S.-S. and F. N. Low. 1977. Scanning electron microscopy of the developing alimentary canal in the chick. *Amer. J. Anat.* 150: 149–174.

Lim, T.-M., K. F. Jaques, C. D. Stern and R. J. Keynes. 1991. An evaluation of myelomeres and segmentation of the chick embryo spinal cord. *Development* 113: 227–238.

Lipton, B. H. and A. G. Jacobson. 1974. Analysis of normal somite development. *Dev. Biol.* 38: 73–90.

Locy, W. A. and O. Larsell. 1916a. The embryology of the bird's lung. I. *Amer. J. Anat.* 20: 1–44.

Locy, W. A. and O. Larsell. 1916b. The embryology of the bird's lung. II. *Amer. J. Anat.* 19: 447–470.

Lumsden, A. and R. Keynes. 1989. Segmental patterns of neuronal development in the chick hindbrain. *Nature* 337: 424–428.

Manasek, F. J., M. B. Burnside and R. E. Waterman. 1972. Myocardial cell shape change as a mechanism of embryonic heart looping. *Dev. Biol.* 29: 349–371.

Manner, J., W. Seidl and G. Steding. 1993. Correlation between the embryonic head flexures and cardiac development. An experimental study in chick embryos. *Anat. Embryol.* 188: 269–285.

Manner, J., W. Seidl and G. Steding. 1995. Formation of the cervical flexure: An experimental study on chick embryos. *Acta Anat.* 152: 1–10.

Maraud, R., O. Vergnaud and M. Rashedi. 1990. New insights on the mechanism of testis differentiation from the morphogenesis of experimentally induced testes in genetically female chick embryos. *Amer. J. Anat.* 188: 429–437.

Markwald, R. R., T. P. Fitzharris, H. Bank and D. H. Bernanke. 1978. Structural analyses on the matrical organization of glycosaminoglycans in developing endocardial cushions. *Dev. Biol.* 62: 292–316.

Markwald, R. R., T. P. Fitzharris, D. L. Bolender and D. H. Bernanke. 1979. Structural analysis of cell:matrix association during the morphogenesis of atrioventricular cushion tissue. *Dev. Biol.* 69: 634–654.

Martin, A. and J. Langman. 1965. The development of the spinal cord examined by autoradiography. *J. Embryol. Exp. Morphol.* 14: 25–35.

Meier, S. 1978a. Development of the embryonic chick otic placode. I. Light microscopic analysis. *Anat. Rec.* 191: 447–458.

Meier, S. 1978b. Development of the embryonic chick otic placode. II. Electron microscopic analysis. *Anat. Rec.* 191: 459–478.

Meier, S. 1979. Development of the chick embryo mesoblast. *Dev. Biol.* 73: 25–45.

Meier, S. 1980. Development of the chick embryo mesoblast: Pronephros, lateral plate, and early vasculature. *J. Embryol. Exp. Morphol.* 55: 291–306.

Meier, S. 1981. Development of the chick embryo mesoblast: Morphogenesis of the prechordal plate and cranial segments. *Dev. Biol.* 83: 49–61.

Meyer, D. B. and R. O'Rahilly. 1959. The development of the cornea in the chick. *J. Embryol. Exp. Morphol.* 7: 303–315.

Miller, S. A. 1982. Differential proliferation in morphogenesis of lateral body folds. *J. Exp. Zool.* 221: 205–211.

Miller, S. A., K. L. Bresee, C. L. Michaelson and D. A. Tyrell. 1994. Domains of differential cell proliferation and formation of the amnion folds in chick embryo ectoderm. *Anat. Rec.* 238: 225–236.

Millet, S. and R.-M. Alvarado-Mallart. 1995. Expression of the homeobox-containing gene *En-2* during the development of the chick central nervous system. *European J. Neuros.* 7: 777–791.

Morse, D. E. and M. J. C. Hendrix. 1980. Atrial septation. II. Formation of the foramina secunda in the chick. *Dev. Biol.* 78: 25–35.

Narbaitz, R. and V. K. Kapal. 1986. Scanning electron microscopical observations on the differentiating mesonephros of the chick embryo. *Acta Anat.* 125: 183–190.

Nicolet, G. 1971. Avian gastrulation. *Adv. Morphogen.* 9: 231–262.

Noden, D. M. 1975. An analysis of the migratory behavior of avian cephalic neural crest cells. *Dev. Biol.* 42: 106–130.

Nogawa, H. 1981. Analysis of elongating morphogenesis of quail anterior submaxillary gland: Absence of localized cell proliferation. *J. Embryol. Exp. Morphol.* 62: 229–239.

Nogawa, H. and T. Mizuno. 1981. Mesenchymal control over elongating and branching morphogenesis in salivary gland development. *J. Embryol. Exp. Morphol.* 66: 209–221.

O'Rahilly, R. and D. B. Meyer. 1959. The early development of the eye in the chick. *Acta Anat.* 36: 20–58.

Oppenheim, R. W. 1966. Amniotic contraction and embryonic motility in the chick embryo. *Science* 152: 528–529.

Oppenheim, R. W. 1972a. Experimental studies on hatching behavior in the chick. 3. The role of the midbrain and forebrain. *J. Comp. Neurol.* 146: 479–506.

Oppenheim, R. W. 1972b. Prehatching and hatching behaviour in birds: A comparative study of altricial and precocial species. *Anim. Behav.* 20: 644–655.

Overton, J. and R. Meyer. 1984. Aspects of liver and gut development in the chick. *Scann. Electr. Microsc.* II: 737–746.

Packard, D. S., Jr. and A. G. Jacobson. 1976. The influence of axial structures on chick somite formation. *Dev. Biol.* 53: 36–48.

Packard, D. S. and A. G. Jacobson. 1979. Analysis of the physical forces that influence the shape of chick somites. *J. Exp. Zool.* 207: 81–91.

Pardanaud, L., C. Altmann, P. Kitos, F. Dieterlen-Lievre and C. A. Buck. 1987. Vasculogenesis in the early quail blastodisc as studied with a monoclonal antibody recognizing endothelial cells. *Development* 100: 339–349.

Pardanaud, L., F. Yassine and F. Dieterlen-Lievre. 1989. Relationship between vasculogenesis, angiogenesis and haemopoiesis during avian ontogeny. *Development* 105: 473–485.

Patten, B. M. 1971. *Early Embryology of the Chick*, 5th Ed. McGraw-Hill, New York.

Patterson, J. T. 1910. Studies on the early development of the hen's egg. I. History of the early cleavage and the accessory cleavage. *J. Morph.* 21: 101–134.

Piatigorsky, J. 1981. Lens differentiation in vertebrates. A review of cellular and molecular features. *Differentiation* 19: 134–153.

Pikalow, A. S., M. E. Flynn and R. L. Searls. 1994. Development of the cranial flexure and Rathke's pouch in the chick embryo. *Anat. Rec.* 238: 407–414.

Pohlman, A. G. 1919. Concerning the causal factor in the hatching of the chick, with particular reference to the Musculus complexus. *Anat. Rec.* 17: 89–104.

Poole, T. J. and J. D. Coffin. 1988. Developmental angiogenesis: Quail embryonic vasculature. *Scan. Elect. Microsc.* 2: 443–448.

Poole, T. J. and J. D. Coffin. 1989. Vasculogenesis and angiogenesis: Two distinct morphogenetic mechanisms establish embryonic vascular pattern. *J. Exp. Zool.* 251: 224–231.

Poole, T. J. and J. D. Coffin. 1991. Morphogenetic mechanisms in avian vascular development. In *The Development of the Vascular System*, R. N. Feinberg, G. K. Sherer and R. Auerbach (eds.), S. Karger, Basel, pp. 25–36.

Poole, T. J. and M. S. Steinberg. 1984. Different modes of pronephric duct origin among vertebrates. *Scan. Electr. Microsc.* I: 475–482.

Revel, J.-P. and S. S. Brown. 1976. Cell junctions in development, with particular reference to the neural tube. *Cold Spring Harbor Symp. Quant. Biol.* 40: 443–455.

Rodemer-Lenz, E. 1989. On cell contribution to gonadal soma formation in quail-chick chimeras during the indifferent stage of gonadal development. *Anat. Embryol.* 179: 237–242.

Romanoff, A. L. 1960. *The Avian Embryo*. Macmillan, New York.

Romanoff, A. L. and A. J. Romanoff. 1949. *The Avian Egg*. John Wiley & Sons, New York.

Rosenquist, G. C. 1966. A radioautographic study of labeled grafts in the chick blastoderm. Development from primitive-streak stages to stage 12. *Carn. Contrib. Embryol.* No. 262, 38: 31–110.

Rychter, Z. 1962. Experimental morphology of the aortic arches and the heart loop in chick embryos. In *Advances in Morphogenesis*, M. Abercrombie and J. Brachet (eds.). Academic Press, New York.

Sausedo, R. A. and G. C. Schoenwolf. 1993. Cell behaviors underlying notochord formation and extension in avian embryos: Quantitative and immunocytochemical studies. *Anat. Rec.* 237: 58–70.

Schoenwolf, G. C. 1991. Neurepithelial cell behavior during avian neurulation. In *Cell–Cell Interactions in Early Development*, J. Gerhart (ed.). Alan R. Liss, New York, pp. 63–78.

Schoenwolf, G. C. 1994. Formation and patterning of the avian neuraxis: One dozen hypotheses. In *Neural Tube Defects*, Wiley, Chichester, pp. 25–50.

Schoenwolf, G. C. 1995. *Laboratory Studies of Vertebrate and Invertebrate Embryos. Guide and Atlas of Descriptive and Experimental Embryology*. Prentice Hall, Englewood Cliffs, New Jersey.

Schoenwolf, G. C. 1997. *Embryo. CD Color Atlas for Developmental Biology*. Prentice Hall, Englewood Cliffs, NJ.

Schoenwolf, G. C. and M. E. Desmond. 1984a. Descriptive studies of occlusion and reopening of the spinal canal of the early chick embryo. *Anat. Rec.* 209: 251–263.

Schoenwolf, G. C. and M. E. Desmond. 1984b. Neural tube occlusion precedes rapid brain enlargement. *J. Exp. Zool.* 230: 405–407.

Schoenwolf, G. C. and J. L. Smith. 1990. Mechanisms of neurulation: Traditional viewpoint and recent advances. *Development* 109: 243–270.

Schoenwolf, G. C., V. Garcia-Martinez and M. S. Dias. 1992. Mesoderm movement and fate during avian gastrulation and neurulation. *Dev. Dynam.* 193: 235–248.

Sechrist, J. and M. Bronner-Fraser. 1991. birth and differentiation of reticular neurons in the chick hindbrain: Ontogeny of the first neuronal population. *Neuron* 7: 947–963.

Sinning, A. R. and M. D. Olsen. 1988. Surface coat material associated with the developing otic placode/vesicle in the chick. *Anat. Rec.* 220: 198–207.

Spratt, N. T., Jr. 1942. Location of organ-specific regions and their relationship to the development of the primitive streak in the early chick blastoderm. *J. Exp. Zool.* 89: 69–101.

Spratt, N. T., Jr. 1952. Localization of the prospective neural plate in the early chick blastoderm. *J. Exp. Zool.* 120: 109–130.

Stalsberg, H. and R. L. DeHaan. 1968. Endodermal movements during foregut formation in the chick embryo. *Dev. Biol.* 18: 198–215.

Stalsberg, H. and R. L. DeHaan. 1969. The precardiac areas and formation of the tubular heart in the chick embryo. *Dev. Biol.* 19: 128–159.

Stern, C. D. 1990. The marginal zone and its contribution to the hypoblast and primitive streak of the chick embryo. *Development* 109: 667–682.

Stern, C. D. and D. R. Canning. 1988. Gastrulation in birds: A model system for the study of animal morphogenesis. *Experentia* 44: 651–657.

Stoll, R., N. Faucounau and R. Maraud. 1990. Action of estradiol on Müllerian duct regression induced by treatment with norethindrone of female chick embryos. *Gen. Comp. Endocrinol.* 80: 101–106.

Swartz, W. J. and L. V. Domm. 1972. Study on division of primordial germ cells of early chick embryo. *Amer. J. Anat.* 135: 51–70.

Tosney, K. W. 1982. The segregation and early migration of cranial neural crest cells in the avian embryo. *Dev. Biol.* 89: 13–24.

Tsai, L.-J. and J. Overton. 1976. The relation between villus formation and the pattern of extracellular fibers as seen by scanning microscopy. *Dev. Biol.* 52: 76–89.

Ukeshima, A. 1992. Scanning electron microscopy of differentiating chick ovaries during embryonic period. *Zool. Sci.* 9: 733–739.

Ukeshima, A. 1994. Abandonment of germ cells in the embryonic chick ovary: TEM and SEM studies. *Anat. Rec.* 240: 261–266.

Ukeshima, A. and T. Fujimoto. 1991. A fine morphological study of germ cells in asymmetrically developing right and left ovaries of the chick. *Anat. Rec.* 230: 378–386.

Ukeshima, A., K. Yoshinaga and T. Fujimoto. 1991. Scanning and transmission electron microscopic observations of chick primordial germ cells with special reference to the extravasation in their migration course. *J. Electron. Microsc.* 40: 124–128.

Urven, L.E., C. A. Erickson, U. K. Abbott and J. R. McCarrey. 1988. Analysis of germ line development in the chick embryo using an antimouse EC cell antibody. *Development* 103: 299–304.

Urven, L.E., U. K. Abbott and C. A. Erickson. 1989. Distribution of extracellular matrix in the migratory pathway of avian primordial germ cells. *Anat. Rec.* 224: 14–21.

Vakaet, L. 1984. Early development of birds. In *Chimeras in Developmental Biology*, N. LeDouarin and A. McLaren (eds.). Academic Press, London, pp. 71–88.

Van De Kamp, M. and S. R. Hilfer. 1985. Cell proliferation in condensing scleral ectomesenchyme associated with the conjunctival papillae in the chick embryo. *J. Embryol. Exp. Morphol.* 88: 25–37.

Van Rybroek, J. J. and M. D. Olsen. 1981. Surface coat material associated with the cells of the developing lens vesicle in the chick embryo. *Anat. Rec.* 201: 261–271.

Watterson, R. L. 1965. Structure and mitotic behavior of the early neural tube. In *Organogenesis*, R. L. DeHaan and H. Ursprung (eds.). Holt, Rinehart and Winston, New York, pp. 129–159.

Weniger, J.-P. 1991a. Embryonic sex hormones in birds. *Int. J. Dev. Biol.* 35: 1–7.

Weniger, J.-P. 1991b. Estrogen secretion by the chick embryo ovary. *Exp. Clin. Endocrinol.* 98: 9–14.

Weniger, J.-P., J. Chouaraqui and A. Zeis. 1991. Effect of partial decapitation (hypophysectomy) on 17 β-estradiol secretion by the chick embryo ovary. *Differentiation* 47: 57–59.

Weston, J. A. 1963. A radioautographic analysis of the migration and localization of trunk neural crest cells in the chick. *Dev. Biol.* 6: 279–310.

Yang, J.-J. and S. R. Hilfer. 1982. The effect of inhibitors of glycoconjugate synthesis on optic cup formation in the chick embryo. *Dev. Biol.* 92: 41–53.

Yntema, C. L. 1964. Procurement and use of turtle embryos for experimental procedures. *Anat. Rec.* 149: 577–586.

Yntema, C. L. 1968. A series of stages in the embryonic development of *Chelydra serpentina*. *J. Morphol.* 125: 219–252.

Yntema, C. L. 1979. Temperature levels and periods of sex determination during incubation of eggs of *Chelydra serpentina*. *J. Morphol.* 159: 17–28.

Yntema, C. L. 1981. Characteristics of gonads and oviducts in hatchlings and young of *Chelydra serpentina* resulting from three incubation temperatures. *J. Morphol.* 167: 297–304.

Yokouchi, Y., J.-I. Sakiyama and A. Kubowa. 1995. Coordinated expression of *Abd-B* subfamily genes of the *HoxA* cluster in the developing digestive tract of chick embryo. *Dev. Biol.* 169: 76–89.

Yoshinaga, K., M. Nakamura and A. Ukeshima. 1993. Ultrastructural characteristics of primordial germ cells in the quail embryo. *Anat. Rec.* 236: 547–552.

Zaccanti, F., M. Vallisneri and A. Quaglia. 1990. Early aspects of sex differentiation in the gonads of chick embryos. *Differentiation* 43: 71–80.

CHAPTER 22

Mammals

Yolanda P. Cruz

THE MAMMALIA TAKE THEIR NAME FROM THE LATIN WORD for breast, *mamma*, because one of the characteristics that distinguishes mammals from other amniotes is the presence of mammary glands. These glands produce milk, which the mostly small, naked, and helpless young require until they are able to fend for themselves. Another readily recognizable distinguishing characteristic of mammals is hairiness, an adaptation for homeothermy. The other distinctive features of mammals relate to internal anatomy and are thus less obvious: the structures of the middle ear and of the nasal passages, the articulation of the jaw and of the cervical vertebrae, the configuration of the pelvic girdle, oxytocin release by the posterior pituitary, two cardiac ventricles, nonnucleated erythrocytes, ureotely, alveolar lungs, a diaphragm, Brunner's glands, circumvallate tongue papillae, and enormous hypertrophy of the brain (Griffiths 1978). The approximately 4500 species of mammals that share these characteristics nevertheless demonstrate remarkable variations in size, gross anatomy, habitat, life history, trophic preferences, reproductive physiology, and, in particular, embryology. We shall see in this chapter how mammalian embryology reflects the diversity represented by, and informs our understanding of the phylogenetic origins of, this group.

Paleontological evidence indicates that mammals arose from reptilian ancestors. This now-extinct group, the therapsid reptiles, almost certainly did not have body hair (although they may have had whisker-like facial hairs), or suckled their young. Yet, their extant descendants all do. Indeed, the combination of these unique and other *ancestral* characteristics suggests a monophyletic origin for present-day mammals (Kemp 1983, but see Lillegraven 1979, Marshall 1979, and Presley 1981 for arguments favoring polyphyly). Extant mammals are divided into three taxonomic groups based largely on divergences in reproductive strategies. One of these strategies is oviparity, known in extant mammals only among the members of the subclass Prototheria (Greek *protos*, first; + *therion*, wild beast): platypi or duckbills, and echidnas. These animals constitute one order, **Monotremata** (Greek *monos*, single; + *trema*, hole [here referring to the presence of a cloaca, into which the reproductive, digestive, and excretory tracts empty their products]). The only other prototherian order, Triconodonta, is extinct. Extant monotremes have a limited geographic distribution, being found only in Australasia. Like nearly all reptiles and their descendants (including birds), the Monotremata lay eggs; unlike these relatives, however (and like all other mammals), they have body hair in addition to skin glands that secrete milk, which their young lap up upon hatching.

The remaining mammals belong to the subclass Theria (Greek *therion*, wild beast). The **Metatheria** (Greek *meta*, next to; + *therion*, wild beast), consisting of the order Marsupialia (Latin *marsupium*, small purse), are viviparous, although the young are born after extremely brief gestation periods. Such neonates require a prolonged period of suckling, usually while they inhabit the maternal pouch, within which the lactogenic organs are found. The Metatheria have a wider geographic range than do the Prototheria, being found in the Americas (mostly in South America) and Australasia. (Only one metatherian species, the didelphid opossum *Didelphis virginiana*, is indigenous to North America.) Fossil remains of metatherians have been found in Europe and in Antarctica, indicating that today's geographic range represents an essentially relic distribution. The **Eutherian** (Greek *eu*, true; + *therion*, wild beast) **mammals**, popularly called **placental** (Latin *placenta*, flat cake), are also viviparous, but have extended gestation periods during which the young are sustained by a

459

transitory organ that forms in the gravid uterus, the so-called **chorioallantoic placenta**. Certain metatherians, it should be noted, also form placentas, but these are of the so-called **yolk-sac** variety. So great is the variability in reproductive strategies demonstrated by mammals that, unlike hairiness or suckling, these must be *derived*, and not ancestral, characteristics.

Mammals exhibit other reproductive specializations. Known gestation periods range from a brief nine days (marsupial mice) to as long as 22 months (elephants). Although not unique in their viviparity, mammals have the distinction of having invented **implantation**, a physiological process in which the embryo establishes histological connections with tissues of a competent uterus for sustenance. We shall see in this chapter the preadaptations that made viviparity possible, the advantages of such a reproductive strategy, and the evolutionary cost of such a biological innovation. We shall also see how fundamental questions regarding embryology in general are being answered with experimental tools and techniques made possible by our increasing ability to manipulate mammalian embryos.

Gametogenesis

Spermatogenesis

The process by which spermatozoa are produced is remarkably uniform among mammals. Spermatogonia divide mitotically and so maintain a supply of male germ cells. Some of these cells differentiate into primary spermatocytes, which undergo a differentiative, meiotic process called spermatogenesis. During this process, the spermatocyte undergoes two successive cell divisions, the first of which is reductional. The products of the first division, therefore, are two haploid cells, called secondary spermatocytes. Each of these haploid cells then undergoes an equational division, thereby producing two spermatids. Each spermatid proceeds through a sequence called **spermiogenesis**, which transforms it into a specialized cell, the mature male gamete, or **spermatozoon** (or, in short, *sperm*). Four such cells are produced from each primary spermatocyte. Because the spermatogonial population is self-replenishing, the ability of a male to remain reproductive is rarely limited by factors other than the length of his lifetime.

The anatomy of mammalian sperm is also remarkably uniform. Except for minor variations in shape, the sperm head consists of a compact conical, rounded, or hooked membrane-bounded structure containing a highly condensed nucleus; a vanishingly small amount of cytoplasm; and a large, specialized lysosomal vesicle, the acrosome. The sperm head tapers into the so-called midpiece, a short cylindrical segment that encloses mitochondria tightly arrayed around the base of a whiplike organelle, the flagellum. The phenomenal motility of mammalian sperm, which must frequently traverse great distances within the female reproductive tract, is due to the beating motion of the flagellum. It is the variable length of this microtubule-laden organelle that accounts for the wide

Figure 22.1 The spermatozoa of several metatherian (*a* to *f*) and eutherian (*g* to *l*) mammals are shown here, magnified, with the acrosome and midpiece shaded in each. The numbers indicate total length to flagellum tip, in μm. Only the hippopotamus sperm is shown totally. The sperm of humans, rabbits, and the common domestic mammals are approximately 50 μm long; those of rodents are much longer, between 150 and 250 μm. The longest mammalian sperm, however, are probably those of a tiny (10–12 grams average adult weight) marsupial, the honey possum *Tarsipes rostratus*. In this nectar-feeding species, the sperm average 310 μm in length. The sperm of echidnas, based on photomicrographs of sections of mature seminiferous tubules (Griffiths 1978), appear to be about 80 μm long. (*a*) Honey possum, *Tarsipes rostratus*; (*b*) marsupial rat, *Dasyuroides byrnei*; (*c*) short-nosed brindled bandicoot, *Isoodon macrourus*; (*d*) tammar wallaby, *Macropus eugenii*; (*e*) brush-tailed possum, *Trichosurus vulpecula*; (*f*) koala, *Phascolarctos cinereus*; (*g*) hippopotamus, *Hippopotamus amphibius*; (*h*) man, *Homo sapiens*; (*i*) rabbit, *Oryctolagus cuniculus*; (*j*) golden hamster; (*k*) laboratory rat, *Rattus rattus norvegicus*; (*l*) Chinese hamster, *Cricetulus griseus*. (After Setchell 1982.)

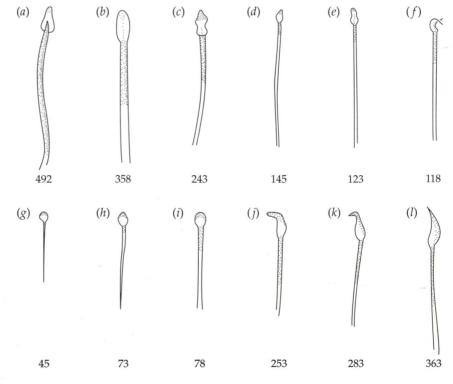

range of mammalian sperm sizes (Figure 22.1).

Perhaps the most noticeable difference among mammalian sperm is the unusual phenomenon of **sperm pairing**, observed exclusively among American marsupials. Here, the plasma membrane overlying the acrosome of one mature sperm is precisely and tightly apposed to the corresponding region on a second sperm (Figure 22.2). The biological significance of this pairing is unclear, but intriguing. Sperm leave the testis unpaired, become coupled in the epididymis, and uncouple in the oviduct (Tyndale-Biscoe and Renfree 1987). It has been suggested that sperm conjugation contributes to efficiency of sperm transport to the vicinity of the egg. The nucleus of the marsupial sperm is asymmetrically positioned with respect to the flagellum; thus pairing may make for greater stability and mobility (Temple-Smith 1994). Indeed, the proportion of sperm that reaches the site of fertilization in the marsupial, the Virginia opossum, *Didelphis virginiana*, for instance, is 5%, nearly 500 times that in the rabbit, a placental mammal. In the opossum, the number of sperm released per insemination, about 13 million, is about ten-fold *fewer* than in rabbits (Bedford et al. 1984).

Oogenesis

The process that generates mature female gametes in mammals is discontinuous. Primordial germ cells, which proliferate in the prenatal conceptus, cease to do so at birth. The neonate thus has differentiated germ cells, or oogonia, which soon after birth begin to increase in size. Many oogonia do not survive this process, perishing at certain times over the life of the female during her juvenile, or sexually immature, stage. Those oogonia that survive become primary oocytes, which at the onset of sexual maturity advance into the first prophase of meiosis. Subsequently and sequentially, in numbers and frequencies characteristic of the species, primary oocytes are released from the ovary into the oviduct during a process called ovulation. Remaining oocytes persist in the ovary, arrested in the dictyotene stage of the first prophase. Thus, in long-lived mammals, later-ovulated oocytes would have been in developmental arrest for several years, or even decades.

Only ovulated oocytes acquire the opportunity to complete meiosis. This is because meiosis resumes after, or is initiated at, fertilization, which occurs in the oviduct. In spermatogenesis, the first meiotic division produces two

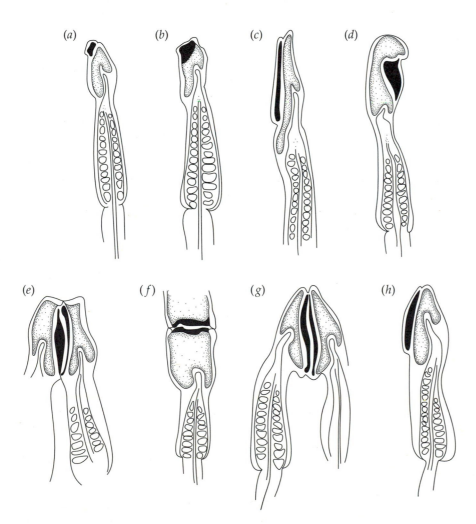

Figure 22.2 Comparison of the basic structural features of the head and midpiece of Australian (*a* to *d*) and American (*e* to *h*) families of metatherians. (*a*) Phalangeridae, Petauridae, Burramyidae, and Macropodidae; (*b*) Peramelidae; (*c*) Dasyuridae; (*d*) Phascolarctidae and Vombatidae; (*e*) Didelphidae; (*f*) Caenolestidae; (*g*) Caluromyidae; (*h*) Microbiotheridae. (After Temple-Smith 1994.)

identical haploid cells (secondary spermatocytes). In oogenesis, the first meiotic division yields likewise haploid but otherwise dissimilar daughter cells: a rather large secondary oocyte and an exceedingly small polar body. This discrepancy in size arises from the eccentric location of the spindle apparatus formed during the first prophase. The completion of the first meiotic division is contingent upon fertilization in most mammals; in exceptional cases such as those of the dog and the fox, the first meiotic division is initiated *upon* fertilization (Baker 1982) (Figure 22.3). Unfertilized oocytes deteriorate, neither completing nor, sometimes even commencing, the first meiotic division.

The fate of the secondary oocyte, now in the oviduct, differs greatly from that of the polar body. In many species, the polar body goes on to divide into two. The secondary oocyte undergoes the second meiotic division and produces an ootid and yet another (a third) polar body. Polar bodies degenerate thereafter and appear to have no developmental significance other than the earlier one of having received, and thus effectively eliminated, a haploid set of chromosomes during the first (reductional) meiotic division. Polar bodies are so named for their function of defining the initial cortical location of the nucleus within the now-voluminous oocyte. They are also, in mammals, nonviable cells, and degenerate within the female reproductive tract in a matter of days. The products of oogenesis are thus one potentially viable, and three nonviable, haploid cells.

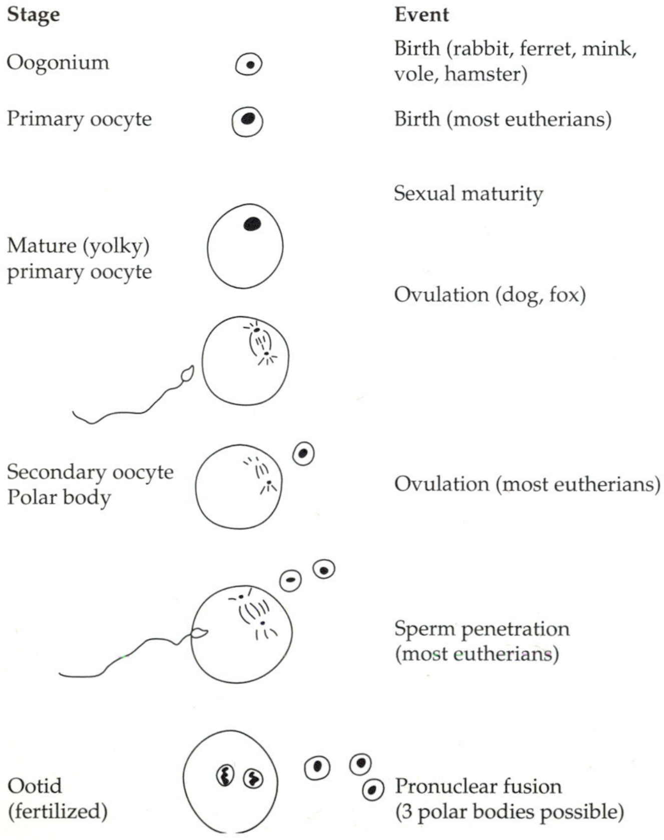

Oogenesis and spermatogenesis would appear to be inefficient or wasteful processes, although in different ways. Spermatogenesis generates vast numbers of sperm, although only one gamete among the millions released per ejaculate is required for fertilization. Oogenesis is inefficient because, unlike spermatogenesis, it yields only one gamete for every meiotic event. Another inefficient feature of oogenesis is **apoptosis** (programmed cell death), whereby numerous progametes perish predictably within, and along with, their follicles (a process called **atresia**) in the prenatal fetus. Follicular atresia occurs over a protracted period during the prenatal, and through the postnatal, period, for reasons that are not clearly understood. Such gradual and relentless depletion of ovarian follicles and oogonia severely erodes the population of cells that become female gametes, as oogonia do not form postnatally. The number of viable oocytes ovulated, much less fertilized, during the lifetime of any given female is thus at best a gross underestimate not only of the number of progametes she was born with, but also of the number of gametes she had the potential to produce.

Oogenesis in monotremes reflects a noteworthy difference between the two families that comprise this mammalian order. Among the Ornithorhynchidae, or the platypi, as in birds to which they bear a superficial resemblance, only the left ovary and oviduct are functional. Oogenesis, therefore, occurs on only that side of the body. The right oviduct, while never attaining as full development as the left, nevertheless shows signs of forming uterine glands during the breeding season (Griffiths 1978). The platypus female reproductive system bears little resemblance to that of birds, however, or other mammals, for that matter. This fact was noted early on by Home (1802), who recognized the close resemblance between the platypus *uterus* and the *oviducts*, not of birds, but of "ovo-viviparous" lizards. Among the monotreme family, Tachyglossidae, or the echidnas, the female reproductive system is bilaterally functional, as is the case in all other mammals. These differences notwithstanding, echidnas and platypi are considered paleontologically more closely related to each other than they are to the Metatheria or Eutheria.

The obvious divergences between the monotremes on the one hand, and the meta- and eutherians on the other, promote periodic debate on the phylogenetic origins of modern-day mammals. Indeed, fossil evidence (Marshall 1979, Tyndale-Biscoe and Renfree 1987) supports the view that the dichotomy between the Eutheria and Metatheria would seem to be secondary and that between these therians and the monotremes, primary. Taken together, these observations argue for a polyphyletic origin for the modern Mammalia (Lillegraven 1975). The opposite view, that

Stage		Event
Oogonium		Birth (rabbit, ferret, mink, vole, hamster)
Primary oocyte		Birth (most eutherians)
		Sexual maturity
Mature (yolky) primary oocyte		
		Ovulation (dog, fox)
Secondary oocyte Polar body		Ovulation (most eutherians)
		Sperm penetration (most eutherians)
Ootid (fertilized)		Pronuclear fusion (3 polar bodies possible)

Figure 22.3 Maturation stages of the eutherian ovum and developmental events in the female eutherian. (After Baker 1982.)

of monophyly, is based on the argument of parsimony (Kemp 1983). The monophyly hypothesis fits in well with the evidence from a comparison of the physiology of reproduction among mammals (Tyndale-Biscoe and Renfree 1987). For example, the cuboidal cells lining the ovarian follicle in modern-day mammals secrete a serum-like fluid, the liquor folliculi. This phenomenon was recognized long ago to occur in monotremes (Flynn and Hill 1939) and more recently among metatherians (reviewed in Tyndale-Biscoe and Renfree 1987) and eutherians (reviewed in Baker 1982). Significantly, the follicles of sauropsid reptiles, the closest extant reptilian relatives of mammals, do not secrete fluid. Parsimony dictates that this shared capability may well be an ancestral characteristic present among the common ancestors of modern mammals, rather than a derived trait evolved independently by different animal groups. We shall see below how embryology reveals other shared characteristics that, taken together, suggest a monophyletic origin for mammals.

Fertilization

Fertilization occurs internally in all known mammals. Gametic union, or **syngamy**, takes place in the oviduct, involving a female gamete enclosed within a glycoprotein matrix, the **zona pellucida**, and a motile sperm propelled by a flagellum. Minor variations on this theme occur among the American metatherians, in which uncoupling of paired sperm in the oviducts precedes syngamy; and among platypi, in which only the left oviduct is functional.

Extracellular Coats

In both metatherians and monotremes, additional extracellular coats are added to the zona pellucida as the oocyte, fertilized or not, courses its way through the oviduct (Figure 22.4). The inner of these coats, in metatherians the so-called **mucolemma** or **mucoid coat** (Hughes 1974, reviewed in Tyndale-Biscoe and Renfree 1987) and in monotremes a two-layered **"albumen" layer** (Caldwell 1887, Hill 1933) slowly thins out (Griffiths 1978) but does not completely disappear (Hughes 1977) during embryogenesis. The obvious implication of such depletion is that the nutrients of this highly hydrated, glycoprotein-rich coat are utilized by the developing embryo. This utilization is patently the case among the monotremes, in which embryonic development occurs autonomously of the mother. The mucoid coat of marsupial embryos apparently serves a similar nutritive function; it, however, disappears early on during the brief gestation period (Tyndale-Biscoe and Renfree 1987).

Both monotreme and metatherian embryos have a third, acellular, shell-like covering. In monotremes, this homogeneous layer appears to be a product of the tubal-gland region of the oviduct. At the time of hatching, the monotreme egg shell has a leathery, not brittle, appearance, attesting perhaps to its keratin content (Griffiths 1978). The metatherian egg shell has been shown to consist of a resilient, proteinaceous, ovokeratin-like substance (Hughes 1977). Unlike the case in monotremes, the metatherian egg shell is shed in utero, persisting for the first two-thirds of gestation (the period preceding organogenesis) in marsupials examined (Hughes 1974; Renfree 1977). Shedding of the metatherian egg shell follows a period of gradual thinning, perhaps driven in part by the expansion of the vesicular conceptus during its blastocyst stage (Tyndale-Biscoe and Renfree 1987).

The biochemical aspects of mammalian fertilization are best known from studies involving eutherians, mostly mice. The zona pellucida assumes two roles during fertilization: first, it binds sperm, and then, it initiates the acrosome reaction after sperm binding has occurred. The sperm-binding component of the mouse zona pellucida is the glycoprotein called ZP3, which appears to be recognized by at least three receptor proteins on the sperm. One of the sperm proteins recognizes galactose residues on ZP3 (Bleil and Wassarman 1990). Another moiety of ZP3, the sugar *N*-acetylglucosamine, is specifically recognized by the sperm cell surface protein, galactosyltransferase (Shur and Neely 1988). Yet another sperm protein that binds to ZP3 is a transmembrane protein with tyrosine kinase activity (Leyton et al. 1992). Such enzymatic activity is known to trigger a multitude of cytoplasmic reactions, which almost certainly includes the acrosome reaction (Leyton et al. 1992).

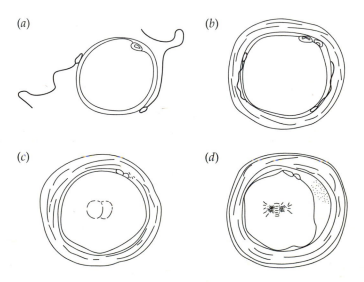

Figure 22.4 Extracellular coats of the metatherian egg. The zona pellucida is formed in the ovarian follicles. (*a*) Fertilization of the zona pellucida-coated secondary oocyte occurs in the oviduct, just before deposition of the mucoid coat (*b*). Further deposition of mucoid layers (*c*) occurs in the uterus where, eventually, the deposition of the shell membrane occurs (*d*). (After Selwood 1982; Selwood and Young 1983.)

Acrosome Reaction

The acrosome reaction mediates the entry of the sperm nucleus into the egg cytoplasm (Figure 22.5). It achieves this, first, by exocytosis of lytic enzymes collectively held in the acrosomal vesicle (Yanagimachi and Noda 1970; Meizel 1984) and second, by local lysis of the zona pellucida so that membrane fusion can occur between the sperm and the egg (Yanagimachi and Noda 1970; Yanagimachi 1988). Membrane fusion is quickly followed in the egg by a rapid and massive release of calcium ions from intracellular stores to the cytoplasm. The resulting ionic shift triggers the cortical granules (lysosomes located within the cortical cytoplasm, or immediately interior to the egg plasma membrane) to undergo exocytosis, releasing their contents during what is called the zona reaction. The lytic activity of released enzymes chemically modifies the zona pellucida sperm receptors so that no further sperm binding occurs (Bleil and Wassarman 1980; Florman and Wassarman 1985; Moller and Wassarman 1989). Thus, the combined effect of the acrosome and zona reactions is the entry of only one, and only exceedingly rarely more than one, sperm nucleus into the egg (Jaffe and Gould 1985).

Sperm and eggs differ in their acquisition of membrane-fusing ability. During sperm maturation, certain membrane proteins require modification before the sperm becomes competent to undergo membrane fusion with the egg (Myles and Primakoff 1992). Acrosomal contents are also required, but not sufficient, for acquisition of sperm

fusibility (Yanagimachi 1988). Indeed, mammalian sperm must be capacitated before the acrosome reaction can occur (reviewed in Gilbert 1994). By contrast, oocytes are competent to undergo membrane fusion even before they are completely mature (Hampl and Eppig 1995, and references therein). Hamster oocytes, for instance, first acquire the capacity for fusion during their growth phase, on attainment of a diameter of a mere 20 μm (Zuccotti et al. 1991), approximately 20% of the final diameter attained (Austin 1982).

Specific membrane domains participate in sperm–egg binding and fusion. The sperm plasma membrane has a specialized anterior head domain, the equatorial region of which overlies the acrosomal membrane (Myles 1993). This region, which usually does not participate in the acrosome reaction and is retained on the head of acrosome-reacted sperm (Myles and Primakoff 1992), is apparently the first region of the sperm plasma membrane to fuse with the egg plasma membrane. Following this fusion, the inner acrosomal membrane becomes contiguous with the sperm surface membrane in metatherians (Rodger and Bedford 1982), but is excluded from the fusion process in eutherians. A similar exclusion is accorded a specialized region on the eutherian egg plasma membrane (Yanagimachi 1988). This region is a cortical-granule-free domain in the region where polar body formation occurs (Ducibella 1991). The molecular basis for membrane-fusion failure to occur in this region has not been determined (Myles 1993).

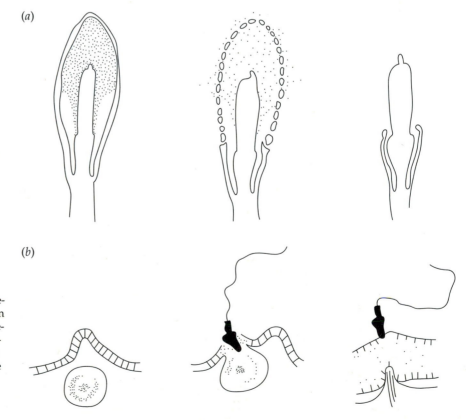

Figure 22.5   The acrosomal (*a*) and zona (*b*) reactions during mammalian fertilization. Fusion of the acrosomal and sperm cell membranes results in exocytosis of acrosomal contents. Likewise, fusion of the cortical-granule and oocyte cell membranes releases lytic enzymes into the perivitelline space. (*a* from Yanagimachi and Noda 1970; Bedford and Cooper 1978; *b* from Austin 1965.)

Early Development

The period of early development commences with the formation of the zygote, followed by cleavage divisions, and then blastulation. In the eutherian, these events occur during the preimplantation period, which ends with the formation of the blastocyst. In the metatherian, in which implantation occurs very late in gestation, if at all, the period of early development proceeds through the formation of a sphere with a single-layered wall, or the unilaminar blastocyst. For prototherian embryos, early development refers to the cleavage stages through hypoblast formation. The common feature of the early developmental stages in mammals, then, is the formation of a cavitated (blastocoel-bearing) embryo.

Monotremes

The three groups of mammals differ greatly in the manner by which their embryos undergo the early cleavage divisions. This difference appears to be the result of the amount of yolk present in the fertilized oocyte. The monotreme egg is telolecithal, and like the telolecithal eggs of reptiles and birds, undergoes meroblastic cleavage (Caldwell 1887). Cleavage results in a sheet of cells, the blastodisc, which, at the 32-cell stage, exhibits regional differentiation. The marginal blastomeres are large and open to the surrounding and underlying yolk; the inner blastomeres are small and delimited on all sides but on their nether surfaces, which face the underlying yolk. This blastomere arrangement resembles that found in sauropsid

reptiles (Griffiths 1978) and in birds at this stage (reviewed in Eyal-Giladi and Kochav 1976). It thus appears that the large volume of yolk in the eggs of these animals, all of which are telolecithal, constrains the nuclear domain not only to the yolk-free cytoplasm, but also to a meroblastic cleavage pattern.

The description of subsequent developmental stages in monotreme eggs is based on the classical work of Wilson and Hill (1908) for the platypus and of Flynn and Hill (1947) for the echidna (Figure 22.6). Further cleavage of the monotreme embryo results in a biconvex blastodisc about five cells thick centrally and one cell thick peripherally. The latter cells produce large numbers of motile cells called **vitellocytes**, which are arranged immediately outside the blastodisc periphery as the **germinal ring** (Flynn and Hill 1939, 1947). As the germinal ring and the blastodisc enlarge, their cells gradually come to cover the surface of the underlying yolk. The blastodisc thins out at its center to the thickness of three cells, and eventually to one cell. Thus the yolk comes to be surrounded by a unilaminar cell sheet, the blastoderm, except for a small area at the pole farthest away from the origin of the blastodisc. Meanwhile, the egg has entered the uterus by this time, and the spherical embryo within, called a **blastocyst**, has grown in diameter. This growth is almost certainly due to both yolk utilization and absorption of uterine secretions. The shell also enlarges, and additionally, its wall thickens (Griffiths 1978). It thus appears that, as is the case in birds, the monotreme uterus secretes substances that contribute to the final stages of formation of the extracellular coats of the egg.

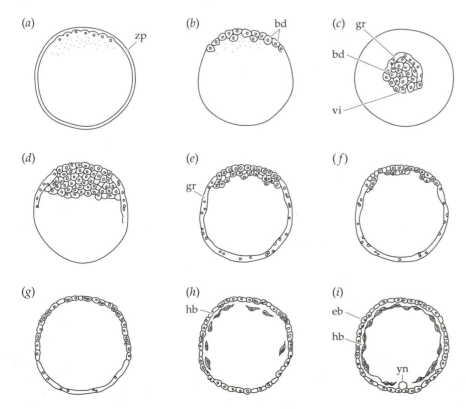

Figure 22.6 Development of the early cell lineages in monotreme embryos. The embryo is shown in sagittal section, with the embryonic area up except in *c*, where a surface view of the embryonic area (or embryoblast) is presented. (*a*) Blastodisc, composed of superficial blastomeres, atop yolk mass. (*b*) Subdisc and marginal cells transform into vitellocytes. (*c*) Surface view of blastodisc showing nascent germ ring. The blastodisc is 6 or 7 cells thick at its center and is surrounded by multinucleated vitellocytes. Eventually, vitellocytes form the syncytial germ ring around the blastodisc. (*d*) The blastodisc thickens some more as the germ ring proceeds to surround the yolk. (*e*) The germ ring completed. (*f*) The blastodisc is reduced to two layers, superficial and deep. Presumptive epiblast and hypoblast (shaded) cells in both layers intermingle as blastodisc expands. (*g*) A unilaminar blastocyst forms. (*h*) The hypoblast cells come to lie in a loose network beneath the epiblast (*i*) The epiblast displaces the germ ring except at the yolk navel. Abbreviations: bd, blastodisc; eb, epiblast; gr, germ ring; hb, hypoblast; vi, vitellocyte; yn, yolk navel; zp, zona pellucida. (After Selwood 1994.)

The unilaminar blastocyst, while still in the oviduct, exhibits signs of cellular differentiation. Its prospective extraembryonic (or primitive) endoderm cells develop pseudopodia, ingress through the underlying (subgerminal) space onto the surface of the yolk, and there form a loose network of cells. Mitotic activity in this network yields cells that eventually transform it into a complete epithelium, the **hypoblast**. Above the hypoblast, therefore, a new flattened space is created, the **blastocoel**. The upper limit of this space is defined by the cell sheet remaining in what used to be the blastodisc; this is the epiblast. At the outer limits of the blastocoel, the hypoblast is pressed against the epiblast. At the opposite side, these two layers come to envelop the yolk, coming together to form a scar-like **yolk navel** at the abembryonic (furthest from the embryo) pole (Griffiths 1978). The monotreme yolk navel is much like that in sauropsidan reptiles, which have telolecithal eggs, but not in fishes, which also have telolecithal eggs. This indicates that a large quantity of yolk is a necessary, but an insufficient, preadaptation for this mode of embryogenesis. Clearly in the case of sauropsids and monotremes, such similarity of embryogenic events must have its origins in close phylogeny. We shall see below that this theme reemerges in the nonyolky eggs of eutherian mammals, which make a yolk sac even without a yolk mass to envelop.

Metatherians

Cleavage in metatherian embryos results in a similarly bilaminar embryo but proceeds in a radically different manner. First, it is holoblastic, producing blastomeres that, during the early divisions, lie free within the space defined by the zona pellucida. Second, cleavage is accompanied not only by yolk extrusion (**deutoplasmolysis**) but also by the secretion of substances that are later deposited as a fibrous matrix onto the inner surface of the zona pellucida (Selwood 1992 and references therein). Third, the cleavage pattern appears to be related to the pattern of yolk elimination, being polar if a large, solitary yolk mass is extruded (Selwood and Sathananthan 1988; Selwood and Smith 1990), and radial if numerous separate yolky vesicles are emitted (Hill 1918; McCrady 1938) (Figure 22.7). Finally, in all studied metatherian species except the brush-tailed possum (Hughes and Hall 1984), cleavages occur in the uterus, because of the rapid passage of the fertilized egg through the oviduct. What is remarkable is that, in spite of these and differences in overall size, metatherian and monotreme blastocysts share strikingly similar features.

Blastocoel formation in the metatherian embryo depends on the presence of an intact zona pellucida. The interior surface of this glycoprotein matrix serves as an adhesive surface for the blastomeres of the 8-celled embryo. The blastomeres do not initially adhere to each other, although they are found close together within the zona pellucida, surrounding the yolk mass or the numerous yolk vesicles extruded during the first three cleavage divisions. Each of the eight blastomeres flattens against the zona pellucida and extends pseudopodia that initiate points of contact with neighboring blastomeres (Selwood and Smith 1990) (Figure 22.8). Cleavage proceeds with the blastomeres in this position, except during cytokinesis when the dividing blastomere momentarily assumes a rounded shape. An epithelium eventually comes to line the zona pellucida, with the cell apices adhering to the zona surface, and the basal lamina facing the cavity that is the blastocoel (Selwood 1992). Within the blastocoel lie ex-

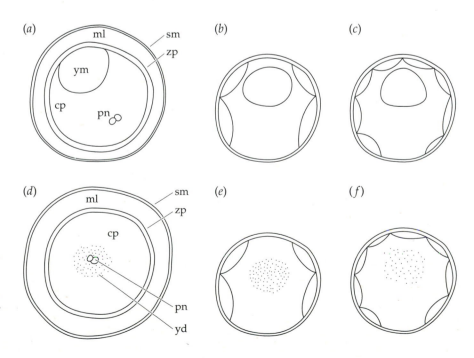

Figure 22.7 Patterns of yolk elimination in dasyurid marsupials (*a* to *c*) and the Virginia opossum (*Didelphis virginiana*), a didelphid (*d* to *f*). Not to scale. All three extracellular coats are shown in *a* and *d*; only the zona pellucida is shown in the remaining figures. (*a*) Yolk is eliminated as a large mass prior to the first cleavage in dasyurid embryos; in the Virginia opossum, yolk droplets are extruded during the first few cleavages. At the 4-cell stage, the blastomeres adhere to the interior of the zona pellucida, leaving the yolk somewhat off-center (*b*), but yolk droplets finely dispersed (*e*) in the nascent blastocoel. *c* and *f* are 8-cell embryos. Abbreviations: cp, cytoplasm; ml, mucoid layer; pn, pronuclei; sm, shell membrane; yd, yolk droplets; ym, yolk mass; zp, zona pellucida. (After McCrady 1938 and Selwood 1994.)

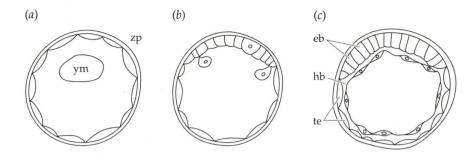

(a) *(b)* *(c)*

zp, ym, eb, hb, te

Figure 22.8 Formation of the bilaminar blastocyst in metatherians. Only the zona pellucida, as an extracellular coat, is shown. (*a*) The yolk mass (shown) or yolk droplets persist until the unilaminar blastocyst is formed by the blastomeres adhering as a monolayer to the interior surface of the zona pellucida. (*b*) On disappearance of the yolk mass or droplets, cells in the presumptive embryonic area (or embryoblast) become squamous. Some of these cells are internalized into the blastocoel by migration or directed (anticlinal) cell division. (*c*) Internalized cells organize into a loose network of hypoblast and primitive endoderm cells, leaving a thickened epiblast consisting of cuboidal cells and an attenuated trophectoderm (or trophoblast) composed of squamous cells. Abbreviations: eb, epiblast; hb, hypoblast; te, trophectoderm; ym, yolk mass; zp, zona pellucida.

truded yolk and other cytoplasmic elements. Thus, the metatherian unilaminar blastocyst resembles its monotreme counterpart, although the role of the its zona pellucida is clearly a less passive one.

Experiments confirm that the metatherian zona pellucida is required for blastocoel formation. Blastomeres of the 8-celled embryo of the dasyurid, *Antechinus stuartii*, denuded of its investments (the shell, mucoid layer, and zona pellucida) and grown in vitro, adhere to the culture dish in much the same way that blastomeres in intact embryos adhere to the interior surface of the zona pellucida (Selwood 1989; Selwood and Smith 1990). These observations imply that without the zona pellucida serving as the common substrate for blastomere adhesion, the metatherian embryo could not form a blastocoel.

The formation of the unilaminar blastocyst signals the establishment of the first two cell lineages in the metatherian embryo. Originally thought to be totipotent (McCrady 1938, 1944), the epithelium of the unilaminar blastocyst instead appears to exhibit regional specialization with respect to developmental potency (Selwood 1992). Specifically, at least in the dasyurid, *Sminthopsis macroura*, only those cells of the unilaminar blastocyst closest to the yolk mass are ever implicated in the formation of the **embryoblast**, the embryo-generating region of the unilaminar blastocyst (Selwood and Woolley 1991). The remaining cells of the unilaminar blastoderm comprise the **trophoblast**, the epithelium that mediates implantation (Sharman 1961; Tyndale-Biscoe and Renfree 1987) and hence, nourishment, of the embryo (Selwood 1992). The delineation between the embryoblast and trophoblast in metatherian embryos that undergo polar yolk emission is easily observed, as the yolk mass serves as a convenient in situ marker. By contrast, in embryos that undergo radial yolk emission, such as those of the didelphid metatherians, the Virginia and the common opossums, the embryolast and trophoblast lineages are apparently set up by the different-sized blastomeres resulting from the first cleavage division. Such blastomere size differences have not been, but could be, attributed to the amount of yolk eliminated by the nascent blastomeres. Some developmental significance has, for example, been suggested for the reported size differences among cleavage-stage blastomeres of certain eutherian embryos (Betteridge and Flechon

1988). Thus, it may be that the yolk plays at least an indirect role in influencing the allocation of blastomeres to different cell lineages in the metatherian embryo.

Eutherians

Early development of mammalian embryos has been best studied in a eutherian, the laboratory mouse. This constrains the following description to use the mouse embryo as a standard, although notable exceptions will be discussed, particularly when they offer insight into the possible significance of otherwise obscure or poorly understood developmental events.

The cleavage-stage eutherian embryo is significantly smaller than its counterpart monotreme or metatherian embryo, owing to the nearly total absence of yolk in the former. Cleavage divisions occur in the oviduct, are holoblastic, and asynchronous (reviewed in McLaren 1982) (Figure 22.9). Moreover, blastomeres not only adhere to each other, but also as early as the first cleavage division (reviewed in Cruz and Pedersen 1991). Cell adhesion molecules have been demonstrated to mediate this process (reviewed in Kemler et al. 1987), which precedes the onset of reiterated cycles of compaction and decompaction in the mouse embryo (Sutherland et al. 1990). **Compaction** is the process by which eutherian blastomeres, commencing at the 8-celled stage, flatten against one another and so maximize cell–cell contact (Lewis and Wright 1935; Ducibella and Anderson 1975). By the 16-celled stage, the embryo resembles a mulberry, for which reason it is referred to as a **morula** (Latin *morus*, mulberry).

Compaction is accompanied by polarization of individual blastomeres (for review, see Johnson and Maro 1986; Fleming and Johnson 1988). Polarization consists of microvilli redistribution (Handyside 1980; Sutherland and Calarco-Gillam 1983), cytoskeletal rearrangements (Sobel et al. 1988), and the reorientation of endocytotic and mem-

Figure 22.9 Preimplantation stages in the mouse embryo. Each cleavage division reduces the size of a blastomere by approximately half, so the total cytoplasmic volume (size) remains the same. Embryos *a* through *e* are shown with their zonae pellucidae and polar bodies. Total diameter of each is approximately 100 μm. (*a*) Fertilized egg; (*b*) 2-cell embryo; (*c*) 4-cell embryo; (*d*) Morula (compacted preimplantation embryo, here 8-cell); (*e*) 16-cell morula with nascent blastocoel; (*f*) Blastocyst, with inner cell mass and trophectoderm (or trophoblast). Abbreviations: bc, blastocoel; im, inner cell mass; pb, polar body; pn, pronucleus; te, trophectoderm; zp, zona pellucida.

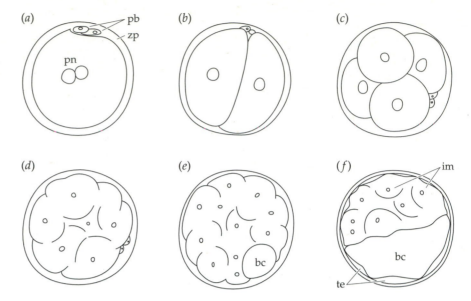

branous elements (Reeve 1981; Maro et al. 1985) so that vesicular traffic is restricted to the free or apical surfaces of exterior blastomeres. In addition, cell–cell adhesion occurs and tight junctions are established between the tightly apposed surfaces of adjacent cells (Johnson and Ziomek 1981; Ziomek and Johnson 1981). The compacted morula acquires smooth surface contours, making individual blastomeres difficult to discern (Figure 22.9).

The morula stage in eutherian embryonic development appears to have no equivalent in other mammals. Both monotreme and metatherian embryos have a unilaminar blastocyst stage, during which the vesicular embryo encloses a fluid-filled cavity, or blastocoel. The eutherian blastocyst is similarly configured, but additionally has the notable feature of a so-called **inner cell mass** (Figure 22.9). This small clump of embryonic cells adheres to the **polar trophectoderm**, its fixed locus of contact with the interior surface of the blastocyst wall, or the **trophoblast**. The cells of the inner cell mass are positionally constrained to their interior location in part by the establishment of cell junctions between adjacent, exterior cells of the 16- to 32-celled morula (Lo and Gilula 1979; Pratt 1989). Cell junction formation, in turn, results in the reconfiguration of the exterior cells into a functional epithelium (Fleming and Johnson 1988), which soon acquires the ability for active ionic transport (Biggers et al. 1988). The resulting osmotic gradient leads to accumulation of fluid in the intercellular spaces, which coalesce into a fluid-filled cavity, the blastocoel. Unlike blastocyst formation in monotremes and metatherians, therefore, the eutherian version of this process requires active transepithelial transport; it relies on neither cellular ingression nor blastomere adhesion to the zona pellucida.

Attainment of the blastocyst stage signals the establishment of the first two cell lineages in the mammalian embryo. One of these lineages is *exclusively* non-embryogenic,

or extraembryonic, and has a trophic function, the *trophoblast* (Johnson 1996). In eutherian embryos and in those metatherian embryos that implant, the trophoblast additionally mediates adhesion by the conceptus to the uterine lining. The other lineage is *only partially* embryogenic, generating both the embryo or fetus, and its extraembryonic tissues (amnion, chorion, yolk sac, and allantois). The term used to designate this tissue in monotreme and metatherian embryos is *epiblast*; in eutherian embryos, it is *inner cell mass*. The developmental potential of the epiblast is arguably identical to that of the inner cell mass (Selwood 1994) (Figure 22.10). Indeed, it has been proposed that the term *pluriblast* be applied to these structures (Johnson and Selwood 1996) in order to eliminate the connotation of a superficial location of the term *epiblast*. By contrast, *pluriblast* acknowledges the pluripotency of the cells that comprise this lineage and would be accurately descriptive of their collective or individual developmental fates, regardless of their initial or final topographical location in or on the conceptus. In this view, the potentially embryogenic tissue of the mammalian blastocyst would be the pluriblast; the definitively nonembryogenic tissue, the trophoblast.

Gastrulation

Gastrulation is the formative process by which the three germinative tissues, or germ layers, of animal embryos are established. These germ layers are the ectoderm (progenitors for integumentary and neural tissue), mesoderm (muscles, bones, connective tissue, circulatory system, urinary system, reproductive organs), and endoderm (digestive system, respiratory system). Gastrulation events are characterized by their dynamism and global conse-

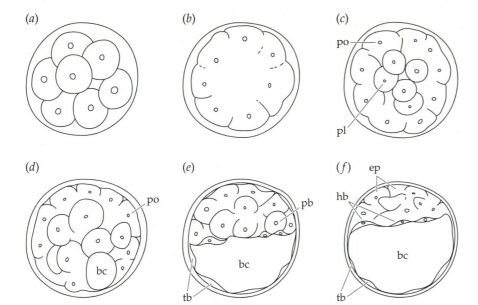

Figure 22.10 New terminology proposed for the eutherian (mouse) embryo. (*a*) The early 8-cell embryo consists of nonpolarized blastomeres. (*b*) The late 8-cell embryo consists of polarized blastomeres, each with a basal and an apical domain. (*c*) The basal domains of these blastomeres are most likely to comprise the inner descendent cells during the ensuing cleavages. These internalized cells are pluriblasts. (*d*) Further pluriblasts are generated during blastocyst formation. (*e*) The early blastocyst consists of pluriblasts surrounded by a differentiated epithelium of trophoblasts. (*f*) The pluriblasts proliferate hypoblasts, which come to line the blastocoel, leaving epiblasts behind. This terminology reflects available information on cell fate in eutherian and metatherian embryos. Abbreviations: bc, blastocoel; eb, epiblast; hb, hypoblast; pl, pluriblast; po, polarblast; tb, trophoblast. (After Johnson 1996.)

quences for the embryo and consist not only of dramatic instances such as cell migration, cell sheet folding, and cell aggregation but also of subtle manifestations such as differential gene activity, cytoskeletal changes, and cellular interactions with the extracellular matrix. So diverse and complex are the phenomena that comprise gastrulation that the process is more easily defined by its results than by its mechanisms.

The manner by which mammalian embryos gastrulate appears to be a direct function of the conformation of the blastocyst. Thus, in the highly yolky eggs of monotremes, moderately yolky eggs of metatherians, and nonyolky but expanding blastocysts of eutherians, topography predicts that the flat, exteriorly situated epiblast (or pluriblast, see foregoing section on Early Development) would differentiate in a manner resembling that of birds and reptiles. The paucity of information concerning these mammalian embryos, however, makes it difficult, albeit logical, to extend such a prediction. The difficulty is confounded by the substantial body of information amassed in the last 20 years concerning gastrulation in the exceptionally configured rodent (notably mouse) embryo. It is important to remember that the mouse blastocyst retains its polar trophectoderm subsequent to hatching from the zona pellucida, and it implants within a few hours of hatching, thus precluding, as it were, blastocyst expansion. Topographic adjustments need to be factored in, therefore, in extending generalizations regarding gastrulation events observed in mouse embryos to parallel events in other mammalian embryos.

Monotremes

Gastrulation in monotreme mammals is poorly understood. Its onset is preceded by the differentiation of blastoderm cells into prospective ectoderm (actually the epi-

blast) and primary endoderm. The primary endoderm cells ingress into the subgerminal space where their pseudopodia fuse together, interconnecting the cells as they undergo repeated mitoses. The primary endoderm thus becomes a complete layer lining the interior of the epiblast; that is, the blastocyst becomes bilaminar. Descriptions of the bilaminar blastocyst are quickly followed by formation in the epiblast area of the so-called **medullary plate** (discussed below), the emergence of a primitive streak within the plate, and organization of the embryonic ectoderm and endoderm, between which the **mesodermal mantle** eventually emerges (reviewed in Griffiths 1978). The edges of the mesodermal layer demarcate the outer limits of the **trilaminar omphalopleure**, that is, the portion of the yolk sac consisting of ectoderm, endoderm, and an interposed layer of mesoderm. Beyond the edges of the mesoderm, the yolk sac remains bilaminar; hence this region is called the **bilaminar omphalopleure**. Formation of the other extraembryonic membranes (amnion, chorion, allantois) in monotremes has not been studied, but is widely believed to be similar to those of other oviparous amniotes, the birds and the reptiles.

Metatherians

Gastrulation in metatherians has been studied not only in several species but also in greater detail. As in monotremes, it commences with the appearance of primary endoderm cells in the unilaminar blastocyst. There are some minor differences, however, in *how* and *when* such cells become distinct from their neighbors. As a result, three types of primary-endoderm formation are recognized (Selwood 1992) (Figure 22.11). These are briefly described below.

PROLIFERATION OF ENDODERM MOTHER CELLS. Primary endoderm cells originate as daughters of ingressed endo-

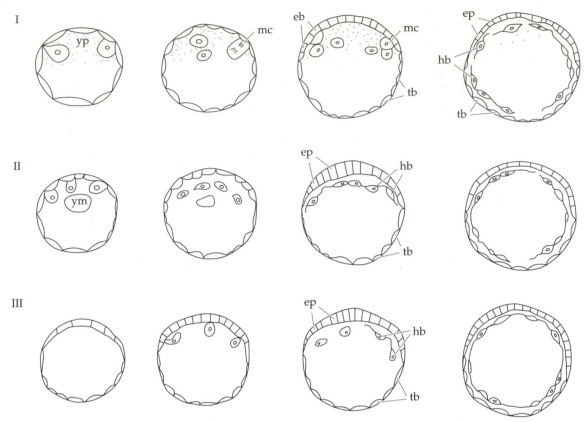

Figure 22.11 Three types (I, II, III) of formation of the primary (or primitive) endoderm (also called the hypoblast) in metatherian embryos. Not to scale. In I, greatly enlarged endoderm mother cells are displaced from the embryoblast into the blastocoel of an otherwise unexpanded unilaminar blastocyst. Once in the blastocoel, these cells proliferate and their descendants come to constitute a loose epithelium, the primary endoderm (hypoblast). In II, an otherwise uniform population of blastocyst cells moves into the blastocoel at the hemisphere defined by the yolk mass. These displaced cells proliferate and later assemble into a loose epithelium (hypoblast) lining the blastocoel. In III, the partially expanded blastocyst reveals a set of blastocyst cells to be cuboidal, rather than typically squamous. These cuboidal cells constitute the epiblast. A subset of epiblast cells moves into the blastocoel, divide, and assemble into the hypoblast. Abbreviations: eb, embryoblast; ep, epiblast; hb, hypoblast; mc, endoderm mother cell; tb, trophoblast; ym, yolk mass; yp, yolk particles. (After Selwood 1992.)

derm mother cells in the Virginia opossum, *Didelphis virginiana* (Selenka 1887; Hartman 1919; McCrady 1938) and in Bennett's wallaby, *Macropus rufogriseus* (Walker and Rose 1983; Selwood 1986). The unilaminar blastocyst in these species is composed of some 60 cells and has a diameter of 0.1 to 0.2 mm before it undergoes a period of rapid expansion. At this stage, certain marginal cells, called the **endoderm mother cells**, increase enormously to 35 × 45 μm. Some of these greatly enlarged cells move gradually into the blastocoel, their ingression being followed by prompt resealing of the blastocyst epithelium. Other endoderm mother cells remain in the epithelium but divide obliquely, such that one daughter cell is internally displaced into the blastocoel. Internalized cells divide several times, decreasing in size as they do. Eventually, a cap of these cells, two or three thick, comes to lie against the interior surface of the unilaminar blastocyst. The area defined on the epithelium by the cap of cells becomes the **embryoblast**. Later, the embryo will form in this hemisphere. The por-

tion of the blastocyst that remains unilaminar at this time retains its apparent trophic function and is thus called the trophoblast. The cap of cells comprises the primitive endoderm, which shortly thereafter proliferates until the trophoblast is lined by a sheet of cells derived from the endoderm mother cells.

The embryoblast appears to differentiate as a result of contact with the primitive endoderm. Embryoblast cells become cuboidal and more basophilic, becoming more clearly recognizable from the neighboring trophoblast cells, which are seen at this time to contain far fewer yolk granules than their neighbors. The thickened embryoblast cells now become the epiblast, or the primitive ectoderm. Meanwhile, the blastocyst itself continues to expand, eventually filling the space within the rapidly thinning zona pellucida. As the zona disappears and the embryoblast contacts the highly attenuated mucoid layer, the shell membrane loses its originally flaccid shape and becomes a tautly inflated sphere. The expanded blastocyst attains a diameter of about 0.4 mm.

DIRECT PROLIFERATION BY CERTAIN BLASTOCYST CELLS. This mode of primary endoderm proliferation occurs in species (such as bandicoots and the stripe-faced dunnart) in which the blastocyst attains a diameter of some 0.3 mm before it makes the unilaminar to bilaminar transition. No visible changes in the unilaminar blastocyst presage the differentiation of the embryoblast from the trophoblast. The unheralded appearance of primary endoderm cells marks the onset of their differentiation from one of the two blastocyst hemispheres. Once primary endoderm cells have appeared, however, the embryoblast becomes distinctly recognizable as highly basophilic cuboidal cells with sharply defined margins. The embryoblast later becomes delineated from the trophoblast by a distinct sutural line (Hill 1910), or transforms into an opaque patch (Hill 1910; Selwood and Woolley 1991) or transparent window (Selwood 1986). In some species, the yolk mass disappears prior to the appearance of the primary endoderm. In those species in which the yolk mass persists past this time, however, it is always located beneath the embryoblast or in the hemisphere containing the embryoblast. Once the primary endoderm is well established, however, the yolk mass disappears (Selwood 1992).

PROLIFERATION OF CELLS IN A DISTINCT EMBRYONIC AREA. In this variation, primary endoderm formation occurs when very little blastocyst expansion has occurred. Reported for only one species, the Tasmanian bettong (*Bettongia cuniculus*), this mode of primary endoderm formation is characterized by the pronounced thickening of certain blastocyst cells into a recognizable embryonic area (Kerr 1935). A proportion of these cells become more rounded, move into the blastocoel, and come to lie on the inner surface of the cells they have left behind on the embryonic area. The embryoblast thus forms. The blastocyst then enlarges from its original diameter of about 0.2 mm to the

expanded dimension of about 0.3 mm. Presently, the internalized cells become the primitive endoderm.

Regardless of the mode of endoderm formation, the metatherian embryo first forms as a unilaminar blastocyst. The transition to bilaminar blastocyst occurs soon after the primitive endoderm is organized, defines the limits of the embryoblast, and later transforms into a cell sheet that comes to line the interior trophoblast surface.

Mesoderm formation in the metatherian embryo begins at various times during the completion of the primitive-endoderm lining of the trophoblast. The first visible indication that mesoderm formation has begun is the proliferation by epiblast (primitive ectoderm) cells on one side of the embryoblast (Hartman 1916; Kerr 1935; McCrady 1938; Lyne and Hollis 1977). The resulting thickened portion of the embryoblast presages the posterior end of the future fetus (Selwood 1992). The embryoblast itself begins to assume a pyriform outline, its long axis situated such that the thickened portion lies at the narrow end. The primitive streak, a thin stream of motile epiblast cells, soon appears along this axis. The anterior end of the primitive streak is indicated by a dimple-like depression, Hensen's node (Figure 22.12).

The mesoderm is first observed as a cell sheet forming within the embryoblast. This sheet is butterfly-shaped and

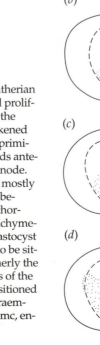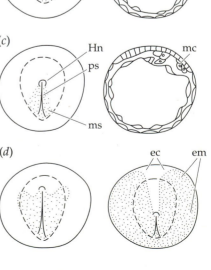

Figure 22.12 Surface (left) and sagittal (right) views (*a* through *c*) of a bilaminar metatherian blastocyst about to become trilaminar. Extracellular coats not shown. (*a*) A flurry of cell proliferative activity occurs on one side of the epiblast. The resulting thickened area defines the posterior end of the future embryo. (*b*) The epiblast elongates in the vicinity of the thickened area, where a faint median furrow, the primitive streak, soon appears. Initiation of the primitive streak may precede the completion of the hypoblast. (*c*) The primitive streak extends anteriorly, through the epiblast, its leading end being defined by the dimple-like Hensen's node. At the streak, cells ingress and come to constitute the mesenchyme, a tissue composed mostly of motile cells contained in the blastocoel. Certain subsets of the mesenchyme, located beneath the epiblast (later the embryonic ectoderm), differentiate into the head process, chordamesoderm, embryonic mesoderm, and embryonic endoderm. (*d*) Eventually, mesenchyme-derived cells migrate beyond the edges of the epiblast, shown here *en face* following blastocyst expansion. The migrating cells produce the extraembryonic mesoderm, which comes to be situated between the extraembryonic ectoderm and the extraembryonic endoderm (formerly the hypoblast). The edge of the extraembryonic mesoderm comes to define the outer limits of the trilaminar yolk sac and is demarcated in the older conceptus by a circumferentially positioned blood vessel, the sinus terminalis (see Figure 22.20). Abbreviations: eb, epiblast; ec, extraembryonic ectoderm; em, extraembryonic mesoderm; hb, hypoblast; Hn, Hensen's node; mc, endoderm mother cell; ms, mesenchyme; ps, primitive streak.

is situated beneath the epiblast, bilaterally symmetrical about the primitive streak. Continued expansion of the mesoderm causes its edges to extend beyond the confines of the embryoblast, which becomes trilaminar as soon as the mesodermal sheet becomes intercalated between the epiblast (primitive ectoderm) and the hypoblast (primitive endoderm). As the mesoderm grows outward from the embryoblast into the trophoblast, the latter region becomes trilaminar as well (Figure 22.12). The edges of the expanded mesoderm come to define the future site of the **sinus terminalis**, a blood vessel that eventually forms, ringlike, around the large vesicle that the conceptus has become.

Eutherians

Gastrulation in the eutherian embryo presumably resembles that in the metatherian. The best-studied eutherian embryo, however, that of the mouse, is much less accessible for study because it has typically invaded the uterine wall and become tightly ensconced in the mucosa by day 5 of gestation. Many of the dynamic stages of gastrulation, such as primitive streak formation, occur at this time. In addition, the murine conceptus is substantially smaller than any metatherian at this stage, having arisen from a nonexpanding blastocyst, which is a mere 0.1 mm in diameter prior to hatching from the zona pellucida. Further confounding anatomical study, the embryonic and outlying extraembryonic areas of the murine conceptus are configured into a steeply concave bowl, much like the closed end of a test tube. The inaccesibility to study of the gastrulating murine conceptus, as well as its exceptional configuration, has traditionally been considered a barrier in the study of gastrulation in this otherwise popular laboratory animal. In recent years, however, significant strides have been made in this area, due to the profitable application of micromanipulation and molecular-biology techniques to the analysis of gastrulation events in the peri-implantation mouse embryo. The findings from these studies will be summarized below.

Primary endoderm formation in the murine embryo commences within 48 hours of the formation of the inner cell mass (Figure 22.13). The embryo, at this time a blastocyst approximately 0.1 mm in diameter, has attained its fully expanded state. The still-present zona pellucida appears to constrain further expansion. Prior to hatching from the zona pellucida on day 5 of gestation, the embryo demonstrates visible cellular differentiation in its inner cell mass. Each of the inner cell mass cells facing the blastocoel rounds up and comes to lie on a basal lamina separating them from the remainder of the inner cell mass. This movement results in the formation of an epithelium, variously referred to as the hypoblast, or primitive endoderm (Boyd and Hamilton 1952; Hafez 1972; Gardner 1985). The remaining cells of the inner cell mass are collectively called the primitive ectoderm (Rossant 1986), or the epiblast. Mitotic divisions within the primitive endoderm

generate motile cells that crawl onto the surface of the mural trophectoderm (that part of the trophectoderm not joined to the inner cell mass). These motile cells later coalesce into an epithelium called the **parietal endoderm** because of its distance from the inner cell mass (Gardner 1982). The portion of this epithelium that remains on the blastocoelic surface of the inner cell mass is called the visceral endoderm (Hogan et al. 1983). At the time of implantation, therefore, the blastocoel has become bounded by the extraembryonic endoderm, in the same way that the monotreme and metatherian yolk masses come to be contained in the yolk sac.

The murine "yolk sac" neither functions as one nor persists intact. It encloses not yolk, but rather the former contents of the blastocoel. Its existence as a two-layered entity is transitory, with the mural trophectoderm cells becoming motile soon after implantation and invading the uterine mucosa. This leaves the primitive endoderm cells, firmly attached to a greatly thickened basal lamina (called **Reichert's membrane**), occupying the position of former mural trophectoderm (Hogan et al. 1980). The postimplantation yolk sac, therefore, consists of an acellular matrix facing the uterine milieu, and the primitive endoderm cells facing the former blastocoel. This gives the murine yolk sac an apparently endodermally derived exterior layer and thus an inside-outside configuration (Morriss 1975; Steven and Morriss 1975; Perry 1981).

The primitive endoderm has an extraembryonic fate in the murine embryo. It contributes solely to the endodermal domain of the yolk sac and the allantois (Gardner and Papaioannou 1975; Gardner and Rossant 1979). It is probably safe to speculate that this conclusion can be extended to other eutherian embryos, and perhaps even metatherian embryos. The unmistakable similarity of embryonic development in the monotreme with that in the avian egg strongly suggests that tissue fates are at least similar, if not identical. Because the chick hypoblast contributes also solely to the extraembryonic endoderm (Vakaet 1984), the monotreme hypoblast very likely has a strictly extraembryonic fate as well.

The remaining tissues in the preimplantation mouse embryo are the polar trophectoderm and the epiblast (that is, the nonhypoblast component of the inner cell mass). The fates of these two tissues will now be described.

The polar trophectoderm has an extraembryonic fate. It contributes descendants to the mural trophectoderm during blastocyst development (Copp 1979; Cruz and Pedersen 1985). Following implantation, however, the polar trophectoderm cells cease to exist as a unilaminar epithelium, but instead accumulate into a compact mass at the embryonic pole. The outermost cells of this mass, like the cells of the mural trophectoderm, invade the uterine surface and there form the ectoplacental cone, a distally tapered plug of embryonic tissue that anchors the embryonic pole of the blastocyst securely into the uterine mucosa. The internally located cells in this mass become the extraembryonic ecto-

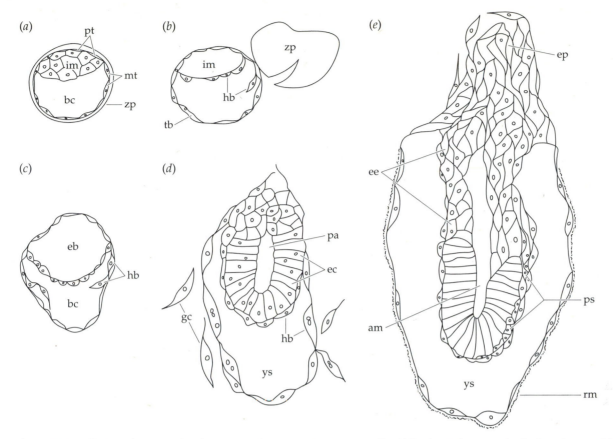

Figure 22.13 Peri-implantation development in the mouse. Not to scale. (*a*) On day 4 of gestation, the blastocyst is free-floating in the uterus. The inner cell mass consists of 10 to 25 cells; the trophoblast, 40 to 55 cells. (*b*) Shedding of the zona pellucida occurs late on day 4, during which the primitive endoderm (hypoblast) cells begin to appear on the blastocoel-facing surface of the inner cell mass and the mural trophoblast. (*c*) At implantation, the hatched blastocyst loses its spherical shape, and its surface becomes adhesive. (*d*) Soon after implantation, the mural trophoblast cells invade the uterine wall as giant trophoblast cells, leaving a fragile network of primitive endoderm cells (hypoblast) on a rigid matrix (Reichert's membrane) to define the cavity of the yolk sac (formerly the blastocoel). The yolk sac has visceral and parietal domains, surrounding exterior of the egg cylinder and lining the far reaches of the yolk sac, respectively. The inner cell mass projects into the yolk sac as a cavitated egg cylinder, its distal end composed of prominently columnar

cells. (*e*) By day 6, the egg cylinder has elongated further, and its unilaminar distal end (or cap) can be seen to constitute the epiblast. Proximally, the egg cylinder consists of inner-cell-mass-derived extraembryonic ectoderm, the cells of which are seamlessly contiguous with those of the polar-trophoblast-derived ectoplacental cone. The primitive streak soon appears (on the right in this diagram), defining the posterior region of the embryo. The cavity of the egg cylinder is the proamnion. The proliferation of mesoderm cells is evident at the site of the primitive streak. Mesoderm cells displaced beyond the epiblast become the extraembryonic mesoderm of the amnion, chorion, and allantois. Abbreviations: am, amnion; bc, blastocoel; eb, epiblast; ec, egg cylinder; ee, extraembryonic ectoderm; ep, ectoplacental cone; gc, giant trophoblast cells; hb, hypoblast; im, inner cell mass; mt, mural trophoblast; pa, proamnion; ps, primitive streak; pt, polar trophoblast; Rm, Reichert's membrane; tb, trophoblast; ys, yolk sac; zp, zona pellucida.

derm. Positioned between the inner cell mass and the ectoplacental cone, the extraembryonic ectoderm later forms the ectodermal portion of the chorion (reviewed in Pedersen 1986) (Figure 22.13).

The fate of the epiblast, or the primitive ectoderm, by contrast, is both embryonic and extraembryonic. All fetal tissues are derived from the epiblast, as are the extraembryonic mesoderm components of the amnion, yolk sac, and chorion (Beddington 1986; Rossant 1986). In this respect, the eutherian epiblast shares a common set of developmental fates with the avian epiblast, and presumably, with those of the monotreme and metatherian as well. No cell lineage analysis has been reported for monotreme embryos thus

far, but a preliminary attempt at fate-map analysis of the epiblast of the stripe-faced dunnart, *Sminthopsis macroura*, suggests that epiblast fate in the marsupial is probably identical to that in eutherians (Cruz et al. 1996).

The murine epiblast forms from the nonhypoblast component of the inner cell mass. On day 6 of gestation, the epiblast can be seen growing into the blastocoel. Continued growth of the epiblast transforms it into a so-called **egg cylinder**, a thickened disc of highly columnar cells seemingly forced to assume the shape of a deep, narrow bowl with steep sides (Figure 22.13). The egg cylinder elongates further for another 24 hours or so, due to proliferation of cells within the distally situated epiblast *and* within the trophectoderm-derived extraembryonic ectoderm of the future chorion. It is worth pointing out that the relative positions in the metatherian embryo of the epiblast- and the trophoblast-derived cells are noticeably the same.

The primitive streak makes its appearance in the epiblast on day 6 of gestation. The epiblast may form initially as a circular disc of cells, but at streak formation, has elongated in the plane of the streak and thus assumes what must be a pyriform outline. This is suggested by the resting position that explanted egg cylinders assume when laid on a flat substratum: the streak is always parallel to the substratum, never perpendicular or oblique. The egg cylinder, therefore, cannot be perfectly cylindrical. It must have a surface flatter in two areas, and it comes to rest on one such surface. The streak must then form along the planes of, and between, the two relatively flat surfaces of the egg cylinder. In this position, the streak would bisect the epiblast into bilaterally symmetrical halves. Thus, were the epiblast to be opened up, its periphery would describe an oval or pyriform shape (Figure 22.14). It is worth pointing out, yet again, that this is precisely the situation in the metatherian conceptus.

Fate-mapping studies of the postimplantation murine egg cylinder reveal even greater similarities among the eutherian, metatherian, and avian epiblasts. When topographic differences are disregarded, the relative locations of the various epiblast domains are seen to be similar, suggesting a fundamentally invariant amniote fate map. Even when nonamniote vertebrates (fish, frog) are considered, a remarkably faithful fate map emerges (Figure 22.15). One is led to conclude that these embryos must follow the same general rules of axis formation and body patterning. Centrally or mediocaudally located on the pregastrulation epiblast is the putative neurectoderm. Beneath this area, the primitive streak forms, displacing the neurectoderm somewhat anteriorly. In this position, the putative neurectoderm is brought into juxtaposition for induction by the chordamesoderm cells, which originally were situated on the exterior of the epiblast, caudal to the neurectoderm itself. Anterior to the putative neurectoderm is a broad tract of putative embryonic endoderm cells. Flanking and somewhat overlapping this domain on each side, the putative mesoderm (embryonic and extraembryonic) stretches outward to the epiblast periphery. The fidelity with which these positions reappear in vertebrate epiblasts is unlikely to be accidental.

Figure 22.14   The mammalian embryo at the primitive streak stage. (*a*) The murine embryo at the egg-cylinder stage exhibits considerable compression of the epiblast, which is seen, when opened up (*b*), to comprise the distal region (or cap) of the egg cylinder. The topology of such an opened-up egg cylinder is remarkably similar to that of a metatherian (*c*) or monotreme (*d*) embryo, here shown *en face*, with anterior to the left. The expanding blastocyst of certain eutherians, such as primates, produces a discoid epiblast (*e*), which has the same topography as its monotreme and metatherian counterparts. Abbreviations: am, amnion; eb, epiblast; ec, egg cylinder; hb, hypoblast; Hn, Hensen's node; ps, primitive streak; ys, yolk sac.

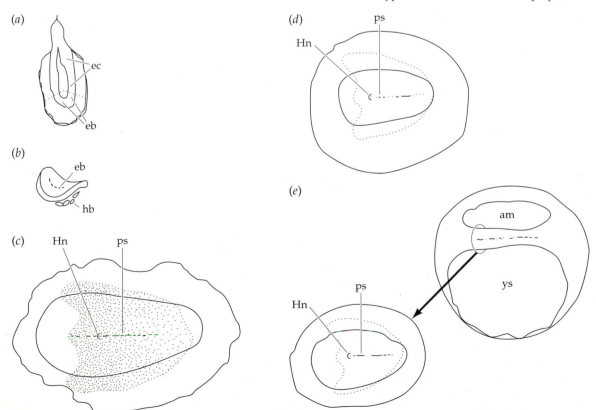

Figure 22.15 Fate map of gastrulating amniote embryos. (*a*) Mouse and (*b*) chick (only embryoblasts shown). Anamniote embryos (*c*) frog and (*d*) fish. The relative positions of major domains are highly conserved, regardless of topology or amount of yolk. Abbreviations: bp, blastopore; ec, ectoderm; en, endoderm; me, mesoderm; ne, neurectoderm; no, notochord; ps, primitive streak; yo, yolk. (After Vakaet 1984; Lawson and Pedersen 1992a; Quinlan and Tam 1996).

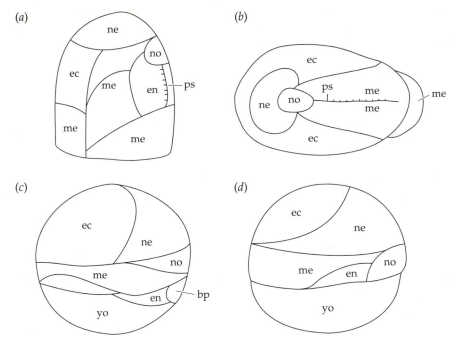

The Medullary Plate

The **medullary plate** consists of the differentiated neurectoderm. This plate has been described in the amphibian neurula and the mammalian embryo (Figure 22.16). It results from the inductive interaction between the chordamesoderm and the overlying ectodermal domain. Because this inductive event requires the prior internalization of the inducer (the chordamesoderm in this case), it can occur *only after* primitive streak formation. The so-called monotreme (Griffiths 1978) and metatherian (McCrady 1938; Tyndale-Biscoe and Renfree 1987) medullary plates, therefore, would appear to be misnamed, as they refer to the surface of the prestreak epiblast. Although the epiblast indeed contains the progenitors of the medullary plate, it also contains the precursors or all other embryonic tissues, and certain extraembryonic tissues. The use of *medullary plate* in these instances is therefore inaccurate and misleading (Selwood 1992; Cruz et al. 1996); the term should be reserved for the differentiated neurectoderm.

Implantation

The embryos of most viviparous mammals undergo some form of implantation and placentation. Although current taxonomic practice would suggest that only eutherian (placental) mammals do so, certain metatherian embryos exhibit unmistakable signs of specific and timely attachment to the uterus (**implantation**) and a definite tendency to form a trophic organ from extraembryonic tissues (**placentation**). Viviparous mammals, of course, resemble their oviparous counterparts (monotremes) and relatives

(reptiles and birds) in their ability to form extraembryonic tissues. We shall see below how this ability is the proximate preadaptation for viviparity, a strategy that has permitted mammals to explore habitats and ecological niches previously unavailable to them.

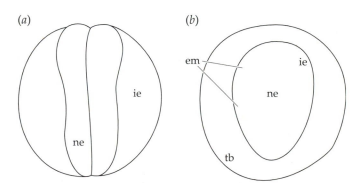

Figure 22.16 The medullary plate in anamniotes and amniotes. Not drawn to scale. (*a*) The visible exterior domains of the amphibian neurula are the neurectoderm and the integumentary ectoderm. The clear separation between these two ectodermal domains is reflected by the use of the term medullary plate interchangeably with neurectoderm. (*b*) The visible exterior domains of the metatherian embryo are the embryoblast and the trophoblast. The embryoblast has been traditionally referred to as the medullary plate although neurulation *has not yet* occurred. This would appear to be a presumptive and thus, inappropriate, use of the term. Furthermore, the embryoblast contains the progenitors of the integumentary ectoderm; thus, it *cannot be merely* a medullary plate. Abbreviations: em, embryoblast; ie, integumentary ectoderm; ne, neurectoderm; tb, trophoblast.

General Considerations

Implantation is a well-orchestrated process that involves a stepwise series of interactions between the hatched embryo and a competent uterus. These interactions occur in two separable phases: adhesion and invasion (Mulholland and Glasser 1991; Damjanov and Wewer 1991; Carson et al. 1991). During adhesion, the embryo aligns with and attaches to the luminal surface of the uterine wall. Adhesion is manifested ultrastructurally by an unusually narrow intercellular space between the luminal epithelium of the uterus and the cell membrane of the trophoblast, formation of junctional complexes between the uterus and embryo, interdigitation of uterine with embryonic microvilli, and a distinctive stickiness between the embryo and the endometrium (reviewed in Parr and Parr 1989) (Figure 22.17). Invasion, on the other hand, consists of tro-

phoblast-originated cytolytic activity that leads to mechanical disruption of uterine epithelial cells. As a result, trophoblast processes often intrude, or insinuate themselves, between uterine cells and later establish contact with the basal lamina and the underlying uterine stroma (Nilsson 1972; Parr 1983; Obrink 1991). Certain uterus-originated changes occur during invasion. These include increased vascular permeability at the implantation sites, which facilitates the establishment of an intimate union between maternal and fetal tissues (Psychoyos 1973). Another such change is the so-called **decidual cell reaction**, which serves to limit the extent of trophoblastic invasion altogether (see below). In sum, implantation is a complex and diverse process that accomplishes a feature unique to mammalian biology: precise and specific attachment of the embryo to the uterine wall.

Blastocyst implantation is preceded by an increase in vascular permeability of the uterine wall. This edematous response is confined to local regions of the uterus, precisely where blastocysts adhere to the epithelial lining (Psychoyos 1973). Its widespread occurrence in studied laboratory mammals and domestic animals suggests that such increase in vascular permeability is important in sustaining both uterus and embryo during the early stages of their interaction (Parr and Parr 1989). Moreover, because the increase in permeability *precedes* implantation and occurs *locally* in the vicinity of the embryo, one is left to speculate that the embryo somehow deploys a signal to the uterus, eliciting an exquisitely timed and precisely positioned response. Indeed, some evidence suggests that the implanting blastocyst may be utilizing prostaglandins or estrogen to help initiate implantation (Dey and Johnson 1986). Such blastocyst-originated estrogen, or similar substance, may in fact be responsible for increasing the number of estrogen and progesterone receptors found in the implantation sites in the rat uterus (De Hertogh et al. 1986).

In many eutherian species, the uterus responds to an implanting blastocyst by mounting the so-called decidual reaction. This histological response transforms the fibroblastic cells of the uterine stroma so they become large, polyploid, and epithelioid. Such transformed cells are believed to constitute a permeability barrier between the blastocyst and the maternal circulation during implantation. Macromolecules and cells prevented from establishing physical contact with the implanting embryo include immunoglobulins, microorganisms, macrophages, and various types of lymphocytes (Tachi et al. 1981; Searle et al. 1983; Bulmer and Johnson 1986). Following implanta-

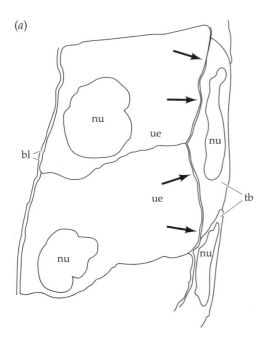

(a)

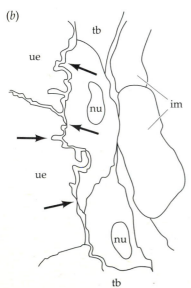

(b)

Figure 22.17 Interpretive sketch of transmission-electron photomicrograph of the close adhesion between the trophoblast and uterine epithelium at the initiation of implantation in (*a*) the rat and (*b*) the mouse. Arrows point to the precise and tight contact between these two epithelia. Abbreviations: bl, basal lamina of uterine epithelium; im, inner cell mass; nu, nucleus; tb, trophoblast; ue, uterine epithelium. (After Kirby 1971; Parr and Parr 1989.)

tion, the decidua loses its barrier function (Welsh and Enders 1987) but at this time, the embryo has become surrounded by a continuous visceral yolk sac, which replaces the decidua as a physical barrier (Parr and Parr 1986). The decidua thus serves a short-lived, but essential, function in protecting the embryo from infections. Perhaps more crucial is its function in deterring maternal immunocytes and immunoglobulins from contacting paternal antigens likely to be present on the surfaces of the embryo (Kirby et al. 1966; Beer and Billingham 1974, 1979). The decidual reaction is clearly of vital significance in ensuring the survival of the embryo.

The decidual response appears to be of benefit to the mother as well. Trophoblast cells are highly invasive (Kirby 1965, 1971), easily and routinely penetrating the uterine epithelium, disrupting the endometrium, and readily establishing contact not only with the underlying basal lamina (Parr et al. 1986; Tung et al. 1986) but also with the endothelia of maternal capillaries (Welsh and Enders 1987) (Figure 22.18). Indeed, uterine epithelial cells die during trophoblast invasion and are phagocytosed by invading trophoblast cells (Tachi et al. 1970). Although the resulting loss of cells is widely construed to be the result of apoptosis, or programmed cell death, on the part of the uterus (Parr et al. 1987), it remains unclear whether the trophoblast cells, or the decidual cells themselves, generate the apoptotic signal (Parr and Parr 1989). What is clear is that decidualization restricts trophoblastic cell invasion, thus preventing excessive damage to the uterus during implantation.

Delayed Implantation and Embryonic Diapause

Implantational delay is a prominent feature in the reproductive biology of many mammals. Here, the cleavage-stage or preimplantation embryo undergoes a quiescent period, during which its normal development is temporarily arrested or slowed down. As a result, blastocyst size increase is temporarily halted, and attachment and implantation are postponed (Given and Enders 1989). Implantational delay occurs naturally among a diverse group of eutherians (armadillos, bats, shrews, seals, otters, bears, ferrets, wolverines, weasels, badgers, skunks, mink, rodents, roe deer) and certain metatherians (kangaroos, wallabees, honey possum) (reviewed in Given and Enders 1989). It is experimentally inducible (by ovariectomy) in some laboratory animals (rat and mouse) but not in others (hamster) (Orsini and Meyer 1962; Weitlauf et al. 1979). Indeed, some mammals (hamster, pig, nonmacropod metatherians, primates) are known not to undergo implantational delay. Nevertheless, because the physiological features of implantational delay across species are similar, and this delay, especially when facultative, is a response to some environmental hardship, this reproductive strategy has received considerable attention.

Implantational delay is manifested in various ways. The hatched or unhatched blastocyst typically fails to undergo expansion or elongation (McRae 1992), and the primitive

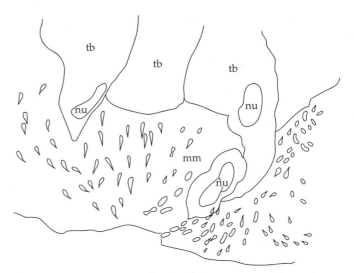

Figure 22.18 Interpretive sketch of light-microscope photomicrograph of mouse trophoblast cells disrupting the uterine myometrium during the invasive phase of implantation. Abbreviations: mm, myometrium; nu, nucleus; tb, trophoblast cell. (After Kirby 1965.)

endoderm may or may not yet have formed (Given and Enders 1989). In marsupials, the delayed blastocyst is typically unexpanded and unilaminar (Tyndale-Biscoe and Renfree 1987). However, in the brown antechinus (*Antechinus stuartii*, a dasyurid marsupial), implantational delay, or **embryonic diapause**, consists of arrested cleavage following the second division. As a result, the embryo remains at the 4-celled stage for approximately five days and at the unilaminar stage for another seven, 20% and 25%, respectively, of the entire gestation period (Selwood 1980). This period of implantational delay coincides with lowered plasma progesterone levels and changes in both uterine histology and corpus luteum development suggestive of conditions in unmated, nonpregnant animals (Cruz and Selwood 1993). It would thus appear that embryonic development in this marsupial is influenced by progesterone levels, and that the uterus mediates progesterone-driven changes in embryonic developmental rate.

Implantational delay strongly indicates that in viviparous mammals, the uterus has a direct role in regulating the pace of embryonic development. Several observations lend considerable support to this view. First, the delay in implantation observed in ovariectomized rats and mice is almost certainly due to progesterone dominance of the uterus in the absence of sufficient amounts of estrogen (Given and Enders 1989). Implantational delay is terminated and implantation is initiated when estrogen is administered to ovariectomized animals (Gidley-Baird 1981). In metatherians, implantational delay is frequently referred to as embryonic diapause because developmental arrest occurs prior to expansion of the unilaminar blastocyst. In kangaroos and wallabies, implantational delay is induced by lactation and terminated when the young leave the pouch. It thus has the effect of spacing out occupancy of the uterus by successively conceived embryos separated in age by the length of the estrous cycle (Tyndale-Biscoe and Renfree 1987). In these animals, implantational delay is experimentally terminated by a pulse of exogenous progesterone but requires a maintenance dose of this hormone for pregnancy to be sustained (reviewed in Tyndale-Biscoe and Renfree 1987). Such a maintenance dose results normally from a fully luteal ovary, not typical of lactating females. Moreover, the uterine histological profile in eutherians undergoing implantational delay is one of quiescence, with moderate glandular activity (Given and Enders 1980; Tachi and Tachi 1986) and, in the luminal epithelium, little or no detectable glycogen and very low alkaline phosphatase activity (Nilsson and Lundkvist 1979). This profile changes dramatically upon estrogen administration, especially in regard to endocytosis, which diminishes greatly (Parr and Parr 1986). In metatherians, abrogation of implantational delay correlates with biochemical changes in the uterine luminal contents, indicating that the effect of the requisite progesterone dose is not directly on the embryo, but on the uterus (Tyndale-Biscoe and Renfree 1987; Given and Enders 1989).

The blastocyst is unlikely to be a passive participant in implantational delay. In experimentally delayed rats, the corpora lutea enlarge when blastocysts are allowed to enter the uterine horns, but not when entry is prevented by ligature (Chatterton et al. 1975). A similarly active role is attributable to the delayed mouse blastocyst, which apparently hastens the loss of implantation-inhibiting factors in uterine fluid (O'Neill and Quinn 1983). Moreover, it is likely that, on reactivation, the delayed mouse blastocyst interacts with the endometrium as normal blastocysts do (Dey and Johnson 1986). Indeed, blastocyst interaction with the endometrium has been suspected to cause the increase in estrogen and progesterone receptors in rat implantation sites (De Hertogh et al. 1986). There is thus evidence of blastocyst-originated changes during implantational delay.

Placentation

The embryos of viviparous mammals receive maternally derived nourishment during their stay in the uterus. This trophic process is mediated by the placenta, a transitory organ that forms during the gestation period from apposed fetal membranes and maternal tissues. Although the term placenta (Latin *placenta*, flat cake, descriptive of the shape of the human placenta), evokes the shape of a disc, mammalian placentas are diverse in morphology, persistence, and tenacity of attachment between the apposed tissues. In fact, the anatomical diversity of placentas stands in stark contrast to the uniformity of placental function: physiological exchange betwen fetus and maternal tissue (Wimsatt 1975; Perry 1981).

Choriovitelline Placenta

Mammalian placentas are classified according to the extraembryonic membrane(s) contributing most prominently to the formed organ. The yolk-sac placenta, for instance, consists of a unilaminar or bilaminar blastocyst wall (also called omphalopleure) fused with, or intimately apposed to, the uterine mucosa. Such a close physical association almost certainly indicates that materials are exchanged between uterus and conceptus. Neither unilaminar nor bilaminar omphalopleure, however, becomes vascularized over the period of its existence during embryonic development. Hence this type of placenta, which is found in bats, mice, armadillos, and certain species of squirrels is more precisely called nonvascular yolk-sac placenta. In certain other squirrel species, and among most insectivores and metatherians, the omphalopleure becomes vascularized and establishes close apposition to, or points of contact with, the uterine mucosa. This constitutes the yolk-sac or choriovitelline placenta.

Placental vascularization is attributed to the presence of the extraembryonic mesoderm. The area vasculosa of the mesoderm frequently extends between the trophoblast and the extraembryonic endoderm of the omphalopleure, which thus becomes trilaminar. An interesting version of the yolk-sac placenta is the inverted yolk-sac placenta, which is found in certain rodents, insectivores, and bats. Here, the vascular mesoderm adheres to the extraembryonic endoderm, and together these closely apposed layers delaminate from a thin layer of nonvascular mesoderm, which continues to adhere to its equally attenuated, overlying trophoblast. A bilaminar, nonvascularized layer thus remains and, for some period of time, persists as the interface between conceptus and uterus. The delaminated layers, on the other hand, either remain vascularized, as occurs among carnivores, or lose their vascular supply and later become wrinkled or shrunken, as occurs among carnivores and many species of bats. In any case, this delamination event creates a space into which the allantois could expand; indeed, the allantois comes to surround the embryo in certain squirrels, pocket gophers, and guinea pigs. Alternatively, the bilaminar remnant (trophoblast and underlying nonvascular mesoderm) could disappear into or invade the uterine mucosa, leaving mostly the extraembryonic endoderm to form a secondary yolk sac. This occurs in many rodents (Mossman 1937).

Chorioallantoic Placenta

The chorioallantoic placenta consists of a portion of the chorion vascularized by allantoic vessels and closely apposed to or fused with the uterine mucosa. It is *the* placenta—the organ referred to by that term. All eutherians have a chorioallantoic placenta, although in some rodents, this placenta forms somewhat late in gestation. Some exceptional metatherians, notably the bandicoot (*Perameles isoodon*), also have a chorioallantoic placenta (Flynn 1923; Padykula and Taylor 1976, 1977), as do some viviparous lizards (Weekes 1928). Vascularization in the chorioallantoic placenta derives from the mesodermal fraction of the allantois; the chorion is never vascularized.

Chorioallantoic placentas are classified by shape, fine structure, or relative proportions of fetal and maternal contributions. Shape is considered a superficial criterion, rather than one based on homology or phylogeny. Thus, depending on the outline of the chorioallantoic area contacting the uterine mucosa, a placenta may be discoid, zonary, cotyledonary, or diffuse (Figure 22.19) (Fabricius 1604, cited in Renfree 1982). The **discoid placenta** consists of a large, flattened, circular structure held in close contact with the uterine mucosa by numerous chorionic villi. Primates and rodents have discoid placentas, as do bats, rabbits, and certain insectivores. The **zonary placenta**, on the other hand, has a circumferential strip or band of chorionic villi, called the hemophagous organ, contacting the surrounding uterine mucosa. This organ may or may not be complete, as in bears, seals, and mustelids like the ferret and mink. In other carnivores, such as the dog and cat, a functional choriovitelline placenta forms early in gestation while the chorioallantoic placenta is developing equatorially (Renfree 1982). There is thus considerable variability within even just the zonary type, which is characteristic also of certain noncarnivorous eutherians such as the elephant, dugong, manatee, and hyrax. In the **cotyledonary placenta**, the villous areas are reduced to small foci scattered over the surface of the chorioallantoic union. These foci, called cotyledons, consist of villi clumped together into well-developed circular tufts, as occur in most ruminants. Cotyledons form adjacent to aglandular endometrial regions or areas called **caruncles**; together, a caruncle and a cotyledon form a **placentome**. The placentome is the actual site of fetal–maternal physiological exchange (Perry 1981; Renfree 1982). The number of caruncles is variable; in most deer species, three or four are found in each uterine horn; but in the goat and giraffe, this number may be as high as 180. The diffuse placenta is found in a wide range of species: pigs, horses, camels, lemurs, whales, dolphins, kangaroos, and possums (Renfree 1982). In this placental type, the villi are more or less evenly distributed across the surface of the chorionic sac, which comes to resemble a towel. The villi interdigitate with depressions in the endometrium; it is here that physiological exchange occurs between mother and fetus.

Villus formation appears to be provoked by the contact established between the maternal and fetal surfaces, and beyond this, by more subtle interactions between these tissues (Mossman 1937). For instance, in horses and whales, in which the (cotyledonary) placenta occupies *both* uterine horns, villus-free areas occur at the termini (which do not contact the mucosa) and over the cervix. It might also be noted here that villus formation is not an exclusive property of the chorioallantoic placenta. The placentas of kangaroos and possums, which are of the diffuse anatomical type, are of the choriovitelline variety (Renfree 1982).

Placentas are sometimes classified by the finer features of the contacting tissues. These contacts may be villous or labyrinthine, depending on whether there is a precise interdigitation of fetal and maternal projections (villi), or

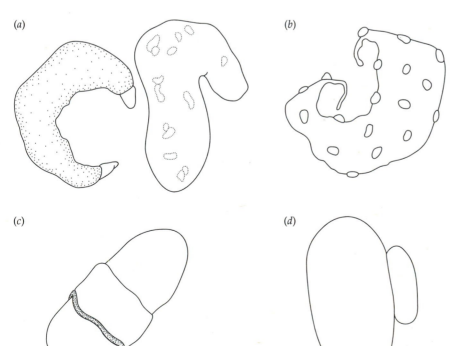

Figure 22.19 Eutherian placental types, not to scale. (*a*) The chorion of the diffuse placenta contains numerous punctate attachment sites (left) or scarlike endometrial cups (right). (*b*) The cotyledonary placenta exhibits multiple caruncles or knoblike protrusions that establish contact with the uterine epithelium. (*c*) The zonary placenta has an equatorial girdle adjacent to a marginal effusion of blood, the hemophagous organ. (*d*) The discoid placenta is positioned on the chorion without an intervening hemophagous structure. (After Renfree 1982.)

TABLE 22.1 *Classification of Placental Types Based on Intimacy of Relation between Chorion and Endometrium*

	Maternal tissue[a]			
Placental type	Endothelium	Connective tissue	Epithelium	Uterine lumen
Epitheliochorial	+	+	+	+
Syndesmochorial	+	+	−	+
Endotheliochorial	+	−	−	−
Hemochorial	−	−	−	−
Hemoendothelial	−	−	−	−

Source: After Mossman 1937.

[a]Tissues separating maternal and fetal blood.

cryptlike mazes of channels at the feto-maternal junction. Further categorization by shape may be superimposed on classification by finer features; thus there are villous zonary placentas, labyrinthine discoid placentas, and so forth (Mossman 1937; Amoroso 1952).

The intimacy of contact between fetal and maternal tissues is another criterion used in the classification of placentas (Grosser 1909). Thus, in the **epitheliochorial** variety, the maternal tissues contribute endothelium, connective tissue, and epithelium, while fetus contributes trophoblast, connective tissue and endothelium. In **syndesmochorial** placentas, the maternal epithelium is missing; and in **endotheliochorial** placentas, the only maternal contribution is the endothelium. **Hemochorial** placentas consist mostly of fetal tissues (trophoblast, connective tissue, and endothelium). Finally, Mossman (1937) adds another category, the **Hemoendothelium**, in which only the fetus has a contribution, the endothelium (Table 22.1).

A classification scheme for placentas based on intimacy of contact incorrectly suggests a physiological advantage when fewer intervening tissues are present at the embryo-maternal interface. If this were so, mammals with greater numbers of placental-tissue layers would be predicted to have less fully developed neonates. This relationship, however, is inconsistent with what is known among eutherians; the newborn foal has far fewer placental-tissue layers than the human neonate, and yet is far more "fully developed" (Renfree 1982). A conservative classification scheme (Steven 1975) has therefore gained currency in recent years (Table 22.2).

The placentas of metatherian mammals have been grouped according to two criteria: the condition of the allantois relative to the chorion (Sharman 1959; Hughes 1984) (Figure 22.20), and the intimacy of the contact between fetal and maternal tissues and circulations *subsequent to* the breakdown of the shell membrane (Hughes 1974). The most common placental type is that found among the known Didelphidae (most American marsupials), Phalangeridae (Australian possums), and Macropodidae (kangaroos and wallabies) (Tyndale-Biscoe and Renfree 1987). Here, the yolk sac is well vascularized and intimately associated with the endometrium. Both vascular and nonvascular portions of the yolk sac, as well as a small area of the chorion, are apposed to the uterine wall.

TABLE 22.2 *Conservative Classification Scheme for Placenta Types*

Histological type	Mammalian order	Common names
Epitheliochorial	Perissodactyla	Horses, rhinos, tapirs
	Cetacea	Whales, dolphins, porpoises
	Lemuridae	Madagascan lemurs
	Artiodactyla	Cloven-hoofed ungulates: cattle, sheep, pigs, deer, antelope, giraffes
Endothelio-chorial	Carnivora (most species)	Felids, mustelids, bears, dogs
	Pinnipedia	Seals
Hemochorial	Insectivora	Shrews
	Rodentia	Rats, mice
	Lagomorpha	Rabbits, hares
	Sirenia	Dugongs, manatees
	Primates (most species)	New and Old World monkeys, apes, humans

Source: After Steven 1975.

TABLE 22.1 *(Continued)*

	Fetal tissue[a]		Placental shape	Examples
Trophoblast	Connective tissue	Endothelium		
+	+	+	Diffuse	Pig, horse
+	+	+	Cotyledonary	Ruminants
+	+	+	Zonary to discoid	Carnivores
+	+	+	Discoid	Insectivores, primates, bats, primitive rodents
−	−	+	Discoid, cup-shaped, spheroidal	Modern rodents

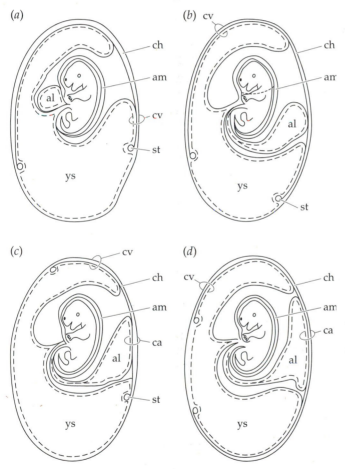

Figure 22.20 Metatherian placental types, not to scale. (*a*) The quokka (*Setonix brachyurus*) yolk-sac placenta has a large trilaminar region that encloses both fetus and allantois in its folds. (*b*) The native cat (*Dasyurus viverrinus*) has a similar placenta and a much larger allantois that approaches, but does not fuse with, the chorion. (*c*) The koala (*Phascolarctos cinereus*) makes both a trilaminar yolk-sac (or choriovitelline) and a chorioallantoic placenta. (*d*) The bandicoot (*Perameles nasuta*) likewise makes both placental types, the chorioallantoic placenta being more extensively attached to the uterine endometrium. Abbreviations: al, allantois; am, amnion; ca, chorioallantoic placenta; ch, chorion; cv, choriovitelline placenta; st, sinus terminalis; ys, yolk sac. (After Tyndale-Biscoe and Renfree 1987.)

The allantois is relatively small and poorly vascularized, does not reach the chorion, becomes a large vessel at the end of gestation, and remains enfolded within the yolk sac (Owen 1834). The second placental type is reported for only two exceptional marsupials, *Dasyurus viverrinus* (Hill 1900) and *Sminthopsis crassicaudata* (Hughes 1974), both dasyurids (marsupial mice). Here, the allantois becomes apposed to the chorion but later retreats; consequently, it does not take part in the formation of the placenta. The allantois subsequently degenerates and lies in the extraembryonic coelom as a vestigial structure (Tyndale-Biscoe and Renfree 1987). The third placental type appears to be restricted to the monotypic family Phascolarctidae. In the koala (*Phascolarctos cinereus*), the allantois becomes apposed to the chorion but does not develop villi, although it is highly vascular in later stages (Caldwell 1884; Semon 1894; Hughes 1974, 1984). The only attachment of the embryo to the uterus is entirely nonvascular and consists of giant cells of the subzonal membrane of the bilaminar yolk sac just exterior to the sinus terminalis (Figure 22.20). The allantois apparently vascularizes the placenta (Amoroso 1952), and likely serves for respiratory exchange (Hughes 1984), although the yolk sac placenta remains as the major organ for nutritive absorption and respiration. Finally, the fourth placental type is truly chorioallantoic and appears restricted to the bandicoots (Peramelidae) (Padykula and Taylor 1976, 1977, 1982). During the later stages of gestation, the chorioallantois forms a discoid placenta, complete with a long, thick umbilicus. It is important to note that although peramelids have the shortest gestation of any marsupial, their neonates are well developed at birth (Sharman 1965) and their postnatal growth is most rapid (Tyndale-Biscoe and Renfree 1987). This suggests that their chorioallantoic placenta is a more efficient organ of exchange than the yolk-sac placenta of other marsupials (Taylor and Padykula 1978; Padykula and Taylor 1982).

The extent of invasiveness of the yolk-sac placenta varies from simple apposition to erosion and penetration of the uterine epithelium (Hughes 1974). In the latter category, attachment to the endometrium is so firm that yolk sacs cannot be pulled away from the endometrium intact (McCrady 1938; New et al. 1977; Krause and Cutts 1984, 1985). In fact, yolk sacs of embryos often fuse together! The placenta of the Virginia opossum (*Didelphis virginiana*) fits this description, as do the placentas of the four-eyed possum (*Philander opossum*) (Enders and Enders 1969), native cat (*Dasyurus viverrinus*) (Hill 1900), and honey possum (*Tarsipes rostratus*) (Tyndale-Biscoe and Renfree 1987). Simple apposition to the uterus by the yolk-sac wall occurs in the bettong (*Bettongia gaimardi*) (Flynn 1930), and is accompanied by attenuation of, but not penetration by, the contacting cells. This arrangment is also true for the quokka (*Setonix brachyurus*), ring-tailed possum (*Pseudocheiris peregrinus*), wallaby (*Macropus rufogriseus*) (Sharman 1961), potoroo (*Potorous tridactyla*) and brush-tailed possum (*Trichosurus vulpecula*) (Hughes 1974). Interdigitation of microvilli, but not endometrial erosion, occurs in the tammar (*Macropus eugenii*) (Walker and Gemmell 1983). No closer contact occurs between maternal and fetal tissues.

It is clear that the wide variety of placental types in eutherian and metatherian mammals reflects the diversity with which fetus and uterus establish physical contact for physiological exchange. The variety thus far discovered correlates only weakly with what we know about mammalian phylogeny; namely, that the choriovitelline placenta is *in general* typical of metatherians, and that the chorioallantoic placenta is *in general* characteristic of eutherian mammals. Enough exceptions exist, it seems, that little gain would be made by establishing finer distinctions.

Organogenesis

Because of the inaccessibility of the developing mammalian embryo, relative to, say, those of other vertebrates, organ formation has been studied little in mammalian embryos. As a result, the best-known textbooks on the subject of vertebrate embryology (Nelsen 1953; Witschi 1956; Patten 1958; Balinsky 1975) are based mostly on what is known from observation and experimentation on avian, reptilian, amphibian, and fish embryos. Only recently has a comprehensive atlas of mouse development, illustrated entirely with photomicrographs of histological preparations, been assembled (Kaufman 1992). Our knowledge of prototherian and metatherian embryology is even less extensive, and is also heavily biased toward general biology (Griffiths 1978), reproductive physiology (Tyndale-Biscoe and Renfree 1987), genetics and evolution (Marshall Graves et al. 1990), and fertilization and early development (CSIRO 1993). The study of noneutherian embryos, of course, is additionally hobbled not only by the relic geographic distribution of these mammals, but also by their relative rarity even within their known ranges.

The advent of modern techniques in cell and developmental biology will almost certainly alter this picture. Already, procedures of cell (and embryo) culture and micromanipulation have become routine (Monk 1987; Hogan et al. 1994), as has electron microscopy (for review, see Cruz 1992). Experimental approaches such as immunohistochemistry, microsurgery, cell labeling, and chimera production have been successfully employed (Le Douarin and McLaren 1984; Rossant and Pedersen 1986; Copp and Cockroft 1990; Beddington et al. 1992). More recently, gene cloning, transgenic technology, culture of embryonic stem cells, homologous recombination, and the polymerase chain reaction have been added to this armory of laboratory approaches (Ciba Foundation Symposium 165 1992; Wassarman and DePamphilis 1993). Reflecting the increased requirement for gauging finely the extent of embryonic development during the postimplantation period, a precise staging standard has recently been devised for estimating the age of gastrulating murine embryos (Downs and Davies 1993). This scheme is intended for the rapid examination of explanted and cultured embryos at magnifications as low as those afforded by a dissecting microscope. The further development of experimental and observational tools such as these will enable mammalian embryologists to make significant progress in our understanding not only of organogenesis, but of embryogenesis in general.

The early events of organogenesis have been studied in streak-stage mouse embryos. During this period, the primitive streak actively advances, and in short order recedes, in the epiblast of the five- to eight-day-old conceptus. This is a dynamic phase in the early postimplantation stages of development, when cells are not only proliferating prodigiously but are also being spatially ordered into the tissues that will become the germ layers, ectoderm, mesoderm, and endoderm (Tam and Beddington 1992). Long before tissue ordering is completed, however, the earliest organs make their appearance: notochord, neural tube, somites, heart. In addition, the body axis lengthens; the extraembryonic membranes enlarge. As in gastrulation, the events that comprise the developmental sequela called organogenesis are far from discrete; they are instead staggered, overlapping, or simultaneous.

Postimplantation Development

Two techniques have been used successfully in recent years to understand how organogenesis occurs in the mouse embryo. The first of these is clonal analysis of embryonic cells labeled in situ with microinjected horseradish peroxidase or fluorescent conjugates. This procedure makes it possible to follow not only the developmental fate of marked cells but also their proliferation rate and migration paths (Lawson et al. 1991). The second technique is even more elegant and provides a counterpoint to the first: grafting of endogenously labeled embryonic cells from a transgenic mouse embryo. The endogenous label that has been used successfully is the bacterial protein

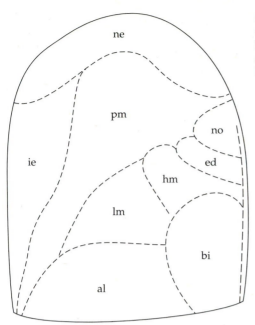

Figure 22.21 Fate map of the early-streak-stage mouse egg cylinder, shown here with the distal cap on top and the primitive streak on the right. Abbreviations: al, allantois; bi, blood islands; ed, endoderm; hm, heart mesoderm; ie, integumentary ectoderm; lm, lateral plate mesoderm; ne, neurectoderm; no, notochord; pm, paraxial mesoderm. (After Lawson and Pedersen 1992b; Quinlan and Tam 1996; Parameswaran and Tam 1995.)

β-galactosidase, the gene for which had been previously incorporated into a known locus on one of the X chromosomes (Tam and Tan 1992). The protein produced is unlike that of mammals and is easily stained. The combination of these two approaches has proved to be a potent approach to charting the fate map of the murine embryo.

CLONAL ANALYSIS. Clonal analysis of streak-stage mouse embryos has demonstrated that lineage restriction of embryonic cells occurs well past the streak stage. The streak itself extends anteriorly by growth of a resident cell population already in the anterior portion of the streak; its posterior half extends by the appropriation of cells from the lateral and anterior epiblast (Lawson and Pedersen 1992a). Mesoderm formation and primitive streak extension occur concurrently in mouse embryos, the posterior half of the streak being formed by the cells that travel through it into extraembryonic and lateral mesoderm. Cells of the axial epiblast immediately anterior to the primitive streak give rise to the cranial neurectoderm; their descendants spread rostrocaudally in the developing neural tube (Lawson and Pedersen 1992b) (Figure 22.21). The notochord appears to be derived from cells initially localized within the so-called node (after the avian Hensen's node), an epiblast feature reputed to be the organizer of the body axis (Beddington 1994). The

predictability of regional fate in the mouse epiblast thus appears to be due to the emphatic alignment of clonal descendants within the epiblast towards the developing primitive streak (Lawson and Pedersen 1992a).

TISSUE GRAFTING. Tissue grafting has yielded results that complement those of clonal analysis. Fate mapping done with carbocyanine dye-labeled and β-galactosidase-expressing cells indicates that the precursor population of the murine neural tube is contained in the distal-cap epiblast of the early-primitive-streak-stage embryo (Figure 22.22). Within this cap, cell fate is further regionalized; cells at the posterior end of the cap, for instance, localize mainly to the caudal parts of the neural tube. Immediately outside the distal regions of the cap, epiblast cells contribute cells to the node, and later, the notochord (Quinlan et al. 1995). Organization of the fetal body plan appears to have been established by the late-primitive-streak stage (Quinlan and Tam, in press). Indeed, it is thought that a preliminary craniocaudal patterning occurs in the neural primordium *before* neurulation (Quinlan et al. 1995). Related to this set of results is the determination that competence for neural induction is restricted to the anterior epiblast of the early-primitive-streak-stage embryo (Ang and Rossant 1993). Further reinforcing this notion of regional specialization are the findings that specific interactions characterize specification and segmentation in the paraxial mesoderm (Tam and Trainor 1994) and that the precursors for the various mesodermal lineages are restricted to the lateral epiblast (Parameswaran and Tam 1996). The inescapable conclusion is that a certain level of topographic specificity characterizes the streak-stage murine embryo.

(a)

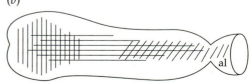

(b)

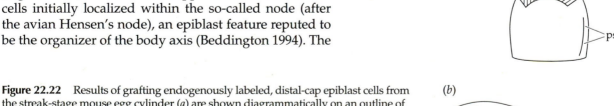

Figure 22.22 Results of grafting endogenously labeled, distal-cap epiblast cells from the streak-stage mouse egg cylinder (*a*) are shown diagrammatically on an outline of the dorsal aspect of the neural tube of the 5-somite-stage embryo obtained following a 24-hour incubation. Grafted Area 1 cells colonized the anterior third of the neural tube; Area 3 cells, the posterior half of the neural tube and the proximal region of the allantois. Area 2 cells, by contrast, contributed to most points along the length of the neural tube. Abbreviations: al, allantois; ps, primitive streak. (After Quinlan et al. 1995.)

Regionalization of cell fate, however, should not be taken to mean fate determination. Descendants of a single clone, for example, seldom remain together (Lawson et al. 1991); indeed, the physical domains occupied by the various germ-layer precursors demonstrate considerable overlap (Tam and Beddington 1992; Tam et al. 1993). Even within such an organ as the embryonic murine brain, cells from different sites intermingle (Morriss-Kay et al. 1994). Taken together, these observations raise the distinct possibility that a number of epiblast cells are not fully committed to a specific germ-layer fate *prior to* ingression (Quinlan and Tam 1996). It thus appears that for an extended period during embryonic development, cell position is at best a tentative predictor of eventual cell fate.

The tentativeness of cell fate assignment is likely to be of considerable adaptive significance to the conceptus as a whole. In a rapidly growing embryo consisting of cells caught in a dynamic flurry of proliferation, migration, and differentiation, it would be desirable for any given cell to retain some measure of developmental flexibility for as long as possible. Such would enable an embryo momentarily disabled by cell cycle delay, for instance, or temporarily compromised by loss of a few cells, to compensate for minor disruptions and resume rather quickly the normal pace of development. It is easy to see how such built-in plasticity could contribute to the wide variety of procedural detail manifested in nearly every phase of mammalian embryogenesis.

Organogenesis in metatherian mammals is unusual in that its duration is remarkably constant for species of very different body sizes. It has been suggested that the length of the organogenetic phase is directly related to the invasiveness of the placenta and most especially to the birth weight of the neonate (Selwood 1980), but more data need to be compiled to see if this prediction holds (Tyndale-Biscoe and Renfree 1987).

The most comprehensive account of organogenesis in marsupials is for *Didelphis virginiana* (McCrady 1938). This scheme recognizes 35 stages of intrauterine development, including the stage of medullary plate extension (see the section on Gastrulation). Because recently, the very existence of a medullary plate (*sensu* McCrady) has been questioned (Cruz et al. 1996), it would be prudent to use another scheme that does not use the medullary plate as a reference point. Partial developmental timetables have recently been published for *Antechinus stuartii* (Selwood 1980), *Monodelphis domestica* (Baggott and Moore 1990), and *Sminthopsis macroura* (Yousef and Selwood 1993), and are likely to prove valuable when completed.

Cell–Matrix Interactions

Organogenesis in mammals has also been studied at the level of cell–cell and cell–matrix interactions. Epitheliomesenchymal interactions are well studied in the development of the murine metanephros and tooth primordium (reviewed in Saxen and Thesleff 1992). An early response of the mesenchyme involved in these interactions is increased cell proliferation in the vicinity of the inducing epithelium (Thesleff and Hurmerinta 1981; Saxen 1987). The resulting cells subsequently aggregate, or condense, into discrete arrays that presently transform into metanephric or tooth primordia.

Metanephrogenesis is described as a series of *permissive* interactions. Here, the induced cell or tissue becomes committed to a certain developmental pathway, but requires exogenous stimuli, made available stepwise, to express its phenotype (Saxen 1977). The formation of kidney tubules is an example of a permissive interaction. The metanephric mesenchyme responds, by tubule formation, to a wide variety of heterotypic tissues (Saxen 1987), but only a predetermined set of cells responds properly; all other mesenchymes tested are unresponsive (Saxen and Thesleff 1992). In the tooth rudiment, on the other hand, the odontogenic potential resides in the epithelium. Such potential thereafter becomes transferred to the mesenchyme, which then begins to act as a *directive* (instructive) inducer (Mina and Kollar 1987; Thesleff et al. 1990). Such induction proceeds as if by fiat; no reiterated series of sequential interactions is necessary for eliciting the final, induced response. Whether permissive or instructive, an inductive sequela clearly alters the competence of the responding tissue.

Postinductive condensation or aggregation of the responding tissue is mediated by largely unknown molecules. Because the extracellular matrix of the metanephric mesenchyme changes following induction, various mesenchymal components have been evaluated for inductive potential. Fibronectin, laminin, collagen, and uvomorulin have been investigated, but changes involving these substances appear to be the result, and not the cause, of aggregate polarization. Only syndecan appears to have a crucial inductive role (Saxen and Thesleff 1992). This heparan-sulfate proteoglycan spans the cell membrane and acts as a matrix receptor (Bernfield and Sanderson 1990). During kidney development, syndecan is present in detectable amounts in the condensed mesenchyme and pretubular aggregates (Vainio et al. 1989). Similar results are obtained in vitro when murine-derived, but not rat-derived, aggregates are examined. Thus, syndecan has demonstrable species-specificity, which makes it a likely candidate as an induction mediator.

Tenascin is another matrix glycoprotein involved in the postinductive condensation of mesenchyme. Expression of this substance in odontogenic mesenchyme, for instance, is induced by dental epithelium (Vainio et al. 1989). In addition, solid-phase binding assays indicate that tenascin is bound by syndecan isolated from dental mesenchyme (Salmivirta et al. 1991). Tenascin, therefore, must help mediate a cell–matrix interaction that promotes mesenchyme aggregation.

Summary

The study of mammalian embryos provides us with an interesting approach to understanding the evolutionary relationships among the members of this diverse group of vertebrates. Of the characteristics that distinguish mammals from other vertebrates, viviparity is perhaps the most salient. This reproductive strategy is made possible by a vast array of physiological and anatomical adaptations. Viviparity has allowed mammals to forego the requirement for yolk-provisioned eggs, but it has also constrained them to develop a means of facilitating the survival of embryos in the female reproductive tract. And because viviparity permits the prolongation of embryonic development in utero, it improves the chances for survival of the neonate. This, in turn, makes it possible to reduce litter size, number of young, and even number of pregnancies. Such a reproductive strategy reduces intraspecific competition while facilitating the occurrence of significant intergenerational interactions. The adaptations and strategies that have made mammals the most successful extant vertebrate group are clearly all linked, and at their crux lies embryogenesis.

Literature Cited

Amoroso, E. C. 1952. Placentation. In *Marshall's Physiology of Reproduction*, 3rd ed., Vol. 2., A. S. Parkes (ed.). Longmans, Green, London, pp. 127–311.

Ang, S.-L. and J. Rossant. 1993. Anterior mesoendoderm induces mouse *engrailed* genes in explant cultures. *Development* 118: 139–149.

Austin, C. R. 1965. *Fertilization*. Prentice-Hall, Inc., Englewood Cliffs, NJ.

Austin, C. R. 1982. The egg. In *Reproduction in Mammals*, vol. I. *Germ Cells and Fertilization*, 2nd ed., C. R. Austin and R. V. Short (eds.). Cambridge University Press, Cambridge, pp. 46–62

Baggott, L. M. and H. D. M. Moore. 1990. Early embryonic development of the grey short-tailed opossum *Monodelphis domestica, in vivo* and *in vitro. J. Zool. (London)* 220: 623–639.

Baker, T. G. 1982. Oogenesis and ovulation. In *Reproduction in Mammals, I. Germ Cells and Fertilization*, 2nd ed., C. R. Austin and R. V. Short (eds.), Cambridge University Press, Cambridge, pp. 17–45.

Balinsky, B. I. 1975. *An Introduction to Embryology*, 4th ed. W. B. Saunders Co., Philadelphia.

Beddington, R. S. P. 1986. Analysis of tissue fate and prospective potency in the egg cylinder. In *Experimental Approaches to Mammalian Development*, J. Rossant and R. A. Pedersen (eds.). Cambridge University Press, Cambridge, pp. 121–147.

Beddington, R. S. P. 1994. Induction of a second neural axis by the mouse node. *Development* 120: 603–612.

Beddington, R. S. P., A. W. Puschel and P. Rashbas. 1992. Use of chimaeras to study gene function in mesodermal tissues during gastrulation and early organogenesis. In *Post-implantation Development in the Mouse* (Ciba Foundation Symposium 165). John Wiley & Sons, Chichester, pp. 61–77.

Bedford, J. M. and G. W. Cooper. 1978. Membrane fusion events in the fertilization of vertebrate eggs. In *Cell Surface Reviews, Vol. 5: Membrane Fusion*, G. Poste and G. L. Nicolson (eds.). North Holland, Amsterdam, pp. 66–125.

Bedford, J. M., J. C. Rodger and W. C. Breed. 1984. Why so many mammalian spermatozoa—a clue from marsupials? *Proc. Roy. Soc. B* 221: 221–233.

Beer, A. E. and R. E. Billingham. 1974. Host responses to intrauterine tissue, cellular and fetal allografts. *J. Reprod. Fertil. (Suppl.)* 21: 59–88.

Beer, A. E. and R. E. Billingham. 1979. Maternal immunological recognition mechanisms during pregnancy. In *Maternal Recognition of Pregnancy* (Ciba Foundation Symposium 64), J. Wheelan (ed.). John Wiley & Sons, Chichester, pp. 293–309.

Bernfield, M. and R. D. Sanderson. 1990. Syndecan, a morphogenetically regulated cell surface proteoglycan that binds extracellular matrix and growth factors. *Philos. Trans. Roy. Soc. London A Math. Phys. Sci.* 327: 171–186.

Betteridge K. J. and J.-E. Flechon. 1988. The anatomy and physiology of pre-attachment bovine embryos. *Theriogenology* 29: 155–161.

Biggers, J. D., J. E. Bell and D. J. Benos. 1988. Mammalian blastocyst: Transport functions in a developing epithelium. *Amer. J. Physiol.* 255: C419–425.

Bleil, J. P. and P. M. Wassarman. 1980. Mammalian sperm-egg interaction: Identification of a glycoprotein in mouse egg zonae pellucidae possessing receptor activity for sperm. *Cell* 20: 873–878.

Bleil, J. P. and P. M. Wassarman. 1990. Identification of a ZP3-binding protein on acrosome-intact mouse sperm by photoaffinity crosslinking. *Proc. Natl. Acad. Sci. U.S.A.* 87: 5563–5567.

Boyd, J. D. and W. J. Hamilton. 1952. Cleavage, early development and implantation of the egg. In *Marshall's Physiology of Reproduction*, 3rd ed., vol 2, A. S. Parkes (ed.). Longmans, Green, London, pp. 1–18.

Bulmer, J. N. and P. M. Johnson. 1986. The T-lymphocyte population in first-trimester human decidua does not express the interleukin-2 receptor. *Immunology* 58: 685–687.

CSIRO. 1993. *Marsupial Reproduction: Gametes, Fertilization and Early Development*. CSIRO, Canberra.

Caldwell, W. H. 1884. On the arrangement of the embryonic membranes in marsupial animals. *Quart. J. Micr. Sci.* 224: 655–658.

Caldwell, W. H. 1887. The embryology of Monotremata and Marsupialia, Pt. 1. *Philos. Trans. Royal Soc. B* 178: 463–486.

Carson, D. D., N. Raboudi and A. L. Jacobs. 1991. Glycoprotein expression and function in embryo-uterine interactions. In *Cellular Signals Controlling Uterine Function*, LA Lavia (ed.). Plenum Press, New York, pp. 107–117.

Chatterton, R. T., Jr., G. J. Macdonald and D. A. Ward. 1975. Effect of blastocysts on rat ovarian steroidogenesis in early pregnancy. *Biol. Reprod.* 13: 77–82.

Ciba Foundation Symposium 165. 1992. *Post-Implantation Development in the Mouse*. John Wiley & Sons, Chichester.

Copp, A. J. 1979. Interaction between inner cell mass and trophectoderm of the mouse blastocyst. II. The fate of the polar trophectoderm. *J. Embryol. Exp. Morphol.* 51: 109–115.

Copp, A. J. and D. C. Cockroft. 1990. *Postimplantation Embryos: A Practical Approach*. IRL Press, Oxford.

Cruz, Y. P. 1992. Role of ultrastructural studies in the analysis of cell lineage in the mammalian pre-implantation embryo. *Micr. Res. Tech.* 22: 103–125.

Cruz, Y. P. and R. A. Pedersen. 1985. Cell fate in the polar trophectoderm of mouse blastocysts as studied by microinjection of cell lineage tracers. *Dev. Biol.* 112: 73–83.

Cruz, Y. P. and R. A. Pedersen. 1991. Origin of embryonic and extraembryonic cell lineages in mammalian embryos. *Curr. Comm. Cell Mol. Biol.* 4: 147–204.

Cruz, Y. P. and L. Selwood. 1993. Uterine histology of the dasyurid marsupial, *Antechinus stuartti*: Relationship with differentiation of the embryo. *J. Reprod. Fertil.* 99: 237–242.

Cruz, Y. P., A. Yousef and L. Selwood 1996. Fate map analysis of the dasyurid marsupial, *Sminthopris macroura* (Gould). *Reprod. Fertil. Dev.* 8: 779–788.

Damjanov, I. and U. M. Wewer. 1991. Uterine extracellular matrix remodeling in pregnancy. In *Cellular Signals Controlling Uterine Function*, L. A. Lavia (ed.). Plenum Press, New York, pp. 99–106.

De Hertogh, R., E. Ekka, I. Vanderheyden and B. Glorieux. 1986. Estrogen and progesterone receptors in the implantation sites and interembryonic segments of rat uterus endometrium and myometrium. Endocrinology 119: 680–684.

Dey, S.K. and D. C. Johnson. 1986. Embryonic signals in pregnancy. *Ann. N.Y. Acad. Sci.* 476: 49–62.

Downs, K. and T. Davies. 1993. Staging of gastrulating mouse embryos by morphological landmarks in the dissecting microscope. *Development* 118: 1255–1266.

Ducibella, T. 1991. Mammalian egg cortical granules and the cortical reaction. In *Elements of Mammalian Fertilization,* Vol. 1., P. M. Wassarman (ed.). CRC Press, Boca Raton, FL, pp. 205–231.

Ducibella, T. and E. Anderson. 1975. Cell shape and membrane changes in the eight-cell mouse embryo: Prerequisites for morphogenesis of of the blastocyst. *Dev. Biol.* 47: 45–48.

Enders, A.C. and R. K. Enders. 1969. The placenta of the four-eyed opossum (*Philander opossum*). *Anat. Rec.* 165: 431–450.

Eyal-Giladi, H. and S. Kochav. 1976. From cleavage to primitive streak formation: A complementary normal table and a new look at the first stages of development in the chick. *Dev. Biol.* 49: 321–327.

Fleming, T. P. and M. H. Johnson. 1988. From egg to epithelium. *Ann. Rev. Cell Biol.* 4: 459–485.

Florman, H. M. and P. M. Wassarman. 1985. O-linked oligosaccharides of mouse egg ZP3 account for its sperm receptor activity. *Cell* 41: 313–324.

Flynn, T. T. 1923. The yolk-sac and allantoic placenta in *Perameles*. *Quart. J. Micr. Sci.* 67: 123–182.

Flynn, T. T. 1930. The uterine cycle of pregnancy and pseudopregnancy as it is in the diprotodont marsupial *Bettongia cuniculus* with notes on other reproductive phenomena in this marsupial. *Proc. Linn. Soc. N.S.W.* 55: 506–531.

Flynn, T. T. and J. P. Hill. 1939. The development of the Monotremata. IV. Growth of the ovarian ovum, fertilization and early cleavage. *Trans. Zool. Soc. London* 24: 445–622.

Flynn, T. T. and J. P. Hill. 1947. The development of the Monotremata. VI. The later stages of cleavage and the formation of the primary germ layers. *Trans. Zool. Soc. London* 26: 1–151.

Gardner, R. L. 1982. Investigation of cell lineage and differentiation in the extraembryonic endoderm of the mouse embryo. *J. Embryol. Exp. Morphol.* 68: 175–198.

Gardner, R. L. 1985. Regeneration of endoderm from primitive ectoderm in the mouse embryo: Fact or artifact? *J. Embryol. Exp. Morphol.* 88: 303–326.

Gardner, R. L. and V. E. Papaioannou. 1975. Differentiataion in the trophectoderm and inner cell mass. In *The Early Development of Mammals*, M. Balls and A. L. Wild (eds.). Cambridge University Press, Cambridge, pp. 107–132.

Gardner, R. L. and J. Rossant. 1979. Investigation of the fate of 4.5-day *post-coitum* mouse inner cell mass cells by blastocyst injection. *J. Embryol. Exp. Morphol.* 52: 141–152.

Gidley-Baird, A. A. 1981. Endocrine control of implantation and delayed implantation in rats and mice. *J. Reprod. Fertil. (Suppl.)* 29: 97–109.

Gilbert, S. F. 1994. *Developmental Biology*, 4th ed. Sinauer Associates, Inc., Sunderland, MA.

Given, R. L. and A. C. Enders. 1980. Mouse uterine glands during the peri-implantation period. I. Fine structure. *Amer. J. Anat.* 157: 169–179.

Given, R. L. and A. C. Enders. 1989. The endometrium of delayed and early implantation. In *Biology of the Uterus*, 2nd ed., R. M. Wynn and W. P. Jollie (eds.). Plenum Medical Book Co., New York, pp. 175-231.

Griffiths, M. 1978. *The Biology of the Monotremes*. Academic Press, New York.

Grosser, O. 1909. Die Wege der fetalen Ernahrung. *Sammlung Anat. u. Physiol. Vortrage u. Aufsatze von Gaup. u. Nagel.* 3: 79–96.

Hafez, E. S. E. 1972. Differentiation of mammalian blastocysts. In *Biology of mammalian Fertilization and Implantation*, K. S. Moghissi and E. S. E. Hafez (eds.). Charles C. Thomas, Springfield, IL, pp. 296–301.

Hampl, A. and J. J. Eppig. 1995. Analysis of the mechanism(s) of metaphase I arrest in maturing mouse oocytes. *Development* 121: 925–933.

Handyside, A. H. 1980. Distribution of antibody and lectin binding sites on dissociated blastomeres from mouse morulae: Evidence for polarization at compaction. *J. Embryol. Exp. Morphol.* 60: 99–116.

Hartman, C. G. 1916. Studies in the development of the opossum *Didelphis virginiana* L. I. History of the early cleavage, II. Formation of the blastocyst. *J. Morphol.* 27: 1–84.

Hartman, C. G. 1919. Studies in the development of the opossum *Didelphis virginiana* L. III. Description of new material on maturation, cleavage and entoderm formation. IV. The bilaminar blastocyst. *J. Morphol.* 32: 1–144.

Hill, C. J. 1933. The development of the Monotremata. I. The histology of the oviduct during gestation. *Trans. Zool. Soc. London* 21: 413–443.

Hill, J. P. 1900. Contributions to the embryology of the Marsupialia. 2. On a further stage of placentation of *Perameles*. 3. On the foetal membranes of *Macropus parma*. *Quart. J. Micr. Sci.* 43: 1–22.

Hill, J. P. 1910. Contributions to the embryology of the Marsupialia. 4. The early development of the Marsupialia with special reference to the native cat (*Dasyurus viverrinus*). *Quart. J. Micr. Sci.* 56: 1–134.

Hill, J. P. 1918. Some observations on the early development of *Didelphys aurita*. *Quart. J. Micr. Sci.* 63: 93–139.

Hogan, B., R. Beddington, F. Costantini and E. Lacy. 1994. *Manipulating the Mouse Embryo, A Laboratory Manual*, 2nd ed., Cold Spring Harbor Laboratory Press, Cold Spring Harbor, NY.

Hogan, B. L. M., A. R. Cooper and M. Kurkinen. 1980. Incorporation into Reichert's membrane of laminin-like extracellular proteins synthesized by parietal endoderm cells of the mouse embryo. *Dev. Biol.* 80: 289–300.

Hogan, B. L. M., D. P. Barlow and R. Tilly. 1983. F9 teratocarcinoma cells as a model for the differentiation of parietal and visceral endoderm in the mouse embryo. *Cancer Surv.* 2: 115–140.

Home, E. 1802. Some observations on the mode of generation of the kangaroo: A particular description of the organs themselves. *Phil. Trans. Roy. Soc.* 85:221–230.

Hughes, R. L. 1974. Morphological studies on implantation in marsupials. *J. Reprod. Fertil.* 39: 173–186.

Hughes, R. L. 1977. Egg membranes and ovarian function during pregnancy in monotremes and marsupials. In *Reproduction and Evolution*, J. H. Calaby and C. H. Tyndale-Biscoe (eds.). Australian Academy of Science, Canberra, pp. 281–291.

Hughes, R. L. 1984. Structural adaptations of the eggs and the fetal membranes of monotremes and marsupials for respiration and metabolic exchange. In *Respiration and Metabolism of Embryonic Vertebrates*, R. Seymour (ed.). Junk, Doordrecht, pp. 389–421.

Hughes, R. L. and L. S. Hall. 1984. Embryonic development in the common brushtail possum *Trichosurus vulpecula*. In *Possums and Gliders*, A. P. Smith and I. D. Hume (eds.). Australian Mammal Society, Sydney, pp. 197–212.

Jaffe, L. A. and M. Gould. 1985. Polyspermy-preventing mechanisms. In *Biology of Fertilization*, vol. 3, C. B. Metz and A. Monroy (eds.). Academic Press, Orlando, pp. 223–250.

Johnson, M. H 1996. Origins of pluriblast and trophoblast in the eutherian conceptus. *Reprod. Fertil. Dev.* 8: 699–709.

Johnson, M. H. and B. Maro. 1986. Time and space in the early mouse embryo: A cell biological approach to cell diversification. In *Experimental Approaches to Mammalian Embryonic Development*, J. Rossant and R. A. Pedersen (eds.). Cambridge University Press, Cambridge, pp. 35–66.

Johnson, M. H. and L. Selwood 1996. Nomenclature of early development in mammals. *Reprod. Fertil. Dev.* 8: 759–764.

Johnson, M. H. and C. A. Ziomek. 1981. The foundation of two distinct cell lineages within the mouse morula. *Cell* 24: 71–80.

Kaufman, M. H. 1992. *The Atlas of Mouse Development.* Academic Press, London.

Kemler, R., B. Babinet, H. Eisen and F. Jacob. 1987. Surface antigen in early differentiation. *Proc. Natl. Acad. Sci. U.S.A.* 74: 4449–4452.

Kemp, T. S. 1983. The relationships of mammals. *Zool. J. Linn. Soc.* 77: 353–384.

Kerr, T. 1935. Blastocysts of *Potorous tridactylus* and *Bettongia cuniculus. Quart. J. Micr. Sci.* 77: 305–315.

Kirby, D. R. S. 1965. The "invasiveness" of the trophoblast. In *The Early Conceptus, Normal and Abnormal*, W. W. Park (ed.). University of St. Andrews Press, Dundee, pp. 68–73.

Kirby, D. R. S. 1971. Blastocyst-uterine relationship before and during implantation. In *The Biology of the Blastocyst*, R. J. Blandau (ed.). The University of Chicago Press, Chicago, pp. 393–411.

Kirby, D. R. S., W. D. Billington and D. A. James. 1966. Transplantation of eggs to the kidney and uterus of immunised mice. *Transplantation* 4: 713–718.

Krause, W. J. and J. H. Cutts. 1984. Scanning electron microscopic observations on the opossum yolk sac chorion immediately prior to attachment. *J. Anat.* 138: 189–191.

Krause, W. J. and J. H. Cutts. 1985. Placentation in the opossum, *Didelphis virginiana. Acta Anat.* 123: 156–171.

Lawson, K. A. and R. A. Pedersen 1992a. Clonal analysis of cell fate during gastrulation and neurulation in the mouse. In *Post-implantation Development in the Mouse* (Ciba Foundation Symposium 165). John Wiley & Sons, Chichester, pp. 3–26.

Lawson, K. A. and R. A. Pedersen. 1992b. Early mesoderm formation in the mouse embryo. In *Formation and Differentiation of Early Embryonic Mesoderm*, R. Bellairs, J. W. Lash and E. J. Sanders (eds.). Plenum Press, New York, pp. 33–46.

Lawson, K. A., J. J. Meneses and R. A. Pedersen. 1991. Clonal analysis of epiblast fate during germ layer formation in the mouse embryo. *Development* 113: 891–911.

Le Douarin, N. and A. McLaren (eds.). 1984. *Chimeras in Developmental Biology.* Academic Press, London.

Lewis, W. H. and E. S. Wright. 1935. On the early development of the mouse egg. *Carnegie Inst. Contrib. Embryol.* 25: 113–143.

Leyton, L., P. Leguen, D. Bunch and P. M. Saling. 1992. Regulation of mouse gametic interaction by sperm tyrosine kinase. *Proc. Natl. Acad. Sci. U.S.A.* 93: 1164–1169.

Lillegraven, J. A. 1975. Biological considerations of the marsupial-placental dichotomy. *Evolution* 29: 707–722.

Lillegraven, J. A. 1979. Reproduction in Mesozoic mammals. In *Mesozoic Mammals: The First Two-thirds of Mammalian History*, J. A. Lillegraven, Z. Kielan-Jaworowska and W. A. Clemens (eds.). University of California Press, Berkeley, pp. 259–276.

Lo, C. W. and N. B. Gilula. 1979. Gap junctional communication in the preimplantation mouse embryo. *Cell* 18: 399–409.

Lyne, A. G. and D. E. Hollis. 1977. The early development of marsupials, with special reference to bandicoots. In *Reproduction and Evolution*, J. H. Calaby and C. H. Tyndale-Biscoe (eds.). Australian Academy of Science, Canberra, pp. 293–302.

Maro, B., M. H. Johnson, S. J. Pickering and D. Louvard. 1985. Changes in the distribution of membranous organelles during mouse early development. *J. Embryol. Exp. Morphol.* 90: 287–309.

Marshall, L. G. 1979. Evolution of metatherian and eutherian (Mammalian) characters: A review based on cladistic methodology. *Zool. J. Linnean Soc.* 66: 369–410.

Marshall Graves J. A., R. M. Hope and D. W. Cooper (eds.). 1990. *Mammals From Pouches and Eggs: Genetics, Breeding and Evolution of Marsupials and Monotremes.* CSIRO, Canberra.

McCrady, E. 1938. The embryology of the opossum. *Amer. Anat. Mem.* 16: 1–233.

McCrady, E. 1944. The evolution and significance of the germ layers. *J. Tenn. Acad. Sci.* 19: 240–251.

McLaren, A. 1982. The embryo. In *Reproduction in Mammals: 2. Embryonic and Fetal Development*, C. R. Austin and R. V. Short (eds.). Cambridge University Press, Cambridge, pp. 1–25.

McRae, A. C. 1992. Effect of ovariectomy on blastocyst expansion and survival in ferrets (*Mustela putorius furo*). *Reprod. Fertil. Dev.* 4: 239–247.

Meizel, S. 1984. The importance of hydrolytic enzymes to an exocytotic event, the mammalian sperm acrosome reaction. *Biol. Rev.* 59: 125–157.

Mina, M. and E. J. Kollar. 1987. The induction of odontogenesis in nondental mesenchyme combined with early murine mandibular arch epithelium. *Arch. Oral Biol.* 32: 123–127.

Moller, C. C. and P. M. Wassarman. 1989. Characterization of a proteinase that cleaves zona pellucida glycoprotein ZP2 following activation of mouse eggs. *Dev. Biol.* 132: 103–112.

Monk, M. (ed.). 1987. *Mammalian Development: A Practical Approach.* IRL Press, Oxford.

Morriss, G. 1975. Placental evolution and embryonic nutrition. In *Comparative Placentation: Essays in Structure and Function.* D. H. Steven (ed.). Academic Press, New York, pp. 87–98.

Morriss-Kay, G., H. Wood and W.-H. Chen. 1994. Normal neurulation in mammals. In *Neural Tube Defects* (Ciba Foundation Symposium 181). John Wiley & Sons, Chichester, pp. 51–69.

Mossman, H. W. 1937. Comparative morphogenesis of the fetal membranes and accessory uterine structures. *Carnegie Inst. Contrib. Embryol.* 26: 129–246.

Mulholland, J. and S. R. Glasser. 1991. Uterine preparation for blastocyst attachment. In *Cellular Signals Controlling Uterine Function*, L. A. Lavia (ed.). Plenum Press, London, pp. 81–97.

Myles, D. G. 1993. Molecular mechanisms of sperm-egg membrane binding and fusion in mammals. *Dev. Biol.* 158: 35–45.

Myles, D. G. and P. Primakoff. 1992. Sperm surface dynamics and fertilization. In *Comparative Spermatology, 20 Years After,* B. Bacetti (ed.). Raven Press, Siena, pp. 671–678.

Nelsen, O. E. 1953. *Comparative Embryology of the Vertebrates.* Blakiston, New York.

New, D. A. T., M. Mizell and D. L. Cockroft. 1977. Growth of opossum embryos *in vitro* during organogenesis. *J. Exp. Morphol.* 41: 111–123.

Nilsson, O. 1972. Ultrastructure of the process of secretion in the rat uterine epithelium at preimplantation. *J. Ultrastruct. Res.* 40: 572–580.

Nilsson, B. O. and O. Lundkvist. 1979. Ultrastructural and histochemical changes of the mouse uterine epithelium on blastocyst activation for implantation. *Anat. Embryol.* 155: 311–321.

Obrink, B. 1991. C-CAM (Cell-CAM 105): A member of the growing immunoglobulin superfamily of cell adhesion proteins. *BioEssays* 13: 227–234.

O'Neill, C. and P. Quinn. 1983. Inhibitory influences of uterine secretions on mouse blastocysts decreases at the time of blastocyst activation. *J. Reprod. Fertil.* 29: 123–126.

Orsini, M. W. and R. K. Meyer. 1962. Effect of varying doses of progesterone on implantation in the ovariectomized hamster. *Proc. Soc. Exp. Biol. Med.* 110: 713–715.

Owen, R. 1834. On the generation of the marsupial animals, with a description of the impregnated uterus of the kangaroo. *Phil. Trans. Roy. Soc.* 1834: 133–164.

Padykula, H. A. and J. M. Taylor. 1976. Ultrastructural evidence for loss of the trophoblastic layer in the chorioallantoic placenta of Australian bandicoots (Marsupialia: Peramelidae). *Anat. Rec.* 186: 357–385.

Padykula, H. A. and J. M. Taylor. 1977. Uniqueness of the bandicoot chorioallantoic placenta (Marsupialia: Peramelidae). Cytological and evolutionary interpretations. In *Reproduction and Evolution*, J. H. Calaby and C. H. Tyndale-Biscoe (eds.). Australian Academy of Science, Canberra, pp. 303–323.

Padykula, H. A. and J. M. Taylor. 1982. Marsupial placentation and its evolutionary significance. *J. Reprod. Fertil. (Suppl.)* 31: 95–104.

Parameswaran, M. and P. P. L. Tam. 1995. Regionalisation of cell fate and morphogenetic movement of the mesoderm during mouse gastrulation. *Dev. Genet.* 17: 16–28

Parr, E. L., H. N. Tung and M. B. Parr. 1987. Apoptosis as the mode of uterine epithelial cell death during embryo implantation in mice and rats. *Biol. Reprod.* 36: 211–225.

Parr, M. B. 1983. Relationship of uterine closure to ovarian hormones and endocytosis in the rat. *J. Reprod. Fertil.* 68: 185–188.

Parr, M. B. and E. L. Parr. 1986. Endocytosis in the rat uterine epithelium at implantation. *Ann. N.Y. Acad. Sci.* 476: 110–121.

Parr, M. B. and E. L. Parr. 1989. The implantation reaction. In *Biology of the Uterus*, R. M. Wynn and W. P. Jolie (eds.). Plenum Medical Book Company, New York, pp. 233–276.

Parr, M. B., H. N. Tung and E. L. Parr. 1986. The ultrastructure of the rat primary decidual zone. *Amer. J. Anat.* 176: 423–436.

Patten, B. M. 1958. *Foundations of Embryology*. McGraw-Hill Book Company, New York.

Pedersen, R. A. 1986. Potency, lineage, and allocation in preimplantation mouse embryos. In *Experimental Approaches to Mammalian Embryonic Development*, J. Rossant and R. A. Pedersen (eds.). Cambridge University Press, Cambridge, pp. 3–34.

Perry, J. A. 1981. The mammalian fetal membranes. *J. Reprod. Fertil.* 62: 321–335.

Pratt, H. P. M. 1989. Marking time and making space: Chronology and topography in the early mouse embryo. *Int. Rev. Cytol.* 117: 99–130.

Presley, R. 1981. Alisphenoid equivalents in placentals, marsupials, monotremes and fossils. *Nature* 294: 668–670.

Psychoyos, A. 1973. Endocrine control of egg implantation. In *Handbook of Physiology*, sect. 7, vol. 2, part 2, R. O. Greep, E. B. Astwood and S. R. Geiger (eds.), Wilkins and Wilkins, Baltimore, pp. 187–215.

Quinlan, G. A. and P. P. L. Tam 1996. Organisation of the body plan: Cell fate and gene activity during gastrulation of the mouse embryo. In *Towards A Molecular Analysis of Mammalian Development*, P. Lonai (ed.). Harwood Academic Press, Chur, pp. 1–27.

Quinlan, G. A., E. A. Williams, S.-S. Tan and P. P. L. Tam. 1995. Neuroectodermal fate of epiblast cells in the distal region of the mouse egg cylinder: Implication for body plan organization during early embryogenesis. *Development* 121: 87–98.

Reeve, W. J. D. 1981. Cytoplasmic polarity develops at compaction in rat and mouse embryos. *J. Embryol. Exp. Morphol.* 62: 351–367.

Renfree, M. B. 1977. Feto-placental influences in marsupial gestation. In *Reproduction and Evolution*, J. H. Calaby and C. H. Tyndale-Biscoe (eds.). Australian Academy of Science, Canberra, pp. 303–311.

Renfree, M. B. 1982. Implantation and placentation. In *Reproduction in Mammals, Book 2. Embryonic and Fetal Development*, 2nd ed. C. R. Austin and R. V. Short (eds.). Cambridge University Press, Cambridge, pp. 26–69.

Rodger, J. C. and J. M. Bedford. 1982. Separation of sperm pairs and sperm-egg interaction in the opossum, *Didelphis virginiana*. *J. Reprod. Fertil.* 64: 171–179.

Rossant, J. 1986. Development of extraembryonic cell lineages in the mouse embryo. In *Experimental Approaches to Mammalian Embryonic Development*, J. Rossant and R. A. Pedersen (eds.). Cambridge University Press, Cambridge, pp. 97–120.

Salmivirta, M., K. Elenius and S. Vainio. 1991. Syndecan from embryonic tooth mesenchyme binds tenascin. *J. Biol. Chem.* 266: 7733–7740.

Saxen, L. 1977. Directive versus permissive induction: A working hypothesis. In *Cell and Tissue Interactions*, J. W. Lash and M. M. Burger (eds.). Raven Press, New York, pp. 1–9.

Saxen, L. 1987. *Organogenesis of the Kidney*. Cambridge University Press, Cambridge.

Saxen, L. and I. Thesleff. 1992. Epithelial-mesenchymal interactions in murine embryogenesis. In *Post-Implantation Development in the Mouse* (Ciba Foundation Symposium 165). John Wiley & Sons, Chichester, pp. 183–198.

Searle, R. F., S. C. Bell and W. D. Billington. 1983. Antigen-bearing decidual cells and macrophages in cultures of mouse decidual tissue. *Placenta* 4: 139–148.

Selenka, E. 1887. Studien uber Entwicklungsgeschichte der Thiere. 4. Das opossum, *Didelphis virginiana*. C. W. Kriedel, Wiesbaden.

Selwood, L. 1980. A timetable of embryonic development of the dasyurid marsupial *Antechinus stuartii* (Macleay). *Aust. J. Zool.* 28: 649–668.

Selwood, L. 1982. A review of maturation and fertization in marsupials with special reference to the dasyurid *Antechinus stuartii*. In *Carnivorous marsupials*, M. Archer (ed.). Royal Zoological Society of New South Wales, Sydney, pp. 65–76.

Selwood, L. 1983. Factors influencing pre-natal fertility in the brown marsupial mouse *Antechinus stuartii*. *J. Reprod. Fertil.* 68: 317–324.

Selwood, L. 1986. Cleavage *in vitro* following destruction of some blastomeres in the marsupial *Antechinus stuartii* (Macleay). *J. Embryol. Exp. Morphol.* 92: 71–84.

Selwood, L. 1989. Development *in vitro* of investment-free marsupial embryos during cleavage and early blastocyst formation. *Gamete Res.* 23: 399–413.

Selwood, L. 1992. Mechanisms underlying the development of pattern in marsupial embryos. In *Current Topics in Developmental Biology*, vol. 27, R. A. Pedersen (ed.). Academic Press, New York, pp. 175–233.

Selwood, L. 1994. Development of early cell lineages in marsupial embryos: An overview. *Reprod. Fertil. Dev.* 6: 507–527.

Selwood, L. and A. H. Sathananthan. 1988. Ultrastructure of early cleavage and yolk extrusion in the marsupial *Antechinus stuartii*. *J. Morphol.* 195: 327–344.

Selwood, L. and D. Smith. 1990. Time-lapse analysis and normal stages of development of cleavage and blastocyst formation in the marsupials the Brown Antechinus and the Stripe-faced Dunnart. *Molec. Reprod. Dev.* 26: 53–62.

Selwood, L. and P. Woolley. 1991. A timetable of embryonic development and ovarian and uterine changes during pregnancy in the stripe-faced dunnart, *Sminthopsis macroura* (Marsupialia: Dasyuridae). *J. Reprod. Fertil.* 91: 213–227.

Semon, R. 1894. Die Embryonalhullen der Monotremen und Marsupialier. In *Zoologische Forschungsreisen in Australien und dem Malayischen Archipel*, Vol. 2, R. Semon (ed.). Gustav Fischer, Jena, pp. 19–58.

Setchell, B. P. 1982. Spermatogenesis and spermatozoa. In *Reproduction in Mammals*, 2nd ed., C. R. Austin and R. V. Short (eds.). Cambridge University Press, Cambridge, pp. 63–101.

Sharman, G. B. 1959. Marsupial reproduction. *Monog. Biol.* 8: 332–368.

Sharman, G. B. 1961. The embryonic membranes and placentation in five genera of diprotodont marsupials. *Proc. Zool. Soc. London* 137: 197–220.

Sharman, B. G. 1965. Marsupials and the evolution of viviparity. *Viewpoints in Biol.* 4: 1–28.

Shur, B. D. and C. A. Neeley. 1988. Plasma membrane association, purification, and partial characterization of mouse sperm β-1,4-galactosyltransferase. *J. Biol. Chem.* 263: 17706–17714.

Sobel, J. S., E. G. Goldstein, J. M. Venuti and M. J. Welsh. 1988. Spectrin and calmodulin in spreading mouse blastomeres. *Dev. Biol.* 126: 47–51.

Steven, D. 1975. Anatomy of the placental barrier. In *Comparative Placentation: Essays in Structure and Function*, D. N. Steven (ed.). Academic Press, New York, pp. 25–36.

Steven, D. and G. Morriss. 1975. Development of fetal membranes. In *Comparative Placentation: Essays in Structure and Function*, D. N. Steven (ed.). Academic Press, New York, pp. 58–67.

Sutherland, A.E. and P. G. Calarco-Gillam. 1983. Analysis of compaction in the preimplantation mouse embryo. *Dev. Biol.* 100: 328–333.

Sutherland, A. E., T. P. Speed and P. G. Calarco. 1990. Inner cell mass allocation in the mouse morula: The role of oriented division during fourth cleavage. *Dev. Biol.* 137: 13–25.

Tachi, S. and C. Tachi. 1986. Macrophages and implantation. *Ann. N.Y. Acad. Sci.* 476: 158–182.

Tachi, S., C. Tachi, A. Knyszynski and H. R. Lindner. 1981. Possible involvement of macrophages in embryo-maternal relationships during ovum implantation in the rat. *J. Exp. Zool.* 217: 81–92.

Tachi, S., C. Tachi and H. R. Lindner. 1970. Ultrastructural features of blastocyst attachment and trophoblastic invasion in rat. *J. Reprod. Fertil.* 21: 37–56.

Tam, P. P. L. and R. S. P. Beddington. 1992. Establishment and organization of germ layers in the gastrulating mouse embryo. In *Post-implantation Development of the Mouse* (Ciba Foundation Symposium 165). John Wiley & Sons, Chichester, pp. 27–49.

Tam, P. P. L. and S.-S. Tan. 1992. The somitogenic potential of cells in the primitive streak and the tail bud of the organogenesis-stage mouse embryo. *Development* 115: 703–715.

Tam, P. P. L. and P. A. Trainor. 1994. Segmentation and specification of the paraxial mesoderm. *Anat. Embryol.* 189: 275–307.

Tam, P. P. L., E. A. Williams and W. Y. Chan. 1993. Gastrulation in the mouse embryo: Ultrastructural and molecular aspects of germ-layer morphogenesis. *Micr. Res. Tech.* 26: 301–328.

Taylor, J. M. and H. A. Padykula. 1978. Marsupial trophoblast and mammalian evolution. *Nature* 271:588.

Temple-Smith, P. 1994. Comparative structure and function of marsupial spermatozoa. *Reprod. Fertil. Dev.* 6: 421–435.

Thesleff, I. and K. Hurmerinta. 1981. Tissue interactions in tooth development. *Differentiation* 18: 75–88.

Thesleff, I., A. Vaahtokari and S. Vainio. 1990. Molecular changes during determination and differentiation of the dental mesenchymal cell lineage. *J. Biol. Buccale* 18: 179–188.

Tung, H. N., Parr, M. B. and E. L. Parr. 1986. The permeability of the primary decidual zone in the rat uterus: An ultrastructural tracer and freeze-fracture study. *Biol. Reprod.* 35: 1045–1058.

Tyndale-Biscoe, H. and M. Renfree. 1987. *Reproductive Physiology of Marsupials*. Cambridge University Press, Cambridge.

Vainio, S., M. Jalkanen and I. Thesleff. 1989. Syndecan and tenascin expression is induced by epithelial-mesenchymal interactions in embryonic tooth mesenchyme. *J. Cell Biol.* 108: 1945–1954.

Vakaet, L. 1984. Early development of birds. In *Chimaeras in Development*, N. Le Douarin and A. McLaren (eds.). Academic Press, London, pp. 71–88.

Walker, M. T. and R. T. Gemmell. 1983. Organogenesis of the pituitary, adrenal, and lung at birth in the wallaby, *Macropus rufogriseus*. *Amer. J. Anat.* 168: 331–344.

Walker, M. T. and R. Rose. 1983. Prenatal development after diapause in the marsupial *Macropus rufogriseus*. *Aust. J. Zool.* 29: 167–187.

Wassarman, P. M. and M. L. DePamphilis (eds.). 1993. *Guide to Techniques in Mouse Development*. Academic Press, San Diego.

Weekes, H. C. 1928. On placentation in reptiles. *Proc. Linn. Soc. N.S.W.* 54:34–60.

Weitlauf, H. M., A. A. Kiessling and R. Buschman. 1979. Comparison of DNA polymerase activity and cell division in normal and delayed implanting mouse embryos. *J. Exp. Zool.* 209: 467–472.

Welsh, A. O. and A. C. Enders. 1987. Trophoblast-decidual cell interactions and establishment of maternal blood circulation in the parietal yolk sac placenta of the rat. *Anat. rec.* 217: 203–219.

Wilson, J. T. and J. P. Hill. 1908. Observations on the development of *Ornithorhynchus*. *Phil. Trans. Roy. Soc. B* 199: 31–168.

Wimsatt, W. A. 1975. Some comparative aspects of implantation. *Biol. Reprod.* 12: 1–40.

Witschi, E. 1956. *Development of Vertebrates*. W. B. Saunders Co., Philadelphia.

Yanagimachi, R. 1988. Mammalian fertilization. In *The Physiology of Reproduction*, E. Knobil and J. Neill (eds.). Raven Press, New York, pp. 135–185.

Yanagimachi, R. and Y. D. Noda. 1970. Electron microscope studies of sperm incorporation into the golden hamster egg. *Amer. J. Anat.* 128: 429–462.

Yousef, A. and L. Selwood. 1993. Embryonic development in culture of the marsupials *Antechinus stuartii* (Macleay) and *Sminthopsis macroura* (Spencer) during preimplantation stages. *Reprod. Fertil. Dev.* 5: 445–458.

Ziomek, C. A. and M. H. Johnson. 1981. Properties of polar and apolar cells from the 16-cell mouse morula. *Roux's Arch. Dev. Biol.* 190: 287–291.

Zuccotti, M., R. Yanagimachi and H. Yanagimachi. 1991. The ability of hamster oolema to fuse with spermatozoa: Its acquisition during oogenesis and loss after fertilization. *Development* 112: 143–152.

SECTION VI

Plants

CHAPTER 23

Plant Life Cycles and Angiosperm Development

Susan R. Singer

THE TERM *PLANT* LOOSELY ENCOMPASSES MANY ORGAN-isms, from algae to angiosperms. In current classification systems, unicellular and multicellular algae are members of the kingdom Protista. This chapter focuses on members of the kingdom Plantae: multicellular, photosynthetic plants that develop from embryos protected by tissues of the parent plant. In this chapter the term *plant* will refer only to members of the Plantae.

Plants have their origins in the green algae, and the transition to land correlates with the evolution of an increasingly protected embryo. Mosses, ferns, gymnosperms (conifers, cycads, and ginkgos), and angiosperms (flowering plants) all develop from protected embryos. Two examples of embryo protection are the seed coat that first appears in the gymnosperms and the fruit that characterizes the angiosperms.

Before we can explore the development of a plant embryo, we must explore the plant life cycle, which alternates between haploid and diploid generations. In alternation of generations, embryo development is seen only in the diploid generation. The embryo, however, is produced by the fusion of gametes that are formed only by the haploid generation. So understanding the relationship between the two generations is important in the study of plant development.

Plant Life Cycles

Unlike animals, plants have multicellular haploid and multicellular diploid stages in their life cycle. Gametes develop from the multicellular haploid **gametophytes** (Greek *phyton*, "plant"). Fertilization gives rise to a multicellular, diploid **sporophyte** that produces haploid spores via meiosis. This type of life cycle is called a **haplodiplontic** life cycle. It differs from our own **diplontic** life cycle,

where only the gametes are in the haploid state (Figure 23.1). Haplodiplontic life cycles are best understood by focusing on where meiosis occurs. Gametes are not the direct result of a meiotic division. Diploid sporophyte cells undergo meiosis to produce haploid **spores.** Each spore goes through mitotic divisions to yield a multicellular, haploid gametophyte. Subsequent mitotic divisions within the gametophyte produce gametes. The diploid sporophyte results from the fusion of two gametes. Among the Plantae, the gametophytes and sporophytes of a species have distinct morphologies (in some algae they look alike). How a single genome can be used to create two unique morphologies is an intriguing puzzle. Aside from the difference in ploidy level between the gametophyte and sporophyte, there may be some interesting parallels to draw with metamorphosis in insects and amphibians.

All plants have alternation of generations. There is an evolutionary trend from a dominant autotrophic (self-feeding) gametophyte and a nutritionally dependent sporophyte to a dependent gametophyte and a dominant autotrophic sporophyte. This is exemplified by exploring the life cycles of a moss, a fern, and an angiosperm (Figures 23.2–23.4). Gymnosperm life cycles bear many similarities to those of angiosperms; the distinctions will be explored in the context of angiosperm development.

The "leafy" moss you walk on in the woods is the gametophyte generation (Figure 23.2). Mosses are **heterosporous**, which means they make two distinct types of spores; these develop into male and female gametophytes. Male gametophytes develop reproductive structures called **antheridia** (singular, antheridium) that produce sperm by mitosis. **Archegonia** (singular, archegonium) develop on female gametophytes and produce eggs by mitosis. Sperm travel to a neighboring plant via a water droplet, are chemically attracted to the entry of the

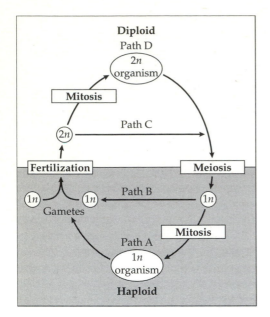

Figure 23.1   Plants have haplodiplontic life cycles that involve mitotic divisions (resulting in multicellularity) in both the haploid and diploid generations (paths A and D). Most animals are diplontic and undergo mitosis only in the diploid generation (paths B and D). Multicellular organisms with haplontic life cycles, such as some species of algae, follow paths A and C.

archegonium, and fertilization results.* The embryonic sporophyte develops within the archegonium and the mature sporophyte stays attached to the gametophyte. The sporophyte is not photosynthetic. Thus both the embryo and the mature sporophyte are nourished by the gametophyte. Meiosis within the capsule of the sporophyte yields haploid spores that are released and eventually germinate to form a male or female gametophyte.

* Have you ever wondered why there are no moss trees? Aside from the fact that the gametophytes of mosses (and other plants) do not have the necessary structural support to attain tree height, it would be very difficult for a sperm to swim up a tree!

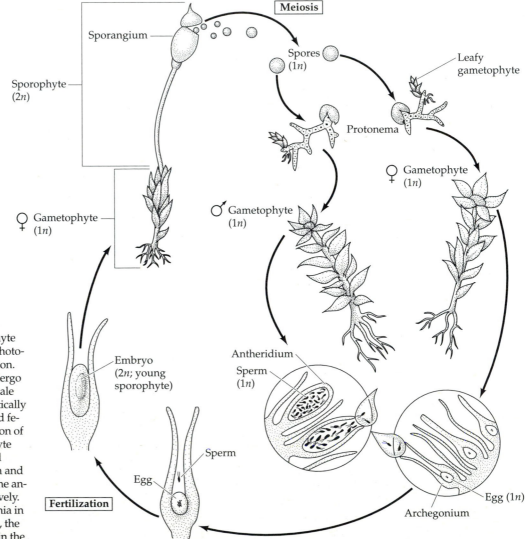

Figure 23.2   Life cycle of a moss (genus *Polytrichum*). The sporophyte generation is dependent on the photosynthetic gametophyte for nutrition. Cells within the sporangium undergo meiosis to produce male and female spores. These spores divide mitotically to produce multicellular male and female gametophytes. Differentiation of the growing tip of the gametophyte produces antheridia in males and archegonia in females. The sperm and egg are produced mitotically in the antheridia and archegonia, respectively. Sperm are carried to the archegonia in water droplets. After fertilization, the sporophyte generation develops in the archegonium and remains attached.

Ferns follow a similar pattern of development to that of mosses, although most (but not all) ferns are **homosporous**. That is, the sporophyte produces only one type of spore within a structure called the **sporangium** (Figure 23.3). Each spore undergoes mitosis, forming a gametophyte that can produce both male and female sex organs. The biggest contrast between the mosses and the ferns is that both the gametophyte and the sporophyte of the fern photosynthesize and are thus autotrophic; the shift to a dominant sporophyte generation is taking place.*

At first glance, angiosperms may appear to have a diplontic life cycle because the gametophyte generation of

the angiosperm is reduced to just a few cells (Figure 23.4). However, mitotic division still follows meiosis, and a multicellular gametophyte produces egg or sperm. Male and female gametophytes have distinct morphologies (i.e., angiosperms are heterosporous), but the gametes they produce no longer rely on water for fertilization. Rather, wind or members of the animal kingdom deliver the male game-

---

* It *is* possible to have tree ferns, for two reasons. First, the gametophyte develops on the ground, where water can facilitate fertilization. Secondly, unlike mosses, the fern sporophyte has vascular tissue, which provides the support and transport system necessary to achieve substantial height.

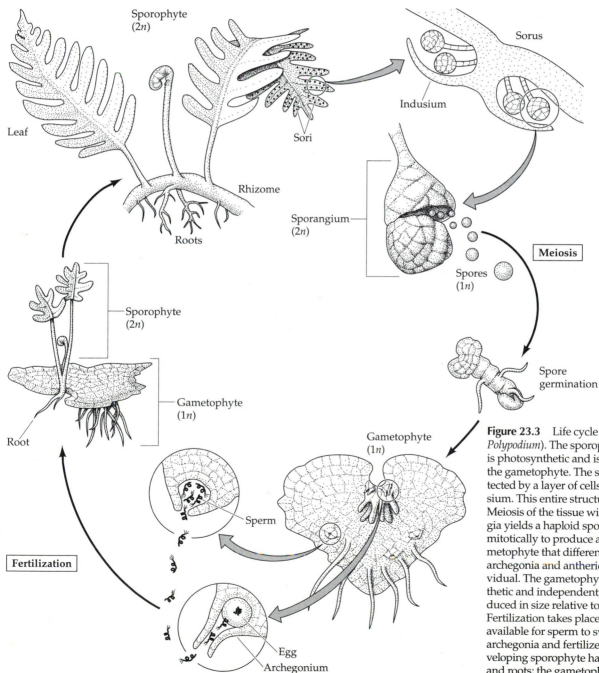

Figure 23.3 Life cycle of a fern (genus *Polypodium*). The sporophyte generation is photosynthetic and is independent of the gametophyte. The sporangia are protected by a layer of cells called the indusium. This entire structure is a sorus. Meiosis of the tissue within the sporangia yields a haploid spore that divides mitotically to produce a heart-shaped gametophyte that differentiates both archegonia and antheridia on one individual. The gametophyte is photosynthetic and independent, although it is reduced in size relative to the sporophyte. Fertilization takes place when water is available for sperm to swim to the archegonia and fertilize the eggs. The developing sporophyte has vascular tissue and roots; the gametophyte does not.

Figure 23.4 Life cycle of an angiosperm, represented here by a pea plant (genus *Pisum*). The sporophyte is the dominant generation, but multicellular male and female gametophytes are produced within the flower of the sporophyte. Cells of the microsporangium within the anther undergo meiosis to produce microspores. Subsequent mitotic divisions are limited, but the end result is multicellular pollen. The megasporangium is protected by two layers of integuments and the ovary wall. Within the megasporangium, meiosis yields four megaspores—three small and one large. Only the large megaspore survives to produce the embryo sac. Fertilization occurs when the pollen germinates and the pollen tube grows toward the embryo sac. The sporophyte generation may be maintained in a dormant state, protected by the seed coat.

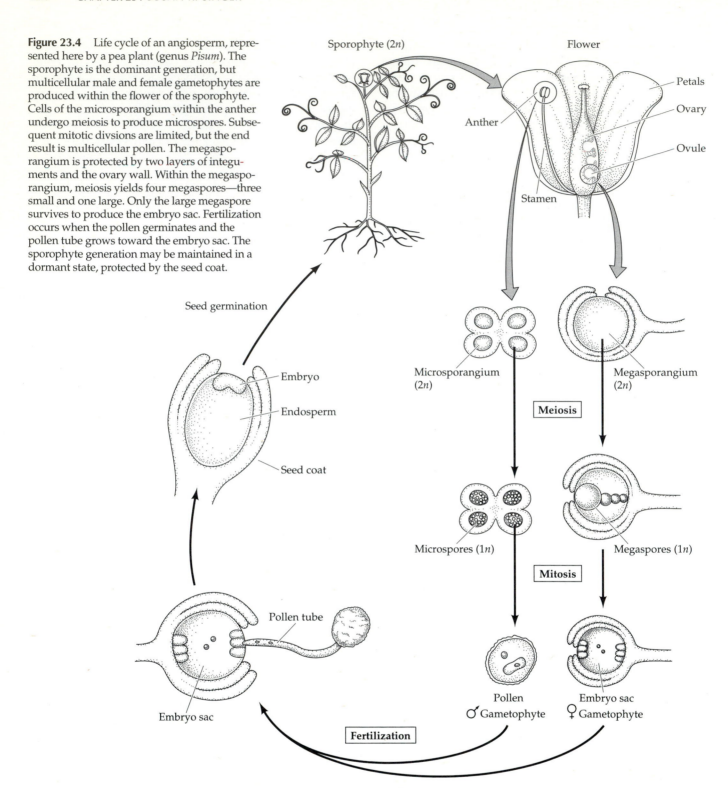

tophyte—**pollen**—to the female parent. Pollen is not merely a cell, but a multicellular complex. Another innovation is the production of a seed coat, which adds an extra layer of protection around the embryo. The seed coat is also found in the gymnosperms. A further protective layer, the fruit, is unique to the angiosperms and aids in dispersal of the enclosed embryos by wind or animals.

What's Unique about Plant Development?

In addition to the haplodiplontic life cycle of plants, plant development differs from animal development in several significant ways.

1. *Plants do not gastrulate.* Plant cells are trapped within rigid cellulose walls that generally prevent cell and tissue migration. Plants develop three basic tissue systems (dermal, ground, and vascular), but do not rely on gastrulation to establish this layered system of tissues. Plant cells divide and differentiate in place.

2. *Plants have sporic meiosis rather than gametic meiosis.* That is, spores, not gametes, are produced by meiosis. Gametes are produced by mitotic divisions following meiosis.

3. *Germ cells are not set aside early in development.* This is also the case in several animal phyla, but it is always true for plants.

4. *Plants have extended embryogenesis.* Clusters of embryonic cells called **meristems** persist long after embryogenesis. Meristems allow for reiterative development and the formation of new structures throughout the life of the plant.

5. *Plants have tremendous plasticity in development.* Individual plant cells are highly plastic. For example, if a shoot is grazed by herbivores, meristems in the leaf axils often grow out to replace the lost part. (This has similarities to regeneration in some animals.) Whole plants can even be regenerated from some single cells. In addition, a plant's form (including branching, height, and relative amounts of vegetative and reproductive structures) is greatly influenced by environmental factors such as light and temperature, and a wide range of morphologies can result from the same genotype. This amazing level of plasticity may help compensate for the plant's lack of mobility.

6. *Plants can tolerate generally higher genetic loads than animals.* For example, half of the maize (corn) genome appears to contain foreign DNA (SanMiguel et al.

1996). Most of it is in the form of retroelements that resemble retroviruses. The maize plant appears to function quite well with all of this "hitchhiking" DNA. Not only can plants compensate for mutations better than animals, but there is evidence that the consequences of both aneuploidy and polyploidy can be adaptive. Many flowers found in the florist shop and the wheat used for bread flour are examples of successful polyploids.

7. *Maternal RNA plays only a minor role in plant embryo development.* This is in contrast to many animal phyla, where maternal RNA stored in the oocyte and used to help pattern the embryo plays a key role in early embryogenesis.

The remainder of this chapter provides a detailed exploration of fertilization and embryo development in angiosperms. Keep in mind that the basic haplodiplontic life cycle seen in the mosses and ferns is also found in the angiosperms, continuing the trend toward increased nourishment and protection of the embryo.

Gamete Production

As with mosses and ferns, angiosperm gametes are produced by the gametophyte generation. These gametes join to form the sporophyte. Embryo development in plants is the study of sporophyte development. In angiosperms, the sporophyte is what is commonly seen as the plant body. The sporophyte produces a series of vegetative structures as a result of meristematic activity. At a certain point in development, internal and external signals trigger a shift to reproductive development (see reviews by McDaniel et al. 1992 and Coupland 1995). Ultimately a meristem initiates floral parts sequentially in whorls of organs modified from leaves (Figure 23.5). The first and second whorls initiate **sepals** and **petals**, respectively; these organs are sterile. The

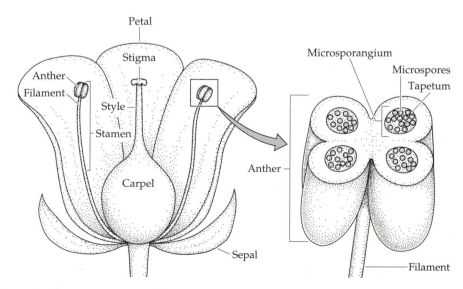

Figure 23.5 Generalized diagram of an angiosperm flower. Tremendous morphological variation is possible in all four organs and appears to be related to reproductive strategies. A detail of the anther shows the four microsporangia. The outer cell layer is the tapetum. Microspores are produced meiotically and will develop into the male gametophyte, pollen.

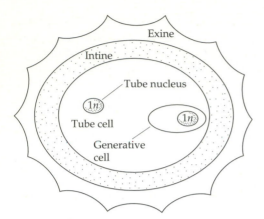

Figure 23.6   Pollen consists of a cell within a cell. The generative cell will undergo division to produce two sperm cells. One will fertilize the egg and the other will join with the polar nuclei, yielding the endosperm.

pollen-producing **stamens** are initiated in the third whorl of the flower and contain four groups of cells, called the **microsporangia** (pollen sacs), within the **anther**. Microsporangia undergo meiosis to produce **microspores**. Unlike ferns, angiosperms are heterosporous, so the prefix *micro* is used to identify the spores that mitotically yield pollen, the male gametophytes. The inner wall of the pollen sac, the **tapetum**, provides nourishment for the developing pollen.

Pollen

Pollen is an extremely simple multicellular structure. The outer wall of the pollen grain, the **exine**, is composed of resistant material provided by both the tapetum (sporophyte generation) and the microspore (gametophyte generation). The inner wall, the **intine**, is produced by the microspore. Mature pollen consists of two cells within a cell (Figure 23.6). The **tube cell** encloses a **generative cell** within it. The generative cell divides to produce two sperm. The tube cell nucleus guides pollen germination and the growth of the pollen tube after the pollen lands on the stigma of the female parent. One of the sperm will fuse with the egg cell to produce the next sporophyte generation. The second sperm will participate in the double fertilization event that provides nourishment for the embryo.

The Ovary

The fourth whorl of organs within the flower form the **carpel**, which gives rise to the female gametophyte (Figure 23.7). The carpel consists of the **stigma** (where the pollen lands), the **style**, and the **ovary**. Following fertilization the ovary wall develops into the **fruit**. This unique angiosperm structure provides further protection for the developing embryo and also enhances seed dispersal by frugivores (fruit-eating animals). Within the ovary are one or more **ovules** attached by a **placenta** to the ovary wall. Fully developed ovules are called **seeds**. The ovule has one or two outer layers of cells called the **integuments**. These enclose the

megasporangium, which contains sporophyte cells that will undergo meiosis to produce **megaspores** (which mitotically produce the female gametophyte). There is a small opening in the integuments, called the **micropyle**, for the pollen tube to grow through. The integuments—an innovation first appearing in the gymnosperms—develop into the seed coat that protects the embryo by providing a waterproof physical barrier. When the mature embryo disperses from the parent plant, diploid sporophyte tissue accompanies the embryo in the form of the seed coat and the fruit.

Within the ovule, meiosis and unequal cytokinesis yield megaspores. The largest of the four megaspores undergoes three mitotic divisions to produce the seven-celled **embyro sac** with eight nuclei (Figure 23.8). The two **synergid cells** surrounding the egg cell may be evolutionary remnants of the archegonium (the female sex organ seen in mosses and ferns). The central cell contains two or more polar nuclei that fuse with the second sperm nucleus and go on to develop the polyploid **endosperm**. Three **antipodal cells** form at the opposite end of the embryo sac from the synergids and degenerate before or during embryo development. There is no known function for the antipodals.

Pollination

Pollination precedes fertilization and refers to the landing and subsequent germination of the pollen on the stigma. Hence it involves an interaction between the gametophytic generation of the male (pollen) and the sporophytic generation of the female (stigmatic surface of the carpel). Pollination can occur within a single flower (self-fertilization), or pollen can land on a different flower on the same

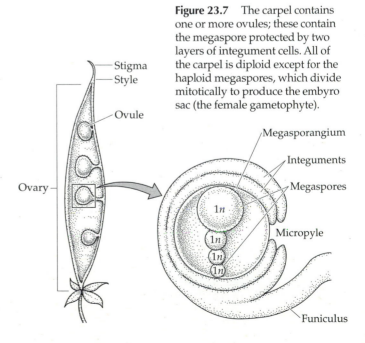

Figure 23.7   The carpel contains one or more ovules; these contain the megaspore protected by two layers of integument cells. All of the carpel is diploid except for the haploid megaspores, which divide mitotically to produce the embyro sac (the female gametophyte).

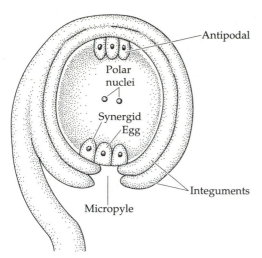

Figure 23.8   The embryo sac is the product of three mitotic divisions of the haploid megaspore. Two of the nuclei are contained within the central cell; the other six cells contain one haploid nucleus each.

or a different plant. In some species, two types of flowers are produced. **Staminate** flowers lack carpels, while **carpellate** flowers lack stamens. Maize plants, for example, have staminate (tassle) and carpellate (ear) flowers on the same plant and are considered to be **monoecious** (Greek *mono*, "one"; *oecos*, "house"). Staminate and carpellate flowers are found on separate willow plants, which are thus considered to be **dioecious** ("two houses"). Only a few plant species have true sex chromosomes. The terms *male* and *female* are most correctly applied to the the gametophyte generations of heterosporous plants, but not to the sporophyte (Cruden and Lloyd 1995).

The arrival of a viable pollen grain on a receptive stigma does not guarantee fertilization. **Interspecific incompatiblity** refers to the failure of pollen from one species to germinate and/or grow on the stigma of another species (for a review, see Taylor 1996). **Intraspecific incompatibility** occurs within a species. **Self-incompatibility** is an example of intraspecific incompatibility; it blocks fertilization between two genetically similar gametes, increasing the probability of new gene combinations by promoting outcrossing (pollination by a different individual of the same species). Not all angiosperms are self-incompatible.

Recognition of self depends on the multiallelic self-incompatibility (*S*) gene (Nasrallah et al. 1994; Dodds et al. 1996). Gametophytic self-incompatibility occurs when the *S* allele of the pollen grain matches either of the *S* alleles of the stigma (remember that the stigma is part of the diploid

sporophyte generation, which has two *S* alleles). In this case, the pollen tube begins developing but stops before reaching the micropyle that leads into the female gametophyte. Sporophytic self-incompatibility occurs when one of the two *S* alleles of the pollen-producing sporophyte (not the gametophyte) matches one of the *S* alleles of the stigma. Most likely, sporophyte contributions to the exine are detected. Figure 23.9 provides examples self-incompatibility interactions.

If the pollen and the stigma are compatible, the pollen tube grows down the style of the carpel towards the micropyle (Figure 23.10). Cell adhesion molecules help move the pollen tube through the style (Lord et al. 1996). The tube nucleus and the sperm cells are kept at the growing tip by bands of callose (a complex carbohydrate). Collectively they are called the male germ unit (MGU). Calcium has long been known to play an essential role in pollen tube growth (Brewbaker and Kwack 1963). Calcium accumulates in the tips of pollen tubes, where open calcium channels are concentrated (Jaffe et al. 1975; Malho et al. 1994). There is direct evidence that pollen tube growth in field poppy is regulated by a slow-moving calcium wave controlled by the phosphoinositide signaling pathway (Franklin-Tong et al. 1996).

(*a*) Gametophytic incompatibility

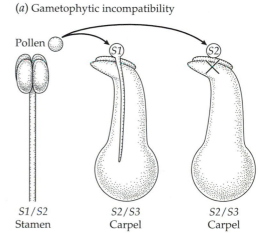

(*b*) Sporophytic incompatibility

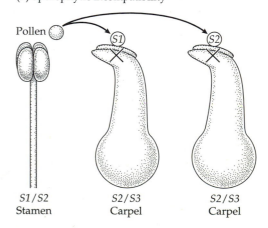

Figure 23.9   Self-incompatibility. *S1*, *S2*, and *S3* are different alleles of the self-incompatibility locus. Plants with gametophytic self-incompatibility reject pollen only when the genotype of the pollen matches one of the carpel's alleles. In sporophytic self-incompatibility, the genotype of the pollen parent, not just the haploid pollen grain itself, can trigger an incompatibility response.

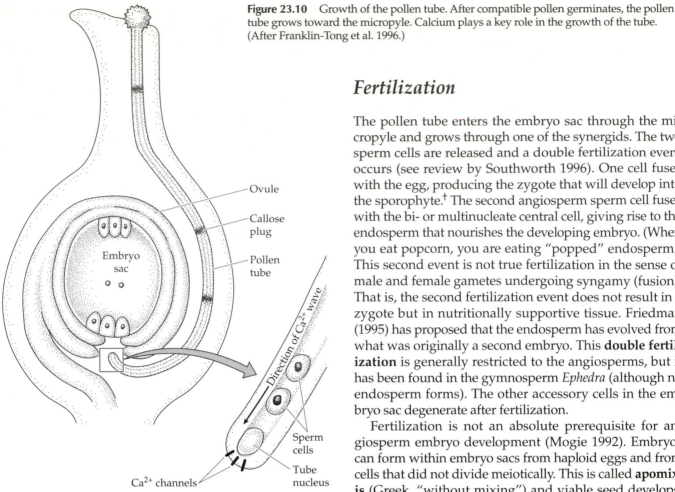

Figure 23.10 Growth of the pollen tube. After compatible pollen germinates, the pollen tube grows toward the micropyle. Calcium plays a key role in the growth of the tube. (After Franklin-Tong et al. 1996.)

Genetic approaches have been useful in investigating how pollen tube growth is guided toward unfertilized ovules. The pollen tube in *Arabidopsis** appears to be guided toward the ovule by a long-distance signal from the ovule (Hulskamp et al. 1995). Analysis of pollen tube growth in ovule mutants of *Arabidopsis* indicates that the haploid embryo sac is particularly important in the long-range guidance of pollen tube growth. Mutants with defective sporophyte tissue in the ovule but a normal haploid embryo sac appear to have normal pollen tube development.

While the evidence points primarily to the role of the gametophyte generation in pollen tube guidance, diploid cells may make some contribution. Two *Arapidopsis* genes, *POP2* and *POP3*, have been identified that specifically guide pollen tubes to the ovule with no other apparent affect on the mutant plants (Wilhelmi and Preuss 1996, 1997). These genes function in both the pollen and the pistil, thus implicating the sporophyte generation in the guidance system. Thus pollen tube growth is a unique example of cell migration in plants that mostly likely relies on the interplay of male and female signals.

---

*A small weed in the mustard family, *Arabidopsis* is used as a model system because of its very small genome.

Fertilization

The pollen tube enters the embryo sac through the micropyle and grows through one of the synergids. The two sperm cells are released and a double fertilization event occurs (see review by Southworth 1996). One cell fuses with the egg, producing the zygote that will develop into the sporophyte.[†] The second angiosperm sperm cell fuses with the bi- or multinucleate central cell, giving rise to the endosperm that nourishes the developing embryo. (When you eat popcorn, you are eating "popped" endosperm.) This second event is not true fertilization in the sense of male and female gametes undergoing syngamy (fusion). That is, the second fertilization event does not result in a zygote but in nutritionally supportive tissue. Friedman (1995) has proposed that the endosperm has evolved from what was originally a second embryo. This **double fertilization** is generally restricted to the angiosperms, but it has been found in the gymnosperm *Ephedra* (although no endosperm forms). The other accessory cells in the embryo sac degenerate after fertilization.

Fertilization is not an absolute prerequisite for angiosperm embryo development (Mogie 1992). Embryos can form within embryo sacs from haploid eggs and from cells that did not divide meiotically. This is called **apomixis** (Greek, "without mixing") and viable seed develops. The viability of haploid sporophytes indicates that ploidy alone does not account for the morphological distinctions between the gametophyte and the sporophyte. Embryos can also develop from cultured sporophytic tissue. These embryos develop with no associated endosperm, and they lack a seed coat.

Embryo Development

Experimental Studies

The angiosperm zygote is embedded within the ovule and ovary and thus is not readily accessible for experimental manipulation. The following approaches, however, have yielded information on the formation of a plant embryo.

- *Histological studies* of embryos at different stages provide useful descriptive information on how carefully regulated cell division results in the construction of an organism, even without the ability to move cells and tissues to shape the embryo (Figure 23.11).

[†] The zygote of the angiosperm produces only a single embryo; the zygote of the gymnosperm, on the other hand, produces two or more embryos after cell division begins, by a process known as cleavage embryony or polyembryony. Usually only one embryo survives.

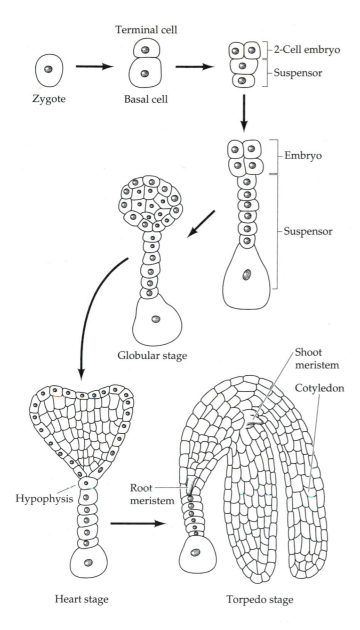

Figure 23.11   Embryogenesis. A representative dicot is shown; a monocot would develop only a single cotyledon. While there are basic patterns of embryogenesis in angiosperms, there is tremendous morphological variation among species.

- *Clonal analysis* involves genetically marking individual cells and following their fate in development (see Poethig 1987 for details on the methodology). For example, seeds heterozygous for a pigmentation gene may be irradiated so that a cell loses the ability to produce pigment. Its descendents will form a colorless sector that can be identified and related to the overall body pattern.

Embryogenesis

Embryogenesis covers development from the time of fertilization until dormancy occurs. The basic body plan of the sporophyte is established during embryogenesis; however, this plan is reiterated and elaborated after dormancy is broken. The major challenges of embryogenesis are:

1. To establish the basic body plan. **Radial patterning** produces three tissue systems, and the **apical-basal** (shoot-root) **axis** is formed.

2. To set aside meristematic tissue for postembryonic elaboration of structure (leaves, roots, flowers, etc.).

3. To establish an accessible food reserve for the germinating embryo until it becomes autotrophic.

Embryogenesis is similar in all angiosperms in terms of the establishment of the basic body plan (Steeves and Sussex 1989). There are differences in pattern elaboration, including differences in the precision of cell division patterns, the extent of endosperm development, cotyledon development, and the extent of shoot meristem development (Esau 1977; Johri et al. 1992).

The basic body plan of the plant laid down during embryogenesis begins with an asymmetrical cell division giving rise to the **terminal cell** and the **basal cell.** The terminal cell gives rise to the **embryo proper.** The basal cell forms closest to the micropyle. It gives rise to the **suspensor.** The distal end of the suspensor is called the **hypophysis** and in many species it gives rise to the root. (The other suspensor cells divide to form a filamentous or spherical organ that degenerates later in embryogenesis.) In both gymnosperms and angiosperms, the suspensor orients the absorptive surface of the embryo toward its food source; in angiosperms it also appears to serve as a nutrient conduit for the developing embryo. Culturing isolated embryos from scarlet runner beans with and without the suspensor has demonstrated the need for a suspensor through the heart-shape stage in dicots (Figure 12.12; Yeung and Sussex 1979). Embryos cultured with a suspensor are twice as likely to survive as embryos cultured without an attached suspensor at the heart-shape

- *Culture experiments* of embryos isolated from ovules and embryos developing *de novo* from cultured sporophytic tissue provide information on the interaction between the embryo and surrounding sporophytic and endosperm tissue.

- *In vitro fertilization experiments* provide information on gamete interactions.

- *Biochemical analysis* of embryos at different stages of development provides information on such things as the stage-specific products necessary for patterning and establishing food reserves.

- *Genetic and molecular analysis of developmental mutants* characterized using the above approaches have greatly enhanced our understanding of embryo development.

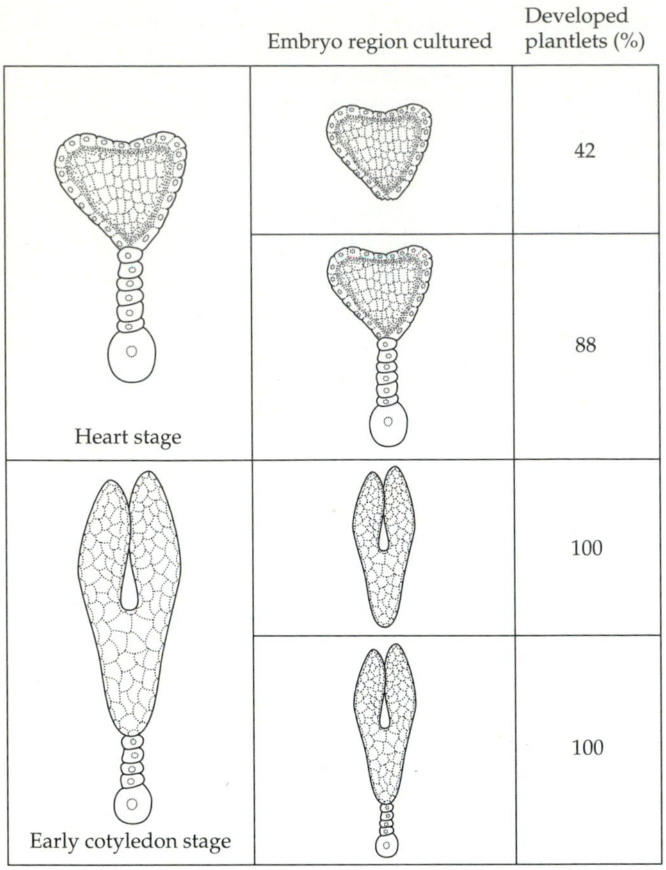

Embryo region cultured		Developed plantlets (%)
Heart stage		42
		88
Early cotyledon stage		100
		100

Figure 23.12 Role of the suspensor in embryogenesis. Culturing embryos with and without their suspensor has demonstrated that the suspensor is essential at the heart stage, but not later. (Data from Yeung and Sussex 1979.)

Axial and radial patterns develop as cell division and differentiation continue (Figure 23.13; also see Bowman 1994 for detailed light micrographs of *Arabidopsis* embryogenesis). The cells of the embryo proper divide in transverse and longitudinal planes to form a globular-stage embryo with tiers of cells. Superficially this stage bears some resemblance to cleavage, but the nuclear/cytoplasmic ratio does not necessarily increase. The emerging shape of the embryo depends on regulation of the patterns of cell division, since the cells are not able to move and reshape the embryo. Cell division planes in the outer layer become restricted and this layer, called the **protoderm**, becomes distinct. Radial patterning emerges in the globular stage as the three tissue systems (dermal, vascular, and ground) of the plant are initiated. The **dermal system** will form from the protoderm and contribute to the outer protective layers of the plant. The **vascular system** functions in support and transport and arises from the procambium cells that differentiate in the center

stage. The suspensor may be a source of hormones. In scarlet runner beans, younger embryos without a suspensor can survive in culture if they are supplemented with gibberellic acid (Cionini et al. 1976).

As establishment of polarity is one of the key achievements of embryogenesis, it is useful to consider why the suspensor and embryo proper develop unique morphologies. Here the study of embryo mutants in maize (corn) and *Arabidopsis* has been particularly helpful. Investigations of suspensor mutants (*sus1*, *sus2*, and *raspberry1*) of *Arabidopsis* have demonstrated that a suspensor can develop embryo-like structures (Schwartz et al. 1994; Yadegari et al. 1994). In these mutants, abnormalities in the embryo proper appear prior to suspensor abnormalities.* Earlier work ablating the embryo proper also demonstrated that suspensors could develop like embryos (Haccius 1963). A signal from the embryo proper to the suspensor may be important in maintaining suspensor identity and blocking the development of the suspensor as an embryo. Molecular analysis of these and other embryonically important genes is providing insight into the mechanisms of communication between the suspensor and embryo proper in the developing embryo.

---

*Another intriguing characteristic of these mutants is that cell differentiation occurs in the absence of morphogenesis. Thus, cell differentiation and morphogenesis can be uncoupled in plant development.

Figure 23.13 Radial and axial patterning. (*a*) Radial patterning in angiosperms begins in the globular stage and results in the establishment of three tissue systems. (*b*) The axial pattern (root-shoot axis) is established by the heart-shape stage.

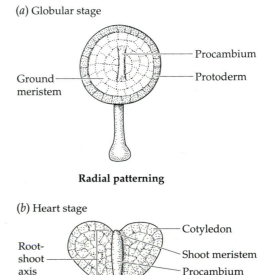

(*a*) Globular stage

Ground meristem — Procambium — Protoderm

Radial patterning

(*b*) Heart stage

Root-shoot axis — Cotyledon — Shoot meristem — Procambium — Root meristem

Axial patterning

of the globular embryo. **Ground tissue** forms from the ground meristem and surrounds the developing vascular tissue; however, the ground and vascular systems form independently. For example, in the *keule* mutant of *Arabidopsis*, the dermal system is defective while the inner tissue systems develop normally (Mayer et al. 1991).

The globular shape of the embryo is lost as **cotyledons** ("first leaves") begin to form. Dicots have two cotyledons, which gives the embryo a heart-shaped appearance as they form. The axial body plan is evident by this heart-shape stage of development. Hormones (specifically, auxins) may mediate the transition from radial to bilateral symmetry (Liu et al. 1993). In monocots such as maize, only a single cotyledon emerges.

Cotyledons may aid in nourishing the plant by becoming photosynthetic after germination (although some never emerge from the ground). In some cases, peas for example, the food reserve in the endosperm is used up before germination and the cotyledons are the nutrient source for the germinating seedling. Mendel's famous wrinkled-seed mutant (the *rugosus* or *r* allele) has a defect in a starch branching enzyme that affects starch, lipid, and protein biosynthesis in the seed and leads to defective cotyledons (Bhattacharyya et al. 1990) Even in the presence of a persistent endosperm (as in maize), the cotyledons store food reserves such as starch, lipids, and proteins. In many monocots the cotyledon becomes a large organ pressed against the endosperm and aids in nutrient transfer to the seedling. Upright cotyledons can give the embryo a torpedo shape, and by this point the suspensor is degenerating. In some plants the cotyledons grow sufficiently long that they must bend to fit within the confines of the ovule. The embryo then looks like a walking stick.

The **shoot apical meristem** and **root apical meristem** are clusters of embryonic cells that persist in postembryonic development and give rise to most of the sporophyte body. The root meristem is partially derived from the hypophysis (the uppermost cell of the suspensor) in some species. All other parts of the sporophyte body are derived from the embryo proper. Genetic evidence indicates that the formation of the shoot and root meristem are regulated independently. This is seen in the *dek23* maize mutant and the *shootmeristemless* (*STM*) mutant in *Arabidopsis*, both of which form a root meristem but fail to initiate a shoot meristem (Clark and Sheridan 1986; Barton and Poethig 1993). The *STM* gene, which has been cloned, is expressed in the late globular stage, before cotyledons form. *STM*'s role appears to be to repress cell differentiation in the shoot apical meristem so that the cells maintain their indeterminate state (Long et al. 1996). Mutants have also been identified that specifically affect development of the root axis in embryogeny (Aeschbacher et al. 1994).

The shoot apical meristem will initiate leaves and ultimately the transition to reproductive development after germination. In *Arabidopsis* the cotyledons are produced from general embryonic tissue, not from the shoot meristem (Barton and Poethig 1993). In many angiosperms, a few leaves are initiated during embryogenesis. In the case of *Arabidopsis*, clonal analysis points to the presence of leaves in the mature embryo even though they are not morphologically well developed (Irish and Sussex 1992). Clonal analysis demonstrated that the cotyledons and first two true leaves of cotton are derived from embryonic tissue rather than an organized meristem (Christianson 1986). Clonal sectors encompassing the first three leaves of cotton do not extend into more apical leaves.

Clonal analysis experiments provide information on cell fate, but do not necessarily indicate whether or not cells are determined for a particular developmental fate. Cells, tissues, and organs are developmentally determined when they have the same fate *in situ*, in isolation, and in or at a new place on the organism (see McDaniel et al. 1992 for more information on developmental states). Clonal analysis has demonstrated that cells that move to a different tissue layer often differentiate according to their new position. Position rather than clonal origin appears to be the critical factor in embryo pattern formation, implicating some type of cell-cell communication (Laux and Jurgens 1994). Microsurgery experiments on somatic carrot embryos demonstrate that isolated pieces of embryo can often replace the missing complement of parts (Schiavone and Racusen 1990). In the shoot apex, a missing cotyledon will be replaced. Isolated embryonic shoots can regenerate a new root; however, isolated root tissue regenerates cotyledons but is less likely to regenerate the shoot axis. Although most embryonic cells are pluripotent and can generate organs such as cotyledons and leaves, only meristems retain this capacity in the postembryonic plant body.

Dormancy

From the earliest stages of embryogenesis there is a high level of zygotic gene expression. As the embryo reaches a maturation phase there is a shift from constructing the basic body plan to creating a food reserve by accumulating storage carbohydrates, proteins, and lipids. Genes coding for seed storage proteins were among the first to be characterized by plant molecular biologists because of the high levels of specific storage protein mRNAs present at different times in embryo development. The high level of metabolic activity in the developing embryo is fueled by continuous input from the parent plant into the ovule. Eventually metabolism slows and the connection of the ovule (seed) to the ovary is severed by the degeneration of the adjacent supporting sporophyte cells. The seed loses water (**dessication**) and the integuments harden to form a tough seed coat. **Dormancy** has occurred, officially ending embryogenesis. The embryo can persist in a dormant state for weeks or years, a fact that affords tremendous survival value. There have even been examples of seeds germinating after thousands of years of dormancy!

Maturation leading to dormancy is the result of a precisely regulated program. The *viviparous* mutation in maize, for example, produces genetic lesions that block dormancy (Steeves and Sussex 1989). The apical meristems of *viviparous* mutants behave like those of ferns, with no pause before producing postembryonic structures. The embryo continues to develop and seedlings emerge from the kernels on the ear attached to the parent plant.

The hormone **abscisic acid** is important in maintaining dormancy in many species. **Gibberellins**, another class of hormones, are important in breaking dormancy.

Germination

The postembryonic phase of plant development begins with **germination**. Some dormant seeds require a period of **after-ripening** during which low level metabolic activities continue to prepare the embryo for germination. Highly evolved interactions between the seed and the environment increase the odds that the germinating seedling will survive to produce another generation. Temperature, water, light, and oxygen are all key in determining the success of germination. **Stratification** is the requirement for chilling (5°C) to break dormancy in some seeds. In a temperate climate, this adaptation prevents germination during the winter months. In addition, seeds have maximum germination rates at moderate temperatures of 25–30°C and often will not germinate at extreme temperatures.

Seeds such as lettuce require light (specifically, the red wavelengths) for germination; thus seeds will not germinate so far below ground that they use up food reserves before photosynthesis is possible. Dessicated seeds may be only 5–20% water. **Imbibition** is the process by which the seed rehydrates by soaking up large volumes of water, swelling to many times its original size. The **radicle** (primary embryonic root) emerges first to enhance water uptake. Water is essential for metabolic activity, but so is oxygen. A seed sitting in a glass of water will not survive. Some species have such hard protective seed coats that they must be **scarified** (scratched or etched) before water and oxygen can cross the barrier. Scarification can occur by the seed being exposed to the weather and other natural elements over time, or by its exposure to acid as the seed passes through the gut of a frugivore. (The frugivores then aid in seed dispersal.)

Most of the sporophyte body plan remains to be elaborated, and the delicate shoot tip must be protected as the shoot emerges through the soil. Three strategies for protecting the shoot tip have evolved (Figure 23.14):

1. Cotyledons protect the shoot tip, often assisted by the **hypocotyl** (the stem below the cotyledons) bending as the shoot emerges through the soil.

2. The **epicotyl** (the stem above the cotyledons) bends so that stem tissue rather than the shoot tip pushes through the soil.

3. In monocots, a special leaflike structure, the **coleoptile**, forms a protective sheath around the shoot tip.

Vegetative Growth

Meristems

As has been mentioned, meristems are clusters of cells that allow the basic body pattern established during embryogeny to be reiterated and extended after germination. Meristems are similar to stem cells in animals. Meristematic cells divide to give rise to one daughter cell that continues to be meristematic and another that differentiates. Meristems fall into three categories: apical, lateral, and intercalary. Figure 23.15 shows the basic parts of the mature sporophyte plant.

Figure 23.14 Germination. Meristems have delicate cells that are susceptible to damage during germination. A variety of strategies have evolved for protecting the shoot meristem during germination. (*a*) Some cotyledons protect the meristem as the shoot emerges from the soil. (*b*) Bent epicotyls (and sometimes hypocotyls; see *a*) push through the soil. (*c*) Monocots utilize the coleoptile, a leaflike structure that sheaths the young shoot tip.

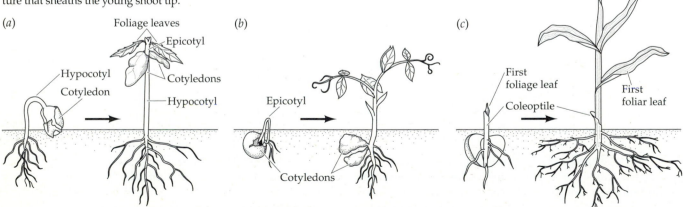

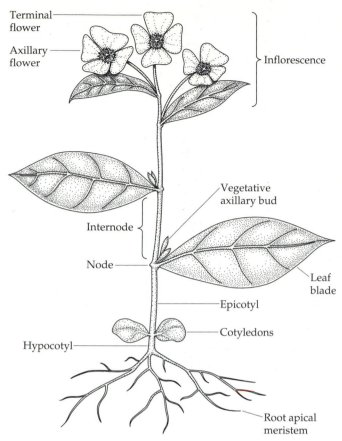

Figure 23.15 Morphology of a generalized angiosperm sporophyte.

Apical meristems occur at the growing shoot and root tips (Figure 23.16). Root apical meristems produce the root cap which consists of lubricated cells that are sloughed off to protect the meristem as it is pushed through the soil by cell division and elongation in more distal cells. It also gives rise to daughter cells that produce the three tissue systems of the root. New root apical meristems are initiated from tissue within the core of the root and emerge through the ground tissue and dermal tissue. Root meristems can also be derived secondarily from the stem; in the case of maize, this is the major source of root mass. The shoot apical meristem produces stems, leaves, and reproductive structures. In addition to the shoot meristem initiated during embryogenesis, axillary shoot apical meristems (axillary buds; see Figure 23.15) derived from the original shoot meristem form in the axils of the leaves. Unlike the new root meristems, these arise from the surface layers of the meristem.

Angiosperm apical meristems are composed of up to three layers of cells (L1, L2, and L3) on the surface (Figure 23.17). One way the contributions of different layers to plant structure are investigated is by constructing chimeras (The fire-breathing chimera from Greek mythology that was a combination of lion, goat, and serpent; botanical chimeras mix tissues from different genotypes together). Plant periclinal chimeras are composed of layers with distinct genotypes with discernible markers. When the L2 has a different genotype than L1 or L3, all pollen will have the L2 genotype indicating that pollen is

Figure 23.16 Shoot and root apical meristems. Both shoots and roots develop from undifferentiated cells clustered in the meristem. In roots, a root cap is also produced, and this protects the meristem as it grows through the soil. Lateral organs (leaves and axillary branches) have a superficial origin in shoot meristems. Root lateral meristems are derived from pericycle cells deep within the root.

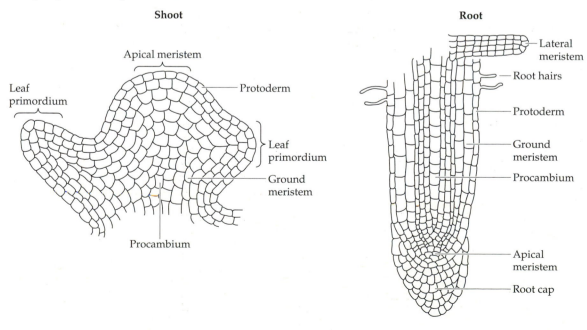

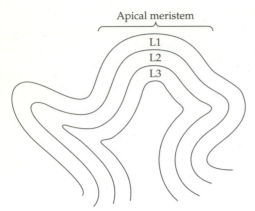

Apical meristem

L1

L2

L3

Figure 23.17   Organization of the shoot apical meristem. Angiosperm meristems have two or three outer layers of cells that are histologically distinct. The distinction is due to repeated anticlinal divisions (the division plane perpendicular to the surface) without periclinal divisions (the division plane parallel to the surface). These layers are labeled L1, L2, and L3. While cells in one of the layers tend to have certain fates, they are not necessarily commited to these fates. If a cell shifts to a new layer because of a periclinal division, it generally develops like other cells in that layer.

derived from the L2. Chimeras have also been used to demonstrate classical induction in plants where, as also occurs in animal development, one layer influences the developmental pathway of the adjacent layer.

Lateral meristems are cylindrical meristems found in shoots and roots that result in secondary growth (the increase in stem and root girth by the production of vascular tissues). Monocot stems do not have lateral meristems, but often have **intercalary meristems** inserted in the stems between mature tissues. The popping sound you can hear in a cornfield on a summer night is actually caused by the rapid increase in stem length due to intercalary meristems.

Phyllotaxy

Branching patterns and leaf development that result in the unique above-ground architecture of different species have their origins in shoot meristems. **Leaf primordia** (clusters of cells that will form the leaf) are initiated at the periphery of the shoot meristem (see Figure 23.16). The union of the leaf and the stem is the **node** and stem tissue between the nodes is called the **internode** (see Figure 23.15). In a simplistic sense, the mature sporophyte is created by stacking these node/internode units together. The positioning of leaves on the stem is called **phyllotaxy**, and it involves communication among existing and newly forming leaf primordia. Leaves may be arranged in various patterns, including a spiral, 180-degree alternation of single leaves, pairs, and whorls of three or more leaves at a node (Jean 1994). Experimentation has generated a number of mechanisms for maintaining geometrically regular spacing of leaves on a plant, including chemical and physical interactions between the new primordia and the shoot apex, and interactions with existing primordia. It is not clear how a specfic pattern of phyllotaxy gets started.

LEAF DEVELOPMENT. Culture experiments have assessed when leaf primordia become determined for leaf development. Research on ferns and angiosperms indicates that the youngest visible leaf primordia are not determined to make a leaf; rather, these youngest primordia can develop

as shoots in culture (Steeves 1966; Smith 1984). The programming for leaf development occurs later. The radial symmetry of the leaf primordium becomes dorsal-ventral, or flattened, in all leaves. The unique shapes of leaves results from regulation of cell division and cell expansion as the leaf blade develops. There are some cases where selective cell death is involved in the shaping of a leaf, but differential cell growth appears to be a more common mechanism (Gifford and Foster 1989). There is much variety in **simple leaf** shape, including deeply lobed oak leaves (Figure 23.18). **Compound leaves** are composed of individual leaflets (and sometimes tendrils) instead of a single leaf blade.

Developmental genetic approaches are being applied to leaf morphogenesis. The maize *knotted1* (*KN1*) gene has been cloned and is expressed in the shoot apical meris-

Figure 23.18   Comparison of simple and compound leaves. Some compound leaves have only leaflets.

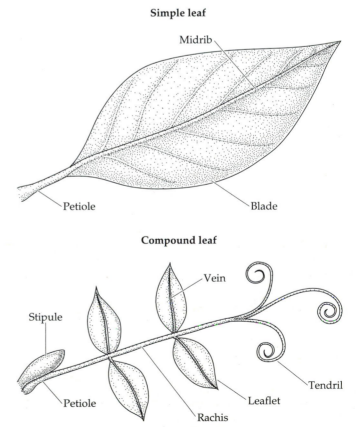

Simple leaf

Midrib

Petiole

Blade

Compound leaf

Vein

Stipule

Petiole

Rachis

Leaflet

Tendril

Wild-type tomato

Tomato leaf
overexpressing
knotted

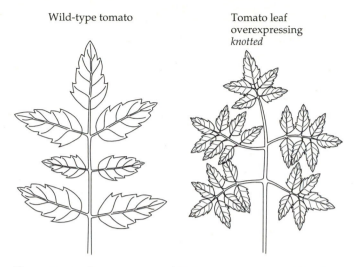

Figure 23.19 Overexpression of *KN1* in tomato. When *KN1*, which was originally cloned in maize, is overexpressed in the tomato, compounding of the leaves is amplified (see Hareven et al. 1996).

tems, but not leaves. It is most likely a homolog of *STM* in Arabidopsis (Long et al. 1996). When *KN1* is inserted into the genome of tomato and overexpressed, the leaves become "super compound" (Figure 23.19; Hareven et al. 1996). Simple leaves become more lobed (but not compound) in response to overexpression of *KN1*, consistent with the hypothesis that compound leaves may be an extreme case of lobing in simple leaves (Jackson 1996). The role of *KN1* in shoot meristem and leaf development is intriguing but does not necessarily imply that leaves are modified shoots.

In some compound leaves, developmental decisions about leaf versus tendril formation are also made. Mutations of two leaf-shape genes can individually and in sum dramatically alter the morphology of the compound pea leaf (Figure 23.20). The *acacia* mutant (*tl*) converts tendrils to leaflets; *afila* (*af*) converts leaflet to tendrils (Gould et al. 1986). The *af tl* double mutant has a complex architecture and resembles a parsley leaf.

BRANCHING. Shoot architecture is also affected by the amount of axillary bud outgrowth. Branching patterns depend on the amount of apical dominance from the shoot tip and plant hormones appear to be the regulators. The hormone **auxin** is produced by young leaves and transported towards the base. It can suppress outgrowth of axillary buds. Grazing and flowering often

Figure 23.20 Leaf morphology mutants in pea. Some compound leaves have tendrils as well as leaflets. Genes have been identified in the pea that convert leaflets to tendrils and tendrils to leaflets. The double mutants have the three-dimensional pattern of tendrils, but terminate with leaflet-like organs.

release buds from apical dominance, at which time branching occurs. **Cytokinins** can release buds from apical dominance. Axillary buds can initiate their own axillary buds, so branching patterns can get quite complex.

Vegetative-to-Reproductive Transition: Plant Metamorphosis

Unlike some animal systems in which the germ line is set aside during early embryogenesis, the germ line in plants is established after the transition from vegetative to reproductive development. The vegetative and reproductive structures of the shoot are all derived from the shoot meristem formed during embryogenesis. Clonal analysis indicates that cells are not set aside in the shoot meristem of the embryo to be used solely in the creation of reproductive structures (McDaniel and Poethig 1988). In maize, irradiated seed gives rise to plants that have visually distinguishable sectors that extend from the vegeta-

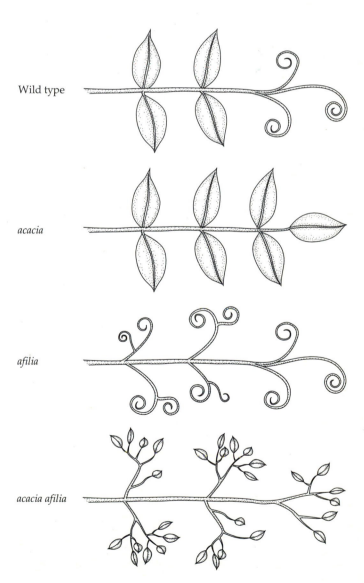

Wild type

acacia

afila

acacia afila

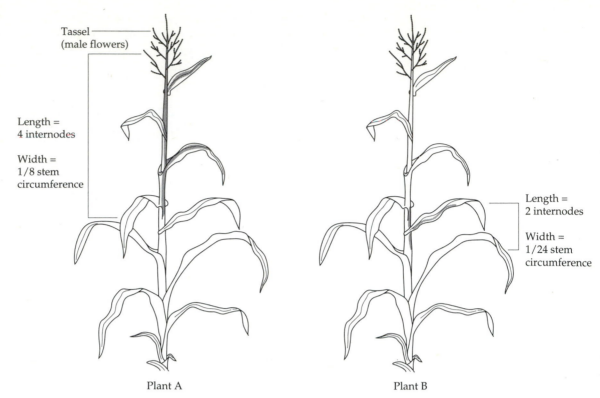

Tassel
(male flowers)

Length =
4 internodes

Width =
1/8 stem
circumference

Length =
2 internodes

Width =
1/24 stem
circumference

Plant A   Plant B

Figure 23.21   Clonal analysis can be used to construct a fate map of a shoot apical meristem in maize seeds. Seeds that are heterozygous for certain pigment genes (anthocyanins) are irradiated so that the dominant allele is lost in a few cells (a chance occurrence). All cells derived from the somatic mutant will be visually distinct from the nonmutant cells. Plants A and B have sectors that reveal the fate of cells in the shoot meristem of the seed. The sector in A includes vegetative and reproductive (tassel) internodes. Thus there is no distinct compartment for the tassel. The sector in A is longer and wider than the sector in B. This indicates that more cells were set aside to contribute to the lower than the upper internodes in the shoot meristem in the seed. The actual number of cells can be calculated by taking the reciprocal of the fraction of the stem circumference the sector occupies. Sector A contributes to 1/8th of the circumference of the stem; thus 8 cells were fated to contribute to these internodes in the seed meristem. Sector B is only 1/24th of the stem circumference; thus 24 cells were fated to contribute to these internodes. In this example, only cells derived from the L1 are being analyzed. It is also important to consider the possible contributions of the L2 and L3 cell layers of the shoot meristem. (Based on the work of McDaniel and Poethig 1988.)

tive portion of the plant into the reproductive regions (Figure 23.21). Sector size and shape are more variable, and these plants do not appear to have distinct developmental compartments.

A simplistic explanation of the flowering process is that a flowering signal from the leaves moves to the shoot apex and induces flowering. The developmental pathway(s) leading to flowering are regulated at numerous control points in different plant organs (roots, cotyledons, leaves, and shoot apices) in various species resulting in a diversity of flowering times and reproductive architectures.

Some plants, especially woody perennials, go through a **juvenile phase** during which the plant cannot produce reproductive structures even if all the appropriate environmental signals are present. The transition from the juvenile to the adult stage may require the acquisition of **competence** by the leaves or meristem to respond to an internal or external signal (McDaniel et al. 1992; Singer et al. 1992; Huala and Sussex 1993). Maximal reproductive success depends on timing flowering, and on balancing

the number of seeds produced with resources allocated to individual seeds. As in animals, different strategies work best for different organisms in different environments. There is a great diversity of flowering patterns among the over 300,000 angiosperms, yet there appears to be an underlying evolutionary conservation of flowering genes and common patterns of floral regulation.

Grafting and organ culture experiments, mutant analyses, and molecular analyses give us a framework for describing the reproductive transition in plants (Figure 23.22). Grafting experiments have identified the sources of signals that promote or inhibit flowering and have provided information on the developmental acquisition of meristem competence to respond to these signals (Lang et al. 1977; Singer et al. 1992; McDaniel et al. 1996; Reid et al. 1996). Analysis of mutants and molecular characterization of these genes are yielding information on the mechanics of the signal/response mechanisms (Coupland 1995; Weigel and Nilsson 1995; Mandel and Yanofsky 1995; Amasino 1996; Beveridge and Murfet 1996).

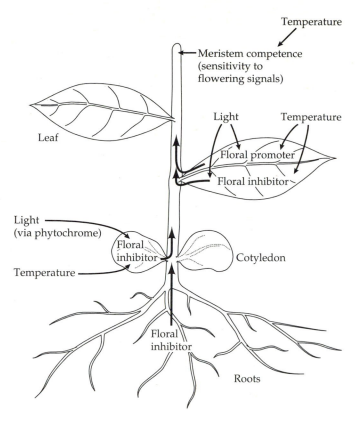

Figure 23.22 Regulation of the vegetative-to-reproductive transition. This model indicates how internal and external factors regulate whether a meristem produces vegetative or reproductive structures. Not all regulatory mechanisms are used in all species, and some species flower independently of external environmental signals. Signals promoting or inhibiting flowering can move from the roots, cotyledons, or leaves to the shoot apex, where meristem competence determines whether or not the plant will respond to the signals. Leaves may also need to develop competence to respond to environmental signals before they can produce floral promoters.

duction of either floral promoter or floral inhibitor. Leaves, cotyledons, and roots have been identified as sources of floral inhibitors in some species (McDaniel et al. 1992; Reid et al. 1996). A critical balance between inhibitor and promoter is needed for the reproductive transition. Not all leaves may be competent to perceive the photoperiodic signal and/or produce such a signal.

In some species, meristems change in their competence to respond to flowering signals during development (Singer et al. 1992). The reproductive transition depends on both meristem competence and signal strength (Figure 23.23). Shoot tip culture experiments in several species (including tobacco, sunflower, and pea) have demonstrated that determination for reproductive development can occur before reproductive morphogenesis (reviewed in

Leaves produce a graft-transmissible substance that induces flowering. In some species, the signal is produced only under specific **photoperiods** (light/dark cycles), while other species are day-neutral and will flower under any photoperiod (Zeevaart 1984). The pigment **phytochrome** is the link to the external environment. Phytochrome, whose structure is modified by red and far-red light, can initiate a cascade of events leading to the pro-

Figure 23.23 Competence and signal strength. *Nicotiana tabacum* (Nt) is a day-neutral plant that flowers when the meristem gains competence to respond to internal signals. *N. silvestris* (Ns) is a long-day plant that flowers when the floral signal(s) reach a critical level. These grafting experiments illustrate that a young Nt shoot is less competent to respond to the Nt flowering signal than an older Nt shoot. Young Nt shoots respond quickly to the floral signal in flowering Ns plants, but flower later when grafted to a young Ns plant with a lower level of signal. (After Singer et al. 1992.)

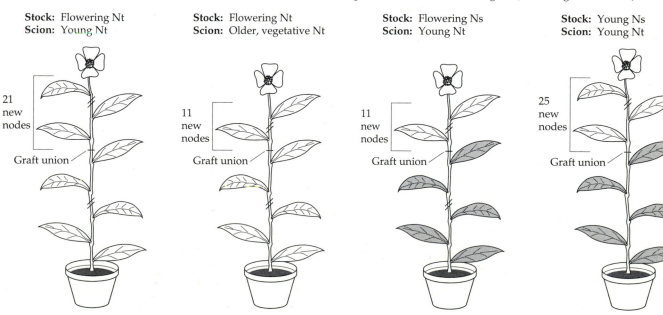

Seed-derived plant

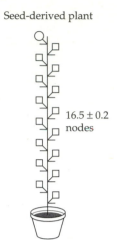

16.5 ± 0.2
nodes

Figure 23.24   Determination for reproductive development. In this experiment, axillary buds three nodes from the base of pea plants were treated when the plants had three expanded leaves. These buds were determined for reproductive development and produced the same number of new nodes (*a*) when they were allowed to grow *in situ* by decapitating the terminal shoot; (*b*) when they were cultured; and (*c*) when just the bud meristem was cultured. If they had not been committed to reproductive development, they would have developed like seed-derived plants and produced many more nodes before flowering.

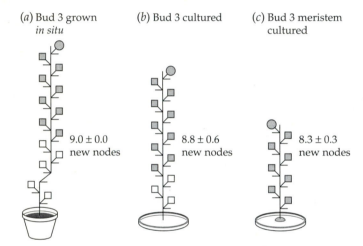

(*a*) Bud 3 grown *in situ*

9.0 ± 0.0 new nodes

(*b*) Bud 3 cultured

8.8 ± 0.6 new nodes

(*c*) Bud 3 meristem cultured

8.3 ± 0.3 new nodes

McDaniel et al. 1992). That is, isolated shoot tips that are determined for reproductive development but are morphologically vegetative will produce the same number of nodes before flowering *in situ* and in culture (Figure 23.24).

The ancestral angiosperm is believed to have formed a terminal flower directly from the terminal shoot apex (Stebbins 1974). A variety of flowering patterns exist where the terminal shoot apex is indeterminate, but axillary buds produce flowers. This introduces an intermediate step in the reproductive process: the transition of a vegetative meristem to an **inflorescence meristem**. The inflorescence meristem initiates axillary meristems that can produce floral organs, but never directly produces floral parts itself. The inflorescence is the reproductive backbone (stem) that displays the flowers (see Figure 23.15). Most likely the production of an inflorescence meristem arose through the action of a gene that supresses terminal flower formation. The recently cloned *centroradialus* (*CEN*) gene in snapdragons supresses terminal flower formation (Bradley et al. 1996). It acts by supressing expression of *floricaula* (*FLO*), which specifies floral meristem identity. Curiously, the expression of *FLO* is necessary for *CEN* to

be turned on. The *Arabidposis* homolog to *CEN* (*TERMINAL FLOWER 1* or *TFL1*) is expressed during the vegetative phase of development as well and has the additional function of delaying the commitment to inflorescence development (Bradley et al. 1997). Garden peas branch one more time than snapdragons do before forming a flower. That is, the axillary meristem does not directly produce a flower, but acts as an infloresecence meristem that initiates floral meristems. Two genes, *Determinate* (*DET*) and *Vegetative 1* (*VEG 1*), are responsible for this more complex inflorescence and only when both are nonfunctional is a terminal flower formed (Figure 23.25; Singer et al. 1996).

The next step in the reproductive process is the specification of floral meristem identity (Weigel 1995). In *Arabidopsis*, *LEAFY* (*LFY*), *APETALA 1* (*AP1*), and *CAULIFLOWER* (*CAL*) are **floral meristem identity genes**. *LFY* is the homolog of *FLO* in snapdragons. Expression of these genes is necessary for the transition from an inflorescence to a floral meristem. Mutants (*lfy*) tend to form leafy shoots in the axils where flowers form in wild type plants; they are unable to make the transition to floral development. If *LFY* is overexpressed, flowering occurs earlier. For

Figure 23.25   Regulation of inflorescence architecture. *Arabidopsis* and snapdragon produce flowers in their axillaries (second-order flowers). In pea, the axillary branches grow out and initiate flowers (third-order flowers) but do not directly produce flowers. Recent evidence suggests that more complex branching patterns in inflorescences are the result of suppressing expression of flowering genes in meristems. Here the stars indicate meristems where floral gene expression is suppressed. In *Arabidopsis* and snapdragon, a single gene suppresses flowering in the first-order meristem. In pea, two genes are necessary to suppress flowering in the first- and second-order meristems.

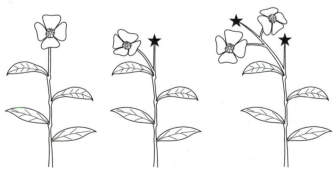

Terminal flower   Second-order flower   Third-order flower

example, aspen was transformed with a *LFY* gene that was expressed throughout the plant and the time to flowering was dramatically shortened from years to months (Weigel and Nilsson 1995).

Floral meristem identity genes initiate a cascade of gene expression that turns on mapping (**cadastral**) genes that further specify pattern and initiate transcription of **floral organ identity genes** (Weigel 1995). *SUPERMAN* (*SUP*)is an example of a cadastral gene in *Arabidopsis* that plays a role in maintaining boundries for organ identity gene expression. Three classes (A, B, and C) of organ identity genes are necessary to specify the four whorls of floral organs (Figure 23.26; Coen and Meyerowitz 1991). These are homeotic genes and include *AP2*, *AGAMOUS* (*AG*), *AP3*, and *PISTILLATA* (*PI*) in *Arabidopsis*. Class A genes (*AP2*) alone specify sepal development. Class A genes and class B genes (*AP3* and *PI*) together specify petals. B and C (*AG*) genes are necessary for stamen formation; class C genes alone specify carpel formation. These genes trigger events resulting in the development of floral organs. When all of these homeotic genes are not expressed in a developing flower, floral parts become leaf like.

Senescence

Flowering and **senescence** (a developmental program leading to death) are closely linked in angiosperms. Individual flower petals in some species senesce following pollination. Fruit ripening (and ultimately over-ripening) is an example of organ senescence. Whole-plant senescence leads to the death of the entire sporophyte generation. **Monocarpic** plants flower once and then senesce. **Polycarpic** plants such as the bristlecone pine can live thousands of years (4900 years is the current record) and flower repeatedly. In polycarpic plants, death is by accident. In monocarpic plants it appears to be genetically programmed. Flowers and/or fruits play a key role in the process and their removal can sometimes delay senescence. It is intriguing that in some legumes senescence can by delayed by removing the developing seed—the embryo may trigger senescence in the parent plant. During flowering and fruit development, nutrients are reallocated from other parts of the plant to support the development of the next generation. That is, the reproductive structures become a nutrient sink, and this can lead to whole-plant senescence.

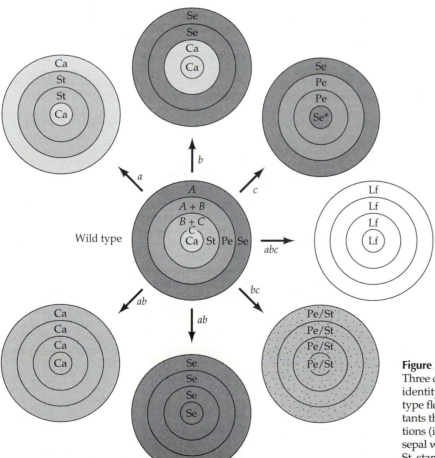

Figure 23.26 ABC model for floral organ specification. Three classes of genes—*A*, *B*, and *C*—regulate organ identity in flowers. The central diagram represents a wild type flower. The surrounding diagrams represent mutants that are missing one or more of these gene functions (indicated by the lowercase *a*, *b*, or *c*). Se, sepal; Se*, sepal with an indeterminate floral meristem; ; Pe, petal; St, stamen; Ca, carpel; Pe/St, a hybrid petal/stamen; Lf, leaf. (Based on the work of Coen and Meyerowitz 1991.)

Summary

Plant embryos develop deeply embedded in parental tissue. The parent tissue provides nutrients but minimal patterning information. Early embryognesis is characterized by the establishment of the root-shoot axis and radial patterning yielding three tissue systems. Pattern emerges by regulation of planes of cell division, since plant cells do not move during development. As the embryo matures, a food reserve is established. Only the rudiments of the basic body plan are established when embryogenesis ceases and the seed enters dormancy. Pattern is elaborated during postembryonic development, when organized clusters of embryonic tissue called meristems construct the reiterative building blocks of the plant. The germ line is not reserved early in development. It is produced by the haploid gametophyte generation which develops from the reproductive structures of the sporophyte generation. Coordination of signaling among leaves, roots, and shoot meristems regulates when a plant transits to the reproductive state. Reproduction may be followed by genetically programmed senescence of the parent plant.

Literature Cited

Aeschbacher, R. A., J. W. Schiefelbein and P. N. Benfey. 1994. The genetic and molecular basis of root development. *Annu. Rev. Plant Physiol. Plant Mol. Biol.* 45: 25–45.

Amasino, R. M. 1996 Control of flowering time in plants. *Curr. Opin. Genet. Dev.* 6: 480–487.

Barton, M. K. and R. S. Poethig. 1993. Formation of the shoot apical meristem in *Arabidopsis thaliana*: An analysis of development in the wild type and in the *shoot meristemless* mutant. *Development* 119: 823–831.

Beveridge, C. A. and I. C. Murfet. 1996. The *gigas* mutant in pea is deficient in the floral stimulus. *Physiol. Plant.* 96: 637–645.

Bhattacharyya, M. K., A. M. Smith, T. H. N. Ellis, C. Hedley and C. Martin. 1990. The wrinkled-seed character of pea described by Mendel is caused by a transposon-like insertion in a gene encoding starch-branching enzyme. *Cell* 60: 115–122.

Bowman, J. 1994. *Arabidopsis: An Atlas of Morphology and Development.* Springer-Verlag, New York.

Bradley, D., R. Carpenter, L. Copsey, C. Vincent, S. Rothstein and E. Coen. 1996. Control of inflorescence architecture in *Antirrhinum*. *Nature* 379: 791–797.

Bradley, D., O. Ratcliffe, C. Vincent, R. Carpenter and E. Coen. 1997. Inflorescence commitment and architecture in *Arabidopsis*. *Science* 275: 80–83.

Brewbaker, J. L. and B. H. Kwack. 1963. The essential role of calcium ions in pollen germination and pollen tube growth. *Amer. J. Bot.* 50: 859–863.

Christianson, M. L. 1986. Fate map of the organizing shoot apex in *Gossypium*. *Amer. J. Bot.* 73: 947–958.

Cionini, P. G., A. Bennici, A. Alpi and F. D'Amato. 1976. Suspensor, gibberellin and *in vitro* development of *Phaseolus coccineus* embryos. *Planta* 131: 115–117.

Clark, J. K. and W. F. Sheridan. 1986. Developmental profiles of the maize embryo-lethal mutants *dek22* and *dek23*. *J. Hered.* 77: 83–92.

Coen, E. S. and E. M. Meyerowitz. 1991. The war of the whorls: Genetic interactions controlling flower development. *Nature* 353: 31–37.

Coupland, G. 1995. Genetic and environmental control of flowering time in *Arabidopsis*. *Trends Genet.* 11: 393–397.

Cruden, R. W. and R. M. Lloyd. 1995. Embryophytes have equivalent sexual phenotypes and breeding sytsems: Why not a common terminology to describe them? *Amer. J. Bot.* 82: 816–825.

Dodds, P. N., A. E. Clarke and E. Newbigin. 1996. A molecular perspective on pollination in flowering plants. *Cell* 85: 141–144.

Esau, K. 1977. *Anatomy of Seed Plants*, 2nd Ed. John Wiley and Sons, New York.

Franklin-Tong, V. E., B. K. Drobak, A. C. Allan, P. A. C. Watkins and A. J. Trewavas. 1996. Growth of pollen tubes in *Papaver rhoeas* is regulated by a slow-moving calcium wave propagated by inositol 1,4,5-trisphosphate. *Plant Cell* 8: 1305–1321.

Friedman, W. E. 1995. Organismal duplication, inclusive fitness theory, and altruism: Understanding the evolution of endosperm and the angiosperm reproductive syndrome. *Proc. Natl. Acad. Sci. USA* 92: 3913–3917.

Gifford, E. M. and A. S. Foster. 1989. *Morphology and Evolution of Vascular Plants.* W. H. Freeman, New York.

Gould, K. S., E. G. Cutter, J. P. W. Young. 1986. Morphogenesis of the compound leaf in three genotypes of the pea, *Pisum sativum*. *Can. J. Bot.* 64: 1268–1276.

Haccius, B. 1963. Restitution in acidity-damaged plant embryos: Regeneration or regulation? *Phytomorphology* 13: 107–115.

Hareven, D, T. Gutfinger, A Pornis, Y. Eshed, and E. Lifschitz. 1996. The making of a compound leaf: genetic manipulation of leaf architecture in tomato. *Cell* 84: 735–744.

Huala, E. and I. M. Sussex. 1993. Determination and cell interactions in reproductive meristems. *Plant Cell* 5: 1157–1165.

Hulskamp, M., K. Schneitz and R. E. Pruitt. 1995. Genetic evidence for a long-range activity that directs pollen tube guidance in *Arabidopsis*. *Plant Cell* 7: 57–64.

Irish, V. F. and I. M. Sussex. 1992. A fate map of the *Arabidopsis* embryonic shoot apical meristem. *Development* 115: 745–754. .

Jackson, D. 1996. Plant morphogenesis: Designing leaves. *Cur. Biol.* 6: 917–919.

Jaffe, L. A., M. H. Weisenseel and L. F. Jaffe. 1975. Calcium accumulation within the growing tips of pollen tubes. *J. Cell Biol.* 67: 488–492.

Jean, R. V. 1994. *Phyllotaxis.* Cambridge University Press, Cambridge.

Johri, B. M., K. B. Ambegaokar and P. S. Srivastava. 1992. *Comparative Embryology of Angiosperms.* Springer-Verlag, New York.

Lang, A., M. K. Chailakhyan and I. A. Frolova. 1977. Promotion and inhibition of flower formation in a day-neutral plant in grafts with short-day and long-day plants. *Proc. Natl. Acad. Sci. USA* 74: 2412–2416.

Laux, T. and G. Jurgens. 1994. Establishing the body plan of the *Arabidopsis* embryo. *Acta Bot. Neer.* 43: 247–260.

Liu, C.-M., Z.-H. Xu and N.-H. Chua. 1993. Auxin polar transport is essential for the establishment of bilateral symmetry during early plant embryogenesis. *Plant Cell* 5: 621–630.

Long, J. A., E. I. Moan, J. I. Medford and M. K. Barton. 1996. A member of the KNOTTED class of homeodomain proteins encoded by the *shootmeristemless* gene of *Arabidopsis*. *Nature* 379: 66–69.

Lord, E. M., L. L. Walling and G. Y. Jauh. 1996. Cell adhesion in plants and its role in pollination. In *Membranes: Specialized Functions in Plants*, M. Smallwood, J. P. Knox and D. J. Bowles (eds.). Bios, Oxford, pp. 21–37.

Malho, R., N. D. Read, M. S. Pais and A. J. Trewavas. 1994. Role of cytosolic free calcium in the reorientation of pollen tube growth. *Plant J.* 5: 331–341.

Mandel, M. A. and M. F. Yanofsky. 1995. A gene triggering flower formation in *Arabidopsis*. *Nature* 377: 522–524.

Mayer, U., R. A. Torres Ruiz, T. Berleth, S. Misera and G. Jurgens. 1991. Mutations affecting body organization in the *Arabidopsis* embryo. *Nature* 353: 402–406.

McDaniel, C. N. and R. S. Poethig. 1988. Cell-lineage patterns in the shoot apical meristem of the germinating maize embryo. *Planta* 175: 13–22.

McDaniel, C. N., S. R. Singer and S. M. E. Smith. 1992. Developmental states associated with the floral transition. *Developmental Biology* 153: 59–69.

McDaniel, C. N., L. K. Hartnett and K. A. Sangrey. 1996. Regulation of node number in day-neutral *Nicotiana tabacum*: A factor in plant size. *Plant J.* 9: 55–61.

Mogie, M. 1992. *The Evolution of Asexual Reproduction in Plants.* Chapman and Hall, New York.

Nasrallah, J. B., J. C. Stein, M. K. Kandasamy and M. E. Nasrallah. 1994. Signaling the arrest of pollen tube development in self-incompatible plants. *Science* 266: 1505–1508.

Poethig, R. S. 1987. Clonal analysis of cell lineage patterns in plant development. *Amer. J. Bot.* 74: 581–594.

Reid, J. B., I. C. Murfet, S. R. Singer, J. L. Weller and S. A. Taylor. 1996. Physiological-genetics of flowering in *Pisum. Seminars Cell Dev. Biol.* 7: 455–463.

SanMiguel, P., A. Tikhonov, Y. Jin, N. Motchoulskaia, D. Zakharov, A. Melake-Berhan, P. Springer, K. J. Edwards, M. Lee and Z. Avramova. 1996. Nested retrotransposons in the intergenic regions of the maize genome. *Science* 274: 765–768.

Schiavone, F. M. and R. H. Racusen. 1990. Microsurgery reveals regional capabilities for pattern reestablishment in somatic carrot embryos. *Dev. Biol.* 141: 211–219.

Schwartz, B. W., E. C. Yeung and D. W. Meinke. 1994. Disruption of morphogenesis and transformation of the suspensor in abnormal suspensor mutants of *Arabidopsis. Development* 120: 3235–3245.

Singer, S. R., C. H. Hannon and S. C. Huber. 1992. Acquisition of competence for floral determination in shoot apices of *Nicotiana. Planta* 188: 546–550.

Singer, S. R., S. L. Maki, J. Sollinger, H. Mullen, J. Fick and A. McCall. 1996. Supression of terminal flower formation in pea: Evolutionary implications. *Dev. Biol.* 175: 377.

Smith, R. H. 1984. Developmental potential of excised primordial and expanding leaves of *Coleus blumei* benth. *Amer. J. Bot.* 71: 114–1120.

Southworth, D. 1996. Gametes and fertilization in flowering plants. *Curr. Top. Dev. Biol.* 34: 259–279.

Stebbins, L. 1974. *Flowering Plants: Evolution Above the Species Level.* Belknap Press, Cambridge, MA.

Steeves, T. A. 1966. On the determination of leaf primordia in ferns. In *Trends in Plant Morphogenesis*, E. G. Cutter (ed.), pp. 200–219. Longman, London.

Steeves, T. A. and I. M. Sussex. 1989. *Patterns in Plant Development,* 2nd Ed. Cambridge University Press, New York.

Taylor, C. B. 1996. More arresting developments: S RNases and interspecific incompatibility. *Plant Cell* 8: 939–941.

Weigel, D. 1995. The genetics of flower development: from floral induction to ovule morphogenesis. *Annu. Rev. Genet.* 29: 19–39.

Weigel, D. and O. Nilsson. 1995. A developmental switch sufficient for flower initiation in diverse plants. *Nature* 377: 495–500.

Wilhelmi, L. K. and D. Preuss. 1996. Self-sterility in *Arabidopsis* due to defective pollen tube guidance. *Science* 274: 1535–1537.

Wilhelmi, L. K. and D. Preuss. 1997. Blazing new trails: Pollen tube guidance in flowering plants. *Plant Physiol.* 113: 307–312.

Yadegari, R., G. R. de Paiva, T. Laux, A. M. Koltunow, N. Apuya, J. L. Zimmerman, R. L. Fischer, J. J. Harada and R. B. Goldberg. 1994. Cell differentiation and morphogenesis are uncoupled in *Arabidopsis raspberry* embryos. *Plant Cell* 6: 1713–1729.

Yeung, E. C. and I. M. Sussex. 1979. Embryogeny of *Phaseolus coccineus*: The suspensor and the growth of the embryo-proper *in vitro. Z. Pflanzenphysiol.* 91: 423–433.

Zeevaart, J. A. D. 1984. Photoperiodic induction, the floral stimulus and floral-promoting substances. In *Light and the Flowering Process*, D. Vince-Prue, B. Thomas and K. E. Cockshull (eds.), pp. 137–142. Academic Press, Orlando.

Index

About the Book

Editor: Andrew D. Sinauer

Project Manager: Carol J. Wigg

Copy Editor: Roberta Lewis

Ink Illustrations: Nancy J. Haver

Electronic Illustrations: Precision Graphics

Production Manager: Christopher Small

Book Design and Layout: Jefferson Johnson

Cover Design: Christine Polaczak

Cover Manufacture: Henry N. Sawyer Company, Inc.

Book Manufacture: Best Book Manufacturers